Springer Monographs in Mathematics

Springer
Berlin
Heidelberg
New York
Barcelona
Budapest
Hong Kong
London
Milan
Paris
Singapore
Tokyo

Andrew Ranicki

High-dimensional Knot Theory

Algebraic Surgery in Codimension 2

with an Appendix
by Elmar Winkelnkemper

Springer

Andrew Ranicki
Department of Mathematics and Statistics
University of Edinburgh
Edinburgh EH9 3JZ, Scotland, UK

Elmar Winkelnkemper
Department of Mathematics
University of Maryland
College Park, MD 20742, USA

The cover design is of a trefoil knot together with a singularity-free spanning surface, that is a Seifert surface. The diagram is taken from the paper of Frankl and Pontrjagin (Math. Ann. 102, 785-789 (1930)), which proved for the first time that every tame knot $k : S^1 \subset S^3$ has a Seifert surface.

Library of Congress Cataloging-in-Publication Data

Ranicki, Andrew, 1948-
High-dimensional knot theory : algebraic surgery in codimension 2
/ Andrew Ranicki ; with an appendix by Elmar Winkelnkemper.
p. cm. -- (Springer monographs in mathematics)
Includes bibliographical references and index.
ISBN 3-540-63389-8 (hardcover : alk. paper)
1. Knot theory 2. Surgery (Topology) 3. Embeddings
(Mathematics) I. Title. II. Series.
QA612.2.R363 1998
514'.224--dc21 98-25080
CIP

Mathematics Subject Classification (1991): 57Q45, 57R67

ISBN 3-540-63389-8 Springer-Verlag Berlin Heidelberg New York

Cover design: *Erich Kirchner, Heidelberg*

Typesetting by the author using a Springer TeX macro package
SPIN 10639194 41/3143-5 4 3 2 1 0 – Printed on acid-free paper

In memory of J. F. Adams

Preface

On my first day as a graduate student at Cambridge in October, 1970 my official Ph. D. supervisor Frank Adams suggested I work on surgery theory. In September he had attended the International Congress at Nice, where Novikov had been awarded the Fields Medal for his work in surgery. Novikov was prevented by the Soviet authorities from going to the Congress himself, and his lecture on hermitian K-theory [219] was delivered by Mishchenko. As usual, Frank had taken meticulous notes, and presented me with a copy. He also suggested I look at 'Novikov's recent paper in *Izvestia*'. This recommendation was quite mysterious to me, since at the time I knew of only one publication called *Izvestia*, and I couldn't imagine that journal publishing an article on topology. I was too shy to ask, but a visit to the library of the Cambridge Philosophical Society soon enlightened me. I started work on Novikov's paper [218] under the actual supervision of Andrew Casson, using a translation kindly provided by Dusa McDuff. (Andrew could not be my official supervisor since he did not have a Ph. D. himself). Ultimately, my reading of [218] became my Ph. D. thesis, which was published as Ranicki [230], [231]. Frank had remained my official supervisor, being ever helpful in answering my many queries on algebraic topology, and generally keeping me under his protective wing. I dedicate this book to his memory, as a token of my gratitude to him.

Edinburgh, June 1998

Table of Contents

Appendix

Introduction

Knot theory high and low

An *n-dimensional knot* (M, N, k) is an embedding of an n-dimensional manifold N^n in an $(n+2)$-dimensional manifold M^{n+2}

$$k : N^n \subset M^{n+2} .$$

An *n-knot* is a knot of the type (S^{n+2}, S^n, k). In general, n can be any positive integer. A *classical knot* is a 1-knot $k : S^1 \subset S^3$.

The homological algebra methods of surgery theory apply to n-dimensional knots for all $n \geq 1$. This book is mainly concerned with knots in the high dimensions $n \geq 4$, for which there is a much closer correspondence between this type of algebra and the topology than in the low dimensions $n = 1, 2, 3$. However, the story begins with the classical case $n = 1$.

The mathematical study of knots started in the 19th century, with the work of Gauss.[1] Towards the end of the century, P. G. Tait (working in Edinburgh) tabulated all the classical knots with ≤ 10 crossings. It was only in the 20th century that a systematic theory of knots was developed. Many algebraic and geometric techniques have been invented and used to deal with classical knots: a large number of invariants is available, including the fundamental group of the knot complement, the Alexander polynomial, the Seifert matrix, assorted signatures, the Jones polynomial, the Vassiliev invariants, ..., although we are still short of a complete classification. The last 20 years have seen a particular flourishing of classical knot theory, involving deep connections with 3-manifold topology, physics and biology. There is a large literature at various levels, including the books by C. Adams [2], Atiyah [12], Burde and Zieschang [34], Crowell and Fox [59], Kauffman [122], Lickorish [167], Livingston [169], Murasugi [207], Reidemeister [250], and Rolfsen [253].

Although high-dimensional knot theory does not have such glamorous applications as classical knot theory, it has many fascinating results of its own, which make use of a wide variety of sophisticated algebraic and geometric methods. This is the first book devoted entirely to high-dimensional knot

[1] See Epple [68], [69], [70] for the history of knot theory.

theory, which previously has been largely confined to research and survey papers. The book actually has two aims:

(i) to serve as an introduction to high-dimensional knot theory, using surgery theory to provide a systematic exposition,
(ii) to serve as an introduction to algebraic surgery theory, using high-dimensional knots as the geometric motivation.

The topological properties of high-dimensional knots are closely related to the algebraic properties of modules and quadratic forms over polynomial extensions. The main theme of the book is the way in which this relationship is essential to both (i) and (ii). High-dimensional knot theory has a somewhat deserved reputation as being an arcane geometric machine – I hope that aim (i) is sufficiently achieved to demystify the geometry, and make it more accessible to algebraic topologists. Likewise, surgery theory has a somewhat deserved reputation as being an arcane algebraic machine – I hope that aim (ii) is sufficiently achieved to demystify the algebra, and make it more accessible to geometric topologists.

Knot theory is a good introduction to surgery since it is easier to visualize knots than manifolds. Many surgery invariants can be viewed as generalizations of (high-dimensional) knot invariants. For example, the self intersection quadratic form μ used by Wall [302] to define the surgery obstruction of a normal map is a generalization of the matrix associated by Seifert [262] to a spanning surface for a classical knot. Moreover, the plumbing construction in [302, Chap. 5] of normal maps with prescribed quadratic form is a generalization of the construction in [262] of classical knots with prescribed matrix.

Artin [8] produced the first non-trivial examples of 2-knots $S^2 \subset S^4$, by spinning classical knots $S^1 \subset S^3$.

The theory of n-knots $S^n \subset S^{n+2}$ and more general n-dimensional knots $N^n \subset M^{n+2}$ evolved with the work of Whitney on removing singularities in the 1940s, Thom's work on transversality and cobordism in the 1950s, the h-cobordism theorem for manifolds of dimension $n \geq 5$ of Smale in 1960 and the consequent surgery theory of high-dimensional manifolds and their submanifolds. The last 35 years have seen the growth of a large body of research literature on codimension 2 embeddings of high-dimensional manifolds in the differentiable, piecewise linear and topological categories. However, it is certainly not the aim of this book to provide a comprehensive account of all the methods and results of high-dimensional knot theory![2] The book has the more limited objective of providing an exposition of the algebraic surgery

[2] In particular, there is very little about links $\cup S^n \subset S^{n+2}$, and not much about the connections between knots and singularities, or about the homotopy-theoretic aspects of high-dimensional knot theory.

method for the construction and classification of high-dimensional knots in the topological category.

The pre-1981 applications of high-dimensional surgery theory to codimension 2 embeddings were considered in Chap. 7 of Ranicki [237] – however, at the time algebraic surgery was not so highly developed, and the treatment still relied on geometric transversality. The current treatment makes full use of the algebraic analogues of transversality obtained by the author since 1981.

Knotty but nice

Two embeddings $k_0, k_1 : N^n \subset M^m$ are *concordant* (or *cobordant*) if there exists an embedding

$$\ell : N \times [0,1] \longrightarrow M \times [0,1] \; ; \; (x,t) \longrightarrow \ell(x,t)$$

such that

$$\ell(x,0) = (k_0(x),0) \; , \; \ell(x,1) = (k_1(x),1) \;\; (x \in N) .$$

Two embeddings $k_0, k_1 : N \subset M$ are *isotopic* if there exists a cobordism ℓ which is level-preserving, with

$$\ell(N \times \{t\}) \subset M \times \{t\} \;\; (t \in [0,1]) .$$

Isotopy is a considerably stronger equivalence relation than concordance – much in the way that the isomorphism of quadratic forms is stronger than stable isomorphism.

Two embeddings $k_0, k_1 : N \subset M$ are *equivalent* if there exists a homeomorphism $h : M \longrightarrow M$ such that

$$hk_0(N) = k_1(N) \subset M .$$

If h is isotopic to the identity then k_0, k_1 are isotopic. Every orientation-preserving homeomorphism $h : S^m \longrightarrow S^m$ is isotopic to the identity, so for embeddings $k : N^n \subset S^m$

$$\text{equivalent} = \text{isotopic} \Longrightarrow \text{cobordant} .$$

An embedding $k : S^n \subset S^m$ is *unknotted* if it is equivalent to the trivial knot $k_0 : S^n \subset S^m$ defined by the standard embedding

$$k_0 : S^n \longrightarrow S^m \; ; \; (x_0, x_1, \ldots, x_n) \longrightarrow (x_0, x_1, \ldots, x_n, 0, \ldots, 0) .$$

These definitions are particularly significant for knots, that is embeddings with codimension $m - n = 2$.

Why 2?

A knot is a codimension 2 embedding. What about embeddings $k : N^n \subset M^m$ with codimension $m - n \neq 2$? Zeeman [312] and Stallings [273] proved that embeddings $k : S^n \subset S^m$ with codimension $m - n \geq 3$ are unknotted in the piecewise linear and topological categories.[3] Thus topological knotting only starts in codimension 2. Many algebraic invariants have been developed to classify the knotting properties of codimension 2 embeddings. Codimension 1 embeddings are almost as interesting as that of codimension 2 embeddings, although they do not have the intuitive appeal of classical knot theory. In fact, codimension 1 embeddings deserve a book of their own! In any case, many of the techniques used in codimension 2 make crucial use of codimension 1 embeddings, such as spanning surfaces.

The fundamental group

The study of knots necessarily involves the fundamental group, as well as the higher homotopy groups and homology. The isomorphism class of the fundamental group $\pi_1(X)$ of the *complement* of a knot (M, N, k)

$$X = M \backslash k(N)$$

is an invariant of the equivalence class. The algebraic theory developed in this book works with modules over an arbitrary ring, reflecting the major role of the fundamental group $\pi_1(X)$ and the group ring $\mathbb{Z}[\pi_1(X)]$ in the classification of knots (M, N, k).

The fundamental group $\pi_1(X)$ of the complement $X = S^3 \backslash k(S^1)$ of a classical knot $k : S^1 \subset S^3$ was the first knot invariant to be studied by the methods of algebraic topology, serving to distinguish many classical knots $k : S^1 \subset S^3$. Dehn's Lemma (formulated in 1910, but only finally proved by Papakyriakopoulos in 1956) states that a classical knot $k : S^1 \subset S^3$ is unknotted if and only if $\pi_1(X) = \mathbb{Z}$. The complement $X = S^{n+2} \backslash k(S^n)$ of any n-knot $k : S^n \subset S^{n+2}$ has the homology of a circle by Alexander duality, $H_*(X) = H_*(S^1)$. Levine [154] proved that for $n \geq 4$ an n-knot $k : S^n \subset S^{n+2}$ is unknotted if and only if the complement X is homotopy equivalent to a circle, i.e. if and only if $\pi_*(X) = \pi_*(S^1)$. This unknotting criterion also holds for $n = 3$ by Levine [159] and Trotter [291], and for $n = 2$ by Freedman [85] (in the topological category). Thus homology does not see knotting, while homotopy detects unknotting.

[3] There is codimension ≥ 3 knotting in the differentiable category. Differentiable embeddings $k : S^n \subset S^m$ with $m - n \geq 3$ were classified by Haefliger [101] and Levine [155].

Local flatness

An n-dimensional knot (M^{n+2}, N^n, k) is *locally flat* at $x \in N$ if x has a neighbourhood in $(M, k(N))$ which is homeomorphic to $(\mathbb{R}^{n+2}, \mathbb{R}^n)$. The knot is *locally flat* if it is locally flat at every $x \in N$.

Given an n-dimensional PL knot (M^{n+2}, N^n, k) there is defined at every point $x \in N$ a PL embedding

$$\mathrm{star}_M(x) \cap N \;=\; \mathrm{star}_N(x) \;=\; D^n \subset \mathrm{star}_M(x) \;=\; D^{n+2}$$

which could be knotted, i.e. not PL equivalent to the standard embedding $D^n \subset D^{n+2}$. The restriction to the links defines a PL $(n-1)$-knot

$$k_x \;:\; \mathrm{link}_M(x) \cap N \;=\; \mathrm{link}_N(x) \;=\; S^{n-1} \subset \mathrm{link}_M(x) \;=\; S^{n+1}\ .$$

The knot type of k_x is a measure of the local singularity of the topology of k at x: if k is locally flat at x then $k_x : S^{n-1} \subset S^{n+1}$ is unknotted (Fox and Milnor [82], with $n = 2$).[4]

Unless specified otherwise, from now on knots (M^{n+2}, N^n, k) will be taken to be locally flat, i.e. only knots which are locally unknotted will be considered.

Surgery theory can also deal with non-locally flat knots, such as arise from singular spaces (Cappell and Shaneson [42], [46], [47]). However, non-locally flat knots require special techniques, such as intersection homology.

For any (locally flat) knot (M^{n+2}, N^n, k) the codimension 2 submanifold $k(N) \subset M$ has a normal bundle, that is a closed regular neighbourhood $(P, \partial P)$ which is the total space of a bundle

$$(D^2, S^1) \longrightarrow (P, \partial P) \longrightarrow k(N)\ .$$

Unless specified otherwise, from now on knots (M^{n+2}, N^n, k) will be taken to be oriented, with M, N compact and the normal bundle of $k : N \subset M$ compatibly oriented.

The *exterior* of a knot (M^{n+2}, N^n, k) is the codimension 0 submanifold of M

$$(X, \partial X) \;=\; (\mathrm{closure}(M \backslash P), \partial P)$$

with a homotopy equivalence to the knot complement

$$X \;\simeq\; M \backslash k(N)\ .$$

A locally flat knot (M^{n+2}, N^n, k) is *homology framed* if

$$k[N] \;=\; 0 \in H_n(M)$$

and the normal bundle of $k(N) \subset M$ is framed

$$(P, \partial P) \;=\; k(N) \times (D^2, S^1)\ ,$$

[4] High-dimensional knots arise in a similar way at isolated singular (i.e. non-manifold) points of a complex hypersurface $V^{2i} \subset S^{2i+1} \subset \mathbb{CP}^{i+1}$ – see the section *Fibred knots and open books* further below in the Introduction.

with a given extension of the projection $\partial X = \partial P \longrightarrow S^1$ to a map $p : X \longrightarrow S^1$. Every n-knot (S^{n+2}, S^n, k) is homology framed, in an essentially unique manner. Homology framed knots are particularly tractable – for example, the knot complement has a canonical infinite cyclic cover, and the knot admits a codimension 1 spanning (= Seifert) surface.

Polynomials

Many knot invariants involve the Laurent polynomial extension ring $A[z, z^{-1}]$ of a ring A, in the first instance for $A = \mathbb{Z}$.

Given a homology framed knot (M, N, k) with exterior $X = \mathrm{cl.}(M \backslash N \times D^2)$ let $\overline{X} = p^*\mathbb{R}$ be the infinite cyclic cover of X classified by the map $p : X \longrightarrow S^1$

$$\begin{array}{ccc} \overline{X} & \longrightarrow & \mathbb{R} \\ \downarrow & & \downarrow \\ X & \xrightarrow{p} & S^1 \end{array}$$

and let $\zeta : \overline{X} \longrightarrow \overline{X}$ be a generating covering translation. The Laurent polynomial ring $\mathbb{Z}[z, z^{-1}]$ acts on the homology groups $H_*(\overline{X})$, with $z = \zeta_* : H_*(\overline{X}) \longrightarrow H_*(\overline{X})$, and also on the fundamental group $\pi_1(\overline{X})$. The fundamental group of X is the ζ_*-twisted extension of $\pi_1(\overline{X})$ by $\mathbb{Z}$

$$\pi_1(X) \;=\; \pi_1(\overline{X}) \times_{\zeta_*} \mathbb{Z} \;=\; \{gz^j \,|\, g \in \pi_1(\overline{X}), j \in \mathbb{Z}\} ,$$

with $\zeta_* : \pi_1(\overline{X}) \longrightarrow \pi_1(\overline{X})$ the induced automorphism and $gz = z\zeta_*(g)$. The group ring of $\pi_1(X)$ is the ζ_*-twisted Laurent polynomial extension ring of the group ring of $\pi_1(\overline{X})$

$$\mathbb{Z}[\pi_1(X)] \;=\; \mathbb{Z}[\pi_1(\overline{X})]_{\zeta_*}[z, z^{-1}] .$$

In particular, if $\zeta_* = 1 : \pi_1(\overline{X}) \longrightarrow \pi_1(\overline{X})$ then

$$\pi_1(X) \;=\; \pi_1(\overline{X}) \times \mathbb{Z} \;,\; \mathbb{Z}[\pi_1(X)] \;=\; \mathbb{Z}[\pi_1(\overline{X})][z, z^{-1}] .$$

The *Alexander polynomial* ([4]) of a classical knot $k : S^1 \subset S^3$

$$\Delta(z) \in \mathbb{Z}[z, z^{-1}]$$

is the basic polynomial invariant of the knot complement $X = S^3 \backslash k(S^1)$, with

$$\Delta(1) \;=\; 1 \;,\; \Delta(z) H_1(\overline{X}) \;=\; 0 .$$

(See Chaps. 17, 33 for the Alexander polynomials of n-knots.) The Alexander polynomial was the first application of polynomial extension rings to knot theory. Many of the abstract algebraic results in this book concern modules and quadratic forms over Laurent polynomial extensions, which are then applied to high-dimensional knots (M, N, k) by considering the algebraic topology of the complement $X = M \backslash k(N)$.

Seifert surfaces

Transversality is a key ingredient of knot theory. For example, every classical knot $k : S^1 \subset \mathbb{R}^3$ (regarding $\mathbb{R}^3$ as a subset of S^3) has plane projections, that is functions $f : \mathbb{R}^3 \longrightarrow \mathbb{R}^2$ such that $fk : S^1 \longrightarrow \mathbb{R}^2$ is an immersion with a finite number of transverse self-intersections. Plane projections are not unique. The minimum number of self-intersections in a plane projection is the *crossing number* of k. This was the first knot invariant, measuring the 'knottedness' of k.

A *Seifert* (or *spanning*) *surface* for a knot (M^{n+2}, N^n, k) is a codimension 1 submanifold $F^{n+1} \subset M^{n+2}$ with boundary $\partial F = k(N)$ and trivial normal bundle $F \times [0,1] \subset M$. Seifert surfaces can be regarded as higher-dimensional analogues of knot projections. Seifert surfaces are in fact the main geometric tool of high-dimensional knot theory, with surgery theory providing the means for transforming one Seifert surface into another.

Codimension 1 transversality guarantees the existence of Seifert surfaces for a homology framed knot (M^{n+2}, N^n, k) – make the map on the knot exterior

$$p : (\mathrm{cl.}(M \backslash (k(N) \times D^2)), k(N) \times S^1) \longrightarrow S^1$$

transverse regular at $1 \in S^1$ and set

$$(F^{n+1}, \partial F) = (p^{-1}(1), k(N)) \subset M^{n+2} .$$

Isolated instances of singular (i.e. immersed) spanning surfaces for classical knots already feature in the work of Tait [282], as in the following example of an 'autotomic' surface spanning the trefoil knot:

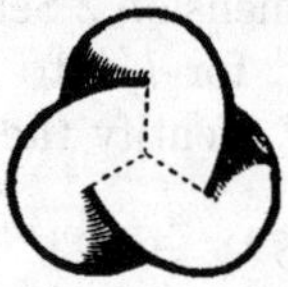

Nonsingular spanning surfaces for classical knots were first obtained by Frankl and Pontrjagin [83]. Seifert [262] obtained a spanning surface of a knot $k : S^1 \subset S^3$ from any knot projection, and defined *genus* of k to be

$$\mathrm{genus}(k) = \min\{\mathrm{genus}(F)\} ,$$

the minimum genus of a spanning surface $F^2 \subset S^3$ with

$$\mathrm{genus}(F) = \frac{1}{2}\,\mathrm{rank}\, H_1(F) \geq 0 .$$

The genus of an alternating knot was shown in [262] to be the genus of the Seifert surface determined by an alternating knot projection.

The intersection properties of $H_1(F)$ were used in [262] to construct the *Seifert matrix* of a Seifert surface F of a classical knot $k : S^1 \subset S^3$, a $2g \times 2g$ matrix V over $\mathbb{Z}$ with $g = \text{genus}(F)$. The difference between V and the transpose V^t is an invertible $2g \times 2g$ matrix $V - V^t$, so that $\det(V - V^t) = \pm 1$, with

$$V - V^t : H_1(F) \xrightarrow{\simeq} H_1(F)^* = \text{Hom}_{\mathbb{Z}}(H_1(F), \mathbb{Z}) = H^1(F)$$

the Poincaré duality isomorphism of F. The Seifert matrix determines the Alexander polynomial by

$$\Delta(z) = \pm\det(V - zV^t) \in \mathbb{Z}[z, z^{-1}] .$$

The *signature* of k

$$\sigma(k) = \text{signature}\,(H_1(F), V + V^t) \in \mathbb{Z}$$

is another classical knot invariant which can be defined using the Seifert form. The Seifert matrix is the most versatile algebraic artefact of a classical knot, although the high degree of non-uniqueness of Seifert surfaces has to be taken into account in the applications.

According to Kervaire and Weber [135] the existence of Seifert surfaces for high-dimensional n-knots 'seems to have become public knowledge during the Morse Symposium at Princeton in 1963. ... It appears in print in Kervaire [132] and Zeeman [312]'. (See also Kervaire [131, Appendix]). Seifert surfaces for arbitrary high-dimensional homology framed knots were obtained by Erle [71].

Every n-knot $k : S^n \subset S^{n+2}$ is of the form $k = \ell_x$ (constructed as in the section *Local flatness* above) for some non-locally flat knot $\ell : G^{n+1} \subset S^{n+3}$ with unique singular point $x \in G$, as follows. Let $i : \widehat{F}^{n+1} \subset D^{n+3}$ be the locally flat embedding of a codimension 2 Seifert surface obtained by pushing a Seifert surface $F^{n+1} \subset S^{n+2}$ for k into D^{n+3} relative to the boundary $\partial F = k(S^n)$. For $x = 0 \in D^{n+3}$ identify the cone $x * k(S^n)$ with D^{n+1}, and let

$$j : x * k(S^n) = D^{n+1} \longrightarrow D^{n+3}$$

be the inclusion. The union

$$\ell = i \cup j : G^{n+1} = \widehat{F}^{n+1} \cup_\partial D^{n+1} \subset D^{n+3} \cup_\partial D^{n+3} = S^{n+3}$$

defines a non-locally flat embedding with $\ell_x = k$.

The results of Levine [154], [157] show that for $n \geq 3$ an n-knot $k : S^n \subset S^{n+2}$ is unknotted (resp. null-cobordant) if and only if k admits a contractible Seifert surface in S^{n+2} (resp. D^{n+3}). One way of applying surgery to knot theory is to start with an arbitrary Seifert surface F for k, and then try to modify the codimension 1 (resp. 2) submanifold $F \subset S^{n+2}$ (resp. D^{n+3}) by surgery[5], making F as contractible as possible.

[5] See the section *Codimension q surgery* further below in the Introduction for the basic definitions.

There are two types of invariants for n-knots $k : S^n \subset S^{n+2}$: the intrinsic invariants of the infinite cyclic cover of the knot complement, and the extrinsic ones associated with the Seifert surfaces. There is a similar distinction for the invariants of arbitrary homology framed knots (M, N, k). The two types of invariants determine each other, although in practice it is not always easy to work out the details of the correspondence. One of the aims of this book is to show how algebraic K- and L-theory can be used to define both types of invariants, and also to relate them to each other.

Fibred knots and open books

A knot (M^{n+2}, N^n, k) is *fibred* if it is homology framed and the canonical projection $p : X \longrightarrow S^1$ on the exterior $X = \mathrm{cl.}(M\backslash(k(N)\times D^2))$ is (homotopic to) a fibre bundle, or equivalently if there exists a Seifert surface $F^{n+1} \subset M$ with a *monodromy* self homeomorphism $h : F \longrightarrow F$ such that :

(i) $h| = 1 : \partial F = k(N) \longrightarrow \partial F$,

(ii) $M = k(N) \times D^2 \cup_{k(N)\times S^1} T(h)$, with

$$T(h) \ = \ (F \times [0,1])/\{(x,0) \sim (h(x),1)\,|\, x \in F\}$$

the *mapping torus* of h, so that $X = T(h)$.

Following Winkelnkemper [308], M is called an *open book* with *page* F and *binding* N.[6]

For the sake of simplicity only fibre bundles over S^1 *and open books with*

$$h_* \ = \ 1 \ : \ \pi_1(F) \longrightarrow \pi_1(F)$$

will be considered, so that

$$\pi_1(M\backslash k(N)) \ = \ \pi_1(X) \ = \ \pi_1(T(h)) \ = \ \pi_1(F) \times \mathbb{Z}\ .$$

The discovery of exotic spheres by Milnor [190], the h-cobordism theorem of Smale, and the surgery classification of exotic spheres in dimensions ≥ 5 by Kervaire and Milnor [134] were powerful incentives to the extension of classical knot theory to knots (S^{n+2}, N^n, k) $(n \geq 1)$, initially for differentiable n-knots by Kervaire [131], [132]. It was proved in [131, Appendix] that an exotic n-sphere Σ^n admits a (differentiable) embedding $k : \Sigma^n \subset S^{n+2}$ if and only if Σ^n is the boundary of a parallelizable $(n+1)$-manifold (e.g. a Seifert surface $F^{n+1} \subset S^{n+2}$), in which case Σ^n represents an element of the group bP_{n+1} of [134].

[6] See the Appendix by Winkelnkemper for the history and applications of open books.

Singular points of complex hypersurfaces (the title of Milnor [196]) provided a large supply of differentiable knots (S^{n+2}, Σ^n, k) with Σ^n an exotic n-sphere: given complex variables $z_0, z_1, \ldots, z_i$ and integers $a_0, a_1, \ldots, a_i \geq 2$ there is defined a fibred knot (S^{2i+1}, N, k) with

$$k : N^{2i-1} = \Sigma(a_0, a_1, \ldots, a_i) = V(f) \cap S^{2i+1} \subset S^{2i+1}$$

where

$$\begin{aligned} f &: \mathbb{C}^{i+1} \longrightarrow \mathbb{C} \; ; \; (z_0, z_1, \ldots, z_i) \longrightarrow z_0^{a_0} + z_1^{a_1} + \ldots + z_i^{a_i} \; , \\ V(f) &= \{(z_0, z_1, \ldots, z_i) \in \mathbb{C}^{i+1} \,|\, f(z_0, z_1, \ldots, z_i) = 0\} \; , \\ S^{2i+1} &= \{(z_0, z_1, \ldots, z_i) \in \mathbb{C}^{i+1} \,|\, |z_0|^2 + |z_1|^2 + \ldots + |z_i|^2 = 1\} \; . \end{aligned}$$

The $(2i-1)$-dimensional manifold N is $(i-2)$-connected, and the fibre of $X = \mathrm{cl.}(S^{2i+1} \backslash (N \times D^2)) \longrightarrow S^1$ is an $(i-1)$-connected $2i$-dimensional Seifert surface $F^{2i} \subset S^{2i+1}$. The precise topological and differentiable nature of $N^{2i-1} \subset S^{2i+1}$ has been the subject of many investigations, ever since Brieskorn [26] identified certain $N = \Sigma(a_0, a_1, \ldots, a_i)$ as exotic spheres.

Fibred knots (M, N, k) play an especially important role in the development of high-dimensional knot theory. On the one hand, the complex hypersurface knots (S^{2i+1}, N^{2i-1}, k) are fibred, and the invariants of fibred knots have many special features, e.g. the fibre is a minimal Seifert surface, with unimodular Seifert matrix, and the extreme coefficients of the Alexander polynomials are $\pm 1 \in \mathbb{Z}$. On the other hand, surgery theory provided ways of recognizing algebraically if a high-dimensional homology framed knot (M^{n+2}, N^n, k) is fibred. In general, the fundamental group $\pi_1(\overline{X}) = \ker(p_* : \pi_1(X) \longrightarrow \mathbb{Z})$ of the infinite cyclic cover $\overline{X}$ of the exterior $X = \mathrm{cl.}(M \backslash (k(N) \times D^2))$ is infinitely generated, but for a fibred knot $\overline{X}$ is homotopy equivalent to the fibre F and $\pi_1(\overline{X}) = \pi_1(F)$ is finitely presented. For an n-knot (S^{n+2}, S^n, k)

$$H_1(X) = H_1(S^1) = \mathbb{Z} \; , \; \pi_1(\overline{X}) = [\pi_1(X), \pi_1(X)] \; .$$

Stallings [272] proved that a classical knot $k : S^1 \subset S^3$ is fibred if and only if $\pi_1(\overline{X})$ is finitely generated. Browder and Levine [31] proved that for $n \geq 4$ a homology framed knot (M^{n+2}, N^n, k) with $\pi_1(X) = \mathbb{Z}$ is fibred if and only if $H_*(\overline{X})$ is finitely generated over $\mathbb{Z}$. Farrell [78] and Siebenmann [266] generalized this result to the non-simply-connected case, using the finiteness obstruction of Wall [300] and Whitehead torsion: the *fibering obstruction* $\Phi(Y) \in Wh(\pi_1(Y))$ is defined for a finite CW complex Y with a finitely dominated infinite cyclic cover $\overline{Y}$ (a *band*), and for $n \geq 4$ a homology framed knot (M^{n+2}, N^n, k) is fibred if and only if the infinite cyclic cover $\overline{X}$ of the exterior X is finitely dominated and $\Phi(X) = 0 \in Wh(\pi_1(P))$.

Winkelnkemper [308] and Quinn [227] applied surgery theory to investigate the existence and uniqueness of open book decompositions for manifolds of dimension ≥ 6, and the closely related bordism groups of diffeomorphisms Δ_*. These results are reproved in Chap. 29 using the algebraic Poincaré complexes of Ranicki [235], [236].

Knot cobordism

The *connected sum* of n-knots $k_1, k_2 : S^n \subset S^{n+2}$ is the n-knot

$$k_1 \# k_2 \ : \ S^n \# S^n \ = \ S^n \subset S^{n+2} \# S^{n+2} \ = \ S^{n+2} \ .$$

An n-knot $k : S^n \subset S^{n+2}$ is *slice* if there exists an $(n+1)$-dimensional knot $\ell : D^{n+1} \subset D^{n+3}$ with $\ell| = k$, or equivalently if k is null-cobordant. The hypothesis of local flatness is crucial here, since the cone on $k(S^n) \subset S^{n+2}$ is a non-locally-flat null-cobordism $D^{n+1} \subset D^{n+3}$.

Fox and Milnor [82] defined the abelian group C_1 of cobordism classes of 1-knots $k : S^1 \subset S^3$, with addition by connected sum, and the trivial knot $k_0 : S^1 \subset S^3$ as the zero element. The group C_1 is countably infinitely generated. The motivation for the definition of C_1 came from the construction of the 1-knots (S^3, S^1, k_x) (already recalled in the section *Locally flat* above) from non-locally flat PL knots (M^4, N^2, κ) $(x \in N)$. It was proved in [82] that the connected sum of 1-knots $k_1, k_2, \ldots, k_r : S^1 \subset S^3$ is slice if and only if there exists a non-locally flat 2-knot $\kappa : S^2 \subset S^4$ with $k_i = k_{x_i}$ $(1 \leq i \leq r)$ the 1-knots defined at the points $x_1, x_2, \ldots, x_r \in S^2$ where κ is not locally flat.

Kervaire [132] defined cobordism for n-knots $k : S^n \subset S^{n+2}$ for all $n \geq 1$. The group of cobordism classes of n-knots with addition by connected sum is denoted by C_n. An n-knot k is such that $[k] = 0 \in C_n$ if and only if k is slice. If $\ell : N^{n+1} \subset M^{n+3}$ is a non-locally flat embedding with a single non-locally flat point $x \in N$ there is defined a (locally flat) n-knot $k = k_x : S^n \subset S^{n+2}$ as before. The n-knot k is slice if and only if the singularity of ℓ at x can be 'resolved', with ℓ replaced near x by a locally flat embedding.

The algebraic determination of the knot cobordism groups C_* was a major preoccupation of high-dimensional knot theorists in the 1960s and 1970s. The algebraic structure of C_n for $n \geq 3$ was worked out in [132], Levine [157], [158] and Stoltzfus [277], with

$$C_n \ = \ C_{n+4} \ , \ \ C_{2i} \ = \ 0$$

and both C_{4*} and C_{4*+2} are countably infinitely generated of the type

$$\bigoplus_{\infty} \mathbb{Z} \oplus \bigoplus_{\infty} \mathbb{Z}_2 \oplus \bigoplus_{\infty} \mathbb{Z}_4 \ .$$

(See Chap. 42 for the structure of C_{2i+1} for $i \geq 1$). The classical knot cobordism group C_1 is still fairly mysterious, with the kernel of the natural surjection $C_1 \longrightarrow C_{4j+1}$ known to be non-trivial, by virtue of the invariants of Casson and Gordon [48].

Simple knots

An n-knot $k : S^n \subset S^{n+2}$ is *simple* if it satisfies one of the following equivalent conditions:

(i) the knot complement $X = S^{n+2} \backslash k(S^n)$ is such that

$$\pi_r(X) = \pi_r(S^1) \text{ for } 1 \leq r \leq (n-1)/2 ,$$

(ii) k admits a Seifert surface $F^{n+1} \subset S^{n+2}$ such that

$$H_r(F) = 0 \text{ for } 1 \leq r \leq (n-1)/2 .$$

Every classical knot $k : S^1 \subset S^3$ is simple. For $n \geq 2$ every n-knot $k : S^n \subset S^{n+2}$ is cobordant to a simple n-knot. The $(2i-1)$-knots $k : S^{2i-1} \subset S^{2i+1}$ constructed by Milnor [196] from singular points of complex hypersurfaces are simple and fibred, with $(i-1)$-connected fibre F^{2i}.

Seifert surfaces and simply-connected surgery theory were used by Kervaire [132] to characterize the homotopy groups $\pi_*(X)$ of the complements $X = S^{n+2} \backslash k(S^n)$ of high-dimensional n-knots $k : S^n \subset S^{n+2}$ (see also Wall [302, p. 18]), and to prove that C_n is isomorphic to the cobordism group of simple n-knots, with $C_{2i} = 0$ $(i \geq 2)$. Levine [156] obtained polynomial invariants of an n-knot k from the homology $H_*(\overline{X})$ of the infinite cyclic cover $\overline{X}$ of X, generalizing the Alexander polynomial. The high-dimensional knot polynomials were used in [156] to characterize the rational homology groups $H_*(\overline{X};\mathbb{Q})$ as $\mathbb{Q}[z,z^{-1}]$-modules, and (working in the differentiable category) were related to the exotic differentiable structures on spheres. Much work was done in the mid-1960s and early 1970s on the isotopy classification of simple n-knots $S^n \subset S^{n+2}$, using the Alexander polynomials, the Seifert matrix and the Blanchfield linking form – see Chaps. 33, 42 for references.

Surgery theory

The Browder–Novikov–Sullivan–Wall surgery theory developed in the 1960s brought a new methodology to high-dimensional knot theory, initially for n-knots $S^n \subset S^{n+2}$ and then for arbitrary codimension 2 embeddings $N^n \subset M^{n+2}$. Surgery theory was then extended to 4-dimensional manifolds with certain fundamental groups by Freedman and Quinn [86], but low-dimensional manifolds have so many distinctive features that the high-dimensional theory is too limited in dimensions 3 and 4 – accordingly, n-dimensional knots $N^n \subset M^{n+2}$ are much harder to classify for $n = 1, 2$ than $n \geq 3$.

It will be assumed that the reader is already familiar with the basics of surgery, at least in the simply-connected case considered by Browder [30].

The basic *surgery* operation on manifolds starts with an n-dimensional manifold N^n and an embedding

$$S^r \times D^{n-r} \subset N .$$

The effect of the surgery is the n-dimensional manifold

$$N'^n = (N \backslash S^r \times D^{n-r}) \cup D^{r+1} \times S^{n-r-1}$$

which is related to N by an elementary cobordism $(W; N, N')$, with

$$W = N \times [0,1] \cup D^{r+1} \times D^{n-r} .$$

Conversely, every cobordism of manifolds is a union of elementary cobordisms (Milnor [192]) – for a closed manifold this is just a handle decomposition. There is a similar decomposition for knot cobordisms.

An *n-dimensional geometric Poincaré complex* X is a finite CW complex with n-dimensional Poincaré duality $H^*(X) \cong H_{n-*}(X)$. The fundamental problem of surgery is to decide if such an X is homotopy equivalent to a compact n-dimensional manifold. The traditional method is to break down the problem into two stages. In the first stage there is a topological K-theory obstruction to the existence of a *normal map*[7] $(f, b) : M \longrightarrow X$, that is a degree 1 map $f : M \longrightarrow X$ from a compact n-dimensional manifold M together with a map $b : \nu_M \longrightarrow \eta$ a map from the stable normal bundle of M to some bundle over X. In the second stage there is an algebraic L-theory obstruction, the *surgery obstruction* of Wall [302]

$$\sigma_*(f, b) \in L_n(\mathbb{Z}[\pi_1(X)])$$

such that $\sigma_*(f, b) = 0$ if (and for $n \geq 5$ only if) $(f, b) : M \longrightarrow X$ can be modified by surgeries on M to a normal bordant homotopy equivalence.[8] The *algebraic L-groups* $L_*(A)$ of [302] are defined for any ring with involution A, and are 4-periodic, $L_n(A) = L_{n+4}(A)$. By construction, $L_{2i}(A)$ is the Witt group of nonsingular $(-)^i$-quadratic forms over A, and $L_{2i+1}(A)$ is a group of automorphisms of nonsingular $(-)^i$-quadratic forms over A.

---

[7] See the section *Algebraic K- and L-theory invariants of knots* further below in the Introduction for the basic constructions of normal maps from n-knots.

[8] See Ranicki [245] for a more streamlined approach, in which the two stages are united in the *total surgery obstruction*.

An n-dimensional *quadratic Poincaré complex* over A is an A-module chain complexes C with a quadratic Poincaré duality, $H^{n-*}(C) \cong H_*(C)$. The algebraic L-groups $L_n(A)$ are the cobordism groups of quadratic Poincaré complexes over A, by the algebraic theory of surgery of Ranicki [235], [236], [237]. The *quadratic kernel* of an n-dimensional normal map $(f,b) : M \longrightarrow X$ is an n-dimensional quadratic Poincaré complex (C,ψ) with $C = \mathcal{C}(f^!)$ the algebraic mapping cone of the Umkehr $\mathbb{Z}[\pi_1(X)]$-module chain map

$$f^! \; : \; C(\widetilde{X}) \;\simeq\; C(\widetilde{X})^{n-*} \xrightarrow{\widetilde{f}^*} C(\widetilde{M})^{n-*} \;\simeq\; C(\widetilde{M})$$

with $\widetilde{X}$ the universal cover of X and $\widetilde{M} = f^*\widetilde{X}$ the pullback cover of M, and with a $\mathbb{Z}[\pi_1(X)]$-module chain equivalence

$$C(\widetilde{M}) \;\simeq\; \mathcal{C}(f^!) \oplus C(\widetilde{X}) \; .$$

The surgery obstruction of (f,b) is the quadratic Poincaré cobordism class

$$\sigma_*(f,b) \;=\; (\mathcal{C}(f^!),\psi) \in L_n(\mathbb{Z}[\pi_1(X)]) \; .$$

There is also the notion of geometric Poincaré pair $(X,\partial X)$, with a rel ∂ surgery obstruction $\sigma_*(f,b) \in L_n(\mathbb{Z}[\pi_1(X)])$ for a normal map $(f,b) : (M,\partial M) \longrightarrow (X,\partial X)$ from a manifold with boundary and with $\partial f = f| : \partial M \longrightarrow \partial X$ a homotopy equivalence.

Algebraic transversality

The main technique used in the book is *algebraic transversality*, an analogue of the geometric transversality construction

$$F^{n+1} \;=\; p^{-1}(1) \subset M^{n+2}$$

of a Seifert surface of a homology framed knot (M^{n+2}, N^n, k) from a map $p : X \longrightarrow S^1$ on the knot exterior $X = \mathrm{cl.}(M \backslash (k(N) \times D^2))$ which is transverse regular at $1 \in S^1$ – cutting X along F results in a fundamental domain $(X_F; F, zF)$ for the infinite cyclic cover $\overline{X} = p^*\mathbb{R}$ of X. The technique extracts finitely generated A-module data from finitely generated $A[z,z^{-1}]$-module data, with $A[z,z^{-1}]$ the Laurent polynomial extension of a ring A.

Algebraic transversality is a direct descendant of the linearization trick of Higman [111] for converting a matrix in $A[z,z^{-1}]$ by stabilization and elementary transformations to a matrix with linear entries $a_0 + a_1 z$ $(a_0, a_1 \in A)$. As explained by Waldhausen [299] and Ranicki [244] linearization corresponds to the geometric transversality construction of a fundamental domain for an infinite cyclic cover of a compact manifold. The algebraic transversality methods used to prove the theorem of Bass, Heller and Swan [14] on the Whitehead group of a polynomial extension

$$Wh(\pi \times \mathbb{Z}) = Wh(\pi) \oplus \widetilde{K}_0(\mathbb{Z}[\pi]) \oplus \widetilde{\mathrm{Nil}}_0(\mathbb{Z}[\pi]) \oplus \widetilde{\mathrm{Nil}}_0(\mathbb{Z}[\pi])$$

and its algebraic L-theory analogues are used in this book to define and relate the invariants of high-dimensional knots associated with both complements and Seifert surfaces.

The splitting theorem for the Wall surgery obstruction groups

$$\begin{aligned} L^s_n(\mathbb{Z}[\pi \times \mathbb{Z}]) &= L^s_n(\mathbb{Z}[\pi]) \oplus L^h_{n-1}(\mathbb{Z}[\pi]) \,, \\ L^h_n(\mathbb{Z}[\pi \times \mathbb{Z}]) &= L^h_n(\mathbb{Z}[\pi]) \oplus L^p_{n-1}(\mathbb{Z}[\pi]) \end{aligned}$$

was obtained by Shaneson [263], Novikov [218] and Ranicki [231], [244], with L^p_* (resp. L^h_*, L^s_*) denoting the quadratic L-groups defined using f.g. projective (resp. f.g. free, based f.g. free) modules. The theorem was motivated by the codimension 1 splitting results of Farrell and Hsiang [80] and Wall [302, 12.6]. The connections between $Wh(\pi \times \mathbb{Z})$ and $\widetilde{K}_0(\mathbb{Z}[\pi])$ in algebraic K-theory and between $L_*(\mathbb{Z}[\pi \times \mathbb{Z}])$ and $L_{*-1}(\mathbb{Z}[\pi])$ in algebraic L-theory are algebraic analogues of the codimension 1 transversality construction of Seifert surfaces for knots.

The topological invariance of rational Pontrjagin classes

Algebraic transversality in L-theory may be regarded as a spinoff from Novikov's proof of the topological invariance of the rational Pontrjagin classes, for which he was awarded the Fields Medal in 1970. The report of Atiyah [10] included:

Undoubtedly the most important single result of Novikov, and one which combines in a remarkable degree both algebraic and geometric methods, is his famous proof of the topological invariance of the (rational) Pontrjagin classes of a differentiable manifold. ...
Perhaps you will understand Novikov's result more easily if I mention a purely geometrical theorem (not involving Pontrjagin classes) which lies at the heart of Novikov's proof. This is as follows:

THEOREM *(formulation due to* L. SIEBENMANN*) If a differentiable manifold X is homeomorphic to a product $M \times \mathbb{R}^n$ (where M is compact, differentiable, simply-connected and has dimension ≥ 5) then X is diffeomorphic to a product $M' \times \mathbb{R}^n$.*

... As is well-known many topological problems are very much easier if one is dealing with simply-connected spaces. Topologists are very happy when

they can get rid of the fundamental group and its algebraic complications. Not so Novikov! Although the theorem above involves only simply-connected spaces, a key step in his proof consists in perversely introducing a fundamental group, rather in the way that (on a much more elementary level) puncturing the plane makes it non-simply-connected. This bold move has the effect of simplifying the geometry at the expense of complicating the algebra, but the complication is just manageable and the trick works beautifully. It is a real master stroke and completely unprecedented.

Novikov's proof of the topological invariance of the rational Pontrjagin classes was published in [217]. (See Ranicki [247, Chap. 4] for an account of some other proofs of the topological invariance of the rational Pontrjagin classes.) The theorem had a tremendous influence on the subsequent development of high-dimensional manifold theory, such as the disproof of the manifold Hauptvermutung by Casson and Sullivan [249], the Kirby–Siebenmann structure theory for high-dimensional topological manifolds and the Chapman–Ferry–Quinn theory of controlled topology, as well as high-dimensional knot theory.[9] Moreover, Novikov himself contributed to further progress, notably the paper [218] already mentioned in the preface, which included the definitive formulation of the 'Novikov conjecture' (reprinted and translated in [81]). The S^1-valued Morse theory of Novikov [220] was yet another contribution, which initiated the topological applications of the 'Novikov rings' of the Laurent polynomial extension $A[z,z^{-1}]$ of a ring A

$$A((z)) = A[[z]][z^{-1}] \ , \ A((z^{-1})) = A[[z^{-1}]][z] \ .$$

The Novikov rings also play a role in the algebraic treatment of high-dimensional knot theory, since a high-dimensional knot is fibred precisely when the knot exterior is acyclic with coefficients in the Novikov rings $\mathbb{Z}((z)),\mathbb{Z}((z^{-1}))$.

Localization

The algebraic properties of high-dimensional knots are best understood in terms of the algebraic K- and L-theory of various localizations $\Sigma^{-1}A[z,z^{-1}]$ of Laurent polynomial extensions $A[z,z^{-1}]$. Traditionally, the localization of a ring A inverting a multiplicative subset $S\subset A$ of central non-zero divisors is the ring $S^{-1}A$ of fractions a/s $(a\in A, s\in S)$ with $a/s=b/t$ if and only if $at=bs\in A$.

The algebraic K-theory localization exact sequence of Bass [13]

$$\dots \longrightarrow K_1(A) \longrightarrow K_1(S^{-1}A) \longrightarrow K_1(A,S) \longrightarrow K_0(A) \longrightarrow \dots$$

[9] For example, Novikov [217] proved that for $n\geq 5$ every n-knot $k: S^n\subset S^{n+2}$ is equivalent to a differentiable embedding.

identifies the relative K-group $K_1(A \longrightarrow S^{-1}A)$ of the inclusion $A \longrightarrow S^{-1}A$ with the class group of the exact category of S-torsion A-modules of homological dimension 1. For $A = \mathbb{Z}$, $S = \mathbb{Z}\backslash\{0\}$ the localization is $S^{-1}A = \mathbb{Q}$, and the relative K-group is the class group of the exact category of finite abelian groups

$$K_1(\mathbb{Z}, S) = \mathbb{Q}^\bullet/\{\pm 1\} = \bigoplus_{p \text{ prime}} K_0(\mathbb{Z}_p) = \bigoplus_{p \text{ prime}} \mathbb{Z}$$

detected by the exponents of the primary cyclic groups.

The algebraic L-theory localization exact sequence of Ranicki [237]

$$\dots \longrightarrow L_n(A) \longrightarrow L_n(S^{-1}A) \longrightarrow L_n(A,S) \longrightarrow L_{n-1}(A) \longrightarrow \dots$$

identifies the relative L-group $L_n(A \longrightarrow S^{-1}A)$ of the inclusion $A \longrightarrow S^{-1}A$ with the cobordism group $L_n(A,S)$ of $(n-1)$-dimensional quadratic Poincaré complexes over A which are $S^{-1}A$-contractible. For $A = \mathbb{Z}$, $S = \mathbb{Z}\backslash\{0\}$, $S^{-1}A = \mathbb{Q}$, $n = 0$ the relative L-group is the Witt group of linking forms on finite abelian groups, with

$$L_0(\mathbb{Z}, S) = \mathbb{Z}_2 \oplus \mathbb{Z}_8 \oplus \bigoplus_{p \neq 2 \text{ prime}} L_0(\mathbb{Z}_p) = \bigoplus_\infty \mathbb{Z}_2 \oplus \bigoplus_\infty \mathbb{Z}_4 \oplus \mathbb{Z}_8$$

detected by the Hasse-Minkowski invariants and the signature mod 8.

The localization of the Laurent polynomial extension ring $\mathbb{Z}[z,z^{-1}]$ inverting the multiplicative subset

$$P = \{p(z)\,|\,p(1) = \pm 1\} \subset \mathbb{Z}[z,z^{-1}]$$

is particularly significant in the theory of spherical knots $k : S^n \subset S^{n+2}$. The localization $P^{-1}\mathbb{Z}[z,z^{-1}]$ has the universal property that a finite f.g. free $\mathbb{Z}[z,z^{-1}]$-module chain complex C is $P^{-1}\mathbb{Z}[z,z^{-1}]$-contractible if and only if C is $\mathbb{Z}$-contractible via the augmentation

$$\mathbb{Z}[z,z^{-1}] \longrightarrow \mathbb{Z}\ ;\ z \longrightarrow 1\,.$$

The groups in the algebraic K-theory localization exact sequence

$$\begin{aligned}\dots \longrightarrow K_1(\mathbb{Z}[z,z^{-1}]) &\longrightarrow K_1(P^{-1}\mathbb{Z}[z,z^{-1}]) \\ &\longrightarrow K_1(\mathbb{Z}[z,z^{-1}],P) \xrightarrow{0} K_0(\mathbb{Z}[z,z^{-1}]) \longrightarrow \dots\end{aligned}$$

can be identified with multiplicative groups of units

$$\begin{aligned}K_1(\mathbb{Z}[z,z^{-1}]) &= \mathbb{Z}[z,z^{-1}]^\bullet = \{\pm z^j\,|\,j \in \mathbb{Z}\}\,,\\ K_1(P^{-1}\mathbb{Z}[z,z^{-1}]) &= (P^{-1}\mathbb{Z}[z,z^{-1}])^\bullet = \{\frac{p(z)}{q(z)}\,|\,p(z),q(z) \in P\}\,,\\ K_1(\mathbb{Z}[z,z^{-1}],P) &= (P^{-1}\mathbb{Z}[z,z^{-1}])^\bullet/\mathbb{Z}[z,z^{-1}]^\bullet\end{aligned}$$

The torsion group $L_n(\mathbb{Z}[z,z^{-1}],P)$ in the algebraic L-theory localization exact sequence

$$\dots \longrightarrow L_n(\mathbb{Z}[z,z^{-1}]) \longrightarrow L_n(P^{-1}\mathbb{Z}[z,z^{-1}]) \\ \longrightarrow L_n(\mathbb{Z}[z,z^{-1}],P) \longrightarrow L_{n-1}(\mathbb{Z}[z,z^{-1}]) \longrightarrow \dots$$

is the cobordism group of $\mathbb{Z}$-contractible $(n-1)$-dimensional quadratic Poincaré complexes over $\mathbb{Z}[z,z^{-1}]$, with involution $\overline{z}=z^{-1}$.

Algebraic *K*- and *L*-theory invariants of knots

The homological invariants of high-dimensional n-knots $k:S^n\subset S^{n+2}$ can be defined using the methods of algebraic K- and L-theory.

Let $k:S^n\subset S^{n+2}$ be an n-knot with exterior

$$(X,\partial X) \;=\; (\mathrm{cl.}(S^{n+2}\backslash(k(S^n)\times D^2)),k(S^n)\times S^1)\ .$$

Let $\overline{X}$ be the infinite cyclic cover of X induced from the universal cover $\mathbb{R}$ of S^1 by pullback along the canonical homology equivalence $p:X\longrightarrow S^1$, with $\mathbb{Z}$-equivariant lift $\overline{p}:\overline{X}\longrightarrow\mathbb{R}$. The kernel $\mathbb{Z}[z,z^{-1}]$-module chain complex

$$C \;=\; \mathcal{C}(\overline{p}:\overline{X}\longrightarrow\mathbb{R})_{*+1}$$

is $\mathbb{Z}$-contractible, with homology finitely generated P-torsion $\mathbb{Z}[z,z^{-1}]$-module

$$H_r(C) \;=\; H_r(\overline{X})\ \ (1\leq r\leq n)\ .$$

The Alexander polynomials $\Delta_r(z)\in P$ of $k:S^n\subset S^{n+2}$ are such that

$$\Delta_r(z)H_r(\overline{X}) \;=\; 0\ \ (1\leq r\leq n)\ .$$

The P-torsion class

$$[H_r(\overline{X})] \;=\; \Delta_r(z)\in K_1(\mathbb{Z}[z,z^{-1}],P) \;=\; (P^{-1}\mathbb{Z}[z,z^{-1}])^\bullet$$

determines $H_r(\overline{X})$ up to extensions. The P-torsion Euler characteristic

$$\chi_P(C) \;=\; \sum_{r=1}^{n}(-)^r[H_r(\overline{X})] \;=\; \prod_{r=1}^{n}\Delta_r(z)^{(-)^r}\in K_1(\mathbb{Z}[z,z^{-1}],P)$$

is the Reidemeister torsion of the knot k. The algebraic K-theoretic interpretation of Reidemeister torsion is due to Milnor [194]. The $\mathbb{Z}[z,z^{-1}]$-coefficient Poincaré duality chain equivalence

$$\phi \;=\; [X]\cap - \;:\; C^{n+2-*} \;=\; \mathrm{Hom}_{\mathbb{Z}[z,z^{-1}]}(C,\mathbb{Z}[z,z^{-1}])_{n+2-*} \xrightarrow{\simeq} C$$

induces the Blanchfield linking pairings

$$H_r(\overline{X}) \times H_{n+1-r}(\overline{X}) \longrightarrow P^{-1}\mathbb{Z}[z,z^{-1}]/\mathbb{Z}[z,z^{-1}] .$$

The pairings are determined by the non-simply connected surgery quadratic kernel of a degree 1 normal map from an $(n+2)$-dimensional manifold with boundary to an $(n+2)$-dimensional geometric Poincaré pair

$$(g,c) \; : \; (X,\partial X) \longrightarrow (D^{n+3},k(S^n)) \times S^1$$

with $g : X \longrightarrow D^{n+3} \times S^1$ a homology equivalence and $\partial g : \partial X \longrightarrow k(S^n) \times S^1$ the identity. For $n \geq 3$ the knot k is unknotted if and only if (g,c) is a homotopy equivalence, and k is null-cobordant if and only if (g,c) is normal bordant by a homology equivalence to a homotopy equivalence.

The high-dimensional knot cobordism groups C_{2*-1} were expressed by Levine [157] as the Witt groups of Seifert matrices in $\mathbb{Z}$, and by Cappell and Shaneson [40] as certain types of algebraic Γ-groups. Pardon [221], Smith [270] and Ranicki [237, Chap. 7.9] expressed the knot cobordism groups as torsion L-groups. The cobordism class of an n-knot $k : S^n \subset S^{n+2}$ is the cobordism class of its Blanchfield complex

$$[k] \;=\; (C,\phi) \in C_n \;=\; L_{n+3}(\mathbb{Z}[z,z^{-1}],P) \;\; (n \geq 3) .$$

(There is no difference between symmetric and quadratic Poincaré structures in this case.)

Given a Seifert surface $F^{n+1} \subset S^{n+2}$ for an n-knot $k : S^n \subset S^{n+2}$ the inclusion defines a degree 1 normal map

$$(f,b) \; : \; (F,\partial F) \longrightarrow (D^{n+3},k(S^n))$$

from an $(n+1)$-dimensional manifold with boundary to an $(n+1)$-dimensional geometric Poincaré pair, with $\partial f = 1 : \partial F \longrightarrow k(S^n)$. For $n = 2i-1$ the Seifert matrix V of F defines a bilinear form on $H_i(F)$ which is a refinement of the simply-connected surgery quadratic kernel of (f,b), and the surgery obstruction of (f,b) is given by

$$\sigma_*(f,b) \;=\; \begin{cases} \frac{1}{8}\,\text{signature}(H_i(F),V+V^t) \\ \text{Arf invariant}(H_i(F),V) \end{cases}$$

$$\in L_{2i}(\mathbb{Z}) = \begin{cases} \mathbb{Z} & \text{if } i \equiv 0 (\text{mod } 2) \\ \mathbb{Z}_2 & \text{if } i \equiv 1 (\text{mod } 2) . \end{cases}$$

For odd i the theorem of Levine [156] expresses the Arf invariant in terms of the Alexander polynomial $\Delta(z) = \det(V - zV^t)$, with

$$\text{Arf invariant}(H_i(F),V) \;=\; \begin{cases} 0 & \text{if } \Delta(-1) \equiv \pm 1 (\text{mod } 8) \\ 1 & \text{if } \Delta(-1) \equiv \pm 3 (\text{mod } 8) . \end{cases}$$

The surgery obstruction $\sigma_*(f,b) \in L_{2i}(\mathbb{Z})$ is detected by just the signature and the Arf invariant. The knot cobordism class $[k] \in C_{2i-1}$ is a refinement of the surgery obstruction – it is the Witt class of any Seifert matrix for k, and is determined by an infinite number of invariants.

Codimension q surgery

The surgery theoretic technique which most directly applies to the classification of high-dimensional knots is the splitting obstruction theory for submanifolds of codimension 2. This is best understood in the general context of codimension q submanifolds for arbitrary $q \geq 2$.

If $N^n \subset M^m$ is a submanifold of codimension $q = m - n$ and the embedding $S^r \times D^{n-r} \subset N$ extends to an embedding $D^{r+1} \times D^{n-r} \subset M$ the effect of the *codimension q* (or *ambient*) *surgery* on N is a submanifold

$$N'^n = (N \backslash S^r \times D^{n-r}) \cup D^{r+1} \times S^{n-r-1} \subset M^m$$

with an elementary codimension q subcobordism

$$(W; N, N') \subset M \times ([0,1]; \{0\}, \{1\}) .$$

Conversely, every codimension q subcobordism is a union of elementary subcobordisms.

Given a homotopy equivalence $h : M' \longrightarrow M$ of m-dimensional manifolds it is possible to make h transverse regular at any submanifold $N \subset M$, so that the restriction $(f, b) = h| : N' = h^{-1}(N) \longrightarrow N$ is a degree 1 normal map of n-dimensional manifolds. The homotopy equivalence h *splits along* $N \subset M$ if it is homotopic to a map (also denoted by h) such that the restrictions $f = h| : N' \longrightarrow N$, $h| : M' \backslash N' \longrightarrow M \backslash N$ are also homotopy equivalences. Wall [302, Chap. 11] defined the LS-groups $LS_*(\Phi)$ to fit into the exact sequence

$$\begin{aligned} \dots \longrightarrow L_{m+1}(\mathbb{Z}[\pi_1(M \backslash N)] \longrightarrow \mathbb{Z}[\pi_1(M)]) \longrightarrow LS_n(\Phi) \\ \longrightarrow L_n(\mathbb{Z}[\pi_1(N)]) \longrightarrow L_m(\mathbb{Z}[\pi_1(M \backslash N)] \longrightarrow \mathbb{Z}[\pi_1(M)]) \longrightarrow \dots \end{aligned}$$

depending only on the system Φ of fundamental groups of M, N and $M \backslash N$. The *splitting obstruction* $s(h) \in LS_n(\Phi)$ of [302, Chap. 12] is such that $s(h) = 0$ if (and for $n \geq 5$ only if) h splits, i.e. if $N' \subset M'$ can be modified by codimension q surgeries until h splits. The splitting obstruction $s(h)$ has image the surgery obstruction $\sigma_*(f, b) \in L_n(\mathbb{Z}[\pi_1(N)])$. If $m - n \geq 3$ then

$$\pi_1(M \backslash N) = \pi_1(M) \ , \ s(h) = \sigma_*(f, b) \in LS_n(\Phi) = L_n(\mathbb{Z}[\pi_1(N)]) .$$

The Browder–Casson–Sullivan–Wall theorem states that for $m - n \geq 3$, $n \geq 5$ a homotopy equivalence $h : M' \longrightarrow M$ splits along $N^n \subset M^m$ if and only if $\sigma_*(f, b) = 0 \in L_n(\mathbb{Z}[\pi_1(N)])$ – again, knotting only starts in codimension $m - n = 2$.

In surgery theory one fixes the homotopy type of a space with Poincaré duality, and then decides if this contains a topological manifold. In the applications of the splitting obstruction theory to codimension 2 embeddings the homology type of the complement is fixed, and the knot is trivial if and only if it contains the homotopy type of the complement of a standard embedding, which is decided by homology surgery theory. This point of view was initiated by López de Medrano [170] in connection with the study of n-knots $k : S^n \subset S^{n+2}$ which are invariant under a fixed point free involution on S^{n+2}, generalizing the work of Browder and Livesay [32].

Cappell and Shaneson ([39]–[47] etc.) extended the obstruction theory of Wall [302] to a homology surgery theory, and used it to obtain many results for codimension 2 embeddings. The homology surgery obstruction groups are the algebraic Γ-groups $\Gamma_*(\mathcal{F})$, the generalizations of the algebraic L-groups defined for any morphism of rings with involution $\mathcal{F} : A \longrightarrow B$. The group $\Gamma_{2i}(\mathcal{F})$ is the Witt group of B-nonsingular $(-)^i$-quadratic forms over A, and $\Gamma_{2i+1}(\mathcal{F}) \subseteq L_{2i+1}(B)$ for surjective $\mathcal{F}$. The high-dimensional knot cobordism groups C_* are particular types of Γ-groups. See Levine and Orr [161] for a survey of the applications of surgery theory to knots (and links).

The surgery treatment of high-dimensional knot theory in Part Two combines the codimension 2 surgery methods of Cappell and Shaneson [40], Freedman [84] and Matsumoto [184] with the author's algebraic methods. In particular, the LS-groups $LS_*(\Phi)$ are generalized in Chap. 22 to the codimension 2 homology splitting obstruction groups $\Gamma S_*(\Phi)$.

Fredholm localization

In order to obtain algebraic expressions for the cobordism groups of high-dimensional knots $N^n \subset M^{n+2}$ more general than the n-knots $S^n \subset S^{n+2}$ it is necessary to work with the algebraic K- and L-theory of the less familiar *noncommutative localization* $\Sigma^{-1}\Lambda$ of Cohn [53], which is defined for any set Σ of square matrices in a ring Λ (see Chap. 9 for the definition). In the applications

$$\Lambda \;=\; \mathbb{Z}[\pi_1(M\backslash N)] \;\;,\;\; H_*(M\backslash N;\Sigma^{-1}\Lambda) \;=\; 0\;.$$

The noncommutative localization $\Omega^{-1}A[z,z^{-1}]$ appropriate to open books is the *Fredholm localization* inverting the set Ω of square matrices ω in a Laurent polynomial extension $A[z,z^{-1}]$ such that coker(ω) is a f.g. projective A-module. The Fredholm localization has the universal property that a finite f.g. free $A[z,z^{-1}]$-module chain complex C is A-module chain equivalent to a finite f.g. projective A-module chain complex if and only if the induced finite f.g. free $\Omega^{-1}A[z,z^{-1}]$-module chain complex $\Omega^{-1}C$ is acyclic, $H_*(\Omega^{-1}C) = 0$. The Fredholm localization $\Omega^{-1}A[z,z^{-1}]$ is closely related to the Novikov rings $A((z))$, $A((z^{-1}))$, with $H_*(\Omega^{-1}C) = 0$ if and only if

$$H_*(A((z))\otimes_{A[z,z^{-1}]} C) \;=\; H_*(A((z^{-1}))\otimes_{A[z,z^{-1}]} C) \;=\; 0\;.$$

The obstruction theory of Quinn [227] for the existence and uniqueness of open book decompositions of high-dimensional manifolds will be given a chain complex formulation in Chap. 28. The *asymmetric signature* of an m-dimensional manifold M is the cobordism class of the $\mathbb{Z}[\pi_1(M)]$-module chain complex $C(\widetilde{M})$ of the universal cover $\widetilde{M}$ and its Poincaré duality $\phi_M : C(\widetilde{M})^{m-*} \simeq C(\widetilde{M})$, regarded as an asymmetric Poincaré complex generalizing the Seifert form of a knot. For even m the asymmetric signature takes value in a group $LAsy^{2*}(\mathbb{Z}[\pi_1(M)])$ which is infinitely generated, with

$$LAsy^{2*}(\mathbb{Z}) = \bigoplus_{\infty}\mathbb{Z} \oplus \bigoplus_{\infty}\mathbb{Z}_2 \oplus \bigoplus_{\infty}\mathbb{Z}_4 .$$

(In the simply-connected case $\pi_1(M) = \{1\}$ the invariant is just the signature of M, the original open book obstruction of Winkelnkemper [308]). The asymmetric signature is 0 for odd m. In Chap. 29 the asymmetric signature will be identified with a generalization of the Blanchfield form of a knot

$$\sigma^*(M;\Omega) = (\Omega^{-1}C(\widetilde{M})[z,z^{-1}],(1-z)\phi_M) \in L_m(\Omega^{-1}\mathbb{Z}[\pi_1(M)][z,z^{-1}]) .$$

The asymmetric signature is such that $\sigma^*(M;\Omega) = 0$ if (and for $m \geq 6$ only if) M admits an open book decomposition. For any knot (M,N,k)

$$\sigma^*(M;\Omega) = (\Omega^{-1}C(\widetilde{X}),(1-z)\phi_X) \in L_m(\Omega^{-1}\mathbb{Z}[\pi_1(M)][z,z^{-1}])$$

with $X = \text{cl.}(M\backslash(k(N) \times D^2))$ the exterior of k, and $\widetilde{X}$ the cover of X induced from $\widetilde{M} \times \mathbb{R}$ by a map $X \longrightarrow M \times S^1$. By definition, M has an open book decomposition with binding N whenever X fibres over S^1. For $m \geq 6$ this is the case if and only if $\pi_1(X) = \pi_1(M) \times \mathbb{Z}$, $H_*(\Omega^{-1}C(\widetilde{X})) = 0$ (i.e. the infinite cyclic cover $\overline{X}$ of X is finitely dominated) and the Farrell–Siebenmann fibering obstruction is $\Phi(X) = 0 \in Wh(\pi_1(X))$. The asymmetric signature is the obstruction to improving the empty knot $(M,\emptyset,k_\emptyset)$ with $p = \text{constant} : X = M \longrightarrow S^1$ by codimension 2 surgeries to a fibred knot (M,N,k), i.e. to M having an open book decomposition.

Structure

The book has two parts and an appendix. Part One deals with the algebraic K-theory aspects of high-dimensional knot theory, with invariants such as the Alexander polynomials. Part Two deals with the algebraic L-theory aspects of high-dimensional knot theory, with invariants such as the Seifert matrix, the Blanchfield pairing, the multisignature and the coupling invariants. The appendix (by Elmar Winkelnkemper) is an account of the history and applications of open books, a field where surgery and knot theory meet in a particularly fruitful way.

I am grateful to Desmond Sheiham for reading a preliminary version of the book, and making valuable suggestions for improvements.

Errata (if any) will be posted on the WWW home page

http://www.maths.ed.ac.uk/~aar

Part One

Algebraic *K*-theory

1. Finite structures

At first sight, Chap. 1 has little to do with high-dimensional knot theory! However, the homological methods used in knot theory frequently involve the finiteness and torsion properties of chain complexes over polynomial extension rings, such as the chain complex of the infinite cyclic cover of a knot complement. In particular, the algebraic L-theory treatment of high-dimensional knot theory developed in Part Two is based on the algebraic K-theory of chain complexes with Poincaré duality.

Chap. 1 brings together the essential definitions from the algebraic theory of finite structures on chain complexes, and the applications to CW complexes. Milnor [194], [199], Bass [13], Cohen [52] and Rosenberg [254] are standard references for algebraic K-theory and the applications to topology. See Ranicki [238], [239], [241] for a fuller account of the K_0- and K_1-groups in terms of chain complexes.

1A. The Wall finiteness obstruction

The Wall finiteness obstruction is an algebraic K-theory invariant which decides if a 'finitely dominated' infinite complex is homotopy equivalent to a finite complex, where complex is understood to be a chain complex in algebra and a CW complex in topology.

Let A be an associative ring with 1. Unless otherwise specified, A-modules are understood to be left A-modules.

An A-module is *projective* if it is a direct summand of a free A-module. The following conditions on an A-module P are equivalent:

(i) P is a f.g. (= finitely generated) projective A-module,
(ii) P is a direct summand of a f.g. free A-module A^n,
(iii) P is isomorphic to the image im(p) of a projection $p = p^2 : A^n \longrightarrow A^n$ of a f.g. free A-module.

The *projective class group* $K_0(A)$ is the abelian group of formal differences $[P] - [Q]$ of isomorphism classes of f.g. projective A-modules P, Q, with

$$[P]-[Q] \;=\; [P']-[Q'] \in K_0(A)$$

if and only if there exists an A-module isomorphism

$$P \oplus Q' \oplus R \;\cong\; P' \oplus Q \oplus R$$

for a f.g. projective A-module R.

The *reduced projective class group* is the quotient of $K_0(A)$

$$\widetilde{K}_0(A) \;=\; \mathrm{coker}(K_0(\mathbb{Z}) \longrightarrow K_0(A))$$

with

$$K_0(\mathbb{Z}) \;=\; \mathbb{Z} \longrightarrow K_0(A)\;;\; n \longrightarrow [A^n]\;.$$

Equivalently, $\widetilde{K}_0(A)$ is the abelian group of stable isomorphism classes $[P]$ of f.g. projective A-modules P, with $[P]=[P'] \in \widetilde{K}_0(A)$ if and only if there exists an A-module isomorphism

$$P \oplus Q \;\cong\; P' \oplus Q$$

for a f.g. projective A-module Q.

A *finite domination* of an A-module chain complex C is a finite f.g. free A-module chain complex D with chain maps $f: C \longrightarrow D$, $g: D \longrightarrow C$ and a chain homotopy $gf \simeq 1 : C \longrightarrow C$. An A-module chain complex C is *finitely dominated* if it admits a finite domination. In [238] it is shown that an A-module chain complex C is finitely dominated if and only if C is chain equivalent to a finite chain complex of f.g. projective A-modules

$$P\;:\;\dots \longrightarrow 0 \longrightarrow \dots \longrightarrow 0 \longrightarrow P_n \longrightarrow P_{n-1} \longrightarrow \dots \longrightarrow P_0\;.$$

The *projective class* of a finitely dominated complex C is defined by

$$[C] \;=\; [P] \;=\; \sum_{i=0}^{\infty}(-)^i[P_i] \in K_0(A)$$

for any such P. An A-module chain complex C is *chain homotopy finite* if it is chain equivalent to a finite f.g. free A-module chain complex. The *reduced projective class* $[C] \in \widetilde{K}_0(A)$ of a finitely dominated A-module chain complex is such that $[C]=0$ if and only if C is chain homotopy finite.

A *finite domination* (K,f,g,h) of a topological space X is a finite CW complex K together with maps $f: X \longrightarrow K$, $g: K \longrightarrow X$ and a homotopy $h: gf \simeq 1 : X \longrightarrow X$. A topological space X is *homotopy finite* if it is homotopy equivalent to a finite CW complex.

The *Wall finiteness obstruction* of a finitely dominated CW complex X is the reduced projective class of the cellular $\mathbb{Z}[\pi_1(X)]$-module chain complex $C(\widetilde{X})$ of the universal cover $\widetilde{X}$

$$[X] \;=\; [C(\widetilde{X})] \in \widetilde{K}_0(\mathbb{Z}[\pi_1(X)])\;.$$

Proposition 1.1 (Wall [300])
A connected CW complex X is finitely dominated if and only if the fundamental group $\pi_1(X)$ is finitely presented and the $\mathbb{Z}[\pi_1(X)]$-module chain complex $C(\widetilde{X})$ is finitely dominated, in which case $[X] = 0 \in \widetilde{K}_0(\mathbb{Z}[\pi_1(X)])$ if and only if X is homotopy finite. □

1B. Whitehead torsion

Whitehead torsion is an algebraic K-theory invariant which is a generalization of the determinant. In the first instance torsion is defined for a homotopy equivalence of finite complexes (in algebra or topology), although in the applications to fibred knots in Chaps. 15, 33 it is necessary to consider a somewhat more general context.

In dealing with Whitehead torsion it will always be assumed that the ground ring A has the invariant basis property, meaning that invertible matrices have to be square, so that f.g. free A-modules have a well-defined dimension.

For example, any ring which admits a morphism $A \longrightarrow F$ to a field F (such as a group ring $\mathbb{Z}[\pi]$, with $F = \mathbb{Z}_2$) has the invariant basis property.

The finite general linear groups of a ring A

$$GL_n(A) \ = \ \mathrm{Aut}_A(A^n) \ \ (n \geq 1)$$

are related by the inclusions

$$GL_n(A) \longrightarrow GL_{n+1}(A) \ ; \ f \longrightarrow f \oplus 1_A \ .$$

The infinite general linear group of A is the union of the finite linear groups

$$GL_\infty(A) \ = \ \bigcup_{n=1}^{\infty} GL_n(A) \ .$$

The *torsion group* of A is the abelianization of $GL_\infty(A)$

$$K_1(A) \ = \ GL_\infty(A)^{ab} \ = \ GL_\infty(A)/[GL_\infty(A), GL_\infty(A)] \ ,$$

i.e. the quotient of $GL_\infty(A)$ by the normal subgroup

$$[GL_\infty(A), GL_\infty(A)] \triangleleft GL_\infty(A)$$

generated by the commutators $xyx^{-1}y^{-1}$ $(x, y \in GL_\infty(A))$.

The *torsion* of an automorphism $\alpha : A^n \longrightarrow A^n$ is the class of the matrix $(\alpha_{ij}) \in GL_n(A)$ of α

$$\tau(\alpha) \ = \ (\alpha_{ij}) \in K_1(A) \ .$$

More generally, the torsion of an automorphism $f : P \longrightarrow P$ of a f.g. projective A-module P is defined by

$$\tau(f) = \tau(h^{-1}(f \oplus 1_Q)h : A^n \longrightarrow A^n) \in K_1(A)$$

for any f.g. projective A-module Q with an isomorphism $h : A^n \cong P \oplus Q$.

In dealing with the torsion groups of rings adopt the following:

Terminology 1.2 The *Whitehead group* of a ring A is the reduced torsion group

$$Wh_1(A) = \mathrm{coker}(K_1(\mathbb{Z}) \longrightarrow K_1(A)) = \widetilde{K}_1(A)$$

in the algebraic context of an arbitrary ring A. In the topological context of a group ring $A = \mathbb{Z}[\pi]$ this is understood to be the Whitehead group of the group π

$$Wh(\pi) = K_1(\mathbb{Z}[\pi])/\{\pm g \,|\, g \in \pi\} .$$

A chain equivalence is *simple* if $\tau = 0 \in Wh$. □

A f.g. free A-module chain complex C is *based* if it is finite and each C_r is a based f.g. free A-module. The *torsion* of a chain equivalence $f : C \longrightarrow D$ of based f.g. free A-module chain complexes is

$$\begin{aligned}\tau(f) &= \tau(\mathcal{C}(f)) \\ &= \tau(d + \Gamma : \mathcal{C}(f)_{odd} \longrightarrow \mathcal{C}(f)_{even}) \in Wh_1(A) ,\end{aligned}$$

with $\mathcal{C}(f)$ the algebraic mapping cone and $\Gamma : 0 \simeq 1 : \mathcal{C}(f) \longrightarrow \mathcal{C}(f)$ any chain contraction. The chain equivalence is *simple* if $\tau(f) = 0 \in Wh_1(A)$.

The *Whitehead torsion* of a homotopy equivalence $f : X \longrightarrow Y$ of finite CW complexes is the torsion of the chain equivalence $\widetilde{f} : C(\widetilde{X}) \longrightarrow C(\widetilde{Y})$ of the cellular based f.g. free $\mathbb{Z}[\pi_1(X)]$-module chain complexes induced by a lift $\widetilde{f} : \widetilde{X} \longrightarrow \widetilde{Y}$ of f to the universal covers $\widetilde{X}, \widetilde{Y}$ of X, Y

$$\tau(f) = \tau(\widetilde{f} : C(\widetilde{X}) \longrightarrow C(\widetilde{Y})) \in Wh(\pi_1(X)) .$$

A *finite structure* on a topological space X is an equivalence class of pairs (K, f) with K a finite CW complex and $f : X \longrightarrow K$ a homotopy equivalence, subject to the equivalence relation

$$(K, f) \sim (K', f') \text{ if } \tau(f'f^{-1} : K \longrightarrow K') = 0 \in Wh(\pi_1(X)) .$$

A *finite structure* on an A-module chain complex C is an equivalence class of pairs (D, f) with D a finite chain complex of based f.g. free A-modules and $f : C \longrightarrow D$ a chain equivalence, subject to the equivalence relation

$$(D, f) \sim (D', f') \text{ if } \tau(f'f^{-1} : D \longrightarrow D') = 0 \in Wh_1(A) .$$

Proposition 1.3 *The finite structures on a connected CW complex X are in one-one correspondence with the finite structures on the cellular* $\mathbb{Z}[\pi_1(X)]$*-module chain complex* $C(\widetilde{X})$ *of the universal cover* $\widetilde{X}$. □

Let $(B, A \subseteq B)$ be a pair of rings. In the applications, B will be one of the polynomial extension rings $A[z], A[z^{-1}], A[z, z^{-1}]$ of A. A B-module chain complex C is *A-finitely dominated* if it is finitely dominated when regarded as an A-module chain complex. An *A-finite structure* on a B-module chain complex C is a finite structure on C when regarded as an A-module chain complex. A B-module chain complex C admits an A-finite structure if and only if C is A-finitely dominated and $[C] = 0 \in \widetilde{K}_0(A)$, in which case there is one A-finite structure for each element of $Wh_1(A)$.

A finite chain complex C of f.g. projective A-modules is *round* if

$$[C] \;=\; 0 \in \operatorname{im}(K_0(\mathbb{Z}) \longrightarrow K_0(A)) \ ,$$

i.e. if C is chain equivalent to a finite chain complex of f.g. free A-modules with Euler characteristic $\chi(C) = 0$. The *absolute torsion* of a chain equivalence $f : C \longrightarrow D$ of round finite chain complexes of based f.g. free A-modules was defined in [239] to be an element $\tau(f) \in K_1(A)$ with image $\tau(f) \in Wh_1(A)$.

A *round A-finite structure* on a B-module chain complex C is an equivalence class of pairs (D, f) with D a round finite chain complex of based f.g. free A-modules and $f : C \longrightarrow D$ a chain equivalence, subject to the equivalence relation

$$(D, f) \sim (D', f') \text{ if } \tau(f'f^{-1} : D \longrightarrow D') = 0 \in K_1(A) \ .$$

A B-module chain complex C admits a round A-finite structure if and only if C is A-finitely dominated and $[C] = 0 \in K_0(A)$, in which case there is one round A-finite structure for each element of $K_1(A)$. The algebraic mapping cone $\mathcal{C}(f)$ of a self chain map $f : C \longrightarrow C$ of an A-finitely dominated B-module chain complex C has a canonical round finite structure. See Ranicki [241, Chap. 6] for a detailed account.

A finite CW complex X is *round* if

$$\chi(X) \;=\; 0 \in K_0(\mathbb{Z}) \;=\; \mathbb{Z} \ .$$

A *round finite structure* on a round finite CW complex X is a round finite structure on the cellular $\mathbb{Z}[\pi_1(X)]$-module chain complex $C(\widetilde{X})$ of the universal cover $\widetilde{X}$.

1C. The mapping torus

The *mapping torus* of a self map $f : X \longrightarrow X$ is the identification space

$$T(f) = X \times [0,1]/\{(x,0) = (f(x),1) \,|\, x \in X\} .$$

For any maps $f : X \longrightarrow Y$, $g : Y \longrightarrow X$ there are defined inverse homotopy equivalences

$$T(gf) \longrightarrow T(fg) \; ; \; (x,s) \longrightarrow (f(x),s) ,$$
$$T(fg) \longrightarrow T(gf) \; ; \; (y,t) \longrightarrow (g(y),t)$$

by Mather [183].

The mapping torus $T(h)$ of a self map $h : X \longrightarrow X$ of a finitely dominated space X has a canonical round finite structure, represented by

$$T(fhg : K \longrightarrow K) \simeq T(gfh) \simeq T(h)$$

for any finite domination

$$(K, f : X \longrightarrow K, g : K \longrightarrow X, gf \simeq 1 : X \longrightarrow X) .$$

See [241, Chap. 6] for a detailed account of the finiteness and torsion properties of the mapping torus.

Given a ring morphism $f : A \longrightarrow B$ regard B as a (B,A)-bimodule by

$$B \times B \times A \longrightarrow B \; ; \; (b,x,a) \longrightarrow b.x.f(a) .$$

An A-module M induces the B-module

$$\begin{aligned} f_! M &= B \otimes_A M \\ &= B \otimes_{\mathbb{Z}} M/\{bf(a) \otimes x - b \otimes ax \,|\, a \in A, b \in B, x \in M\} . \end{aligned}$$

Let $f : X \longrightarrow X$ be a self map of a connected CW complex X. Given a base point $x_0 \in X$ and a path $\omega : I \longrightarrow X$ from $\omega(0) = x_0 \in X$ to $\omega(1) = f(x_0) \in X$ define an endomorphism of the fundamental group at x_0

$$f_* : \pi_1(X) = \pi_1(X,x_0) \xrightarrow{f_\#} \pi_1(X,f(x_0)) \xrightarrow{\omega_\#^{-1}} \pi_1(X,x_0) .$$

Let $f^*\widetilde{X}$ be the pullback along $f : X \longrightarrow X$ of the universal cover $\widetilde{X}$ of X. The lift of $f : X \longrightarrow X$ to a $\pi_1(X)$-equivariant map $\widetilde{f} : f^*\widetilde{X} \longrightarrow \widetilde{X}$ induces a chain map of the cellular $\mathbb{Z}[\pi_1(X)]$-module chain complexes

$$\widetilde{f} : C(f^*\widetilde{X}) = (f_*)_! C(\widetilde{X}) \longrightarrow C(\widetilde{X}) .$$

The fundamental group of the mapping torus $T(f)$ is the amalgamation

$$\begin{aligned}\pi_1(T(f)) &= \pi_1(X) *_{f_*} \{z\} \\ &= \Big(\pi_1(X) * \{z\}\Big) \Big/ \{gz = zf_*(g) \mid g \in \pi_1(X)\}\ .\end{aligned}$$

The natural map $\pi_1(X) \longrightarrow \pi_1(T(f))$ need not be injective. However, if $f_* : \pi_1(X) \longrightarrow \pi_1(X)$ is an automorphism then

$$\pi_1(T(f)) = \pi_1(X) \times_{f_*} \mathbb{Z}$$

is the f_*-twisted extension of $\pi_1(X)$ by $\mathbb{Z}$, with an exact sequence of groups

$$\{1\} \longrightarrow \pi_1(X) \longrightarrow \pi_1(T(f)) \longrightarrow \mathbb{Z} \longrightarrow \{1\}\ ,$$

and the fundamental group ring is the f_*-twisted Laurent polynomial extension of $\mathbb{Z}[\pi_1(X)]$

$$\mathbb{Z}[\pi_1(T(f))] = \mathbb{Z}[\pi_1(X)]_{f_*}[z, z^{-1}]\ .$$

Definition 1.4 A self map $f : X \longrightarrow X$ is *untwisted* if $f_* : \pi_1(X) \longrightarrow \pi_1(X)$ is an inner automorphism. □

Suppose that $f : X \longrightarrow X$ is untwisted, with

$$f_*(x) = gxg^{-1} \in \pi_1(X) \quad (x \in \pi_1(X))$$

for some $g \in \pi_1(X)$. Let $\omega' : I \longrightarrow X$ be a path such that

$$\omega(i) = \omega'(i) \in X \ (i = 0, 1)\ ,\ g = \omega'^{-1}\omega \in \pi_1(X)\ ,$$

so that

$$f_\# = \omega'_\# : \pi_1(X, x_0) \longrightarrow \pi_1(X, f(x_0))\ .$$

Replacing ω by ω' in the construction of f_* gives

$$\begin{aligned} f_* &= 1 : \pi_1(X) \longrightarrow \pi_1(X)\ , \\ \pi_1(T(f)) &= \pi_1(X) \times \mathbb{Z}\ , \\ \mathbb{Z}[\pi_1(T(f))] &= \mathbb{Z}[\pi_1(X)][z, z^{-1}]\end{aligned}$$

with a $\pi_1(X)$-equivariant lift $\widetilde{f} : \widetilde{X} \longrightarrow \widetilde{X}$ of $f : X \longrightarrow X$ inducing the $\mathbb{Z}[\pi_1(X)]$-module chain map $\widetilde{f} : C(\widetilde{X}) \longrightarrow C(\widetilde{X})$.

Now consider a connected space X with a connected infinite cyclic cover $\overline{X}$. The fibration $\overline{X} \longrightarrow X \longrightarrow S^1$ induces an exact sequence of fundamental groups

$$\{1\} \longrightarrow \pi_1(\overline{X}) \longrightarrow \pi_1(X) \longrightarrow \pi_1(S^1) \longrightarrow \{1\}$$

and $\pi_1(X) = \pi_1(\overline{X}) \times_\alpha \mathbb{Z}$ with $\alpha = \zeta_* : \pi_1(\overline{X}) \longrightarrow \pi_1(\overline{X})$ the automorphism induced by a generating covering translation $\zeta : \overline{X} \longrightarrow \overline{X}$.

Definition 1.5 An infinite cyclic cover $\overline{X}$ of a CW complex X is *untwisted* if a generating covering translation $\zeta : \overline{X} \longrightarrow \overline{X}$ is untwisted. □

Proposition 1.6 *A connected infinite cyclic cover $\overline{X}$ of a connected CW complex X is untwisted if and only if the group extension*

$$\{1\} \longrightarrow \pi_1(\overline{X}) \longrightarrow \pi_1(X) \longrightarrow \pi_1(S^1) \longrightarrow \{1\}$$

is trivial, in which case

$$\pi_1(X) = \pi_1(\overline{X}) \times \mathbb{Z} \ , \ \ \mathbb{Z}[\pi_1(X)] = \mathbb{Z}[\pi_1(\overline{X})][z, z^{-1}]$$

and a generating covering translation $\zeta : \overline{X} \longrightarrow \overline{X}$ has a $\pi_1(X)$-equivariant lift $\widetilde{\zeta} : \widetilde{X} \longrightarrow \widetilde{X}$ inducing a $\mathbb{Z}[\pi_1(X)]$-module chain equivalence $\widetilde{\zeta} : C(\widetilde{X}) \longrightarrow C(\widetilde{X})$, with $\widetilde{X}$ the universal cover of X. □

The mapping torus $T(f)$ of a map $f : X \longrightarrow X$ is equipped with a canonical infinite cyclic cover

$$\overline{T}(f) = \Big(\coprod_{n=-\infty}^{\infty} X \times \{n\} \times [0,1] \Big) / (x, n, 1) \sim (f(x), 0, n+1) \ ,$$

which is the pullback $\overline{T}(f) = p^* \mathbb{R}$ of the universal cover $\mathbb{R}$ of S^1 along the projection

$$p : T(f) \longrightarrow I/(0 \sim 1) = S^1 \ ; \ (x, s) \longrightarrow [s] \ .$$

Proposition 1.7 *A map $f : X \longrightarrow X$ inducing an automorphism $f_* : \pi_1(X) \longrightarrow \pi_1(X)$ is untwisted if and only if the infinite cyclic cover $\overline{T}(f)$ of $T(f)$ is untwisted.* □

2. Geometric bands

A band is a compact space W with an infinite cyclic cover $\overline{W}$ which is finitely dominated. Bands occur naturally in the classification of manifolds that fibre over the circle S^1 such as fibred knot complements, the bordism of diffeomorphisms, and open book decompositions. A manifold which fibres over S^1

$$M = T(h : F \longrightarrow F)$$

is a band, with $\overline{M} = F \times \mathbb{R} \simeq F$ homotopy finite. Conversely, if M is an n-dimensional manifold band then the infinite cyclic cover $\overline{M}$ is a finitely dominated $(n-1)$-dimensional geometric Poincaré complex with a homotopy equivalence

$$M \simeq T(\zeta : \overline{M} \longrightarrow \overline{M})$$

(ζ = generating covering translation), so that M has the homotopy theoretic properties of a fibre bundle over S^1 with 'fibre' $\overline{M}$, but in general there are obstructions to M actually fibring over S^1.

The problem of deciding if a manifold band W fibres over S^1 was first studied by Stallings [272] in dimension 3, and then by Browder and Levine [31] in dimensions ≥ 5 with $\pi_1(W) = \mathbb{Z}$. In the non-simply-connected high-dimensional case Farrell [78], [79] and Siebenmann [266] obtained a Whitehead group obstruction for a manifold band to fibre over S^1.

Much of the progress of high-dimensional compact topological manifolds achieved in the last 30 years depends on non-compact manifolds with tame ends, starting with Novikov's proof of the topological invariance of the rational Pontrjagin classes. See Hughes and Ranicki [113] for an account of tame ends, bands, and some of these applications. In [113, 17.11] it is proved that every tame end of dimension ≥ 6 has an open neighbourhood which is an infinite cyclic cover of a compact manifold band.

Here is the formal definition in the CW category:

Definition 2.1 (i) A CW *band* is a finite CW complex X with a finitely dominated infinite cyclic cover $\overline{X}$.
(ii) A CW band is *untwisted* if $\overline{X}$ is untwisted (1.5). □

Remark 2.2 The band terminology was introduced by Siebenmann [265]. □

Example 2.3 The mapping torus $T(h)$ of a homotopy equivalence $h : F \longrightarrow F$ of a finite CW complex F is a CW band, with infinite cyclic cover $\overline{T}(h) \simeq F$ homotopy finite. The mapping torus $T(h)$ is untwisted if and only if h is untwisted. □

Example 2.4 Every finitely dominated CW complex X is homotopy equivalent to the infinite cyclic cover $\overline{Y}$ of an untwisted CW band Y. For any finite domination $(K, f : X \longrightarrow K, g : K \longrightarrow X, gf \simeq 1 : X \longrightarrow X)$ the mapping torus $Y = T(fg : K \longrightarrow K)$ is a CW band which is homotopy equivalent to $T(gf) \simeq X \times S^1$, with the infinite cyclic cover $\overline{Y} = \overline{T}(fg)$ homotopy equivalent to $\overline{T}(gf) \simeq X \times \mathbb{R} \simeq X$. □

The obstruction theory for fibering manifold bands over S^1 will be described in Chap. 16. It is a special case of the obstruction theory for fibering CW bands over S^1.

Definition 2.5 A CW band X *fibres over* S^1 if there exists a simple homotopy equivalence $X \simeq T(h)$ to the mapping torus $T(h)$ of a simple homotopy self equivalence $h : F \longrightarrow F$ of a finite CW complex F and the diagram

$$\begin{array}{ccc} X & \xrightarrow{\simeq} & T(h) \\ & \searrow \quad \swarrow & \\ & S^1 & \end{array}$$

is homotopy commutative, so that the infinite cyclic cover $\overline{X}$ of X is the pullback of the canonical infinite cyclic cover $\overline{T}(h)$ of $T(h)$. □

The fibering obstructions $\Phi^+(X), \Phi^-(X) \in Wh(\pi_1(X))$ of a CW band X are defined as follows. The mapping torus $T(\zeta)$ of the generating covering translation $\zeta : \overline{X} \longrightarrow \overline{X}$ of the infinite cyclic cover $\overline{X}$ of a space X is such that the projection

$$q^+ \ : \ T(\zeta) \longrightarrow X \ ; \ (x, s) \longrightarrow p(x)$$

is a homotopy equivalence, with $p : \overline{X} \longrightarrow X$ the covering projection. Define similarly the homotopy equivalence

$$q^- \ : \ T(\zeta^{-1}) \longrightarrow X \ ; \ (x, s) \longrightarrow p(x) \ .$$

Definition 2.6 The *fibering obstructions* of a CW band X with respect to a choice of generating covering translation $\zeta : \overline{X} \longrightarrow \overline{X}$ are the Whitehead torsions

$$\begin{aligned} \Phi^+(X) &= \tau(q^+ : T(\zeta) \longrightarrow X) \ , \\ \Phi^-(X) &= \tau(q^- : T(\zeta^{-1}) \longrightarrow X) \in Wh(\pi_1(X)) \end{aligned}$$

defined using the canonical finite structures on $T(\zeta)$, $T(\zeta^{-1})$. □

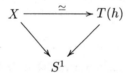

The fibering obstructions $\Phi^+(X), \Phi^-(X)$ of a CW band X measure the difference between the intrinsic finite structure of X determined by the cell decomposition and the extrinsic finite structures determined by the canonical finite structures on the mapping tori $T(\zeta), T(\zeta^{-1})$.

Proposition 2.7 (i) *The torsion of a homotopy equivalence $f: X \longrightarrow Y$ of CW bands is the difference of the fibering obstructions*

$$\begin{aligned}\tau(f) &= \Phi^+(Y) - \Phi^+(X) \\ &= \Phi^-(Y) - \Phi^-(X) \in Wh(\pi_1(X)) .\end{aligned}$$

In particular, $\Phi^+(X)$ and $\Phi^-(X)$ are simple homotopy invariants of X.
(ii) *The difference*

$$\begin{aligned}\Phi^-(X) - \Phi^+(X) &= \tau(-z\zeta^{-1} : C(\widetilde{X})_\alpha[z,z^{-1}] \longrightarrow C(\widetilde{X})_\alpha[z,z^{-1}]) \\ &\in Wh(\pi_1(X))\end{aligned}$$

is a homotopy invariant of a CW band X, such that $\Phi^-(X) - \Phi^+(X) = 0$ if and only if there exists a finite structure $(K, f: \overline{X} \longrightarrow K)$ on $\overline{X}$ such that

$$\tau(f\zeta f^{-1} : K \longrightarrow K) = 0 \in Wh(\pi_1(\overline{X})) ,$$

in which case

$$\tau(T(f\zeta f^{-1}) \longrightarrow X) = \Phi^+(X) = \Phi^-(X) \in Wh(\pi_1(X)) .$$

(iii) *A CW band X fibres over S^1 if and only if*

$$\Phi^+(X) = \Phi^-(X) = 0 \in Wh(\pi_1(X)) .$$

Proof See Ranicki [244, Chap. 20]. □

Example 2.8 Let $f: X \longrightarrow X$ be a self homotopy equivalence of a connected finitely dominated CW complex X. The mapping torus $Y = T(f)$ has a canonical finite structure, with infinite cyclic cover $\overline{Y} = \overline{T}(f) \simeq X$. The fibering obstructions of any CW band Z in the canonical finite structure of Y are given by

$$\begin{aligned}\Phi^+(Z) &= 0 , \\ \Phi^-(Z) &= \tau(-z\widetilde{f}^{-1} : C(\widetilde{X})_\alpha[z,z^{-1}] \longrightarrow C(\widetilde{X})_\alpha[z,z^{-1}]) \in Wh(\pi_1(Y)) ,\end{aligned}$$

with $C(\widetilde{X})$ the cellular chain complex of the universal cover $\widetilde{X}$ of X and

$$\alpha = f_* : \pi_1(X) \longrightarrow \pi_1(X) \ , \ \pi_1(Y) = \pi_1(X) \times_\alpha \mathbb{Z} . \qquad \square$$

3. Algebraic bands

Algebraic bands are the chain complex analogues of the geometric bands of Chap. 2. An algebraic band is a finite chain complex C of finitely generated free modules over the Laurent polynomial extension $A[z,z^{-1}]$ of a ring A such that C is finitely dominated over A, i.e. A-module chain equivalent to a finite finitely generated projective A-module chain complex. The most obvious application of algebraic bands to high-dimensional knot theory is via fibred knots, but the related algebra is useful in the study of all knots. For example, the results of Milnor [195] that for any field F the F-coefficient homology of the infinite cyclic cover $\overline{X}$ of a finite CW complex X with $H_*(X) = H_*(S^1)$ is finite-dimensional

$$\dim_F H_*(\overline{X};F) < \infty ,$$

and that if X is an $(n+2)$-dimensional manifold with boundary $\partial X = S^n \times S^1$ (e.g. the exterior of an n-knot $k : S^n \subset S^{n+2}$) there are Poincaré duality isomorphisms

$$H^{n+1-*}(\overline{X};F) \cong H_*(\overline{X},S^n;F)$$

show that every n-knot has the F-coefficient homological properties of a fibred n-knot, with fibre $\overline{X}$.

Definition 3.1 (i) The *polynomial extension* $A[z]$ of a ring A is the ring consisting of the polynomials $\sum\limits_{j=0}^{\infty} a_j z^j$ with coefficients $a_j \in A$ such that $\{j \geq 0 \,|\, a_j \neq 0\}$ is finite.

(ii) The *Laurent polynomial extension* $A[z,z^{-1}]$ is the ring consisting of the polynomials $\sum\limits_{j=-\infty}^{\infty} a_j z^j$ with coefficients $a_j \in A$ such that $\{j \in \mathbb{Z} \,|\, a_j \neq 0\}$ is finite. □

Definition 3.2 A *chain complex band over* $A[z,z^{-1}]$ is a based f.g. free $A[z,z^{-1}]$-module chain complex which is A-finitely dominated. □

Example 3.3 If A is a Dedekind ring then an A-module chain complex C is finitely dominated if and only if the homology $H_*(C)$ is finitely generated. Thus a based f.g. free $A[z,z^{-1}]$-module chain complex C is a band if and only if the homology $H_*(C)$ is finitely generated as an A-module. □

Example 3.4 An infinite cyclic cover $\overline{X} = \widetilde{X}/\pi$ of a finite CW complex X with $\pi_1(X) = \pi \times \mathbb{Z}$ is finitely dominated if and only if $C(\widetilde{X})$ is finitely dominated over $\mathbb{Z}[\pi]$, by Wall [300] (cf. 1.1). Thus X is a CW complex band if and only if $C(\widetilde{X})$ is a chain complex band. See 3.5 for further details. □

Remark 3.5 Let X be a connected CW complex with universal cover $\widetilde{X}$ and fundamental group The connected infinite cyclic covers $\overline{X}$ of X correspond to the normal subgroups $\pi \lhd \pi_1(X)$ such that $\pi_1(X)/\pi = \mathbb{Z}$, with

$$\overline{X} = \widetilde{X}/\pi \ , \ \pi_1(\overline{X}) = \pi \ .$$

Given any expression of $\pi_1(X)$ as a group extension

$$\{1\} \longrightarrow \pi \longrightarrow \pi_1(X) \longrightarrow \mathbb{Z} \longrightarrow \{1\}$$

let $z \in \pi_1(X)$ be a lift of $1 \in \mathbb{Z}$. Conjugation by z defines an automorphism

$$\alpha : \pi \longrightarrow \pi \ ; \ x \longrightarrow z^{-1}xz$$

such that

$$\pi_1(X) = \pi \times_\alpha \mathbb{Z} \ , \ \mathbb{Z}[\pi_1(X)] = \mathbb{Z}[\pi]_\alpha[z, z^{-1}] \ .$$

An infinite cyclic cover $\overline{X}$ of X is finitely dominated if and only if $\pi_1(\overline{X}) = \pi$ is finitely presented and the cellular $\mathbb{Z}[\pi_1(X)]$-module chain complex $C(\widetilde{X})$ is $\mathbb{Z}[\pi]$-finitely dominated (1.1). If $\overline{X}$ is untwisted (1.5) then

$$\alpha = 1 \ , \ \pi_1(X) = \pi \times \mathbb{Z} \ , \ \mathbb{Z}[\pi_1(X)] = \mathbb{Z}[\pi][z, z^{-1}] \ .$$

(i) A connected finite CW complex X is an untwisted band if and only if $\pi_1(X) = \pi \times \mathbb{Z}$ and $C(\widetilde{X})$ is a chain complex band over $\mathbb{Z}[\pi][z, z^{-1}]$ (as in 3.4). If X is a has a finite 2-skeleton then π is finitely presented, and if $\overline{X}$ is untwisted then $\overline{\pi}$ is finitely presented also.
(ii) In the twisted case $\pi_1(X) = \pi \times_\alpha \mathbb{Z}$ with $\alpha \neq 1$ it may be that $\pi_1(X)$ is finitely presented but π is not finitely presented: for example, if $X = T(2 : S^1 \longrightarrow S^1)$ then the canonical infinite cyclic cover $\overline{X} = \overline{T}(2)$ is the dyadic solenoid, and

$$\begin{aligned} \pi_1(X) &= \pi \times_\alpha \mathbb{Z} = \{y, z \,|\, zyz^{-1} = y^2\} \ , \\ \pi_1(\overline{X}) &= \pi = [\pi_1(X), \pi_1(X)] = \mathbb{Z}[1/2] \ , \ \alpha = 2 : \pi \longrightarrow \pi \end{aligned}$$

with $\pi_1(X)$ finitely presented and π not finitely generated, let alone finitely presented. (I am indebted to C.J.B.Brookes for this example.) □

Definition 3.6 Given an A-module chain complex C and a chain map $f : C \longrightarrow C$ define the *algebraic mapping tori* $T^+(f), T^-(f)$ to be the $A[z, z^{-1}]$-module chain complexes

$$\begin{aligned} T^+(f) &= \mathcal{C}(1 - zf : C[z, z^{-1}] \longrightarrow C[z, z^{-1}]) \ , \\ T^-(f) &= \mathcal{C}(1 - z^{-1}f : C[z, z^{-1}] \longrightarrow C[z, z^{-1}]) \ . \end{aligned}$$ □

If C is finitely dominated then $T^+(f)$ an $T^-(f)$ have canonical round finite structures. If $f : C \longrightarrow C$ is an isomorphism then $T^+(f)$ is related to $T^-(f^{-1})$ by the isomorphism

$$T^-(f^{-1}) \longrightarrow T^+(f) \; ; \; (x,y) \longrightarrow (-zf(x),y) \; .$$

If $f : C \longrightarrow C$ is a chain equivalence then $T^+(f)$ is chain equivalent to $T^-(f^{-1})$.

The (geometric) mapping torus $T(f)$ of an untwisted map of spaces $f : X \longrightarrow X$ is such that

$$\pi_1(T(f)) \;=\; \pi_1(X) \times \mathbb{Z} \; , \; \mathbb{Z}[\pi_1(T(f))] \;=\; \mathbb{Z}[\pi_1(X)][z,z^{-1}] \; .$$

The cellular chain complex $C(\widetilde{T}(f))$ of the universal cover $\widetilde{T}(f)$ of $T(f)$ is an algebraic mapping torus of the induced $\mathbb{Z}[\pi_1(X)]$-module chain map $\widetilde{f} : C(\widetilde{X}) \longrightarrow C(\widetilde{X})$ with $C(\widetilde{X})$ the cellular $\mathbb{Z}[\pi_1(X)]$-module chain complex of the universal cover $\widetilde{X}$ of X

$$C(\widetilde{T}(f)) \;=\; T^+(\widetilde{f}) \;=\; \mathcal{C}(1 - z\widetilde{f} : C(\widetilde{X})[z,z^{-1}] \longrightarrow C(\widetilde{X})[z,z^{-1}]) \; .$$

If X is a finitely dominated CW complex then $C(\widetilde{X})$ is a finitely dominated $\mathbb{Z}[\pi_1(X)]$-module chain complex and $C(\widetilde{T}(f))$ has a canonical round finite structure.

Example 3.7 If C is a based f.g. free A-module chain complex and $f : C \longrightarrow C$ is a chain equivalence then the algebraic mapping tori $T^+(f), T^-(f)$ are chain complex bands which are A-module chain equivalent to C. □

By analogy with the convention for $Wh_1(A)$ (1.4):

Convention 3.8 (i) The *Whitehead group* of the Laurent polynomial extension $A[z,z^{-1}]$ is

$$\begin{aligned} Wh_1(A[z,z^{-1}]) &= \text{coker}(K_1(\mathbb{Z}[z,z^{-1}]) \longrightarrow K_1(A[z,z^{-1}])) \\ &= K_1(A[z,z^{-1}])/\{\pm z^j \,|\, j \in \mathbb{Z}\} \; . \end{aligned}$$

in the algebraic context of an arbitrary ring A.
(ii) In the topological context of a group ring $A = \mathbb{Z}[\pi]$ this is understood to be the Whitehead group of the group $\pi \times \mathbb{Z}$

$$Wh(\pi \times \mathbb{Z}) \;=\; K_1(\mathbb{Z}[\pi \times \mathbb{Z}])/\{\pm z^j g \,|\, j \in \mathbb{Z}, g \in \pi\} \; .$$

(iii) There are corresponding conventions for a *simple chain equivalence* $f : C \longrightarrow D$ of based f.g. free $A[z,z^{-1}]$-module chain complexes, with $\tau(f) = 0 \in Wh_1(A[z,z^{-1}])$ in an algebraic context, and $\tau(f) = 0 \in Wh(\pi \times \mathbb{Z})$ in a topological context with $A = \mathbb{Z}[\pi]$. □

Definition 3.9 A chain complex band C *fibres* if it is simple chain equivalent to the algebraic mapping torus

$$T^+(h) \;=\; \mathcal{C}(1 - zh : C[z,z^{-1}] \longrightarrow C[z,z^{-1}])$$

of a simple chain equivalence $h : C \longrightarrow C$. □

Definition 3.10 The *fibering obstructions* of an $A[z,z^{-1}]$-module chain complex band C are the Whitehead torsions

$$\begin{aligned}\Phi^+(C) &= \tau(q^+ : T^+(\zeta^{-1})\longrightarrow C) ,\\ \Phi^-(C) &= \tau(q^- : T^-(\zeta)\longrightarrow C) \in Wh_1(A[z,z^{-1}]) ,\end{aligned}$$

defined using the canonical finite structures on $T^+(\zeta^{-1})$, $T^-(\zeta^{-1})$ with

$$\begin{aligned}\zeta &: C \longrightarrow C \; ; \; x \longrightarrow zx ,\\ \zeta^{-1} &: C \longrightarrow C \; ; \; x \longrightarrow z^{-1}x .\end{aligned}$$

□

By analogy with 2.7:

Proposition 3.11 (i) *A chain complex band C fibres if and only if*

$$\Phi^+(C) = \Phi^-(C) = 0 \in Wh_1(A[z,z^{-1}]) .$$

(ii) *The torsion of a chain equivalence $f : C\longrightarrow D$ of chain complex bands is the difference of the fibering obstructions*

$$\begin{aligned}\tau(f) &= \Phi^+(D) - \Phi^+(C)\\ &= \Phi^-(D) - \Phi^-(C) \in Wh_1(A[z,z^{-1}]) .\end{aligned}$$

In particular, $\Phi^+(C)$ and $\Phi^-(C)$ are simple chain homotopy invariants of C.
(ii) *The difference*

$$\begin{aligned}&\Phi^-(C) - \Phi^+(C)\\ &= \tau(-z\zeta^{-1} : C[z,z^{-1}]\longrightarrow C[z,z^{-1}]) \in Wh_1(A[z,z^{-1}])\end{aligned}$$

is a chain homotopy invariant of a chain complex band C, such that $\Phi^-(C) - \Phi^+(C) = 0$ if and only if there exists an A-finite structure $(B, f : B\longrightarrow C)$ on C such that

$$\tau(f^{-1}\zeta f : B\longrightarrow B) = 0 \in Wh_1(A) ,$$

in which case

$$\tau(T^+(f^{-1}\zeta^{-1}f)\longrightarrow C) = \Phi^+(C) = \Phi^-(C) \in Wh_1(A[z,z^{-1}]) .$$

Proof See Ranicki [244, Chap. 20]. □

By analogy with 2.8:

Example 3.12 Let P be a finitely dominated A-module chain complex, and let $f : P\longrightarrow P$ be a chain equivalence. The fibering obstructions of any based f.g. free $A[z,z^{-1}]$-module chain complex C in the canonical round finite structure of $T^+(f)$ are given by

$$\begin{aligned}\Phi^+(C) &= 0 ,\\ \Phi^-(C) &= \tau(-zf : P[z,z^{-1}]\longrightarrow P[z,z^{-1}]) \in Wh_1(A[z,z^{-1}]) ,\end{aligned}$$

so that C fibres if and only if P is chain homotopy finite and f is simple. Similarly for $T^-(f)$. □

4. Localization and completion in K-theory

As already recalled in the Introduction, localization and completion are the basic algebraic techniques for computing the algebraic K- and L-groups, by reducing the computation for a complicated ring to simpler rings (e.g. fields). The classic example of localization and completion is the Hasse-Minkowski principle by which quadratic forms over $\mathbb{Z}$ are related to quadratic forms over $\mathbb{Q}$ and the finite fields $\mathbb{F}_p$ and the p-adic completions $\widehat{\mathbb{Z}}_p$, $\widehat{\mathbb{Q}}_p$ of $\mathbb{Z}, \mathbb{Q}$ (p prime). The localization of polynomial rings is particularly relevant to knot theory, starting with the way in which the Blanchfield form takes its values in the localization of $\mathbb{Z}[z, z^{-1}]$ inverting the Alexander polynomials.

For any ring morphism $f : A \longrightarrow B$ the algebraic K-groups of A and B are related by a long exact sequence

$$\ldots \longrightarrow K_n(A) \xrightarrow{f_!} K_n(B) \longrightarrow K_n(f) \longrightarrow K_{n-1}(A) \longrightarrow \ldots$$

with relative K-groups $K_*(f)$. In particular, $K_1(f)$ is the abelian group of equivalence classes of triples (P, Q, g) given by f.g. projective A-modules P, Q and a B-module isomorphism

$$g : f_!P = B \otimes_A P \longrightarrow f_!Q = B \otimes_A Q ,$$

subject to the equivalence relation defined by:

$$\begin{aligned} (P, Q, g) \sim (P', Q', g') \ & \text{if there exist a f.g. projective } A\text{-module } R \\ & \text{and an } A\text{-module isomorphism} \\ & h : P \oplus Q' \oplus R \cong P' \oplus Q \oplus R \\ & \text{such that} \end{aligned}$$

$$\tau((g^{-1} \oplus g' \oplus 1_{f_!R})h : f_!(P \oplus Q' \oplus R) \xrightarrow{\simeq} f_!(P \oplus Q' \oplus R)) = 0 \in K_1(B) .$$

In the special case when $f : A \longrightarrow B = S^{-1}A$ is the inclusion of A in the localization inverting a multiplicative subset $S \subset A$ the relative K-groups $K_*(f)$ are identified with the K-groups of the exact category of homological dimension 1 S-torsion A-modules.

4A. Commutative localization

This section only deals with commutative localization, in which only central elements of a ring are inverted – see Chap. 9 below for noncommutative localization

Definition 4.1 (i) A *multiplicative subset* $S \subset A$ is a subset of central elements which is closed under multiplication, with $st \in S$ for each $s, t \in S$, and such that $1 \in S$.
(ii) The *localization* $S^{-1}A$ is the ring obtained from A by inverting every $s \in S$, with elements the equivalence classes a/s of pairs $(a, s) \in A \times S$ subject to the equivalence relation

$$(a, s) \sim (b, t) \text{ if } (at - bs)u = 0 \in A \text{ for some } u \in S .$$

Addition and multiplication are by

$$a/s + b/t = (at + bs)/st \ , \ (a/s)(b/t) = ab/st . \qquad \square$$

Proposition 4.2 *The ring morphism*

$$i : A \longrightarrow S^{-1}A \ ; \ a \longrightarrow a/1$$

has the universal property that for any ring morphism $f : A \longrightarrow B$ with $f(s) \in B$ invertible for every $s \in S$ there is a unique morphism $F : S^{-1}A \longrightarrow B$ with

$$f = Fi : A \longrightarrow S^{-1}A \longrightarrow B . \qquad \square$$

The morphism $i : A \longrightarrow S^{-1}A$ is not injective in general. However, it is injective if $S \subset A$ consists of non-zero divisors.

Example 4.3 Given a ring A and a central element $s \in A$ define the multiplicative subset

$$S = (s)^\infty = \{s^k \,|\, k \geq 0\} \subset A .$$

The localization of A inverting S is written

$$S^{-1}A = A[1/s] .$$

(i) If $s \in A$ is a non-zero divisor every non-zero element $x \in S^{-1}A$ has a unique expression as $x = a/s^k$ $(a \in A, k \geq 0)$.
(ii) If $s = s^2 \in A$ is an idempotent then

$$S = \{1, s\} \ , \ A = s(A) \oplus (1 - s)(A) \ , \ A[1/s] = s(A) . \qquad \square$$

Given an A-module M write the induced $S^{-1}A$-module as

$$S^{-1}M = S^{-1}A \otimes_A M .$$

Localization is exact, so that for any A-module chain complex C there is a natural identification

$$H_*(S^{-1}C) = S^{-1}H_*(C) .$$

Definition 4.4 (i) An A-module M is *S-torsion* if $S^{-1}M = 0$, or equivalently if for every $x \in M$ there exists $s \in S$ such that $sx = 0 \in M$. (ii) An *(A,S)-module* M is an S-torsion A-module of h.d. (= homological dimension) 1, that is an A-module with a f.g. projective resolution

$$0 \longrightarrow P_1 \xrightarrow{d} P_0 \longrightarrow M \longrightarrow 0$$

such that there exist $s \in S$ and $\Gamma \in \mathrm{Hom}_A(P_0, P_1)$ with

$$d\Gamma = s : P_0 \longrightarrow P_0 \ , \ \Gamma d = s : P_1 \longrightarrow P_1 \ .$$

Let $\mathbb{H}(A,S)$ be the category of (A,S)-modules. The *S-torsion class* of an (A,S)-module M is the stable isomorphism class

$$\tau_S(M) = [M] \in K_0(\mathbb{H}(A,S)) \ . \qquad \square$$

In an exact sequence of S-torsion A-modules

$$0 \longrightarrow M' \longrightarrow M \longrightarrow M'' \longrightarrow 0$$

M is an (A,S)-module if and only if M' and M'' are (A,S)-modules. Thus $\mathbb{H}(A,S)$ is an exact category in the sense of Quillen [225], and the algebraic K-groups $K_*(\mathbb{H}(A,S))$ are defined.

4B. The algebraic K-theory localization exact sequence

Proposition 4.5 (Bass [13, XII], Quillen [225])
Let $S \subset A$ be a multiplicative subset in a ring A such that the natural map $i : A \longrightarrow S^{-1}A$ is injective. The algebraic K-groups of A and $S^{-1}A$ are related by the localization exact sequence

$$\dots \longrightarrow K_n(A) \xrightarrow{i} K_n(S^{-1}A) \xrightarrow{\partial} K_n(A,S) \xrightarrow{j} K_{n-1}(A) \longrightarrow \dots$$

with i induced by the inclusion $A \longrightarrow S^{-1}A$ and

$$K_n(A,S) = K_{n-1}(\mathbb{H}(A,S)) \ \ (n \in \mathbb{Z}) \ . \qquad \square$$

In the case $n = 1$

$$\begin{aligned} j \ : \ K_1(A,S) = K_0(\mathbb{H}(A,S)) &\longrightarrow K_0(A) \ ; \\ \tau_S(M) &\longrightarrow [M] = [P_0] - [P_1] \ , \end{aligned}$$

$$\begin{aligned} \partial \ : \ K_1(S^{-1}A) \longrightarrow K_1(A,S) \ ; \ &\tau(a/s : A^n \longrightarrow A^n) \\ &\longrightarrow \tau_S(\mathrm{coker}(a : A^n \longrightarrow A^n)) - \tau_S(\mathrm{coker}(s : A^n \longrightarrow A^n)) \ . \end{aligned}$$

Proposition 4.6 *The following conditions on a finite f.g. projective A-module chain complex C are equivalent:*

(i) *the homology A-modules $H_*(C)$ are S-torsion, with $S^{-1}H_*(C) = 0$,*
(ii) *there exists $s \in S$ such that $sH_*(C) = 0$,*
(iii) *there exist A-module morphisms $\Gamma : C_r \longrightarrow C_{r+1}$ and an element $s \in S$ such that*

$$d\Gamma + \Gamma d = s : C_r \longrightarrow C_r \quad (r \in \mathbb{Z}) ,$$

(iv) *C is $S^{-1}A$-contractible, i.e. the induced finite f.g. projective $S^{-1}A$-module chain complex $S^{-1}C$ is chain contractible,*
(v) *C is homology equivalent to a finite chain complex D of (A, S)-modules.*

Proof (i) $\Longleftrightarrow$ (ii) Trivial.
(i) $\Longleftrightarrow$ (iii) For any ring R a finite projective R-module chain complex P is chain contractible if and only if $H_*(P) = 0$.
(iii) $\Longrightarrow$ (i) It follows from the chain homotopy $\Gamma : s \simeq 0 : C \longrightarrow C$ that $s = 0 : H_*(C) \longrightarrow H_*(C)$, and hence that $S^{-1}H_*(C) = 0$.
(i) $\Longrightarrow$ (v) By Ranicki [237, 6.4.2] there exists a finite chain complex D of (A, S)-modules with a chain map $C \longrightarrow D$ inducing isomorphisms

$$H_*(C) \cong H_*(D) .$$

(v) $\Longrightarrow$ (i) For each D_r there exists $s_r \in S$ such that $s_r D_r = 0$. The product $s = \prod_r s_r \in S$ is such that

$$sH_*(D) = sH_*(C) = 0$$

and there exists a chain homotopy $\Gamma : s \simeq 0 : C \longrightarrow C$.
(iii) $\Longleftrightarrow$ (iv) Every chain contraction of $S^{-1}C$ is of the form Γ/s, with

$$d\Gamma + \Gamma d = s : C_r \longrightarrow C_r . \qquad \square$$

Definition 4.7 An $S^{-1}A$-contractible finitely dominated A-module chain complex C has an *S-torsion class invariant*

$$\tau_S(C) = \sum_{i=0}^{\infty} (-)^i \tau_S(D_i) \in K_1(A, S) = K_0(\mathbb{H}(A, S)) ,$$

with D any finite (A, S)-module chain complex homology equivalent to C.

$\square$

Proposition 4.8 (i) *The morphism j in the localization exact sequence*

$$\ldots \longrightarrow K_1(A) \xrightarrow{i} K_1(S^{-1}A) \xrightarrow{\partial} K_1(A, S) \xrightarrow{j} K_0(A) \longrightarrow \ldots$$

sends the S-torsion class of C to the projective class

$$j : K_1(A,S) \longrightarrow K_0(A) \; ; \; \tau_S(C) \longrightarrow [C] \; .$$

(ii) *The connecting map* $\partial : K_1(S^{-1}A) \longrightarrow K_1(A,S)$ *sends the torsion* $\tau(S^{-1}C)$ *of the contractible based f.g. free* $S^{-1}A$*-module chain complex* $S^{-1}C$ *induced from a based f.g. free* A*-module chain complex* C *with* $S^{-1}H_*(C) = 0$ *to the* S*-torsion class*

$$\partial\tau(S^{-1}C) \;=\; \tau_S(C) \in K_1(A,S) \; .$$

(iii) *If each* $H_r(C)$ $(r \geq 0)$ *is an* (A,S)*-module*

$$\tau_S(C) \;=\; \sum_{r=0}^{\infty}(-)^r \tau_S(H_r(C)) \in K_1(A,S) \; . \qquad \square$$

Definition 4.9 (i) For any commutative ring A let $\max(A)$ denote the set of maximal ideals in A.
(ii) For any field F let $\mathcal{M}(F)$ be the set of irreducible monic polynomials

$$p(z) \;=\; a_0 + a_1 z + a_2 z^2 + \ldots + a_d z^d \in F[z] \;\; (a_d = 1) \; . \qquad \square$$

Remark 4.10 (i) The maximal ideals of a Dedekind ring A are the non-zero prime ideals $\mathcal{P} \lhd A$. For $S = A\backslash\{0\} \subset A$ a standard devissage argument identifies

$$K_*(A,S) \;=\; \bigoplus_{\mathcal{P}\in\max(A)} K_{*-1}(A/\mathcal{P}) \; ,$$

with $S^{-1}A = F$ the quotient field of A. See Chap. 18 for more on the K-theory of Dedekind rings.
(ii) The polynomial extension of a field F is a principal ideal domain $F[z]$ with maximal ideals the principal ideals generated by the monic irreducible polynomials, so that there is defined a bijection

$$\mathcal{M}(F) \xrightarrow{\;\simeq\;} \max(F[z]) \; ; \; p(z) \longrightarrow (p(z)) \; .$$

The quotient field of $F[z]$ is the function field

$$F(z) \;=\; S^{-1}F[z] \;\; (S = F[z]\backslash\{0\})$$

and the localization exact sequence breaks up into short exact sequences

$$0 \longrightarrow K_n(F[z]) \longrightarrow K_n(F(z)) \longrightarrow K_n(F[z],S) \longrightarrow 0 \; .$$

The computation

$$K_1(F[z],S) \;=\; \mathbb{Z}[\mathcal{M}(F)]$$

and its L-theoretic analogues will play an important role in the computation of the high-dimensional knot cobordism groups in Part Two. $\square$

Definition 4.11 A *cartesian morphism* of rings with multiplicative subsets

$$f : (A,S) \longrightarrow (B,T)$$

is a morphism of rings $f : A \longrightarrow B$ such that $f| : S \longrightarrow T$ is a bijection and such that each

$$[f] : A/(s) \longrightarrow B/(f(s)) \quad (s \in S)$$

is an isomorphism of A-modules. □

For any multiplicative subset $S \subset A$ there is defined an A-module isomorphism

$$\varinjlim_{s\in S} A/(s) \xrightarrow{\simeq} S^{-1}A/A \ ; \ \{a_s \mid s \in S\} \longrightarrow a_s/s$$

so that a cartesian morphism induces a cartesian square of rings

$$\begin{array}{ccc} A & \longrightarrow & S^{-1}A \\ \downarrow & & \downarrow \\ B & \longrightarrow & T^{-1}B \end{array}$$

with an exact sequence of A-modules

$$0 \longrightarrow A \longrightarrow S^{-1}A \oplus B \longrightarrow T^{-1}B \longrightarrow 0 \ .$$

Proposition 4.12 *A cartesian morphism $f : (A,S) \longrightarrow (B,T)$ determines an isomorphism of exact categories*

$$\mathbb{H}(A,S) \xrightarrow{\simeq} \mathbb{H}(B,T) \ ; \ M \longrightarrow B \otimes_A T \ ,$$

so that

$$K_*(A,S) = K_*(B,T) \ ,$$

and there are defined a morphism of localization exact sequences

$$\begin{array}{ccccccccc} \dots \longrightarrow & K_n(A) & \longrightarrow & K_n(S^{-1}A) & \longrightarrow & K_n(A,S) & \longrightarrow & K_{n-1}(A) & \longrightarrow \dots \\ & \downarrow & & \downarrow & & \downarrow & & \downarrow & \\ \dots \longrightarrow & K_n(B) & \longrightarrow & K_n(T^{-1}B) & \longrightarrow & K_n(B,T) & \longrightarrow & K_{n-1}(B) & \longrightarrow \dots \end{array}$$

and a Mayer–Vietoris exact sequence

$$\begin{aligned} \dots \longrightarrow K_n(A) \longrightarrow & K_n(S^{-1}A) \oplus K_n(B) \\ & \longrightarrow K_n(T^{-1}B) \longrightarrow K_{n-1}(A) \longrightarrow \dots \ . \end{aligned}$$

Proof See Karoubi [116, App. 5] for a proof that $\mathbb{H}(A,S) \longrightarrow \mathbb{H}(B,T)$ is an isomorphism of exact categories, and Weibel [305, 4.2] for a proof of the naturality. (It is not assumed that B is a flat A-module, so the localization theorem of Quillen [225, Chap. 8] does not give the naturality directly). □

Definition 4.13 The multiplicative subsets $S, T \subset A$ are *coprime* if for every $s \in S$, $t \in T$ the ideals $(s), (t) \triangleleft A$ are coprime, i.e. if there exist $a, b \in A$ such that

$$as + bt = 1 \in A . \qquad \square$$

The localizations inverting coprime multiplicative subsets $S, T \subset A$ and the product multiplicative subset

$$ST = \{st \mid s \in S, t \in T\} \subset A$$

fit into a cartesian square of rings

$$\begin{array}{ccc} A & \longrightarrow & S^{-1}A \\ \big\downarrow & & \big\downarrow \\ T^{-1}A & \longrightarrow & (ST)^{-1}A \end{array}$$

since inclusion defines a cartesian morphism $(A, S) \longrightarrow (T^{-1}A, S)$ and

$$(ST)^{-1}A = S^{-1}(T^{-1}A) .$$

Proposition 4.14 *Let $S, T \subset A$ be coprime multiplicative subsets.*
(i) *The relative algebraic K-groups are such that*

$$K_*(A, ST) = K_*(A, S) \oplus K_*(A, T)$$

and there is defined a Mayer–Vietoris exact sequence

$$\dots \longrightarrow K_n(A) \longrightarrow K_n(S^{-1}A) \oplus K_n(T^{-1}A) \longrightarrow K_n((ST)^{-1}A) \longrightarrow K_{n-1}(A) \longrightarrow \dots .$$

(ii) *A finitely dominated A-module chain complex C is $(ST)^{-1}A$-contractible if and only if it is chain equivalent to the sum $C' \oplus C''$ of a finitely dominated $S^{-1}A$-contractible A-module chain complex C' and a finitely dominated $T^{-1}A$-contractible A-module chain complex C'', in which case $C' \simeq T^{-1}C$ and $C'' \simeq S^{-1}C$.*
(iii) *The ST-torsion class of an $(ST)^{-1}A$-contractible finitely dominated A-module chain complex C is*

$$\tau_{ST}(C) = (\tau_S(T^{-1}C), \tau_T(S^{-1}C)) \in K_1(A, ST) = K_1(A, S) \oplus K_1(A, T) .$$

Proof (i) Apply 4.12 to the cartesian morphism $(A, S) \longrightarrow (T^{-1}A, S)$. Alternatively, note that every (A, ST)-module M has a canonical direct sums splitting

$$M = M' \oplus M''$$

with

$$M' = \{x \in M \mid sx = 0 \in M \text{ for some } s \in S\} ,$$
$$M'' = \{x \in M \mid tx = 0 \in M \text{ for some } t \in T\} ,$$

defining an isomorphism of exact categories

$$\mathbb{H}(A, ST) \longrightarrow \mathbb{H}(A, S) \oplus \mathbb{H}(A, T) \ ; \ M \longrightarrow (M', M'') .$$

(ii) If C is chain equivalent to $C' \oplus C''$ then

$$(ST)^{-1}C \simeq T^{-1}(S^{-1}C') \oplus S^{-1}(T^{-1}C'') \simeq 0 .$$

Conversely, suppose that C is $(ST)^{-1}A$-contractible. Replacing C by a chain equivalent finite f.g. projective A-module chain complex (if necessary) let

$$C' = T^{-1}C \ , \ C'' = S^{-1}C \ ,$$

so that there is defined a short exact sequence of A-module chain complexes

$$0 \longrightarrow C \longrightarrow C' \oplus C'' \longrightarrow (ST)^{-1}C \longrightarrow 0 .$$

Since $(ST)^{-1}C$ is contractible both C' and C'' are finitely dominated A-module chain complexes, such that $C \simeq C' \oplus C''$ with C' $S^{-1}A$-contractible and C'' $T^{-1}A$-contractible.

(iii) Immediate from (ii). □

Definition 4.15 Let A be a ring with a multiplicative subset $S \subset A$. The *S-adic completion of A* is the ring defined by the inverse limit

$$\widehat{A}_S = \varprojlim_{s \in S} A/(s) .$$

Let $\widehat{S} \subset \widehat{A}_S$ be the multiplicative subset defined by the image of $S \subset A$ under the canonical inclusion $A \longrightarrow \widehat{A}_S$. □

Example 4.16 For $S = (s)^\infty \subset A$ as in 4.3

$$\widehat{A}_S = \widehat{A}_s = \ker(I - T : \prod_{k=1}^{\infty} A/(s^k) \longrightarrow \prod_{k=1}^{\infty} A/(s^k))$$

with

$$I - T \ : \ \prod_{k=1}^{\infty} A/(s^k) \longrightarrow \prod_{k=1}^{\infty} A/(s^k) \ ; \ \prod_{k=1}^{\infty} a_k \longrightarrow \prod_{k=1}^{\infty} (a_k - a_{k-1}) . \ \square$$

Proposition 4.17 *For any ring with multiplicative subset (A, S) the localization $S^{-1}A$ and completion $\widehat{A}_S$ are such that the inclusion $A \longrightarrow \widehat{A}_S$ defines a cartesian morphism*

$$(A, S) \longrightarrow (\widehat{A}_S, \widehat{S}) ,$$

inducing a cartesian square of rings

$$\begin{array}{ccc} A & \longrightarrow & S^{-1}A \\ \downarrow & & \downarrow \\ \widehat{A}_S & \longrightarrow & \widehat{S}^{-1}\widehat{A}_S \end{array}$$

and a Mayer–Vietoris exact sequence

$$\dots \longrightarrow K_n(A) \longrightarrow K_n(S^{-1}A) \oplus K_n(\widehat{A}_S) \longrightarrow K_n(\widehat{S}^{-1}\widehat{A}_S) \longrightarrow K_{n-1}(A) \longrightarrow \dots . \ \square$$

5. K-theory of polynomial extensions

The algebraic K-theory of the polynomial extensions $A[z]$, $A[z,z^{-1}]$ enters high-dimensional knot theory via the action of the group of covering translations on the infinite cyclic cover of a knot complement, and a related endomorphism of the chain complex of a Seifert surface.

The exterior of an n-knot $k : S^n \subset S^{n+2}$ is a space

$$X \ = \ \mathrm{cl.}(S^{n+2}\backslash(S^n \times D^2))$$

with a canonical infinite cyclic cover $\overline{X}$, so that $C(\overline{X})$ is a finite f.g. free $\mathbb{Z}[z,z^{-1}]$-module chain complex. A choice of Seifert surface $F^{n+1} \subset S^{n+2}$ for k determines a $\mathbb{Z}$-module chain complex $\dot{C}(F)$ with a chain map $f : \dot{C}(F) \longrightarrow \dot{C}(F)$, such that

$$\mathcal{C}(1 - f + zf : \dot{C}(F)[z,z^{-1}] \longrightarrow \dot{C}(F)[z,z^{-1}]) \simeq C(\overline{X}) \ .$$

In this connection it is convenient to introduce another indeterminate s over $\mathbb{Z}$ (related to z by $s = (1-z)^{-1}$), and to regard $\dot{C}(F)$ as a $\mathbb{Z}[s]$-module chain complex via $s = f$ – this point of view will become particularly useful in the treatment of Seifert and Blanchfield complexes in Chap. 32. In any case, Chap. 5 is devoted to the algebraic K-theory of $A[z]$, $A[z,z^{-1}]$ for an arbitrary ring A.

The direct summand $NK_1(A)$ in the Bass–Heller–Swan ([13], [14]) direct sum decompositions

$$\begin{aligned} K_1(A[z]) \ &= \ K_1(A) \oplus NK_1(A) \ , \\ K_1(A[z,z^{-1}]) \ &= \ K_1(A) \oplus K_0(A) \oplus NK_1(A) \oplus NK_1(A) \end{aligned}$$

is the reduced nilpotent class group $\widetilde{\mathrm{Nil}}_0(A)$. This identification will now be recalled and extended, using the chain complex interpretation of Ranicki [244].

The Laurent polynomial extension $A[z,z^{-1}]$ of a ring A is the localization of $A[z]$ inverting the multiplicative subset

$$Z \ = \ (z)^\infty \ = \ \{z^k \,|\, k \geq 0\} \subset A[z] \ ,$$

that is

$$Z^{-1}A[z] \;=\; A[z,z^{-1}] \;.$$

The Laurent polynomial extension ring can also be expressed as

$$A[z,z^{-1}] \;=\; Z'^{-1}A[z^{-1}] \;,$$

with $Z' = \{z^{-k} \mid k \geq 0\} \subset A[z^{-1}]$.

Let $A^\bullet$ be the multiplicative group of units of A

Terminology 5.1 (i) The $\left\{\begin{array}{l} \textit{leading} \\ \textit{trailing} \end{array}\right.$ *coefficient* of a non-zero polynomial $\sum\limits_{j=-\infty}^{\infty} a_j z^j \in A[z,z^{-1}]$ is the coefficient $a_j \neq 0 \in A$ with the $\left\{\begin{array}{l} \text{largest} \\ \text{smallest} \end{array}\right.$ possible $j \in \mathbb{Z}$.

(ii) The *extreme coefficients* of a non-zero polynomial $\sum\limits_{j=-\infty}^{\infty} a_j z^j \in A[z,z^{-1}]$ are the leading and trailing coefficients $a_m, a_n \neq 0 \in A$ (which may be the same). □

Let

$$i \,:\, A \longrightarrow A[z,z^{-1}] \;,\; i_\pm \,:\, A[z^{\pm 1}] \longrightarrow A[z,z^{-1}] \;,\; \widetilde{i}_\pm \,:\, A \longrightarrow A[z^{\pm 1}]$$

denote the inclusions, so that there is defined a commutative diagram

$$\begin{array}{ccc} A & \xrightarrow{\;\widetilde{i}_+\;} & A[z] \\ {\scriptstyle \widetilde{i}_-}\big\downarrow & \searrow{\scriptstyle i} & \big\downarrow{\scriptstyle i_+} \\ A[z^{-1}] & \xrightarrow{\;i_-\;} & A[z,z^{-1}] \end{array}$$

Given an A-module M let

$$M[z] \;=\; A[z] \otimes_A M \;=\; \sum_{j=0}^{\infty} z^j M$$

be the induced $A[z]$-module. Similarly for the induced $A[z,z^{-1}]$-module

$$M[z,z^{-1}] \;=\; A[z,z^{-1}] \otimes_A M \;=\; \sum_{j=-\infty}^{\infty} z^j M \;.$$

Proposition 5.2 (i) *An $A[z]$-module M is an A-module with an endomorphism*

$$\zeta \,:\, M \longrightarrow M \;;\; x \longrightarrow zx \;.$$

For any such M there is defined an exact sequence of $A[z]$-modules

$$0 \longrightarrow M[z] \xrightarrow{z-\zeta} M[z] \longrightarrow M \longrightarrow 0$$

with

$$M[z] \longrightarrow M \; ; \; \sum_{j=0}^{\infty} z^j x_j \longrightarrow \sum_{j=0}^{\infty} \zeta^j(x_j) \; .$$

(ii) *For any A-modules L, M there is defined an injection*

$$\mathrm{Hom}_A(L,M)[z] \longrightarrow \mathrm{Hom}_{A[z]}(L[z],M[z]) \; ;$$

$$\sum_{j=0}^{\infty} z^j f_j \longrightarrow \left(\sum_{k=0}^{\infty} z^k x_k \longrightarrow \sum_{j=0}^{\infty}\sum_{k=0}^{\infty} z^{j+k} f_j(x_k)\right) \; .$$

If L is f.g. projective this is an isomorphism, allowing the identification

$$\mathrm{Hom}_A(L,M)[z] \;=\; \mathrm{Hom}_{A[z]}(L[z],M[z]) \; . \qquad \square$$

Remark 5.3 If L is not f.g. projective then in general

$$\mathrm{Hom}_A(L,M)[z] \;\neq\; \mathrm{Hom}_{A[z]}(L[z],M[z]) \; .$$

Specifically, consider the special case

$$L \;=\; \sum_{0}^{\infty} A \quad , \quad M = A$$

and the $A[z]$-module morphism

$$f \; : \; L[z] \longrightarrow M[z] \; ; \; \sum_{j=0}^{\infty} z^j(x_0, x_1, \dots) \longrightarrow \sum_{j=0}^{\infty} z^j x_j \; .$$

Then f cannot be expressed as $\sum_{j=0}^{\infty} z^j f_j$ for for some set of A-module morphisms $\{f_j \in \mathrm{Hom}_A(L,M) \,|\, j \geq 0\}$ with $\{j \geq 0 \,|\, f_j \neq 0\}$ finite. $\square$

Similarly for $A[z,z^{-1}]$-modules:

Proposition 5.4 (i) *An $A[z,z^{-1}]$-module M is an A-module with an automorphism*

$$\zeta \; : \; M \longrightarrow M \; ; \; x \longrightarrow zx \; .$$

For any such M there is defined an exact sequence of $A[z,z^{-1}]$-modules

$$0 \longrightarrow M[z,z^{-1}] \xrightarrow{z-\zeta} M[z,z^{-1}] \longrightarrow M \longrightarrow 0$$

with

$$M[z,z^{-1}] \longrightarrow M \; ; \; \sum_{j=-\infty}^{\infty} z^j x_j \longrightarrow \sum_{j=-\infty}^{\infty} \zeta^j(x_j) \; .$$

(ii) *For any A-modules L, M there is defined an injection*

$$\mathrm{Hom}_A(L,M)[z,z^{-1}] \longrightarrow \mathrm{Hom}_{A[z,z^{-1}]}(L[z,z^{-1}],M[z,z^{-1}]) ;$$

$$\sum_{j=-\infty}^{\infty} z^j f_j \longrightarrow \left(\sum_{k=-\infty}^{\infty} z^k x_k \longrightarrow \sum_{j=-\infty}^{\infty}\sum_{k=-\infty}^{\infty} z^{j+k} f_j(x_k)\right) .$$

If L is f.g. projective this is an isomorphism, allowing the identification

$$\mathrm{Hom}_A(L,M)[z,z^{-1}] = \mathrm{Hom}_{A[z,z^{-1}]}(L[z,z^{-1}],M[z,z^{-1}]) . \qquad \square$$

Let $\mathbb{P}(A)$ be the exact category of f.g. projective A-modules, so that

$$K_*(A) = K_*(\mathbb{P}(A)) .$$

An element $\nu \in A$ is *nilpotent* if $\nu^N = 0 \in A$ for some $N \geq 0$. An endomorphism $\nu : P \longrightarrow P$ of an A-module P is nilpotent if $\nu \in \mathrm{Hom}_A(P,P)$ is nilpotent.

Definition 5.5 (i) The *nilpotent category* $\mathrm{Nil}(A)$ is the exact category in which an object is a pair (P,ν) with P a f.g. projective A-module and $\nu : P \longrightarrow P$ a nilpotent endomorphism. A morphism in $\mathrm{Nil}(A)$

$$f : (P,\nu) \longrightarrow (P',\nu')$$

is an A-module morphism $f : P \longrightarrow P'$ such that

$$f\nu = \nu' f : P \longrightarrow P' .$$

A sequence of objects and morphisms in $\mathrm{Nil}(A)$

$$0 \longrightarrow (P,\nu) \longrightarrow (P',\nu') \longrightarrow (P'',\nu'') \longrightarrow 0$$

is exact if $0 \longrightarrow P \longrightarrow P' \longrightarrow P'' \longrightarrow 0$ is an exact sequence of the underlying f.g. projective A-modules.
(ii) The *nilpotent K-groups of A* are given by

$$\mathrm{Nil}_*(A) = K_*(\mathrm{Nil}(A)) .$$

(iii) The *reduced nilpotent K-groups of A* are given by

$$\widetilde{\mathrm{Nil}}_*(A) = \mathrm{coker}(K_*(\mathbb{P}(A)) \longrightarrow K_*(\mathrm{Nil}(A)))$$

with

$$\mathbb{P}(A) \longrightarrow \mathrm{Nil}(A) \; ; \; P \longrightarrow (P,0) ,$$

such that

$$\mathrm{Nil}_*(A) = K_*(A) \oplus \widetilde{\mathrm{Nil}}_*(A) .$$

(iv) The *nilpotent class group* is

$$\mathrm{Nil}_0(A) = K_0(\mathrm{Nil}(A)) ,$$

the group of equivalence classes of objects (P,ν) in $\mathrm{Nil}(A)$, subject to the relation: $(P',\nu') \sim (P,\nu) + (P'',\nu'')$ if there exists a short exact sequence

$$0 \longrightarrow (P,\nu) \longrightarrow (P',\nu') \longrightarrow (P'',\nu'') \longrightarrow 0 .$$

The *reduced nilpotent class group of* A is

$$\widetilde{\mathrm{Nil}}_0(A) = \mathrm{coker}(K_0(A) \longrightarrow \mathrm{Nil}_0(A)) ,$$

the group of equivalence classes of objects (P,ν) in $\mathrm{Nil}(A)$, subject to the relations :

(a) $(P',\nu') \sim (P,\nu) + (P'',\nu'')$ if there exists a short exact sequence

$$0 \longrightarrow (P,\nu) \longrightarrow (P',\nu') \longrightarrow (P'',\nu'') \longrightarrow 0 ,$$

(b) $(P,0) \sim 0$ for any f.g. projective A-module P. □

Proposition 5.6 (Bass [13, XII.9.4], Quillen [225], Grayson [96])
(i) *The relative groups* $K_*(A[z],Z)$ *in the localization exact sequence*

$$\ldots \longrightarrow K_n(A[z]) \xrightarrow{i_+} K_n(A[z,z^{-1}]) \xrightarrow{\partial_+} K_n(A[z],Z) \longrightarrow K_{n-1}(A[z]) \longrightarrow \ldots$$

are the algebraic K*-groups*

$$K_*(A[z],Z) = K_{*-1}(\mathbb{H}(A[z],Z))$$

of the exact category $\mathbb{H}(A[z],Z)$ *of* $(A[z],Z)$*-modules. Use the isomorphism of exact categories*

$$\mathrm{Nil}(A) \xrightarrow{\simeq} \mathbb{H}(A[z],Z) \; ; \; (P,\nu) \longrightarrow \mathrm{coker}(z-\nu : P[z] \longrightarrow P[z])$$

to identify

$$\begin{aligned} K_*(A[z],Z) &= K_{*-1}(\mathbb{H}(A[z],Z)) \\ &= \mathrm{Nil}_{*-1}(A) = K_{*-1}(A) \oplus \widetilde{\mathrm{Nil}}_{*-1}(A) . \end{aligned}$$

(ii) *The localization exact sequence breaks up into split exact sequences*

$$0 \longrightarrow K_n(A[z]) \xrightarrow{i_+} K_n(A[z,z^{-1}]) \xrightarrow{\partial_+} \mathrm{Nil}_{n-1}(A) \longrightarrow 0 ,$$

so that

$$K_n(A[z,z^{-1}]) = K_n(A[z]) \oplus \mathrm{Nil}_{n-1}(A) \quad (n \in \mathbb{Z}) .$$

(iii) *The algebraic* K*-groups of* $A[z]$ *fit into split exact sequences*

$$0 \longrightarrow K_n(A) \xrightarrow{\widetilde{i}_+} K_n(A[z]) \xrightarrow{\widetilde{\partial}_+} \widetilde{\mathrm{Nil}}_{n-1}(A) \longrightarrow 0$$

with

$$\begin{aligned}\widetilde{\partial}_+ \ : \ K_n(A[z]) &\xrightarrow{i_+} K_n(A[z,z^{-1}]) \\ &\xrightarrow{\partial_-} K_n(A[z^{-1}],Z')/K_{n-1}(A) \ = \ \widetilde{\mathrm{Nil}}_{n-1}(A) \ ,\end{aligned}$$

so that

$$K_n(A[z]) \ = \ K_n(A) \oplus \widetilde{\mathrm{Nil}}_{n-1}(A) \ \ (n \in \mathbb{Z}) \ . \qquad \square$$

Remark 5.7 Let P be a f.g. projective A-module, and let $R = \mathrm{Hom}_A(P,P)$ be the endomorphism ring.
(i) A linear element

$$a \ = \ a_0 + a_1 z \in \mathrm{Hom}_{A[z]}(P[z],P[z]) \ = \ R[z]$$

is a unit $a \in R[z]^\bullet$ (i.e. an automorphism of $P[z]$) if and only if $a_0 : P {\longrightarrow} P$ is an automorphism and $\nu = a_1(a_0)^{-1} \in R$ is nilpotent, in which case

$$(a_0 + a_1 z)^{-1} \ = \ (a_0)^{-1}(\sum_{j=0}^{\infty}(-\nu)^j z^j) \ : \ P[z] \longrightarrow P[z] \ .$$

(ii) A linear element

$$a \ = \ a_0 + a_1 z \in \mathrm{Hom}_{A[z,z^{-1}]}(P[z,z^{-1}],P[z,z^{-1}]) \ = \ R[z,z^{-1}]$$

is a unit $a \in R[z,z^{-1}]^\bullet$ if and only if $a_0 + a_1 \in R$ is a unit and

$$f \ = \ (a_0 + a_1)^{-1} a_0 \in R$$

is a *near-projection* in the sense of Lück and Ranicki [176], i.e. such that $f(1-f) \in R$ is nilpotent. If $k \geq 0$ is so large that $(f(1-f))^k = 0 \in R$ then $f^k + (1-f)^k \in R^\bullet$, and

$$f_\omega \ = \ (f^k + (1-f)^k)^{-1} f^k \in R$$

is a projection, $(f_\omega)^2 = f_\omega$. The f.g. projective A-modules

$$M \ = \ f_\omega(P) \ , \ N \ = \ (1 - f_\omega)(P)$$

are such that $P = M \oplus N$ and

$$\begin{aligned} a_0 + a_1 z \ &= \ (b_0 + b_1 z) \oplus (c_0 + c_1 z) \ : \\ P[z,z^{-1}] \ &= \ M[z,z^{-1}] \oplus N[z,z^{-1}] \longrightarrow \\ &\qquad\qquad P[z,z^{-1}] \ = \ M[z,z^{-1}] \oplus N[z,z^{-1}] \end{aligned}$$

with $b_0 : M {\longrightarrow} M$, $c_1 : N {\longrightarrow} N$ isomorphisms and $(b_0)^{-1}b_1 : M {\longrightarrow} M$, $(c_1)^{-1}c_0 : N {\longrightarrow} N$ nilpotent. Thus $b_0 + b_1 z : M[z] {\longrightarrow} M[z]$ is an $A[z]$-module automorphism and $c_0 z^{-1} + c_1 : N[z^{-1}] {\longrightarrow} N[z^{-1}]$ is an $A[z^{-1}]$-module automorphism. $\square$

The split exact sequences

$$0 \longrightarrow K_1(A[z]) \longrightarrow K_1(A[z,z^{-1}]) \longrightarrow \mathrm{Nil}_0(A) \longrightarrow 0\,,$$

$$0 \longrightarrow K_1(A) \longrightarrow K_1(A[z]) \longrightarrow \widetilde{\mathrm{Nil}}_0(A) \longrightarrow 0$$

will now be interpreted in terms of chain homotopy nilpotent chain maps of finitely dominated A-module chain complexes, generalizing the treatment of Ranicki [244].

Definition 5.8 An A-module chain map $\nu : C \longrightarrow C$ is *chain homotopy nilpotent* if for some $k \geq 0$ there exists a chain homotopy $\nu^k \simeq 0 : C \longrightarrow C$, or equivalently if $\nu \in H_0(\mathrm{Hom}_A(C,C))$ is nilpotent. □

See Ranicki [244, Chap. 11] for the definition of the nilpotent class

$$[C,\nu] \in \mathrm{Nil}_0(A)$$

of a finitely dominated A-module chain complex C with a chain homotopy nilpotent map $\nu : C \longrightarrow C$.

Proposition 5.9 *For a finitely dominated A-module chain complex P the following conditions on a chain map $\nu : P \longrightarrow P$ are equivalent:*

(i) *ν is chain homotopy nilpotent,*

(ii) *the $A[z]$-module chain map $1 - z\nu : P[z] \longrightarrow P[z]$ is a chain equivalence, with chain homotopy inverse*

$$(1 - z\nu)^{-1} = \sum_{j=0}^{\infty} z^j \nu^j : P[z] \xrightarrow{\simeq} P[z]\,,$$

(iii) *the $A[z^{-1}]$-module chain map $1 - z^{-1}\nu : P[z^{-1}] \longrightarrow P[z^{-1}]$ is a chain equivalence, with chain homotopy inverse*

$$(1 - z^{-1}\nu)^{-1} = \sum_{j=-\infty}^{0} z^j \nu^{-j} : P[z^{-1}] \longrightarrow P[z^{-1}]\,.$$

Proof (i) $\Longleftrightarrow$ (ii) The $A[z]$-module chain map $1 - z\nu : P[z] \longrightarrow P[z]$ is a chain equivalence if and only if the element

$$1 - z\nu \in H_0(\mathrm{Hom}_{A[z]}(P[z],P[z])) = H_0(\mathrm{Hom}_A(P,P))[z]$$

is a unit, which by 5.7 is the case if and only if $\nu \in H_0(\mathrm{Hom}_A(P,P))$ is nilpotent.
(ii) $\Longleftrightarrow$ (iii) Trivial. □

Proposition 5.10 *The following conditions on a finite f.g. free $A[z]$-module chain complex E^+ are equivalent:*

(i) *E^+ is A-finitely dominated and the A-module chain map*

$$\zeta \; : \; E^+ \longrightarrow E^+ \; ; \; x \longrightarrow zx$$

is chain homotopy nilpotent,

(ii) $H_*(A[z,z^{-1}] \otimes_{A[z]} E^+) = 0$,

(iii) E^+ *is homology equivalent to a finite complex of* $(A[z], Z)$*-modules.*

Proof Immediate from 5.5, 4.5 (i). □

If E^+ is an $A[z,z^{-1}]$-contractible finite f.g. free $A[z]$-module chain complex the nilpotent class of (E^+, ζ) is given by

$$[E^+, \zeta] \;=\; \sum_{r=0}^{\infty}(-)^r[P_r, \nu_r] \in \mathrm{Nil}_0(A)$$

for any finite chain complex in $\mathrm{Nil}(A)$

$$(P,\nu) \;:\; \dots \longrightarrow (P_n, \nu_n) \longrightarrow (P_{n-1}, \nu_{n-1}) \longrightarrow \dots \longrightarrow (P_0, \nu_0)$$

such that E^+ is chain equivalent to $\mathcal{C}(z - \nu : P[z] \longrightarrow P[z])$.

Proposition 5.10′ *The following conditions on a finite f.g. free* $A[z^{-1}]$*-module chain complex* E^- *are equivalent:*

(i) E^- *is* A*-finitely dominated and the* A*-module chain map*

$$\zeta^{-1} \; : \; E^- \longrightarrow E^- \; ; \; x \longrightarrow z^{-1}x$$

is chain homotopy nilpotent,

(ii) $H_*(A[z,z^{-1}] \otimes_{A[z^{-1}]} E^-) = 0$,

(iii) E^- *is homology equivalent to a finite complex of* $(A[z^{-1}], Z')$*-modules.*

Proof As for 5.11, but with E^+, z replaced by E^-, z^{-1}. □

For any $A[z]$-module chain complex E^+ there is defined a short exact sequence of $A[z]$-module chain complexes

$$1 \longrightarrow E^+[z] \xrightarrow{z-\zeta} E^+[z] \xrightarrow{p^+} E^+ \longrightarrow 1$$

with

$$\zeta \; : \; E^+ \longrightarrow E^+ \; ; \; x \longrightarrow zx \; ,$$

$$p^+ \; : \; E^+[z] \;=\; A[z] \otimes_A E^+ \longrightarrow E^+ \; ; \; \sum_{j=1}^{\infty} a_j z^j \otimes x_j \longrightarrow \sum_{j=1}^{\infty} a_j \zeta^j(x_j) \; .$$

The $A[z]$-module chain map

$$r^+ = (p^+ \;\; 1) \; : \; \mathcal{C}(z - \zeta : E^+[z] \longrightarrow E^+[z]) \longrightarrow E^+$$

is a homology equivalence. If E^+ is a projective $A[z]$-module chain complex then r^+ is a chain equivalence.

Definition 5.11 The *fibering obstruction* of an A-finitely dominated based f.g. free $A[z]$-module chain complex E^+ is

$$\Phi(E^+) \;=\; \tau(r^+ : \mathcal{C}(z-\zeta : E^+[z] \longrightarrow E^+[z]) \longrightarrow E^+) \in K_1(A[z]) \;. \qquad \square$$

By analogy with the properties (3.12) of the fibering obstructions $\Phi^+(E)$, $\Phi^-(E)$ of an $A[z,z^{-1}]$-module chain complex band E:

Proposition 5.12 (i) *The fibering obstruction of an A-finitely dominated based f.g. free $A[z]$-module chain complex E^+ is such that $\Phi(E^+) = 1 \in K_1(A[z])$ if and only if E^+ is simple chain equivalent to $\mathcal{C}(z-h : P[z] \longrightarrow P[z])$ for some finite f.g. projective A-module chain complex P and self chain map $h : P \longrightarrow P$.*
(ii) *The torsion of a chain equivalence $f : C \longrightarrow D$ of A-finitely dominated based f.g. free $A[z]$-module chain complexes is the difference of the fibering obstructions*

$$\tau(f) \;=\; \Phi(D) - \Phi(C) \in K_1(A[z]) \;. \qquad \square$$

Remark 5.13 In Chap. 8 it will be proved that if E^+ is a based f.g. free $A[z]$-module chain complex such that the induced $A[z,z^{-1}]$-module chain complex $E = A[z,z^{-1}] \otimes_A E^+$ is a A-finitely dominated then E^+ is A-finitely dominated. See 5.16 (ii) below for the relationship between the fibering obstructions $\Phi^+(E) \in K_1(A[z,z^{-1}])$, $\Phi(E^+) \in K_1(A[z])$ of 3.10, 5.11 in the case when E^+ and E are both A-finitely dominated. $\square$

Proposition 5.14 (i) *The torsion group of $A[z,z^{-1}]$ fits into the direct sum system*

$$K_1(A[z]) \underset{j_+}{\overset{i_+}{\rightleftarrows}} K_1(A[z,z^{-1}]) \underset{\Delta_+}{\overset{\partial_+}{\rightleftarrows}} \mathrm{Nil}_1(A)$$

with

$$\begin{aligned}
i_+ \;&:\; K_1(A[z]) \longrightarrow K_1(A[z,z^{-1}]) \;; \\
&\qquad \tau(E^+) \longrightarrow \tau(A[z,z^{-1}] \otimes_{A[z]} E^+) \;, \\
j_+ \;&:\; K_1(A[z,z^{-1}]) \longrightarrow K_1(A[z]) \;; \\
&\qquad \tau(A[z,z^{-1}] \otimes_{A[z]} E^+) \longrightarrow \Phi(E^+), \\
\partial_+ \;&:\; K_1(A[z,z^{-1}]) \longrightarrow \mathrm{Nil}_0(A) \;; \\
&\qquad \tau(A[z,z^{-1}] \otimes_{A[z]} E^+) \longrightarrow [E^+,\zeta] \;, \\
\Delta_+ \;&:\; \mathrm{Nil}_0(A) \longrightarrow K_1(A[z,z^{-1}]) \;; \\
&\qquad [P,\nu] \longrightarrow \tau(z-\nu : P[z,z^{-1}] \longrightarrow P[z,z^{-1}]) \;.
\end{aligned}$$

(ii) *The torsion group of $A[z]$ fits into the direct sum system*

$$K_1(A) \underset{\widetilde{j}_+}{\overset{\widetilde{i}_+}{\rightleftarrows}} K_1(A[z]) \underset{\widetilde{\Delta}_+}{\overset{\widetilde{\partial}_+}{\rightleftarrows}} \widetilde{\mathrm{Nil}}_0(A)$$

with

$$\widetilde{i}_+ \ : \ K_1(A) \longrightarrow K_1(A[z]) \ ; \ \tau(E) \longrightarrow \tau(E[z]) \ ,$$

$$\widetilde{j}_+ \ : \ K_1(A[z]) \longrightarrow K_1(A) \ ; \ \tau(E^+) \longrightarrow \tau(A \otimes_{A[z]} E^+) \ ,$$

$$\widetilde{\partial}_+ \ : \ K_1(A[z]) \longrightarrow \widetilde{\mathrm{Nil}}_0(A) \ ; \ \tau(E^+) \longrightarrow [E^-, \zeta^{-1}]$$

$$(A[z,z^{-1}] \otimes_{A[z]} E^+ \ = \ A[z,z^{-1}] \otimes_{A[z^{-1}]} E^-) \ ,$$

$$\widetilde{\Delta}_+ \ : \ \widetilde{\mathrm{Nil}}_0(A) \longrightarrow K_1(A[z]) \ ;$$

$$[P, \nu] \longrightarrow \tau(1 - z\nu : P[z] \longrightarrow P[z]) \ . \qquad \square$$

Similarly:

Proposition 5.14′ (i) *The torsion group of* $A[z,z^{-1}]$ *fits into the direct sum system*

$$K_1(A[z^{-1}]) \underset{j_-}{\overset{i_-}{\rightleftarrows}} K_1(A[z,z^{-1}]) \underset{\Delta_-}{\overset{\partial_-}{\rightleftarrows}} \mathrm{Nil}_0(A)$$

with

$$i_- \ : \ K_1(A[z^{-1}]) \longrightarrow K_1(A[z,z^{-1}]) \ ;$$

$$\tau(E^-) \longrightarrow \tau(A[z,z^{-1}] \otimes_{A[z^{-1}]} E^-) \ ,$$

$$j_- \ : \ K_1(A[z,z^{-1}]) \longrightarrow K_1(A[z^{-1}]) \ ;$$

$$\tau(A[z,z^{-1}] \otimes_{A[z^{-1}]} E^-) \longrightarrow$$

$$\Phi(E^-) \ = \ \tau(r^- : \mathcal{C}(z^{-1} - \zeta^{-1} : E^-[z^{-1}] \longrightarrow E^-[z^{-1}]) \longrightarrow E^-) \ ,$$

$$\partial_- \ : \ K_1(A[z,z^{-1}]) \longrightarrow \mathrm{Nil}_0(A) \ ;$$

$$\tau(A[z,z^{-1}] \otimes_{A[z^{-1}]} E^-) \longrightarrow [E^-, \zeta^{-1}] \ ,$$

$$\Delta_- \ : \ \mathrm{Nil}_0(A) \longrightarrow K_1(A[z,z^{-1}]) \ ;$$

$$[P, \nu] \longrightarrow \tau(z^{-1} - \nu : P[z,z^{-1}] \longrightarrow P[z,z^{-1}]) \ .$$

(ii) *The torsion group of* $A[z^{-1}]$ *fits into the direct sum system*

$$K_1(A) \underset{\widetilde{j}_-}{\overset{\widetilde{i}_-}{\rightleftarrows}} K_1(A[z^{-1}]) \underset{\widetilde{\Delta}_-}{\overset{\widetilde{\partial}_-}{\rightleftarrows}} \widetilde{\mathrm{Nil}}_0(A)$$

with

$$\widetilde{i}_- \ : \ K_1(A) \longrightarrow K_1(A[z^{-1}]) \ ; \ \tau(E) \longrightarrow \tau(E[z^{-1}]) \ ,$$

$$\widetilde{j}_- \ : \ K_1(A[z^{-1}]) \longrightarrow K_1(A) \ ; \ \tau(E^-) \longrightarrow \tau(A \otimes_{A[z^{-1}]} E^-) \ ,$$

$$\widetilde{\partial}_- \ : \ K_1(A[z^{-1}]) \longrightarrow \widetilde{\mathrm{Nil}}_0(A) \ ; \ \tau(E^-) \longrightarrow [E^+, \zeta]$$

$$(A[z,z^{-1}] \otimes_{A[z^{-1}]} E^- \ = \ A[z,z^{-1}] \otimes_{A[z]} E^+) \ ,$$

$$\widetilde{\Delta}_- \ : \ \widetilde{\mathrm{Nil}}_0(A) \longrightarrow K_1(A[z^{-1}]) \ ;$$

$$[P, \nu] \longrightarrow \tau(1 - z^{-1}\nu : P[z^{-1}] \longrightarrow P[z^{-1}]) \ . \qquad \square$$

Example 5.15 (i) If A is a Dedekind ring every object (P,f) in $\mathrm{Nil}(A)$ has a finite filtration

$$(\ker(f),0)\subset(\ker(f^2),f|)\subset\ldots\subset(\ker(f^N),f|)\;=\;(P,f)\;\;(f^N=0)$$

such that

$$f(\ker(f^{r+1}))\subseteq\ker(f^r)\;\;(r\geq 0)$$

with A-module isomorphisms

$$\ker(f^{r+1})/\ker(f^r)\xrightarrow{\simeq}f^r(\ker(f^{r+1}))\;;\;x\longrightarrow f^r(x)\;.$$

The filtration quotients $\ker(f^{r+1})/\ker(f^r)$ are f.g. torsion-free A-modules, and hence f.g. projective, with

$$(P,f)\;=\;\sum_{r=0}^{\infty}(\ker(f^{r+1})/\ker(f^r),0)\;=\;(P,0)\in\mathrm{Nil}_0(A)\;.$$

It follows that

$$\mathrm{Nil}_0(A)\;=\;K_0(A)\;\;,\;\;\widetilde{\mathrm{Nil}}_0(A)\;=\;0\;.$$

(In fact, $\widetilde{\mathrm{Nil}}_0(A)=0$ for any regular ring A, by Bass, Heller and Swan [14].)
(ii) Let $A=F$ be a field. Dimension and determinant define isomorphisms

$$\dim\;:\;K_0(F)\xrightarrow{\simeq}\mathbb{Z}\;;\;[P]\longrightarrow\dim_F(P)\;,$$

$$\det\;:\;K_1(F)\xrightarrow{\simeq}F^\bullet\;;\;\tau\longrightarrow\det(\tau)\;,$$

with $F^\bullet=F\backslash\{0\}$, and

$$\begin{aligned}\mathrm{Nil}_0(F)&=K_0(F)=\mathbb{Z}\;,\\ K_1(F[z,z^{-1}])&=K_1(F)\oplus K_0(F)\;=\;F^\bullet\oplus\mathbb{Z}\;.\end{aligned}$$ □

Proposition 5.16 *Let E^+ be a based f.g. free $A[z]$-module chain complex, and let*

$$E\;=\;A[z,z^{-1}]\otimes_{A[z]}E^+\;=\;A[z,z^{-1}]\otimes_{A[z^{-1}]}E^-\;.$$

(i) *If E^+ is A-finitely dominated then $\zeta^{-1}:E/E^-\longrightarrow E/E^-$ is chain homotopy nilpotent and*

$$\widetilde{\partial}_+\Phi(E^+)\;=\;-[E/E^-,\zeta^{-1}]\in\widetilde{\mathrm{Nil}}_0(A)$$

for any compatibly based f.g. free $A[z^{-1}]$-module chain complex E^- with

$$E\;=\;A[z,z^{-1}]\otimes_{A[z]}E^+\;=\;A[z,z^{-1}]\otimes_{A[z^{-1}]}E^-\;.$$

(ii) *If E^+ and E are A-finitely dominated then $\zeta/\zeta:E/E^+\longrightarrow E/E^+$ is chain homotopy nilpotent and*

$$\begin{aligned} i_+\Phi^+(E) - \Phi(E) &= \tau(-\zeta : E[z,z^{-1}] \longrightarrow E[z,z^{-1}]) + \Delta_+(E/E^+,\zeta) \\ &\in K_1(A[z,z^{-1}]) . \end{aligned}$$

Proof (i) The $A[z^{-1}]$-module chain map defined by

$$\begin{aligned} s^- \ :\ E^- &\longrightarrow \mathcal{C}(1-z^{-1}\zeta : E^+[z^{-1}] \longrightarrow E^+[z^{-1}]) ; \\ \sum_{j=-\infty}^{0} \zeta^j(x_j) &\longrightarrow \sum_{j=-\infty}^{0} z^j x_j \ \ (x_j \in E^+) \end{aligned}$$

induces an $A[z,z^{-1}]$-module chain equivalence

$$\begin{aligned} 1\otimes s^- \ :\ E \ =\ & A[z,z^{-1}] \otimes_{A[z^{-1}]} E^- \\ & \xrightarrow{\simeq} \mathcal{C}(1-z^{-1}\zeta : E^+[z,z^{-1}] \longrightarrow E^+[z,z^{-1}]) . \end{aligned}$$

The composite of $1\otimes s^-$ and the $A[z,z^{-1}]$-module chain isomorphism

$$\begin{aligned} t \ :\ & \mathcal{C}(1-z^{-1}\zeta : E^+[z,z^{-1}] \longrightarrow E^+[z,z^{-1}]) \\ & \xrightarrow{\simeq} \mathcal{C}(z-\zeta : E^+[z,z^{-1}] \longrightarrow E^+[z,z^{-1}]) \ ;\ (x,y) \longrightarrow (zx,y) \end{aligned}$$

is a chain homotopy inverse for $1\otimes r^+$

$$\begin{aligned} & t(1\otimes s^-) \ =\ (1\otimes r^+)^{-1} \ : \\ & E \ =\ A[z,z^{-1}] \otimes_{A[z^{-1}]} E^- \xrightarrow{\simeq} \mathcal{C}(z-\zeta : E^+[z,z^{-1}] \longrightarrow E^+[z,z^{-1}]) , \end{aligned}$$

with torsion

$$\tau(t) \ =\ \tau(z : E^+[z,z^{-1}] \longrightarrow E^+[z,z^{-1}]) \in K_1(A[z,z^{-1}]) .$$

The $A[z^{-1}]$-module chain map

$$\begin{aligned} \mathcal{C}(1-z^{-1}\zeta : E^+[z^{-1}] \longrightarrow E^+[z^{-1}]) &\longrightarrow E ; \\ (\sum_{j=-\infty}^{0} z^j x_j, \sum_{j=-\infty}^{0} z^j y_j) &\longrightarrow \sum_{j=-\infty}^{0} \zeta^j(x_j) \end{aligned}$$

is a chain equivalence, and

$$\partial_-\tau(1\otimes s^-) \ =\ [E/E^-,\zeta^{-1}] + [E^+\cap E^-,0] \in \mathrm{Nil}_0(A) .$$

The fibering obstruction $\Phi(E^+) = \tau(r^+) \in K_1(A[z])$ is thus such that

$$\begin{aligned} \widetilde{\partial}_+\tau(r^+) &= \partial_-\tau(1\otimes r^+) \ =\ -\partial_-\tau(1\otimes s^-) - \partial_-\tau(t) \\ &= -[E/E^-,\zeta^{-1}] - [E^+\cap E^-,0] + [E^+,0] \\ &= -[E/E^-,\zeta^{-1}] \in \widetilde{\mathrm{Nil}}_0(A) \subseteq \mathrm{Nil}_0(A) . \end{aligned}$$

(ii) Immediate from (i) and 5.10. □

Remark 5.17 The chain complexes E, E^- in 5.16 (i) need not be A-finitely dominated, although the quotient

$$E/E^- = E^+/E^+ \cap E^-$$

is A-finitely dominated. For example, if $s \in A$ is a central non-unit (e.g. $s = 2 \in A = \mathbb{Z}$) then the chain complexes

$$\begin{aligned} E^+ &= \mathcal{C}(z - s : A[z] \longrightarrow A[z]) , \\ E &= \mathcal{C}(z - s : A[z, z^{-1}] \longrightarrow A[z, z^{-1}]) , \\ E^- &= \mathcal{C}(1 - sz^{-1} : A[z^{-1}] \longrightarrow A[z^{-1}]) \end{aligned}$$

are such that E^+ is A-finitely dominated but E and E^- are not A-finitely dominated, with

$$H_0(E^+) = A \ , \ H_0(E) = H_0(E^-) = A[1/s] . \qquad \square$$

Example 5.18 Let E^+ be the 1-dimensional based f.g. free $A[z]$-module chain complex

$$E^+ : \ldots \longrightarrow 0 \longrightarrow E_1^+ = F[z] \xrightarrow{d^+} E_0^+ = G[z]$$

with F, G based f.g. free A-modules and

$$d^+ = d_0 + d_1 z : F[z] \longrightarrow G[z]$$

for some $d_0, d_1 \in \mathrm{Hom}_A(F, G)$. The 1-dimensional f.g. free $A[z^{-1}]$-module chain complex

$$E^- : \ldots \longrightarrow 0 \longrightarrow E_1^- = F[z^{-1}] \xrightarrow{d^-} E_0^- = G[z^{-1}]$$

defined by

$$d^- = d_1 + d_0 z^{-1} : F[z] \longrightarrow G[z]$$

is such that

$$A[z, z^{-1}] \otimes_{A[z]} E^+ = A[z, z^{-1}] \otimes_{A[z^{-1}]} E^- .$$

The induced A-module chain complex

$$A \otimes_{A[z]} E^+ : \ldots \longrightarrow A[z] \otimes_A E_1^+ = F \xrightarrow{d_0} A \otimes_{A[z]} E_0^+ = G$$

is such that $H_*(A \otimes_{A[z]} E^+) = 0$ if and only if $d_0 : F \longrightarrow G$ is an isomorphism, in which case the inclusion $G \longrightarrow E^-$ is an A-module chain equivalence with

$$\zeta^{-1} = -d_1(d_0)^{-1} : H_0(E^-) = G \longrightarrow H_0(E^-) = G .$$

The $A[z]$-module chain complex E^+ is contractible if and only if $d_0 : F \longrightarrow G$ is an isomorphism and $d_1(d_0)^{-1} : G \longrightarrow G$ is nilpotent, in which case

$$\begin{aligned}\tau(r^+) \;&=\; \tau(E^+) \;=\; (\tau(d_0 : F \longrightarrow G), [G, -d_1(d_0)^{-1}]) \\ &\in K_1(A[z]) \;=\; K_1(A) \oplus \widetilde{\mathrm{Nil}}_0(A) \;.\end{aligned}$$

If $d_1 : F \longrightarrow G$ is an isomorphism the A-module morphism

$$f \;:\; E_0^+ \;=\; G[z] \longrightarrow G \;;\; \sum_{j=0}^{\infty} z^j x_j \longrightarrow \sum_{j=0}^{\infty} (-d_0(d_1)^{-1})^j (x_j)$$

defines an A-module chain equivalence $f : E^+ \longrightarrow G$ such that

$$f\zeta f^{-1} \;=\; -d_0(d_1)^{-1} \;:\; G \longrightarrow G \;,$$

and E^+ is A-finitely dominated with

$$\begin{aligned}\tau(r^+ : \mathcal{C}(\zeta - z : E^+[z] \longrightarrow E^+[z]) \longrightarrow E^+) \;&=\; (\tau(-d_1 : F \longrightarrow G), 0) \\ \in K_1(A[z]) \;&=\; K_1(A) \oplus \widetilde{\mathrm{Nil}}_0(A) \;. \qquad \square\end{aligned}$$

As in Ranicki [240], [244] it is possible to consider both the original "algebraically significant" Bass–Heller–Swan type direct sum decomposition of $K_1(A[z, z^{-1}])$ and $Wh_1(A[z, z^{-1}])$ and the transfer-invariant "geometrically significant" direct sum decomposition. The latter decomposition will play a greater role in this book.

Definition 5.19 The *geometrically significant decomposition* of the Whitehead group of $A[z, z^{-1}]$

$$Wh_1(A[z, z^{-1}]) \;=\; Wh_1(A) \oplus \widetilde{K}_0(A) \oplus \widetilde{\mathrm{Nil}}_0(A) \oplus \widetilde{\mathrm{Nil}}_0(A)$$

is the decomposition determined by the isomorphism

$$\begin{aligned}&Wh_1(A) \oplus \widetilde{K}_0(A) \oplus \widetilde{\mathrm{Nil}}_0(A) \oplus \widetilde{\mathrm{Nil}}_0(A) \xrightarrow{\simeq} Wh_1(A[z, z^{-1}]) \;; \\ &(\tau(f : M \longrightarrow M), [N], [P^+, \nu^+], [P^-, \nu^-]) \longrightarrow \\ &\tau(f : M[z, z^{-1}] \longrightarrow M[z, z^{-1}]) + \tau(-z : N[z, z^{-1}] \longrightarrow N[z, z^{-1}]) \\ &\qquad + \tau(1 - z^{-1}\nu^+ : P^+[z, z^{-1}] \longrightarrow P^+[z, z^{-1}]) \\ &\qquad + \tau(1 - z\nu^- : P^-[z, z^{-1}] \longrightarrow P^-[z, z^{-1}]) \;. \qquad \square\end{aligned}$$

6. *K*-theory of formal power series

The results of Chap. 5 on the K-theory of the polynomial extensions $A[z], A[z,z^{-1}]$ are extended to the K-theory of the formal power series ring $A[[z]]$ and the Novikov ring $A((z))$. This extension will be used in Chap. 8 to show that a finite f.g. free $A[z,z^{-1}]$-module chain complex is A-finitely dominated (i.e. a band) if and only if

$$H_*(A((z))\otimes_{A[z,z^{-1}]} C) \ = \ H_*(A((z^{-1}))\otimes_{A[z,z^{-1}]} C) \ = \ 0 \ .$$

In particular, if $\Delta(z)H_*(C) = 0$ for some $\Delta(z) = \sum\limits_{j=0}^{d} a_j z^j \in A[z,z^{-1}]$ with $a_0, a_d \in A^\bullet$ (e.g. the product of the Alexander polynomials for a fibred knot) then C is a chain complex band.

As in Chap. 5 let z be an indeterminate over a ring A, and let

$$Z \ = \ \{z^k \,|\, k \geq 0\} \subset A[z] \ .$$

Definition 6.1 The *formal power series ring* of A is the Z-adic completion of $A[z]$

$$\widehat{A[z]}_Z \ = \ A[[z]] = \varprojlim_k A[z]/(z^k)$$

consisting of all the formal power series $\sum\limits_{j=0}^{\infty} a_j z^j$ with coefficients $a_j \in A$. □

Given an A-module M let

$$M[[z]] \ = \ A[[z]]\otimes_A M \ = \ \prod_{j=0}^{\infty} z^j M$$

be the induced $A[[z]]$-module.

Proposition 6.2 *For any A-modules L, M there is defined an isomorphism*

$$\mathrm{Hom}_{A[[z]]}(L[[z]], M[[z]]) \longrightarrow \mathrm{Hom}_A(L,M)[[z]] \ ; \ f \longrightarrow \sum_{j=0}^{\infty} z^j f_j$$

with $f_j \in \mathrm{Hom}_A(L,M)$ given by

$$f(x) \ = \ \sum_{j=0}^{\infty} z^j f_j(x) \in M[[z]] \ \ (x \in L) \ . \qquad □$$

Proposition 6.3 (i) *Let P be an A-module, and let*

$$R = \mathrm{Hom}_A(P,P)$$

be the endomorphism ring. An $A[[z]]$-module endomorphism

$$a = \sum_{j=0}^{\infty} a_j z^j \in \mathrm{Hom}_{A[[z]]}(P[[z]],P[[z]])) = R[[z]] \quad (a_j \in R)$$

is an automorphism if and only if $a_0 : P \longrightarrow P$ is an automorphism, that is

$$R[[z]]^\bullet = R^\bullet(1+zR[[z]]) .$$

In particular, the case $P=A$ gives the units of $A[[z]]$ to be

$$A[[z]]^\bullet = A^\bullet(1+zA[[z]]) .$$

(ii) *A finite f.g. free $A[z]$-module chain complex E^+ is such that*

$$H_*(A[[z]] \otimes_{A[z]} E^+) = 0$$

if and only if the A-module chain map

$$\zeta : E^+ \longrightarrow E^+ ; x \longrightarrow zx$$

is an A-module chain equivalence.

Proof (i) The augmentation

$$\widetilde{j}_+ : R[[z]] \longrightarrow R ; \sum_{j=0}^{\infty} a_j z^j \longrightarrow a_0$$

is a ring morphism, so that if $a \in R[[z]]^\bullet$ then $\widetilde{j}_+(a) = a_0 \in R^\bullet$. Conversely, if $a_0 \in R^\bullet$ then $a \in R[[z]]^\bullet$ with inverse

$$a^{-1} = \Big(1 + \sum_{j=1}^{\infty}\big(-(a_0)^{-1}\sum_{k=1}^{\infty} a_k z^k\big)^j\Big)(a_0)^{-1} \in R[[z]]^\bullet .$$

(ii) The exact sequence of $A[z]$-module chain complexes

$$0 \longrightarrow A[z]\otimes_A E^+ \xrightarrow{z-\zeta} A[z]\otimes_A E^+ \longrightarrow E^+ \longrightarrow 0$$

induces an exact sequence of $A[[z]]$-module chain complexes

$$0 \longrightarrow A[[z]]\otimes_A E^+ \xrightarrow{z-\zeta} A[[z]]\otimes_A E^+ \longrightarrow A[[z]]\otimes_{A[z]} E^+ \longrightarrow 0 .$$

Thus $H_*(A[[z]]\otimes_{A[z]}E^+) = 0$ if and only if $z-\zeta : A[[z]]\otimes_A E^+ \longrightarrow A[[z]]\otimes_A E^+$ is an $A[[z]]$-module chain equivalence. This is the case if and only if $\zeta : E^+ \longrightarrow E^+$ is an A-module chain equivalence, with chain homotopy inverse

$$(z-\zeta)^{-1} = -\sum_{j=0}^{\infty} z^j(\zeta)^{-j-1} : A[[z]]\otimes_A E^+ \longrightarrow A[[z]]\otimes_A E^+ . \qquad \square$$

Remark 6.4 The ideal $J = zA[[z]] \triangleleft A[[z]]$ is such that $1 + J \subseteq A[[z]]^\bullet$, so that the projection $A[[z]] \longrightarrow A[[z]]/J = A$ induces an isomorphism

$$K_0(A[[z]]) \xrightarrow{\simeq} K_0(A) \; ; \; [P] \longrightarrow [A \otimes_{A[[z]]} P]$$

(Silvester [267, p. 21]). However, the functor

$$\mathbb{P}(A[[z]]) \longrightarrow \mathbb{P}(A) \; ; \; P \longrightarrow A \otimes_{A[[z]]} P$$

is not an equivalence of categories, and the split surjection $K_1(A[[z]]) \longrightarrow K_1(A)$ is not in general an isomorphism. In Chap. 14 the direct summand $\widehat{NK}_1(A)$ in

$$K_1(A[[z]]) \;=\; K_1(A) \oplus \widehat{NK}_1(A)$$

will be identified with the abelianization $\widehat{W}(A)^{ab}$ of the multiplicative group of Witt vectors

$$\widehat{W}(A) \;=\; 1 + zA[[z]] \subset A[[z]]^\bullet \, . \qquad \square$$

Definition 6.5 (i) The *Novikov ring* of A is the localization of $A[[z]]$ inverting $\widehat{Z} = \{z^k \,|\, k \geq 0\}$

$$A((z)) \;=\; \widehat{Z}^{-1} A[[z]]$$

consisting of the formal power series $\sum\limits_{j=-\infty}^{\infty} a_j z^j$ with coefficients $a_j \in A$ such that $\{j \leq 0 \,|\, a_j \neq 0\}$ is finite.
(ii) The *Novikov homology* of an $A[z, z^{-1}]$-module chain complex C is the homology of the induced $A((z))$-module chain complex

$$H_*(A((z)) \otimes_{A[z,z^{-1}]} C) \, . \qquad \square$$

Proposition 6.6 *The inclusion of the polynomial ring $A[z]$ in the power series ring $A[[z]]$ defines a cartesian morphism of rings with multiplicative subsets*

$$(A[z], Z) \longrightarrow (A[[z]], \widehat{Z})$$

with a cartesian square of rings

$$\begin{array}{ccc} A[z] & \longrightarrow & A[z, z^{-1}] \\ \big\downarrow & & \big\downarrow \\ A[[z]] & \longrightarrow & A((z)) \end{array}$$

Proof Immediate from the definition (4.11), since for every $k \geq 1$

$$A[z]/(z^k) \;=\; A[[z]]/(z^k) \;=\; \sum_{j=0}^{k-1} z^j A \, . \qquad \square$$

By analogy with 5.6 :

Proposition 6.7 (i) *The relative groups $K_*(A[[z]],\widehat{Z})$ in the localization exact sequence*

$$\dots \longrightarrow K_n(A[[z]]) \xrightarrow{\widetilde{i}_+} K_n\big(A((z))\big) \xrightarrow{\partial_+} K_n(A[[z]],\widehat{Z}) \longrightarrow K_{n-1}(A[[z]]) \longrightarrow \dots$$

are the algebraic K-groups $K_{*-1}(\mathbb{H}(A[[z]],\widehat{Z}))$ of the exact category $\mathbb{H}(A[[z]],\widehat{Z})$ of $(A[[z]],\widehat{Z})$-modules. The isomorphism of exact categories

$$\mathrm{Nil}(A) \xrightarrow{\simeq} \mathbb{H}(A[[z]],\widehat{Z}) \ ; \ (P,\nu) \longrightarrow \mathrm{coker}(z-\nu : P[[z]]\longrightarrow P[[z]])$$

gives the identifications

$$K_*(A[[z]],\widehat{Z}) \ = \ K_{*-1}(\mathbb{H}(A[[z]],\widehat{Z})) \ = \ \mathrm{Nil}_{*-1}(A) \ .$$

(ii) *The localization exact sequence for $A[[z]]\longrightarrow A((z))$ breaks up into split short exact sequences*

$$0 \longrightarrow K_n(A[[z]]) \longrightarrow K_n\big(A((z))\big) \xrightarrow{\partial_+} K_n(A[[z]],\widehat{Z}) \longrightarrow 0 \ ,$$

so that

$$K_n\big(A((z))\big) \ = \ K_n(A[[z]]) \oplus \mathrm{Nil}_{n-1}(A) \quad (n \in \mathbb{Z}) \ .$$

Proof (i) The functor

$$\mathbb{H}(A[z],Z) \longrightarrow \mathbb{H}(A[[z]],\widehat{Z}) \ ; \ M \longrightarrow A[[z]] \otimes_{A[z]} M$$

is an equivalence by 4.12, and $\mathbb{H}(A[z],Z)$ is equivalent to $\mathrm{Nil}(A)$ by 5.6 (i).
(ii) The map of localization exact sequences

$$\begin{array}{ccccccccc}
\cdots \longrightarrow & K_n(A[z]) & \longrightarrow & K_n(A[z,z^{-1}]) & \xrightarrow{\partial_+} & K_n(A[z],Z) & \longrightarrow \cdots \\
 & \downarrow & & \downarrow & & \downarrow & \\
\cdots \longrightarrow & K_n(A[[z]]) & \longrightarrow & K_n(A((z))) & \xrightarrow{\partial_+} & K_n(A[[z]],\widehat{Z}) & \longrightarrow \cdots
\end{array}$$

includes isomorphisms $K_*(A[z],Z) \cong K_*(A[[z]],\widehat{Z})$. Use the splitting maps

$$\Delta_+ \ : \ K_*(A[z],Z) \ = \ \mathrm{Nil}_{*-1}(A) \longrightarrow K_*(A[z,z^{-1}])$$

given by 5.6 for the connecting maps $\partial_+ : K_*(A[z,z^{-1}])\longrightarrow K_*(A[z],Z)$ to define splitting maps

$$\Delta_+ \ : \ K_*(A[[z]],\widehat{Z}) \ = \ \mathrm{Nil}_{*-1}(A) \longrightarrow K_*(A[z,z^{-1}]) \longrightarrow K_*\big(A((z))\big)$$

for $\partial_+ : K_*\big(A((z))\big)\longrightarrow K_*(A[[z]],\widehat{Z})$. For $* = 1$

$$\Delta_+ \ : \ K_1(A[[z]],\widehat{Z}) \ = \ \mathrm{Nil}_0(A) \longrightarrow K_1\big(A((z))\big) \ ;$$
$$(P,\nu) \longrightarrow \tau(z-\nu : P((z))\longrightarrow P((z))) \ . \qquad \square$$

For any free $A[[z]]$-module chain complex E^+ there is defined a short exact sequence of $A[[z]]$-module chain complexes

$$0 \longrightarrow E^+[[z]] \xrightarrow{z-\zeta} E^+[[z]] \xrightarrow{p^+} E^+ \longrightarrow 0$$

with

$$\zeta \, : \, E^+ \longrightarrow E^+ \; ; \; x \longrightarrow zx \, ,$$

$$p^+ \, : \, E^+[[z]] \; = \; A[[z]] \otimes_A E^+ \longrightarrow E^+ \; ; \sum_{j=0}^{\infty} a_j z^j \otimes x_j \longrightarrow \sum_{j=0}^{\infty} a_j \zeta^j(x_j) \, .$$

The $A[[z]]$-module chain map

$$(p^+ \;\; 0) \; : \; \mathcal{C}(z-\zeta : E^+[[z]] \longrightarrow E^+[[z]]) \longrightarrow E^+$$

is a homology equivalence. If E^+ is A-finitely dominated the chain complex $T(\zeta)$ has a canonical round $A[[z]]$-finite structure, and q^+ is a chain equivalence.

By analogy with 5.14 (i):

Proposition 6.8 *The torsion group of the Novikov ring $A((z))$ fits into the direct sum system*

$$K_1(A[[z]]) \underset{j_+}{\overset{i_+}{\rightleftarrows}} K_1\big(A((z))\big) \underset{\Delta_+}{\overset{\partial_+}{\rightleftarrows}} \mathrm{Nil}_0(A)$$

with

$$\begin{aligned}
i_+ \; &: \; K_1(A[[z]]) \longrightarrow K_1\big(A((z))\big) \; ; \; \tau(E^+) \longrightarrow \tau(A((z)) \otimes_{A[[z]]} E^+) \, , \\
j_+ \; &: \; K_1\big(A((z))\big) \longrightarrow K_1(A[[z]]) \; ; \; \tau(A((z)) \otimes_{A[[z]]} E^+) \longrightarrow \\
&\qquad\qquad \tau(\mathcal{C}(z-\zeta : E^+[[z]] \longrightarrow E^+[[z]]) \longrightarrow E^+) \, , \\
\partial_+ \; &: \; K_1\big(A((z))\big) \longrightarrow \mathrm{Nil}_0(A) \; ; \; \tau(A((z)) \otimes_{A[[z]]} E^+) \longrightarrow [E^+, \zeta] \, , \\
\Delta_+ \; &: \; \mathrm{Nil}_0(A) \longrightarrow K_1\big(A((z))\big) \; ; \; [P, \nu] \longrightarrow \tau(z - \nu : P((z)) \longrightarrow P((z))) \, .
\end{aligned}$$

□

The Novikov ring $A((z^{-1}))$ is defined by analogy with $A((z))$ using z^{-1} in place of z. The multiplicative subset

$$Z' \; = \; \{z^{-k} \,|\, k \geq 0\} \subset A[z^{-1}]$$

is such that

$$Z'^{-1}A[z^{-1}] \; = \; A[z, z^{-1}] \;\; , \;\; \widehat{A[z^{-1}]}_{Z'} \; = \; A[[z^{-1}]] \; = \; \varprojlim_k A[z^{-1}]/(z^{-k}) \, .$$

The inclusion of $A[z^{-1}]$ in the completion $A[[z^{-1}]]$ defines a cartesian morphism of rings with multiplicative subsets

$$(A[z^{-1}], Z') \longrightarrow (A[[z^{-1}]], \widehat{Z}') \, ,$$

with a cartesian square of rings

$$\begin{array}{ccc} A[z^{-1}] & \longrightarrow & A[z,z^{-1}] \\ \big\downarrow & & \big\downarrow \\ A[[z^{-1}]] & \longrightarrow & A((z^{-1})) \end{array}$$

Proposition 6.7′ (i) *The exact category* $\mathbb{H}(A[[z^{-1}]],\widehat{Z}')$ *of* $(A[[z^{-1}]],\widehat{Z}')$*-modules is such that there are natural equivalences*

$$\mathbb{H}(A[[z^{-1}]],\widehat{Z}') \approx \mathbb{H}(A[z^{-1}],Z') \approx \mathrm{Nil}(A)\ ,$$

and

$$K_*(A[[z^{-1}]],\widehat{Z}') = K_*(A[z^{-1}],Z') = \mathrm{Nil}_{*-1}(A)\ .$$

(ii) *The localization exact sequence for* $A[[z^{-1}]]\longrightarrow A((z^{-1}))$

$$\dots \longrightarrow K_n(A[[z^{-1}]]) \longrightarrow K_n\big(A((z^{-1}))\big) \xrightarrow{\partial_+} K_n(A[[z^{-1}]],\widehat{Z}') \longrightarrow K_{n-1}(A[[z^{-1}]]) \longrightarrow \dots$$

breaks up into split short exact sequences

$$0 \longrightarrow K_n(A[[z^{-1}]]) \longrightarrow K_n\big(A((z^{-1}))\big) \xrightarrow{\partial_+} K_n(A[[z^{-1}]],\widehat{Z}') \longrightarrow 0\ ,$$

and up to isomorphism

$$K_n\big(A((z^{-1}))\big) = K_n(A[[z^{-1}]]) \oplus \mathrm{Nil}_{n-1}(A)\ . \qquad \square$$

Proposition 6.8′ *The torsion group of* $A((z^{-1}))$ *fits into the direct sum system*

$$K_1(A[[z^{-1}]]) \underset{j_-}{\overset{i_-}{\rightleftarrows}} K_1\big(A((z^{-1}))\big) \underset{\Delta_-}{\overset{\partial_-}{\rightleftarrows}} \mathrm{Nil}_0(A)$$

with

$$\begin{aligned}
i_- &: K_1(A[[z^{-1}]]) \longrightarrow K_1\big(A((z^{-1}))\big)\ ;\\
&\qquad \tau(E^-) \longrightarrow \tau(A((z^{-1}))\otimes_{A[[z^{-1}]]} E^-)\ ,\\
j_- &: K_1\big(A((z^{-1}))\big) \longrightarrow K_1(A[[z^{-1}]])\ ;\ \tau(A((z^{-1}))\otimes_{A[[z^{-1}]]} E^-) \longrightarrow\\
&\qquad \tau(\mathcal{C}(z^{-1}-\zeta^{-1} : E^-[[z^{-1}]]\longrightarrow E^-[[z^{-1}]])\longrightarrow E^-)\ ,\\
\partial_- &: K_1\big(A((z^{-1}))\big) \longrightarrow \mathrm{Nil}_0(A)\ ;\\
&\qquad \tau(A((z^{-1}))\otimes_{A[[z^{-1}]]} E^-) \longrightarrow [E^-,\zeta^{-1}]\ ,\\
\Delta_- &: \mathrm{Nil}_0(A) \longrightarrow K_1\big(A((z^{-1}))\big)\ ;\\
&\qquad [P,\nu] \longrightarrow \tau(z^{-1}-\nu : P((z^{-1}))\longrightarrow P((z^{-1})))\ .
\end{aligned}$$

$\square$

7. Algebraic transversality

Algebraic transversality is the chain complex analogue of the geometric transversality technique used to construct fundamental domains for infinite cyclic covers of compact manifolds and finite CW complexes. Refer to Ranicki [244, Chap. 4] for a previous account of algebraic transversality: here, only the additional results required for the new applications are proved. The construction in Part Two of the algebraic invariants of knots will make use of the L-theory version of algebraic transversality for chain complexes with Poincaré duality, the analogue of the geometric transversality construction of a Seifert surface fundamental domain for the infinite cyclic cover of a knot complement.

A Mayer–Vietoris presentation of a finite f.g. free $A[z,z^{-1}]$-module chain complex is the chain complex analogue of the cutting of an infinite cyclic cover of a connected finite CW complex into two halves by means of a separating finite subcomplex. Mayer–Vietoris presentations will be used in Chap. 8 to prove that a finite f.g. free $A[z,z^{-1}]$-module chain complex E is A-finitely dominated if and only if

$$H_*(A((z))\otimes_{A[z,z^{-1}]} E) = H_*(A((z^{-1}))\otimes_{A[z,z^{-1}]} E) = 0 .$$

Definition 7.1 Let M be a based f.g. free $A[z,z^{-1}]$-module, with basis elements $\{b_1,b_2,\dots,b_r\}$. A f.g free $A[z]$-submodule $M^+\subset M$ is *compatibly based* if it is such that

$$A[z,z^{-1}]\otimes_{A[z]} M^+ = M$$

with basis elements $\{z^{-N_1^+}b_1, z^{-N_1^+}b_2,\dots,z^{-N_r^+}b_r\}$ for some $N_j^+\geq 0$, in which case

$$\begin{aligned}\tau(1: A[z,z^{-1}]\otimes_{A[z]} M^+\longrightarrow M) &= \tau(z^{N^+}: A[z,z^{-1}]\longrightarrow A[z,z^{-1}]) = 0\\ \in Wh_1(A[z,z^{-1}]) &= \operatorname{coker}(K_1(\mathbb{Z}[z,z^{-1}])\longrightarrow K_1(A[z,z^{-1}]))\end{aligned}$$

$$(N^+ = \sum_{j=0}^{r} N_j^+ \geq 0) .$$

Similarly for f.g free $A[z^{-1}]$-submodules $M^-\subset M$ such that

$$A[z,z^{-1}]\otimes_{A[z^{-1}]} M^- \;=\; M$$

with basis elements $\{z^{N_1^-}b_1, z^{N_2^-}b_2, \dots, z^{N_r^-}b_r\}$ for some $N_j^- \geq 0$. □

Definition 7.2 A *Mayer–Vietoris presentation* (E^+, E^-) of a based f.g. free $A[z,z^{-1}]$-module chain complex E is a choice of a compatibly based f.g. free $A[z]$-module subcomplex $E^+ \subset E$ and a compatibly based f.g. free $A[z^{-1}]$-module subcomplex $E^- \subset E$. □

Proposition 7.3 (i) *Every based f.g. free $A[z,z^{-1}]$-module chain complex E admits Mayer–Vietoris presentations (E^+, E^-).*
(ii) *For any Mayer–Vietoris presentation (E^+, E^-) of a based f.g. free $A[z,z^{-1}]$-module chain complex E there is defined an A-module exact sequence*

$$0 \longrightarrow E^+\cap E^- \longrightarrow E^+\oplus E^- \longrightarrow E \longrightarrow 0 \;.$$

Also, the subcomplexes

$$C \;=\; E^+\cap E^- \;,\; D \;=\; E^+\cap zE^- \subset E$$

are based f.g. free A-module subcomplexes, and there is defined an exact sequence of $A[z,z^{-1}]$-module chain complexes and chain maps

$$0 \longrightarrow C[z,z^{-1}] \xrightarrow{f-zg} D[z,z^{-1}] \xrightarrow{h} E \longrightarrow 0$$

with

$$f \;:\; C \longrightarrow D \;;\; x \longrightarrow \zeta x \;,$$

$$g \;:\; C \longrightarrow D \;;\; x \longrightarrow x \;,$$

$$h \;:\; D[z,z^{-1}] \longrightarrow E \;;\; \sum_{j=-\infty}^{\infty} a_j z^j \otimes x \longrightarrow \sum_{j=-\infty}^{\infty} a_j z^j x \;,$$

and such that the chain equivalence

$$q \;=\; (h \;\; 0) \;:\; \mathcal{C}(f-zg : C[z,z^{-1}] \longrightarrow D[z,z^{-1}]) \longrightarrow E$$

has $\tau(q) = 0 \in Wh_1(A[z,z^{-1}])$, i.e.

$$\tau(q) \in \operatorname{im}(K_1(\mathbb{Z}[z,z^{-1}]) \longrightarrow K_1(A[z,z^{-1}])) \;.$$

Proof (i) Let E be n-dimensional

$$E \;:\; \longrightarrow 0 \longrightarrow E_n \longrightarrow E_{n-1} \longrightarrow \dots \longrightarrow E_1 \longrightarrow E_0 \longrightarrow 0 \dots \;,$$

and let F_r be the f.g. free A-module generated by the basis of E_r

$$E_r \;=\; F_r[z,z^{-1}] \;\; (0 \leq r \leq n) \;.$$

Now

$$d_E = \sum_{j=-N_r^+}^{N_r^-} z^j d_j : E_r = \sum_{j=-\infty}^{\infty} z^j F_r \longrightarrow E_{r-1} = \sum_{j=-\infty}^{\infty} z^j F_{r-1}$$

for some A-module morphisms $d_j : F_r \longrightarrow F_{r-1}$ and some integers $N_r^+, N_r^- \geq 0$. Let E^+ be the n-dimensional compatibly based f.g. free $A[z]$-module chain complex defined by

$$E_r^+ = \sum_{j=-N_r^+}^{\infty} z^j F_r \subset E_r \ (0 \leq r \leq n) .$$

Similarly, let E^- be the n-dimensional f.g. free $A[z^{-1}]$-module chain complex defined by

$$E_r^- = \sum_{j=-\infty}^{N_r^-} z^j F_r \subset E_r \ \ (0 \leq r \leq n) .$$

(ii) Let F be a based f.g. free A-module, and let $M = F[z, z^{-1}]$ be the induced based f.g. free $A[z, z^{-1}]$-module. The Mayer–Vietoris presentation (M^+, M^-) of M defined for any $N^+, N^- \geq 0$ by

$$M^+ = \sum_{j=-N^+}^{\infty} z^j F \ , \ M^- = \sum_{j=-\infty}^{N^-} z^j F \subset M$$

determines an exact sequence of A-modules

$$0 \longrightarrow M^+ \cap M^- \longrightarrow M^+ \oplus M^- \longrightarrow M \longrightarrow 0$$

with

$$M^+ \cap M^- = \sum_{j=-N^+}^{N^-} z^j F \ , \ M^+ \cap \zeta M^- = \sum_{j=-N^+}^{N^-+1} z^j F$$

based f.g. free A-modules. The A-module morphisms

$$\begin{aligned} f &: M^+ \cap M^- \longrightarrow M^+ \cap \zeta M^- \ ; \ x \longrightarrow \zeta x \ , \\ g &: M^+ \cap M^- \longrightarrow M^+ \cap \zeta M^- \ ; \ x \longrightarrow x \ , \\ h &: (M^+ \cap \zeta M^-)[z, z^{-1}] \longrightarrow M \ ; \ \sum_{j=-\infty}^{\infty} a_j z^j \otimes x \longrightarrow \sum_{j=-\infty}^{\infty} a_j z^j x \ , \end{aligned}$$

are such that the chain equivalence

$$q = (h \ 0) : \mathcal{C}(f - zg : (M^+ \cap M^-)[z, z^{-1}] \longrightarrow (M^+ \cap \zeta M^-)[z, z^{-1}]) \longrightarrow M$$

has torsion $\tau(q) = 0 \in Wh_1(A[z, z^{-1}])$. □

Proposition 7.4 *A based f.g. free $A[z,z^{-1}]$-module chain complex E is a band (i.e A-finitely dominated) if and only if E^+ and E^- are A-finitely dominated, for any Mayer–Vietoris presentation (E^+,E^-) of E with respect to any choice of base, in which case the A-module chain maps*

$$\begin{aligned} \zeta \,:\, E/E^+ &\longrightarrow E/E^+ \;;\; x \longrightarrow zx \;, \\ \zeta^{-1} \,:\, E/E^- &\longrightarrow E/E^- \;;\; x \longrightarrow z^{-1}x \end{aligned}$$

are chain homotopy nilpotent. The fibering obstructions of a band E $\Phi^{\pm}(E) \in Wh_1(A[z,z^{-1}])$ have the form

$$\begin{aligned} \Phi^+(E) &= (\phi^+,-[E^-],-[E/E^+,\zeta],-[E/E^-,\zeta^{-1}]) \;, \\ \Phi^-(E) &= (\phi^-,[E^+],-[E/E^+,\zeta],-[E/E^-,\zeta^{-1}]) \\ \in Wh_1(A[z,z^{-1}]) &= Wh_1(A)\oplus \widetilde{K}_0(A)\oplus \widetilde{\mathrm{Nil}}_0(A)\oplus \widetilde{\mathrm{Nil}}_0(A) \end{aligned}$$

with respect to the geometrically significant direct sum decomposition of $Wh_1(A[z,z^{-1}])$ (5.19), for some $\phi^+,\phi^- \in Wh_1(A)$ such that

$$\phi^+ - \phi^- \;=\; \tau(\zeta : E \longrightarrow E) \in Wh_1(A) \;.$$

Proof The $A[z]$-module chain equivalence of 5.16

$$r^+ \;:\; \mathcal{C}(z-\zeta : E^+[z] \longrightarrow E^+[z]) \longrightarrow E^+$$

has torsion of the form

$$\tau(r^+) \;=\; (\phi^-,[E/E^-,\zeta^{-1}]) \in K_1(A[z]) \;=\; K_1(A)\oplus \widetilde{\mathrm{Nil}}_0(A)$$

inducing

$$\begin{aligned} \tau(1\otimes r^+) \;&=\; (\phi^-,0,0,[E/E^-,\zeta^{-1}]) \\ &\in Wh_1(A[z,z^{-1}]) \;=\; Wh_1(A)\oplus \widetilde{K}_0(A)\oplus \widetilde{\mathrm{Nil}}_0(A)\oplus \widetilde{\mathrm{Nil}}_0(A) \;. \end{aligned}$$

The $A[z,z^{-1}]$-module chain equivalence

$$\begin{aligned} h \;:\; A[z,z^{-1}]\otimes_{A[z]} T(\zeta) \;&=\; \mathcal{C}(z-\zeta : E^+[z,z^{-1}] \longrightarrow E^+[z,z^{-1}]) \\ \longrightarrow T^+(\zeta^{-1}) \;&=\; \mathcal{C}(1-z\zeta^{-1} : E[z,z^{-1}] \longrightarrow E[z,z^{-1}]) \;; \\ &(x,y) \longrightarrow (\zeta^{-1}(x),y) \end{aligned}$$

has torsion

$$\begin{aligned} \tau(h) \;&=\; \tau(z-\zeta : (E/E^+)[z,z^{-1}] \longrightarrow (E/E^+)[z,z^{-1}]) \\ &\qquad + \tau(\zeta^{-1} : E[z,z^{-1}] \longrightarrow E[z,z^{-1}]) \\ &=\; (-\tau(\zeta),[E/E^+],\,[E/E^+,\zeta],0) \;, \\ &\in Wh_1(A[z,z^{-1}]) \;=\; Wh_1(A)\oplus \widetilde{K}_0(A)\oplus \widetilde{\mathrm{Nil}}_0(A)\oplus \widetilde{\mathrm{Nil}}_0(A) \end{aligned}$$

It now follows from the factorization

$$1 \otimes q^+ \;=\; q^+h \;:\; A[z,z^{-1}] \otimes_{A[z]} T(\zeta) \longrightarrow$$
$$T^+(\zeta^{-1}) \longrightarrow A[z,z^{-1}] \otimes_{A[z]} E^+ \;=\; E$$

that

$$\begin{aligned}\Phi^+(E) &= \tau(q^+) \;=\; \tau(1 \otimes q^+) - \tau(h)\\ &= (\phi^+, -[E^-], -[E/E^+, \zeta], -[E/E^-, \zeta^{-1}])\\ &\in Wh_1(A[z,z^{-1}]) \;=\; Wh_1(A) \oplus \widetilde{K}_0(A) \oplus \widetilde{\mathrm{Nil}}_0(A) \oplus \widetilde{\mathrm{Nil}}_0(A)\ ,\end{aligned}$$

with $\phi^+ = \phi^- + \tau(\zeta) \in Wh_1(A)$. Similarly for $\Phi^-(E)$. □

Example 7.5 Let X be a finite CW complex with fundamental group $\pi_1(X) = \pi \times \mathbb{Z}$. The cellular chain complex of the universal cover $\widetilde{X}$ of X is a finite chain complex $C(\widetilde{X})$ of based f.g. free $\mathbb{Z}[\pi_1(X)]$-modules, with

$$\mathbb{Z}[\pi_1(X)] \;=\; \mathbb{Z}[\pi][z,z^{-1}]\ .$$

Let $\overline{X} = \widetilde{X}/\pi$ be the untwisted infinite cyclic cover of X classified by $\pi_1(\overline{X}) = \pi \subset \pi_1(X)$, and let $\zeta : \overline{X} \longrightarrow \overline{X}$ be the covering translation. A map $c : X \longrightarrow S^1$ inducing

$$c_* \;=\; \text{projection} \;:\; \pi_1(X) \;=\; \pi \times \mathbb{Z} \longrightarrow \pi_1(S^1) \;=\; \mathbb{Z}$$

is a classifying map for $\overline{X} = c^*\mathbb{R}$. Replacing X by a simple homotopy equivalent finite CW complex (e.g. a closed regular neighbourhood in some high-dimensional Euclidean space) it is possible to choose $c : X \longrightarrow S^1$ transverse regular at a point $* \in S^1$, so that $U = c^{-1}(\{*\}) \subset X$ is a subcomplex with an embedding $U \times (0,1) \subset X$. Cutting X along U gives a compact fundamental domain $(V; U, \zeta U)$ for $\overline{X}$, with

$$\overline{X} \;=\; \bigcup_{j=-\infty}^{\infty} \zeta^j(V; U, \zeta U)\ .$$

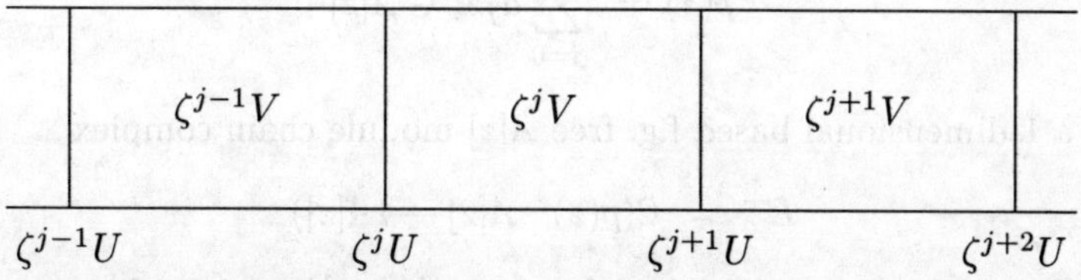

The subcomplexes

$$\overline{X}^+ \;=\; \bigcup_{j=0}^{\infty} \zeta^j V\ ,\ \overline{X}^- \;=\; \bigcup_{j=-\infty}^{-1} \zeta^j V \subset \overline{X}$$

are such that

$$\overline{X}^+ \cap \overline{X}^- = U \ , \ \overline{X}^+ \cap \zeta \overline{X}^- = V \ .$$

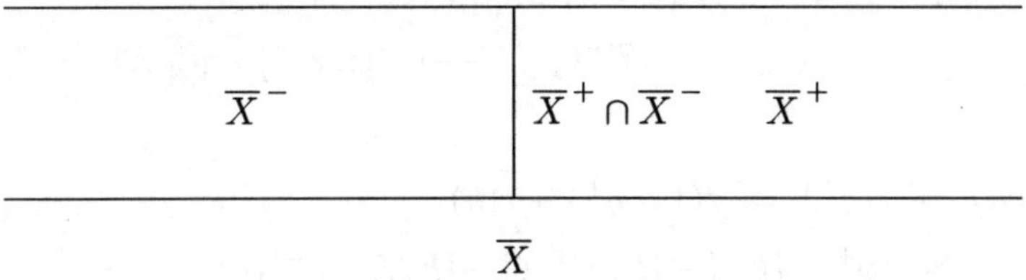

Codimension 1 manifold transversality thus determines a finite Mayer-Vietoris presentation $(E^+, E^-) = (C(\widetilde{X}^+), C(\widetilde{X}^-))$ of $E = C(\widetilde{X})$ with

$$C = E^+ \cap E^- = C(\widetilde{U}) \ , \ D = E^+ \cap zE^- = C(\widetilde{V}) \ .$$

If X is a band the fibering obstructions $\Phi^\pm(X) \in Wh(\pi_1(X))$ are of the form

$$\begin{aligned}
&[\Phi^+(X)] = \\
&\quad (\phi^+, -[\overline{X}^-], -[C(\widetilde{X})/C(\widetilde{X}^+), \zeta], -[C(\widetilde{X})/C(\widetilde{X}^-), \zeta^{-1}]) \ , \\
&[\Phi^-(X)] = \\
&\quad (\phi^-, [\overline{X}^+], -[C(\widetilde{X})/C(\widetilde{X}^+), \zeta], -[C(\widetilde{X})/C(\widetilde{X}^-), \zeta^{-1}]) \ , \\
&\in Wh((\pi_1(X))) \\
&\quad = Wh(\pi_1(\overline{X})) \oplus \widetilde{K}_0(\mathbb{Z}[\pi_1(\overline{X})]) \oplus \widetilde{\mathrm{Nil}}_0(\mathbb{Z}[\pi_1(\overline{X})]) \oplus \widetilde{\mathrm{Nil}}_0(\mathbb{Z}[\pi_1(\overline{X})])
\end{aligned}$$

with $\phi^+ = \phi^- + \tau(\zeta : \overline{X} \longrightarrow \overline{X}) \in Wh(\pi_1(\overline{X}))$. See Ranicki [244, 8.15] for a discussion of codimension 1 transversality for CW complexes with maps $X \longrightarrow S^1$ and chain complexes over polynomial extension rings $A[z, z^{-1}]$. It is possible to avoid the use of manifolds and obtain codimension 1 CW complex transversality by purely combinatorial methods. □

Example 7.6 Given a polynomial

$$p(z) = \sum_{j=0}^{d} a_j z^j \in A[z]$$

define a 1-dimensional based f.g. free $A[z]$-module chain complex

$$E^+ = \mathcal{C}(p(z) : A[z] \longrightarrow A[z]) \ .$$

The induced 1-dimensional based f.g. free $A[z, z^{-1}]$-module chain complex

$$E = A[z, z^{-1}] \otimes_{A[z]} E^+ = \mathcal{C}(p(z) : A[z, z^{-1}] \longrightarrow A[z, z^{-1}])$$

has a Mayer–Vietoris presentation (E^+, E^-) with

$$E^- = \mathcal{C}(p(z) : \sum_{j=-\infty}^{-1} z^j A \longrightarrow \sum_{j=-\infty}^{d-1} z^j A) ,$$

$$C = E^+ \cap E^- = \sum_{j=0}^{d-1} z^j A ,$$

$$D = E^+ \cap \zeta E^- = \mathcal{C}(p(z) : A \longrightarrow \sum_{j=0}^{d} z^j A) ,$$

$$f : C \longrightarrow D ; \sum_{j=0}^{d-1} z^j x_j \longrightarrow \sum_{j=1}^{d} z^j x_{j-1} ,$$

$$g : C \longrightarrow D ; \sum_{j=0}^{d-1} z^j x_j \longrightarrow \sum_{j=0}^{d-1} z^j x_j$$

and an exact sequence

$$0 \longrightarrow C[z] \xrightarrow{f-zg} D[z] \longrightarrow E^+ \longrightarrow 0 .$$

The following conditions are equivalent :

(i) $p(z) \in A[z]^\bullet$,
(ii) $H_*(E^+) = 0$,
(iii) the A-module chain map $f : C \longrightarrow D$ is a chain equivalence, and $f^{-1}g : C \longrightarrow C$ is chain homotopy nilpotent,
(iv) $a_0 \in A^\bullet$ and the $d \times d$ matrix with entries in A

$$\begin{pmatrix} -a_1(a_0)^{-1} & 1 & 0 & \dots & 0 \\ -a_2(a_0)^{-1} & 0 & 1 & \dots & 0 \\ -a_3(a_0)^{-1} & 0 & 0 & \dots & 0 \\ \vdots & \vdots & \vdots & \ddots & \vdots \\ -a_d(a_0)^{-1} & 0 & 0 & \dots & 0 \end{pmatrix}$$

is nilpotent.

□

Proposition 7.7 *The following conditions on a finite f.g. free $A[z,z^{-1}]$-module chain complex E are equivalent :*

(i) $H_*(E) = 0$,
(ii) *for any Mayer–Vietoris presentation (E^+, E^-) of E with respect to any choice of basis for E the $A[z]$-module chain complex E^+ is A-finitely dominated and the A-module chain map*

$$\zeta : E^+ \longrightarrow E^+ ; x \longrightarrow zx$$

is chain homotopy nilpotent,

(iii) *for any Mayer–Vietoris presentation* (E^+, E^-) *of* E *the* $A[z^{-1}]$*-module chain complex* E^- *is* A*-finitely dominated and the* A*-module chain map*

$$\zeta^{-1} \,:\, E^- \longrightarrow E^- \;;\; x \longrightarrow z^{-1}x$$

is chain homotopy nilpotent.

Proof (i) $\Longrightarrow$ (ii)+(iii) For any Mayer–Vietoris presentation (E^+, E^-) of a finite f.g. free $A[z, z^{-1}]$-module chain complex E there is defined a commutative diagram of $A[z, z^{-1}]$-module chain complexes and chain maps

$$\begin{array}{ccccc} & & E^+[z,z^{-1}] \oplus E^-[z,z^{-1}] & & \\ & \nearrow \begin{pmatrix} -z & 0 \\ 0 & 1 \end{pmatrix} & & \searrow \begin{pmatrix} 1-z^{-1}\zeta & 0 \\ 0 & 1-z\zeta^{-1} \end{pmatrix} & \\ E^+[z,z^{-1}] \oplus E^-[z,z^{-1}] & & & & E^+[z,z^{-1}] \oplus E^-[z,z^{-1}] \\ & \nwarrow i = \begin{pmatrix} i^+ \\ i^- \end{pmatrix} & & \nearrow j = \begin{pmatrix} j^+ \\ j^- \end{pmatrix} & \\ & C[z,z^{-1}] & \xrightarrow{\;f-zg\;} & D[z,z^{-1}] & \end{array}$$

with

$$\begin{aligned} f \,&:\, C \,=\, E^+ \cap E^- \longrightarrow D \,=\, E^+ \cap \zeta E^- \;;\; x \longrightarrow \zeta x \,, \\ g \,&:\, C \,=\, E^+ \cap E^- \longrightarrow D \,=\, E^+ \cap \zeta E^- \;;\; x \longrightarrow x \,, \\ i^{\pm} \,&=\, \text{inclusion} \,:\, C \,=\, E^+ \cap E^- \longrightarrow E^{\pm} \,, \\ j^+ \,&=\, \text{inclusion} \,:\, D \,=\, E^+ \cap \zeta E^- \longrightarrow E^+ \,, \\ j^- \,&:\, D \,=\, E^+ \cap \zeta E^- \longrightarrow E^- \;;\; x \longrightarrow \zeta^{-1}x \,. \end{aligned}$$

(The commutative diagram of Ranicki [244, 10.13] is of this type). If E is contractible each of the chain maps in the diagram is a chain equivalence, the chain complexes E^+, E^- are A-finitely dominated, the $A[z, z^{-1}]$-module chain map $f - zg : C[z, z^{-1}] \longrightarrow D[z, z^{-1}]$ is a chain equivalence, the A-module chain maps $i : C \longrightarrow E^+ \oplus E^-$, $j : E \longrightarrow E^+ \oplus E^-$ are chain equivalences, and the A-module chain maps $\zeta : E^+ \longrightarrow E^+$, $\zeta^{-1} : E^- \longrightarrow E^-$ are chain homotopy nilpotent.

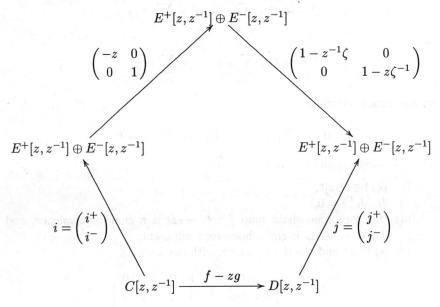

(ii) $\Longrightarrow$ (i) For A-finitely dominated E^+ the $A[z]$-module chain equivalence $q^+ : T(\zeta) \longrightarrow E^+$ of 5.16 induces an $A[z,z^{-1}]$-module chain equivalence

$$\begin{aligned} &1 \otimes q^+ : \\ &A[z,z^{-1}] \otimes_{A[z]} T(\zeta) \;=\; \mathcal{C}(\zeta - z : E^+[z,z^{-1}] \longrightarrow E^+[z,z^{-1}]) \longrightarrow E \,, \end{aligned}$$

with

$$E^+[z,z^{-1}] \;=\; A[z,z^{-1}] \otimes_A E^+ \;\;,\;\; E \;=\; A[z,z^{-1}] \otimes_{A[z]} E^+ \,.$$

If also $\zeta : E^+ \longrightarrow E^+$ is chain homotopy nilpotent the $A[z,z^{-1}]$-module chain map $\zeta - z : E^+[z,z^{-1}] \longrightarrow E^+[z,z^{-1}]$ is a chain equivalence, with chain homotopy inverse

$$(\zeta - z)^{-1} \;=\; -\sum_{j=0}^{\infty} z^{-j-1}\zeta^j \;:\; E^+[z,z^{-1}] \longrightarrow E^+[z,z^{-1}] \,,$$

so that $A[z,z^{-1}] \otimes_{A[z]} T(\zeta)$ is a contractible $A[z,z^{-1}]$-module chain complex and hence so is E.
(iii) $\Longrightarrow$ (i) As for (ii) $\Longrightarrow$ (i), but with E^+, z replaced by E^-, z^{-1}. □

Example 7.8 Given a polynomial $p(z) = \sum\limits_{j=0}^{d} a_j z^j \in A[z]$ let $E, E^+, E^-, C,$ f, g be as in 7.6, with

$$E \;=\; \mathcal{C}(p(z) : A[z,z^{-1}] \longrightarrow A[z,z^{-1}]) \;\;,\;\; C \;=\; \sum_{j=0}^{d-1} z^j A \;\;\text{etc.}$$

The following conditions are equivalent:

(i) $p(z) \in A[z,z^{-1}]^\bullet$,
(ii) $H_*(E) = 0$,
(iii) the A-module chain map $f - g : C \longrightarrow E$ is a chain equivalence, and

$$(f-g)^{-1} f (f-g)^{-1} g \;:\; C \longrightarrow C$$

is chain homotopy nilpotent,
(iv) the $(d+1) \times (d+1)$ matrices with entries in A defined by

$$M_1 \;=\; \begin{pmatrix} a_0 & 0 & 0 & \dots & 0 \\ a_1 & 1 & 0 & \dots & 0 \\ a_2 & 0 & 1 & \dots & 0 \\ \vdots & \vdots & \vdots & \ddots & \vdots \\ a_d & 0 & 0 & \dots & 1 \end{pmatrix} \;,\; M_2 \;=\; \begin{pmatrix} 0 & 1 & 0 & \dots & 0 \\ 0 & 0 & 1 & \dots & 0 \\ 0 & 0 & 0 & \dots & 0 \\ \vdots & \vdots & \vdots & \ddots & \vdots \\ 0 & 0 & 0 & \dots & 0 \end{pmatrix}$$

are such that

$$M_1 - M_2 = \begin{pmatrix} a_0 & -1 & 0 & \dots & 0 \\ a_1 & 1 & -1 & \dots & 0 \\ a_2 & 0 & 1 & \dots & 0 \\ \vdots & \vdots & \vdots & \ddots & \vdots \\ a_d & 0 & 0 & \dots & 1 \end{pmatrix}$$

is invertible (which happens if and only if $\sum_{j=0}^{d} a_j \in A^{\bullet}$) and the product

$$M_1(M_1 - M_2)^{-1} M_2 (M_1 - M_2)^{-1}$$

is nilpotent.

□

8. Finite domination and Novikov homology

Finite domination for chain complexes was already defined in Chap. 1. Following Ranicki [246], algebraic transversality will now be used to prove that a finite f.g. free $A[z,z^{-1}]$-module chain complex E is A-finitely dominated if and only if the Novikov homology vanishes

$$H_*(A((z))\otimes_{A[z,z^{-1}]}E) \;=\; H_*(A((z^{-1}))\otimes_{A[z,z^{-1}]}E) \;=\; 0\ .$$

The special case $A = \mathbb{Z}$ will be used in Chap. 33 to prove that a high-dimensional n-knot $k : S^n \subset S^{n+2}$ fibres if and only if the extreme coefficients of the Alexander polynomials are units.

The *suspension* of an A-module chain complex E is the A-module chain complex SE with

$$d_{SE} \;=\; d_E \;:\; (SE)_r \;=\; E_{r-1} \longrightarrow (SE)_{r-1} \;=\; E_{r-2}\ .$$

Proposition 8.1 *Let E be a based f.g. free $A[z,z^{-1}]$-module chain complex with a Mayer–Vietoris presentation (E^+,E^-). The following conditions are equivalent:*

(i) *E^+ is A-finitely dominated,*
(ii) *$H_*(A((z^{-1}))\otimes_{A[z]}E^+)=0$,*
(iii) *$H_*(A((z^{-1}))\otimes_{A[z^{-1}]}E^-)=0$,*
(iv) *$H_*(A((z^{-1}))\otimes_{A[z,z^{-1}]}E)=0$,*
(v) *$A[[z^{-1}]]\otimes_{A[z^{-1}]}E^-$ is A-finitely dominated, and the A-module chain map*

$$1\otimes\zeta^{-1} \;:\; A[[z^{-1}]]\otimes_{A[z^{-1}]}E^- \longrightarrow A[[z^{-1}]]\otimes_{A[z^{-1}]}E^-$$

is chain homotopy nilpotent,

(vi) *E/E^- is A-finitely dominated, and the A-module chain map*

$$\zeta^{-1} \;:\; E/E^- \longrightarrow E/E^-\ ;\ x \longrightarrow z^{-1}x$$

is chain homotopy nilpotent.

Moreover, if these conditions are satisfied there is defined a chain equivalence

$$(E/E^-,\zeta^{-1}) \;\simeq\; S(A[[z^{-1}]]\otimes_{A[z^{-1}]}E^-,1\otimes\zeta^{-1})\ ,$$

so that

$$[E/E^-,\zeta^{-1}] = -[A[[z^{-1}]]\otimes_{A[z^{-1}]}E^-,1\otimes\zeta^{-1}]\in \mathrm{Nil}_0(A)\ .$$

Proof (ii) $\Longleftrightarrow$ (iii) $\Longleftrightarrow$ (iv) Applying $A((z^{-1}))\otimes_{A[z,z^{-1}]}-$ to the identities

$$A[z,z^{-1}]\otimes_{A[z]}E^+ = A[z,z^{-1}]\otimes_{A[z^{-1}]}E^- = E$$

gives the identities

$$A((z^{-1}))\otimes_{A[z]}E^+ = A((z^{-1}))\otimes_{A[z^{-1}]}E^- = A((z^{-1}))\otimes_{A[z,z^{-1}]}E\ .$$

(iii) $\Longleftrightarrow$ (v) This follows from the equivalences of categories

$$\mathbb{H}(A[[z^{-1}]],\widehat{Z}')\approx\mathbb{H}(A[z^{-1}],Z')\approx \mathrm{Nil}(A)$$

given by 6.7′.
(iv) $\Longrightarrow$ (i) The cartesian square of rings

$$\begin{array}{ccc} A[z^{-1}] & \longrightarrow & A[z,z^{-1}] \\ \big\downarrow & & \big\downarrow \\ A[[z^{-1}]] & \longrightarrow & A((z^{-1})) \end{array}$$

gives an exact sequence of $A[z]$-modules

$$0\longrightarrow A[z^{-1}]\longrightarrow A[z,z^{-1}]\oplus A[[z^{-1}]]\longrightarrow A((z^{-1}))\longrightarrow 0\ .$$

Write the induced f.g. free $A((z^{-1}))$-module chain complex as

$$\widehat{E} = A((z^{-1}))\otimes_{A[z,z^{-1}]}E = A((z^{-1}))\otimes_{A[z^{-1}]}E^-\ .$$

The f.g. free $A[z^{-1}]$-module chain complex E^- fits into an exact sequence of $A[z^{-1}]$-module chain complexes

$$0\longrightarrow E^-\xrightarrow{\ i\ }E^-[z,z^{-1}]\oplus E^-[[z^{-1}]]\longrightarrow\widehat{E}\longrightarrow 0\ ,$$

with

$$\begin{aligned} E^-[z,z^{-1}] &= A[z,z^{-1}]\otimes_{A[z^{-1}]}E^- = E\ ,\\ E^-[[z^{-1}]] &= A[[z^{-1}]]\otimes_{A[z^{-1}]}E^-\ . \end{aligned}$$

By hypothesis $H_*(\widehat{E})=0$, so that i is an $A[z^{-1}]$-module chain equivalence. Let

$$j\ :\ E^+\cap E^-\longrightarrow E^+\oplus E^-[[z^{-1}]]$$

be the A-module chain map defined by inclusions in each component. The algebraic mapping cones of i and j are chain equivalent A-module chain complexes, since $E/E^-\cong E^+/(E^+\cap E^-)$. But i is a chain equivalence, so that j is also a chain equivalence. Since $E^+ \cap E^-$ is a finite f.g. free A-module chain complex this shows that both E^+ and $E^-[[z^{-1}]]$ are A-finitely dominated.

(i) $\Longrightarrow$ (ii) As already noted in Chap. 5 the A-finite domination of E^+ implies that there is defined an $A[z]$-module chain equivalence

$$r^+ = (p^+ \ \ 0) \ : \ \mathcal{C}(\zeta - z : E^+[z] \longrightarrow E^+[z]) \longrightarrow E^+ \ .$$

The $A((z^{-1}))$-module chain map

$$\zeta - z \ : \ A((z^{-1})) \otimes_A E^+ = A((z^{-1})) \otimes_{A[z]} E^+[z] \longrightarrow A((z^{-1})) \otimes_{A[z]} E^+[z]$$

is an automorphism, with inverse

$$(\zeta - z)^{-1} = -\sum_{j=-\infty}^{0} (\zeta)^j z^{j-1} \ :$$

$$A((z^{-1})) \otimes_A E^+ = A((z^{-1})) \otimes_{A[z]} E^+[z] \longrightarrow A((z^{-1})) \otimes_{A[z]} E^+[z] \ ,$$

so that $A((z^{-1})) \otimes_{A[z]} E^+$ is a contractible $A((z^{-1}))$-module chain complex.

(i) $\Longleftrightarrow$ (vi) It is clear from the isomorphism

$$E/E^- \cong E^+/E^+ \cap E^-$$

that E^+ is A-finitely dominated if and only if E/E^- is A-finitely dominated. If this is the case the inclusion

$$\mathcal{C}(z^{-1} - \zeta^{-1} : E^-[z^{-1}] \longrightarrow E^-[z^{-1}])$$

$$\longrightarrow \mathcal{C}(z^{-1} - \zeta^{-1} : E[z^{-1}] \longrightarrow E[z^{-1}]) \ (\simeq E)$$

is a chain equivalence, so that the $A[z^{-1}]$-module chain map

$$z^{-1} - \zeta^{-1} \ : \ (E/E^-)[z^{-1}] \longrightarrow (E/E^-)[z^{-1}]$$

is a chain equivalence and $\zeta^{-1} : E/E^- \longrightarrow E/E^-$ is chain homotopy nilpotent. □

Proposition 8.1′ *Let E be a based f.g. free $A[z, z^{-1}]$-module chain complex with a Mayer–Vietoris presentation (E^+, E^-). The following conditions are equivalent:*

(i) *E^- is A-finitely dominated,*
(ii) $H_*(A((z)) \otimes_{A[z^{-1}]} E^-) = 0,$
(iii) $H_*(A((z)) \otimes_{A[z]} E^+) = 0,$
(iv) $H_*(A((z)) \otimes_{A[z, z^{-1}]} E) = 0,$
(v) *$A[[z]] \otimes_{A[z]} E^+$ is A-finitely dominated, and the A-module chain map*

$$1 \otimes \zeta \ : \ A[[z]] \otimes_{A[z]} E^+ \longrightarrow A[[z]] \otimes_{A[z]} E^+$$

is chain homotopy nilpotent,

(vi) *E/E^+ is A-finitely dominated, and the A-module chain map*

$$\zeta \,:\, E/E^+ \longrightarrow E/E^+ \,;\, x \longrightarrow zx$$

is chain homotopy nilpotent.

Moreover, if these conditions are satisfied there is defined a chain equivalence

$$(E/E^+,\zeta) \simeq S(A[[z]]\otimes_{A[z]} E^+, 1\otimes\zeta) \;,$$

so that

$$[E/E^+,\zeta] = -[A[[z]]\otimes_{A[z]} E^+, 1\otimes\zeta] \in \mathrm{Nil}_0(A) \;.$$

Proof As for 8.1, but with E^+, E^-, z replaced by E^-, E^+, z^{-1}. □

Proposition 8.2 *A finite f.g. free $A[z]$-module chain complex E^+ is A-finitely dominated if and only if $H_*(A((z^{-1}))\otimes_{A[z]} E^+) = 0$.*

Proof Apply 8.1 to the induced finite f.g. free $A[z,z^{-1}]$-module chain complex $E = A[z,z^{-1}]\otimes_{A[z]} E^+$. □

Similarly:

Proposition 8.2′ *A finite f.g. free $A[z^{-1}]$-module chain complex E^- is A-finitely dominated if and only if $H_*(A((z))\otimes_{A[z^{-1}]} E^-) = 0$.* □

Proposition 8.3 *The following conditions on a finite f.g. free $A[z,z^{-1}]$-module chain complex E are equivalent:*

(i) *E is A-finitely dominated,*
(ii) *$H_*(A((z))\otimes_{A[z,z^{-1}]} E) = H_*(A((z^{-1}))\otimes_{A[z,z^{-1}]} E) = 0$.*

Proof Let (E^+,E^-) be a Mayer–Vietoris presentation of E, for any choice of base. By 7.4 E is A-finitely dominated if and only if E^+ and E^- are A-finitely dominated. The equivalence (i) $\Longleftrightarrow$ (ii) is immediate from 8.1 and 8.1′, which give that E^+ is A-finitely dominated if and only if

$$H_*(A((z^{-1}))\otimes_{A[z,z^{-1}]} E) = 0 \;,$$

and that E^- is A-finitely dominated if and only if

$$H_*(A((z))\otimes_{A[z,z^{-1}]} E) = 0 \;.$$ □

Remark 8.4 Given a space W let

$$W^\infty = W\cup\{\infty\}$$

be the one-point compactification. The *homotopy link of* ∞ $e(W)$ of W in the sense of Quinn [228] is the space of paths

$$\omega \,:\, ([0,1],\{0\}) \longrightarrow (W^\infty,\{\infty\})$$

such that $\omega^{-1}(\infty) = \{0\}$. If W is *tame at* ∞ in the sense of [228] the homotopy link is such that there is defined a homotopy pushout

$$\begin{array}{ccc} e(W) & \longrightarrow & \{\infty\} \\ {\scriptstyle p_W}\downarrow & & \downarrow \\ W & \xrightarrow{\ i\ } & W^\infty \end{array}$$

with

$$p_W \ : \ e(W) \longrightarrow W \ ; \ \omega \longrightarrow \omega(1) \ .$$

Thus for tame W the homology groups of $e(W)$ are such that there is defined an exact sequence

$$\dots \longrightarrow H_n(e(W)) \longrightarrow H_n(W) \longrightarrow H_n^{lf}(W) \longrightarrow H_{n-1}(e(W)) \longrightarrow \dots ,$$

with $H_*^{lf}(W) = H_*(W^\infty, \{\infty\})$ the locally finite homology groups of W. The chain level properties of the homotopy link construction are investigated in Hughes and Ranicki [113], including the definition of the *end complex* of a based free A-module chain complex E with

$$E_r \ = \ \sum_{I_r} A \ \ (r \in \mathbb{Z})$$

as the A-module chain complex

$$e(E) \ = \ \mathcal{C}(i : E \longrightarrow E^{lf})_{*+1} \ ,$$

with

$$E_r^{lf} \ = \ \prod_{I_r} A \ \ (r \in \mathbb{Z})$$

and $i : E \longrightarrow E^{lf}$ the inclusion. It is proved in [113] that if W is tame at ∞, and $\widetilde{W}$ is the universal cover of W, then the pullback cover $\widetilde{e(W)}$ of $e(W)$ is such that

$$H_*(\widetilde{e(W)}) \ = \ H_*(e(C(\widetilde{W})))$$

with $\widetilde{p}_W : H_*(\widetilde{e(W)}) \longrightarrow H_*(\widetilde{W})$ induced by the projection

$$e(C(\widetilde{W})) \longrightarrow C(\widetilde{W}) \ .$$

As in 7.5 let X be a finite CW complex with an infinite cyclic cover $W = \overline{X}$ classified by

$$p \ = \ \text{projection} \ : \ \pi_1(X) \ = \ \pi \times \mathbb{Z} \longrightarrow \mathbb{Z} \ ,$$

with a finite fundamental domain $(V; U, \zeta U)$ such that

$$W \ = \ W^+ \cup W^- \ = \ \bigcup_{j=-\infty}^{\infty} \zeta^j(V; U, \zeta U) \ ,$$

with

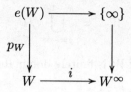

$$W^+ = \bigcup_{j=0}^{\infty} \zeta^j V \ , \ W^- = \bigcup_{j=-\infty}^{-1} \zeta^j V \ , \ W^+ \cap W^- = U \ .$$

It is proved in [113] that if W is finitely dominated then W^+, W^-, W are all tame, and the composites

$$e(W^+) \xrightarrow{p_{W^+}} W^+ \longrightarrow W \ ,$$
$$e(W^-) \xrightarrow{p_{W^-}} W^- \longrightarrow W \ ,$$
$$e(W^+) \sqcup e(W^-) \longrightarrow e(W)$$

are homotopy equivalences. Assume that

$$\pi_1(U) \ = \ \pi_1(V) \ = \ \pi_1(W) \ = \ \pi \ ,$$

and let $\widetilde{U}, \widetilde{V}, \widetilde{W}$ be the universal covers of U, V, W, such that

$$\widetilde{W} \ = \ \widetilde{W}^+ \cup \widetilde{W}^- \ = \ \bigcup_{j=-\infty}^{\infty} \zeta^j(\widetilde{V}; \widetilde{U}, \zeta\widetilde{U}) \ ,$$

with

$$\widetilde{W}^+ = \bigcup_{j=0}^{\infty} \zeta^j \widetilde{V} \ , \ \widetilde{W}^- = \bigcup_{j=-\infty}^{-1} \zeta^j \widetilde{V} \ , \ \widetilde{W}^+ \cap \widetilde{W}^- = \widetilde{U} \ .$$

Let $A = \mathbb{Z}[\pi]$, so that

$$A[z, z^{-1}] \ = \ \mathbb{Z}[\pi \times \mathbb{Z}] \ .$$

The cellular chain complex of $\widetilde{W}^+$ is a based f.g. free $A[z]$-module chain complex

$$E^+ \ = \ C(\widetilde{W}^+) \ .$$

The induced finite f.g. free $A[z, z^{-1}]$-module chain complex is the cellular chain complex of $\widetilde{W}$

$$A[z, z^{-1}] \otimes_{A[z]} E^+ \ = \ C(\widetilde{W}) \ ,$$

and the induced based f.g. free $A[[z]]$-module chain complex is the locally finite cellular chain complex of $\widetilde{W}^+$

$$A[[z]] \otimes_{A[z]} E^+ \ = \ C(\widetilde{W}^+)^{lf} \ = \ C^{lf}(\widetilde{W}^+) \ ,$$

so that the end complex of E^+ is given by

$$e(E^+) \ = \ \mathcal{C}(i : E^+ \longrightarrow A[[z]] \otimes_{A[z]} E^+)_{*+1} \ .$$

The condition $H_*(A((z^{-1})) \otimes_{A[z]} E^+) = 0$ of 8.1 for a finite f.g. free $A[z]$-module chain complex E^+ to be A-finitely dominated is just that the composite

$$e(E^+) \xrightarrow{\text{proj.}} E^+ \xrightarrow{\text{incl.}} E$$

be a homology equivalence. The condition $H_*(A((z^{-1})) \otimes_{A[z^{-1}]} E^-) = 0$ of 8.1′ for a finite f.g. free $A[z^{-1}]$-module chain complex E^- to be A-finitely dominated is just that the composite

$$e(E^-) \xrightarrow{\text{proj.}} E^- \xrightarrow{\text{incl.}} E$$

be a homology equivalence. The two conditions of 8.3 for a finite f.g. free $A[z, z^{-1}]$-module chain complex E to be A-finitely dominated

$$H_*(A((z)) \otimes_{A[z,z^{-1}]} E) = 0 \ , \ H_*(A((z^{-1})) \otimes_{A[z,z^{-1}]} E) = 0$$

are just that the chain maps

$$e(E^+) \longrightarrow E \ , \ e(E^-) \longrightarrow E$$

be homology equivalences, for any Mayer–Vietoris presentation (E^+, E^-). □

Example 8.5 Let F be a field. An F-module chain complex E is finitely dominated if and only if the homology F-modules $H_*(E)$ are finitely generated (i.e. finite dimensional vector spaces of F). In (i) (resp. (ii)) below E will be the F-module chain complex defined by a finite f.g. free $F[z]$- (resp. $F[z, z^{-1}]$-) module chain complex, and the finite dimensionality of $H_*(E)$ will be related to the localization of $F[z]$ inverting the multiplicative subset $S = F[z]\backslash\{0\} \subset F[z]$ of all the non-zero polynomials, which is the quotient field of $F[z]$

$$S^{-1}F[z] = F(z) ,$$

the function field in one variable.
(i) If E^+ is a finite f.g. free $F[z]$-module chain complex which is F-finitely dominated then each $H_r(E^+)$ $(r \geq 0)$ is a finite dimensional F-vector space. The characteristic polynomial

$$g_r(z) = \det(z - \zeta : H_r(E^+)[z] \longrightarrow H_r(E^+)[z]) \in F[z]$$

is such that $g_r(z)H_r(E) = 0$, by the Cayley-Hamilton theorem. The product of characteristic polynomials is an element

$$g(z) = \prod_{r=0}^{\infty} g_r(z) \in S$$

such that $g(z)H_*(E^+) = 0$, so that E^+ is $F(z)$-contractible. The localization of $F[[z]]$ inverting $Z = \{z^k \mid k \geq 0\}$

$$Z^{-1}F[[z]] = F((z))$$

is the quotient field of $F[[z]]$, with inclusions

$$F(z) \longrightarrow F((z)) \ , \ F(z) \longrightarrow F((z^{-1})) \ .$$

If E^+ is a finite f.g. free $F[z]$-module chain complex which is $F(z)$-contractible then it is $F((z^{-1}))$-contractible, so that E^+ is F-finitely dominated by 8.2. Thus a finite f.g. free $F[z]$-module chain complex E^+ is F-finitely dominated if and only if E^+ is $F(z)$-contractible. For 1-dimensional E^+ this is the case if and only if the dimensions of E_0^+, E_1^+ are equal, and

$$\det(d : E_1^+ \longrightarrow E_0^+) \neq 0 \in F[z]$$

with respect to any choice of bases.
(ii) If E is a finite f.g. free $F[z,z^{-1}]$-module chain complex which is F-finitely dominated, then as in (i) there exists an element $g(z) \in S$ such that $g(z)H_*(E) = 0$ and E is $F(z)$-contractible. Conversely, if E is a finite f.g. free $F[z,z^{-1}]$-module chain complex which is $F(z)$-contractible then it is both $F((z))$- and $F((z^{-1}))$-contractible, so that E is F-finitely dominated by 8.3. (Alternatively, apply (i) to a finite f.g. free $F[z]$-module chain complex E^+ such that $E = F[z,z^{-1}] \otimes_{F[z]} E^+$, and use the identity $H_*(E) = \text{coker}(\zeta : H_*(E^+) \longrightarrow H_*(E^+))$.) Thus a finite f.g. free $F[z,z^{-1}]$-module chain complex E is F-finitely dominated if and only if E is $F(z)$-contractible (Milnor [194], [195]). As in (i), for 1-dimensional E this is the case if and only if the dimensions of E_0, E_1 are equal and

$$\det(d : E_1 \longrightarrow E_0) \neq 0 \in F[z,z^{-1}]$$

with respect to any choice of bases. □

Proposition 8.6 *Let*

$$p(z) = \sum_{j=m}^{n} a_j z^j \in A[z,z^{-1}]$$

be a Laurent polynomial with extreme coefficients $a_m, a_n \neq 0 \in A$.
(i) *If the leading coefficient is a unit $a_n \in A^\bullet$ then $p(z) \in A((z^{-1}))^\bullet$. In addition, if $n = 0$ then $p(z) \in A[z^{-1}] \cap A[[z^{-1}]]^\bullet$.*
(ii) *If the trailing coefficient is a unit $a_m \in A^\bullet$ then $p(z) \in A((z))^\bullet$. In addition, if $m = 0$ then $p(z) \in A[z] \cap A[[z]]^\bullet$.*
(iii) *If the extreme coefficients are units $a_m, a_n \in A^\bullet$ then*

$$p(z) \in A((z))^\bullet \cap A((z^{-1}))^\bullet \ .$$

Proof (i) Use 6.3 to identify

$$A[[z^{-1}]]^\bullet = A^\bullet(1 + z^{-1}A[[z^{-1}]]) \ .$$

A Laurent polynomial with leading coefficient a unit $a_n \in A^\bullet$ is a product

$$p(z) = z^n(\sum_{j=m-n}^{0} a_{j+n}z^j) \in A((z^{-1}))$$

of a unit $z^n \in A[z,z^{-1}]^\bullet$ and a unit $\sum\limits_{j=m-n}^{0} a_{j+n}z^j \in A[[z^{-1}]]^\bullet$.
(ii) As for (i), reversing the roles of (i) and (ii).
(iii) Immediate from (i) and (ii). □

Example 8.7 (i) If A is an integral domain

$$A[z] \cap A((z^{-1}))^\bullet = \{\sum_{j=0}^{d} a_j z^j \,|\, a_d \in A^\bullet\} ,$$

$$A[z^{-1}] \cap A((z))^\bullet = \{\sum_{j=-d}^{0} a_j z^j \,|\, a_{-d} \in A^\bullet\} ,$$

$$A[z,z^{-1}] \cap A((z))^\bullet \cap A((z^{-1}))^\bullet = \{\sum_{j=m}^{n} a_j z^j \,|\, a_m, a_n \in A^\bullet\} .$$

(ii) If $e = e^2 \in A$ is an idempotent then

$$1 - e + ez \in A[z] \cap A[z,z^{-1}]^\bullet \subseteq A[z,z^{-1}] \cap A((z))^\bullet \cap A((z^{-1}))^\bullet$$

with

$$\begin{aligned}(1-e+ez)^{-1} &= 1 - e + ez^{-1} \\ &\in A[z^{-1}] \cap A[z,z^{-1}]^\bullet \subseteq A[z,z^{-1}] \cap A((z))^\bullet \cap A((z^{-1}))^\bullet .\end{aligned}$$

□

Example 8.8 Write the centre of A as

$$Z(A) = \{a \in A \,|\, ab = ba \in A \text{ for all } b \in A\} ,$$

and define the multiplicative subsets

$$S = \{\sum_{j=0}^{d} a_j z^j \,|\, a_j \in Z(A), a_d = 1\} ,$$

$$\widetilde{S} = \{\sum_{j=0}^{d} a_j z^j \,|\, a_j \in Z(A), a_0 = 1\} \subset A[z]$$

of all the central polynomials with leading/constant coefficient 1. It is immediate from 8.6 that the inclusions

$$A[z] \longrightarrow A((z^{-1})) \ , \ A[z] \longrightarrow A[[z]]$$

factor through embeddings

$$S^{-1}A[z] \longrightarrow A((z^{-1})) \ , \ \widetilde{S}^{-1}A[z] \longrightarrow A[[z]] .$$

(If $A = F$ is a field then $S^{-1}F[z] = F(z)$, $\widetilde{S}^{-1}F[z] = F[z]_{(z)}$ as in 8.5). If E^+ is a finite f.g. free $A[z]$-module chain complex such that $p(z)H_*(E^+) = 0$ for some $p(z) \in S$ then

$$H_*(S^{-1}E^+) = 0 \ , \ H_*(A((z^{-1})) \otimes_{A[z]} E^+) = 0 \ ,$$

and by 8.2 E^+ is A-finitely dominated.

(ii) Every element $\widetilde{p}(z) = \sum\limits_{j=0}^{d} a_j z^j \in \widetilde{S}$ has an inverse

$$\widetilde{p}(z)^{-1} = 1 + \sum_{j=1}^{\infty} \Big(- \sum_{k=1}^{d} a_k z^k \Big)^j \in A[[z]] \ ,$$

defining an embedding

$$\widetilde{S}^{-1}A[z] \longrightarrow A[[z]] \ .$$

If E is a finite f.g. free $A[z,z^{-1}]$-module chain complex with a Mayer–Vietoris presentation (E^+,E^-) such that $\widetilde{p}(z)H_*(E^+) = 0$ for some $\widetilde{p}(z) \in \widetilde{S}$ then

$$H_*(\widetilde{S}^{-1}E^+) = 0 \ , \ H_*(A((z)) \otimes_{A[z^{-1}]} E^-) = 0 \ ,$$

and by 8.2′ E^- is A-finitely dominated. If A is a commutative ring a finite f.g. free $A[z]$-module chain complex E^+ is $\widetilde{S}^{-1}A[z]$-contractible if and only if E^+ is A-contractible via $A[z] \longrightarrow A; z \longrightarrow 0$. If $a \neq 0 \in A$ is not a unit (e.g. $a = 2 \in A = \mathbb{Z}$) then $E^+ = \mathcal{C}(1 - za : A[z] \longrightarrow A[z])$ is A-contractible but not A-finitely dominated. If A is a field then the leading coefficient $a_d \neq 0 \in A$ of every $\widetilde{p}(z) \in \widetilde{S}$ is a unit, so that $p(z)$ is invertible in $A((z^{-1}))$ and there is defined an embedding

$$\widetilde{S}^{-1}A[z] \longrightarrow A((z^{-1})) \ .$$

Thus every A-contractible complex E^+ is $A((z^{-1}))$-contractible, and hence A-finitely dominated.

(iii) If E is a finite f.g. free $A[z,z^{-1}]$-module chain complex such that

$$p(z)H_*(E) = 0 \ , \ \widetilde{q}(z)H_*(E) = 0$$

for some $p(z) \in S$, $\widetilde{q}(z) \in \widetilde{S}$ then

$$\begin{aligned} &H_*(S^{-1}E) = 0 \ , \ H_*(\widetilde{S}^{-1}E) = 0 \ , \\ &H_*(A((z^{-1})) \otimes_{A[z,z^{-1}]} E) = 0 \ , \ H_*(A((z)) \otimes_{A[z,z^{-1}]} E) = 0 \end{aligned}$$

and by 8.3 E is A-finitely dominated. □

Example 8.9 As in 4.2 let

$$S = (s)^\infty = \{s^k \,|\, k \geq 0\} \subset A$$

be the multiplicative subset defined by the powers of a central non-zero divisor $s \in A$, and as in Chap. 5 let $Z = \{z^k \,|\, k \geq 0\} \subset A[z]$. There is a natural identification

$$(A, S) \;=\; (A[z]/(z-s), Z) \longrightarrow (\widehat{A}_s, \widehat{S}) \;=\; (A[[z]]/(z-s), \widehat{Z}) \;,$$

allowing the cartesian square

$$\begin{array}{ccc} A[z]/(z-s) & \longrightarrow & A[z,z^{-1}]/(z-s) \\ \big\downarrow & & \big\downarrow \\ A[[z]]/(z-s) & \longrightarrow & A((z))/(z-s) \end{array}$$

to be identified with the cartesian square

$$\begin{array}{ccc} A & \longrightarrow & A[1/s] \\ \big\downarrow & & \big\downarrow \\ \widehat{A}_s & \longrightarrow & \widehat{A}_s[1/s] \end{array}$$

Define the based f.g. free $A[z,z^{-1}]$-module chain complex

$$E \;=\; \mathcal{C}(z-s : A[z,z^{-1}] \longrightarrow A[z,z^{-1}])$$

and consider the Mayer–Vietoris presentation (E^+, E^-) of E defined by

$$\begin{aligned} E^+ &= \mathcal{C}(z-s : A[z] \longrightarrow A[z]) \;, \\ E^- &= \mathcal{C}(z-s : z^{-1}A[z^{-1}] \longrightarrow A[z^{-1}]) \;\cong\; \mathcal{C}(1-sz^{-1} : A[z^{-1}] \longrightarrow A[z^{-1}]) \end{aligned}$$

with

$$E^+ \cap E^- \;=\; A \;\;,\;\; E \;=\; A[z,z^{-1}] \otimes_{A[z]} E^+ \;=\; A[z,z^{-1}] \otimes_{A[z^{-1}]} E^- \,.$$

The homology A-modules are such that

$$\begin{aligned} H_0(E^+) &= A[z]/(z-s) \;=\; A \;, \\ H_0(E^-) &= H_0(E) \;=\; H_0(A[z,z^{-1}] \otimes_{A[z]} E^+) \\ &= A[z^{-1}]/(1-sz^{-1}) \;=\; A[z,z^{-1}]/(z-s) \;=\; A[1/s] \;, \\ H_0(A[[z]] \otimes_{A[z]} E^+) &= A[[z]]/(z-s) \;=\; \widehat{A}_s \;, \\ H_0(A((z)) \otimes_{A[z]} E^+) &= H_0(A((z)) \otimes_{A[z,z^{-1}]} E) \\ &= H_0(A[[z]] \otimes_{A[z]} E^+) \;=\; A((z))/(z-s) \;=\; \widehat{A}_s[1/s] \;, \\ H_0(A[[z^{-1}]] \otimes_{A[z^{-1}]} E^-) &= H_0(A((z^{-1})) \otimes_{A[z]} E^+) \\ &= H_0(A((z^{-1})) \otimes_{A[z,z^{-1}]} E) \;=\; 0 \end{aligned}$$

and

$$\begin{aligned} \zeta &= s : H_0(E^+) = A \longrightarrow H_0(E^+) = A \\ \zeta^{-1} &= s^{-1} : H_0(E^-) = A[1/s] \longrightarrow H_0(E^-) = A[1/s] . \end{aligned}$$

If $s \in A$ is not a unit then E^- is not finitely dominated. Let (A, s) denote the homological dimension 1 $A[z]$-module defined by $H_0(E^+) = A$ with z acting by $z = s : A \longrightarrow A$, with f.g. free resolution E^+. In terms of homological algebra

$$\begin{aligned} H_0(A[[z]] \otimes_{A[z]} E^+) &= \mathrm{Tor}_0^{A[z]}((A,s), A[[z]]) = \widehat{A}_s , \\ H_0(A[z,z^{-1}] \otimes_{A[z]} E^+) &= \mathrm{Tor}_0^{A[z]}((A,s), A[z,z^{-1}]) = A[1/s] , \\ H_0(A((z)) \otimes_{A[z]} E^+) &= \mathrm{Tor}_0^{A[z]}((A,s), A((z))) = \widehat{A}_s[1/s] . \end{aligned}$$

The inclusion $A \longrightarrow E^+$ is an A-module chain equivalence, so E^+ is A-finitely dominated. The inclusion $E^- \longrightarrow E$ is an A-module chain equivalence. In accordance with 8.1, 8.1′, 8.1, 8.5 the following conditions on s, E, E^+, E^- are equivalent:

(i) $s \in A$ is a unit,
(ii) $A[1/s] = A$,
(iii) $\widehat{A}_s = 0$,
(iv) $z - s \in A[[z]]$ is a unit,
(v) E is A-finitely dominated,
(vi) the inclusion $A \longrightarrow E$ is an A-module chain equivalence,
(vii) E^- is A-finitely dominated,
(viii) the inclusion $A \longrightarrow E^-$ is an A-module chain equivalence,
(ix) $z = s : H_0(E^+) = A \longrightarrow H_0(E^+) = A$ is an isomorphism. □

Example 8.10 (i) Given a central element $a \in A$ define the multiplicative subset

$$(z-a)^\infty = \{(z-a)^k \,|\, k \geq 0\} \subset A[z]$$

with

$$((z-a)^\infty)^{-1} A[z] = A[z, (z-a)^{-1}] .$$

A finite f.g. free $A[z]$-module chain complex E^+ is $A[z,(z-a)^{-1}]$-contractible if and only if E^+ is A-finitely dominated and $\zeta - a : E^+ \longrightarrow E^+$ is chain homotopy nilpotent, and

$$\begin{aligned} K_*(A[z], (z-a)^\infty) &= \mathrm{Nil}_{*-1}(A) , \\ K_*(A[z,(z-a)^{-1}]) &= K_*(A[z]) \oplus \mathrm{Nil}_{*-1}(A) . \end{aligned}$$

(ii) Given a central unit $a \in A^\bullet$ define the multiplicative subset

$$(z-a)^\infty = \{(z-a)^k \,|\, k \geq 0\} \subset A[z, z^{-1}]$$

as in (i), with

$$((z-a)^\infty)^{-1}A[z,z^{-1}] = A[z,z^{-1},(z-a)^{-1}] .$$

A finite f.g. free $A[z,z^{-1}]$-module chain complex E is $A[z,z^{-1},(z-a)^{-1}]$-contractible if and only if E is A-finitely dominated and $\zeta - a : E \longrightarrow E$ is chain homotopy nilpotent, and

$$K_*(A[z,z^{-1}],(z-a)^\infty) = \mathrm{Nil}_{*-1}(A) ,$$
$$K_*(A[z,z^{-1},(z-a)^{-1}]) = K_*(A[z,z^{-1}]) \oplus \mathrm{Nil}_{*-1}(A) . \qquad \square$$

By analogy with 5.16:

Proposition 8.11 *The connecting map*

$$\partial : K_1(A((z))) \longrightarrow K_1(A[[z]],\widehat{Z}) = \mathrm{Nil}_0(A)$$

is such that for any based f.g. free $A[z]$-module chain complex E^+ with

$$H_*(A((z)) \otimes_{A[z]} E^+) = 0$$

the torsion $\tau(A((z)) \otimes_{A[z]} E^+) \in K_1(A((z)))$ has image

$$\begin{aligned}\partial\tau(A((z)) \otimes_{A[z]} E^+) &= [A[[z]] \otimes_{A[z]} E^+, 1 \otimes \zeta] \\ &= -[E/E^+, \nu^+] \in \mathrm{Nil}_0(A) .\end{aligned}$$

∂ is split by the 'algebraically significant' split injection

$$\begin{aligned}\overline{B} : \mathrm{Nil}_0(A) &\longrightarrow K_1(A((z))) ; \\ (P,\nu) &\longrightarrow \tau((1-\nu)^{-1}(z-\nu) : P((z)) \longrightarrow P((z)))\end{aligned}$$

and also by the 'geometrically significant' split injection

$$\begin{aligned}\overline{B}' : \mathrm{Nil}_0(A) &\longrightarrow K_1(A((z))) ; \\ (P,\nu) &\longrightarrow \tau(\nu - z : P((z)) \longrightarrow P((z))) .\end{aligned}$$

Proof If E is a based f.g. free $A[z,z^{-1}]$-module chain complex such that

$$H_*(A((z)) \otimes_{A[z,z^{-1}]} E) = 0$$

(as in 8.1′) then for any Mayer–Vietoris presentation (E^+,E^-)

$$[E/E^+,\nu^+] = -[A[[z]] \otimes_{A[z]} E^+, 1 \otimes \zeta] \in \mathrm{Nil}_0(A) . \qquad \square$$

Remark 8.12 (i) Bieri and Eckmann [19] and Brown [33] have obtained a necessary and sufficient homological criterion for a projective A-module chain complex C to be finitely dominated, namely that the functor

$$\begin{aligned} H_*(C,-) \ &: \ \{\text{right } A\text{-modules}\} \longrightarrow \{\text{graded } \mathbb{Z}\text{-modules}\} \ ; \\ M &\longrightarrow H_*(C;M) \ = \ H_*(M \otimes_A C) \end{aligned}$$

preserve products.
(ii) A nilpotent space X is finitely dominated if and only if $H_*(X)$ is finitely generated, by Mislin [202]. In the simply-connected case this was first observed by Milnor: in this case finite domination is the same as homotopy finiteness, since $\widetilde{K}_0(\mathbb{Z}) = 0$. □

9. Noncommutative localization

Traditionally, localization is defined in the context of commutative algebra. However, ever since the work of Ore it has also been possible to localize noncommutative rings. High-dimensional knot theory requires the noncommutative localization matrix inversion method of Cohn [53], [54]. The algebraic K- and L-theory invariants of codimension 2 embeddings frequently involve this type of localization of a polynomial ring, as will become apparent in Part Two. (This is particularly the case for links, although links are beyond the scope of the book). The localization exact sequences of algebraic K- and L-theory also hold in the noncommutative case. This chapter deals with K-theory, and L-theory will be considered in Chap. 25.

In dealing with noncommutative localization it will always be assumed that the rings involved have the invariant basis property, so that invertible matrices are square.

Definition 9.1 (Cohn [53, Chap. 7])
For any ring A and any set Σ of square matrices in A let $\Sigma^{-1}A$ be the *localization of A inverting Σ*, the ring obtained from A by taking a ring presentation with generators all the elements of A and all the entries m'_{ij} in formal inverses $M' = (m'_{ij})$ of the matrices $M \in \Sigma$, subject to all the relations holding in A as well as

$$MM' = M'M = I \ (M \in \Sigma) . \qquad \square$$

By analogy with 4.2:

Proposition 9.2 *The ring morphism*

$$i : A \longrightarrow \Sigma^{-1}A ; a \longrightarrow a/1$$

has the universal property that for any ring morphism $f : A \longrightarrow B$ such that $f(M)$ is invertible for every $M \in \Sigma$ there is a unique morphism $F : \Sigma^{-1}A \longrightarrow B$ such that

$$f = Fi : A \longrightarrow \Sigma^{-1}A \longrightarrow B . \qquad \square$$

Example 9.3 (i) Let A be a commutative ring. Given a multiplicative subset $S \subset A$ (4.1) let

$$\langle S\rangle \;=\; A\cap(S^{-1}A)^{\bullet} \;=\; \{a\in A\,|\,ab\in S \text{ for some } b\in A\}\ .$$

(Note that $S\subseteq\langle S\rangle$.) A square matrix M in A becomes invertible in $S^{-1}A$ if and only if $\det(M)\in\langle S\rangle$, in which case

$$M^{-1} \;=\; \det(M)^{-1}\mathrm{adj}(M)$$

with the adjoint matrix defined in A. The localization of A inverting the set Σ of square matrices in A with $\det(M)\in S$ is

$$\Sigma^{-1}A \;=\; S^{-1}A \;=\; \langle S\rangle^{-1}A\ .$$

(ii) Let Σ be any set of square matrices in a ring A such that the entries are central in A. The inverses M^{-1} $(M\in\Sigma)$ in $\Sigma^{-1}A$ commute with the images $i(a)\in\Sigma^{-1}A$ $(a\in A)$, since $aM=Ma$ implies $M^{-1}i(a)=i(a)M^{-1}$. In order to invert $M\in\Sigma$ it thus suffices to invert $\det(M)\in A$, and the localization of A inverting Σ is the localization

$$\Sigma^{-1}A \;=\; S^{-1}A$$

inverting the multiplicative subset $S\subset A$ generated by the determinants $\det(M)\in A$ $(M\in\Sigma)$. In particular, if A is commutative then so is $\Sigma^{-1}A$. □

Example 9.4 Given a ring A and an element $s\in A$ let $\Sigma=\{(s)\}$, the set consisting of the 1×1 matrix (s). The localization of A inverting Σ is

$$\Sigma^{-1}A \;=\; A*\mathbb{Z}[z]/(1-zs,1-sz)$$

with $(1-zs,1-sz)\triangleleft A*\mathbb{Z}[z]$ the two-sided ideal in the free product of A and $\mathbb{Z}[z]$ generated by $1-zs$ and $1-sz$. If $s\in A$ is central then

$$\Sigma^{-1}A \;=\; A[\frac{1}{s}] \;=\; A[z]/(1-zs)$$

and the natural map $A\longrightarrow\Sigma^{-1}A$ is injective if and only if $s\in A$ is a non-zero divisor. □

Given an A-module M write the induced $\Sigma^{-1}A$-module as

$$\Sigma^{-1}M \;=\; \Sigma^{-1}A\otimes_A M\ .$$

Definition 9.5 (i) An (A,Σ)*-module* is an A-module M with a f.g. projective A-module resolution

$$0\longrightarrow P_1\xrightarrow{\;d\;}P_0\longrightarrow M\longrightarrow 0$$

such that $d \in \mathrm{Hom}_A(P_1, P_0)$ is a $\Sigma^{-1}A$-isomorphism, i.e. such that the induced $\Sigma^{-1}A$-module morphism $d : \Sigma^{-1}P_1 \longrightarrow \Sigma^{-1}P_0$ is an isomorphism.
(ii) Let $\mathbb{H}(A, \Sigma)$ be the exact category of (A, Σ)-modules. □

Example 9.6 Given a multiplicative subset $S \subset A$ as in 4.1 (i) let Σ be the set of 1×1 matrices (s) $(s \in S)$. In this case

$$\Sigma^{-1}A = S^{-1}A \ , \ \mathbb{H}(A, \Sigma) = \mathbb{H}(A, S) \ .$$ □

Proposition 9.7 *The natural map $i : A \longrightarrow \Sigma^{-1}A$ is injective if and only if there exists an injective ring morphism $j : A \longrightarrow \Lambda$ such that the matrices in Σ are Λ-invertible.*
Proof If i is injective take

$$j = i : A \longrightarrow \Lambda = \Sigma^{-1}A \ .$$

Conversely, given an injection $j : A \longrightarrow \Lambda$ as in the statement there is a factorization

$$j : A \xrightarrow{i} \Sigma^{-1}A \longrightarrow \Lambda$$

by 9.2, from which it is clear that i is injective. □

The algebraic K-theory localization exact sequence (4.5) has the following generalization to the noncommutative case (cf. Vogel [297, 9.14], Schofield [257, 4.16]) :

Proposition 9.8 *Let Σ be a set of square matrices in a ring A such that the natural map $i : A \longrightarrow \Sigma^{-1}A$ is injective. The relative K-group $K_1(i)$ in the exact sequence*

$$\dots \longrightarrow K_1(A) \xrightarrow{i} K_1(\Sigma^{-1}A) \xrightarrow{\partial} K_1(i) \xrightarrow{j} K_0(A) \longrightarrow \dots$$

is isomorphic to the abelian group

$$K_1(A, \Sigma) = K_0(\mathbb{H}(A, \Sigma))$$

of equivalence classes of triples (P, Q, f) given by f.g. projective A-modules P, Q together with a $\Sigma^{-1}A$-isomorphism $f : P \longrightarrow Q$ subject to the equivalence relation defined by:

$$(P, Q, f) \sim (P', Q', f') \ \text{if there exist a f.g. projective } A\text{-module } R \text{ and an } A\text{-module isomorphism } g : P \oplus Q' \oplus R \cong P' \oplus Q \oplus R \text{ such that}$$

$$\tau((f^{-1} \oplus f' \oplus 1_{\Sigma^{-1}})g : \Sigma^{-1}(P \oplus Q' \oplus R) \xrightarrow{\cong} \Sigma^{-1}(P \oplus Q' \oplus R)) = 0 \in K_1(\Sigma^{-1}A) \ .$$

Proof Define a natural map

$$K_1(A,\Sigma) \longrightarrow K_1(i) \; ; \; (P,Q,f) \longrightarrow (P,Q,f) \; .$$

For f.g. projective A-modules P,Q every $\Sigma^{-1}A$-module isomorphism h : $\Sigma^{-1}P \longrightarrow \Sigma^{-1}Q$ can be expressed as the composite

$$h \; = \; g^{-1}f \; : \; \Sigma^{-1}P \longrightarrow \Sigma^{-1}Q \longrightarrow \Sigma^{-1}Q$$

for some $\Sigma^{-1}A$-isomorphisms $f : P \longrightarrow Q$, $g : Q \longrightarrow Q$, allowing the definition of the inverse isomorphism

$$K_1(i) \longrightarrow K_1(A,\Sigma) \; ; \; (P,Q,h) \longrightarrow (P,Q,f) - (Q,Q,g) \; . \qquad \square$$

Example 9.9 Let $S \subset A$ be a subset which is a *right denominator set* in the sense of Stenström [274, p. 52], satisfying the conditions:

(i) $st \in S$ for all $s,t \in S$,
(ii) if $sa = 0 \in A$ for some $s \in S, a \in A$ then $a = 0 \in A$,
(iii) for all $a \in A, s \in S$ there exist $b,b' \in A$, $t,t' \in S$ such that

$$at \; = \; sb \; , \; t'a \; = \; b's \in A \; ,$$

(iv) $1 \in S$.

As in 9.6 let Σ be the set of 1×1 matrices (s) $(s \in S)$. The localization of A inverting Σ is the ring of fractions

$$\Sigma^{-1}A \; = \; S^{-1}A \; ,$$

the ring of equivalence classes a/s of pairs $(a,s) \in A \times S$ subject to the equivalence relation

$$(a,s) \; \sim \; (b,t) \text{ if } \; ca \; = \; db \in A \; , \; cs \; = \; dt \in S \text{ for some } c,d \in A \; .$$

The natural map $i : A \longrightarrow \Sigma^{-1}A = S^{-1}A$ is injective, so that 9.8 applies. Grayson [98] extended the algebraic K-theory localization exact sequence of Bass and Quillen to this type of noncommutative localization, identifying

$$K_*(A \longrightarrow S^{-1}A) \; = \; K_{*-1}(\mathbb{H}(A,S)) \; ,$$

as in the case of commutative localization recalled in Chap. 4. $\square$

Definition 9.10 (Cappell and Shaneson [40])
A ring morphism $f : A \longrightarrow B$ is *locally epic* if for every finite subset $B_0 \subseteq B$ there exists a unit $u \in B^\bullet$ such that $uB_0 \subseteq f(A)$. $\square$

Example 9.11 (i) A surjective ring morphism $f : A \longrightarrow B$ is locally epic.
(ii) A localization map $i : A \longrightarrow \Sigma^{-1}A$ is locally epic. $\square$

Proposition 9.12 *Let $f : A \longrightarrow B$ be a locally epic morphism of rings, and let Σ be the set of B-invertible square matrices in A.*
(i) *The factorization*

$$f \, : \, A \xrightarrow{\ i\ } \Sigma^{-1}A \longrightarrow B$$

has the universal property that a finite f.g. free A-module chain complex C is B-contractible if and only if C is $\Sigma^{-1}A$-contractible.
(ii) *The natural map i is injective if and only if there exists an injective ring morphism $j : A \longrightarrow \Lambda$ such that a square matrix in A is B-invertible if and only if it is Λ-invertible.*
(iii) *If $f : A \longrightarrow B$ is a surjection with kernel ideal*

$$I \;=\; \ker(f : A \longrightarrow B) \lhd A$$

and $\Sigma_1 \subseteq \Sigma$ is the set of B-invertible square matrices M in A with $f(M)$ an identity matrix then

$$\Sigma^{-1}A \;=\; \Sigma_1^{-1}A \,,$$

and the natural map from A to the I-adic completion

$$j \, : \, A \longrightarrow \widehat{A}_I \;=\; \varprojlim_k A/I^k$$

factorizes as

$$j \, : \, A \xrightarrow{\ i\ } \Sigma_1^{-1}A \longrightarrow \widehat{A}_I$$

with $\Sigma_1^{-1}A \longrightarrow \widehat{A}_I$ sending $(1+x)^{-1}$ $(x \in I)$ to the element

$$\varprojlim_k (1 - x + x^2 - \ldots + (-)^{k-1}x^{k-1}) \in \widehat{A}_I \;=\; \varprojlim_k A/I^k \,.$$

The morphism $j : A \longrightarrow \widehat{A}_I$ is injective if and only if

$$\bigcap_{k=1}^{\infty} I^k \;=\; \{0\} \,,$$

in which case $i : A \longrightarrow \Sigma^{-1}A$ is also injective.

Proof (i) Note first that an A-module chain complex C is chain contractible if and only if there exist A-module morphisms $\Gamma : C_r \longrightarrow C_{r+1}$ $(r \in \mathbb{Z})$ such that the A-module morphisms

$$\Delta \;=\; d\Gamma + \Gamma d \; : \; C_r \longrightarrow C_r$$

are isomorphisms. For if

$$\Gamma \; : \; 0 \simeq 1 \; : \; C \longrightarrow C$$

is a chain contraction then $\Delta = 1$. Conversely, given Γ with Δ an isomorphism the A-module morphisms

$$\Gamma' = \Delta^{-1}\Gamma : C_r \longrightarrow C_{r+1}$$

define a chain contraction $\Gamma' : 0 \simeq 1 : C \longrightarrow C$.

Now suppose that C is finite f.g. free and B-contractible. Since $f : A \longrightarrow B$ is locally epic it is possible to lift a chain contraction

$$0 \simeq 1 \ : \ B \otimes_A C \longrightarrow B \otimes_A C$$

to A-module morphisms $\Gamma : C_r \longrightarrow C_{r+1}$. The corresponding A-module morphisms $\Delta = d\Gamma + \Gamma d$ are B-isomorphisms, hence (by 9.2) $\Sigma^{-1}A$-isomorphisms, and C is $\Sigma^{-1}A$-contractible.

(ii) Immediate from 9.7.

(iii) In order to identify $\Sigma^{-1}A = \Sigma_1^{-1}A$ it suffices to show that if M is a B-invertible $k \times k$ matrix in A then M is $\Sigma_1^{-1}A$-invertible. Since f is locally epic it is possible to lift $f(M)^{-1}$ to a $k \times k$ matrix L in A such that $LM \in \Sigma_1$, so that M is $\Sigma_1^{-1}A$-invertible, with $M^{-1} = (LM)^{-1}L$. □

Example 9.13 Let A be a commutative ring.

(i) If $f : A \longrightarrow B$, I are as in 9.12 then a square matrix M in A is B-invertible if and only if $\det(M) \in (1+I)A^\bullet$, so that

$$\Sigma^{-1}A = \Sigma_1^{-1}A = (1+I)^{-1}A \, .$$

(ii) Let $I = (s) \triangleleft A$ be the principal ideal generated by an element $s \in A$, and let $f : A \longrightarrow B = A/I$ be the projection. By (i) the localization of A inverting the set Σ of B-invertible square matrices in A is the (commutative) localization $\Sigma^{-1}A = (1+I)^{-1}A$, which is the localization of A inverting all the elements $t \in A$ coprime to s (i.e. such that $as + bt = 1$ for some $a, b \in A$). The natural map $g : A \longrightarrow \widehat{A}_I$ is injective if and only if $\bigcap\limits_{k=1}^{\infty} s^k A = \{0\}$. □

Note that if $f : A \longrightarrow B$ is not injective then in general the natural map $i : A \longrightarrow \Sigma^{-1}A$ in 9.12 is not injective, so 9.8 does not apply:

Example 9.14 (i) If $f : A \longrightarrow B$ and $x, y \in A$ are such that

$$xy = 0 \in A \ , \ f(x) = 0 \in B \ , \ f(y) \in B^\bullet$$

then

$$i(x)i(y) = 0 \in \Sigma^{-1}A \ , \ i(y) \in \Sigma \subseteq (\Sigma^{-1}A)^\bullet$$

and $i(x) = 0 \in \Sigma^{-1}A$.

(ii) Let

$$f : A = \mathbb{R}[\mathbb{Z}_2] = \mathbb{R}[T]/(1-T^2) \longrightarrow B = \mathbb{R} \ ; \ p + qT \longrightarrow p + q$$

and take $x = 1 - T$, $y = 1 + T \in \mathbb{R}[\mathbb{Z}_2]$ in (i). The kernel ideal

$$I = \ker(f) = (1-T) \triangleleft A$$

is such that

$$I = I^2 = I^3 = \dots .$$

The natural maps in this case

$$i = f = g : A = \mathbb{R}[\mathbb{Z}_2] \longrightarrow \Sigma^{-1}A = (1+I)^{-1}A = \widehat{A}_I = B = \mathbb{R}$$

are not injective, with $x \neq 0$, $i(x) = 0$. □

Example 9.15 Given a ring B let

$$f : A = B[z] \longrightarrow B ; z \longrightarrow 0$$

be the augmentation map, with kernel ideal

$$I = \ker(f) = zB[z] \triangleleft B[z] .$$

The I-adic completion of A is the ring of formal power series of B

$$\widehat{A}_I = B[[z]]$$

and the natural map $g : B[z] \longrightarrow B[[z]]$ is an injection. Every square matrix M in A can be expressed as $\sum\limits_{j=0}^{\infty} M_j z^j$ with each M_j a square matrix in B, and M is B-invertible if and only if M_0 is invertible. The localization of A inverting the set Σ of B-invertible square matrices in A is thus

$$\Sigma^{-1}A = (1+zB[z])^{-1}B[z] ,$$

and the natural map $i : B[z] \longrightarrow (1+zB[z])^{-1}B[z]$ is an injection by 9.12 (ii). □

The noncommutative versions of 4.13, 4.14 are given by:

Definition 9.16 Two sets Σ_1, Σ_2 of square matrices in a ring A are *coprime* if the natural maps between the localizations of A inverting Σ_1, Σ_2 and $\Sigma_1 \cup \Sigma_2$ fit into a cartesian square

$$\begin{array}{ccc} A & \longrightarrow & (\Sigma_1)^{-1}A \\ \big\downarrow & & \big\downarrow \\ (\Sigma_2)^{-1}A & \longrightarrow & (\Sigma_1 \cup \Sigma_2)^{-1}A \end{array}$$

i.e. if the sequence of additive groups

$$0 \longrightarrow A \longrightarrow (\Sigma_1)^{-1}A \oplus (\Sigma_2)^{-1}A \longrightarrow (\Sigma_1 \cup \Sigma_2)^{-1}A \longrightarrow 0$$

is exact. □

Proposition 9.17 *If Σ_1, Σ_2 are coprime sets of square matrices in a ring A such that the natural maps*

$$i_1 \,:\, A \longrightarrow (\Sigma_1)^{-1}A \ , \ i_2 \,:\, A \longrightarrow (\Sigma_2)^{-1}A$$

are injective then the natural map

$$i_{12} \,:\, A \longrightarrow (\Sigma_1 \cup \Sigma_2)^{-1}A$$

is injective, and the relative algebraic K-groups are such that

$$K_*(A, \Sigma_1 \cup \Sigma_2) \ = \ K_*(A, \Sigma_1) \oplus K_*(A, \Sigma_2)$$

with a Mayer–Vietoris exact sequence

$$\begin{aligned} \dots \longrightarrow K_n(A) \longrightarrow K_n((\Sigma_1)^{-1}A) \oplus K_n((\Sigma_2)^{-1}A) \\ \longrightarrow K_n((\Sigma_1 \cup \Sigma_2)^{-1}A) \longrightarrow K_{n-1}(A) \longrightarrow \dots . \end{aligned}$$

Proof Identify

$$\mathbb{H}(A, \Sigma_1 \cup \Sigma_2) \ = \ \mathbb{H}(A, \Sigma_1) \times \mathbb{H}(A, \Sigma_2)$$

and apply 9.8. □

10. Endomorphism K-theory

Endomorphism K-theory is the algebraic K-theory of modules with an endomorphism, such as arise in knot theory via the Seifert matrix.[10] An A-module P with an endomorphism $f : P \longrightarrow P$ is essentially the same as a module (P, f) over the polynomial ring $A[z]$, with the indeterminate z acting on P by f. This correspondence will be used to relate the algebraic K-groups $K_*(\Sigma^{-1}A[z])$ of the localizations $\Sigma^{-1}A[z]$ of $A[z]$ to the K-groups of pairs (P, f) with P a f.g. projective A-module.

Two localizations of polynomial extension rings are especially relevant to knot theory:

(i) the 'Fredholm' localization $\Omega_+^{-1}A[z]$ of $A[z]$ inverting the set Ω_+ of square matrices in $A[z]$ with cokernel a f.g. projective A-module (10.4),

(ii) the localization $\Pi^{-1}A[z, z^{-1}]$ of $A[z, z^{-1}]$ inverting the set Π of square matrices in $A[z, z^{-1}]$ which become invertible over A under the augmentation $z \longrightarrow 1$ (10.17).

The algebraic description in Part 2 of the high-dimensional knot cobordism groups will involve algebraic L-theory analogues of the endomorphism class groups, defined using these localizations.

The splitting theorems of Chap. 5

$$K_1(A[z, z^{-1}]) = K_1(A[z]) \oplus \mathrm{Nil}_0(A) ,$$

$$K_1(A[z]) = K_1(A) \oplus \widetilde{\mathrm{Nil}}_0(A)$$

will now be extended to splitting theorems

$$K_1(\Omega_+^{-1}A[z]) = K_1(A[z]) \oplus \mathrm{End}_0(A) ,$$

$$K_1(\widetilde{\Omega}_+^{-1}A[z]) = K_1(\Pi^{-1}A[z, z^{-1}]) = K_1(A) \oplus \widetilde{\mathrm{End}}_0(A)$$

with $\mathrm{End}_0(A)$ and $\widetilde{\mathrm{End}}_0(A)$ the absolute and reduced endomorphism class groups of Almkvist [5] and Grayson [96] such that

$$\mathrm{End}_0(A) = K_0(A) \oplus \widetilde{\mathrm{End}}_0(A) .$$

[10] In the original treatment of Seifert [262] this was the matrix of an endomorphism, whereas nowadays it is viewed as the matrix of a bilinear form.

10A. The endomorphism category

Definition 10.1 Let A be any ring.
(i) The *endomorphism category* $\mathrm{End}(A)$ is the exact category in which an object is a pair (P,f) with P a f.g. projective A-module and $f : P \longrightarrow P$ an endomorphism. A morphism in $\mathrm{End}(A)$

$$g : (P,f) \longrightarrow (P',f')$$

is an A-module morphism $g : P \longrightarrow P'$ such that

$$gf = f'g : P \longrightarrow P' .$$

A sequence of objects and morphisms in $\mathrm{End}(A)$

$$0 \longrightarrow (P,f) \longrightarrow (P',f') \longrightarrow (P'',f'') \longrightarrow 0$$

is exact if $0 \longrightarrow P \longrightarrow P' \longrightarrow P'' \longrightarrow 0$ is an exact sequence of the underlying f.g. projective A-modules.
(ii) The *endomorphism K-groups of A* are given by

$$\mathrm{End}_*(A) = K_*(\mathrm{End}(A)) .$$

(iii) The *reduced endomorphism K-groups of A* are given by

$$\widetilde{\mathrm{End}}_*(A) = \mathrm{coker}\big(K_*(\mathbb{P}(A)) \longrightarrow K_*(\mathrm{End}(A))\big)$$

with

$$\mathbb{P}(A) \longrightarrow \mathrm{End}(A) \; ; \; P \longrightarrow (P,0) ,$$

such that

$$\mathrm{End}_*(A) = K_*(A) \oplus \widetilde{\mathrm{End}}_*(A) .$$

(iv) The *endomorphism class group of A* is

$$\mathrm{End}_0(A) = K_0(\mathrm{End}(A)) ,$$

the group of equivalence classes of objects (P,f) in $\mathrm{End}(A)$, subject to the relation:

$(P',f') \sim (P,f) + (P'',f'')$ if there exists a short exact sequence

$$0 \longrightarrow (P,f) \longrightarrow (P',f') \longrightarrow (P'',f'') \longrightarrow 0 .$$

The *reduced endomorphism class group of A* is the quotient group

$$\widetilde{\mathrm{End}}_0(A) = \mathrm{coker}(K_0(A) \longrightarrow \mathrm{End}_0(A)) ,$$

with

$$K_0(A) \longrightarrow \mathrm{End}_0(A) \; ; \; [P] \longrightarrow [P,0] . \qquad \square$$

Remark 10.2 (i) The endomorphism class group of a field F is given by

$$\mathrm{End}_0(F) \;=\; \mathbb{Z}[\mathcal{M}(F)]$$

with $\mathcal{M}(F)$ the set of irreducible monic polynomials $p(z) \in F[z]$ (4.9) – the class $[P, f] \in \mathrm{End}_0(F)$ of an endomorphism $f : P \longrightarrow P$ of a finite dimensional F-vector space P is determined by the factorization of the characteristic polynomial

$$\mathrm{ch}_z(P, f) \;=\; \det(z - f : P[z] \longrightarrow P[z]) \in F[z]$$

as a product of irreducible monic polynomials. See Chaps. 14, 18 below for more detailed accounts.
(ii) The class $[P, f] \in \mathrm{End}_0(A)$ is determined by the characteristic polynomial of $f : P \longrightarrow P$ for any ring A (including the noncommutative case!) – see Chaps. 11, 14 below for more detailed accounts. If A is not a field the relationship between the factorization properties of polynomials in $A[z]$ and the structure of $\mathrm{End}_0(A)$ is necessarily more complicated than in (i), since a factorization of the characteristic polynomial of an object (P, f) in $\mathrm{End}(A)$ may not be realized by an exact sequence

$$0 \longrightarrow (P_1, f_1) \longrightarrow (P, f) \longrightarrow (P_2, f_2) \longrightarrow 0\,. \qquad \square$$

The endomorphism K-theory of A is related to the K-theory of the localization of $A[z]$ inverting matrices of the following type:

Proposition 10.3 *The following conditions on a $k \times k$ matrix $\omega = (\omega_{ij})$ in $A[z]$ are equivalent:*

(i) *the $A[z]$-module morphism*

$$\omega \;:\; A[z]^k \longrightarrow A[z]^k\;;$$

$$(x_1, x_2, \dots, x_k) \longrightarrow \Big(\sum_{j=1}^{k} x_j\omega_{1j}, \sum_{j=1}^{k} x_j\omega_{2j}, \dots, \sum_{j=1}^{k} x_j\omega_{kj}\Big)$$

is injective and the cokernel is a f.g. projective A-module,

(ii) *the 1-dimensional f.g. free $A[z]$-module chain complex*

$$E^+ \;:\; \dots \longrightarrow 0 \longrightarrow A[z]^k \xrightarrow{\;\omega\;} A[z]^k$$

is A-finitely dominated,

(iii) *ω is invertible in $A((z^{-1}))$.*

Proof (i) $\Longrightarrow$ (ii) E^+ is A-module chain equivalent to the 0-dimensional f.g. projective A-module chain complex P defined by $P_0 = \mathrm{coker}(\omega)$.
(ii) $\Longleftrightarrow$ (iii) Immediate from 8.2.
(iii) $\Longrightarrow$ (i) The $A[z]$-module morphism $\omega : A[z]^k \longrightarrow A[z]^k$ is injective (i.e.

$H_1(E^+) = 0$) since $A[z] \longrightarrow A((z^{-1}))$ is injective. It has to be shown that $H_0(E^+) = \mathrm{coker}(\omega)$ is a f.g. projective A-module. Let

$$\omega = \sum_{j=0}^{N} \omega_j z^j$$

with each ω_j a $k \times k$ matrix in A, and $N \geq 0$. Let E be the $A[z^{-1}]$-module subcomplex of the induced $A[z,z^{-1}]$-module chain complex

$$E = A[z,z^{-1}] \otimes_{A[z]} E^+$$

and let $E^- \subset E$ be the $A[z^{-1}]$-module subcomplex defined by

$$d^- = \omega| : E_1^- = \sum_{i=-\infty}^{-1} z^i A^k \longrightarrow E_0^- = \sum_{j=-\infty}^{N-1} z^j A^k .$$

The intersection $E^+ \cap E^-$ is the 0-dimensional f.g. free A-module chain complex with

$$(E^+ \cap E^-)_0 = \sum_{j=0}^{N-1} z^j A^k .$$

As in the proof of 8.1 there is defined a short exact sequence

$$0 \longrightarrow E^+ \cap E^- \longrightarrow E^+ \oplus (A[[z^{-1}]] \otimes_{A[z^{-1}]} E^-) \longrightarrow A((z^{-1})) \otimes_{A[z]} E^+ \longrightarrow 0 .$$

By hypothesis

$$H_*(A((z^{-1})) \otimes_{A[z]} E^+) = 0 ,$$

so that there is defined an A-module isomorphism

$$H_0(E^+ \cap E^-) \cong H_0(E^+) \oplus H_0(A[[z^{-1}]] \otimes_{A[z^{-1}]} E^-) .$$

Thus $H_0(E^+)$ is (isomorphic to) a direct summand of the f.g. free A-module $H_0(E^+ \cap E^-)$, and $H_0(E^+)$ is a f.g. projective A-module. □

10B. The Fredholm localizations $\Omega_+^{-1}A[z]$, $\widetilde{\Omega}_+^{-1}A[z]$

By analogy with the definition of a Fredholm operator on Hilbert space as one with finite dimensional kernel and cokernel:

Definition 10.4 (i) A matrix ω in $A[z]$ is *Fredholm* if it is square and satisfies the equivalent conditions of 10.3.
(ii) Let Ω_+ be the set of Fredholm matrices ω in $A[z]$.
(iii) The *Fredholm localization* of $A[z]$ is the ring $\Omega_+^{-1}A[z]$ obtained from $A[z]$ by inverting Ω_+. □

Remark 10.5 It follows from 10.3 that $\Omega_+^{-1}A[z]$ is the localization of $A[z]$ in the sense of Schofield [257, 4.1] inverting the morphisms $f : P \longrightarrow Q$ of f.g. projective $A[z,z^{-1}]$-modules which are injective and such that coker(f) is a f.g. projective A-module. □

Every matrix ω in $A[z]$ can be expressed as

$$\omega = \sum_{j=0}^{d} \omega_j z^j$$

with each ω_j a matrix with entries in A (of the same size as ω), for some $d \geq 0$.

Definition 10.6 (i) A matrix $\omega = \sum_{j=0}^{d} \omega_j z^j$ with entries in $A[z]$ is *monic* if ω is square and the leading coefficient ω_d is the identity matrix in A.
(ii) Let $\Omega_{+,mon}$ be the set of monic matrices ω in $A[z]$, so that the localization $\Omega_{+,mon}^{-1}A[z]$ is defined as in Chap. 9. □

Proposition 10.7 (i) *The natural map* $A[z] \longrightarrow \Omega_+^{-1}A[z]$ *is injective.*
(ii) *Every monic matrix in* $A[z]$ *is Fredholm, so that* $\Omega_{+,mon} \subset \Omega_+$.
(iii) *The localizations of* $A[z]$ *inverting* $\Omega_{+,mon}$ *and* Ω_+ *coincide*

$$\Omega_{+,mon}^{-1}A[z] = \Omega_+^{-1}A[z] .$$

Proof (i) Immediate from the factorization

$$A[z] \longrightarrow \Omega_+^{-1}A[z] \longrightarrow A((z^{-1}))$$

of the injection $A[z] \longrightarrow A((z^{-1}))$.
(ii) If $\omega = \sum_{j=0}^{d} \omega_j z^j$ is a monic $k \times k$ matrix define an A-module isomorphism

$$A^{kd} \xrightarrow{\simeq} \text{coker}(\omega) \; ; \; (a_i)_{0 \leq i \leq kd-1} \longrightarrow \sum_{j=0}^{d-1} (a_{jk}, a_{jk+1}, \dots, a_{jk+k-1}) z^j .$$

Thus coker$(\omega) = A^{kd}$ is a f.g. free A-module of rank kd, and ω is Fredholm.
(iii) Given a Fredholm matrix ω in $A[z]$ let

$$P = \text{coker}(\omega) \; , \; \zeta : P \longrightarrow P \; ; \; x \longrightarrow zx .$$

Let Q be a f.g. projective A-module such that $P \oplus Q$ is a f.g. free A-module, say $P \oplus Q = A^d$. The $A[z]$-module morphism

$$\eta = 1 \oplus z : A[z]^d = (P \oplus Q)[z] \longrightarrow A[z]^d = (P \oplus Q)[z]$$

with coker$(\eta) = Q$ has a Fredholm matrix in $A[z]$. Let ω' be the matrix of the $A[z]$-module morphism

$$z-(\zeta\oplus 0)\ :\ A[z]^d\ =\ (P\oplus Q)[z]\longrightarrow A[z]^d\ =\ (P\oplus Q)[z]\ .$$

Now $(A^d,\zeta\oplus 0)$ is an $A[z]$-module of homological dimension 1, with a resolution by both $\omega\oplus\eta$ and ω'. In order to invert ω it therefore suffices to invert the monic matrix ω'. □

Definition 10.8 Let $\widetilde{\Omega}_+$ be the set of square matrices ω in $A[z]$ with ω_0 invertible. □

Proposition 10.9 *The localization of $A[z]$ inverting $\widetilde{\Omega}_+$ is the localization of $A[z]$ inverting all the elements in $1+zA[z]$ (regarded as 1×1 matrices)*

$$\widetilde{\Omega}_+^{-1}A[z]\ =\ (1+zA[z])^{-1}A[z]$$

with an injection

$$\widetilde{\Omega}_+^{-1}A[z]\longrightarrow A[[z]]\ ;\ (1+zp(z))^{-1}\longrightarrow\sum_{k=0}^{\infty}(-)^k z^k p(z)^k$$

and the natural map $A[z]\longrightarrow\widetilde{\Omega}_+^{-1}A[z]$ is injective.
Proof As in 9.15. □

Example 10.10 (i) The indeterminate z is a monic 1×1 matrix in $A[z]$, so that z is monic (as a matrix) and

$$A[z,z^{-1}]\subset\Omega_+^{-1}A[z]\ .$$

(ii) If $P=P^2$ is a projection matrix in A then $M=I-P+zP\in\Omega_+$, since $M^{-1}=I-P+z^{-1}P$ is defined in $A[z,z^{-1}]\subset A((z^{-1}))$.
(iii) If A is commutative then the multiplicative subsets

$$S\ =\ A[z]\cap A((z^{-1}))^\bullet\ ,$$

$$S_{mon}\ =\ \{\sum_{j=0}^{d}a_jz^j\,|\,a_d=1\in A\}\subseteq S\ ,$$

$$T\ =\ A[z]\cap A[[z]]^\bullet\ =\ \{\sum_{j=0}^{d}a_jz^j\,|\,a_0\in A^\bullet\}\subset A[z]\ ,$$

$$\widetilde{S}\ =\ \{\sum_{j=0}^{d}a_jz^j\,|\,a_0=1\in A^\bullet\}\subseteq T$$

are such that

$$\begin{aligned}\Omega_+\ &=\ \{M\,|\,\det(M)\in S\}\ ,\ \Omega_+^{-1}A[z]\ =\ S^{-1}A[z]\ =\ S_{mon}^{-1}A[z]\ ,\\ \widetilde{\Omega}_+\ &=\ \{M\,|\,\det(M)\in T\}\ ,\ \widetilde{\Omega}_+^{-1}A[z]\ =\ \widetilde{S}^{-1}A[z]\ =\ T^{-1}A[z]\ .\end{aligned}$$

(iv) If A is an integral domain then (as in 8.7)

$$S = \{\sum_{j=0}^{d} a_j z^j \mid a_d \in A^\bullet\} \subset A[z] .$$

It follows from 8.7 and 10.3 that a polynomial $p(z) = \sum\limits_{j=0}^{d} a_j z^j \in A[z]$ is such that $A[z]/(p(z))$ is a f.g. projective A-module if and only if the leading coefficient $a_d \in A$ is a unit, in which case $A[z]/(p(z))$ is a f.g. free A-module of rank d.
(v) If $A = \mathbb{Z}$ then $\Omega_+^{-1}A[z] = S^{-1}\mathbb{Z}[z, z^{-1}]$ is a principal ideal domain (Farber [75, 2.5]).
(vi) If $A = F$ is a field then

$$\begin{aligned}\Omega_+^{-1}F[z] &= \{p(z)/q(z) \mid q(z) \neq 0\} \\ &= F(z)\end{aligned}$$

is the function field of F (= the quotient field of $F[z]$), and

$$\begin{aligned}\widetilde{\Omega}_+^{-1}F[z] &= \{p(z)/q(z) \mid q(0) \neq 0\} \\ &= F[z]_{(z)} \subset F(z)\end{aligned}$$

is the localization of $F[z]$ at the maximal ideal $(z) \triangleleft F[z]$. The Laurent polynomial extension $F[z, z^{-1}]$ is the localization of $F[z]$ away from (z), and the various localizations fit into a cartesian square of rings

$$\begin{array}{ccc} F[z] & \longrightarrow & F[z, z^{-1}] \\ \big\downarrow & & \big\downarrow \\ F[z]_{(z)} & \longrightarrow & F(z) \end{array}$$

□

Proposition 10.11 *The following conditions on a finite f.g. free $A[z]$-module chain complex E^+ are equivalent:*

(i) $H_*(\Omega_+^{-1}E^+) = 0$,
(ii) $H_*(A((z^{-1})) \otimes_{A[z]} E^+) = 0$,
(iii) E^+ *is A-finitely dominated.*

Proof (i) $\iff$ (ii) A square matrix ω in $A[z]$ becomes invertible in $\Omega_+^{-1}A[z]$ if and only if it becomes invertible in $A((z^{-1}))$.
(ii) $\iff$ (iii) By 8.1. □

Example 10.12 If E^+ is a finite f.g. free $A[z]$-module chain complex such that $p(z)H_*(E^+) = 0$ for some $p(z) \in A[z]$ with $\text{coker}(p(z) : A[z] \longrightarrow A[z])$ a f.g. projective A-module (i.e. such that $(p(z))$ is a 1×1 Fredholm matrix in $A[z]$) then E^+ is A-finitely dominated. □

Proposition 10.13 *The following conditions on a finite f.g. free $A[z]$-module chain complex E^+ are equivalent:*

(i) $H_*(\widetilde{\Omega}_+^{-1}E^+) = 0$,
(ii) $H_*(A[[z]] \otimes_{A[z]} E^+) = 0$,
(iii) $H_*(A \otimes_{A[z]} E^+) = 0$ *via* $A[z] \longrightarrow A; z \longrightarrow 0$,
(iv) $\zeta : E^+ \longrightarrow E^+$ *is an A-module chain equivalence.*

Proof (i) $\Longleftrightarrow$ (ii) A square matrix ω in $A[z]$ becomes invertible in $\widetilde{\Omega}_+^{-1}A[z]$ if and only if it becomes invertible in $A[[z]]$.
(ii) $\Longleftrightarrow$ (iii) An element $\sum\limits_{j=0}^{\infty} a_j z^j \in A[[z]]$ is a unit if and only if $a_0 \in A$ is a unit.
(iii) $\Longleftrightarrow$ (iv) By the exact sequence

$$0 \longrightarrow E^+ \xrightarrow{\zeta} E^+ \longrightarrow A \otimes_{A[z]} E^+ \longrightarrow 0 . \qquad \square$$

The nilpotent category Nil(A) is the full subcategory of End(A) with objects (P, ν) such that $\nu : P \longrightarrow P$ is nilpotent. The inclusion

$$\mathrm{Nil}(A) \longrightarrow \mathrm{End}(A) \ ; \ (P, \nu) \longrightarrow (P, \nu)$$

induces natural morphisms

$$\mathrm{Nil}_*(A) \longrightarrow \mathrm{End}_*(A) \ , \ \widetilde{\mathrm{Nil}}_*(A) \longrightarrow \widetilde{\mathrm{End}}_*(A) \ .$$

The relative groups in the localization exact sequence

$$\begin{aligned} \dots \longrightarrow K_n(A[z]) \xrightarrow{i_+} K_n(\Omega_+^{-1}A[z]) \xrightarrow{\partial_+} & K_n(A[z], \Omega_+) \\ & \longrightarrow K_{n-1}(A[z]) \longrightarrow \dots \end{aligned}$$

are the algebraic K-groups

$$K_*(A[z], \Omega_+) \ = \ K_{*-1}(\mathbb{H}(A[z], \Omega_+))$$

of the exact category $\mathbb{H}(A[z], \Omega_+)$ of $(A[z], \Omega_+)$-modules.

By analogy with 5.6 and 5.14:

Proposition 10.14 (i) *The functor*

$$\begin{aligned} & \mathrm{End}(A) \longrightarrow \mathbb{H}(A[z], \Omega_+) \ ; \\ & (P, f) \longrightarrow \mathrm{coker}(z - f : P[z] \longrightarrow P[z]) \ = \ P \textit{ with } z = f \end{aligned}$$

is an isomorphism of exact categories, so that

$$\begin{aligned} K_*(A[z], \Omega_+) \ &= \ K_{*-1}(\mathbb{H}(A[z], \Omega_+)) \\ &= \ \mathrm{End}_{*-1}(A) \ = \ K_{*-1}(A) \oplus \widetilde{\mathrm{End}}_{*-1}(A) \ . \end{aligned}$$

(ii) *The localization exact sequence breaks up into split exact sequences*

$$0 \longrightarrow K_n(A[z]) \xrightarrow{i_+} K_n(\Omega_+^{-1}A[z]) \xrightarrow{\partial_+} \mathrm{End}_{n-1}(A) \longrightarrow 0 ,$$

so that

$$K_n(\Omega_+^{-1}A[z]) = K_n(A[z]) \oplus \mathrm{End}_{n-1}(A) \quad (n \in \mathbb{Z}) .$$

In particular, for $n = 1$ there is defined a direct sum system

$$K_1(A[z]) \underset{j_+}{\overset{i_+}{\rightleftarrows}} K_1(\Omega_+^{-1}A[z]) \underset{\Delta_+}{\overset{\partial_+}{\rightleftarrows}} \mathrm{End}_0(A)$$

with

$$\begin{aligned}
i_+ &: K_1(A[z]) \longrightarrow K_1(\Omega_+^{-1}A[z]) ; \ \tau(E^+) \longrightarrow \tau(\Omega_+^{-1}E^+) , \\
j_+ &: K_1(\Omega_+^{-1}A[z]) \longrightarrow K_1(A[z]) ; \\
&\qquad \tau(\Omega_+^{-1}E^+) \longrightarrow \tau(r^+ : \mathcal{C}(z-\zeta : E^+[z] \longrightarrow E^+[z]) \longrightarrow E^+) , \\
\partial_+ &: K_1(\Omega_+^{-1}A[z]) \longrightarrow \mathrm{End}_0(A) ; \ \tau(\Omega_+^{-1}E^+) \longrightarrow [E^+, \zeta] , \\
\Delta_+ &: \mathrm{End}_0(A) \longrightarrow K_1(\Omega_+^{-1}A[z]) ; \\
&\qquad [P, f] \longrightarrow \tau(z - f : \Omega_+^{-1}P[z] \longrightarrow \Omega_+^{-1}P[z]) . \qquad \square
\end{aligned}$$

Example 10.15 Let X be a finitely dominated CW complex, and let $f : X \longrightarrow X$ be a map which induces $f_* = 1 : \pi_1(X) \longrightarrow \pi_1(X)$. Write

$$A = \mathbb{Z}[\pi_1(X)] \ , \ \Lambda = \Omega_+^{-1}A[z] .$$

The mapping torus $T(f)$ is Λ-contractible, with

$$\tau(T(f); \Lambda) = (0, [X, f]) \in K_1(\Lambda) = K_1(A[z]) \oplus \mathrm{End}_0(A) .$$

(The endomorphism class $[X, f] \in \mathrm{End}_0(A)$ agrees with the universal functorial Lefschetz invariant of Lück [174].) The condition $f_* = 1$ can be dropped, at the expense of dealing with the ring

$$A_{f_*}[z] = A * \mathbb{Z}[z]/\{az = zf_*(a)\} \quad (a \in A)$$

instead of $A_1[z] = A[z]$, such that

$$\mathbb{Z}[\pi_1(T(f))] = \mathbb{Z}[\pi_1(X)]_{f_*}[z, z^{-1}] . \qquad \square$$

Proposition 10.16 *The torsion group of $\widetilde{\Omega}_+^{-1}A[z]$ fits into the direct sum system*

$$K_1(A) \underset{\widetilde{j}_+}{\overset{\widetilde{i}_+}{\rightleftarrows}} K_1(\widetilde{\Omega}_+^{-1}A[z]) \underset{\widetilde{\Delta}_+}{\overset{\widetilde{\partial}_+}{\rightleftarrows}} \widetilde{\mathrm{End}}_0(A)$$

with

$$\begin{aligned}
\widetilde{i}_+ \ &: \ K_1(A) \longrightarrow K_1(\widetilde{\Omega}_+^{-1}A[z]) \ ; \ \tau(E) \longrightarrow \tau(\widetilde{\Omega}_+^{-1}E[z]) \ , \\
\widetilde{\partial}_+ \ &: \ K_1(\widetilde{\Omega}_+^{-1}A[z]) \longrightarrow \widetilde{\mathrm{End}}_0(A) \ ; \\
&\qquad \tau(\widetilde{\Omega}_+^{-1}\mathcal{C}(f_0 + zf_1 : P[z] \longrightarrow P[z])) \longrightarrow [P, -f_0^{-1}f_1] \ , \\
\widetilde{j}+ \ &: \ K_1(\widetilde{\Omega}_+^{-1}A[z]) \longrightarrow K_1(A) \ ; \ \tau(\widetilde{\Omega}_+^{-1}E^+) \longrightarrow \tau(A \otimes_{A[z]} E^+) \ , \\
\widetilde{\Delta}_+ \ &: \ \widetilde{\mathrm{End}}_0(A) \longrightarrow K_1(\widetilde{\Omega}_+^{-1}A[z]) \ ; \\
&\qquad [P, f] \longrightarrow \tau(1 - zf : \widetilde{\Omega}_+^{-1}P[z] \longrightarrow \widetilde{\Omega}_+^{-1}P[z]) \ .
\end{aligned}$$

Proof Let $\Omega_-^{-1}A[z^{-1}]$ be the localization of $A[z^{-1}]$ inverting the set Ω_- of Fredholm matrices in $A[z^{-1}]$ (i.e. the $k \times k$ matrices ω such that $\mathrm{coker}(\omega : A[z^{-1}]^k \longrightarrow A[z^{-1}]^k)$ is a f.g. projective A-module, for all $k \geq 1$). By 10.14

$$K_1(\Omega_-^{-1}A[z^{-1}]) \ = \ K_1(A) \oplus K_0(A) \oplus \widetilde{\mathrm{Nil}}_0(A) \oplus \widetilde{\mathrm{End}}_0(A) \ .$$

As in Chap. 5 let $Z = \{z^k \,|\, k \geq 0\} \subset A[z]$, so that $Z^{-1}A[z] = A[z, z^{-1}]$. The cartesian square of rings

$$\begin{array}{ccc}
A[z] & \longrightarrow & \widetilde{\Omega}_+^{-1}A[z] \\
\big\downarrow & & \big\downarrow \\
A[z, z^{-1}] & \longrightarrow & \Omega_-^{-1}A[z^{-1}]
\end{array}$$

induces excision isomorphisms

$$K_*(A[z], Z) \ \cong \ K_*(\widetilde{\Omega}_+^{-1}A[z], \Omega_-) \ .$$

Thus

$$K_1(\widetilde{\Omega}_+^{-1}A[z], \Omega_-) \ = \ K_1(A[z], Z) \ = \ K_0(A) \oplus \widetilde{\mathrm{Nil}}_0(A) \ ,$$

and

$$K_1(\widetilde{\Omega}_+^{-1}A[z]) \ = \ K_1(A) \oplus \widetilde{\mathrm{End}}_0(A) \ . \qquad \square$$

10C. The A-contractible localization $\Pi^{-1}A[z, z^{-1}]$

The A-contractible localization $\Pi^{-1}A[z, z^{-1}]$ has the universal property that a finite f.g. free $A[z, z^{-1}]$-module chain complex C is A-contractible if and only if C is $\Pi^{-1}A[z, z^{-1}]$-contractible. This is the localization of $A[z, z^{-1}]$ of greatest relevance to knot theory, as will be spelled out in Chaps. 17, 32, 33. The results of 10A. are now used to identify

$$\begin{aligned}
K_1(\Pi^{-1}A[z, z^{-1}]) \ &= \ K_1(\widetilde{\Omega}_+^{-1}A[z]) \\
&= \ K_1(A) \oplus \widetilde{\mathrm{End}}_0(A) \ .
\end{aligned}$$

Definition 10.17 (i) An $A[z,z^{-1}]$-module chain complex C is *A-contractible* if $A\otimes_{A[z,z^{-1}]}C$ is a contractible A-module chain complex, regarding A as an $A[z,z^{-1}]$-module via the augmentation map

$$p : A[z,z^{-1}] \longrightarrow A ; \sum_{j=-\infty}^{\infty} a_j z^j \longrightarrow \sum_{j=-\infty}^{\infty} a_j .$$

(ii) A $k\times k$ matrix $\omega=(\omega_{ij})$ in $A[z,z^{-1}]$ is *A-invertible* if $p(\omega)=(p(\omega_{ij}))$ is an invertible $k\times k$ matrix in A.
(iii) Let Π be the set of A-invertible matrices in $A[z,z^{-1}]$, so that the localization $\Pi^{-1}A[z,z^{-1}]$ is defined as in 9.1. □

Proposition 10.18 (i) *The augmentation* $p : A[z,z^{-1}]\longrightarrow A; z\longrightarrow 1$ *has a canonical factorization*

$$p : A[z,z^{-1}] \longrightarrow \Pi^{-1}A[z,z^{-1}] \longrightarrow A .$$

(ii) *The natural map* $A[z,z^{-1}]\longrightarrow\Pi^{-1}A[z,z^{-1}]$ *is an injection.*
(iii) *The following conditions on a finite f.g. free* $A[z,z^{-1}]$*-module chain complex* E *are equivalent:*

(a) E *is* A*-contractible,*
(b) $H_*(\Pi^{-1}E)=0$,
(c) $H_*(A\otimes_{A[z,z^{-1}]}E)=0$,
(d) $\zeta-1 : E\longrightarrow E$ *is a chain equivalence.*

Proof Apply 9.12, noting that the natural map into the $(z-1)$-adic completion

$$A[z,z^{-1}] \longrightarrow \varprojlim_k A[z,z^{-1}]/(z-1)^k = A[[z-1]][z^{-1}]$$

is an injection. □

Example 10.19 (i) If A is a commutative ring then

$$\Pi^{-1}A[z,z^{-1}] = P^{-1}A[z,z^{-1}]$$

with

$$P = \{\sum_{j=-\infty}^{\infty} a_j z^j \mid \sum_{j=-\infty}^{\infty} a_j \in A^\bullet\} \subset A[z,z^{-1}] .$$

A finite f.g. free $A[z,z^{-1}]$-module chain complex E is A-contractible if and only if

$$p(z)H_*(E) = 0$$

for some $p(z)\in P$.
(ii) If $A=F$ is a field then $P^{-1}F[z,z^{-1}]\subset F(z)$, so that an F-contractible finite f.g. free $F[z,z^{-1}]$-module chain complex E is $F(z)$-contractible, and hence by 8.3 E is F-finitely dominated. □

Proposition 10.20 *Let E be a finite based f.g. free $A[z,z^{-1}]$-module chain complex.*
(i) *E is A-contractible if and only if it is chain equivalent to the algebraic mapping cone $\mathcal{C}(1-f+zf : P[z,z^{-1}]\longrightarrow P[z,z^{-1}])$ for some finite f.g. projective A-module chain complex P with a chain map $f : P\longrightarrow P$, in which case it is possible to choose a chain equivalence such that*

$$\tau(E \simeq \mathcal{C}(1-f+zf)) \;=\; \tau(A\otimes_{A[z,z^{-1}]} E) \in \mathrm{im}(K_1(A)\longrightarrow K_1(A[z,z^{-1}])) \ .$$

(ii) *If E is A-contractible then E is A-finitely dominated if and only if $f, 1-f : P\longrightarrow P$ are chain equivalences for any (P,f) as in* (i), *in which case E is A-module chain equivalent to P and*

$$f \simeq (1-\zeta)^{-1} \;:\; P \simeq E \longrightarrow P \simeq E \ .$$

Proof (i) The algebraic transversality of Ranicki [240, Chap. 8] shows that for every based f.g. free $A[z,z^{-1}]$-module chain complex E there is a simple chain equivalence

$$i \;:\; E \xrightarrow{\simeq} \mathcal{C}(g+zh : C[z,z^{-1}]\longrightarrow D[z,z^{-1}])$$

for some based f.g. free A-module chain complexes C, D and chain maps $g,h : C\longrightarrow D$. Applying $A\otimes_{A[z,z^{-1}]} -$ there is obtained a simple A-module chain equivalence

$$1\otimes i \;:\; A\otimes_{A[z,z^{-1}]} E \longrightarrow \mathcal{C}(g+h : C\longrightarrow D) \ .$$

Thus E is A-contractible if and only if $g+h : C\longrightarrow D$ is a chain equivalence, in which case the chain map

$$f \;=\; (g+h)^{-1}h \;:\; P \;=\; C \longrightarrow P$$

is such that there is defined an $A[z,z^{-1}]$-module chain equivalence

$$\begin{aligned} e \;:\; \mathcal{C}(1-f+zf : P[z]\longrightarrow P[z]) &\xrightarrow{(g+h\ \ 0)} \\ \mathcal{C}(g+zh : C[z]\longrightarrow D[z]) &\xrightarrow{i^{-1}} E \end{aligned}$$

with torsion

$$\begin{aligned} \tau(e) \;&=\; \tau(g+h : C\longrightarrow D) \\ &=\; \tau(A\otimes_{A[z,z^{-1}]} E) \in \mathrm{im}(K_1(A)\longrightarrow K_1(A[z,z^{-1}])) \ . \end{aligned}$$

Alternatively, use the following construction of (P,f) for an A-contractible based f.g. free $A[z,z^{-1}]$-module chain E. Since E is A-contractible the chain map $\zeta - 1 : E\longrightarrow E$ is an A-module chain equivalence (10.18). Let (E^+, E^-) be a compatibly based Mayer–Vietoris presentation of E, so that

$$E \;=\; A[z,z^{-1}]\otimes_{A[z]} E^+ \;=\; A[z,z^{-1}]\otimes_{A[z^{-1}]} E^- \ ,$$

with

$$\begin{aligned}\tau(1:A[z,z^{-1}]\otimes_{A[z^{\pm 1}]}E^{\pm}\longrightarrow E) &= \tau(z^{N^{\pm}}:A[z,z^{-1}]\longrightarrow A[z,z^{-1}]) \\ &\in K_1(A[z,z^{-1}])\end{aligned}$$

for some $N^{\pm}\in\mathbb{Z}$. Now E^+ is $A[[z]]$-contractible, so that

$$A((z))\otimes_{A[z^{-1}]}E^- = A((z))\otimes_{A[z,z^{-1}]}E = A((z))\otimes_{A[z]}E^+ \simeq 0\,.$$

Thus E^- is $A((z))$-contractible, and by 8.1′ E^- is A-finitely dominated. The projection $E^-[z]\longrightarrow E$ defines an $A[z]$-module chain equivalence

$$j\,:\,\mathcal{C}(1-z\zeta^{-1}:E^-[z]\longrightarrow E^-[z])\xrightarrow{\simeq}E$$

such that the $A[z]$-module chain map

$$k\,:\,E^+\longrightarrow E\xrightarrow{j^{-1}}\mathcal{C}(1-z\zeta^{-1}:E^-[z]\longrightarrow E^-[z])$$

is an $A[z]$-module chain equivalence with torsion

$$\tau(k) = \tau(A\otimes_{A[z]}E^+)\in \mathrm{im}(K_1(A)\longrightarrow K_1(A[z]))\,.$$

(ii) For any A-finitely dominated finite f.g. free $A[z,z^{-1}]$-module chain complex E there is defined an $A[z,z^{-1}]$-module chain equivalence

$$e\,:\,\mathcal{C}(1-z\zeta^{-1}\;:E[z,z^{-1}]\longrightarrow E[z,z^{-1}])\longrightarrow E\,,$$

inducing an A-module chain equivalence

$$1\otimes e\,:\,\mathcal{C}(1-\zeta^{-1}\;:E\longrightarrow E)\longrightarrow A\otimes_{A[z,z^{-1}]}E\,.$$

Thus if E is A-contractible the A-module chain map $1-\zeta^{-1}:E\longrightarrow E$ is a chain equivalence, and the chain equivalence

$$f = (1-\zeta)^{-1}\;:\,E\longrightarrow E$$

is such that

$$1-f = -\zeta(1-\zeta)^{-1}\;:\,E\longrightarrow E$$

is also a chain equivalence, with an $A[z,z^{-1}]$-module chain equivalence

$$\begin{aligned}e\,:\,&\mathcal{C}(1-f+zf:E[z,z^{-1}]\longrightarrow E[z,z^{-1}])\xrightarrow{\simeq}\\ &\mathcal{C}((1-\zeta^{-1})^{-1}(1-z\zeta^{-1}):E[z,z^{-1}]\longrightarrow E[z,z^{-1}])\longrightarrow E\end{aligned}$$

such that $\tau(f)\in\mathrm{im}(K_1(A)\longrightarrow K_1(A[z,z^{-1}]))$ with respect to an appropriate choice of $A[z,z^{-1}]$-module basis for E.

Conversely, if E is an A-contractible based f.g. free $A[z,z^{-1}]$-module chain complex with $E\simeq\mathcal{C}(1-f+zf:P[z,z^{-1}]\longrightarrow P[z,z^{-1}])$ and $f,1-f:P\longrightarrow P$ A-module chain equivalences there are defined A-module chain equivalences

$$\begin{aligned} E &\simeq \mathcal{C}(1 - f + zf : P[z, z^{-1}] \longrightarrow P[z, z^{-1}]) \\ &\simeq \mathcal{C}(1 + z(1-f)^{-1}f : P[z, z^{-1}] \longrightarrow P[z, z^{-1}]) \simeq P . \end{aligned}$$ □

Proposition 10.21 (i) *The torsion group of $\Pi^{-1}A[z, z^{-1}]$ fits into the direct sum system*

$$K_1(A) \underset{\widetilde{j}}{\overset{\widetilde{i}}{\rightleftarrows}} K_1(\Pi^{-1}A[z, z^{-1}]) \underset{\widetilde{\Delta}}{\overset{\widetilde{\partial}}{\rightleftarrows}} \widetilde{\mathrm{End}}_0(A)$$

with

$$\begin{aligned} \widetilde{i} &: K_1(A) \longrightarrow K_1(\Pi^{-1}A[z, z^{-1}]) \; ; \; \tau(E) \longrightarrow \tau(\Pi^{-1}E[z, z^{-1}]) , \\ \widetilde{\partial} &: K_1(\Pi^{-1}A[z, z^{-1}]) \longrightarrow \widetilde{\mathrm{End}}_0(A) \; ; \; \tau(\Pi^{-1}E) \longrightarrow \widetilde{\partial}(E) = [P, f] \\ &\qquad\qquad (\text{with } (P, f) \text{ as in 10.20}) , \\ \widetilde{j} &: K_1(\Pi^{-1}A[z, z^{-1}]) \longrightarrow K_1(A) \; ; \; \tau(\Pi^{-1}E) \longrightarrow \tau(A \otimes_{A[z, z^{-1}]} E) , \\ \widetilde{\Delta} &: \widetilde{\mathrm{End}}_0(A) \longrightarrow K_1(\Pi^{-1}A[z, z^{-1}]) \; ; \\ &\qquad [P, f] \longrightarrow \tau(1 - f + zf : \Pi^{-1}F[z, z^{-1}] \longrightarrow \Pi^{-1}F[z, z^{-1}]) . \end{aligned}$$

(ii) *The torsion group of $\Pi^{-1}A[z, z^{-1}, (1-z)^{-1}]$ fits into the direct sum system*

$$K_1(\Pi^{-1}A[z, z^{-1}]) \underset{j}{\overset{i}{\rightleftarrows}} K_1(\Pi^{-1}A[z, z^{-1}, (1-z)^{-1}]) \underset{\Delta}{\overset{\partial}{\rightleftarrows}} \mathrm{Nil}_0(A)$$

with

$$\begin{aligned} i &: K_1(\Pi^{-1}A[z, z^{-1}]) \longrightarrow K_1(\Pi^{-1}A[z, z^{-1}, (1-z)^{-1}]) \; ; \\ &\qquad\qquad \tau(E) \longrightarrow \tau((1-z)^{-1}E) , \\ \partial &: K_1(\Pi^{-1}A[z, z^{-1}, (1-z)^{-1}]) \longrightarrow \mathrm{Nil}_0(A) \; ; \\ &\qquad\qquad \tau((1-z)^{-1}E) \longrightarrow [E^!, 1 - \zeta] , \\ \Delta &: \mathrm{Nil}_0(A) \longrightarrow K_1(\Pi^{-1}A[z, z^{-1}, (1-z)^{-1}]) \; ; \; [Q, \nu] \longrightarrow \\ &\quad \tau(1 - (1-z)^{-1}\nu : \Pi^{-1}Q[z, z^{-1}, (1-z)^{-1}] \longrightarrow \Pi^{-1}Q[z, z^{-1}, (1-z)^{-1}]) . \end{aligned}$$

(iii) *The torsion groups are such that*

$$\begin{aligned} K_1(\Pi^{-1}A[z, z^{-1}]) &= K_1(A) \oplus \widetilde{\mathrm{End}}_0(A) , \\ K_1(\Pi^{-1}A[z, z^{-1}, (1-z)^{-1}]) &= K_1(A) \oplus K_0(A) \oplus \widetilde{\mathrm{Nil}}_0(A) \oplus \widetilde{\mathrm{End}}_0(A) . \end{aligned}$$

(iv) *An* $(A[z,z^{-1}],\Pi)$*-module is an h.d. 1* $A[z,z^{-1}]$*-module* M *such that* $1-\zeta : M \longrightarrow M$ *is an automorphism. Every* $(A[z,z^{-1}],\Pi)$*-module* M *admits a resolution*

$$0 \longrightarrow P[z,z^{-1}] \xrightarrow{1-f+zf} P[z,z^{-1}] \longrightarrow M \longrightarrow 0$$

for some (P,f) *in* $\mathrm{End}_0(A)$*, and there is defined an exact sequence*

$$\mathrm{Nil}_0(A) \oplus \mathrm{Nil}_0(A) \xrightarrow{j} \mathrm{End}_0(A) \xrightarrow{k} K_1(A[z,z^{-1}],\Pi) \longrightarrow 0$$

with

$$\begin{aligned} j([M,\mu],[N,\nu]) &= [M\oplus N, \mu\oplus(1-\nu)] ,\\ k[P,f] &= [\mathrm{coker}(1-f+zf : P[z,z^{-1}] \longrightarrow P[z,z^{-1}])] . \end{aligned}$$

Proof (i) Let Π_+ be the set of square matrices in $A[z]$ which are A-invertible via $z \longrightarrow 1$. The ring isomorphism

$$\rho : A[z] \xrightarrow{\simeq} A[z] ;\ z \longrightarrow 1-z$$

induces a ring isomorphism

$$\rho : (\Pi_+)^{-1}A[z] = \Pi^{-1}A[z,z^{-1}] \xrightarrow{\simeq} \widetilde{\Omega}_+^{-1}A[z] ;\ z \longrightarrow 1-z .$$

Now apply 10.16 to identify

$$K_1(\Pi^{-1}A[z,z^{-1}]) = K_1(\widetilde{\Omega}_+^{-1}A[z]) = K_1(A) \oplus \widetilde{\mathrm{End}_0}(A) .$$

(ii) For any A-invertible $n\times n$ matrix M in $A[z,z^{-1}]$ there exist an invertible $n\times n$ matrix L in A and an $n\times n$ matrix K in $A[z,z^{-1}]$ such that

$$LM + (1-z)K = I_n .$$

Thus the commutative square of rings

$$\begin{array}{ccc} A[z,z^{-1}] & \longrightarrow & \Pi^{-1}A[z,z^{-1}] \\ \big\downarrow & & \big\downarrow \\ A[z,z^{-1},(1-z)^{-1}] & \longrightarrow & \Pi^{-1}A[z,z^{-1},(1-z)^{-1}] \end{array}$$

is cartesian, and there is a Mayer–Vietoris exact sequence in algebraic K-theory

$$\begin{aligned} \dots \longrightarrow K_1(A[z,z^{-1}]) &\longrightarrow K_1(\Pi^{-1}A[z,z^{-1}]) \oplus K_1(A[z,z^{-1},(1-z)^{-1}]) \\ &\longrightarrow K_1(\Pi^{-1}A[z,z^{-1},(1-z)^{-1}]) \longrightarrow K_0(A[z,z^{-1}]) \longrightarrow \dots . \end{aligned}$$

The decomposition of $K_1(\Pi^{-1}A[z,z^{-1},(1-z)^{-1}])$ now follows from (i).
(iii) Immediate from (i) and (ii).
(iv) It is convenient to introduce another indeterminate over A, called s, and to let Ω_+ denote the set of Fredholm matrices in $A[s]$. From (i), (ii) and (iii) there is defined a commutative braid of exact sequences

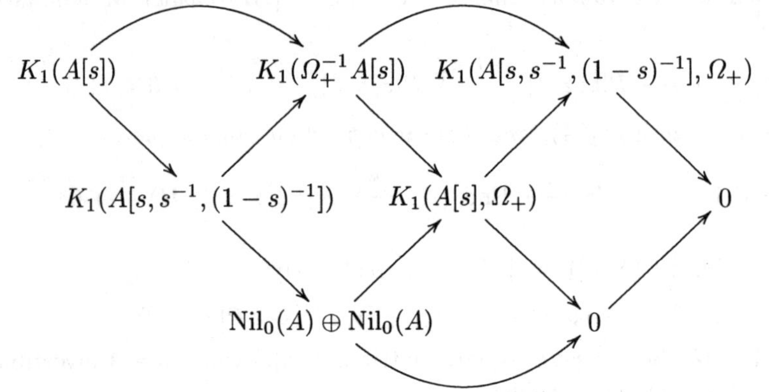

with

$$K_1(A[s],\Omega_+) \;=\; \mathrm{End}_0(A)\ .$$

The isomorphism of rings

$$A[s,s^{-1},(1-s)^{-1}] \xrightarrow{\simeq} A[z,z^{-1},(1-z)^{-1}]\ ;\ s \longrightarrow (1-z)^{-1}$$

induces an isomorphism of exact categories

$$\mathbb{H}(A[s,s^{-1},(1-s)^{-1}],\Omega_+) \;\cong\; \mathbb{H}(A[z,z^{-1},(1-z)^{-1}],\Pi)\ .$$

The inclusion $A[z,z^{-1}]\longrightarrow A[z,z^{-1},(1-z)^{-1}]$ induces an isomorphism of exact categories

$$\mathbb{H}(A[z,z^{-1}],\Pi) \;\cong\; \mathbb{H}(A[z,z^{-1},(1-z)^{-1}],\Pi)\ ,$$

since Π is coprime to $(1-z)$. It follows that

$$\begin{aligned}
K_1(A[z,z^{-1}],\Pi) \;&=\; K_1(A[z,z^{-1},(1-z)^{-1}],\Pi)\\
&=\; K_1(A[s,s^{-1},(1-s)^{-1}],\Omega_+)\\
&=\; \mathrm{coker}(K_1(A[s,s^{-1},(1-s)^{-1}])\longrightarrow K_1(\Omega_+^{-1}A[s]))\\
&=\; \mathrm{coker}(K_1(A[s],(s(1-s))^\infty)\longrightarrow K_1(A[s],\Omega_+))\\
&=\; \mathrm{coker}(\mathrm{Nil}_0(A)\oplus\mathrm{Nil}_0(A)\longrightarrow\mathrm{End}_0(A))\ .
\end{aligned}$$

(More conceptually, note that an object (P,f) in $\mathrm{End}_0(A)$ is such that

$$1-f+zf\ :\ P[z,z^{-1}]\longrightarrow P[z,z^{-1}]$$

is an $A[z,z^{-1}]$-module isomorphism if and only if $f:P\longrightarrow P$ is a near-projection (i.e. $f(1-f):P\longrightarrow P$ is nilpotent), if and only if (P,f) is isomorphic to $(M,\mu)\oplus(N,1-\nu)$ for some objects (M,μ), (N,ν) in the nilpotent category $\mathrm{Nil}(A)$. See 5.7 for near-projections.) □

Example 10.22 Let F be a field. If E is an F-contractible based f.g. free $F[z,z^{-1}]$-module chain complex then

$$\widetilde{\partial}(E) = [H_*(E),(1-\zeta)^{-1}] + d[F,1] \in \widetilde{\mathrm{End}}_0(F)$$

with

$$d = \chi(H_*(E^+)) - \chi(H_*(E)) \in \mathbb{Z}$$

for any finite f.g. free $F[z]$-module chain complex E^+ such that

$$F[z,z^{-1}] \otimes_{F[z]} E^+ = E ,$$

interpreting negative values of d using chain complex suspension. □

Example 10.23 (i) Given a group π let

$$\Lambda = \Pi^{-1}\mathbb{Z}[\pi][z,z^{-1}] \ , \ \Lambda_1 = \Pi^{-1}\mathbb{Z}[\pi][z,z^{-1},(1-z)^{-1}]$$

with Π the set of $\mathbb{Z}[\pi]$-invertible square matrices in $\mathbb{Z}[\pi][z,z^{-1}]$, and define Whitehead groups

$$Wh_1(\Lambda) = K_1(\Lambda)/\{\pm gz^j \mid g \in \pi, j \in \mathbb{Z}\} ,$$

$$Wh_1(\Lambda_1) = K_1(\Lambda_1)/\{\pm gz^j(1-z)^k \mid g \in \pi, j,k \in \mathbb{Z}\} .$$

By 10.21

$$Wh_1(\Lambda) = Wh(\pi) \oplus \widetilde{\mathrm{End}}_0(\mathbb{Z}[\pi]) ,$$

$$Wh_1(\Lambda_1) = Wh(\pi) \oplus \widetilde{K}_0(\mathbb{Z}[\pi]) \oplus \widetilde{\mathrm{End}}_0(\mathbb{Z}[\pi]) \oplus \widetilde{\mathrm{Nil}}_0(\mathbb{Z}[\pi]) .$$

Assume that π is finitely presented, and let K be a connected finite CW complex with $\pi_1(K) = \pi$. Given a finite CW complex X and a cellular map $h : X \longrightarrow K \times S^1$ let $\widetilde{X}, \widetilde{K}$ be the universal covers of X, K, so that there is induced a $\mathbb{Z}[\pi][z,z^{-1}]$-module chain map $\widetilde{h} : C(\widetilde{X}) \longrightarrow C(\widetilde{K} \times \mathbb{R})$. By 10.18 $\widetilde{h}$ is a $\mathbb{Z}[\pi]$-homology equivalence if and only if it is a Λ-homology equivalence. From now on, assume this is indeed the case, so that there is defined a Λ-coefficient Whitehead torsion

$$\tau(h;\Lambda) = (\tau_1,\tau_2) \in Wh_1(\Lambda) = Wh(\pi) \oplus \widetilde{\mathrm{End}}_0(\mathbb{Z}[\pi]) .$$

(See (ii) below for the application to knot complements). The first component of $\tau(h;\Lambda)$ is the $\mathbb{Z}[\pi]$-coefficient Whitehead torsion

$$\tau_1 = \tau(h;\mathbb{Z}[\pi]) = \tau(1 \otimes \widetilde{h} : \mathbb{Z}[\pi] \otimes_{\mathbb{Z}[\pi][z,z^{-1}]} C(\widetilde{X}) \longrightarrow C(\widetilde{K} \times S^1)) \in Wh(\pi) .$$

The second component of $\tau(h;\Lambda)$

$$\tau_2 = [C,f] \in \widetilde{\mathrm{End}}_0(\mathbb{Z}[\pi])$$

is constructed as follows. Let $\overline{X} = h^*(K \times \mathbb{R})$ be the induced infinite cyclic cover of X, so that h lifts to a $\mathbb{Z}$-equivariant map $\overline{h} : \overline{X} \longrightarrow K \times \mathbb{R}$. Making h CW transverse at $K \times \{*\} \subset K \times S^1$ (up to simple homotopy equivalence, as in 7.5) it is possible to obtain a fundamental domain $(V;U,\zeta U)$ for $\overline{X}$ with a π_1-isomorphism map

$$(b;a,\zeta a) : (V;U,\zeta U) \longrightarrow K \times ([0,1];\{0\},\{1\}) .$$

Let

$$\begin{aligned} i_+ &: U = V \cap \zeta^{-1}V \longrightarrow V ; x \longrightarrow \zeta x \\ i_- &: U = V \cap \zeta^{-1}V \longrightarrow V ; x \longrightarrow x \end{aligned}$$

and let $\widetilde{U}, \widetilde{V}$ be the universal covers over U, V, so that there is defined a morphism of short exact sequences of finite f.g. free $\mathbb{Z}[\pi][z, z^{-1}]$-module chain complexes

$$\begin{array}{ccccccccc} 0 & \longrightarrow & C(\widetilde{U})[z,z^{-1}] & \xrightarrow{i_+ - zi_-} & C(\widetilde{V})[z,z^{-1}] & \longrightarrow & C(\widetilde{X}) & \longrightarrow & 0 \\ & & \big\downarrow \widetilde{a} & & \big\downarrow \widetilde{b} & & \big\downarrow \widetilde{h} & & \\ 0 & \longrightarrow & C(\widetilde{K})[z,z^{-1}] & \xrightarrow{1-z} & C(\widetilde{K} \times I)[z,z^{-1}] & \longrightarrow & C(\widetilde{K} \times \mathbb{R}) & \longrightarrow & 0 \end{array}$$

The $\mathbb{Z}[\pi]$-module chain maps induced on the chain complex kernels

$$i_+ , i_- : C = \mathcal{C}(\widetilde{a} : C(\widetilde{U}) \longrightarrow C(\widetilde{K})) \longrightarrow D = \mathcal{C}(\widetilde{b} : C(\widetilde{V}) \longrightarrow C(\widetilde{K}))$$

are such that $i_+ - i_- : C \longrightarrow D$ is a $\mathbb{Z}[\pi]$-module chain equivalence. The second component of $\tau(h; \Lambda)$ is the endomorphism class $\tau_2 = [C, f] \in \widetilde{\mathrm{End}}_0(\mathbb{Z}[\pi])$ of the $\mathbb{Z}[\pi]$-module chain map

$$f = -(i_+ - i_-)^{-1} i_- : C \longrightarrow C ,$$

for which

$$1 - f = (i_+ - i_-)^{-1} i_+ : C \longrightarrow C .$$

The $\mathbb{Z}[\pi][z, z^{-1}]$-module chain complexes $C(\widetilde{X}), C(\widetilde{K} \times \mathbb{R})$ are Λ_1-contractible, so the split injection $Wh_1(\Lambda) \longrightarrow Wh_1(\Lambda_1)$ sends $\tau(h; \Lambda)$ to a difference of absolute Reidemeister torsion type invariants

$$\tau(h; \Lambda_1) = \tau(X; \Lambda_1) - \tau(K \times S^1; \Lambda_1) \in Wh_1(\Lambda_1) ,$$

with

$$\tau(K \times S^1; \Lambda_1) = \tau((1-z)^{\chi(K)}) = 0 \in Wh_1(\Lambda_1) .$$

(See Chap. 16 below for a brief account of Reidemeister torsion).

(ii) Let $k : S^n \subset S^{n+2}$ be a knot, and let

$$X = \mathrm{cl.}(S^{n+2} \backslash (k(S^n) \times D^2)) \subset S^{n+2}$$

be the knot exterior. The generator $1 \in H^1(X) = \mathbb{Z}$ can be represented by a map $h : X \longrightarrow S^1$ which is a $\mathbb{Z}$-homology equivalence, and which is transverse at a point $* \in S^1$. Thus $F^{n+1} = h^{-1}(*) \subset X$ is a Seifert surface for k, with $\partial F = k(S^n) \subset S^{n+2}$. The $\Pi^{-1}\mathbb{Z}[z, z^{-1}]$-coefficient Whitehead torsion of h is then defined as in (i)

$$\tau(h; \Pi^{-1}\mathbb{Z}[z, z^{-1}]) = [C(F), f] \in Wh(\Pi^{-1}\mathbb{Z}[z, z^{-1}]) = \widetilde{\mathrm{End}}_0(\mathbb{Z}) ,$$

with $K = \{\mathrm{pt.}\}$ and $f = -(i_+ - i_-)^{-1} i_- : C(F) \longrightarrow C(F)$. See Chaps. 17, 33 for the expression of this invariant in terms of the Alexander polynomials of k. □

11. The characteristic polynomial

The characteristic polynomial is the basic invariant of an endomorphism of a f.g. free module over a commutative ring A. The Alexander polynomials of knots (Chaps. 17, 33) are scaled characteristic polynomials.

The following treatment of determinants and characteristic polynomials for endomorphisms of f.g. projective A-modules is an adaptation of Goldman [90] and Almkvist [5], where further details may be found. In Chap. 14 the endomorphism class will be shown to be determined by the characteristic polynomial.

In the following definition the ground ring A is not assumed to be commutative.

Definition 11.1 The *reverse* of a polynomial of degree d

$$p(z) \ = \ \sum_{j=0}^{d} a_j z^j \in A[z]$$

with leading coefficient a unit $a_d \in A^\bullet$ is the polynomial

$$\begin{aligned}\widetilde{p}(z) \ &= \ (a_d)^{-1} z^d p(z^{-1}) \\ &= \ (a_d)^{-1}\Big(\sum_{j=0}^{d} a_{d-j} z^j\Big) \in A[z]\end{aligned}$$

with constant coefficient $\widetilde{p}(0) = 1 \in A$. □

The reverse of a monic polynomial

$$p(z) \ = \ a_0 + a_1 z + \ldots + a_{d-1} z^{d-1} + z^d$$

is the polynomial

$$\widetilde{p}(z) \ = \ z^d p(z^{-1}) \ = \ 1 + a_{d-1} z + \ldots + a_0 z^d$$

with the same coefficients but in reverse order.

For the remainder of Chap. 11 the ground ring A will be assumed to be commutative.

Definition 11.2 Let M be a $d \times d$ matrix with entries in A.
(i) The *characteristic polynomial* of M is

$$\mathrm{ch}_z(M) \;=\; \det(zI_d - M) \in A[z] \;,$$

a monic polynomial of degree d.
(ii) The *reverse characteristic polynomial* of M is

$$\widetilde{\mathrm{ch}}_z(M) \;=\; \det(I_d - zM) \in A[z] \;,$$

a polynomial of degree $\leq d$ with constant coefficient 1. □

The characteristic polynomials in 11.2 are related by

$$\widetilde{\mathrm{ch}}_z(M) \;=\; z^d \mathrm{ch}_{z^{-1}}(M) \in A[z] \;,$$

i.e. $\widetilde{\mathrm{ch}}_z(M)$ is the reverse of $\mathrm{ch}_z(M)$ in the sense of 11.1.

Definition 11.3 The *trace* and *determinant* of an endomorphism $f : P \longrightarrow P$ of a f.g. projective A-module P are

$$\begin{aligned} \mathrm{tr}(f) \;&=\; \mathrm{tr}(f \oplus 0 : P \oplus Q \longrightarrow P \oplus Q) \;, \\ \det(f) \;&=\; \det(f \oplus 1 : P \oplus Q \longrightarrow P \oplus Q) \in A \end{aligned}$$

for any f.g. projective A-module Q such that $P \oplus Q$ is a f.g. free A-module. □

Example 11.4 The trace and determinant of a $d \times d$ matrix M in A are (of course) the trace and determinant of the endomorphism $M : A^d \longrightarrow A^d$. If

$$\mathrm{ch}_z(M) \;=\; \det(zI_d - M) \;=\; p(z) \;=\; \sum_{j=0}^{d} a_j z^j \in A[z]$$

then

$$\widetilde{\mathrm{ch}}_z(M) \;=\; \det(I_d - zM) \;=\; \widetilde{p}(z) \;=\; \sum_{j=0}^{d} \widetilde{a}_j z^j \in A[z] \;.$$

The coefficients $\widetilde{a}_j \in A$ $(0 \leq j \leq d)$ are such that

$$\begin{aligned} \widetilde{a}_j \;&=\; a_{d-j} \;\;,\;\; \widetilde{a}_0 \;=\; a_d \;=\; 1 \;\;, \\ \widetilde{a}_d \;&=\; a_0 \;=\; (-1)^d \det(M) \in A \;\;, \\ \widetilde{a}_1 \;&=\; a_{d-1} \;=\; -\mathrm{tr}(M) \in A \;, \\ \widetilde{a}_j \;&=\; a_{d-j} \;=\; (-1)^j \mathrm{tr}(\Lambda^j(M) : \Lambda^j(A^d) \longrightarrow \Lambda^j(A^d)) \in A \;, \end{aligned}$$

with $\Lambda^*(A^d)$ the exterior algebra on A^d. □

Example 11.5 For any monic polynomial of degree d

$$p(z) = \sum_{j=0}^{d} a_j z^j \in A[z] \quad (a_d = 1)$$

the quotient A-module $A[z]/(p(z))$ is a d-dimensional f.g. free A-module, with

$$\begin{aligned} \mathrm{ch}_z(A[z]/(p(z)),\zeta) &= p(z) \,, \\ \widetilde{\mathrm{ch}}_z(A[z]/(p(z)),\zeta) &= \widetilde{p}(z) = \sum_{j=0}^{d} a_{d-j} z^j \in A[z] \,. \end{aligned}$$

□

Proposition 11.6 (i) *For any endomorphisms $f : P \longrightarrow P$, $g : Q \longrightarrow Q$ of f.g. projective A-modules*

$$\det(f \oplus g : P \oplus Q \longrightarrow P \oplus Q) = \det(f : P \longrightarrow P)\det(g : Q \longrightarrow Q) \in A \,,$$

and if $P = Q$

$$\det(fg : P \longrightarrow P) = \det(f : P \longrightarrow P)\det(g : P \longrightarrow P) \in A \,,$$

with

$$\det(1 : P \longrightarrow P) = 1 \in A \,.$$

(ii) *An endomorphism $f : P \longrightarrow P$ of a f.g. projective A-module P is an automorphism if and only if* $\det(f) \in A$ *is a unit.*
(iii) *If A is an integral domain with quotient field F then*

$$\det(f : P \longrightarrow P) = \det(1 \otimes f : F \otimes_A P \longrightarrow F \otimes_A P) \in A \subseteq F \,,$$

with $\det(f) \neq 0$ *if and only if $f : P \longrightarrow P$ is injective, if and only if $1 \otimes f : F \otimes_A P \longrightarrow F \otimes_A P$ is an automorphism.* □

Remark 11.7 If F is a f.g. free A-module of rank $d \geq 1$ then

$$\det(0 : F \longrightarrow F) = 0 \,, \quad \det(z : F[z] \longrightarrow F[z]) = z^d \,,$$

as usual. But in general for a f.g. projective A-module P

$$\det(0 : P \longrightarrow P) \neq 0 \in A \,, \quad \det(z : P[z] \longrightarrow P[z]) \neq z^d \,.$$

For example, if $P = e(A)$ for some idempotent $e = e^2 \in A$ then

$$\begin{aligned} \det(0 : P \longrightarrow P) &= 1 - e \in A \,, \\ \det(z : P[z] \longrightarrow P[z]) &= 1 - e + ze \in A \,. \end{aligned}$$

□

Definition 11.8 Let $f : P \longrightarrow P$ be an endomorphism of a f.g. projective A-module P.
(i) The *characteristic polynomial* of (P, f) is

$$\mathrm{ch}_z(P, f) = \det(z - f : P[z] \longrightarrow P[z]) \in A[z] .$$

(ii) The *reverse characteristic polynomial* of (P, f) is

$$\widetilde{\mathrm{ch}}_z(P, f) = \det(1 - zf : P[z] \longrightarrow P[z]) \in A[z] . \qquad \square$$

For f.g. free P the characteristic polynomials of 11.8 agree with the definitions in 11.2.

Example 11.9 If $P = e(A)$ for some idempotent $e = e^2 \in A$ there is an evident identification

$$\mathrm{Hom}_A(P, P) = \{f \in A \,|\, efe = f \in A\} ,$$

and for any such f

$$\begin{aligned} \mathrm{ch}_z(P, f) &= \det(z - f : P[z] \longrightarrow P[z]) = ze + 1 - e - f , \\ \widetilde{\mathrm{ch}}_z(P, f) &= \det(1 - zf : P[z] \longrightarrow P[z]) = 1 - zf \in A[z] . \qquad \square \end{aligned}$$

Proposition 11.10 (i) *The characteristic and reverse characteristic polynomials are related by*

$$\widetilde{\mathrm{ch}}_z(P, f) = \mathrm{ch}_z(P, 0)\, \mathrm{ch}_{z^{-1}}(P, f) \in A[z] ,$$

with $\mathrm{ch}_z(P, 0) \in A[z]$ *a unit in* $A[z, z^{-1}]$.
(ii) *Every characteristic polynomial* $\mathrm{ch}_z(P, f) \in A[z]$ *is a factor of a monic polynomial.*
(iii) *The reverse characteristic polynomial is of the form*

$$\widetilde{\mathrm{ch}}_z(P, f) = \sum_{j=0}^{d} (-1)^j \mathrm{tr}(\Lambda^j f : \Lambda^j P \longrightarrow \Lambda^j P) z^j \in A[z]$$

with $\Lambda^j P$ *the* j*th exterior product of* P.
(iv) *If* A *is an integral domain with quotient field* F *then*

$$\begin{aligned} \mathrm{ch}_z(P, f) &= \mathrm{ch}_z(F \otimes_A (P, f)) , \\ \widetilde{\mathrm{ch}}_z(P, f) &= \widetilde{\mathrm{ch}}_z(F \otimes_A (P, f)) \in A[z] \subseteq F[z] . \end{aligned}$$

In particular, $\mathrm{ch}_z(P, f)$ *is a monic polynomial of degree* $\dim_F(F \otimes_A P)$.
Proof (i) Let $P = \mathrm{im}(p = p^2 : A^k \longrightarrow A^k)$, so that

$$\begin{aligned} \mathrm{ch}_z(P, 0) &= \det(z : P[z] \longrightarrow P[z]) \\ &= \det(1 - p + pz : A[z]^k \longrightarrow A[z]^k) \in A[z] \cap A[z, z^{-1}]^\bullet \end{aligned}$$

with inverse

$$\begin{aligned} \mathrm{ch}_z(P,0)^{-1} &= \det(z^{-1} : P[z^{-1}] \longrightarrow P[z^{-1}]) \\ &= \det(1-p+pz^{-1} : A[z^{-1}]^k \longrightarrow A[z^{-1}]^k) \\ &\in A[z^{-1}] \cap A[z,z^{-1}]^\bullet . \end{aligned}$$

By the result of Goldman [87, 2.3]

$$\mathrm{ch}_z(P,0) = \sum_{i=0}^{n} e_i z^i \in A[z]$$

for a complete set of orthogonal idempotents $e_0, e_1, \ldots, e_n \in A$ with

$$\begin{aligned} e_i e_j &= 0 \ (i \neq j) \ , \ (e_i)^2 = e_i \ , \ \sum_{i=0}^{n} e_i = 1 \in A \\ \mathrm{ann}\,(\Lambda^i P) &= (e_0 + e_1 + \ldots + e_{i-1})(A) \ , \\ \det(z : P[z] \longrightarrow P[z])^{-1} &= \sum_{i=0}^{n} e_i z^{-i} \in A[z,z^{-1}] \ . \end{aligned}$$

The identity $\widetilde{\mathrm{ch}}_z(P,f) = \mathrm{ch}_z(P,0)\,\mathrm{ch}_{z^{-1}}(P,f)$ is obtained by applying the multiplicative property of the determinant to the identity

$$1 - zf = z(z^{-1} - f) \ : \ P[z,z^{-1}] \longrightarrow P[z,z^{-1}] \ .$$

(ii) Given (P,f) let Q be a f.g. projective A-module such that $P \oplus Q$ is f.g. free, and note that

$$\mathrm{ch}_z(P,f)\mathrm{ch}_z(Q,0) = \mathrm{ch}_z(P \oplus Q, f \oplus 0) \in A[z]$$

is a monic polynomial – this is a special case of 10.7 (ii). (Grayson [96, p. 440] has a different proof: "the characteristic polynomial is monic locally on spec(A), and thus divides a monic polynomial.")
(iii) See Almkvist [5, 2.3].
(iv) Immediate from 11.6 (iii). □

Example 11.11 (i) If P is f.g. free of rank d then

$$\mathrm{ch}_z(P,0) = z^d \ , \ \mathrm{ch}_z(P,f) = z^d \widetilde{\mathrm{ch}}_{z^{-1}}(P,f) \in A[z] \ ,$$

so $p(z) = \mathrm{ch}_z(P,f)$ and $\widetilde{p}(z) = \widetilde{\mathrm{ch}}_z(P,f)$ are related as in 11.1.
(ii) If $P = e(A)$ for some idempotent $e = e^2 \in A$ then

$$\begin{aligned} \mathrm{ch}_z(P,0) &= \det(z : P[z] \longrightarrow P[z]) = 1 - e + ze \ , \\ \widetilde{\mathrm{ch}}_z(P,0) &= \det(1 : P[z] \longrightarrow P[z]) = 1 \in A[z] \ . \end{aligned}$$

□

The Cayley-Hamilton theorem extends to endomorphisms of f.g. projective A-modules:

Proposition 11.12 (Goldman [87, 2.1], Almkvist [5, 1.6])
If the characteristic polynomial and the reverse characteristic polynomial of (P, f) are given by

$$\mathrm{ch}_z(P,f) \;=\; p(z) \;\;,\;\; \widetilde{\mathrm{ch}}_z(P,f) \;=\; q(z) \in A[z]$$

then

$$p(f) \;=\; q(f) \;=\; 0 \;:\; P \longrightarrow P \;. \qquad \square$$

Definition 11.13 Let $W(A)$ be the abelian group of *rational Witt vectors*, the subgroup $W(A) \subset A[[z]]^\bullet$ consisting of the formal power series of the type

$$\sum_{j=0}^{\infty} a_j z^j \;=\; \frac{p(z)}{q(z)} \in 1 + zA[[z]] \subset A[[z]]^\bullet$$

with $p(z), q(z) \in A[z]$ such that $a_0 = p(0) = q(0) = 1 \in A$. $\square$

Proposition 11.14 (Almkvist [6])
The reverse characteristic polynomial defines an isomorphism

$$\widetilde{\mathrm{End}}_0(A) \xrightarrow{\;\simeq\;} W(A) \;;\; (P,f) \longrightarrow \widetilde{\mathrm{ch}}_z(P,f)$$

with inverse

$$W(A) \xrightarrow{\;\simeq\;} \widetilde{\mathrm{End}}_0(A) \;;\; p(z) \longrightarrow [A[z]/(p(z))] \;.$$

Proof See 14.12 below (which also works in the noncommutative case). $\square$

Example 11.15 (Almkvist [5, 1.12])
For any endomorphism $f : P \longrightarrow P$ of a f.g. projective A-module P the logarithmic derivative of $\widetilde{\mathrm{ch}}_z(P,f)$ is such that

$$-z\,\widetilde{\mathrm{ch}}_z(P,f)^{-1}\widetilde{\mathrm{ch}}_z'(P,f) \;=\; \sum_{n=1}^{\infty} \mathrm{tr}(f^n : P \longrightarrow P) z^n \in A[[z]] \;.$$

If the additive group of A is torsion-free the reverse characteristic polynomial is given by the exponential trace formula

$$\widetilde{\mathrm{ch}}_z(P,f) \;=\; \exp\Big(-\sum_{n=1}^{\infty} \frac{\mathrm{tr}(f^n)}{n} z^n \Big)$$
$$\in 1 + zA[[z]] \subset W(A) \;.$$

$\square$

Proposition 11.16 *The following conditions on a polynomial*

$$p(z) = \sum_{j=0}^{d} a_j z^j \in A[z]$$

are equivalent:

(i) $p(z) \in A((z^{-1}))^{\bullet}$,
(ii) $p(z) \in A[z]$ *is a non-zero divisor and* $A[z]/(p(z))$ *is a f.g. projective* A*-module,*
(iii) $p(z)$ *is a factor of a monic polynomial in* $A[z]$.

Proof (i) $\iff$ (ii) By 8.2 applied to $E^+ = \mathcal{C}(p(z) : A[z] \longrightarrow A[z])$.
(ii) $\Longrightarrow$ (iii) Define the multiplicative subset

$$S = (p(z))^{\infty} = \{p(z)^k \mid k \geq 0\} \subset A[z] .$$

The $(A[z], S)$-module $P = A[z]/(p(z))$ has a f.g. free $A[z]$-module resolution

$$0 \longrightarrow A[z] \xrightarrow{p(z)} A[z] \longrightarrow P \longrightarrow 0$$

and also the f.g. projective $A[z]$-module resolution

$$0 \longrightarrow P[z] \xrightarrow{z-\zeta} P[z] \longrightarrow P \longrightarrow 0 ,$$

which are related by a chain equivalence. The characteristic polynomial of the endomorphism $\zeta \in \mathrm{Hom}_A(P, P)$ is thus of the form

$$\mathrm{ch}_z(P, \zeta) = p(z)t(z) \in A[z]$$

for some unit $t(z) \in A[z]^{\bullet}$. (If $a_d = 1 \in A$ then P is f.g. free of rank d and (P, ζ) has characteristic polynomial $\mathrm{ch}_z(P, \zeta) = p(z)$ – see 12.18 below). Now $\mathrm{ch}_z(P, \zeta)$ is a factor of a monic polynomial by 11.10 (ii), and hence so is $p(z)$.
(iii) $\Longrightarrow$ (ii) Every monic polynomial in $A[z]$ is a unit in $A((z^{-1}))$. □

Similarly:

Proposition 11.17 *The following conditions on a Laurent polynomial*

$$p(z) = \sum_{j=m}^{n} a_j z^j \in A[z, z^{-1}]$$

are equivalent:

(i) $p(z) \in A((z))^{\bullet} \cap A((z^{-1}))^{\bullet}$,
(ii) $p(z) \in A[z, z^{-1}]$ *is a non-zero divisor and* $A[z, z^{-1}]/(p(z))$ *is a f.g. projective* A*-module,*
(iii) $p(z)$ *is a factor of a polynomial in* $A[z, z^{-1}]$ *with extreme coefficients units.*

Proof (i) $\iff$ (ii) By 8.3 applied to $E = \mathcal{C}(p(z) : A[z,z^{-1}] \longrightarrow A[z,z^{-1}])$.
(ii) $\Longrightarrow$ (iii) Define the multiplicative subset

$$S = (p(z))^\infty = \{p(z)^k \mid k \geq 0\} \subset A[z,z^{-1}] .$$

The $(A[z,z^{-1}],S)$-module $P = A[z,z^{-1}]/(p(z))$ has a f.g. free $A[z,z^{-1}]$-module resolution

$$0 \longrightarrow A[z,z^{-1}] \xrightarrow{p(z)} A[z,z^{-1}] \longrightarrow P \longrightarrow 0$$

and also the f.g. projective $A[z,z^{-1}]$-module resolution

$$0 \longrightarrow P[z,z^{-1}] \xrightarrow{z-\zeta} P[z,z^{-1}] \longrightarrow P \longrightarrow 0 ,$$

which are related by a chain equivalence. The characteristic polynomial of the automorphism $\zeta \in \mathrm{Hom}_A(P,P)$ is thus of the form

$$\mathrm{ch}_z(P,\zeta) = p(z)t(z) \in A[z,z^{-1}]$$

for some unit $t(z) \in A[z,z^{-1}]^\bullet$, with

$$\mathrm{ch}_0(P,\zeta) = \det(-\zeta : P \longrightarrow P) \in A^\bullet .$$

(If $a_m \in A^\bullet$, $a_n = 1 \in A$ then P is f.g. free of rank $n-m$ and (P,ζ) has characteristic polynomial $\mathrm{ch}_z(P,\zeta) = z^{-m}p(z)$). Now $\mathrm{ch}_z(P,\zeta)$ is a factor of a monic polynomial by 11.10 (ii), and hence so is $p(z)$.
(iii) $\Longrightarrow$ (ii) Every monic polynomial in $A[z,z^{-1}]$ with constant coefficient a unit in A is a unit in both $A((z))$ and $A((z^{-1}))$. $\square$

12. Primary K-theory

In the applications of endomorphism K- (and L-) theory to the computation of the high-dimensional knot cobordism groups it is necessary to restrict attention to endomorphisms $f : P \longrightarrow P$ of f.g. projective A-modules such that $p(f) = 0$ for a particular type of polynomial $p(z) \in A[z]$, e.g. an (Alexander) polynomial $p(z) \in A[z]$ with $p(1) \in A^\bullet$.

This chapter studies the endomorphism class groups $\text{End}_0^S(A)$, $\widetilde{\text{End}}_0^S(A)$ of pairs (P, f) with $p(f) = 0$ for some $p(z) \in S$, with $S \subset A[z]$ a multiplicative subset. The splitting theorems of Chap. 10 will now be specialized to the splitting theorems of Grayson [96]

$$\begin{aligned} K_1(S^{-1}A[z]) &= K_1(A[z]) \oplus \text{End}_0^S(A) , \\ K_1(\widetilde{S}^{-1}A[z]) &= K_1(A) \oplus \widetilde{\text{End}}_0^S(A) \end{aligned}$$

for appropriate $S, \widetilde{S} \subset A[z]$.

Definition 12.1 Let $S \subset A[z]$ be a multiplicative subset.
(i) An object (P, f) in $\text{End}(A)$ is *S-primary* if

$$z - f \,:\, S^{-1}P[z] \longrightarrow S^{-1}P[z]$$

is an $S^{-1}A[z]$-module automorphism.
(ii) The *S-primary endomorphism category* $\text{End}^S(A)$ is the full subcategory of $\text{End}(A)$ with S-primary objects (P, f). The *S-primary endomorphism K-groups* of A are defined by

$$\text{End}_*^S(A) \;=\; K_*(\text{End}^S(A)) .$$

(iii) If $z \in S$ use the embedding

$$\mathbb{P}(A) \longrightarrow \text{End}(A) \,;\; P \longrightarrow (P, 0)$$

to define the *reduced S-primary endomorphism K-groups* of A by

$$\widetilde{\text{End}}_*^S(A) \;=\; \text{coker}(K_*(A) \longrightarrow K_*(\text{End}^S(A))) ,$$

such that

$$\mathrm{End}_*^S(A) = K_*(A) \oplus \widetilde{\mathrm{End}}_*^S(A) . \qquad \square$$

Example 12.2 Suppose that A is commutative, and that $S \subset A[z]$ is a multiplicative subset. An object $(P, f) \in \mathrm{End}(A)$ with characteristic polynomial

$$\mathrm{ch}_z(P, f) = \det(z - f : P[z] \longrightarrow P[z]) \in S \subset A[z]$$

is S-primary. (See 12.6 below for a generalization). $\square$

Example 12.3 Let A, S be as in 12.1. The $A[z]$-module given for any $p(z) \in S$ by

$$P = A[z]/(p(z))$$

determines an object (P, ζ) in $\mathrm{End}^S(A)$ if and only if $p(z) \in A((z^{-1}))^\bullet$.

(i) If $p(z) = \sum\limits_{j=0}^{d} a_j z^j \in S$ has leading coefficient $a_d \in A^\bullet$ then $p(z) \in A((z^{-1}))^\bullet$ and P is a f.g. free A-module of rank d. The A-module isomorphism

$$f : A^d \xrightarrow{\simeq} P ; (b_1, b_2, \dots, b_d) \longrightarrow b_1 + b_2 z + \dots + b_d z^{d-1}$$

is such that

$$f^{-1}\zeta f = \begin{pmatrix} 0 & 0 & 0 & \dots & -a_0 a_d^{-1} \\ 1 & 0 & 0 & \dots & -a_1 a_d^{-1} \\ 0 & 1 & 0 & \dots & -a_2 a_d^{-1} \\ \vdots & \vdots & \vdots & \ddots & \vdots \\ 0 & 0 & 0 & \dots & -a_{d-1} a_d^{-1} \end{pmatrix} : A^d = A \oplus A \oplus \dots \oplus A \longrightarrow A^d .$$

(This is the *companion matrix* of linear algebra). If A is commutative then

$$\begin{aligned} \mathrm{ch}_z(P, \zeta) &= \mathrm{ch}_z(A^d, f^{-1}\zeta f) \\ &= (a_d)^{-1}\Big(\sum_{j=0}^{d} a_j z^j\Big) = (a_d)^{-1} p(z) \in S \subset A[z] . \end{aligned}$$

(ii) If $p(z) = 1 - e + ez \in S$ for an idempotent $e = e^2 \in A$ then

$$(P, \zeta) = (e(A), 0)$$

is an object in $\mathrm{End}^S(A)$. $\square$

Proposition 12.4 *Let $S \subset A[z]$ be a multiplicative subset. The following conditions on an object (P, f) in* $\mathrm{End}(A)$ *are equivalent:*

(i) *(P, f) is S-primary,*
(ii) $S^{-1}(P, f) = 0$,
(iii) $p(f) = 0 : P \longrightarrow P$ *for some $p(z) \in S$.*

Proof (i) $\Longleftrightarrow$ (ii) The exact sequence

$$0 \longrightarrow P[z] \xrightarrow{z-f} P[z] \longrightarrow (P,f) \longrightarrow 0$$

defines a f.g. projective $A[z]$-module resolution of the $A[z]$-module (P,f) (= A-module P with $z = f : P \longrightarrow P$). Localization is exact, so there is induced an exact sequence of $S^{-1}A[z]$-modules

$$0 \longrightarrow S^{-1}P[z] \xrightarrow{z-f} S^{-1}P[z] \longrightarrow S^{-1}(P,f) \longrightarrow 0$$

and $z-f$ is an isomorphism if and only if $S^{-1}(P,f) = 0$.
(i) $\Longrightarrow$ (iii) If (P,f) is S-primary then the inverse of $z-f$ is of the form

$$(z-f)^{-1} = g(z)/p(z) \ : \ S^{-1}P[z] \longrightarrow S^{-1}P[z]$$

for some

$$p(z) \in S \ , \ g(z) \in \mathrm{Hom}_{A[z]}(P[z],P[z]) = \mathrm{Hom}_A(P,P)[z] \ ,$$

so that

$$(z-f)g(z) = p(z) \ : \ P[z] \longrightarrow P[z]$$

and $p(f) = 0 : P \longrightarrow P$.
(iii) $\Longrightarrow$ (i) Let $p(f) = 0 : P \longrightarrow P$ for some

$$p(z) = \sum_{j=0}^{d} a_j z^j \in S \ .$$

The $A[z]$-module morphism

$$\begin{aligned} g(z) &= (p(z)-p(f))/(z-f) \\ &= \sum_{j=1}^{d}\sum_{k=0}^{j-1} a_j f^{j-1-k} z^k \ : \ P[z] \longrightarrow P[z] \end{aligned}$$

is such that

$$(z-f)^{-1} = g(z)/p(z) \ : \ S^{-1}P[z] \longrightarrow S^{-1}P[z] \ . \qquad \square$$

Terminology 12.5 A polynomial $p(z) \in A[z]$ with central coefficients generates a principal ideal

$$\mathcal{P} = (p(z)) \triangleleft A[z]$$

and a multiplicative subset

$$(p(z))^\infty = \{up(z)^k \mid u \in Z(A^\bullet), k \geq 0\} \subset A[z] \ ,$$

where $Z(A^\bullet) \subset A$ is the subset of central units. Abbreviate $(p(z))^\infty$-primary to $\mathcal{P}$-*primary*, in line with the usual terminology. $\square$

Example 12.6 Suppose that A is commutative, and that $S \subset A[z]$ is a multiplicative subset. The following conditions on an object $(P, f) \in \text{End}(A)$ are equivalent:

(i) (P, f) is S-primary,
(ii) $\text{ch}_z(P, f) \in A[z]$ is a unit in $S^{-1}A[z]$,
(iii) $\text{ch}_z(P, f)$ divides an element $p(z) \in S$.

These conditions are satisfied if $\text{ch}_z(P, f) \in S$ (12.2). In particular, if $S = (p(z))^\infty$ (as in 12.4) for a monic polynomial $p(z) \in A[z]$ the conditions are satisfied if $\text{ch}_z(P, f) = p(z)^k$ for some $k \geq 1$. □

Recall from 4.13 that multiplicative subsets $S, T \subset A[z]$ are coprime if for all $p(z) \in S$, $q(z) \in T$ there exist $a(z), b(z) \in A[z]$ such that

$$a(z)p(z) + b(z)q(z) \ = \ 1 \in A[z] \ .$$

Proposition 12.7 *If $S, T \subset A[z]$ are coprime central multiplicative subsets the multiplicative subset*

$$ST \ = \ \{p(z)q(z) \,|\, p(z) \in S, q(z) \in T\} \subset A[z]$$

is such that

$$\begin{aligned} \text{End}^{ST}(A) \ &= \ \text{End}^S(A) \times \text{End}^T(A) \ , \\ \text{End}_*^{ST}(A) \ &= \ \text{End}_*^S(A) \oplus \text{End}_*^T(A) \ . \end{aligned}$$

Proof By 12.4, for any object (P, f) in $\text{End}^{ST}(A)$ there exist $p(z) \in S$, $q(z) \in T$ such that

$$p(f)q(f) \ = \ 0 \ : \ P \longrightarrow P \ .$$

If $a(z), b(z) \in A[z]$ are such that $a(z)p(z) + b(z)q(z) = 1$ then the endomorphisms $a(f)p(f), b(f)q(f) : P \longrightarrow P$ are idempotents such that

$$a(f)p(f) + b(f)q(f) \ = \ 1 \ : \ P \longrightarrow P \ .$$

The A-modules

$$P_S \ = \ S^{-1}P \ = \ a(f)p(f)(P) \ \ , \ \ P_T \ = \ T^{-1}P \ = \ b(f)q(f)(P)$$

are such that

$$(P, f) \ = \ (P_S, f_S) \oplus (P_T, f_T)$$

with (P_S, f_S) S-primary and (P_T, f_T) T-primary. □

Remark 12.8 If F is a field then $F[z]$ is a unique factorization domain, and every monic polynomial $p(z) \in F[z]$ has a unique factorization as a product of powers of irreducible monic polynomials $p(z)$, so that

$$\text{End}_*(F) \ = \ \bigoplus_{p(z) \in \max(F[z])} \text{End}_*^{p(z)^\infty}(F)$$

according to the factorization of the characteristic polynomial. See Chap. 18 for a more detailed account of $\mathrm{End}_0(F)$. □

Definition 12.9 Let $S \subset A[z]$ be a multiplicative subset.

(i) S has *leading units* if for each $\sum\limits_{j=0}^{d} a_j z^j \in S$ the leading coefficient is a unit, i.e. if $a_d \in A^\bullet$.

(ii) S has *constant units* if for each $\sum\limits_{j=0}^{d} a_j z^j \in S$ the constant coefficient is a unit, i.e. if $a_0 \in A^\bullet$. □

Remark 12.10 (i) If $S \subset A[z]$ has leading units then $S \subseteq A[z] \cap A((z^{-1}))^\bullet$. If A is an integral domain then $S \subset A[z]$ has leading units if and only if $S \subseteq A[z] \cap A((z^{-1}))^\bullet$.
(ii) If $S \subset A[z]$ has constant units then $S \subseteq A[z] \cap A[[z]]^\bullet$. If A is an integral domain then $S \subset A[z]$ has constant units if and only if $S \subseteq A[z] \cap A[[z]]^\bullet$. □

A chain complex (C, f) in $\mathrm{End}(A)$ is a f.g. projective A-module chain complex C together with a chain map $f : C \longrightarrow C$.

Definition 12.11 Let $S \subset A[z]$ be a multiplicative subset. A finite chain complex (C, f) in $\mathrm{End}(A)$ is *S-primary* if

$$z - f \; : \; S^{-1}C[z] \longrightarrow S^{-1}C[z]$$

is an $S^{-1}A[z]$-module chain equivalence. The chain map $f : C \longrightarrow C$ is said to be *S-primary.* □

By analogy with 12.4:

Proposition 12.12 *The following conditions on a finite chain complex* (C, f) *in* $\mathrm{End}(A)$ *are equivalent:*

(i) (C, f) *is S-primary,*
(ii) $S^{-1}(H_*(C), f_*) = 0,$
(iii) $p(f) \simeq 0 : C \longrightarrow C$ *for some* $p(z) \in S$. □

By analogy with 10.10:

Proposition 12.13 *Let $S \subset A[z]$ be a multiplicative subset such that*

$$S \subseteq A[z] \cap A((z^{-1}))^\bullet .$$

(i) *The inclusion $A[z] \longrightarrow S^{-1}A[z]$ has the universal property that a finite f.g. free $A[z]$-module chain complex E^+ is such that $H_*(S^{-1}E^+) = 0$ if and only if E^+ is A-finitely dominated and $\zeta : E^+ \longrightarrow E^+$ is S-primary.*
(ii) *The functor*

$$\mathbb{H}(A[z], S) \longrightarrow \mathrm{End}^S(A) \; ; \; M \longrightarrow (M, \zeta)$$

is an isomorphism of categories. The relative K-groups

$$K_*(A[z],S) = K_{*-1}(\mathbb{H}(A[z],S))$$

in the localization exact sequence

$$\dots \longrightarrow K_n(A[z]) \longrightarrow K_n(S^{-1}A[z]) \longrightarrow K_n(A[z],S) \longrightarrow K_{n-1}(A[z]) \longrightarrow \dots$$

can be expressed as

$$K_*(A[z],S) = \mathrm{End}^S_{*-1}(A) .$$

The exact sequence breaks up into split exact sequences

$$0 \longrightarrow K_n(A[z]) \longrightarrow K_n(S^{-1}A[z]) \longrightarrow K_n(A[z],S) \longrightarrow 0 .$$

(iii) *The torsion group of $S^{-1}A[z]$ fits into the direct sum system*

$$K_1(A[z]) \underset{j_+}{\overset{i_+}{\rightleftarrows}} K_1(S^{-1}A[z]) \underset{\Delta_+}{\overset{\partial_+}{\rightleftarrows}} \mathrm{End}^S_0(A)$$

with

$$\begin{aligned}
i_+ &: K_1(A[z]) \longrightarrow K_1(S^{-1}A[z]) \ ; \ \tau(E^+) \longrightarrow \tau(S^{-1}E^+) , \\
j_+ &: K_1(S^{-1}A[z]) \longrightarrow K_1(A[z]) \ ; \\
&\quad \tau(S^{-1}E^+) \longrightarrow \tau(r^+ : \mathcal{C}(z-\zeta : E^+[z] \longrightarrow E^+[z]) \longrightarrow E^+) , \\
\partial_+ &: K_1(S^{-1}A[z]) \longrightarrow \mathrm{End}^S_0(A) \ ; \ \tau(S^{-1}E^+) \longrightarrow [E^+,\zeta] , \\
\Delta_+ &: \mathrm{End}^S_0(A) \longrightarrow K_1(S^{-1}A[z]) \ ; \\
&\quad [P,f] \longrightarrow \tau(z-f : S^{-1}P[z] \longrightarrow S^{-1}P[z]) .
\end{aligned}$$

□

Example 12.14 If A is commutative then by 11.16 $S = A[z] \cap A((z^{-1}))^\bullet$ is the multiplicative subset of all the polynomials in $A[z]$ which divide monic polynomials, with

$$S^{-1}A[z] = \Omega_+^{-1}A[z] \ , \ \mathrm{End}^S(A) = \mathrm{End}(A)$$

and 12.13 (iii) is just the splitting of 10.14. Moreover, $S^{-1}A[z]$ is the localization of $A[z]$ inverting all the monic polynomials. □

As in 11.1, given a polynomial $p(z) = \sum\limits_{j=0}^{d} a_j z^j \in A[z]$ with leading unit $a_d \in A^\bullet$ define the reverse polynomial

$$\widetilde{p}(z) = (a_d)^{-1} z^d p(z^{-1}) = (a_d)^{-1}\Big(\sum_{j=0}^{d} a_{d-j} z^j\Big) \in A[z]$$

with constant unit $\widetilde{p}(0) = 1$. If $q(z) = \sum\limits_{k=0}^{e} b_k z^k \in A[z]$ has leading unit $b_e \in A$ then

$$\widetilde{pq}(z) = \widetilde{p}(z)\widetilde{q}(z) \in A[z] .$$

Definition 12.15 Given a multiplicative subset $S \subset A[z]$ with leading units let $\widetilde{S} \subset A[z]$ be the *reverse* multiplicative subset with constant units consisting of the reverse elements $\widetilde{p}(z)$ of $p(z) \in S$. □

Proposition 12.16 *Let $S \subset A[z]$ be a multiplicative subset with leading units and $z \in S$. An object (P, f) in* End(A) *is (P, f) is S-primary if and only if $1 - zf : \widetilde{S}^{-1}P[z] \longrightarrow \widetilde{S}^{-1}P[z]$ is an $\widetilde{S}^{-1}A[z]$-module automorphism.*

Proof If (P, f) is S-primary there exist an A-module morphism

$$g(z) = \sum_{j=0}^{d} g_j z^j \ : \ P[z] \longrightarrow P[z] \ \ (g_j \in \mathrm{Hom}_A(P, P))$$

with $g_d : P \longrightarrow P$ an automorphism and $p(z) \in S$ such that

$$(z - f)g(z) = p(z) \ : \ P[z] \longrightarrow P[z] .$$

The $A[z]$-module morphism defined by

$$\widetilde{g}(z) = \sum_{j=0}^{d} (g_d)^{-1} g_{d-j} z^j \ : \ P[z] \longrightarrow P[z]$$

is such that

$$(1 - zf)\widetilde{g}(z) = \widetilde{p}(z) \ : \ P[z] \longrightarrow P[z],$$

so that $1 - zf : P[z] \longrightarrow P[z]$ is an $\widetilde{S}^{-1}A[z]$-module isomorphism.

Conversely, suppose $1 - zf : P[z] \longrightarrow P[z]$ is an $\widetilde{S}^{-1}A[z]$-equivalence, so that there exist an $A[z]$-module morphism

$$\widetilde{g}(z) = \sum_{j=0}^{d} \widetilde{g}_j z^j \ : P[z] \longrightarrow P[z] \ \ (\widetilde{g}_j \in \mathrm{Hom}_A(P, P))$$

with $\widetilde{g}_0 = 1 : P \longrightarrow P$ and

$$p(z) = \sum_{j=m}^{n} a_j z^j \in S$$

such that

$$(1 - zf)\widetilde{g}(z) = \widetilde{p}(z) \ : \ P[z] \longrightarrow P[z] ,$$

with

$$\widetilde{p}(z) = (a_n)^{-1}(\sum_{j=m}^{n} a_{m+n-j} z^j) \in \widetilde{S} .$$

The $A[z]$-module morphism defined by

$$g(z) = \sum_{j=0}^{d} \widetilde{g}_{d-j} z^j : P[z] \longrightarrow P[z]$$

is such that

$$\begin{aligned}(z-f)g(z) &= (a_n)^{-1}(\sum_{j=m}^{n} a_j z^{j-m}) \\ &= z^{-m}(a_n)^{-1}p(z) : P[z] \longrightarrow P[z] ,\end{aligned}$$

so that $z - f : P[z] \longrightarrow P[z]$ is an $S^{-1}A[z]$-module automorphism. □

Example 12.17 If $S = (z)^\infty$ then $\widetilde{S} = \{1\}$, and 12.16 identifies the (z)-primary (= nilpotent) category $\mathrm{End}^{(z)^\infty}(A) = \mathrm{Nil}(A)$ with the subcategory of $\mathrm{End}(A)$ consisting of the objects (P, f) such that $1 - zf : P[z] \longrightarrow P[z]$ is an automorphism. □

Example 12.18 If A is commutative and $S \subset A[z]$ is a multiplicative subset with leading units and $z \in S$ then 12.16 shows that an object (P, f) in $\mathrm{End}(A)$ is S-primary if and only if the reverse characteristic polynomial $\widetilde{\mathrm{ch}}_z(P, f) \in A[z]$ is a unit in $\widetilde{S}^{-1}A[z]$. In particular, this is the case if $\mathrm{ch}_z(P, f)$ divides an element of S, or if $\widetilde{\mathrm{ch}}_z(P, f) \in \widetilde{S}$. □

Proposition 12.19 *Let $S \subset A[z]$ be a multiplicative subset with leading units and $z \in S$, with reverse multiplicative subset $\widetilde{S} \subset A[z]$. A chain complex (C, f) in $\mathrm{End}(A)$ is S-primary if and only if the $\widetilde{S}^{-1}A[z]$-module chain map $1 - zf : \widetilde{S}^{-1}C[z] \longrightarrow \widetilde{S}^{-1}C[z]$ is a chain equivalence.*
Proof The chain homotopy class of an S-primary chain equivalence $f : C \longrightarrow C$ is an object $(H_0(\mathrm{Hom}_A(C, C)), f)$ in $\mathrm{End}^S(H_0(\mathrm{Hom}_A(C, C)))$. Now apply 12.4 and 12.15. □

An S-primary chain equivalence $f : C \longrightarrow C$ of a finitely dominated A-module chain complex C has an S-primary class invariant

$$[C, f] \in \mathrm{End}_0^S(A) .$$

By analogy with 10.16:

Proposition 12.20 *If $S \subset A[z]$ is a multiplicative subset with leading units and $z \in S$ the torsion group of $\widetilde{S}^{-1}A[z]$ fits into the direct sum system*

$$K_1(A) \underset{\widetilde{j}_+}{\overset{\widetilde{i}_+}{\rightleftarrows}} K_1(\widetilde{S}^{-1}A[z]) \underset{\widetilde{\Delta}_+}{\overset{\widetilde{\partial}_+}{\rightleftarrows}} \widetilde{\mathrm{End}}_0^S(A)$$

with

$$\begin{aligned}
\widetilde{i}_+ \ &: \ K_1(A) \longrightarrow K_1(\widetilde{S}^{-1}A[z]) \ ; \ \tau(E) \longrightarrow \tau(\widetilde{S}^{-1}E[z]) \ , \\
\widetilde{j}_+ \ &: \ K_1(\widetilde{S}^{-1}A[z]) \longrightarrow K_1(A) \ ; \ \tau(\widetilde{S}^{-1}E^+) \longrightarrow \tau(A \otimes_{A[z]} E^+) \ , \\
\widetilde{\partial}_+ \ &: \ K_1(\widetilde{S}^{-1}A[z]) \longrightarrow \widetilde{\mathrm{End}}_0^S(A) \ ; \\
&\qquad \tau(\widetilde{S}^{-1}\mathcal{C}(f_0 + zf_1 : P[z] \longrightarrow P[z])) \longrightarrow [P, -f_0^{-1}f_1] \ , \\
\widetilde{\Delta}_+ \ &: \ \widetilde{\mathrm{End}}_0^S(A) \longrightarrow K_1(\widetilde{S}^{-1}A[z]) \ ; \\
&\qquad [P, f] \longrightarrow \tau(1 - zf : \widetilde{S}^{-1}P[z] \longrightarrow \widetilde{S}^{-1}P[z]) \ .
\end{aligned}$$

Proof Let $S_- \subset A[z^{-1}]$ be the image of $S \subset A[z]$ under the isomorphism

$$A[z] \xrightarrow{\simeq} A[z^{-1}] \ ; \ z \longrightarrow z^{-1} \ .$$

This is the product

$$S_- \ = \ Z\widetilde{S}$$

of the coprime multiplicative subsets $Z = \{z^k\}$, $\widetilde{S} \subset A[z]$, so that there is defined a cartesian square of rings

$$\begin{array}{ccc}
A[z] & \longrightarrow & \widetilde{S}^{-1}A[z] \\
\big\downarrow & & \big\downarrow \\
A[z, z^{-1}] & \longrightarrow & S_-^{-1}A[z^{-1}]
\end{array}$$

with excision isomorphisms

$$K_*(A[z], Z) \ \cong \ K_*(\widetilde{S}^{-1}A[z], S_-) \ .$$

Combining the special case $* = 1$ with

$$\begin{aligned}
K_1(A[z], Z) \ &= \ K_0(A) \oplus \widetilde{\mathrm{Nil}}_0(A) \ , \\
K_1(S_-^{-1}A[z^{-1}]) \ &= \ K_1(S^{-1}A[z]) \\
&= \ K_1(A) \oplus K_0(A) \oplus \widetilde{\mathrm{Nil}}_0(A) \oplus \widetilde{\mathrm{End}}_0^S(A)
\end{aligned}$$

gives

$$K_1(\widetilde{S}^{-1}A[z]) \ = \ K_1(A) \oplus \widetilde{\mathrm{End}}_0^S(A) \ . \qquad \square$$

Example 12.21 If A is commutative and $S \subset A[z]$ is the multiplicative subset of all the monic polynomials then every object (P, f) in $\mathrm{End}(A)$ is S-primary, since $\mathrm{ch}_z(P, f) \in A[z]$ divides a monic polynomial (11.10) and so is a unit in $S^{-1}A[z]$. The reverse multiplicative subset is

$$\widetilde{S} \ = \ 1 + zA[z] \subset A[z] \ ,$$

and

$$\mathrm{End}^S(A) = \mathrm{End}(A) , \; \mathrm{End}_*^S(A) = \mathrm{End}_*(A) , \; \widetilde{\mathrm{End}}_*^S(A) = \widetilde{\mathrm{End}}_*(A) .$$

The splitting theorem of 12.20 coincides in this case with the splitting theorem of 10.16, identifying

$$\Omega_+^{-1}A[z] = S^{-1}A[z] , \; \widetilde{\Omega}_+^{-1}A[z] = \widetilde{S}^{-1}A[z] ,$$

so that

$$\begin{aligned} K_*(S^{-1}A[z]) &= K_*(A[z]) \oplus \mathrm{End}_{*-1}(A) , \\ K_*(\widetilde{S}^{-1}A[z]) &= \mathrm{End}_*(A) . \end{aligned}$$

□

Example 12.22 The multiplicative subset with leading units

$$Z = \{z^j \,|\, j \geq 0\} \subset A[z]$$

is such that

$$\widetilde{Z} = \{1\} , \; Z^{-1}A[z] = A[z,z^{-1}] , \; \widetilde{Z}^{-1}A[z] = A[z] .$$

An object (P,f) in $\mathrm{End}(A)$ is Z-primary if and only if f is nilpotent, that is

$$f^N = 0 : P \longrightarrow P$$

for some $N \geq 0$. Thus

$$\mathrm{End}^Z(A) = \mathrm{Nil}(A) , \; \mathrm{End}_*^Z(A) = \mathrm{Nil}_*(A)$$

and 12.13, 12.20 give the direct sum decompositions of Chap. 5

$$\begin{aligned} K_1(A[z,z^{-1}]) &= K_1(A[z]) \oplus \mathrm{Nil}_0(A) , \\ K_1(A[z]) &= K_1(A) \oplus \widetilde{\mathrm{Nil}}_0(A) . \end{aligned}$$

□

Example 12.23 The multiplicative subset with leading units

$$S = z^\infty(1-z)^\infty = \{z^j(1-z)^k \,|\, j,k \geq 0\} \subset A[z]$$

is such that

$$\begin{aligned} &\widetilde{S} = \{(1-z)^k \,|\, k \geq 0\} , \; S^{-1}A[z] = A[z,z^{-1},(1-z)^{-1}] , \\ &\widetilde{S}^{-1}A[z] = A[z,(1-z)^{-1}] \cong A[z,z^{-1}] . \end{aligned}$$

An object (P,f) in $\mathrm{End}(A)$ is S-primary if and only if f is a near-projection (5.7), that is

$$(f-f^2)^N = 0 : P \longrightarrow P$$

for some $N \geq 0$. For any such N the endomorphism

$$f_\omega = (f^N + (1-f)^N)^{-1} f^N = f_\omega^2 : P \longrightarrow P$$

is the unique projection commuting with f such that $f - f_\omega : P \longrightarrow P$ is nilpotent. Use f_ω and the nilpotent endomorphisms

$$\begin{aligned}\nu_0 &= f| : P_0 = (1-f_\omega)(P) \longrightarrow P_0 ,\\ \nu_1 &= (1-f)| : P_1 = f_\omega(P) \longrightarrow P_1\end{aligned}$$

to define isomorphisms

$$\begin{aligned}\mathrm{End}_0^S(A) &\xrightarrow{\simeq} K_0(A) \oplus K_0(A) \oplus \widetilde{\mathrm{Nil}}_0(A) \oplus \widetilde{\mathrm{Nil}}_0(A) ;\\ &\qquad [P,f] \longrightarrow ([P_0],[P_1],[P_0,\nu_0],[P_1,\nu_1]) ,\\ \widetilde{\mathrm{End}}_0^S(A) &\xrightarrow{\simeq} K_0(A) \oplus \mathrm{Nil}_0(A) \oplus \mathrm{Nil}_0(A) ;\\ &\qquad [P,f] \longrightarrow ([P_0]-[P_1],[P_0,\nu_0],[P_1,\nu_1]) .\end{aligned}$$

Use 12.13 and 12.20 to identify

$$\begin{aligned}&K_1(S^{-1}A[z]) = \\ &\qquad K_1(A) \oplus K_0(A) \oplus K_0(A) \oplus \widetilde{\mathrm{Nil}}_0(A) \oplus \widetilde{\mathrm{Nil}}_0(A) \oplus \widetilde{\mathrm{Nil}}_0(A) ,\\ &K_1(\widetilde{S}^{-1}A[z]) = K_1(A) \oplus K_0(A) \oplus \widetilde{\mathrm{Nil}}_0(A) \oplus \widetilde{\mathrm{Nil}}_0(A) .\end{aligned}$$

□

Example 12.24 Given a monic central polynomial of degree n

$$p(z) = \sum_{j=0}^{d} a_j z^j \in A[z] \quad (a_d = 1)$$

define the multiplicative subset

$$S = (p(z))^\infty = \{p(z)^k \mid k \geq 0\} \subset A[z]$$

with leading units. Identify the $(A[z],S)$-module $A[z]/(p(z))$ with the object of $\mathrm{End}^S(A)$ defined by

$$(P,f) = (A[z]/(p(z)),\zeta) ,$$

with P a f.g. free A-module of rank n and $f : P \longrightarrow P$ an S-primary endomorphism such that

$$\begin{aligned}&p(f) = 0 : P \longrightarrow P ,\\ &\partial_+\tau(p(z) : S^{-1}A[z] \longrightarrow S^{-1}A[z]) = (P,f) \in \mathrm{End}_0^S(A) ,\end{aligned}$$

As in 12.3 the A-module isomorphism

$$g : A^d \xrightarrow{\simeq} P ; (b_1,b_2,\dots,b_d) \longrightarrow b_1 + b_2 z + \dots + b_d z^{d-1}$$

is such that

$$g^{-1}fg = \begin{pmatrix} 0 & 0 & 0 & \dots & -a_0 \\ 1 & 0 & 0 & \dots & -a_1 \\ 0 & 1 & 0 & \dots & -a_2 \\ \vdots & \vdots & \vdots & \ddots & \vdots \\ 0 & 0 & 0 & \dots & -a_{d-1} \end{pmatrix} : A^d = A \oplus A \oplus \dots \oplus A \longrightarrow A^d .$$

If A is commutative the characteristic polynomials are given by

$$\begin{aligned} \mathrm{ch}_z(P,f) &= p(z) \in S \subset A[z] , \\ \widetilde{\mathrm{ch}}_z(P,f) &= \widetilde{p}(z) \in \widetilde{S} \subset A[z] . \end{aligned}$$

If $A = F$ is a field and $p(z) \in F[z]$ is irreducible then

$$\partial_+ \tau(p(z)) = [F[z]/(p(z)), \zeta] = 1 \in \mathrm{End}_0^S(F) = \mathbb{Z}$$

is a generator. □

Remark 12.25 Let A be a commutative ring, and let $S \subset A[z]$ be a multiplicative subset with leading units.
(i) The composite

$$\mathrm{End}_0^S(A) \xrightarrow{\Delta_+} K_1(S^{-1}A[z]) \xrightarrow{\det} (S^{-1}A[z])^\bullet$$

sends an S-primary endomorphism class (P,f) to the characteristic polynomial

$$\begin{aligned} \det \Delta_+[P,f] &= \det(z - f : P[z] \longrightarrow P[z]) \\ &= \mathrm{ch}_z(P,f) \in (S^{-1}A[z])^\bullet . \end{aligned}$$

(ii) If $z \in S$ the composite

$$\widetilde{\mathrm{End}}_0^S(A) \xrightarrow{\widetilde{\Delta}_+} K_1(\widetilde{S}^{-1}A[z]) \xrightarrow{\det} (\widetilde{S}^{-1}A[z])^\bullet$$

sends a reverse S-primary endomorphism class (P,f) to the reverse characteristic polynomial

$$\begin{aligned} \det \widetilde{\Delta}_+[P,f] &= \det(1 - zf : P[z] \longrightarrow P[z]) \\ &= \widetilde{\mathrm{ch}}_z(P,f) \in (\widetilde{S}^{-1}A[z])^\bullet . \end{aligned}$$

□

Example 12.26 Let $A = F$ be a field, and as in 12.21 let $S \subset F[z]$ be the subset of all the monic polynomials, so that $\widetilde{S} \subset F[z]$ is the subset of all the polynomials with constant coefficient 1. Thus

$$S^{-1}F[z] = (F[z]\backslash\{0\})^{-1}F[z] = F(z)$$

is the quotient field of $F[z]$ (cf. 8.5), and

$$\widetilde{S}^{-1}F[z] = F[z]_{(z)} = \{p(z)/q(z) \,|\, q(0) \in F^\bullet\} \subset F(z) .$$

The multiplicative subsets $\widetilde{S}, \{z^k \mid k \geq 0\} \subset F[z]$ are coprime, so that there is defined a cartesian square

$$\begin{array}{ccc} F[z] & \longrightarrow & F[z,z^{-1}] \\ \downarrow & & \downarrow \\ F[z]_{(z)} & \longrightarrow & F(z) \end{array}$$

For each of the rings A in the square the determinant defines an isomorphism

$$\det\ :\ K_1(A) \xrightarrow{\simeq} A^\bullet\ ,$$

and

$$\begin{aligned}
&\widetilde{\mathrm{Nil}}_0(F)\ =\ 0\ ,\ \ K_1(F[z])\ =\ K_1(F)\ =\ F[z]^\bullet\ =\ F^\bullet\ ,\\
&K_0(F)\ =\ \mathbb{Z}\ ,\ \ \mathrm{End}_0(F)\ =\ \mathbb{Z}\oplus\widetilde{\mathrm{End}}_0(F)\ ,\\
&K_1(F[z]_{(z)})\ =\ K_1(F[z])\oplus\widetilde{\mathrm{End}}_0(F)\ =\ F[z]^\bullet_{(z)}\ ,\\
&K_1(F(z))\ =\ K_1(F[z])\oplus\mathrm{End}_0(F)\ =\ F(z)^\bullet\ .
\end{aligned}$$

The characteristic polynomial defines an isomorphism

$$\mathrm{End}_0(F) \xrightarrow{\simeq} F(z)^\bullet/F^\bullet\ ;\ [P,f] \longrightarrow \mathrm{ch}_z(P,f)\ .$$

The reverse characteristic polynomial defines an isomorphism

$$\begin{aligned}
\widetilde{\mathrm{End}}_0(F) \xrightarrow{\simeq} F[z]^\bullet_{(z)}/F[z]^\bullet\ &=\ F(z)^\bullet/(F^\bullet\oplus\{z^n \mid n\in\mathbb{Z}\})\ ;\\
[P,f] &\longrightarrow \widetilde{\mathrm{ch}}_z(P,f)\ .
\end{aligned}$$

By 12.13 there are defined isomorphisms

$$\begin{aligned}
(i_+\ \ \Delta_+)\ :\ K_1(F)\oplus\mathrm{End}_0(F) &\xrightarrow{\simeq} K_1(F(z))\ ;\\
(\tau(a),[P,f]) &\longrightarrow \tau(a)+\tau(z-f: P(z)\longrightarrow P(z))\ ,\\
(\widetilde{i}_+\ \ \widetilde{\Delta}_+)\ :\ K_1(F)\oplus\widetilde{\mathrm{End}}_0(F) &\xrightarrow{\simeq} K_1(F[z]_{(z)})\ ;\\
(\tau(a),[P,f]) &\longrightarrow \tau(a)+\tau(1-zf: P[z]_{(z)}\longrightarrow P[z]_{(z)})\ .
\end{aligned}$$

The inclusion $F[z]_{(z)}\longrightarrow F(z)$ induces the injection

$$\begin{aligned}
K_1(F[z]_{(z)})\ =\ K_1(F)\oplus\widetilde{\mathrm{End}}_0(F) &\longrightarrow K_1(F(z))\ =\ K_1(F)\oplus\mathrm{End}_0(F)\ ;\\
(\tau(a),[P,f]) &\longrightarrow (\tau(a)-\tau(g),[Q,g])\ ,
\end{aligned}$$

with

$$\begin{aligned}
(Q,g)\ &=\ (\mathrm{coker}(1-zf: P[z]\longrightarrow P[z]),z)\\
&=\ (\mathrm{coker}(1-zf: P[z,z^{-1}]\longrightarrow P[z,z^{-1}]),z)
\end{aligned}$$

such that $g : Q \longrightarrow Q$ is an automorphism. See 13.29 below for a proof that

$$\mathrm{ch}_z(P,f) = a_d z^d \widetilde{\mathrm{ch}}_z(Q,g) \in F[z]$$

with

$$d = \dim_F(P) - \dim_F(Q) \ , \ a_d = \det(-g : Q \longrightarrow Q) \in F^{\bullet} \ . \qquad \square$$

13. Automorphism K-theory

Automorphism K-theory is the algebraic K-theory of modules with an automorphism, such as arise from fibred knots. An A-module M with an automorphism $h : M \longrightarrow M$ is essentially the same as a module M over the Laurent polynomial ring $A[z,z^{-1}]$, with the invertible indeterminate z acting on M by h. This correspondence will be used to express the various automorphism K-groups of A in terms of the algebraic K-groups $K_*(\Sigma^{-1}A[z,z^{-1}])$ of the localizations $\Sigma^{-1}A[z,z^{-1}]$ inverting appropriate sets Σ of square matrices in $A[z,z^{-1}]$.

The automorphism class group $\mathrm{Aut}_0(A)$ of a ring A is defined by analogy with the endomorphism class group $\mathrm{End}_0(A)$, using automorphisms $h : P \longrightarrow P$ of f.g. projective A-modules. By definition, an automorphism $h : P \longrightarrow P$ is fibred if $h - 1 : P \longrightarrow P$ is also an automorphism, and there is also a fibred automorphism class group $\mathrm{Aut}_0^{fib}(A)$. The algebraic description in Part Two of the high-dimensional fibred knot cobordism groups and the bordism groups of automorphisms of manifolds will involve the algebraic L-theory analogues of the automorphism class groups (for $A = \mathbb{Z}, \mathbb{Z}[\pi_1]$ respectively).

The algebraic K-theory splitting theorems of Chaps. 10, 11, 12 are now extended to

$$\begin{aligned} K_1(\Omega^{-1}A[z,z^{-1}]) &= K_1(A[z,z^{-1}]) \oplus \mathrm{Aut}_0(A) , \\ K_1(\Omega_{fib}^{-1}A[z,z^{-1}]) &= K_1(A[z,z^{-1}]) \oplus \mathrm{Aut}_0^{fib}(A) \end{aligned}$$

for the appropriate localizations of $A[z,z^{-1}]$.

13A. The Fredholm localization $\Omega^{-1}A[z,z^{-1}]$

By analogy with the 10.3:

Proposition 13.1 (Ranicki [248])
The following conditions on a $k \times k$ matrix $\omega = (\omega_{ij})$ in $A[z,z^{-1}]$ are equivalent:

(i) *the $A[z,z^{-1}]$-module morphism*

$$\omega : A[z,z^{-1}]^k \longrightarrow A[z,z^{-1}]^k ;$$
$$(x_1,x_2,\dots,x_k) \longrightarrow (\sum_{i=1}^{k} x_i\omega_{1i}, \sum_{i=1}^{k} x_i\omega_{2i},\dots,\sum_{i=1}^{k} x_i\omega_{ki})$$

is injective and the cokernel is a f.g. projective A-module,
(ii) *the 1-dimensional f.g. free $A[z,z^{-1}]$-module chain complex*

$$E : \dots \longrightarrow 0 \longrightarrow A[z,z^{-1}]^k \xrightarrow{\omega} A[z,z^{-1}]^k$$

is A-finitely dominated,
(iii) *ω is invertible in $A((z)) \times A((z^{-1}))$.* □

Remark 13.2 13.1 can be used to prove that a based f.g. free $A[z,z^{-1}]$-module chain complex C is A-finitely dominated (i.e. a chain complex band) if and only if C is chain equivalent to the algebraic mapping torus

$$T^-(h) = \mathcal{C}(1 - z^{-1}h : P[z,z^{-1}] \longrightarrow P[z,z^{-1}])$$

of an automorphism $h : P \longrightarrow P$ of a finite f.g. projective A-module chain complex P. See Ranicki [248, 1.9]. □

By analogy with the definition in 10.4 of Fredholm matrices in $A[z]$:

Definition 13.3 (i) A matrix ω in $A[z,z^{-1}]$ is *Fredholm* if it is square and satisfies the equivalent conditions of 13.2.
(ii) Let Ω be the set of Fredholm matrices ω in $A[z,z^{-1}]$.
(iii) The *Fredholm localization* of $A[z,z^{-1}]$ is the ring $\Omega^{-1}A[z,z^{-1}]$ obtained from $A[z,z^{-1}]$ by inverting Ω. □

A matrix ω with entries in $A[z,z^{-1}]$ can be expressed as

$$\omega = \sum_{j=m}^{n} \omega_j z^j$$

with each ω_j a matrix with entries in A (of the same size as ω), for some $m \leq n$.

Definition 13.4 (i) A matrix $\omega = \sum\limits_{j=m}^{n} \omega_j z^j$ with entries in $A[z,z^{-1}]$ is *bionic* if ω is square, ω_n is the identity matrix and ω_m is invertible in A.
(ii) Let Ω_{bio} be the set of bionic matrices ω in $A[z,z^{-1}]$. □

By analogy with 10.7:

Proposition 13.5 (i) *The natural map* $A[z,z^{-1}] \longrightarrow \Omega^{-1}A[z]$ *is injective.*
(ii) *Every bionic matrix in* $A[z,z^{-1}]$ *is Fredholm, so that* $\Omega_{bio} \subset \Omega$.
(iii) *The localizations of* $A[z,z^{-1}]$ *inverting* Ω_{bio} *and* Ω *coincide*

$$\Omega_{bio}^{-1}A[z,z^{-1}] \;=\; \Omega^{-1}A[z,z^{-1}] \;.$$

Proof (i) Immediate from the factorization

$$A[z,z^{-1}] \longrightarrow \Omega^{-1}A[z] \longrightarrow A((z))$$

of the injection $A[z,z^{-1}] \longrightarrow A((z))$.
(ii) If $\omega = \sum\limits_{j=m}^{n} \omega_j z^j$ is a bionic $k \times k$ matrix define an A-module isomorphism

$$A^{k(n-m)} \xrightarrow{\simeq} \mathrm{coker}(\omega) \;;$$

$$(a_i)_{0 \le i \le k(n-m)-1} \longrightarrow \sum_{j=0}^{n-m-1} (a_{jk}, a_{jk+1}, \dots, a_{jk+k-1}) z^j \;.$$

Thus $\mathrm{coker}(\omega) = A^{k(n-m)}$ is a f.g. free A-module of rank $k(n-m)$, and ω is Fredholm.
(iii) Given a Fredholm matrix ω in $A[z,z^{-1}]$ let

$$P \;=\; \mathrm{coker}(\omega) \;\;,\;\; \zeta \;:\; P \longrightarrow P \;;\; x \longrightarrow zx \;.$$

Let Q be a f.g. projective A-module such that $P \oplus Q$ is a f.g. free A-module, say $P \oplus Q = A^d$. The $A[z,z^{-1}]$-module morphism

$$\begin{aligned}\eta \;=\; 1 \oplus (z-1) \;:\; A[z,z^{-1}]^d \;&=\; (P \oplus Q)[z,z^{-1}] \\ &\longrightarrow A[z,z^{-1}]^d \;=\; (P \oplus Q)[z,z^{-1}]\end{aligned}$$

with $\mathrm{coker}(\eta) = Q$ has a Fredholm matrix in $A[z,z^{-1}]$. Let ω' be the matrix of the $A[z,z^{-1}]$-module morphism

$$z-(\zeta \oplus 1) \;:\; A[z,z^{-1}]^d \;=\; (P \oplus Q)[z,z^{-1}] \longrightarrow A[z,z^{-1}]^d \;=\; (P \oplus Q)[z,z^{-1}] \;.$$

Now $(P \oplus Q, \zeta \oplus 1)$ is an $A[z,z^{-1}]$ of homological dimension 1, with a resolution by both $\omega \oplus \eta$ and ω'. In order to invert ω it therefore suffices to invert the bionic matrix ω'. □

Definition 13.6 (i) An automorphism $h : P \longrightarrow P$ of a module P is *fibred* if $h - 1 : P \longrightarrow P$ is also an automorphism.
(ii) A Fredholm matrix $\omega = \sum_{j=-\infty}^{\infty} \omega_j z^j$ in $A[z, z^{-1}]$ is *fibred* if $\sum_{j=-\infty}^{\infty} \omega_j$ is an invertible matrix in A.
(iii) Let $\Omega_{fib} \subset \Omega$ be the set of fibred Fredholm matrices in $A[z, z^{-1}]$. □

Example 13.7 (i) If H is an invertible matrix in A such that $I - H$ is also invertible then $I - H + zH$ is a fibred Fredholm matrix in $A[z, z^{-1}]$.
(ii) If $P = P^2$ is a projection matrix in A then $I - P + zP$ is a fibred Fredholm matrix in $A[z, z^{-1}]$.
(iii) If Q is a nilpotent square matrix in A then $I - zQ$ is a fibred Fredholm matrix in $A[z, z^{-1}]$. □

Proposition 13.8 (i) *A Fredholm matrix ω in $A[z, z^{-1}]$ is fibred if and only if it is A-invertible.*
(ii) *The localization $\Omega_{fib}^{-1} A[z, z^{-1}]$ is a subring of $\Pi^{-1} A[z, z^{-1}]$ with canonical factorizations*

$$A[z, z^{-1}] \longrightarrow \Omega_{fib}^{-1} A[z, z^{-1}] \longrightarrow \Pi^{-1} A[z, z^{-1}] \longrightarrow A$$

with Π the set of A-invertible matrices in $A[z, z^{-1}]$ (10.17). □

The inclusions

$$A[z, z^{-1}] \longrightarrow A((z)) \ , \ A[z, z^{-1}] \longrightarrow A((z^{-1}))$$

have canonical factorizations

$$A[z, z^{-1}] \longrightarrow \Omega^{-1} A[z, z^{-1}] \longrightarrow A((z)) \ ,$$
$$A[z, z^{-1}] \longrightarrow \Omega^{-1} A[z, z^{-1}] \longrightarrow A((z^{-1}))$$

with $A[z, z^{-1}] \longrightarrow \Omega^{-1} A[z, z^{-1}]$ injective. The factorization

$$A[z, z^{-1}] \longrightarrow \Omega^{-1} A[z, z^{-1}] \longrightarrow A((z)) \times A((z^{-1}))$$

is such that the localization $A[z, z^{-1}] \longrightarrow \Omega^{-1} A[z, z^{-1}]$ has the same universal property as the completion $A[z, z^{-1}] \longrightarrow A((z)) \times A((z^{-1}))$, namely:

Proposition 13.9 *The following conditions on a finite f.g. free $A[z, z^{-1}]$-module chain complex E are equivalent:*

(i) $H_*(\Omega^{-1} E) = 0$,
(ii) $H_*(A((z)) \otimes_{A[z,z^{-1}]} E) = 0$, $H_*(A((z^{-1})) \otimes_{A[z,z^{-1}]} E) = 0$,
(iii) *E is A-finitely dominated.*

Proof By 8.3. □

Similarly, $A[z, z^{-1}] \longrightarrow \Omega_{fib}^{-1} A[z, z^{-1}]$ is injective and such that:

Proposition 13.10 *The following conditions on a finite f.g. free $A[z,z^{-1}]$-module chain complex E are equivalent:*

(i) $H_*(\Omega_{fib}^{-1}E)=0$,

(ii) $H_*(A((z))\otimes_{A[z,z^{-1}]}E)=0$, $H_*(A((z^{-1}))\otimes_{A[z,z^{-1}]}E)=0$, $H_*(A\otimes_{A[z,z^{-1}]}E)=0$,

(iii) *E is A-finitely dominated and A-contractible,*

(iv) *E is A-finitely dominated and $\zeta-1:E\longrightarrow E$ is a chain equivalence.*

□

Example 13.11 (i) If E is a finite f.g. free $A[z,z^{-1}]$-module chain complex such that $p(z)H_*(E)=0$ for some $p(z)\in A[z,z^{-1}]$ with coker$(p(z):A[z,z^{-1}]\longrightarrow A[z,z^{-1}])$ a f.g. projective A-module (i.e. such that $(p(z))$ is a 1×1 Fredholm matrix in $A[z,z^{-1}]$) then E is A-finitely dominated.
(ii) As for (i), but with $p(1)\in A^\bullet$, in which case $(p(z))$ is a 1×1 fibred Fredholm matrix in $A[z,z^{-1}]$, and E is A-finitely dominated and A-contractible.

□

Example 13.12 (i) For a commutative ring A define the multiplicative subsets

$$\begin{aligned}
P &= \{\sum_{j=m}^{n} a_j z^j \in A[z,z^{-1}] \mid \sum_{j=m}^{n} a_j \in A^\bullet\}\ ,\\
Q &= A[z,z^{-1}]\cap A((z))^\bullet \cap A((z^{-1}))^\bullet\ ,\\
Q_{bio} &= \{\sum_{j=m}^{n} a_j z^j \in A[z,z^{-1}] \mid a_n=1, a_m\in A^\bullet\}\ ,\\
R &= P\cap Q = \{\sum_{j=m}^{n} a_j z^j \in Q \mid \sum_{j=m}^{n} a_j \in A^\bullet\}\subset A[z,z^{-1}]\ .
\end{aligned}$$

The localization $\Pi^{-1}A[z,z^{-1}]$ of $A[z,z^{-1}]$ is the one defined in Chap. 10, such that a square matrix ω in $A[z,z^{-1}]$ is invertible in $\Pi^{-1}A[z,z^{-1}]$ if and only if it is A-invertible (via $z\longrightarrow 1$). The (fibred) Fredholm localizations of $A[z,z^{-1}]$ are the localizations inverting Q and R are

$$\begin{aligned}
\Omega^{-1}A[z,z^{-1}] &= Q^{-1}A[z,z^{-1}] = Q_{bio}^{-1}A[z,z^{-1}]\ ,\\
\Omega_{fib}^{-1}A[z,z^{-1}] &= R^{-1}A[z,z^{-1}]\ .
\end{aligned}$$

A finite f.g. free $A[z,z^{-1}]$-module chain complex E is A-finitely dominated if and only if $p(z)H_*(E)=0$ for some $p(z)\in Q$. A finite f.g. free $A[z,z^{-1}]$-module chain complex E is A-finitely dominated and A-contractible if and only if $q(z)H_*(E)=0$ for some $q(z)\in R$.
(ii) If A is an integral domain then the multiplicative subsets Q,R in (i) are given by

$$Q = \{\sum_{j=m}^{n} a_j z^j \,|\, a_m, a_n \in A^\bullet\} ,$$

$$R = \{\sum_{j=m}^{n} a_j z^j \,|\, a_m, a_n, \sum_{j=m}^{n} a_j \in A^\bullet\} \subset A[z, z^{-1}] .$$

Thus Q consists of all the Laurent polynomials with the extreme coefficients units in A, and $R = P \cap Q$ is the subset of those which project via $z \longrightarrow 1$ to a unit in A. It follows from 13.1 that a polynomial $p(z) = \sum\limits_{j=m}^{n} a_j z^j \in A[z, z^{-1}]$ is such that $A[z, z^{-1}]/(p(z))$ is a f.g. projective A-module if and only if the extreme coefficients $a_m, a_n \in A$ are units, in which case $A[z, z^{-1}]/(p(z))$ is a f.g. free A-module of rank $n - m$.
(iii) If $A = F$ is a field then the multiplicative subsets P, Q, R in (i) are such that

$$\begin{aligned} P \;=\; R \;&=\; \{\sum_{j=-\infty}^{\infty} a_j z^j \,|\, \sum_{j=-\infty}^{\infty} a_j \in F^\bullet\} \\ &\subset Q \;=\; F[z, z^{-1}] \backslash \{0\} \subset F[z, z^{-1}] \end{aligned}$$

and

$$\begin{aligned} \Omega^{-1} F[z, z^{-1}] \;&=\; Q^{-1} F[z, z^{-1}] \;=\; F(z) , \\ \Omega_{fib}^{-1} F[z, z^{-1}] \;&=\; P^{-1} F[z, z^{-1}] \;=\; R^{-1} F[z, z^{-1}] \\ &=\; F[z]_{(1-z)} \;=\; \{p(z)/q(z) \,|\, q(1) \in F^\bullet\} \subset F(z) . \end{aligned}$$

The multiplicative subset

$$S \;=\; (1 - z)^\infty \;=\; \{(1 - z)^k \,|\, k \geq 0\} \subset F[z, z^{-1}]$$

is coprime to P and such that

$$Q \;=\; PS \;,\quad S^{-1} F[z, z^{-1}] \;=\; F[z, z^{-1}, (1 - z)^{-1}] ,$$

so that there is defined a cartesian square

$$\begin{array}{ccc} F[z, z^{-1}] & \longrightarrow & P^{-1} F[z, z^{-1}] \\ \big\downarrow & & \big\downarrow \\ F[z, z^{-1}, (1 - z)^{-1}] & \longrightarrow & F(z) \end{array}$$

□

13B. The automorphism category

Definition 13.13 (i) The *automorphism category* $\mathrm{Aut}(A)$ is the full subcategory of $\mathrm{End}(A)$ with objects (P,h) such that $h : P \longrightarrow P$ is an automorphism. The *automorphism K-groups of A* are given by

$$\mathrm{Aut}_*(A) = K_*(\mathrm{Aut}(A)) ,$$

with $\mathrm{Aut}_0(A)$ the *automorphism class group.*
(ii) The *fibred automorphism category* $\mathrm{Aut}^{fib}(A)$ is the full subcategory of $\mathrm{Aut}(A)$ with objects (P,h) such that $h-1 : P \longrightarrow P$ is an automorphism. The *fibred automorphism K-groups of A* are given by

$$\mathrm{Aut}^{fib}_*(A) = K_*(\mathrm{Aut}^{fib}(A)) . \qquad \square$$

Remark 13.14 The automorphism class group $\mathrm{Aut}_0(A)$ is the abelian group with one generator (P,h) for each isomorphism class in $\mathrm{Aut}(A)$ and relations

$$(P,h) + (P',h') - (P \oplus P', h \oplus h') .$$

The torsion group $K_1(A)$ is a quotient of $\mathrm{Aut}_0(A)$

$$K_1(A) = \mathrm{Aut}_0(A)/\{(P,h) + (P,k) - (P,hk)\} . \qquad \square$$

Example 13.15 Let A be a commutative ring.
(i) Let (P,h) be an object in $\mathrm{End}(A)$. The characteristic polynomial

$$\mathrm{ch}_z(P,h) = \det(z-h : P[z] \longrightarrow P[z]) = \sum_{j=0}^{d} a_j z^j \in A[z]$$

is such that

$$\mathrm{ch}_0(P,h) = \det(-h : P \longrightarrow P) = a_0 ,$$

$$\mathrm{ch}_1(P,h) = \det(1-h : P \longrightarrow P) = \sum_{j=0}^{d} a_j \in A .$$

Thus (P,h) is in $\mathrm{Aut}(A)$ if and only if $a_0 \in A^\bullet$, and (P,h) is in $\mathrm{Aut}^{fib}(A)$ if and only if $a_0, \sum\limits_{j=0}^{d} a_j \in A^\bullet$.
(ii) The multiplicative subsets $S, T \subset A[z]$ defined by

$$S = \{\sum_{j=0}^{d} a_j z^j \,|\, a_0 \in A^\bullet\} \cap A((z^{-1}))^\bullet ,$$

$$T = \{\sum_{j=0}^{d} a_j z^j \,|\, a_0, \sum_{j=0}^{d} a_j \in A^\bullet\} \cap A((z^{-1}))^\bullet$$

are such that

$$A[z]\cap(S^{-1}A[z])^{\bullet} = S \ , \ A[z]\cap(T^{-1}A[z])^{\bullet} = T$$

and it follows from 12.4 and (i) that

$$\mathrm{End}^S(A) = \mathrm{Aut}(A) \ , \ \mathrm{End}^T(A) = \mathrm{Aut}^{fib}(A) \ .$$

If A is an integral domain then

$$S = \{\sum_{j=0}^{d} a_j z^j \,|\, a_0, a_d \in A^{\bullet}\} \ ,$$

$$T = \{\sum_{j=0}^{d} a_j z^j \,|\, a_0, a_d, \sum_{j=0}^{d} a_j \in A^{\bullet}\} \subset A[z] \ .$$

□

By analogy with 10.14:

Proposition 13.16 (i) *The relative groups in the localization exact sequence for the Fredholm localization $\Omega^{-1}A[z,z^{-1}]$ (13.3)*

$$\dots \longrightarrow K_n(A[z,z^{-1}]) \xrightarrow{i} K_n(\Omega^{-1}A[z,z^{-1}]) \xrightarrow{\partial} K_n(A[z,z^{-1}],\Omega) \longrightarrow K_{n-1}(A[z,z^{-1}]) \longrightarrow \dots$$

are the algebraic K-groups

$$K_*(A[z,z^{-1}],\Omega) = K_{*-1}(\mathbb{H}(A[z,z^{-1}],\Omega))$$

of the exact category $\mathbb{H}(A[z,z^{-1}],\Omega)$ of $(A[z,z^{-1}],\Omega)$-modules. The functor

$$\mathrm{Aut}(A) \longrightarrow \mathbb{H}(A[z,z^{-1}],\Omega) \ ;$$
$$(P,h) \longrightarrow \mathrm{coker}(z-h : P[z,z^{-1}] \longrightarrow P[z,z^{-1}]) = P \textit{ with } z=h$$

is an isomorphism of exact categories, and

$$\begin{aligned} K_*(A[z,z^{-1}],\Omega) &= K_{*-1}(\mathbb{H}(A[z,z^{-1}],\Omega)) \\ &= \mathrm{Aut}_{*-1}(A) \ . \end{aligned}$$

(ii) *The localization exact sequence breaks up into split exact sequences*

$$0 \longrightarrow K_n(A[z,z^{-1}]) \xrightarrow{i} K_n(\Omega^{-1}A[z,z^{-1}]) \xrightarrow{\partial} \mathrm{Aut}_{n-1}(A) \longrightarrow 0 \ ,$$

so that

$$K_n(\Omega^{-1}A[z,z^{-1}]) = K_n(A[z,z^{-1}]) \oplus \mathrm{Aut}_{n-1}(A) \quad (n \in \mathbb{Z}) \ .$$

Similarly for $(\Omega^{fib})^{-1}[z,z^{-1}]$ and $\mathrm{Aut}^{fib}(A)$. □

In particular, for $n = 1$:

Proposition 13.17 (i) *The torsion group of $\Omega^{-1}A[z,z^{-1}]$ fits into the direct sum system*

$$K_1(A[z,z^{-1}]) \underset{j}{\overset{i}{\rightleftarrows}} K_1(\Omega^{-1}A[z,z^{-1}]) \underset{\Delta}{\overset{\partial}{\rightleftarrows}} \mathrm{Aut}_0(A)$$

with

$$\begin{aligned}
i\ &:\ K_1(A[z,z^{-1}]) \longrightarrow K_1(\Omega^{-1}A[z,z^{-1}])\ ;\ \tau(E) \longrightarrow \tau(\Omega^{-1}E)\ ,\\
j\ &:\ K_1(\Omega^{-1}A[z,z^{-1}]) \longrightarrow K_1(A[z,z^{-1}])\ ;\\
&\qquad \tau(\Omega^{-1}E) \longrightarrow \Phi^+(E)\\
&\qquad\quad = \tau(q^+ : \mathcal{C}(1 - z\zeta^{-1} : E[z,z^{-1}] \longrightarrow E[z,z^{-1}]) \longrightarrow E)\ ,\\
\partial\ &:\ K_1(\Omega^{-1}A[z,z^{-1}]) \longrightarrow \mathrm{Aut}_0(A)\ ;\ \tau(\Omega^{-1}E) \longrightarrow [E,\zeta]\ ,\\
\Delta\ &:\ \mathrm{Aut}_0(A) \longrightarrow K_1(\Omega^{-1}A[z,z^{-1}])\ ;\\
&\qquad [P,h] \longrightarrow \tau(1 - zh^{-1} : \Omega^{-1}P[z,z^{-1}] \longrightarrow \Omega^{-1}P[z,z^{-1}])\ .
\end{aligned}$$

(iii) *The torsion group of $\Omega^{-1}_{fib}A[z,z^{-1}]$ fits into the direct sum system*

$$K_1(A[z,z^{-1}]) \underset{\widetilde{j}}{\overset{\widetilde{i}}{\rightleftarrows}} K_1(\Omega^{-1}_{fib}A[z,z^{-1}]) \underset{\widetilde{\Delta}}{\overset{\widetilde{\partial}}{\rightleftarrows}} \mathrm{Aut}_0^{fib}(A)$$

with $\widetilde{i},\widetilde{j},\widetilde{\partial},\widetilde{\Delta}$ defined in the same way as i,j,∂,Δ in (i).
(iv) *The inclusion $\Omega^{-1}_{fib}A[z,z^{-1}] \longrightarrow \Pi^{-1}A[z,z^{-1}]$ induces*

$$\begin{aligned}
K_1(\Omega^{-1}_{fib}A[z,z^{-1}]) &= K_1(A[z,z^{-1}]) \oplus \mathrm{Aut}_0^{fib}(A)\\
\longrightarrow K_1(\Pi^{-1}A[z,z^{-1}]) &= K_1(A) \oplus \widetilde{\mathrm{End}}_0(A)\ ;\\
(\tau(\Omega^{-1}_{fib}E),[P,h]) &\longrightarrow (\tau(A \otimes_{A[z,z^{-1}]} E),[E,(1-\zeta)^{-1}] + [P,h])
\end{aligned}$$

with $K_1(\Pi^{-1}A[z,z^{-1}]) = K_1(A) \oplus \widetilde{\mathrm{End}}_0(A)$ as in 10.21. □

Example 13.18 Let X be an untwisted CW complex band with

$$\pi_1(X) = \pi \times \mathbb{Z}\ ,\ \pi_1(\overline{X}) = \pi$$

and write

$$\Lambda = \Omega^{-1}\mathbb{Z}[\pi][z,z^{-1}]\ .$$

Now X is Λ-contractible, with Λ-coefficient torsion

$$\begin{aligned}
\tau(X;\Lambda) &= (\Phi^+(X),[\overline{X},\zeta])\\
&\in K_1(\Lambda)/\{\pm\pi \times \mathbb{Z}\} = Wh(\pi \times \mathbb{Z}) \oplus \mathrm{Aut}_0(\mathbb{Z}[\pi])\ .
\end{aligned}$$

The fibering obstruction $\Phi^+(X) \in Wh(\pi \times \mathbb{Z})$ is a simple homotopy invariant, and the automorphism class $[\overline{X}, \zeta] \in \mathrm{Aut}_0(\mathbb{Z}[\pi])$ is a homotopy invariant. The Λ-coefficient torsion of a homotopy equivalence $f : X \longrightarrow Y$ of untwisted bands is given by

$$\begin{aligned} \tau(f;\Lambda) &= \tau(Y;\Lambda) - \tau(X;\Lambda) \\ &= (\Phi^+(Y) - \Phi^+(X), [\overline{Y}, \zeta_Y] - [\overline{X}, \zeta_X]) \\ &= (\tau(f), 0) \in Wh(\pi \times \mathbb{Z}) \oplus \mathrm{Aut}_0(\mathbb{Z}[\pi]) . \end{aligned}$$

The Λ-coefficient torsion of the mapping torus $T(h)$ of an untwisted self homotopy equivalence $h : P \longrightarrow P$ of a finite CW complex P with $\pi_1(P) = \pi$ is

$$\tau(T(h);\Lambda) = (0, [P, h]) \in Wh(\pi \times \mathbb{Z}) \oplus \mathrm{Aut}_0(\mathbb{Z}[\pi]) . \qquad \square$$

Definition 13.19 Let $S \subset A[z, z^{-1}]$ be a multiplicative subset.
(i) An object (P, h) in $\mathrm{Aut}(A)$ is *S-primary* if it is defined in $\mathrm{End}^S(A)$, i.e. if

$$z - h \; : \; S^{-1}P[z, z^{-1}] \longrightarrow S^{-1}P[z, z^{-1}]$$

is an $S^{-1}A[z, z^{-1}]$-module automorphism. The *S-primary automorphism category*

$$\mathrm{Aut}^S(A) = \mathrm{End}^S(A) \cap \mathrm{Aut}(A)$$

is the full subcategory of $\mathrm{Aut}(A)$ with S-primary objects (P, h). The *S-primary automorphism K-groups* of A are defined by

$$\mathrm{Aut}^S_*(A) = K_*(\mathrm{Aut}^S(A)) .$$

(ii) The *fibred S-primary automorphism category*

$$\mathrm{Aut}^{fib,S}(A) = \mathrm{Aut}^S(A) \cap \mathrm{Aut}^{fib}(A)$$

and the *fibred S-primary automorphism K-groups of A* are defined by

$$\mathrm{Aut}^{fib,S}_*(A) = K_*(\mathrm{Aut}^{fib,S}(A)) .$$

(iii) S is *bionic* if for each $\sum\limits_{j=m}^{n} a_j z^j \in S$ the extreme non-zero coefficients are $a_n = 1 \in A$ and a unit $a_m \in A^\bullet$. $\square$

Example 13.20 If A is commutative and $S \subset A[z, z^{-1}]$ is any multiplicative subset then an object (P, h) in $\mathrm{Aut}(A)$ is S-primary if and only if $\mathrm{ch}_z(P, h)$ is a factor of an element $p(z) \in S$. In particular, this is the case if $\mathrm{ch}_z(P, h) \in S$. $\square$

Example 13.21 If a multiplicative subset $S \subset A[z, z^{-1}]$ is such that

$$S \subset A((z))^\bullet \cap A((z^{-1}))^\bullet$$

(e.g. if S has extreme units) then the inclusions of $A[z, z^{-1}]$ factor as

$$A[z,z^{-1}] \longrightarrow S^{-1}A[z,z^{-1}] \longrightarrow \Omega^{-1}A[z,z^{-1}] \longrightarrow A((z)) ,$$

$$A[z,z^{-1}] \longrightarrow S^{-1}A[z,z^{-1}] \longrightarrow \Omega^{-1}A[z,z^{-1}] \longrightarrow A((z^{-1})) ,$$

and there is defined an isomorphism of categories.

$$\mathbb{H}(A[z,z^{-1}],S) \xrightarrow{\simeq} \mathrm{Aut}^S(A) \; ; \; P \longrightarrow (P,\zeta) .$$

The relative K-groups in the localization exact sequence

$$\begin{aligned}\dots \longrightarrow K_n(A[z,z^{-1}]) \longrightarrow K_n(S^{-1}A[z,z^{-1}]) \longrightarrow K_n(A[z,z^{-1}],S) \\ \longrightarrow K_{n-1}(A[z,z^{-1}]) \longrightarrow \dots\end{aligned}$$

can thus be expressed as

$$K_*(A[z,z^{-1}],S) \;=\; K_{*-1}(\mathbb{H}(A[z,z^{-1}],S)) \;=\; \mathrm{Aut}^S_{*-1}(A) .$$

In fact, this exact sequence breaks up into split exact sequences

$$0 \longrightarrow K_n(A[z,z^{-1}]) \longrightarrow K_n(S^{-1}A[z,z^{-1}]) \longrightarrow K_n(A[z,z^{-1}],S) \longrightarrow 0 . \; \square$$

By analogy with 12.7:

Proposition 13.22 *If $S,T \subset A[z,z^{-1}]$ are coprime central multiplicative subsets then*

$$\begin{aligned}\mathrm{Aut}^{ST}(A) &= \mathrm{Aut}^S(A) \times \mathrm{Aut}^T(A) ,\\ \mathrm{Aut}^{ST}_*(A) &= \mathrm{Aut}^S_*(A) \oplus \mathrm{Aut}^T_*(A)\end{aligned}$$

and similarly for fibred automorphisms.

Proof As for 12.7. □

By analogy with 13.17:

Proposition 13.23 *For any multiplicative subset $S \subset A[z,z^{-1}]$ such that $S \subseteq A((z))^\bullet \cap A((z^{-1}))^\bullet$ the torsion group of $S^{-1}A[z,z^{-1}]$ fits into the direct sum system*

$$K_1(A[z,z^{-1}]) \underset{j}{\overset{i}{\rightleftarrows}} K_1(S^{-1}A[z,z^{-1}]) \underset{\Delta}{\overset{\partial}{\rightleftarrows}} \mathrm{Aut}^S_0(A)$$

with

$$\begin{aligned}
i \; &: \; K_1(A[z,z^{-1}]) \longrightarrow K_1(S^{-1}A[z,z^{-1}]) \; ; \; \tau(E) \longrightarrow \tau(S^{-1}E) ,\\
j \; &: \; K_1(S^{-1}A[z,z^{-1}]) \longrightarrow K_1(A[z,z^{-1}]) \; ;\\
&\quad \tau(S^{-1}E) \longrightarrow \Phi^+(E)\\
&\qquad = \tau(q^+ : \mathcal{C}(1-z\zeta^{-1} : E[z,z^{-1}] \longrightarrow E[z,z^{-1}]) \longrightarrow E) ,\\
\partial \; &: \; K_1(S^{-1}A[z,z^{-1}]) \longrightarrow \mathrm{Aut}^S_0(A) \; ; \; \tau(S^{-1}E) \longrightarrow [E,\zeta] ,\\
\Delta \; &: \; \mathrm{Aut}^S_0(A) \longrightarrow K_1(S^{-1}A[z,z^{-1}]) \; ;\\
&\quad [P,h] \longrightarrow \tau(1-zh^{-1} : S^{-1}P[z,z^{-1}] \longrightarrow S^{-1}P[z,z^{-1}]) . \qquad \square
\end{aligned}$$

Remark 13.24 For a commutative ring A the composite

$$\mathrm{Aut}_0^S(A) \xrightarrow{\Delta} K_1(S^{-1}A[z,z^{-1}]) \xrightarrow{\det} (S^{-1}A[z,z^{-1}])^\bullet$$

sends an S-primary automorphism class (P,h) to

$$\begin{aligned}\det \Delta[P,h] &= \det(1 - zh^{-1} : P[z]\longrightarrow P[z]) = \widetilde{\mathrm{ch}}_z(P,h^{-1}) \\ &= \mathrm{ch}_z(P,h)/\det(-h : P\longrightarrow P) \in (S^{-1}A[z,z^{-1}])^\bullet .\end{aligned}$$

□

Proposition 13.25 *Let $S \subset A[z]$ be a multiplicative subset with constant units, and let*

$$T = (z)^\infty S = \{z^k p(z) \mid k \geq 0, p(z) \in S\} \subset A[z] .$$

(i) *The S- and T-primary endomorphism categories and K-groups are such that*

$$\begin{aligned}\mathrm{End}^S(A) &= \mathrm{Aut}^S(A) \ , \quad \mathrm{End}_*^S(A) = \mathrm{Aut}_*^S(A) \ , \\ \mathrm{End}^T(A) &= \mathrm{End}^{(z)^\infty}(A) \times \mathrm{End}^S(A) \\ &= \mathrm{Nil}(A) \times \mathrm{Aut}^S(A) \\ \mathrm{End}_*^T(A) &= \mathrm{Nil}_*(A) \oplus \mathrm{Aut}_*^S(A) .\end{aligned}$$

(ii) *If for every $p(z) = \sum\limits_{j=0}^{d} a_j z^j \in S$ it is the case that $p(1) = \sum\limits_{j=0}^{d} a_j \in A^\bullet$ then*

$$\begin{aligned}\mathrm{End}^S(A) &= \mathrm{Aut}^S(A) = \mathrm{Aut}^{fib,S}(A) \ , \\ \mathrm{End}_*^S(A) &= \mathrm{Aut}^S(A) = \mathrm{Aut}_*^{fib,S}(A) .\end{aligned}$$

Proof (i) For any object (P,f) in $\mathrm{End}^S(A)$ there exists an element

$$p(z) = \sum_{j=0}^{d} a_j z^j \in S \subset A[z]$$

such that $p(f) = 0 : P\longrightarrow P$, by 12.4. Now $a_0 \in A^\bullet$ (by hypothesis), so that (P,f) is defined in $\mathrm{Aut}^S(A)$, with the inverse defined by

$$f^{-1} = -(a_0)^{-1} \sum_{j=1}^{d} a_j f^{j-1} : P \longrightarrow P .$$

The multiplicative subsets $(z)^\infty, S \subset A[z]$ are coprime, so that 12.7 applies.
(ii) Given (P,f) in $\mathrm{End}^S(A)$ let $p(z) \in S$ be as in the proof of (i). Since $\sum\limits_{j=0}^{d} a_j \in A^\bullet$ (P,f) is defined in $\mathrm{Aut}^{fib,S}(A)$, with the inverse of $1-f : P\longrightarrow P$ given by

$$(1-f)^{-1} = (\sum_{i=0}^{d} a_i)^{-1} \sum_{j=1}^{d} a_j (\sum_{k=0}^{j-1} f^k) : P \longrightarrow P .$$

□

The multiplicative subset with extreme units

$$(1-z)^\infty = \{(1-z)^j \mid j \geq 0\} \subset A[z,z^{-1}]$$

has localization

$$((1-z)^\infty)^{-1} A[z,z^{-1}] = A[z,z^{-1},(1-z)^{-1}] .$$

In dealing with $(1-z)$-primary objects adopt the following terminology.

Definition 13.26 (i) An A-module endomorphism $f : P \longrightarrow P$ is *unipotent* if $1 - f : P \longrightarrow P$ is nilpotent, in which case f is an automorphism.
(ii) A chain map $h : C \longrightarrow C$ is *chain homotopy unipotent* if $h - 1 : C \longrightarrow C$ is chain homotopy nilpotent, in which case $h : C \longrightarrow C$ is a chain equivalence.
(iii) The *unipotent category*

$$\mathrm{Aut}^{uni}(A) = \mathrm{End}^{(1-z)^\infty}(A) = \mathrm{Aut}^{(1-z)^\infty}(A)$$

is the exact category of pairs (P,f) with P a f.g. projective A-module and $f : P \longrightarrow P$ an unipotent automorphism.
(iv) The *unipotent automorphism K-groups of A* are given by

$$\mathrm{Aut}^{uni}_*(A) = K_*(\mathrm{Aut}^{uni}(A)) .$$

□

Proposition 13.27 (i) *An A-module endomorphism $f : P \longrightarrow P$ is $(1-z)$-primary if and only if f is a unipotent automorphism.*
(ii) *A finite f.g. free $A[z,z^{-1}]$-module chain complex C is $A[z,z^{-1},(1-z)^{-1}]$-contractible if and only if C is A-finitely dominated and $\zeta : C \longrightarrow C$ is chain homotopy unipotent.*
(iii) *The functor*

$$\mathrm{Nil}(A) \longrightarrow \mathrm{Aut}^{uni}(A) \; ; \; (P,\nu) \longrightarrow (P,1+\nu)$$

is an isomorphism of categories, and

$$\begin{aligned} K_1(A[z,z^{-1},(1-z)^{-1}]) &= K_1(A[z,z^{-1}]) \oplus \mathrm{Nil}_0(A) \\ &= K_1(A) \oplus K_0(A) \oplus K_0(A) \oplus \widetilde{\mathrm{Nil}}_0(A) \oplus \widetilde{\mathrm{Nil}}_0(A) \oplus \widetilde{\mathrm{Nil}}_0(A) , \end{aligned}$$

$$\mathrm{End}_0^{z^\infty (1-z)^\infty}(A) = \mathrm{Nil}_0(A) \oplus \mathrm{Aut}_0^{uni}(A) = \mathrm{Nil}_0(A) \oplus \mathrm{Nil}_0(A) ,$$

in accordance with 12.23 and 13.25. □

Example 13.28 Let X be a finite CW complex, with fundamental group $\pi_1(X) = \pi$ and universal cover $\widetilde{X}$. Write

$$\Lambda = \mathbb{Z}[\pi][z,z^{-1},(1-z)^{-1}] .$$

The product $X \times S^1$ is Λ-contractible, with Λ-coefficient torsion

$$\begin{aligned}
&\tau(X \times S^1; \Lambda) \\
&= \tau(1 - z : C(\widetilde{X})[z, z^{-1}, (1-z)^{-1}] \longrightarrow C(\widetilde{X})[z, z^{-1}, (1-z)^{-1}]) \\
&= (0, (C(\widetilde{X}), 0)) \in Wh_1(\Lambda) = Wh_1(\mathbb{Z}[\pi][z, z^{-1}]) \oplus \mathrm{Nil}_0(\mathbb{Z}[\pi]) . \qquad \square
\end{aligned}$$

Example 13.29 For a field F the multiplicative subsets

$$S = \{\sum_{j=0}^{\infty} a_j z^j \mid a_0 \in F^\bullet\} \ , \ T = (z)^\infty S \subset F[z]$$

are such that

$$S^{-1}F[z] = F[z]_{(z)} \ , \ T^{-1}F[z] = F(z)$$

with a cartesian square

$$\begin{array}{ccc}
F[z] & \longrightarrow & F[z, z^{-1}] \\
\downarrow & & \downarrow \\
F[z]_{(z)} & \longrightarrow & F(z)
\end{array}$$

(as in 12.26) and identifications

$$\begin{aligned}
\mathbb{H}(F[z], T) &= \mathbb{H}(F[z], (z)^\infty) \times \mathbb{H}(F[z], S) \\
&= \mathrm{End}(F) = \mathrm{Nil}(F) \times \mathrm{Aut}(F) , \\
\mathrm{End}_0^T(F) &= \mathrm{End}_0(F) \\
&= \mathrm{End}_0^{(z)^\infty}(F) \oplus \mathrm{End}_0^S(F) \\
&= \mathrm{Nil}_0(F) \oplus \mathrm{Aut}_0(F) .
\end{aligned}$$

From 5.15

$$\widetilde{\mathrm{Nil}}_0(F) = 0 \ , \ \mathrm{Nil}_0(F) = K_0(F) = \mathbb{Z} ,$$

so that

$$\begin{aligned}
\mathrm{End}_0(F) &= \mathrm{Nil}_0(F) \oplus \mathrm{Aut}_0(F) = \mathbb{Z} \oplus \mathrm{Aut}_0(F) , \\
\widetilde{\mathrm{End}}_0(F) &= \mathrm{Aut}_0(F) .
\end{aligned}$$

Every object (P, f) in $\mathrm{End}(F)$ can be expressed as a direct sum

$$(P, f) = (P_1, f_1) \oplus (P_2, f_2)$$

with

$$\begin{aligned}(P_1, f_1) &= (\mathrm{coker}(z - f : S^{-1}P[z] \longrightarrow S^{-1}P[z]), \zeta) \\ &= (\bigcup_{r=0}^{\infty} \ker(f^r : P \longrightarrow P), f|) \in \mathrm{Nil}(F) \ , \\ (P_2, f_2) &= (\mathrm{coker}(z - f : P[z, z^{-1}] \longrightarrow P[z, z^{-1}]), \zeta) \\ &= (\bigcap_{r=0}^{\infty} \mathrm{im}(f^r : P \longrightarrow P), f|) \in \mathrm{Aut}(F) \ .\end{aligned}$$

If

$$\mathrm{ch}_z(P, f) = \sum_{j=m}^{n} a_j z^j \in F[z] \ \ (a_m \in F^{\bullet}, a_n = 1)$$

then

$$\begin{aligned}&\dim_F(P) = n \ , \ \dim_F(P_1) = m \ , \ \dim_F(P_2) = n - m \ , \\ &\mathrm{ch}_z(P_1, f_1) = z^m \ , \ \mathrm{ch}_z(P_2, f_2) = \sum_{j=0}^{n-m} a_{j+m} z^j \in F[z] \ , \\ &a_m = (-1)^{n-m} \det(f_2) \in F^{\bullet} \ .\end{aligned}$$

The automorphism object constructed in 12.26

$$(Q, g) = (\mathrm{coker}(1 - zf : P[z, z^{-1}] \longrightarrow P[z, z^{-1}]), z) = (P_2, (f_2)^{-1})$$

is such that

$$\widetilde{\mathrm{ch}}_z(Q, g) = (a_m)^{-1} \left(\sum_{j=0}^{n-m} a_{j+m} z^j \right) \in F[z] \ .$$

The multiplicative subsets

$$S^{fib} = \{ \sum_{j=0}^{\infty} a_j z^j \,|\, a_0, \sum_{j=0}^{\infty} a_j \in F^{\bullet} \} \ , \ (1 - z)^{\infty} \subset F[z]$$

are coprime, and such that

$$S = (1 - z)^{\infty} S^{fib} \subset F[z] \ ,$$

so that

$$\begin{aligned}\mathrm{Aut}(F) = \mathrm{End}^S(F) &= \mathrm{End}^{(1-z)^{\infty}}(F) \times \mathrm{End}^{S^{fib}}(F) \\ &= \mathrm{Nil}(F) \times \mathrm{Aut}^{fib}(F)\end{aligned}$$

(up to isomorphism) and

$$\mathrm{Aut}_0(F) = \mathrm{Nil}_0(F) \oplus \mathrm{Aut}_0^{fib}(F) = \mathbb{Z} \oplus \mathrm{Aut}_0^{fib}(F) \ . \qquad \square$$

Example 13.30 For any field F the multiplicative subsets P, Q of $F[z, z^{-1}]$ defined by

$$P = \{ \sum_{j=-\infty}^{\infty} a_j z^j \mid \sum_{j=-\infty}^{\infty} a_j \in F^\bullet \} \ , \ Q = F[z, z^{-1}] \backslash \{0\}$$

are such that $(1-z)^\infty, P$ are coprime, $(1-z)^\infty P = Q$, there is defined a cartesian square of rings

$$\begin{array}{ccc} F[z,z^{-1}] & \longrightarrow & P^{-1}F[z,z^{-1}] \\ \downarrow & & \downarrow \\ F[z,z^{-1},(1-z)^{-1}] & \longrightarrow & F(z) \end{array}$$

and there are identifications

$$\begin{aligned}
\Omega_{fib}^{-1}F[z,z^{-1}] &= P^{-1}F[z,z^{-1}] = F[z,z^{-1}]_{(1-z)} \ , \\
\mathbb{H}(F[z,z^{-1}],P) &= \mathrm{Aut}^P(F) = \mathrm{Aut}^{fib}(F) \ , \\
\mathbb{H}(F[z,z^{-1}],(1-z)^\infty) &= \mathrm{Aut}^{(1-z)^\infty}(F) = \mathrm{Nil}(F) \ , \\
\Omega^{-1}F[z,z^{-1}] &= Q^{-1}F[z,z^{-1}] = F(z) \ , \\
\mathbb{H}(F[z,z^{-1}],Q) &= \mathrm{Aut}(F) \\
&= \mathbb{H}(F[z,z^{-1}],(1-z)^\infty) \times \mathbb{H}(F[z,z^{-1}],P) \\
&= \mathrm{Nil}(F) \times \mathrm{Aut}^{fib}(F) \ .
\end{aligned}$$

□

Remark 13.31 If A is a ring which is not a field then the natural map

$$\mathrm{Aut}_0^{uni}(A) \oplus \mathrm{Aut}_0^{fib}(A) \longrightarrow \mathrm{Aut}_0(A)$$

need not be an isomorphism, since in general it is not possible to express every object (P,h) in $\mathrm{Aut}(A)$ as a sum

$$(P,h) = (P_1,h_1) \oplus (P_2,h_2)$$

with (P_1,h_1) unipotent and (P_2,h_2) fibred. For example, if $A = \mathbb{Z}$ and

$$h = \begin{pmatrix} 1 & 2 \\ 1 & 3 \end{pmatrix} : P = \mathbb{Z} \oplus \mathbb{Z} \longrightarrow P = \mathbb{Z} \oplus \mathbb{Z}$$

the characteristic polynomial of (P,h)

$$\det(z - h : P[z] \longrightarrow P[z]) = z^2 - 4z + 1 \in \mathbb{Z}[z]$$

is irreducible, and

$$\det(1 - h : P \longrightarrow P) = -2 \neq 0, \pm 1 \in \mathbb{Z} \ .$$

Thus (P,h) is not the sum of a unipotent object and a fibred object in $\mathrm{Aut}(\mathbb{Z})$.

□

14. Witt vectors

A Witt vector is a formal power series with constant coefficient 1. The reverse characteristic polynomial of an endomorphism is a Witt vector which determines the endomorphism K-theory class. In Chap. 17 the Reidemeister torsion of an A-contractible finite f.g. $A[z, z^{-1}]$-module chain complex E will be identified with the Witt vector determined by the Alexander polynomials. In the applications to knot theory in Chap. 33 E will be the cellular chain complex of the infinite cyclic cover of the knot complement.

The endomorphism K-theory class of an endomorphism $f : P \longrightarrow P$ of a f.g. free $\mathbb{Z}$-module P is entirely determined by the rank of P and the characteristic polynomial of f, and a factorization of the polynomial determines a primary decomposition of the class. The situation is considerably complicated for L-theory. The high-dimensional knot cobordism groups C_* are the algebraic L-groups of quadratic forms over $\mathbb{Z}$ with certain types of endomorphisms (Chap. 31). The actual computation of C_* is complicated by the possibility that for an endomorphism of a quadratic form a factorization of the characteristic polynomial in $\mathbb{Z}[z, z^{-1}]$ as a product of powers of coprime irreducible polynomials does not in general lead to a primary decomposition of the endomorphism L-theory class, with *coupling invariants* obstructing such an expression – see Chap. 40 for more details.

Almkvist [5], [6] and Grayson [97] proved that the reduced endomorphism class group $\widetilde{\mathrm{End}}_0(A)$ of a commutative ring A is isomorphic to the group $W(A)$ of rational Witt vectors. There will now be obtained such an isomorphism

$$\widetilde{\mathrm{ch}} \; : \; \widetilde{\mathrm{End}}_0(A) \xrightarrow{\simeq} W(A)^{ab} \; ; \; (P, f) \longrightarrow \widetilde{\mathrm{ch}}_z(P, f)$$

for any ring A, with $W(A)^{ab}$ the abelianization of the group $W(A)$ of rational Witt vectors. In the first instance, the summand $\widehat{NK}_1(A)$ in the direct sum decomposition

$$K_1(A[[z]]) \; = \; K_1(A) \oplus \widehat{NK}_1(A)$$

will be identified with the abelianization $\widehat{W}(A)^{ab}$ of the group of Witt vectors $\widehat{W}(A)$.

Definition 14.1 (i) A *Witt vector* over A is an infinite sequence

$$a = (a_1, a_2, a_3, \ldots) \in \prod_1^\infty A \quad (a_i \in A)$$

which can be regarded as a formal power series with constant coefficient 1

$$1 + \sum_{i=1}^\infty a_i z^i \in A[[z]] .$$

(ii) The *Witt vector group* $\widehat{W}(A)$ is the group of Witt vectors over A, with multiplication by

$$(1 + \sum_{i=1}^\infty a_i z^i)(1 + \sum_{j=1}^\infty b_j z^j) = 1 + \sum_{k=1}^\infty (\sum_{i+j=k} a_i b_j) z^k . \qquad \square$$

By 6.3 the units of $A[[z]]$ are given by

$$\begin{aligned} A[[z]]^\bullet &= A^\bullet(1 + zA[[z]]) \\ &= \{\sum_{i=0}^\infty a_i z^i \in A[[z]] \,|\, a_0 \in A^\bullet\} , \end{aligned}$$

so that

$$\widehat{W}(A) = 1 + zA[[z]] \subset A[[z]]^\bullet .$$

The abelianization

$$\begin{aligned} \widehat{W}(A)^{ab} &= \widehat{W}(A)/[\widehat{W}(A), \widehat{W}(A)] \\ &= \widehat{W}(A)/\{xyx^{-1}y^{-1} \,|\, x, y \in \widehat{W}(A)\} \end{aligned}$$

is an abelian group (!).

Example 14.2 (Almkvist [5, 6.13])
If A is a commutative ring which is additively torsion-free (so that $\mathbb{Q} \subseteq A$) there is defined an isomorphism of abelian groups

$$\prod_{j=1}^\infty A \xrightarrow{\simeq} \widehat{W}(A) \; ; \; (a_1, a_2, a_3, \ldots) \longrightarrow \exp\Big(\int_0^z (a_1 - a_2 s + a_3 s^2 - \ldots) ds\Big)$$

with $\int_0^z s^k ds = z^{k+1}/(k+1)$. $\square$

Witt vectors arise as the noncommutative determinants of invertible matrices in $A[[z]]$, using the following generalization of the Dieudonné determinant of an invertible matrix in a local ring (cf. Rosenberg [254, 2.3.5]).

A $k \times k$ matrix $M = \sum_{r=0}^{\infty} M_r z^r$ in $A[[z]]$ is invertible if and only if M_0 is an invertible $k \times k$ matrix in A. Given such an M define the invertible $k \times k$ matrix in $A[[z]]$

$$B = (M_0)^{-1} M = (b_{ij})_{1 \le i,j \le k} \in GL_k(A[[z]])$$

in which the diagonal entries are Witt vectors

$$b_{ii} \in \widehat{W}(A) = 1 + zA[[z]] \subset A[[z]]^{\bullet} .$$

Use Gaussian elimination (i.e. elementary row operations) to express B as a product

$$B = LU$$

with $L = (\ell_{ij})$ a $k \times k$ lower triangular matrix in $A[[z]]$ and $U = (u_{ij})$ a $k \times k$ upper triangular matrix in $A[[z]]$ with diagonal entries

$$\ell_{ii} = 1 \ , \ u_{ii} \in \widehat{W}(A) = 1 + zA[[z]] \subset A[[z]]^{\bullet} .$$

Definition 14.3 The *Dieudonné determinant Witt vector* of an invertible matrix $M \in GL_k(A[[z]])$ is the abelianized Witt vector

$$DW(M) = u_{11}u_{22}\dots u_{kk} \in \widehat{W}(A)^{ab} . \qquad \square$$

Example 14.4 If A is commutative then

$$DW(M) = \det(B) = \det((M_0)^{-1}M) \in \widehat{W}(A) = 1 + zA[[z]] . \qquad \square$$

Example 14.5 For an invertible 2×2 matrix $M \in GL_2(A[[z]])$ with

$$B = M_0^{-1} M = \begin{pmatrix} b_{11} & b_{12} \\ b_{21} & b_{22} \end{pmatrix}$$

take

$$L = \begin{pmatrix} 1 & 0 \\ b_{21}(b_{11})^{-1} & 1 \end{pmatrix} , \ U = \begin{pmatrix} b_{11} & b_{12} \\ 0 & b_{22} - b_{21}(b_{11})^{-1}b_{12} \end{pmatrix}$$

so that $B = LU$ and

$$DW(M) = b_{11}(b_{22} - b_{21}(b_{11})^{-1}b_{12}) \in \widehat{W}(A)^{ab} . \qquad \square$$

By analogy with 5.14 (ii):

Proposition 14.6 *The torsion group of $A[[z]]$ fits into the direct sum system*

$$K_1(A) \underset{\widetilde{j}_+}{\overset{\widetilde{i}_+}{\rightleftarrows}} K_1(A[[z]]) \underset{\widetilde{\Delta}_+}{\overset{\widetilde{\partial}_+}{\rightleftarrows}} \widehat{W}(A)^{ab}$$

with

$$\widetilde{i}_+ : K_1(A) \longrightarrow K_1(A[[z]]) \ ; \ \tau(E) \longrightarrow \tau(E[[z]]) \ ,$$

$$\widetilde{j}_+ : K_1(A[[z]]) \longrightarrow K_1(A) \ ; \ \tau(E^+) \longrightarrow \tau(A \otimes_{A[[z]]} E^+) \ ,$$

$$\widetilde{\partial}_+ : K_1(A[[z]]) \longrightarrow \widehat{W}(A)^{ab} \ ; \ \tau(M : A[[z]]^k \longrightarrow A[[z]]^k) \longrightarrow DW(M) \ ,$$

$$\widetilde{\Delta}_+ \; : \; \widehat{W}(A)^{ab} \longrightarrow K_1(A[[z]]) \; ;$$

$$(a_1, a_2, \dots) \longrightarrow \tau(1 + \sum_{j=1}^{\infty} a_j z^j : A[[z]] \longrightarrow A[[z]]) \; .$$

Proof The morphisms

$$\widetilde{i}_+ \; : \; K_1(A) \longrightarrow K_1(A[[z]]) \; , \; \widetilde{j}_+ \; : \; K_1(A[[z]]) \longrightarrow K_1(A)$$

are such that $\widetilde{j}_+ \widetilde{i}_+ = 1$, since they are induced by the ring morphisms

$$\begin{aligned} \widetilde{i}_+ &= \text{inclusion} \; : \; A \longrightarrow A[[z]] \; , \\ \widetilde{j}_+ &= \text{projection} \; : \; A[[z]] \longrightarrow A \; ; \; z \longrightarrow 0 \; . \end{aligned}$$

It is clear from the definition of $DW(M)$ that $\widetilde{\partial}_+ \widetilde{\Delta}_+ = 1$. The exactness of

$$\widehat{W}(A)^{ab} \xrightarrow{\widetilde{\Delta}_+} K_1(A[[z]]) \xrightarrow{\widetilde{j}_+} K_1(A)$$

is given by the following argument of Suslin (Pazhitnov [223, 8.2]) : for any invertible $k \times k$ matrix $M \in GL_k(A[[z]])$ such that $\widetilde{j}_+(M) \in E_k(A)$ let

$$M' \in \ker(GL_k(A[[z]]) \longrightarrow GL_k(A))$$

be the result of applying to M the elementary operations which reduce $\widetilde{j}_+(M)$ to $I_k \in E_k(A)$. The diagonal entries in M' are units in $A[[z]]$, and repeated application of the matrix identity

$$\begin{pmatrix} a & b \\ c & d \end{pmatrix} = \begin{pmatrix} 1 & 0 \\ ca^{-1} & 1 \end{pmatrix} \begin{pmatrix} a & 0 \\ 0 & d - ca^{-1}b \end{pmatrix} \begin{pmatrix} 1 & a^{-1}b \\ 0 & 1 \end{pmatrix}$$

allows M' to be reduced by further elementary operations to the stabilization of a 1×1 matrix in $\ker(\widetilde{j}_+ : GL_1(A[[z]]) \longrightarrow GL_1(A)) = \widehat{W}(A)$. □

The subset

$$\widehat{W}(A) \cap A[z] \; = \; 1 + zA[z] \subset \widehat{W}(A)$$

is closed under multiplication, but not under inverses : for example

$$(1+z)^{-1} \; = \; \prod_{j=0}^{\infty} (-z)^j \in \widehat{W}(A) \backslash (\widehat{W}(A) \cap A[z]) \; .$$

Definition 14.7 (i) The *rational Witt vector group* $W(A)$ is the subgroup of $\widehat{W}(A)$ generated by the subset $\widehat{W}(A) \cap A[z] \subset \widehat{W}(A)$ of all the polynomials $p(z) = \sum\limits_{j=0}^{\infty} a_j z^j \in A[z]$ with $a_0 = 1 \in A$.

(ii) Let A be a commutative ring. The *reduced rational Witt vector group*

$\widetilde{W}(A)$ is the quotient of the rational Witt vector group $W(A)$

$$\widetilde{W}(A) = \operatorname{coker}(K_0(A) \longrightarrow W(A))$$

with

$$K_0(A) \longrightarrow W(A) \; ; \; [P] \longrightarrow \det(1 - z : P[z] \longrightarrow P[z]) \; . \qquad \square$$

In other words, $W(A)$ is the subgroup of $\widehat{W}(A)$ consisting of the elements

$$a = (a_1, a_2, a_3, \ldots) \in \prod_1^\infty A$$

such that

$$1 + \sum_{j=1}^{\infty} a_j z^j = \frac{p(z)}{q(z)} \in A[[z]]$$

for some polynomials $p(z), q(z) \in A[z]$ with $p(0) = q(0) = 1 \in A$. ($W(A)$ was already defined in 11.13 for commutative A).

Proposition 14.8 *Let A be an integral domain with quotient field F. Let $\mathcal{M}(F)$ be the set of irreducible monic polynomials in $F[z]$, and let $\mathcal{M}_A(F) \subseteq \mathcal{M}(F)$ be the subset of the F-irreducible monic polynomials in $A[z]$.*
(i) *The rational Witt vector group $W(F)$ is (isomorphic to) the free abelian group on $\mathcal{M}(F)\backslash\{z\}$, with an isomorphism*

$$\mathbb{Z}[\mathcal{M}(F)\backslash\{z\}] \xrightarrow{\simeq} W(F) \; ;$$

$$p(z) = \sum_{j=0}^{d} a_j z^j \longrightarrow \widetilde{p}(z) = z^d p(z^{-1}) = \sum_{j=0}^{d} a_{d-j} z^j \quad (a_d = 1) \; .$$

(In the terminology of 11.1 $\widetilde{p}(z)$ is the reverse polynomial of $p(z)$.)
(ii) *The rational Witt vector group $W(A)$ is a subgroup of $W(F)$.*
(iii) *If A is integrally closed (e.g. a Dedekind ring) $W(A)$ is the subgroup of $W(F)$ given by*

$$W(A) = \mathbb{Z}[\mathcal{M}_A(F)\backslash\{z\}] \subseteq W(F) = \mathbb{Z}[\mathcal{M}(F)\backslash\{z\}] \; .$$

Proof (i) The polynomial ring $F[z]$ is a principal ideal domain, and every element $q(z) \in F[z]$ with $q(0) = 1$ has a unique expression as a product of irreducible polynomials

$$q(z) = q_1(z)^{n_1} q_2(z)^{n_2} \ldots q_k(z)^{n_k} \in F[z] \quad (n_1, n_2, \ldots, n_k \geq 1)$$

with $q_1(z), q_2(z), \ldots, q_k(z) \in F[z]$ coprime and such that $q_j(0) = 1$. Moreover, $q(z) = \widetilde{p}(z)$ with $p(z) \in \mathcal{M}(F)\backslash\{z\}$.
(ii) The composite

$$W(A) \longrightarrow W(F) \longrightarrow \widehat{W}(F)$$

is the composite of the injections $W(A)\longrightarrow\widehat{W}(A)$, $\widehat{W}(A)\longrightarrow\widehat{W}(F)$, so that $W(A)\longrightarrow W(F)$ is also an injection.
(iii) Every monic polynomial $p(z) \in A[z]$ has a unique factorization as a product

$$p(z) \;=\; p_1(z)p_2(z)\dots p_k(z) \in F[z]$$

of irreducible monic polynomials $p_j(z) \in \mathcal{M}(F)$ $(1 \leq j \leq k)$. Since A is integrally closed the factors are actually defined in $A[z]$ (Eisenbud [67, Prop. 4.12]), so that $p_j(z) \in \mathcal{M}_A(F)$. Passing to the reverse polynomials gives a factorization

$$\widetilde{p}(z) \;=\; \widetilde{p}_1(z)\widetilde{p}_2(z)\dots\widetilde{p}_k(z) \in A[z]\ .$$

Every polynomial $q(z) \in A[z]$ with $q(0) = 1$ thus has a unique factorization

$$q(z) \;=\; q_1(z)q_2(z)\dots q_k(z) \in A[z]$$

with $q_j(z) = \widetilde{p}_j(z)$ $(1 \leq j \leq k)$ the reverse of an F-irreducible monic polynomial $p_j(z) \in A[z]$. □

Example 14.9 (Almkvist [7, 3.5])
(i) If F is an algebraically closed field

$$\mathcal{M}(F) \;=\; \{z+\lambda \,|\, \lambda \in F\}$$

and there is defined an isomorphism

$$\mathbb{Z}[F^\bullet] \xrightarrow{\simeq} W(F)\ ;\ \sum_{j=1}^{k} n_j\lambda_j \longrightarrow \prod_{j=1}^{k}(1+\lambda_j z)^{n_j}\ .$$

(ii) For $F = \mathbb{R}$

$$\mathcal{M}(\mathbb{R}) \;=\; \{z+\lambda \,|\, \lambda \in \mathbb{R}\} \cup \{(z+\mu)(z+\overline{\mu}) \,|\, \mu \in \mathbb{H}\backslash\mathbb{R}\}$$

with $\mathbb{H} \subset \mathbb{C}$ the upper half-plane, and there is defined an isomorphism

$$\mathbb{Z}[\mathbb{H}] \xrightarrow{\simeq} W(\mathbb{R})\ ;\ \lambda \longrightarrow \begin{cases} 1+\lambda z & \text{if } \lambda \in \mathbb{R} \\ (1+\lambda z)(1+\overline{\lambda} z) & \text{otherwise}\ . \end{cases}$$

□

Definition 14.10 The *(noncommutative) reverse characteristic polynomial* of an endomorphism $f : P\longrightarrow P$ of a f.g. projective A-module P is the Dieudonné determinant Witt vector (14.3)

$$\widetilde{\mathrm{ch}}_z(P,f) \;=\; DW(1-z(f\oplus 0) : (P\oplus Q)[[z]]\longrightarrow(P\oplus Q)[[z]]) \in W(A)^{ab}$$

for any f.g. projective A-module Q such that $P\oplus Q$ is a f.g. free A-module. □

Example 14.11 If A is a commutative ring then $W(A)$ is abelian, and

$$W(A)^{ab} = W(A) = \{\widetilde{p}(z)/\widetilde{q}(z) \mid p(z), q(z) \in S\} \subset \widehat{W}(A)$$

with $S \subset A[z]$ the multiplicative subset of all monic polynomials, where

$$\widetilde{p}(z) = z^{\deg(p(z))} p(z^{-1}) \in F[z]$$

is the reverse polynomial with constant coefficient 1 determined by $p(z) \in S$. The reverse characteristic polynomial $\widetilde{\mathrm{ch}}_z(P, f) \in W(A)$ of 14.10 is just the reverse characteristic polynomial in the terminology of Chap. 11, with

$$\widetilde{\mathrm{ch}}_z(P, f) = \det(1 - z(f \oplus 0) : (P \oplus Q)[z] \longrightarrow (P \oplus Q)[z]) \in W(A)$$

the ordinary determinant. □

Proposition 14.12 (Almkvist [6], Grayson [97] for commutative A.)
(i) *The reverse characteristic polynomial defines a natural isomorphism*

$$\begin{aligned} \widetilde{\mathrm{ch}} \;:\; \widetilde{\mathrm{End}_0}(A) &\xrightarrow{\simeq} W(A)^{ab} ; \\ [P, f] &\longrightarrow \widetilde{\mathrm{ch}}_z(P, f) = \det(1 - zf : P[z] \longrightarrow P[z]) , \end{aligned}$$

(ii) *The projective class and reverse characteristic polynomial define a natural isomorphism*

$$\mathrm{End}_0(A) \xrightarrow{\simeq} K_0(A) \oplus W(A)^{ab} ; \; [P, f] \longrightarrow ([P], \widetilde{\mathrm{ch}}_z(P, f)) .$$

Proof (i) As in Chap. 10 let $\widetilde{\Omega}_+$ be the set of square matrices ω in $A[z]$ such that the square matrix $\omega(0)$ in A is invertible, with

$$\widetilde{\Omega}_+^{-1} A[z] = (1 + zA[z])^{-1} A[z]$$

by 10.9. Now 10.16 gives a direct sum decomposition

$$K_1(A) \underset{\widetilde{j}_+}{\overset{\widetilde{i}_+}{\rightleftarrows}} K_1(\widetilde{\Omega}_+^{-1} A[z]) \underset{\widetilde{\Delta}_+}{\overset{\widetilde{\partial}_+}{\rightleftarrows}} \widetilde{\mathrm{End}_0}(A) .$$

The proof of 14.6 gives a direct sum system

$$K_1(A) \underset{\widetilde{j}_+}{\overset{\widetilde{i}_+}{\rightleftarrows}} K_1(\widetilde{\Omega}_+^{-1} A[z]) \underset{\widetilde{\Delta}_+}{\overset{\widetilde{\partial}_+}{\rightleftarrows}} W(A)^{ab} .$$

It now follows from the factorization

$$\widetilde{\mathrm{ch}} \;:\; \widetilde{\mathrm{End}_0}(A) \xrightarrow{\widetilde{\Delta}_+} K_1(\widetilde{\Omega}_+^{-1} A[z]) \xrightarrow{\widetilde{\partial}_+} W(A)^{ab}$$

that $\widetilde{\mathrm{ch}} : \widetilde{\mathrm{End}_0}(A) \longrightarrow W(A)^{ab}$ is an isomorphism, with inverse

$$\widetilde{\mathrm{ch}}^{-1} : W(A)^{ab} \longrightarrow \widetilde{\mathrm{End}}_0(A) \; ; \; p(z) \longrightarrow [A[z]/(\widetilde{p}(z)), z]$$

$$(p(z) = \sum_{j=0}^{d} a_j z^j \ , \ \widetilde{p}(z) = z^d p(z^{-1}) \ , \ a_0 = 1, a_d \neq 0 \in A) \ .$$

(ii) Immediate from (i). □

Example 14.13 Let X be a connected finitely dominated CW complex with a regular cover $\widetilde{X}$ with group of covering translations π, and let $A = \mathbb{Z}[\pi]$. Let $f : X \longrightarrow X$ be a π-untwisted map, so that there is induced an A-module chain map $\widetilde{f} : C(\widetilde{X}) \longrightarrow C(\widetilde{X})$. The invariant

$$\begin{aligned}[X, f] &= [C(\widetilde{X}), \widetilde{f}] = ([X], [X, f]) \\ &\in \mathrm{End}_0(A) = K_0(A) \oplus \widetilde{\mathrm{End}}_0(A)\end{aligned}$$

has first component the A-coefficient projective class of X

$$[X] = [C(\widetilde{X})] \in K_0(A) \ .$$

The isomorphism $\widetilde{\mathrm{ch}} : \widetilde{\mathrm{End}}_0(A) \xrightarrow{\simeq} W(A)^{ab}$ of 14.12 sends the second component to the *ζ-function*

$$\zeta(X, f) = \widetilde{\mathrm{ch}}_z(P, \widetilde{f}) = \prod_{r=0}^{\infty} \widetilde{\mathrm{ch}}_z(\widetilde{f} : P_r \longrightarrow P_r)^{(-)^r} \in W(A)^{ab} \ ,$$

for any finite f.g. projective A-module chain complex P chain equivalent to $C(\widetilde{X})$. The mapping torus of f

$$T(f) = X \times [0,1]/\{(x,0) = (f(x),1) \mid x \in X\}$$

has a regular cover

$$T(\widetilde{f}) = \widetilde{X} \times [0,1]/\{(\widetilde{x},0) = (\widetilde{f}(\widetilde{x}),1) \mid \widetilde{x} \in \widetilde{X}\}$$

with group of covering translations π. The canonical infinite cyclic cover $\overline{T}(\widetilde{f})$ of $T(\widetilde{f})$ is a regular cover of $T(f)$ with group of covering translations $\pi \times \mathbb{Z}$, with

$$\mathbb{Z}[\pi \times \mathbb{Z}] = A[z, z^{-1}] \ .$$

The cellular $A[z, z^{-1}]$-module chain complex of $\overline{T}(\widetilde{f})$ is the algebraic mapping torus of $\widetilde{f} : C(\widetilde{X}) \longrightarrow C(\widetilde{X})$

$$C(\overline{T}(\widetilde{f})) = \mathcal{C}(\widetilde{f} - z : C(\widetilde{X})[z, z^{-1}] \longrightarrow C(\widetilde{X})[z, z^{-1}]) \ ,$$

with a canonical round finite structure. As in Chap. 10 let Ω_+ the set of Fredholm matrices in $A[z]$, with localization $\Omega_+^{-1}A[z]$. The $\Omega_+^{-1}A[z]$-contractible A-finitely dominated $A[z]$-module chain complex

$$E^+ = \mathcal{C}(\widetilde{f} - z : C(\widetilde{X})[z] \longrightarrow C(\widetilde{X})[z])$$

is A-chain equivalent to $C(\widetilde{X})$, and is such that

$$A[z, z^{-1}] \otimes_{A[z]} E^+ = C(\overline{T}(\widetilde{f})) .$$

The $\Omega_+^{-1}A[z]$-coefficient torsion of $T(f)$ is given by

$$\begin{aligned} \tau(T(f); \Omega_+^{-1}A[z]) &= \tau(\Omega_+^{-1}E^+) = (0, [X, f]) \\ &\in K_1(\Omega_+^{-1}A[z]) = K_1(A[z]) \oplus \mathrm{End}_0(A) \end{aligned}$$

with

$$[X, f] = ([X], \zeta(X, f)) \in \mathrm{End}_0(A) = K_0(A) \oplus W(A)^{ab} .$$

A morphism $A \longrightarrow B$ to a commutative ring B extends to a morphism $\Omega_+^{-1}A[z] \longrightarrow S^{-1}B[z]$, with $S \subset B[z]$ the multiplicative subset of all monic polynomials. The mapping torus $T(f)$ is $S^{-1}B[z]$-contractible, and the image of $\tau(\Omega_+^{-1}E^+) \in K_1(\Omega_+^{-1}A[z])$ is the $S^{-1}B[z]$-coefficient (Reidemeister) torsion of $T(f)$

$$\begin{aligned} \tau(T(f); S^{-1}B[z]) &= S^{-1}B[z] \otimes_{\Omega_+^{-1}A[z]} \tau(\Omega_+^{-1}E^+) \\ &\in K_1(S^{-1}B[z]) = K_1(B[z]) \oplus \mathrm{End}_0(B) , \end{aligned}$$

with component

$$\begin{aligned} B \otimes_A [X, f] &= ([C(\widetilde{X}; B)], \zeta(C(\widetilde{X}; B), \widetilde{f})) \\ &\in \mathrm{End}_0(B) = K_0(B) \oplus W(B) . \end{aligned}$$

A morphism $A \longrightarrow R$ to a semi-simple ring R sends $\zeta(X, f) \in W(A)^{ab}$ to

$$\begin{aligned} \zeta(C(\widetilde{X}; R), \widetilde{f}) &= \widetilde{\mathrm{ch}}_z(H_*(\widetilde{X}; R), \widetilde{f}_*) \\ &= \prod_{r=0}^{\infty} \widetilde{\mathrm{ch}}_z(\widetilde{f}_* : H_r(\widetilde{X}; R) \longrightarrow H_r(\widetilde{X}; R))^{(-)^r} \in W(R)^{ab} . \end{aligned}$$

See Chap. 16 below for further discussion of the case when R is a field. □

Definition 14.14 For any ring A let $T_A \subset A[z]$ be the set of monic polynomials. □

Proposition 14.15 *The endomorphism class group of a field F is such that*

$$\begin{aligned} \mathrm{End}_0(F) &= \mathbb{Z}[\max(F[z])] = \mathbb{Z}[\mathcal{M}(F)] \\ &= K_1(F[z], T_F) = K_0(F) \oplus W(F) = \mathbb{Z} \oplus \left(F(z)^\bullet / F^\bullet\right) , \end{aligned}$$

with inverse isomorphisms

$$\mathbb{Z}[\max(F[z])] \xrightarrow{\simeq} \mathrm{End}_0(F) \; ; \; (p(z)) \longrightarrow [F[z]/(p(z)), z] \; ,$$

$$\mathrm{End}_0(F) \xrightarrow{\simeq} \mathbb{Z}[\max(F[z])] \; ; \; [P, f] \longrightarrow \sum_{j=1}^{k} n_j(p_j(z))$$

$$(\mathrm{ch}_z(P, f) = \prod_{j=1}^{k} p_j(z)^{n_j} \; , \; p_j(z) \in \mathcal{M}(F)) \; .$$

There is also defined an isomorphism

$$\mathrm{End}_0(F) \xrightarrow{\simeq} \mathbb{Z} \oplus \left(F(z)^\bullet / F^\bullet\right) \; ;$$
$$[P, f] \longrightarrow (\dim_F(P), \det(1 - zf : P[z] \longrightarrow P[z])) \; .$$

The reduced endomorphism class group of F is such that

$$\begin{aligned} \widetilde{\mathrm{End}}_0(F) \;&=\; \mathbb{Z}[\max(F[z]) \backslash \{(z)\}] \;=\; \mathbb{Z}[\mathcal{M}(F) \backslash \{z\}] \\ &=\; W(F) \;=\; K_1(F[z, z^{-1}], T_F) \;=\; F(z)^\bullet / F^\bullet \; . \end{aligned}$$ □

See Chap. 19 for a more detailed exposition of the algebraic K-theory of function fields.

Proposition 14.16 *Let A be an integrally closed integral domain with quotient field*

$$F \;=\; S^{-1}A \quad (S = A \backslash \{0\}) \; .$$

and let $\max_A(F[z])$ be the set of maximal ideals $(p(z)) \triangleleft F[z]$ generated by the F-irreducible monic polynomials with coefficients in $A \subset F$

$$p(z) \;=\; a_0 + a_1 z + \ldots \; + a_{d-1} z^{d-1} + a_d z^d \in T_A \subset F[z] \quad (a_d = 1)$$

with coefficients $a_j \in A$.
(i) *The localization map of reduced endomorphism class groups*

$$\widetilde{\mathrm{End}}_0(A) \longrightarrow \widetilde{\mathrm{End}}_0(F) \;=\; \mathbb{Z}[\max(F[z]) \backslash \{z\}] \; ;$$
$$(P, f) \longrightarrow S^{-1}(P, f) \;=\; \mathrm{ch}_z(P, f)$$

is injective, with image $\mathbb{Z}[\max_A(F[z]) \backslash \{z\}]$.
(ii) *The localization map of reduced endomorphism class groups*

$$\widetilde{\mathrm{End}}_0(A) \longrightarrow \widetilde{\mathrm{End}}_0^{T_A}(F) \; ; \; (P, f) \longrightarrow S^{-1}(P, f)$$

is an isomorphism, with image the subgroup of $\mathrm{End}_0(F)$ of endomorphisms $g : Q \longrightarrow Q$ of finite-dimensional F-vector spaces with characteristic polynomial

$$\mathrm{ch}_z(Q, g) \in T_A \subset F[z] \; .$$

The endomorphism class groups of A are given by

$$\begin{aligned}\widetilde{\mathrm{End}}_0(A) &= \widetilde{\mathrm{End}}_0^{T_A}(F) = W(A) = \bigoplus_{p(z)\in \max_A(F[z])\backslash\{(z)\}} \mathrm{End}_0^{p(z)^\infty}(F) \\ &= \mathbb{Z}[\max_A(F[z])\backslash\{(z)\}] \subseteq \widetilde{\mathrm{End}}_0(F) = \mathbb{Z}[\max(F[z])\backslash\{(z)\}] , \\ \mathrm{End}_0(A) &= K_0(A) \oplus \widetilde{\mathrm{End}}_0(A) \\ &= K_0(A) \oplus W(A) = K_0(A) \oplus \mathbb{Z}[\max_A(F[z])\backslash\{(z)\}] .\end{aligned}$$

(iii) *The natural map*

$$\widetilde{\mathrm{Nil}}_0(A) \longrightarrow \widetilde{\mathrm{End}}_0(A) \ ; \ (P,\nu) \longrightarrow (P,\nu)$$

is 0.
(iv) *Let* $P = \{p(z) \,|\, p(1) \in A^\bullet\} \subset A[z,z^{-1}]$ *be the multiplicative subset of 10.19. The* P*-torsion group of* $A[z,z^{-1}]$ *is given by*

$$K_1(A[z,z^{-1}],P) = \widetilde{W}(A)$$

with $\widetilde{W}(A) = \mathrm{coker}(K_0(A)\longrightarrow W(A))$ *the reduced rational Witt group (14.7 (ii)), the cokernel of*

$$K_0(A) \longrightarrow W(A) \ ; \ [L] \longrightarrow \det(1-z : L[z]\longrightarrow L[z]) .$$

Every $(A[z,z^{-1}],P)$*-module* K *has a presentation of the type*

$$0 \longrightarrow L[z,z^{-1}] \xrightarrow{1-f+zf} L[z,z^{-1}] \longrightarrow K \longrightarrow 0$$

for some object (L,f) *in* $\mathrm{End}(A)$*, and the* P*-torsion class of* K *is given by*

$$\tau_P(K) = \det(1-zf : L[z]\longrightarrow L[z]) \in K_1(A[z,z^{-1}],P) = \widetilde{W}(A) .$$

(v) *If* $(L,f),(M,g)$ *are objects in* $\mathrm{End}(A)$ *which are related by a morphism* $i : (L,f)\longrightarrow(M,g)$ *such that* $S^{-1}i : S^{-1}L\longrightarrow S^{-1}M$ *is an isomorphism of* F*-vector spaces then*

$$[L,f] = [M,g] \in \widetilde{\mathrm{End}}_0(A) .$$

(vi) *If* A *is a Dedekind ring then*

$$\begin{aligned}\widetilde{\mathrm{End}}_0(A) &= \bigoplus_{p(z)\in \max_A(F[z])\backslash\{(z)\}} \mathrm{End}_0^{p(z)^\infty}(A) \\ &= \bigoplus_{p(z)\in \max_A(F[z])\backslash\{(z)\}} \mathrm{End}_0^{p(z)^\infty}(F) = \mathbb{Z}[\max_A(F[z])\backslash\{(z)\}] .\end{aligned}$$

Proof (i)+(ii) Apply 14.8 and 14.12.
(iii) This follows from the commutative square

$$\begin{array}{ccc} \widetilde{\mathrm{Nil}}_0(A) & \longrightarrow & \widetilde{\mathrm{Nil}}_0(F) \\ \big\downarrow & & \big\downarrow \\ \widetilde{\mathrm{End}}_0(A) & \longrightarrow & \widetilde{\mathrm{End}}_0(F) \end{array}$$

on noting that $\widetilde{\mathrm{End}}_0(A) \longrightarrow \widetilde{\mathrm{End}}_0(F)$ is an injection by (ii), and $\widetilde{\mathrm{Nil}}_0(F) = 0$ (5.15 (i)).
(iv) Combine (ii), (iii) and the exact sequence of 10.21 (ii)

$$\mathrm{Nil}_0(A) \oplus \mathrm{Nil}_0(A) \longrightarrow \mathrm{End}_0(A) \longrightarrow K_1(A[z,z^{-1}],P) \longrightarrow 0 .$$

(v) Immediate from 14.15, since the characteristic polynomials are such that

$$\begin{aligned} \mathrm{ch}_z(L,f) - \mathrm{ch}_z(M,g) &= \mathrm{ch}_z(S^{-1}(L,f)) - \mathrm{ch}_z(S^{-1}(M,g)) \\ &= 0 \in A[z] \subset F[z] . \end{aligned}$$

(vi) Let (L,f) be an object in $\mathrm{End}(A)$ with L f.g. free, so that

$$\mathrm{ch}_z(L,f) = \det(z - f : L[z] \longrightarrow L[z]) \in A[z]$$

is a monic polynomial in $A[z]$. Write the characteristic polynomial as

$$\mathrm{ch}_z(L,f) = p(z) \in T_A \subset A[z] ,$$

and let

$$p(z) = p_1(z)p_2(z)\dots p_k(z) \in F[z]$$

be a factorization as a product of monic polynomials $p_1(z), p_2(z), \dots, p_k(z) \in T_A \subset A[z]$ which are coprime in $F[z]$. Write

$$\alpha_j(z) = p_1(z)\dots p_{j-1}(z)p_{j+1}(z)\dots p_k(z) \in A[z] \quad (1 \le j \le k) .$$

There exist $\beta_1(z), \beta_2(z), \dots, \beta_k(z) \in A[z]$, $s \in S$ such that

$$\sum_{j=1}^{k} \alpha_j(z)\beta_j(z) = s \in A[z] .$$

The f.g. A-modules

$$L_j = \alpha_j(f)(L) \subseteq L \quad (1 \le j \le k)$$

are torsion-free. Since A is Dedekind each L_j is a f.g. projective A-module, and there are defined objects (L_j, f_j) in $\mathrm{End}^{(p_j(z))^\infty}(A)$ with

$$f_j = f| : L_j \longrightarrow L_j \ , \ \mathrm{ch}_z(L_j, f_j) = p_j(z) \in A[z] .$$

The morphisms defined in $\mathrm{End}(A)$ by

$$\alpha = \bigoplus_{j=1}^{k} \alpha_j(f) : (L,f) \longrightarrow \bigoplus_{j=1}^{k} (L_j, f_j) ,$$

$$\beta = \bigoplus_{j=1}^{k} \beta_j(f) : \bigoplus_{j=1}^{k} (L_j, f_j) \longrightarrow (L,f)$$

are such that $\alpha\beta = s$, $\beta\alpha = s$, so that α becomes an isomorphism in $\mathrm{End}(F)$. It follows from (v) that

$$[L,f] = \sum_{j=1}^{k} [L_j, f_j] \in \widetilde{\mathrm{End}}_0(A) ,$$

and hence that the natural map

$$\bigoplus_{p(z)\in \max_A(F[z])\backslash\{(z)\}} \mathrm{End}_0^{p(z)^\infty}(A) \longrightarrow \widetilde{\mathrm{End}}_0(A)$$

is an isomorphism. On the other hand, the natural map

$$\widetilde{\mathrm{End}}_0(A) \longrightarrow \widetilde{\mathrm{End}}_0^{T_A}(F) = \bigoplus_{p(z)\in \max_A(F[z])\backslash\{(z)\}} \mathrm{End}_0^{p(z)^\infty}(F)$$

is an isomorphism by (ii). The natural map

$$\mathrm{End}_0^{p(z)^\infty}(A) \longrightarrow \mathrm{End}_0^{p(z)^\infty}(F) = \mathbb{Z}$$

is an isomorphism, and $\mathrm{End}_0^{p(z)^\infty}(A)$ is the infinite cyclic group generated by $[A[z]/(p(z)), z]$. □

Remark 14.17 The isomorphism $\mathrm{End}_0(F) \cong \mathbb{Z}[\max(F[z])]$ was first obtained by Kelley and Spanier [130]. □

Example 14.18 A monic polynomial $p(z) \in \mathbb{Z}[z]$ is irreducible in $\mathbb{Z}[z]$ if and only if it is irreducible in $\mathbb{Q}[z]$ (Gauss lemma), so that

$$\max_{\mathbb{Z}}(\mathbb{Q}[z]) = \{\text{irreducible monic polynomials } p(z) \in \mathbb{Z}[z]\}$$

and 14.16 gives an isomorphism

$$\mathbb{Z}[\max_{\mathbb{Z}}(\mathbb{Q}[z])] \xrightarrow{\simeq} \mathrm{End}_0(\mathbb{Z}) ; \; p(z) \longrightarrow [\mathbb{Z}[z]/(p(z)), z] . \quad □$$

Proposition 14.19 (i) *If (P,f) is a finite f.g. projective A-module chain complex P with a chain map $f : P \longrightarrow P$ the isomorphism*

$$\widetilde{\mathrm{ch}} : \widetilde{\mathrm{End}}_0(A) \xrightarrow{\simeq} W(A)^{ab}$$

sends the class $[P,f] \in \widetilde{\mathrm{End}}_0(A)$ to

$$\widetilde{\mathrm{ch}}_z(P,f) \;=\; \prod_{r=0}^{\infty} \widetilde{\mathrm{ch}}_z(f : P_r \longrightarrow P_r)^{(-)^r} \in W(A)^{ab} \;.$$

(ii) *If E^+ is an A-finitely dominated finite f.g. free $A[z]$-module chain complex the isomorphism $\widetilde{\mathrm{ch}} : \widetilde{\mathrm{End}}_0(A) \xrightarrow{\simeq} W(A)^{ab}$ sends the class $[E^+,\zeta] \in \widetilde{\mathrm{End}}_0(A)$ to*

$$\widetilde{\mathrm{ch}}_z(E^+) \;=\; \widetilde{\mathrm{ch}}_z(P,f) \;=\; \prod_{r=0}^{\infty} \widetilde{\mathrm{ch}}_z(f : P_r \longrightarrow P_r)^{(-)^r} \in W(A)^{ab}$$

for any finite f.g. projective A-module chain complex P A-module chain equivalent to E^+, with

$$f \;\simeq\; \zeta \;:\; P \;\simeq\; E^+ \longrightarrow P \;\simeq\; E^+ \;.$$

(iii) *The split surjections $\partial_+, \widetilde{\partial}_+$ in the direct sum systems of 10.14, 10.16 are given by*

$$\begin{aligned} \partial_+ \;:\; K_1(\Omega_+^{-1}A[z]) &\longrightarrow \mathrm{End}_0(A) \;=\; K_0(A) \oplus W(A)^{ab} \;; \\ \tau(\Omega_+^{-1}E^+) &\longrightarrow ([E^+], \widetilde{\mathrm{ch}}_z(E^+,\zeta)) \;, \\ \widetilde{\partial}_+ \;:\; K_1(\widetilde{\Omega}_+^{-1}A[z]) &\longrightarrow \widetilde{\mathrm{End}}_0(A) \;=\; W(A)^{ab} \;; \\ \tau(\widetilde{\Omega}_+^{-1}E^+) &\longrightarrow \widetilde{\mathrm{ch}}_z(E^-,\zeta^{-1}) \;. \end{aligned}$$

(iv) *The natural map $\widetilde{\Omega}_+^{-1}A[z] \longrightarrow A[[z]]$ induces the map*

$$K_1(\widetilde{\Omega}_+^{-1}A[z]) \;=\; K_1(A) \oplus W(A)^{ab} \longrightarrow K_1(A[[z]]) \;=\; K_1(A) \oplus \widehat{W}(A)^{ab}$$

given by the sum of the identity on $K_1(A)$ and the map of abelianizations induced by the inclusion $W(A) \longrightarrow \widehat{W}(A)$.

Proof (i) Immediate from 14.12 and the identity

$$[P,f] \;=\; \sum_{r=0}^{\infty} (-)^r [P_r, f] \in \widetilde{\mathrm{End}}_0(A) \;.$$

(ii) Immediate from (i).
(iii) Immediate from (ii) and 10.14.
(iv) Immediate from 14.6 and 14.12. □

Example 14.20 Let A be a semi-simple ring (e.g. a field), so that every A-module is projective. A finite A-module chain complex P is finitely dominated if and only if the A-modules $H_*(P)$ are finitely generated. Every finitely dominated A-module chain complex P is chain equivalent to its homology $H_*(P)$ (regarded as a chain complex with 0 differentials), and for any chain map $f : P \longrightarrow P$

$$\begin{aligned}\widetilde{\mathrm{ch}}_z(P,f) &= \widetilde{\mathrm{ch}}_z(H_*(P),f_*)\\ &= \prod_{r=0}^{\infty} \widetilde{\mathrm{ch}}_z(H_r(P),f_*)^{(-)^r} \in W(A)^{ab} .\end{aligned}$$

□

Definition 14.21 Let A be a commutative ring. Given a multiplicative subset $S \subset A[z]$ with leading units and $z \in S$ define the *S-primary Witt vector group*

$$\begin{aligned}W^S(A) &= \{\widetilde{p}(z)/\widetilde{q}(z) \,|\, p(z), q(z) \text{ monics dividing some element of } S\}\\ &\subseteq W(A) .\end{aligned}$$

□

Proposition 14.22 (Stienstra [276, p. 60])
For a commutative ring A and a multiplicative subset $S \subset A[z]$ with leading units and $z \in S$ the characteristic polynomial defines a surjection

$$\begin{aligned}\widetilde{\mathrm{ch}} \,:\, \widetilde{\mathrm{End}}_0^S(A) &\longrightarrow W^S(A) \,;\\ [P,f] \longrightarrow \widetilde{\mathrm{ch}}_z(P,f) &= \det(1 - zf : P[z] \longrightarrow P[z])\end{aligned}$$

which is split by

$$W^S(A) \longrightarrow \widetilde{\mathrm{End}}_0^S(A) \,;\, p(z) \longrightarrow [A[z]/p(z)] .$$

Proof The torsion group of any commutative ring R splits as

$$K_1(R) = R^\bullet \oplus SK_1(R) ,$$

with $SK_1(R)$ the kernel of the split surjection

$$\det \,:\, K_1(R) \longrightarrow R^\bullet \,;\, \tau(f) \longrightarrow \det(f) .$$

Thus

$$\begin{aligned}\widetilde{\mathrm{End}}_0^S(A) &= \mathrm{coker}(K_1(A) \longrightarrow K_1(\widetilde{S}^{-1}A[z]))\\ &= \mathrm{coker}(A^\bullet \longrightarrow (\widetilde{S}^{-1}A[z])^\bullet) \oplus \mathrm{coker}(SK_1(A) \longrightarrow SK_1(\widetilde{S}^{-1}A[z]))\\ &= W^S(A) \oplus \mathrm{coker}(SK_1(A) \longrightarrow SK_1(\widetilde{S}^{-1}A[z])) .\end{aligned}$$

□

Remark 14.23 (i) If $S \subset A[z]$ is the multiplicative subset of all monic polynomials in A then the split surjection of 14.21 is the isomorphism of 14.12

$$\widetilde{\mathrm{ch}} \,:\, \widetilde{\mathrm{End}}_0^S(A) = \widetilde{\mathrm{End}}_0(A) \xrightarrow{\simeq} W^S(A) = W(A) .$$

(ii) For a field F and a multiplicative subset $S \subset F[z]$ such that $z \in S$ the split surjection of 14.21 is an isomorphism $\widetilde{\mathrm{ch}} : \widetilde{\mathrm{End}}_0^S(F) \xrightarrow{\simeq} W^S(F)$. See Chap. 18 for further details.

(iii) If A is an integral domain with quotient field F, and $T_A \subset F[z]$ is the multiplicative subset of monic polynomials (14.16) then

$$\widetilde{\mathrm{End}}_0(A) \;=\; W(A) \;=\; \widetilde{\mathrm{End}}_0^{T_A}(F) \;=\; W^{T_A}(F)\;.$$

(iv) If $S = \{z^k \mid k \geq 0\} \subset A[z]$ then

$$\begin{aligned}\widetilde{\mathrm{End}}_0^S(A) &= \widetilde{\mathrm{Nil}}_0(A)\;,\\ W^S(A) &= \{1 + a_1 z + \ldots + a_d z^d \mid a_i \in A \text{ nilpotent}\} \subseteq W(A)\end{aligned}$$

(Almkvist [7, 4.11]). □

Example 14.24 Let A be a commutative ring, and let $S \subset A[z]$ be the multiplicative subset consisting of all the polynomials $p(z) = \sum\limits_{j=m}^{n} a_j z^j$ with $a_m, a_n \in A^\bullet$, so that by 13.5

$$S^{-1}A[z] \;=\; \Omega^{-1}A[z, z^{-1}]\;\;,\;\; \mathrm{End}^S(A) \;=\; \mathrm{End}^\Omega(A)$$

with Ω the set of Fredholm matrices in $A[z, z^{-1}]$. The corresponding multiplicative subset of the reverse polynomials

$$\widetilde{S} \;=\; \{\widetilde{p}(z) \mid p(z) \in S\} \subset A[z]$$

is the subset of S consisting of all the polynomials $\widetilde{p}(z) = \sum\limits_{j=0}^{d} b_j z^j$ with $b_0 = 1, b_d \in A^\bullet$. There are identifications

$$\begin{aligned}\mathrm{End}^{\widetilde{S}}(A) &= \mathrm{Aut}(A)\;,\\ \mathrm{End}^S(A) &= \mathrm{Nil}(A) \times \mathrm{Aut}(A)\;,\\ K_*(S^{-1}A[z]) &= K_*(\Omega^{-1}A[z, z^{-1}])\\ &= K_*(A[z]) \oplus \mathrm{End}_*^S(A)\\ &= K_*(A[z, z^{-1}]) \oplus \mathrm{Aut}_*(A)\;.\end{aligned}$$

By 14.22 the characteristic polynomial defines a split surjection

$$\widetilde{\mathrm{ch}} \;:\; \widetilde{\mathrm{End}}_0^S(A) \;=\; \widetilde{\mathrm{Nil}}_0(A) \oplus \mathrm{Aut}_0(A) \longrightarrow W^S(A)\;.$$ □

15. The fibering obstruction

The fibering obstruction is the Whitehead torsion invariant whose vanishing is the necessary and sufficient condition for a manifold of dimension ≥ 6 with finitely dominated infinite cyclic cover (i.e. a band) to fibre over a circle. Given a codimension 2 framed submanifold $N^n \subset M^{n+2}$ it is natural to ask if the exterior $M \backslash (N \times D^2)$ is actually a fibre bundle over S^1 – see Chap. 33 for more on fibred knots. See Hughes and Ranicki [113] for a general account of the fibering obstruction.

The splitting theorem of Chap. 14

$$K_1(A[[z]]) \;=\; K_1(A) \oplus \widehat{W}(A)^{ab}$$

extends to splittings

$$\begin{aligned} Wh_1\big(A((z))\big) &= Wh_1(A) \oplus \widetilde{K}_0(A) \oplus \widetilde{\mathrm{Nil}}_0(A) \oplus \widehat{W}(A)^{ab} \\ Wh_1\big(A((z^{-1}))\big) &= Wh_1(A) \oplus \widetilde{K}_0(A) \oplus \widehat{W}(A)^{ab} \oplus \widetilde{\mathrm{Nil}}_0(A) \end{aligned}$$

which have an application to fibering obstruction theory.

The most important sources of the algebraic and geometric bands of Chaps. 1,2 are the manifold bands:

Definition 15.1 A *manifold band* is a compact manifold M with a finitely dominated infinite cyclic cover $\overline{M}$. □

A manifold band is a CW band, with the CW structure given by the handlebody decomposition.

Remark 15.2 A manifold band is a 'candidate for fibering' in the terminology of Siebenmann [265], [266]. □

Example 15.3 If F is a compact $(n-1)$-dimensional manifold and $h : F \longrightarrow F$ is a homeomorphism the mapping torus $T(h)$ is a compact n-dimensional manifold band. The canonical map $c : T(h) \longrightarrow S^1$ is the projection of a fibre bundle with fibre F and generating covering translation

$$\zeta \,:\, \overline{T}(h) \;=\; F \times \mathbb{R} \longrightarrow \overline{T}(h) \;=\; F \times \mathbb{R}\;;\; (x,t) \longrightarrow (h(x), t+1)\,. \quad \square$$

Example 15.4 (Siebenmann [264])
If ϵ is a tame end of a non-compact n-dimensional manifold W with $n \geq 6$ there exists an open neighbourhood $\overline{M} \subset W$ of ϵ which is the infinite cyclic cover of an untwisted n-dimensional manifold band M with

$$\pi_1(M) = \pi_1(\epsilon) \times \mathbb{Z} \ , \ \pi_1(\overline{M}) = \pi_1(\epsilon) \ .$$

See Hughes and Ranicki [113, Chaps. 15–17] for an account of the structure theory of tame ends. □

Definition 15.5 A closed n-dimensional manifold M *fibres over* S^1 if M is isomorphic to the mapping torus $T(h)$ of an automorphism $h : F \longrightarrow F$ of a closed $(n-1)$-dimensional manifold F. □

In order for a manifold to fibre over S^1 it is necessary (but not in general sufficient) for it to be a band.

Let M be an n-dimensional manifold band. The classifying map $c : M \longrightarrow S^1$ can be made transverse regular at a point in S^1 with inverse image a codimension 1 framed submanifold

$$F^{n-1} = c^{-1}(\{\text{pt.}\}) \subset M^n$$

with a codimension 0 embedding $F \times (0,1) \subset M$. Cutting M along F there is obtained a fundamental domain

$$(M_F; F, \zeta F) = (M \backslash F \times (0,1); F \times \{0\}, F \times \{1\})$$

for the infinite cyclic cover

$$\overline{M} = \bigcup_{j=-\infty}^{\infty} \zeta^j M_F \ .$$

The manifold band M fibres over S^1 if and only if $\overline{M}$ admits a fundamental domain homeomorphic to a product $F \times (I; \{0\}, \{1\})$. It is thus a direct consequence of the s-cobordism theorem that there exists a fundamental domain $(M_F; F, \zeta F)$ which is an s-cobordism if (and for $n \geq 6$ only if) the manifold band M fibres over S^1.

Stallings [272] proved that every irreducible 3-dimensional manifold band M^3 with $\pi_1(\overline{M}) \neq \mathbb{Z}_2$ fibres over S^1, by considering surgery on the surface

$$F^2 = c^{-1}(\{\text{pt.}\}) \subset M^3 \ ,$$

with $c : M \longrightarrow S^1$ the classifying map for the infinite cyclic cover $\overline{M}$. Browder and Levine [31] proved that for $n \geq 6$ every n-dimensional manifold band M^n with $\pi_1(\overline{M}) = \{1\}$ (or equivalently $c_* : \pi_1(M) \cong \pi_1(S^1) = \mathbb{Z}$) fibres over S^1, by considering surgery on the simply-connected submanifold

$$F^{n-1} = c^{-1}(\{\text{pt.}\}) \subset M^n \ .$$

Farrell [78], [79] and Siebenmann [265], [266] defined a Whitehead torsion obstruction

$$\Phi^+(M) \in Wh(\pi_1(M))$$

for any n-dimensional manifold band M, such that $\Phi^+(M) = 0$ if (and for $n \geq 6$ only if) M fibres over S^1.

Proposition 15.6 *For $n \geq 6$ an n-dimensional manifold band M fibres over S^1 if and only if it fibres as a CW complex band.*

Proof The manifold fibering obstruction $\Phi^+(M)$ is one of the CW complex fibering obstructions $\Phi^+(M), \Phi^-(M)$ (2.9). For a manifold band M the fibering obstructions are dual to each other

$$\Phi^+(M) = (-)^{n-1}\Phi^-(M)^* \in Wh(\pi_1(M)) ,$$

so that $\Phi^+(M) = 0$ if and only if $\Phi^-(M) = 0$. □

For $n = 4, 5$ there are examples of n-dimensional manifold bands M with $\Phi^+(M) = 0$ which do not fibre over S^1 (Kearton [129], Weinberger [272]).

From now on, only untwisted manifold bands M will be considered, so that

$$\pi_1(M) = \pi_1(\overline{M}) \times \mathbb{Z} \ , \ \mathbb{Z}[\pi_1(M)] = \mathbb{Z}[\pi_1(\overline{M})][z, z^{-1}] .$$

This is a convenient simplifying assumption, which avoids the need for twisted coefficients.

Novikov [220] initiated the development of S^1-valued Morse theory, using the power series rings $A((z)), A((z^{-1}))$ to count critical points. Farber [75] used S^1-valued Morse theory to recover the result of Browder and Levine [31] for fibering manifold bands M^n with $n \geq 6$ and $\pi_1(M) = \mathbb{Z}$, using the Fredholm localization for noetherian rings – the result of [31] appears as a special case of a theorem realizing the Novikov inequalities on a manifold M with $\pi_1(M) = \mathbb{Z}$ by Morse maps $M \longrightarrow S^1$ with a minimal number of critical points of each index (namely 0 for the fibering case). See Lazarev [153] for the computation of the Novikov homology of a knot complement in terms of the Alexander polynomials. Pazhitnov [222], [223] used S^1-valued Morse theory to prove that for $n \geq 6$ a compact n-dimensional manifold band M fibres over S^1 if and only if the $A((z))$-coefficient Reidemeister torsion

$$\widehat{\tau}^+(M) = \tau(A((z)) \otimes_{A[z,z^{-1}]} C(\widetilde{M})) \in Wh_1(A((z)))/\widehat{W}(A)^{ab}$$

is such that $\widehat{\tau}^+(M) = 0$, with $C(\widetilde{M})$ the cellular chain complex of the universal cover $\widetilde{M}$ and

$$A = \mathbb{Z}[\pi_1(\overline{M})] \ , \ A[z, z^{-1}] = \mathbb{Z}[\pi_1(M)] .$$

A similar result was obtained by Latour [148]. The geometrically significant splittings of $Wh_1(A((z)))$ and $Wh_1(A((z^{-1})))$ allow this result to be deduced from the original fibering obstruction theory of Farrell [78], [79] and Siebenmann [265], [266] using the algebraic fibering obstruction theory of Ranicki [244, Chap. 21] recalled in Chap. 3.

By analogy with the conventions regarding the Whitehead groups $Wh_1(A)$ (1.4) and $Wh_1(A[z, z^{-1}])$ (3.8) :

Convention 15.7 The *Whitehead group* of the formal power series ring $A((z))$ is

$$\begin{aligned} Wh_1(A((z))) &= \operatorname{coker}(K_1(\mathbb{Z}[z,z^{-1}])\longrightarrow K_1(A((z)))) \\ &= K_1(A((z)))/\{\pm z^j \,|\, j\in\mathbb{Z}\} \end{aligned}$$

in the algebraic context of an arbitrary ring A. In the topological context of a group ring $A=\mathbb{Z}[\pi]$ the Whitehead group is to be understood to be

$$\widehat{Wh}^+(\pi\times\mathbb{Z}) = K_1(\mathbb{Z}[\pi]((z)))/\{\pm z^j g \,|\, j\in\mathbb{Z}, g\in\pi\} .$$

Similarly for

$$\begin{aligned} Wh_1(A((z^{-1}))) &= \operatorname{coker}(K_1(\mathbb{Z}[z,z^{-1}])\longrightarrow K_1(A((z^{-1})))) \\ &= K_1(A((z^{-1})))/\{\pm z^j \,|\, j\in\mathbb{Z}\} \end{aligned}$$

and

$$\widehat{Wh}^-(\pi\times\mathbb{Z}) = K_1(\mathbb{Z}[\pi]((z^{-1})))/\{\pm z^j g \,|\, j\in\mathbb{Z}, g\in\pi\} . \qquad \square$$

By analogy with the geometrically significant decomposition

$$Wh_1(A[z,z^{-1}]) = Wh_1(A)\oplus\widetilde{K}_0(A)\oplus\widetilde{\mathrm{Nil}}_0(A)\oplus\widetilde{\mathrm{Nil}}_0(A)$$

given by 5.19:

Definition 15.8 The *geometrically significant decompositions* of the Whitehead groups of $A((z)), A((z^{-1}))$

$$\begin{aligned} Wh_1(A((z))) &= Wh_1(A)\oplus\widetilde{K}_0(A)\oplus\widetilde{\mathrm{Nil}}_0(A)\oplus\widehat{W}(A)^{ab} \\ Wh_1(A((z^{-1}))) &= Wh_1(A)\oplus\widetilde{K}_0(A)\oplus\widehat{W}(A)^{ab}\oplus\widetilde{\mathrm{Nil}}_0(A) \end{aligned}$$

are the decompositions determined by the isomorphisms

$$\begin{aligned} & Wh_1(A)\oplus\widetilde{K}_0(A)\oplus\widetilde{\mathrm{Nil}}_0(A)\oplus\widehat{W}(A)^{ab} \xrightarrow{\simeq} Wh_1(A((z))) ; \\ & (\tau(f:M\longrightarrow M),[N],[P^+,\nu^+],(a_1,a_2,\dots)) \longrightarrow \\ & \tau(f:M((z))\longrightarrow M((z))) + \tau(-z:N((z))\longrightarrow N((z))) \\ & \qquad + \tau(1-z^{-1}\nu^+ : P^+((z))\longrightarrow P^+((z))) \\ & \qquad + \tau(1+\sum_{j=1}^{\infty} a_j z^j : A((z))\longrightarrow A((z))) , \end{aligned}$$

$$\begin{aligned} & Wh_1(A)\oplus\widetilde{K}_0(A)\oplus\widehat{W}(A)^{ab}\oplus\widetilde{\mathrm{Nil}}_0(A) \xrightarrow{\simeq} Wh_1(A((z^{-1}))) ; \\ & (\tau(f:M\longrightarrow M),[N],(a_1,a_2,\dots),[P^-,\nu^-]) \longrightarrow \\ & \tau(f:M((z^{-1}))\longrightarrow M((z^{-1}))) + \tau(-z:N((z^{-1}))\longrightarrow N((z^{-1}))) \\ & \qquad + \tau(1+\sum_{j=1}^{\infty} a_j z^{-j} : A((z^{-1}))\longrightarrow A((z^{-1}))) \\ & \qquad + \tau(1-z\nu^- : P^-((z^{-1}))\longrightarrow P^-((z^{-1}))) . \end{aligned}$$

$\square$

Recall from 8.3 that a based f.g. free $A[z,z^{-1}]$-module chain complex E is a band if and only if it is both $A((z))$-contractible and $A((z^{-1}))$-contractible.

Definition 15.9 For any chain complex band E over $A[z,z^{-1}]$ let

$$\begin{aligned}\widehat{\tau}^+(E) &= \tau(A((z))\otimes_{A[z,z^{-1}]} E) \in Wh_1(A((z)))\ ,\\ \widehat{\tau}^-(E) &= \tau(A((z^{-1}))\otimes_{A[z,z^{-1}]} E) \in Wh_1(A((z^{-1})))\ .\end{aligned}$$ □

The algebraic mapping tori of an A-module chain map $f : P\longrightarrow P$ are the $A[z,z^{-1}]$-module chain complexes

$$\begin{aligned}T^+(f) &= \mathcal{C}(1-zf : P[z,z^{-1}]\longrightarrow P[z,z^{-1}])\ ,\\ T^-(f) &= \mathcal{C}(1-z^{-1}f : P[z,z^{-1}]\longrightarrow P[z,z^{-1}])\ .\end{aligned}$$

Proposition 15.10 *Let P be a finitely dominated A-module chain complex.*
(i) *For any chain map $f : P\longrightarrow P$*

$$\begin{aligned}\widehat{\tau}^+(T^+(f)) &= (0,0,0,\widetilde{\mathrm{ch}}_z(P,f))\\ &\in Wh_1(A((z))) = Wh_1(A)\oplus\widetilde{K}_0(A)\oplus\widetilde{\mathrm{Nil}}_0(A)\oplus\widehat{W}(A)^{ab}\\ \widehat{\tau}^-(T^-(f)) &= (0,0,\widetilde{\mathrm{ch}}_z(P,f),0)\\ &\in Wh_1(A((z^{-1}))) = Wh_1(A)\oplus\widetilde{K}_0(A)\oplus\widehat{W}(A)^{ab}\oplus\widetilde{\mathrm{Nil}}_0(A)\ .\end{aligned}$$

(ii) *For any chain equivalence $f : P\longrightarrow P$*

$$\begin{aligned}\widehat{\tau}^+(T^-(f)) &= (\tau(f),-[P],0,\widetilde{\mathrm{ch}}_z(P,f^{-1}))\\ &\in Wh_1(A((z))) = Wh_1(A)\oplus\widetilde{K}_0(A)\oplus\widetilde{\mathrm{Nil}}_0(A)\oplus\widehat{W}(A)^{ab}\\ \widehat{\tau}^-(T^+(f)) &= (\tau(f),[P],\widetilde{\mathrm{ch}}_z(P,f^{-1}),0)\\ &\in Wh_1(A((z^{-1}))) = Wh_1(A)\oplus\widetilde{K}_0(A)\oplus\widehat{W}(A)^{ab}\oplus\widetilde{\mathrm{Nil}}_0(A)\ .\end{aligned}$$ □

Proposition 15.11 *For any chain complex band E over $A[z,z^{-1}]$*

$$\begin{aligned}\widehat{\tau}^+(E) &= (\phi^+,-[E^-],-[E/E^+,\zeta],\widetilde{\mathrm{ch}}_z(E^-,\zeta^{-1}))\\ &\in Wh_1(A((z))) = Wh_1(A)\oplus\widetilde{K}_0(A)\oplus\widetilde{\mathrm{Nil}}_0(A)\oplus\widehat{W}(A)^{ab}\\ \widehat{\tau}^-(E) &= (\phi^-,[E^+],\widetilde{\mathrm{ch}}_z(E^+,\zeta),-[E/E^-,\zeta^{-1}])\\ &\in Wh_1(A((z^{-1}))) = Wh_1(A)\oplus\widetilde{K}_0(A)\oplus\widehat{W}(A)^{ab}\oplus\widetilde{\mathrm{Nil}}_0(A)\end{aligned}$$

for some $\phi^+,\phi^-\in Wh_1(A)$ such that

$$\phi^+-\phi^- = \tau(\zeta : E\longrightarrow E)\in Wh_1(A)\ .$$

Proof By 7.4 the fibering obstructions of E are of the form

$$\begin{aligned}
\Phi^+(E) &= \tau(q^+ : T^+(\zeta^{-1}) \longrightarrow E) \\
&= (\phi^+, -[E^-], -[E/E^+, \zeta], -[E/E^-, \zeta^{-1}]) \\
\Phi^-(E) &= \tau(q^- : T^-(\zeta) \longrightarrow E) \\
&= (\phi^-, [E^+], -[E/E^+, \zeta], -[E/E^-, \zeta^{-1}]) \\
&\in Wh_1(A[z, z^{-1}]) \ = \ Wh_1(A) \oplus \widetilde{K}_0(A) \oplus \widetilde{\mathrm{Nil}}_0(A) \oplus \widetilde{\mathrm{Nil}}_0(A) \ .
\end{aligned}$$

The $A[z, z^{-1}]$-module chain equivalence $q^+ : T^+(\zeta^{-1}) \longrightarrow E$ induces an $A((z))$-module chain equivalence

$$1 \otimes q^+ \ : \ A((z)) \otimes_{A[z,z^{-1}]} T^+(\zeta^{-1}) \longrightarrow A((z)) \otimes_{A[z,z^{-1}]} E$$

of contractible complexes. Applying 15.10 (i) gives

$$\begin{aligned}
\widehat{\tau}^+(E) &= \tau(1 \otimes q^+) + \widehat{\tau}^+(T^+(\zeta^{-1})) \\
&= (\phi^+, -[E^-], -[E/E^+, \zeta], -\widetilde{\mathrm{ch}}_z(E/E^-, \zeta^{-1})) \\
&\qquad\qquad + (0, 0, 0, \widetilde{\mathrm{ch}}_z(E, \zeta^{-1})) \\
&= (\phi^+, -[E^-], -[E/E^+, \zeta], \widetilde{\mathrm{ch}}_z(E^-, \zeta^{-1})) \\
&\in Wh_1\big(A((z))\big) \ = \ Wh_1(A) \oplus \widetilde{K}_0(A) \oplus \widetilde{\mathrm{Nil}}_0(A) \oplus \widehat{W}(A)^{ab} \ .
\end{aligned}$$

Similarly for $\widehat{\tau}^-(E) \in Wh_1\big(A((z^{-1}))\big)$. □

Proposition 15.12 *The fibering obstructions of a chain complex band E over $A[z, z^{-1}]$ are such that $\Phi^+(E) = \Phi^-(E) = 0 \in Wh_1(A[z, z^{-1}])$ if and only if*

$$\begin{aligned}
\widehat{\tau}^+(E) &= 0 \in Wh_1\big(A((z))\big)/\widehat{W}(A)^{ab} \ , \\
\widehat{\tau}^-(E) &= 0 \in Wh_1\big(A((z^{-1}))\big)/\widehat{W}(A)^{ab} \ .
\end{aligned}$$

Proof Immediate from 15.11, since the components of

$$\begin{aligned}
\Phi^+(E) &= (\phi^+, -[E^-], -[E/E^+, \zeta], -[E/E^-, \zeta^{-1}]) \\
\Phi^-(E) &= (\phi^-, [E^+], -[E/E^+, \zeta], -[E/E^-, \zeta^{-1}]) \\
&\in Wh_1(A[z, z^{-1}]) \ = \ Wh_1(A) \oplus \widetilde{K}_0(A) \oplus \widetilde{\mathrm{Nil}}_0(A) \oplus \widetilde{\mathrm{Nil}}_0(A)
\end{aligned}$$

are determined by the components of

$$\begin{aligned}
\widehat{\tau}^+(E) &= (\phi^+, -[E^-], -[E/E^+, \zeta]) \\
&\in Wh_1\big(A((z))\big)/\widehat{W}(A)^{ab} \ = \ Wh_1(A) \oplus \widetilde{K}_0(A) \oplus \widetilde{\mathrm{Nil}}_0(A) \ , \\
\widehat{\tau}^-(E) &= (\phi^-, [E^+], -[E/E^-, \zeta^{-1}]) \\
&\in Wh_1\big(A((z^{-1}))\big)/\widehat{W}(A)^{ab} \ = \ Wh_1(A) \oplus \widetilde{K}_0(A) \oplus \widetilde{\mathrm{Nil}}_0(A) \ .
\end{aligned}$$

□

Definition 15.13 A *geometric Poincaré band* is a finite n-dimensional geometric Poincaré complex X with

$$\pi_1(X) = \pi \times \mathbb{Z} \ , \ \mathbb{Z}[\pi_1(X)] \ = \ \mathbb{Z}[\pi][z, z^{-1}]$$

satisfying any one of the following equivalent conditions:

(i) X is a CW band, i.e. the infinite cyclic cover $\overline{X} = \widetilde{X}/\pi$ of X is finitely dominated,

(ii) the cellular $\mathbb{Z}[\pi][z, z^{-1}]$-module chain complex $C(\widetilde{X})$ of the universal cover $\widetilde{X}$ of X is finitely dominated,

(iii) $H_*(X; \Omega^{-1}\mathbb{Z}[\pi][z, z^{-1}]) = 0$,

in which case $\overline{X}$ is a finitely dominated $(n-1)$-dimensional geometric Poincaré complex. □

Example 15.14 A manifold band (15.1) is a geometric Poincaré band. □

Proposition 15.15 (Ranicki [244, p. 163])
The torsion $\tau(X) \in Wh(\pi \times \mathbb{Z})$ of an n-dimensional geometric Poincaré band X and the Farrell–Siebenmann fibering obstructions

$$\Phi^+(X) \ = \ (\phi^+, -[E^-], \nu^+, \nu^-) \ , \ \Phi^-(X) \ = \ (\phi^-, [E^+], \nu^+, \nu^-)$$

are such that

$$\begin{aligned}\tau(X) \ &= \ \Phi^+(X) + (-)^n \Phi^-(X)^* \\ &= \ (\phi^+ + (-)^n(\phi^-)^*, (-)^{n+1}[E^+]^* - [E^-], \\ &\qquad \nu^+ + (-)^{n+1}(\nu^-)^*, \nu^- + (-)^{n+1}(\nu^+)^*) \\ &\in Wh(\pi \times \mathbb{Z}) \ = \ Wh(\pi) \oplus \widetilde{K}_0(\mathbb{Z}[\pi]) \oplus \widetilde{\mathrm{Nil}}_0(\mathbb{Z}[\pi]) \oplus \widetilde{\mathrm{Nil}}_0(\mathbb{Z}[\pi])\end{aligned}$$

with $E = C(\widetilde{X})$ and

$$\begin{aligned}\phi^+ - \phi^- \ &= \ \tau(\zeta : \overline{X} \longrightarrow \overline{X}) \in Wh(\pi) \ , \\ [E^+] + [E^-] \ &= \ [E] \ = \ [\overline{X}] \in \widetilde{K}_0(\mathbb{Z}[\pi]) \ .\end{aligned}$$

Also, 15.11 gives

$$\begin{aligned}&\widehat{\tau}^+(X) \ = \ (\phi^+, -[E^-], \nu^+) \\ &\quad \in Wh_1(\mathbb{Z}[\pi]((z)))/\widehat{W}(\mathbb{Z}[\pi])^{ab} \ = \ Wh_1(\mathbb{Z}[\pi]) \oplus \widetilde{K}_0(\mathbb{Z}[\pi]) \oplus \widetilde{\mathrm{Nil}}_0(\mathbb{Z}[\pi]) \\ &\widehat{\tau}^-(X) \ = \ (\phi^-, [E^+], \nu^-) \\ &\quad \in Wh_1(\mathbb{Z}[\pi]((z^{-1})))/\widehat{W}(\mathbb{Z}[\pi])^{ab} \ = \ Wh_1(\mathbb{Z}[\pi]) \oplus \widetilde{K}_0(\mathbb{Z}[\pi]) \oplus \widetilde{\mathrm{Nil}}_0(\mathbb{Z}[\pi]) \ .\end{aligned}$$

In particular, for a simple Poincaré band $\tau(X) = 0 \in Wh(\pi \times \mathbb{Z})$ (e.g. for a manifold band), so that

$$\begin{aligned}
\phi^- &= (-)^{n+1}(\phi^+)^* \in Wh(\pi) , \\
[E^-] &= (-)^{n+1}[E^+]^* \in \widetilde{K}_0(\mathbb{Z}[\pi]) , \\
\nu^- &= (-)^n(\nu^+)^* \in \widetilde{\mathrm{Nil}}_0(\mathbb{Z}[\pi]) .
\end{aligned}$$

□

Proposition 15.16 *The Farrell–Siebenmann fibering obstruction*

$$\Phi^+(M) \;=\; \tau(T(\zeta)\longrightarrow M) \in Wh(\pi \times \mathbb{Z})$$

of an n-dimensional manifold band M with $\pi_1(M) = \pi\times\mathbb{Z}$ and the Pazhitnov fibering obstruction

$$\widehat{\tau}^+(M) \;=\; \tau(\mathbb{Z}[\pi]((z)) \otimes_{\mathbb{Z}[\pi][z,z^{-1}]} C(\widetilde{M})) \in Wh_1\big(\mathbb{Z}[\pi]((z))\big)/\widehat{W}(\mathbb{Z}[\pi])^{ab}$$

are such that $\Phi^+(M) = 0$ if and only if $\widehat{\tau}^+(M) = 0$.

Proof M is a simple Poincaré band, so that by 15.15 the Farrell–Siebenmann fibering obstruction

$$\begin{aligned}
\Phi^+(M) &= (\phi^+, -[E^-], \nu^+, (-)^n(\nu^+)^*) \\
&\in Wh(\pi \times \mathbb{Z}) \;=\; Wh(\pi) \oplus \widetilde{K}_0(\mathbb{Z}[\pi]) \oplus \widetilde{\mathrm{Nil}}_0(\mathbb{Z}[\pi]) \oplus \widetilde{\mathrm{Nil}}_0(\mathbb{Z}[\pi])
\end{aligned}$$

determines and is determined by the Pazhitnov fibering obstruction

$$\begin{aligned}
\widehat{\tau}^+(M) &= (\phi^+, -[E^-], \nu^+) \\
&\in Wh_1\big(\mathbb{Z}[\pi]((z))\big)/\widehat{W}(\mathbb{Z}[\pi])^{ab} \;=\; Wh(\pi) \oplus \widetilde{K}_0(\mathbb{Z}[\pi]) \oplus \widetilde{\mathrm{Nil}}_0(\mathbb{Z}[\pi]) .
\end{aligned}$$

□

Remark 15.17 The fibering obstructions can also be described using the Fredholm localization $\Omega^{-1}A[z,z^{-1}]$ of 13.3. A finite based f.g. free $A[z,z^{-1}]$-module chain complex E is A-finitely dominated if and only if it is $\Omega^{-1}A[z,z^{-1}]$-contractible, in which case it has $\Omega^{-1}A[z,z^{-1}]$-coefficient torsion

$$\begin{aligned}
\tau(\Omega^{-1}E) &= (\Phi^+(E), [E,\zeta]) \\
&\in K_1(\Omega^{-1}A[z,z^{-1}]) \;=\; K_1(A[z,z^{-1}]) \oplus \mathrm{Aut}_0(A)
\end{aligned}$$

by the splitting theorem of 13.17 (cf. Ranicki [248]). □

Example 15.18 Let X be an untwisted CW band with fundamental group $\pi_1(X) = \pi \times \mathbb{Z}$, universal cover $\widetilde{X}$ and infinite cyclic cover $\overline{X} = \widetilde{X}/\pi$, and write

$$\Lambda \;=\; \Omega^{-1}\mathbb{Z}[\pi][z,z^{-1}] .$$

Now X is Λ-acyclic, with Λ-coefficient torsion

$$\begin{aligned}
\tau(X;\Lambda) &= (\Phi^+(X), [\overline{X},\zeta]) \\
&\in Wh_1(\Lambda) \;=\; Wh_1(\mathbb{Z}[\pi][z,z^{-1}]) \oplus \mathrm{Aut}_0(\mathbb{Z}[\pi]) .
\end{aligned}$$

□

16. Reidemeister torsion

Reidemeister torsion is an invariant of simple homotopy type, a precursor of Whitehead torsion, which was originally used in the combinatorial classification of lens spaces. There are many connections between Reidemeister torsion and knots, particularly the Alexander polynomials of knots – see Milnor [193], [194], [195] and Turaev [293], as well as Chap. 17 below. In the current chapter the treatment of Reidemeister torsion in [194] will be generalized to define a relative K-theory invariant for chain complexes.

For any ring morphism $f : A \longrightarrow B$ there is defined an exact sequence of algebraic K-groups

$$K_1(A) \xrightarrow{f} K_1(B) \longrightarrow K_1(f) \longrightarrow K_0(A) \xrightarrow{f} K_0(B) ,$$

with $K_1(f)$ the algebraic K-group of triples (P, Q, g) with P, Q f.g. projective A-modules and $g : B \otimes_A P \longrightarrow B \otimes_A Q$ a B-module isomorphism. A finitely dominated A-module chain complex C with a round finite structure ϕ_B on $B \otimes_A C$ has a relative K-theory invariant

$$[C, \phi_B] \in K_1(f)$$

with image the projective class $[C] \in K_0(A)$ (Ranicki [241]). The projective class of a finitely dominated B-contractible A-module chain complex C is an element

$$[C] \in \ker(f : K_0(A) \longrightarrow K_0(B)) = \operatorname{im}(K_1(f) \longrightarrow K_0(A)) ,$$

the image of $[C, \phi_B] \in K_1(f)$ for any choice of round finite structure ϕ_B on $B \otimes_A C$. The projective class is $[C] = 0$ if and only if C admits a round A-finite structure ϕ_A, in which case

$$[C, \phi_B] \in \ker(K_1(f) \longrightarrow K_0(A)) = \operatorname{im}(K_1(B) \longrightarrow K_1(f))$$

is the image of $\tau(B \otimes_A C, 1 \otimes \phi_A) \in K_1(B)$.

Definition 16.1 (i) The *absolute Reidemeister torsion* of a B-contractible A-module chain complex C with a round finite structure ϕ_A is

$$\Delta(C, \phi_A) = \tau(B \otimes_A C, 1 \otimes \phi_A) \in K_1(B) .$$

(ii) The *relative Reidemeister torsion* of a B-contractible f.g. free A-module chain complex C is

$$\begin{aligned}\Delta(C) &= [\Delta(C,\phi_A)] \\ &\in \text{coker}(f : K_1(A)\longrightarrow K_1(B)) = \ker(K_1(f)\longrightarrow K_0(A))\end{aligned}$$

using any round finite structure ϕ_A on C. □

Terminology 16.2 If C is a B-contractible based f.g. free A-module chain complex the bases determine a round finite structure ϕ_A on C, and the Reidemeister torsion $\Delta(C,\phi_A) = \tau(B\otimes_A C, 1\otimes\phi_A)$ is written

$$\Delta(C) = \tau(B\otimes_A C) \in K_1(B) .$$ □

Proposition 16.3 (i) *If C is a B-contractible f.g. free A-module chain complex then $B\otimes_A C$ has a canonical round finite structure ϕ_B with*

$$\Delta(C,\phi_A) = \tau(1 : (B\otimes_A C, 1\otimes\phi_A)\longrightarrow(B\otimes_A C,\phi_B)) \in K_1(B)$$

for any round finite structure ϕ_A on C.
(ii) *If C is a B-contractible based f.g. free A-module chain complex then*

$$\Delta(C) = \sum_{r=0}^{\infty}(-)^r\tau(1 : B\otimes_A C_r\longrightarrow(B\otimes_A C_r,\phi_r)) \in K_1(B)$$

with ϕ_r the (stable) basis of $B\otimes_A C_r$ determined by the canonical round finite structure ϕ_B on B.
Proof (i) Define

$$D_r = B\otimes_A C_r \ , \ E_r = \ker(d : D_r\longrightarrow D_{r-1}) \ (r\geq 0)$$

and use a chain contraction $\Gamma : 0 \simeq 1 : D\longrightarrow D$ to define an isomorphism of B-module chain complexes

$$(i\ \Gamma|) : \mathcal{C}(1 : E\longrightarrow E) \longrightarrow D$$

with $i : E_r\longrightarrow D_r$ the inclusion. For each $r\geq 0$ there is defined a contractible chain complex

$$0 \longrightarrow E_r \longrightarrow D_r \longrightarrow \dots \longrightarrow D_1 \longrightarrow D_0 \longrightarrow 0 ,$$

so that there is an isomorphism

$$E_r\oplus D_{r-1}\oplus D_{r-3}\oplus\dots \cong D_r\oplus D_{r-2}\oplus\dots$$

and E_r is (stably) f.g. free. Choosing an arbitrary (stable) basis for each E_r determines the canonical round finite structure ϕ_B on $D = B\otimes_A C$.
(ii) Immediate from (i). □

Example 16.4 (i) Let X be a finite CW complex with finite fundamental group $\pi_1(X) = \pi$ and Euler characteristic $\chi(X) = 0$. Also, let

$$f : A = \mathbb{Z}[\pi] \longrightarrow B = \mathbb{Q}[\pi]/(\sum_{g \in \pi} g)$$

be the natural ring morphism, and let $C = C(\widetilde{X})$ be the cellular A-module chain complex of the universal cover $\widetilde{X}$, with the bases determined by the cell structure up to $\pm\pi$. It follows from $\mathbb{Q}[\pi] = \mathbb{Q} \oplus B$ that

$$H_*(\widetilde{X};\mathbb{Q}) = H_*(\mathbb{Q}[\pi] \otimes_A C) = H_*(\mathbb{Q} \otimes_A C) \oplus H_*(B \otimes_A C) .$$

If π acts trivially on $H_*(\widetilde{X};\mathbb{Q})$ then

$$H_*(\widetilde{X};\mathbb{Q}) = H_*(\mathbb{Q} \otimes_A C) , \quad H_*(B \otimes_A C) = 0 ,$$

and X has an absolute Reidemeister torsion invariant as in Milnor [194, 12.4]

$$\Delta(X) = \Delta(C) \in Wh(B) = K_1(B)/\{\pm\pi\} .$$

If $h : Y \longrightarrow X$ is a homotopy equivalence of finite CW complexes for which τ is defined

$$h_*\Delta(Y) - \Delta(X) = [\tau(h)] \in \mathrm{im}(Wh(\pi) \longrightarrow Wh(B)) ,$$

with $\tau(h) \in Wh(\pi)$ the Whitehead torsion. Thus $\Delta(X)$ is a simple homotopy invariant, while the relative Reidemeister torsion

$$[\Delta(X)] \in \mathrm{coker}(f : Wh(\pi) \longrightarrow Wh(B))$$

is a homotopy invariant of X.
(ii) In the applications of absolute Reidemeister torsion to the classification of lens spaces (due to Reidemeister, Franz and deRham) $\pi = \mathbb{Z}_m$ is a cyclic group, with

$$\mathbb{Q}[\mathbb{Z}_m] = \bigoplus_{d|m} \mathbb{Q}(\zeta_d) , \quad B = \bigoplus_{d|m, d \neq 1} \mathbb{Q}(\zeta_d)$$

products of cyclotomic fields $\mathbb{Q}(\zeta_d)$ ($\zeta_d = e^{2\pi i/d}$), and

$$K_1(\mathbb{Q}[\mathbb{Z}_m]) = \bigoplus_{d|m} \mathbb{Q}(\zeta_d)^\bullet , \quad K_1(B) = \bigoplus_{d|m, d \neq 1} \mathbb{Q}(\zeta_d)^\bullet .$$

By results of Higman and Bass

$$Wh(\mathbb{Z}_m) = \mathbb{Z}^{[m/2]+1-\delta(m)}$$

with $\delta(m)$ the number of divisors of m. □

Proposition 16.5 *Let*

$$i \,:\, A \longrightarrow B \,=\, S^{-1}A$$

be the inclusion of a ring in the localization inverting a multiplicative subset $S \subset A$.
(i) *An* $S^{-1}A$*-contractible finite f.g. projective* A*-module chain complex has an* S*-torsion class* $\tau_S(C) \in K_1(A,S)$ *(4.7) with image the projective class* $[C] \in K_0(A)$.
(ii) *If* $[C] = 0 \in K_0(A)$ *a choice of round finite structure on* C *determines a lift of the* S*-torsion class* $\tau_S(C) \in K_1(A,S)$ *to the element* $\tau(S^{-1}C) \in K_1(S^{-1}A)$, *and the relative Reidemeister torsion of* C *agrees with the* S*-torsion class*

$$\begin{aligned}\Delta(C) \,&=\, [\tau(S^{-1}C)] \,=\, \tau_S(C) \\ &\in \operatorname{coker}(i : K_1(A) \longrightarrow K_1(S^{-1}A)) \,=\, \ker(K_1(A,S) \longrightarrow K_0(A)) \ . \qquad \square\end{aligned}$$

Example 16.6 Let A be an integral domain with quotient field F. Let

$$i \,:\, A \longrightarrow F \,=\, S^{-1}A$$

be the inclusion, with $S = A\backslash\{0\} \subset A$, so that

$$i \,:\, K_1(A) \longrightarrow K_1(F) \,=\, F^\bullet \ ; \ (a_{jk}) \longrightarrow \det(a_{jk}) \ .$$

If C is a finite f.g. free A-module chain complex then $H_*(F \otimes_A C) = 0$ if and only if $H_*(C)$ is S-torsion, that is $sH_*(C) = 0$ for some $s \in S$. A choice of $s \in S$ such that $sH_*(C) = 0$ determines a round finite structure on C. As in 16.5 the relative Reidemeister torsion of C is an element

$$\Delta(C) \in \operatorname{coker}(i : K_1(A) \longrightarrow F^\bullet)$$

which (at least for A a unique factorization ring) may be identified with the class of $\dfrac{1}{s}$ for some element $s \in S$ such that $sH_*(C) = 0$, as follows.
(i) The *annihilator* of an A-module M is the ideal

$$\operatorname{ann}(M) \,=\, \{a \in A \,|\, aM = 0\} \triangleleft A \ .$$

Every f.g. A-module M admits a presentation of the type

$$A^m \xrightarrow{\ d\ } A^n \longrightarrow M \longrightarrow 0$$

with $m \leq \infty$. The *order ideal* $o(M) \triangleleft A$ of M (or 0*th Fitting ideal*) is the ideal generated by the $n \times n$ minors of the matrix of d, such that

$$o(M) \subseteq \operatorname{ann}(M) \ .$$

If $m < n$ then $o(M) = \operatorname{ann}(M) = \{0\}$.
If $m = n$ then $o(M) = (\det(d))$ and the annihilator of M is

$$\text{ann}(M) = \{a \in A \,|\, ao_1(M) \subseteq o(M)\} \lhd A$$

with $o_1(M)$ the Fitting ideal generated by the $(n-1) \times (n-1)$ minors of the matrix of d – see Eisenbud [67, p. 511].
(ii) Suppose now that A is a unique factorization domain. The *order* of a f.g. A-module M is

$$\begin{aligned}\text{ord}(M) &= \{\text{greatest common divisor of } o(M)\} \\ &\in \text{coker}(i : K_1(A) \longrightarrow F^\bullet) .\end{aligned}$$

If M has a presentation (as in (i)) with $m = n$ and $o(M)$ is a principal ideal then

$$\text{ord}(M) = \det(d) \in \text{coker}(i : K_1(A) \longrightarrow F^\bullet) .$$

If A is a principal ideal domain then every f.g. S-torsion A-module M is a finite direct sum of cyclic S-torsion A-modules

$$M = \bigoplus_{j=1}^{k} A/s_jA \quad (s_j \in S\,,\, s_{j+1} \,|\, s_j) ,$$

and

$$\text{ann}(M) = (s_1)\,,\; o(M) = \Big(\prod_{j=1}^{k} s_j\Big)\,,\; o_1(M) = \Big(\prod_{j=2}^{k} s_j\Big) \lhd A$$

with

$$\text{ord}(M) = \prod_{j=1}^{k} s_j \in \text{coker}(i : K_1(A) \longrightarrow F^\bullet) .$$

(iii) If A is noetherian ring the homology modules $H_r(C)$ $(r \geq 0)$ of a finite f.g. free A-module chain complex C are f.g., and $H_*(F \otimes_A C) = 0$ if and only if each $H_r(C)$ is an S-torsion A-module. If A is a noetherian unique factorization domain the relative Reidemeister torsion of a finite f.g. free A-module chain complex C with $H_*(F \otimes_A C) = 0$ is given by the formula

$$\Delta(C) = \prod_{r \geq 0} \text{ord}(H_r(C))^{(-)^r} \in \text{coker}(i : K_1(A) \longrightarrow F^\bullet) . \qquad \square$$

Example 16.7 For any ring A let

$$i : A[z, z^{-1}] \longrightarrow B = \Omega^{-1}A[z, z^{-1}]$$

be the inclusion of $A[z, z^{-1}]$ in the Fredholm localization (13.3). A finitely dominated $A[z, z^{-1}]$-module chain complex C is $\Omega^{-1}A[z, z^{-1}]$-acyclic if and only if it is A-finitely dominated, in which case the relative Reidemeister torsion of C is the automorphism class of Chap. 13

$$\begin{aligned}\Delta(C) &= [C, \zeta] \\ &\in \operatorname{coker}(i : K_1(A[z,z^{-1}]) \longrightarrow K_1(\Omega^{-1}A[z,z^{-1}])) = \operatorname{Aut}_0(A) .\end{aligned}$$

(This is also a special case of 16.5). □

Example 16.8 The Laurent polynomial extension $F[z,z^{-1}]$ of a field F is a principal ideal domain with quotient $F(z)$ the field of rational functions. A f.g. $F[z,z^{-1}]$-module M is torsion if and only if $\dim_F(M) < \infty$, in which case M is a finite direct sum of cyclic $F[z,z^{-1}]$-modules

$$M = \bigoplus_{j=1}^{k} F[z,z^{-1}]/(s_j(z))$$

with

$$s_j(z) \neq 0 \in F[z,z^{-1}] \ , \ s_{j+1}(z) \,|\, s_j(z) \ ,$$

$$\dim_F(M) = \sum_{j=1}^{k} \deg(s_j(z)) < \infty \ ,$$

$$\operatorname{ord}(M) = \prod_{j=1}^{k} s_j(z) = \operatorname{ch}_z(M, \zeta) \in F[z,z^{-1}] \ .$$

A finite f.g. free $F[z,z^{-1}]$-module chain complex C is $F(z)$-contractible if and only if the homology F-vector spaces $H_r(C)$ $(r \geq 0)$ are finite dimensional, in which case the relative Reidemeister torsion is given by

$$\begin{aligned}\Delta(C) &= \prod_{r\geq 0} \operatorname{ord}(H_r(C))^{(-)^r} = \prod_{r\geq 0} \operatorname{ch}_z(H_r(C), z)^{(-)^r} \\ &\in \operatorname{coker}(K_1(F[z,z^{-1}]) \longrightarrow K_1(F(z))) \\ &= F(z)^\bullet/\{uz^n \,|\, u \in F^\bullet, n \in \mathbb{Z}\}\end{aligned}$$

as in Milnor [195] and Turaev [293]. This is also the automorphism torsion of C (16.7). See Chap. 19 below for an elaboration of this example. □

17. Alexander polynomials

In Chap. 17 it is assumed that the ground ring A is an integral domain, with quotient field F.

This chapter only deals with the algebraic properties of Alexander polynomials – see Chap. 33E for the applications to knot theory.

The Alexander polynomials of an n-dimensional A-contractible finite f.g. projective $A[z,z^{-1}]$-module chain complex E

$$\Delta_r(z) \in A[z] \quad (r \geq 0)$$

are chain homotopy invariants such that

$$\Delta_r(1) = 1 \ , \ \Delta_r(0) \neq 0 \ , \ \Delta_r(z)H_r(E) = 0$$

with $\Delta_r(z) = 1$ for $r \geq n$. The multiplicative subset

$$P = \{p(z) \in A[z,z^{-1}] \,|\, p(1) \in A^\bullet\} \subset A[z,z^{-1}]$$

has the universal property that a finite f.g. projective $A[z,z^{-1}]$-module chain complex E is A-contractible if and only if E is $P^{-1}A[z,z^{-1}]$-contractible. Recall that the Witt vector groups $W(A), \widetilde{W}(A)$ of Chap. 14 are defined by

$$W(A) = \{\frac{p(z)}{q(z)} \,|\, p(z), q(z) \in A[z], p(0) = q(0) = 1\} \subset F(z)^\bullet ,$$

$$\widetilde{W}(A) = W(A)/\{\det(1 - z : P[z] \longrightarrow P[z])\}$$

for f.g. projective A-modules P. The two main results of Chap. 17 are that for an A-contractible E:

(i) the Alexander polynomials of E determine the P-torsion K-theory class

$$\tau_P(E) = \prod_{r=0}^{\infty} \Delta_r(1-z)^{(-)^r} \in K_1(A[z,z^{-1}],P) = \widetilde{W}(A) ,$$

which is sent by the injection $\widetilde{W}(A) \subseteq \widetilde{W}(F)$ to the Reidemeister torsion

$$[\tau_P(E)] = \Delta(F \otimes_A E) \in \widetilde{W}(F) = (F(z)^\bullet/F^\bullet)/\{(1-z)\} .$$

(ii) E is A-finitely dominated if and only if the extreme coefficients of the Alexander polynomials $\Delta_r(z)$ are units in A.

The Alexander polynomials of knot theory arise in the case $A = \mathbb{Z}$. The criterion for finite domination is an abstraction of the result that a high-dimensional knot k is fibred if and only if the extreme coefficients of the Alexander polynomials of k are units in $\mathbb{Z}$ (i.e. ± 1).

A finite f.g. projective $A[z,z^{-1}]$-module chain complex E is A-contractible if and only if the A-module morphisms $1-\zeta : H_*(E) \longrightarrow H_*(E)$ are isomorphisms, in which case the induced F-module morphisms are also isomorphisms

$$1-\zeta \; : \; H_*(E;F) \xrightarrow{\simeq} H_*(E;F) \; .$$

Definition 17.1 The *Alexander polynomials* of an A-contractible finite f.g. projective $A[z,z^{-1}]$-module chain complex E are

$$\begin{aligned}\Delta_r(z) \;&=\; \det((z-\zeta)(1-\zeta)^{-1} : H_r(E;F)[z] \longrightarrow H_r(E;F)[z]) \\ &\in F[z] \;\; (r \geq 0) \; ,\end{aligned}$$

with

$$H_*(E;F) \;=\; H_*(F \otimes_A E) \;=\; H_*(F[z,z^{-1}] \otimes_{A[z,z^{-1}]} E) \; . \qquad \square$$

The Alexander polynomials $\Delta_r(z)$ are in fact defined in $A[z] \subset F[z]$:

Proposition 17.2 *Let $Q \subset A[z]$ be the multiplicative subset of monic polynomials, so that as in 10.10 (iii) the localization of $A[z]$ inverting Q is the Fredholm localization of $A[z]$*

$$Q^{-1}A[z] \;=\; \Omega_+^{-1}A[z] \; .$$

(i) *The Q-torsion class of a finite chain complex (D,f) in* $\mathrm{End}(A)$ *is given by*

$$\begin{aligned}\tau_Q(D,f) \;&=\; \Big(\sum_{r=0}^{\infty} (-)^r [D_r] \, , \, \prod_{r=0}^{\infty} \det(1 - zf : D_r[z] \longrightarrow D_r[z])^{(-)^r} \Big) \\ &\in K_1(A[z],Q) \;=\; \mathrm{End}_0(A) \;=\; K_0(A) \oplus W(A) \; .\end{aligned}$$

(ii) *The ring morphism*

$$A[z] \longrightarrow A[z,z^{-1},(1-z)^{-1}] \; ; \; z \longrightarrow (1-z)^{-1}$$

induces an exact functor

$$\begin{aligned}\mathrm{End}(A) \;&=\; \mathbb{H}(A[z],Q) \longrightarrow \mathbb{H}(A[z,z^{-1},(1-z)^{-1}],P) \;=\; \mathbb{H}(A[z,z^{-1}],P) \; ; \\ &(D,f) \longrightarrow \mathrm{coker}(1 - f + zf : D[z,z^{-1}] \longrightarrow D[z,z^{-1}])\end{aligned}$$

and a morphism of torsion K-groups

$$\begin{aligned}&K_1(A[z],Q) \;=\; \mathrm{End}_0(A) \;=\; K_0(A) \oplus W(A) \longrightarrow \\ &K_1(A[z,z^{-1},(1-z)^{-1}],P) \;=\; K_1(A[z,z^{-1}],P) \;=\; \widetilde{W}(A) \; ; \\ &\qquad \tau_Q(D,f) \longrightarrow \tau_P(E)\end{aligned}$$

with

$$E = \mathcal{C}(1-f+zf : D[z,z^{-1}] \longrightarrow D[z,z^{-1}])$$

an A-contractible finite f.g. projective $A[z,z^{-1}]$-module chain complex and

$$\begin{aligned}\tau_P(E) &= \prod_{r=0}^{\infty} \det(1-zf : D_r[z] \longrightarrow D_r[z])^{(-)^r} \\ &= \prod_{r=0}^{\infty} \det(1-zf : H_r(D;F)[z] \longrightarrow H_r(D;F)[z])^{(-)^r} \\ &= \prod_{r=0}^{\infty} \det(1-z(1-\zeta)^{-1} : H_r(E;F)[z] \longrightarrow H_r(E;F)[z])^{(-)^r} \\ &= \prod_{r=0}^{\infty} \Delta_r(1-z)^{(-)^r} \in \widetilde{W}(A) .\end{aligned}$$

(iii) *Every A-contractible finite f.g. projective $A[z,z^{-1}]$-module chain complex E is chain equivalent to $\mathcal{C}(1-f+zf : D[z,z^{-1}] \longrightarrow D[z,z^{-1}])$ for some finite chain complex (D,f) in* End(A) *(10.20). If*

$$\det(1-f+zf : H_r(D;F)[z] \longrightarrow H_r(D;F)[z]) = z^{n_r}(\sum_{j=0}^{m_r} a_{j,r}z^j) \in A[z]$$

$$(a_{0,r} \neq 0 \ , \ \sum_{j=0}^{m_r} a_{j,r} = 1 \in A)$$

then the Alexander polynomials of E are given by

$$\Delta_r(z) = \sum_{j=0}^{m_r} a_{j,r}z^j \in A[z] .$$

(iv) *If A is a Dedekind ring then for (D,f), E as in (iii) the A-modules $H_*(D)$ are finitely generated, and the torsion-free quotients*

$$L_r = H_r(D)/\text{torsion}$$

are f.g. projective A-modules, with the Alexander polynomials of E given by

$$\Delta_r(z) = z^{-n_r}\det(1-f+zf : L_r[z] \longrightarrow L_r[z]) \in A[z] \subset F[z] \quad (r \geq 0)$$

(n_r as in (iii)*).*

(v) *If A is a Dedekind ring and the A-modules $H_*(E)$ are finitely generated*[11] *then the torsion-free quotients*

$$M_r = H_r(E)/\text{torsion}$$

[11]if and only if E is A-finitely dominated (17.8)

are f.g. projective A-modules, with the Alexander polynomials of E given by

$$\begin{aligned} \Delta_r(z) &= \det((z-\zeta)(1-\zeta)^{-1} : M_r[z] \longrightarrow M_r[z]) \\ &\in A[z] \subset F[z] \quad (r \geq 0) \ . \end{aligned}$$

Proof (i)+(ii) Immediate from 14.16 (iv).
(iii) The polynomial

$$p_r(z) = \det(1 - f + zf : H_r(D;F)[z] \longrightarrow H_r(D;F)[z]) \in F[z]$$

is such that $p_r(1) = 1$, so that $p_r(z) \neq 0$. Moreover, $p_r(z) \in A[z] \subset F[z]$, since it is possible to choose a basis for $H_r(D;F)$ in $\mathrm{im}(H_r(D) \longrightarrow H_r(D;F))$, and the matrix of $1-f+zf$ with respect to the corresponding basis for $H_r(D;F)[z]$ has entries in $A[z]$. The connecting maps in the homology exact sequence

$$\begin{aligned} \dots \longrightarrow H_r(D;F)[z,z^{-1}] &\xrightarrow{1-f+zf} H_r(D;F)[z,z^{-1}] \\ &\longrightarrow H_r(E;F) \xrightarrow{\partial} H_{r-1}(D;F)[z,z^{-1}] \longrightarrow \dots \end{aligned}$$

are $\partial = 0$, and there is defined an exact sequence in $\mathrm{End}(F)$

$$0 \longrightarrow (C_r, g) \longrightarrow (H_r(D;F), f) \longrightarrow (H_r(E;F), (1-\zeta)^{-1}) \longrightarrow 0$$

with

$$C_r = \{x \in H_r(D;F) \,|\, (f^2 - f)^k(x) = 0 \text{ for some } k \geq 0\} \ , \quad g = f| \ .$$

For any indeterminate s over A

$$\mathrm{ch}_s(C_r, g) = s^{\ell_r - m_r - n_r}(s-1)^{n_r} \in A[s] \ ,$$

with

$$\begin{aligned} \ell_r &= \dim_F H_r(D;F) \ , \quad m_r = \dim_F H_r(E;F) \ , \\ n_r &= \dim_F\Big(\bigcup_{k=0}^{\infty} \ker((1-f)^k : H_r(D;F) \longrightarrow H_r(D;F))\Big) \end{aligned}$$

and

$$\mathrm{ch}_s(H_r(D;F), f) = \mathrm{ch}_s(C_r, g)\mathrm{ch}_s(H_r(E;F), (1-\zeta)^{-1}) \in A[s] \ .$$

Substituting $s = (1-z)^{-1}$ and multiplying by $(1-z)^{\ell_r}$ gives

$$\begin{aligned} &\det(1 - f + zf : H_r(D;F)[z] \longrightarrow H_r(D;F)[z]) \\ &\quad = \det(1 - g + zg : C_r[z] \longrightarrow C_r[z]) \\ &\qquad\quad \det((z-\zeta)(1-\zeta)^{-1} : H_r(E;F)[z] \longrightarrow H_r(E;F)[z]) \ , \\ &\det(1 - g + zg : C_r[z] \longrightarrow C_r[z]) = z^{n_r} \in A[z] \ . \end{aligned}$$

(iv) By the universal coefficient theorem

$$H_r(D;F) = F \otimes_A L_r ,$$

and by 11.6 (iii)

$$\begin{aligned}\det(1-f+zf : H_r(D;F)[z] \longrightarrow H_r(D;F)[z]) \\ = \det(1-f+zf : L_r[z] \longrightarrow L_r[z]) \in A[z] \subset F[z] .\end{aligned}$$

(v) By the universal coefficient theorem

$$H_r(E;F) = F \otimes_A M_r ,$$

and by 11.6 (iii)

$$\begin{aligned}\det((z-\zeta)(1-\zeta)^{-1} : H_r(E;F)[z] \longrightarrow H_r(E;F)[z]) \\ = \det((z-\zeta)(1-\zeta)^{-1} : M_r[z] \longrightarrow M_r[z]) \in A[z] \subset F[z] .\end{aligned}$$ □

Remark 17.3 Let E be an A-contractible finite f.g. projective $A[z,z^{-1}]$-module chain complex, as in 17.2.
(i) The Laurent polynomial extension $F[z,z^{-1}]$ is a principal ideal domain and each $H_r(E;F)$ is a f.g. $F[z,z^{-1}]$-module such that

$$H_r(E;F) = \sum_{j=1}^{k_r} F[z,z^{-1}]/(\lambda_{j,r}(z))$$

for some polynomials $\lambda_{j,r}(z) \in F[z,z^{-1}]$ with $\lambda_{j,r}(1) = 1 \in F$. The Alexander polynomial $\Delta_r(z) \in A[z,z^{-1}]$ is a generator of the order ideal (16.6)

$$\begin{aligned} o(H_r(E;F)) &= \{p(z) \in F[z,z^{-1}] \,|\, p(z)H_r(E;F) = 0\} \\ &= \prod_{j=1}^{k_r}(\lambda_{j,r}(z)) = (\prod_{j=1}^{k_r}\lambda_{j,r}(z)) \lhd F[z,z^{-1}] \end{aligned}$$

(Levine [156], Milnor [195]).
(ii) It is necessary to define the Alexander polynomials $\Delta_*(z)$ using $H_*(E;F)$ instead of $H_*(E)$, since the F-modules $H_*(E;F)$ are finitely generated and in general the A-modules $H_*(E)$ are not finitely generated – see (iii) below for an explicit example.
(iii) Let $A = \mathbb{Z}$, $F = \mathbb{Q}$ and let E be the $\mathbb{Z}$-contractible finite f.g. free $\mathbb{Z}[z,z^{-1}]$-module chain complex

$$E : \ldots \longrightarrow 0 \longrightarrow \mathbb{Z}[z,z^{-1}] \xrightarrow{2z-1} \mathbb{Z}[z,z^{-1}] .$$

In this case $H_0(E) = \mathbb{Z}[1/2]$ is an infinitely generated free $\mathbb{Z}$-module. The $\mathbb{Q}$-coefficient homology

$$H_0(E;\mathbb{Q}) = \mathbb{Q}[z,z^{-1}]/(2z-1) = \mathbb{Q}$$

is a 1-dimensional $\mathbb{Q}$-vector space with $\zeta = 1/2 : \mathbb{Q} \longrightarrow \mathbb{Q}$, and the Alexander polynomial is $\Delta_0(z) = 2z - 1$. □

Proposition 17.4 *The Alexander polynomials of an A-contractible finite f.g. projective $A[z,z^{-1}]$-module chain complex E*

$$\Delta_r(z) = \sum_{j=0}^{m_r} a_{j,r} z^j \in A[z] \quad (r \geq 0)$$

are such that

$$\begin{aligned}
m_r &= \dim_F H_r(E;F)\ ,\\
\mathrm{ch}_z(H_r(E;F),\zeta) &= \det(1-\zeta)\Delta_r(z) \in A[z]\ ,\\
\mathrm{ch}_z(H_r(E;F),(1-\zeta)^{-1}) &= z^{m_r}\Delta_r(1-z^{-1})\\
&= \sum_{j=0}^{m_r} a_{j,r}(z-1)^j z^{m_r-j} \in A[z]\ ,\\
\mathrm{ch}_z(H_r(E;F),-\zeta(1-\zeta)^{-1}) &= (z-1)^{m_r}\Delta_r(z(z-1)^{-1})\\
&= \sum_{j=0}^{m_r} a_{j,r} z^j (z-1)^{m_r-j} \in A[z]\ ,\\
\Delta_r(0) &= a_{0,r} = \det(-\zeta(1-\zeta)^{-1} : H_r(E;F) \longrightarrow H_r(E;F)) \neq 0 \in A\ ,\\
a_{m_r,r} &= \det((1-\zeta)^{-1} : H_r(E;F) \longrightarrow H_r(E;F)) \neq 0 \in A\ ,\\
\Delta_r(z)H_r(E) &= 0\ ,\ \Delta_r(1) = \sum_{j=0}^{m_r} a_{j,r} = 1 \in A\ .
\end{aligned}$$

Proof Let (D,f) be a finite chain complex in $\mathrm{End}(A)$ with an $A[z,z^{-1}]$-module chain equivalence

$$\mathcal{C}(1-f+zf : D[z,z^{-1}] \longrightarrow D[z,z^{-1}]) \simeq E\ .$$

Define the finite f.g. projective $A[z]$-module chain complex

$$E^+ = \mathcal{C}(1-f+zf : D[z] \longrightarrow D[z])$$

such that

$$A[z,z^{-1}] \otimes_{A[z]} E^+ \simeq E\ .$$

Now

$$\det(1-f+zf : H_r(D;F)[z] \longrightarrow H_r(D;F)[z]) = z^{n_r} \sum_{j=0}^{m_r} a_{j,r} z^j \in A[z]$$

with

$$\begin{aligned} n_r &= \dim_F\big(\ker(\zeta_* : H_r(E^+;F)\longrightarrow H_r(E^+;F))\big) \\ &= \dim_F H_r(E^+;F) - \dim_F H_r(E;F) \geq 0 \end{aligned}$$

as in 14.19 (i), and as in 17.2 the Alexander polynomials of E are given by

$$\begin{aligned} \Delta_r(z) &= z^{-n_r}\det(1 - f + zf : H_r(D;F)[z]\longrightarrow H_r(D;F)[z]) \\ &= \sum_{j=0}^{m_r} a_{j,r} z^j \in A[z] \ . \end{aligned}$$

□

Proposition 17.5 *Let E be a 1-dimensional A-contractible f.g. free $A[z,z^{-1}]$-module chain complex*

$$E \ : \ \ldots \longrightarrow 0 \longrightarrow E_1 \xrightarrow{\ d\ } E_0$$

with

$$d \ = \ \sum_{j=0}^{m} d_j z^j \ : \ E_1 \ = \ A[z,z^{-1}]^k \longrightarrow E_0 \ = \ A[z,z^{-1}]^k$$

and

$$\det(d) \ = \ \sum_{j=n}^{mk} a_j z^j \in A[z] \ \ (a_n \neq 0 \in A) \ .$$

The Alexander polynomial of E is given by

$$\Delta_0(z) \ = \ z^{-n}\det(d(\sum_{j=0}^{m} d_j)^{-1} : E_0\longrightarrow E_0) \in A[z] \ .$$

Proof Define the A-module

$$D \ = \ \sum_0^m A^k \ .$$

The A-module endomorphisms

$$g \ = \ \begin{pmatrix} d_0 & 0 & 0 & \ldots & 0 \\ -d_1 & 1 & 0 & \ldots & 0 \\ d_2 & 0 & 1 & \ldots & 0 \\ \vdots & \vdots & \vdots & \ddots & \vdots \\ (-)^m d_m & 0 & 0 & \ldots & 1 \end{pmatrix} \ : \ D \longrightarrow D \ ,$$

$$h \ = \ \begin{pmatrix} 0 & 1 & 0 & \ldots & 0 \\ 0 & 0 & 1 & \ldots & 0 \\ 0 & 0 & 0 & \ldots & 0 \\ \vdots & \vdots & \vdots & \ddots & \vdots \\ 0 & 0 & 0 & \ldots & 0 \end{pmatrix} \ : \ D \longrightarrow D$$

are such that

$$g+zh = \begin{pmatrix} 1 & -z & z^2 & \dots & (-)^m z^m \\ 0 & 1 & 0 & \dots & 0 \\ 0 & 0 & 1 & \dots & 0 \\ \vdots & \vdots & \vdots & \ddots & \vdots \\ 0 & 0 & 0 & \dots & 1 \end{pmatrix}^{-1} \begin{pmatrix} d & 0 & 0 & \dots & 0 \\ 0 & 1 & 0 & \dots & 0 \\ 0 & 0 & 1 & \dots & 0 \\ \vdots & \vdots & \vdots & \ddots & \vdots \\ 0 & 0 & 0 & \dots & 1 \end{pmatrix}$$

$$\begin{pmatrix} 1 & 0 & 0 & \dots & 0 \\ d_1 & 1 & 0 & \dots & 0 \\ -d_2 & 0 & 1 & \dots & 0 \\ \vdots & \vdots & \vdots & \ddots & \vdots \\ (-)^{m+1}d_m & 0 & 0 & \dots & 1 \end{pmatrix}^{-1} : D \longrightarrow D .$$

(The matrix identity comes from the algebraic transversality of the Higman linearization trick – see Chap. 7 above, and Ranicki [244, 10.6]). Now

$$\det(g+h) = \det(\sum_{j=0}^{m} d_j) \in A^\bullet ,$$

so that $g+h : D \longrightarrow D$ is an automorphism. The endomorphism

$$f = (g+h)^{-1}h : D \longrightarrow D$$

is such that

$$\begin{aligned} &\mathcal{C}(1-f+zf : D[z,z^{-1}] \longrightarrow D[z,z^{-1}]) \\ &= \mathcal{C}((g+h)^{-1}(g+zh) : D[z,z^{-1}] \longrightarrow D[z,z^{-1}]) \\ &\simeq \mathcal{C}(d : A[z,z^{-1}]^k \longrightarrow A[z,z^{-1}]^k) = E . \end{aligned}$$

By 17.2 the Alexander polynomial of E is

$$\begin{aligned} \Delta_0(z) &= z^{-n}\det(1-f+zf : D[z] \longrightarrow D[z]) \\ &= z^{-n}\det(d(\sum_{j=0}^{m} d_j)^{-1} : E_0 \longrightarrow E_0) \in A[z] . \end{aligned}$$

□

Example 17.6 Given a polynomial

$$p(z) = z^n \sum_{j=0}^{m} a_j z^j \in A[z]$$

with $p(1) = 1 \in A$, $a_0, a_m \neq 0 \in A$, define

$$\begin{aligned} E^+ &= \mathcal{C}(p(z) : A[z] \longrightarrow A[z]) , \\ E &= A[z,z^{-1}] \otimes_{A[z]} E^+ = \mathcal{C}(p(z) : A[z,z^{-1}] \longrightarrow A[z,z^{-1}]) \end{aligned}$$

such that

$$\dim_F H_0(E^+;F) \;=\; m+n \;,\; \dim_F H_0(E;F) \;=\; m \;.$$

The Alexander polynomial of E is

$$\Delta_0(z) \;=\; z^{-n}p(z) \;=\; \sum_{j=0}^{m} a_j z^j \in A[z] \;. \qquad \square$$

In fact, 17.5 is the 1-dimensional case of:

Proposition 17.7 *Let E be a finite f.g. projective $A[z,z^{-1}]$-module chain complex which is A-contractible, so that E is $P^{-1}A[z,z^{-1}]$-contractible, with*

$$P \;=\; \{p(z) \in A[z,z^{-1}] \,|\, p(1) \in A^\bullet\} \subset A[z,z^{-1}] \;.$$

(i) *The boundary map ∂ in the localization exact sequence*

$$\dots \longrightarrow K_1(A[z,z^{-1}]) \longrightarrow K_1(P^{-1}A[z,z^{-1}]) \xrightarrow{\partial} K_1(A[z,z^{-1}],P) \longrightarrow K_0(A[z,z^{-1}]) \longrightarrow \dots$$

sends the torsion $\tau(P^{-1}E) \in K_1(P^{-1}A[z,z^{-1}])$ (with respect to arbitrary choice of bases for each E_r) to the P-torsion class (= Reidemeister torsion, by 16.5)

$$\tau_P(E) \;=\; \Delta(E) \;=\; \prod_{r=0}^{\infty} \Delta_r(1-z)^{(-)^r} \in K_1(A[z,z^{-1}],P) \;=\; \widetilde{W}(A) \;,$$

with

$$\begin{aligned}\Delta_r(1-z) \;&=\; \widetilde{\mathrm{ch}}_z(H_r(D;F),f) \\ &=\; \det(1-zf : H_r(D;F)[z] \longrightarrow H_r(D;F)[z]) \in W(A)\end{aligned}$$

for any finite chain complex (D,f) in $\mathrm{End}_0(A)$ such that

$$\mathcal{C}(1-f+zf : D[z,z^{-1}] \longrightarrow D[z,z^{-1}]) \;\simeq\; E \;.$$

(ii) *There exist $A[z,z^{-1}]$-module morphisms $\Gamma : E_r \longrightarrow E_{r+1}$ such that*

$$(d\Gamma+\Gamma d-1)(E_r) \subseteq (1-z)(E_r) \quad (r \geq 0)$$

and $d+\Gamma : E_{odd} \longrightarrow E_{even}$ is an A-isomorphism. If bases are chosen for each E_r such that

$$\det(1 \otimes (d+\Gamma) : A \otimes_{A[z,z^{-1}]} E_{odd} \longrightarrow A \otimes_{A[z,z^{-1}]} E_{even}) \;=\; 1 \in A$$

then for some $n \in \mathbb{Z}$

$$\begin{aligned}\tau(P^{-1}E) &= \tau(d+\Gamma : P^{-1}E_{odd} \longrightarrow P^{-1}E_{even}) \in K_1(P^{-1}A[z,z^{-1}]) \ ,\\ \det(d+\Gamma : E_{odd} \longrightarrow E_{even}) &= z^n \prod_{r=0}^{\infty} \Delta_r(z)^{(-)^r} \in P \subset A[z,z^{-1}] \ .\end{aligned}$$

□

The extreme coefficients of the Alexander polynomials determine if a chain complex is a band :

Proposition 17.8 *Let E be an A-contractible finite f.g. projective $A[z,z^{-1}]$-module chain complex with Alexander polynomials*

$$\Delta_r(z) = \sum_{j=0}^{m_r} a_{j,r} z^j \in A[z] \quad (a_{0,r}, a_{m_r,r} \neq 0 \in A, r \geq 0) \ .$$

(i) *The chain complex E is A-finitely dominated (i.e. E is a band) if and only if the extreme coefficients $a_{0,r}, a_{m_r,r} \in A$ are units.*
(ii) *If A is a Dedekind ring then E is A-finitely dominated if and only if the homology groups $H_r(E)$ $(r \geq 0)$ are finitely generated as A-modules.*
Proof (i) Define the multiplicative subset

$$S = \{\sum_{j=m}^{n} a_j z^j \,|\, a_m, a_n \in A \text{ are units}\} \subset A[z,z^{-1}] \ ,$$

so that as in 13.12 (i)

$$S^{-1}A[z,z^{-1}] = \Omega^{-1}A[z,z^{-1}] \ .$$

By 13.8 E is A-finitely dominated if and only if it is $S^{-1}A[z,z^{-1}]$-contractible. Thus E is A-finitely dominated if and only if $\Delta_r(z) \in S$ $(r \geq 0)$.
(ii) For any noetherian ring A the homology of a f.g. A-module chain complex consists of f.g. A-modules. Conversely, if A is Dedekind and the homology groups $H_r(E)$ are f.g. A-modules then (as in 17.3 (iii)) the Alexander polynomials are given by

$$\Delta_r(z) = \det((z-\zeta)(1-\zeta)^{-1} : M_r[z] \longrightarrow M_r[z]) \in A[z] \quad (r \geq 0)$$

with each $M_r = H_r(E)/\text{torsion}$ a f.g. projective A-module. The extreme coefficients of $\Delta_r(z)$ are units

$$\Delta_r(0) = \det(-\zeta(1-\zeta)^{-1} : M_r \longrightarrow M_r) \ , \ \det((1-\zeta)^{-1} : M_r \longrightarrow M_r) \in A^{\bullet}$$

so that (i) applies. □

Example 17.9 The 1-dimensional A-contractible f.g. free $A[z,z^{-1}]$-module chain complex of 17.6

$$E = \mathcal{C}(z^n \sum_{j=0}^{m} a_j z^j : A[z,z^{-1}] \longrightarrow A[z,z^{-1}])$$

is A-finitely dominated if and only if the coefficients $a_0, a_m \in A$ are units, in which case $E \simeq A^m$. □

Remark 17.10 Suppose that $A = \mathbb{Z}$. For any object (L, f) in $\mathrm{End}(\mathbb{Z})$ the $(\mathbb{Z}[z, z^{-1}], P)$-module

$$K = \mathrm{coker}(1 - f + zf : L[z, z^{-1}] \longrightarrow L[z, z^{-1}])$$

is a torsion-free abelian group (Crowell [58], Trotter [291, Lemma 2.1]). By 17.8 (ii) K is a finitely generated abelian group if and only if the extreme coefficients of the Alexander polynomial

$$\Delta(z) = \det(1 - f + zf : L[z] \longrightarrow L[z]) = \sum_{j=0}^{m} a_j z^j \in \mathbb{Z}[z]$$

are units $a_0, a_m \in \mathbb{Z}^\bullet = \{\pm 1\}$, in which case K is a f.g. free abelian group of rank m. □

The classic application of Alexander polynomials to high-dimensional knot complements is given by:

Example 17.11 (Milnor [193, Chap. 2], [194], [195])
(i) Let X be a connected finite CW complex with a homology equivalence $p : X \longrightarrow S^1$ (as in 10.23 (ii)). Let $\overline{X} = p^*\mathbb{R}$ be the pullback infinite cyclic cover, and let $\overline{p} : \overline{X} \longrightarrow \mathbb{R}$ be a lift of p to a $\mathbb{Z}$-equivariant map. The finite f.g. free $\mathbb{Z}[z, z^{-1}]$-module chain complex

$$E = \mathcal{C}(\overline{f} : C(\overline{X}) \longrightarrow C(\mathbb{R}))_{*+1}$$

is $\mathbb{Z}$-contractible, so that the Alexander polynomials of E are defined

$$\Delta_r(z) \in \mathbb{Z}[z] \ \ (r \geq 0)$$

with $\Delta_0(z) = 1$. If $(X_F; F, \zeta F)$ is a fundamental domain for $\overline{X}$ then the inclusions

$$g : \dot{C}(F) \longrightarrow \dot{C}(X_F) \ , \ h : \dot{C}(F) = \dot{C}(\zeta F) \longrightarrow \dot{C}(X_F)$$

are such that $g - h : \dot{C}(F) \longrightarrow \dot{C}(X_F)$ is a chain equivalence, with

$$\dot{C}(F) = \mathcal{C}(C(F) \longrightarrow C(\{0\}))_{*+1} \ ,$$
$$\dot{C}(X_F) = \mathcal{C}(C(X_F) \longrightarrow C([0,1]))_{*+1}$$

the reverse chain complexes of F and X_F. (If $\overline{X}$ is finitely dominated, e.g. if $p : X \longrightarrow S^1$ is the projection of a fibre bundle, and $F \simeq X_F \simeq \overline{X}$, then g, h are chain equivalences with $g^{-1}h \simeq \zeta : \dot{C}(F) \longrightarrow \dot{C}(F)$.) The chain map

$$f = (g - h)^{-1}(-h) : D = \dot{C}(F) \longrightarrow D = \dot{C}(F)$$

is such that

$$E \simeq \mathcal{C}(1-f+zf : D[z,z^{-1}] \longrightarrow D[z,z^{-1}])$$

as in 17.2. The abelian groups defined by

$$L_r = H_r(D)/\text{torsion} \quad (r \geq 1)$$

are f.g. free, with

$$\mathbb{Q} \otimes_{\mathbb{Z}} L_r = H_r(D;\mathbb{Q}) .$$

Let

$$P = \{p(z) \mid p(1) = \pm 1\} \subset \mathbb{Z}[z,z^{-1}] .$$

The Alexander polynomials of E are given by

$$\Delta_r(z) = \det(1-f+zf : L_r[z] \longrightarrow L_r[z]) \in P \quad (r \geq 1)$$

and the P-primary torsion is given by

$$\tau_P(E) = \sum_{r=1}^{\infty}(-)^r[L_r,f] \in K_1(\mathbb{Z}[z,z^{-1}],P) = \text{End}_0(\mathbb{Z})/\{(\mathbb{Z},0),(\mathbb{Z},1)\}$$

with $P = \{p(z) \in \mathbb{Z}[z,z^{-1}] \mid p(1) = \pm 1 \in \mathbb{Z}\}$. The isomorphism of 17.7

$$K_1(\mathbb{Z}[z,z^{-1}],P) \cong \widetilde{W}(\mathbb{Z}) = W(\mathbb{Z})/\{(1-z)\}$$

sends the P-primary torsion $\tau_P(E)$ to the Reidemeister torsion (16.5)

$$\begin{aligned}\Delta(X) &= \prod_{r=1}^{\infty} \Delta_r(1-z)^{(-)^r} \\ &= \prod_{r=1}^{\infty} \det(1-zf : L_r[z] \longrightarrow L_r[z])^{(-)^r} \\ &= \prod_{r=1}^{\infty} \det(1-zf : D_r[z] \longrightarrow D_r[z])^{(-)^r} \in \widetilde{W}(\mathbb{Z}) ,\end{aligned}$$

which is sent by the injection $\widetilde{W}(\mathbb{Z}) \subset \widetilde{W}(\mathbb{Q})$ to

$$\begin{aligned}\Delta(X;\mathbb{Q}) &= \prod_{r=1}^{\infty} \det(1-zf : H_r(X;\mathbb{Q})[z] \longrightarrow H_r(X;\mathbb{Q})[z])^{(-)^r} \\ &\in \widetilde{W}(\mathbb{Q}) = (\mathbb{Q}(z)^{\bullet}/\mathbb{Q}^{\bullet})/\{(1-z)\} .\end{aligned}$$

(ii) If $k : S^n \subset S^{n+2}$ is an n-knot with exterior

$$X = \text{cl.}(S^{n+2} \backslash k(S^n) \times D^2)$$

then any representative $p : X \longrightarrow S^1$ of the generator

$$1 \in [X,S^1] = H^1(X) = \mathbb{Z}$$

is a homology equivalence, so that (i) applies, with $F^{n+1} = p^{-1}(\text{pt.}) \subset S^{n+2}$ a Seifert surface. □

18. K-theory of Dedekind rings

The construction of Reidemeister torsion using relative K-groups in Chap. 16 is now combined with the algebraic K-theory exact sequence for a Dedekind ring A. In particular, the torsion projective class of a f.g. free A-module chain complex C is expressed in terms of the torsion of the homology A-modules $H_*(C)$ and the maximal ideal structure of A. The corresponding algebraic L-theoretic expression for a chain complex with Poincaré duality will be used in Chaps. 39–42 in the computation of the high-dimensional knot cobordism groups C_*.

Let A be a Dedekind ring. Write the quotient field of A as

$$F = S^{-1}A ,$$

with $S = A\backslash\{0\}$.

Remark 18.1 The following properties of a Dedekind ring A are well-known – see e.g. Milnor [199].
(i) An *ideal class* of A is an equivalence class of ideals $\mathfrak{I} \lhd A$, with

$$\mathfrak{I} \sim \mathfrak{J} \text{ if } x\mathfrak{I} = y\mathfrak{J} \text{ for some } x, y \in A .$$

The *ideal class group* $C(A)$ is the set of ideal classes of A, with the multiplication of ideals as the abelian group law. Every ideal is a product of maximal ideals (= non-zero prime ideals for Dedekind A), so that the ideal class group is generated by the ideal classes of maximal ideals. For every $\mathfrak{P} \in \max(A)$ there exist $r \in \mathfrak{P}$, $\mathfrak{Q} \in \max(A)$ such that

$$(r) = \mathfrak{P}\mathfrak{Q} \lhd A ,$$

with generators $p_1, p_2 \in \mathfrak{P}$ and $q_1, q_2 \in \mathfrak{Q}$ such that

$$p_1q_1 + p_2q_2 = r \in A .$$

Define

$$a = \frac{p_1q_1}{r} , \quad b = \frac{p_2q_1}{r} , \quad c = \frac{p_1q_2}{r} \in A$$

such that

$$a(1-a) = bc \in A .$$

The projections

$$p = \begin{pmatrix} a & b \\ c & 1-a \end{pmatrix}, \ 1-p = \begin{pmatrix} 1-a & -b \\ -c & a \end{pmatrix} : A \oplus A \longrightarrow A \oplus A$$

are such that there are defined A-module isomorphisms

$$\begin{aligned} \mathrm{im}(p) &\xrightarrow{\simeq} \mathcal{P} ; \ p(x,y) \longrightarrow p_1 x + p_2 y , \\ \mathrm{im}(1-p) &\xrightarrow{\simeq} \mathcal{Q} ; \ (1-p)(x,y) \longrightarrow q_1 x + q_2 y . \end{aligned}$$

Thus $\mathcal{P}$ is a f.g. projective A-module, with an A-module isomorphism

$$\mathcal{P} \oplus \mathcal{Q} \cong A \oplus A .$$

Moreover, if $F = (A\backslash\{0\})^{-1}A$ is the quotient field of A, there is defined an A-module isomorphism

$$\mathcal{Q} \xrightarrow{\simeq} \{x \in F \,|\, x\mathcal{P} \subseteq A\} ; \ q \longrightarrow \frac{q}{r} .$$

For a principal ideal $\mathcal{P} = (r)$ take $\mathcal{Q} = A$, $(p_1, p_2) = (r, 0)$, $(q_1, q_2) = (1, 0)$, so that

$$p = \begin{pmatrix} 1 & 0 \\ 0 & 0 \end{pmatrix}, \ 1-p = \begin{pmatrix} 0 & 0 \\ 0 & 1 \end{pmatrix} : A \oplus A \longrightarrow A \oplus A$$

with an A-module isomorphism

$$A \xrightarrow{\simeq} \mathcal{P} ; \ x \longrightarrow rx .$$

(ii) Every f.g. A-module M has homological dimension 1, with a f.g. projective A-module resolution of length ≤ 1

$$0 \longrightarrow P_1 \longrightarrow P_0 \longrightarrow M \longrightarrow 0 ,$$

and M is isomorphic to the direct sum of a finite number of copies of A, $A/\mathcal{P}^k$ ($\mathcal{P} \in \max(A)$, $k \geq 1$) and possibly also a single $\mathcal{Q} \in \max(A)$. M is projective if there are no summands of type $A/\mathcal{P}^k$. M is S-torsion if there are no summands of type $A, \mathcal{Q}$. The ideal class group of A is isomorphic to the reduced projective class group of A, with an isomorphism

$$C(A) \xrightarrow{\simeq} \widetilde{K}_0(A) ; \ [\mathcal{P}] \longrightarrow [\mathcal{P}] .$$

The category of (A, S)-modules is

$$\mathbb{H}(A, S) = \{\text{f.g. } S\text{-torsion } A\text{-modules}\} ,$$

and there is defined an isomorphism

$$\begin{aligned}\mathbb{Z}[\max(A)] &\xrightarrow{\simeq} K_1(A,S) \;=\; K_0(\mathbb{H}(A,S))\;;\\ \sum_{\mathcal{P}} n_{\mathcal{P}}[\mathcal{P}] &\longrightarrow \sum_{\mathcal{P}}(\mathrm{sign}\, n_{\mathcal{P}})\big[(A/\mathcal{P})^{|n_{\mathcal{P}}|}\big] \;=\; \sum_{\mathcal{P}}(\mathrm{sign}\, n_{\mathcal{P}})\big[A/(\mathcal{P}^{|n_{\mathcal{P}}|})\big]\;.\end{aligned}$$

The ideal class group fits into an exact sequence

$$F^{\bullet} \longrightarrow \mathbb{Z}[\max(A)] \longrightarrow C(A) \longrightarrow 0$$

with

$$\begin{aligned}&\mathbb{Z}[\max(A)] \longrightarrow C(A)\;;\; m[\mathcal{P}] \longrightarrow [\mathcal{P}^m]\;,\\ &F^{\bullet} \longrightarrow K_0(\mathbb{H}(A,S)) \;=\; \mathbb{Z}[\max(A)]\;;\; a/s \longrightarrow [A/(a)] - [A/(s)]\;.\end{aligned}$$

(iii) For each $\mathcal{P} \in \max(A)$ let $\mathbb{H}(A,\mathcal{P}^{\infty})$ be the full subcategory of $\mathbb{H}(A,S)$ consisting of the $\mathcal{P}$-primary f.g. torsion A-modules. Let $\pi \in \mathcal{P}\backslash\mathcal{P}^2$ be a uniformizer, and define the multiplicative subset

$$\mathcal{P}^{\infty} \;=\; \{\pi^k \,|\, k \geq 0\} \subset A\;.$$

The localization $(\mathcal{P}^{\infty})^{-1}A = A[1/\pi]$ fits into a cartesian square

$$\begin{array}{ccc} A & \longrightarrow & A[1/\pi] \\ \big\downarrow & & \big\downarrow \\ A_{\mathcal{P}} & \longrightarrow & F \end{array}$$

with $A_{\mathcal{P}} = (A\backslash\mathcal{P})^{-1}A$ the local ring of A at $\mathcal{P}$. The following conditions on an (A,S)-module M are equivalent:

(a) M is $\mathcal{P}$-primary, i.e. an $(A,\mathcal{P}^{\infty})$-module,
(b) $M[1/\pi] \;=\; 0$,
(c) $M = A_{\mathcal{P}} \otimes_A M$,
(d) the annihilator of M is a power of $\mathcal{P}$

$$\mathrm{ann}(M) \;=\; \{a \in A \,|\, ax = 0 \in M \text{ for all } x \in M\} \;=\; \mathcal{P}^k$$

for some $k \geq 0$.

(iv) The ring of algebraic integers A in an algebraic number field F is a Dedekind ring with finite ideal class group $C(A) = \widetilde{K}_0(A)$. □

Proposition 18.2 *Let A be a Dedekind ring, and let $\mathcal{P} \lhd A$ be a maximal ideal with uniformizer π.*
(i) *The relative K-groups in the localization exact sequence*

$$\ldots \longrightarrow K_n(A) \longrightarrow K_n(A[1/\pi]) \longrightarrow K_n(A,\mathcal{P}^{\infty}) \longrightarrow K_{n-1}(A) \longrightarrow \ldots$$

are such that

$$K_*(A,\mathcal{P}^{\infty}) \;=\; K_{*-1}(A/\mathcal{P})\;,$$

with $A/\mathcal{P}$ the residue class field.
(ii) *Every $(A,\mathcal{P}^\infty)$-module M has a direct sum decomposition*

$$M = \sum_{k=1}^{\infty} M_k$$

such that $\operatorname{ann}(M_k) = \mathcal{P}^k$ *and M_k is a f.g. free $A/\mathcal{P}^k$-module. The $\mathcal{P}$-primary class invariant of M is given by*

$$\begin{aligned}[M] &= \sum_{j=0}^{\infty} \dim_{A/\mathcal{P}}(\mathcal{P}^j M/\mathcal{P}^{j+1}M) \\ &= \sum_{k=1}^{\infty} k \dim_{A/\mathcal{P}^k}(M_k) \in K_1(A,\mathcal{P}^\infty) = K_0(A/\mathcal{P}) = \mathbb{Z} .\end{aligned}$$

(iii) *The projective class $[C] \in K_0(A)$ of a finite f.g. projective A-module chain complex C with $\mathcal{P}$-primary homology $H_*(C)$ is the image of the $\mathcal{P}^\infty$-torsion class invariant*

$$\chi_{\mathcal{P}}(C) = \sum_{r=0}^{\infty} (-)^r [H_r(C)]_{\mathcal{P}} \in K_1(A,\mathcal{P}) = K_0(A/\mathcal{P}) = \mathbb{Z} .$$

Proof Localization at $\mathcal{P}$ defines an isomorphism of exact categories

$$\mathbb{H}(A,\mathcal{P}^\infty) \xrightarrow{\simeq} \mathbb{H}(A_{\mathcal{P}},\mathcal{P}^\infty) ; M \longrightarrow M = A_{\mathcal{P}} \otimes_A M ,$$

allowing every f.g. $\mathcal{P}$-primary A-module M to be regarded as a module over the local ring $A_{\mathcal{P}}$. Every f.g. projective module over a local ring is free, and matrices can be diagonalized, so that M has a f.g. free $A_{\mathcal{P}}$-module resolution

$$0 \longrightarrow A_{\mathcal{P}}^n \xrightarrow{d} A_{\mathcal{P}}^n \longrightarrow M \longrightarrow 0$$

with

$$d = \begin{pmatrix} \pi^{i_1} & 0 & \dots & 0 \\ 0 & \pi^{i_2} & \dots & 0 \\ \vdots & \vdots & \ddots & \vdots \\ 0 & 0 & \dots & \pi^{i_n} \end{pmatrix} : A_{\mathcal{P}}^n \longrightarrow A_{\mathcal{P}}^n$$

for some $i_1, i_2, \dots, i_n \geq 1$. Thus M has a direct sum decomposition of the form

$$M = \sum_{k=1}^{\infty} M_k$$

with $\operatorname{ann}(M_k) = \mathcal{P}^k$ and M_k a f.g. free $A/\mathcal{P}^k$-module. Every $(A,\mathcal{P}^\infty)$-module M has a finite filtration

$$\{0\} = \mathcal{P}^k M \subset \mathcal{P}^{k-1} M \subset \dots \subset \mathcal{P}M \subset M$$

such that the successive quotients $\mathcal{P}^j M/\mathcal{P}^{j+1}M$ are finite dimensional vector spaces over $A/\mathcal{P}$, with

$$\begin{aligned}\dim_{A/\mathcal{P}}(\mathcal{P}^j M/\mathcal{P}^{j+1}M) &= \sum_{k=j+1}^{\infty} \dim_{A/\mathcal{P}}(\mathcal{P}^j M_k/\mathcal{P}^{j+1}M_k) \\ &= \sum_{k=j+1}^{\infty} \dim_{A/\mathcal{P}^k}(M_k) .\end{aligned}$$

The embedding

$$\mathbb{P}(A/\mathcal{P}) \longrightarrow \mathbb{H}(A,\mathcal{P}^\infty) \; ; \; M_1 \longrightarrow M_1$$

induces devissage isomorphisms in algebraic K-theory, with

$$K_*(A,\mathcal{P}^\infty) \;=\; K_{*-1}(A/\mathcal{P})$$

(Bass [13, pp. 405,702] for $* = 1$, Quillen [225, p. 105] in general). □

Proposition 18.3 *Given a Dedekind ring A and a subset $J \subseteq \max(A)$ let $\mathcal{P}_j$ be the maximal ideal indexed by $j \in J$, with uniformizer $\pi_j \in \mathcal{P}_j$, and define the multiplicative subset*

$$S_J \;=\; \{\pi_1^{k_1}\pi_2^{k_2}\dots\pi_m^{k_m} \,|\, k_1,k_2,\dots,k_m \geq 1\} \subset A .$$

(i) *The relative terms in the algebraic K-theory localization exact sequence*

$$\dots \longrightarrow K_n(A) \longrightarrow K_n(S_J^{-1}A) \longrightarrow K_n(A,S_J) \longrightarrow K_{n-1}(A) \longrightarrow \dots$$

are such that

$$K_n(A,S_J) \;=\; \sum_{j\in J} K_{n-1}(A/\mathcal{P}_j) \;\; (n \geq \mathbb{Z}) .$$

(ii) *An $S_J^{-1}A$-contractible finite f.g. projective A-module chain complex C has an S_J-torsion class invariant*

$$\chi_J(C) \;=\; \sum_{j\in J}\chi_j(C)[j] \in K_1(A,S_J) \;=\; \sum_{j\in J} K_0(A/\mathcal{P}_j) \;=\; \mathbb{Z}[J] ,$$

with

$$\chi_j(C) \;=\; \chi_{\mathcal{P}_j}(H_*(C)_{\mathcal{P}_j}) \in K_1(A,\mathcal{P}_j^\infty) \;=\; K_0(A/\mathcal{P}_j) \;=\; \mathbb{Z}$$

the $\mathcal{P}_j^\infty$-torsion class invariant. The projective class of C is the image of the S_J-torsion class invariant

$$[C] \;=\; \sum_{j\in J}\chi_j(C) \in \operatorname{im}(\sum_{j\in J} K_1(A,\mathcal{P}_j^\infty) \longrightarrow K_0(A)) .$$

(iii) *For an $S_J^{-1}A$-contractible finite f.g. projective A-module chain complex C with $[C] = 0 \in K_0(A)$ a choice of round finite structure on C determines a lift of $\chi_J(C) \in K_1(A, S_J)$ to $\tau(S_J^{-1}C) \in K_1(S_J^{-1}A)$, so that as in 16.3 the Reidemeister torsion of C is given by*

$$\begin{aligned}\Delta(C) &= [\tau(S_J^{-1}C)] = \chi_J(C)\\ &\in \mathrm{coker}(K_1(A)\longrightarrow K_1(S_J^{-1}A)) = \ker(K_1(A,S_J)\longrightarrow K_0(A)) .\end{aligned}$$

Proof (i) Every (A, S_J)-module M has a $\mathcal{P}$-primary decomposition

$$M = \sum_{j\in J} A_{\mathcal{P}_j} \otimes_A M ,$$

so that

$$\mathbb{H}(A, S_J) = \coprod_{j\in J} \mathbb{H}(A, \mathcal{P}_j^\infty) ,$$

$$K_*(A, S_J) = K_{*-1}(\mathbb{H}(A, S_J)) = \sum_{j\in J} K_{*-1}(A, \mathcal{P}_j^\infty) = \sum_{j\in J} K_{*-1}(A/\mathcal{P}_j) .$$

The relative K-groups $K_*(A, S_J)$ are such that

$$K_*(A, S_J) = K_{*-1}(\mathbb{H}(A, S_J)) = \sum_{j\in J} K_{*-1}(A/\mathcal{P}_j) .$$

(ii)+(iii) Apply 18.2 at each of the maximal ideals $\mathcal{P}_j \lhd A$ ($j \in J$), summing all the contributions. □

Example 18.4 In the maximal case $J = \max(A)$ the localization

$$S_J^{-1}A = (A\backslash\{0\})^{-1}A = F$$

is the quotient field of A, and the localization exact sequence is

$$\dots \longrightarrow K_n(A) \longrightarrow K_n(F) \longrightarrow \sum_{\mathcal{P}\in\max(A)} K_{n-1}(A/\mathcal{P}) \longrightarrow K_{n-1}(A) \longrightarrow \dots .$$

□

Proposition 18.5 *Let A be a Dedekind ring, and let $T_A \subset A[z]$ be the multiplicative subset of monic polynomials*

$$p(z) = \sum_{j=0}^{d} a_j z^j \in A[z] \quad (a_d = 1) .$$

The reverse multiplicative subset $\widetilde{T}_A \subset A[z]$ consists of the polynomials

$$\widetilde{p}(z) = \sum_{j=0}^{d} \widetilde{a}_j z^j \in A[z]$$

with constant coefficient $\widetilde{a}_0 = 1 \in A$ (i.e. the reverse polynomials).

(i) *The localization* $T_A^{-1}A[z]$ *is a Dedekind ring with*

$$\begin{aligned}K_{n+1}(T_A^{-1}A[z]) &= K_{n+1}(A[z]) \oplus \mathrm{End}_n(A)\ ,\\ K_{n+1}(\widetilde{T}_A^{-1}A[z]) &= K_{n+1}(A) \oplus \widetilde{\mathrm{End}}_n(A)\ .\end{aligned}$$

(ii) *For any multiplicative subset* $S \subset A$ *the exact functor*

$$S^{-1}\ :\ \mathrm{End}(A) \longrightarrow \mathrm{End}^{T_A}(S^{-1}A)\ ;\ (P,f) \longrightarrow S^{-1}(P,f)$$

is an embedding of the endomorphism category of A *as a subcategory of the* T_A*-primary endomorphism category of* $S^{-1}A$*. An object* (Q,g) *in* $\mathrm{End}(S^{-1}A)$ *is isomorphic to* $S^{-1}(P,f)$ *for an object* (P,f) *in* $\mathrm{End}(A)$ *if and only if* $[Q] \in \mathrm{im}(\widetilde{K}_0(A)\longrightarrow\widetilde{K}_0(S^{-1}A))$ *and* (Q,g) *is* T_A*-primary, i.e. such that* $Q = S^{-1}P$ *for some f.g. projective* A*-module and*

$$\mathrm{ch}_z(Q,g) \in A[z] \subset S^{-1}A[z]\ .$$

The localization map of reduced endomorphism class groups

$$S^{-1}\ :\ \widetilde{\mathrm{End}}_0(A) \longrightarrow \widetilde{\mathrm{End}}_0^{T_A}(S^{-1}A)\ ;\ [P,f] \longrightarrow S^{-1}[P,f]$$

is an isomorphism.

Proof (i) See Lam [143, IV.1.3] for a proof that $T_A^{-1}A[z]$ is a Dedekind ring. The expressions for $K_*(T_A^{-1}A[z])$, $K_*(\widetilde{T}_A^{-1}A[z])$ are given by 12.13, 12.20.
(ii) The characteristic polynomial of an object (P,f) in $\mathrm{End}(A)$ is a monic polynomial $p(z) = \mathrm{ch}_z(P,f) \in T_A$ such that $p(f) = 0$ (11.12), so that $S^{-1}(P,f)$ is an object in $\mathrm{End}^{T_A}(S^{-1}A)$.

Conversely, given an object (Q,g) in $\mathrm{End}^{T_A}(S^{-1}A)$ let $p(z) \in T_A$ be such that $p(g) = 0 : Q\longrightarrow Q$. For any f.g. projective A-module L such that $Q = S^{-1}L$ define a g-invariant A-submodule containing L

$$P\ =\ L + g(L) + \ldots + g^{d-1}(L) \subset Q\ =\ S^{-1}L$$

and let $f = g| : P\longrightarrow P$. Now P is a f.g. A-module which is torsion-free, and thus f.g. projective, so that (P,f) is an object in $\mathrm{End}(A)$ such that $S^{-1}(P,f) = (Q,g)$. The localization map $S^{-1} : \widetilde{\mathrm{End}}_0(A)\longrightarrow\widetilde{\mathrm{End}}_0^{T_A}(S^{-1}A)$ is an isomorphism by 14.16, with inverse given by

$$\widetilde{\mathrm{End}}_0^{T_A}(S^{-1}A) \longrightarrow \widetilde{\mathrm{End}}_0(A)\ ;\ [Q,g] \longrightarrow [P,f]\ .$$ □

Remark 18.6 The localization maps in reduced endomorphism K-theory

$$S^{-1}\ :\ \widetilde{\mathrm{End}}_*(A) \longrightarrow \widetilde{\mathrm{End}}_*^{T_A}(S^{-1}A)$$

are not isomorphisms in general. The localization exact sequence of 18.2

$$\begin{aligned}\ldots \longrightarrow K_{n+1}(T_A^{-1}A[z]) \longrightarrow K_{n+1}(S^{-1}T_A^{-1}A[z]) \longrightarrow K_{n+1}(T_A^{-1}A[z],S)\\ \longrightarrow K_n(T_A^{-1}A[z]) \longrightarrow \ldots\end{aligned}$$

is a direct sum of the localization exact sequences

$$\begin{aligned}
\ldots \longrightarrow K_{n+1}(A[z]) \longrightarrow K_{n+1}(S^{-1}A[z]) \longrightarrow K_{n+1}(A[z],S) \\
\longrightarrow K_n(A[z]) \longrightarrow \ldots ,
\end{aligned}$$

$$\begin{aligned}
\ldots \longrightarrow K_n(A) \longrightarrow K_n(S^{-1}A) \longrightarrow \bigoplus_{\mathcal{P}\in \max_S(A)} K_{n-1}(A/\mathcal{P}A) \\
\longrightarrow K_{n-1}(A) \longrightarrow \ldots ,
\end{aligned}$$

$$\begin{aligned}
\ldots \longrightarrow \widetilde{\mathrm{End}}_n(A) \longrightarrow \widetilde{\mathrm{End}}_n^{T_A}(S^{-1}A) \longrightarrow \bigoplus_{\mathcal{P}\in \max_S(A)} \widetilde{\mathrm{End}}_{n-1}(A/\mathcal{P}A) \\
\longrightarrow \widetilde{\mathrm{End}}_{n-1}(A) \longrightarrow \ldots
\end{aligned}$$

with $\max_S(A) \subseteq \max(A)$ the set of maximal ideals $\mathcal{P} \lhd A$ with uniformizers $\pi \in \mathcal{P}$ such that $\pi \in S$. In general, $\widetilde{\mathrm{End}}_*(A/\mathcal{P}A) \neq 0$. For example, if $* = 0$, $A = \mathbb{Z}$, $S = \mathbb{Z}\backslash\{0\}$ and $\mathcal{P} = (p)$ then $\widetilde{\mathrm{End}}_0(\mathbb{Z}_p) = W(\mathbb{Z}_p) \neq 0$ by 14.8, 14.12. □

19. K-theory of function fields

The computation of the knot cobordism groups C_* in Chaps. 40–42 will make use of the algebraic K- and L-theory of the function field $F(z)$ of a field F. The function field is the quotient field of $F[z]$, which is a Dedekind ring. The results of Chap. 18 are now specialized to describe the K-theory of $F(z)$ – see Chap. 39 for the L-theory of $F(z)$.

The polynomial extension of a field F is a principal ideal domain $F[z]$, and hence a Dedekind ring with quotient the function field

$$S^{-1}F[z] = F(z) ,$$

where $S = F[z]\backslash\{0\}$. The algebraic K-theory localization exact sequence breaks up into split exact sequences

$$0 \longrightarrow K_n(F[z]) \longrightarrow K_n(F(z)) \longrightarrow K_n(F[z],S) \longrightarrow 0$$

with

$$K_n(F[z],S) = \mathrm{End}_{n-1}(F) = \bigoplus_{\mathcal{P}\in\max(F[z])} K_{n-1}(F[z]/\mathcal{P}) .$$

The maximal ideals of $F[z]$ are the principal ideals

$$\mathcal{P} = (p(z)) \triangleleft F[z]$$

generated by the irreducible monic polynomials $p(z) \in \mathcal{M}(F)$ (4.9), allowing the identification

$$\max(F[z]) = \mathcal{M}(F) .$$

For any such $\mathcal{P}$ there is defined a multiplicative subset

$$\mathcal{P}^\infty = \{p(z)^k \,|\, k \geq 0\} \subset F[z]$$

such that the localization

$$(\mathcal{P}^\infty)^{-1}F[z] = F[z,p(z)^{-1}]$$

fits into a cartesian square

$$\begin{array}{ccc} F[z] & \longrightarrow & F[z,p(z)^{-1}] \\ \downarrow & & \downarrow \\ F[z]_{p(z)} & \longrightarrow & F(z) \end{array}$$

with $F[z]_{(p(z))}$ the local ring of $F[z]$ at $\mathcal{P}$, obtained by inverting all the polynomials coprime to $p(z)$. An endomorphism $f : V \longrightarrow V$ of a finite dimensional F-vector space V is $\mathcal{P}$-primary in the sense of 12.5 if the induced $F[z,p(z)^{-1}]$-module endomorphism

$$z - f \ : \ V[z,p(z)^{-1}] \longrightarrow V[z,p(z)^{-1}]$$

is an isomorphism. As in 12.5 adopt the usual terminology, calling such endomorphisms (V,f) *$\mathcal{P}$-primary.*

Proposition 19.1 *The following conditions on an endomorphism $f : V \longrightarrow V$ of a finite dimensional F-vector space V are equivalent:*

(i) (V,f) *is $\mathcal{P}$-primary,*
(ii) $\mathrm{ch}_z(V,f) = p(z)^k$ *for some* $k \geq 1$,
(iii) $p(f)^k = 0 : V \longrightarrow V$ *for some* $k \geq 1$. □

Similarly for the Laurent polynomial extension $F[z,z^{-1}]$, and automorphism K-theory $\mathrm{Aut}_*(F)$, with

$$\max(F[z,z^{-1}]) \ = \ \mathcal{M}(F)\backslash\{(z)\} \ ,$$

and $\mathcal{P}$-primary automorphisms.

Proposition 19.2 (i) *The endomorphism category of a field F is the disjoint union of the $(p(z))$-primary endomorphism categories of F*

$$\mathrm{End}(F) \ = \ \coprod_{p(z)\in\mathcal{M}(F)} \mathrm{End}^{p(z)^\infty}(F) \ .$$

The endomorphism K-groups are direct sums of the ordinary K-groups of the residue class fields $F[z]/(p(z))$

$$\begin{aligned}\mathrm{End}_*(F) \ &= \ \sum_{p(z)\in\mathcal{M}(F)} \mathrm{End}_*^{p(z)^\infty}(F) \\ &= \ \sum_{p(z)\in\mathcal{M}(F)} K_*(F[z]/(p(z))) \ .\end{aligned}$$

(ii) *The automorphism category of a field F is the disjoint union*

$$\begin{aligned}\mathrm{Aut}(F) \ &= \ \coprod_{p(z)\neq z\in\mathcal{M}(F)} \mathrm{Aut}^{p(z)^\infty}(F) \\ &= \ \coprod_{p(z)\neq z\in\mathcal{M}(F)} \mathrm{End}^{p(z)^\infty}(F) \ .\end{aligned}$$

The automorphism K-groups are direct sums

$$\begin{aligned}\mathrm{Aut}_*(F) &= \widetilde{\mathrm{End}}_*(F)\\ &= \sum_{p(z)\neq z\in\mathcal{M}(F)} \mathrm{Aut}_*^{p(z)^\infty}(F)\\ &= \sum_{p(z)\neq z\in\mathcal{M}(F)} K_*(F[z]/(p(z)))\ .\end{aligned}$$

(iii) *The algebraic K-groups of the function field $F(z)$ are such that*

$$\begin{aligned}K_*(F(z)) &= K_*(F[z])\oplus \mathrm{End}_{*-1}(F)\\ &= K_*(F[z,z^{-1}])\oplus \mathrm{Aut}_{*-1}(F)\\ &= K_*(F)\oplus K_{*-1}(F)\oplus \mathrm{Aut}_{*-1}(F)\ .\end{aligned}$$

Proof (i) Let $S\subset F[z]$ be the multiplicative subset consisting of the monic polynomials, so that $S^{-1}F[z]=F(z)$. An $(F[z],S)$-module is a finite dimensional F-vector space V together with an endomorphism $f:V\longrightarrow V$, so that

$$\mathbb{H}(F[z],S) = \mathrm{End}(F)\ .$$

Factorize the characteristic polynomial of (V,f) as a product

$$\begin{aligned}\mathrm{ch}_z(V,f) &= \det(z-f:V[z]\longrightarrow V[z])\\ &= \prod_{j=1}^{k} p_j(z)^{m_j}\in S\subset F[z]\quad (m_j\geq 0)\end{aligned}$$

of powers of distinct irreducible monic polynomials $p_j(z)\in\mathcal{M}(F)$. The polynomials

$$g_j(z) = p_1(z)^{m_1}\dots p_{j-1}(z)^{m_{j-1}}p_{j+1}(z)^{m_{j+1}}\dots p_k(z)^{m_k}\in F[z]$$

are such that there exist $a_1(z),a_2(z),\dots,a_k(z)\in F[z]$ with

$$\sum_{j=1}^{k} a_j(z)g_j(z) = 1\in F\ .$$

The subspaces

$$V_j = a_j(f)g_j(f)(V)\subseteq V$$

are such that there is a direct sum decomposition

$$(V,f) = \bigoplus_{j=1}^{k}(V_j,f_j)$$

with (V_j,f_j) $(p_j(z))$-primary, and

$$\mathrm{ch}_z(V_j, f_j) = p_j(z)^{m_j} \in F[z] .$$

The expression of S as a product

$$S = \prod_{p(z)\in \mathcal{M}(F)} p(z)^\infty$$

of the coprime multiplicative subsets

$$p(z)^\infty = \{p(z)^k \mid k \geq 0\} \subset F[z]$$

thus determines identifications

$$\begin{aligned} \mathrm{End}(F) &= \mathbb{H}(F[z], S) \\ &= \coprod_{p(z)\in \mathcal{M}(F)} \mathbb{H}(F[z], p(z)^\infty) = \coprod_{p(z)\in \mathcal{M}(F)} \mathrm{End}^{p(z)^\infty}(F) . \end{aligned}$$

Let $\mathbb{P}(F)$ be the additive category of finite dimensional F-vector spaces. The inclusion

$$\mathbb{P}(F[z]/(p(z))) \longrightarrow \mathrm{End}^{p(z)^\infty}(F) \; ; \; F[z]/(p(z)) \longrightarrow (F[z]/(p(z)), \zeta)$$

induces isomorphisms in algebraic K-theory by devissage (18.2), so that

$$\mathrm{End}_*^{p(z)^\infty}(F) = K_*(F[z]/(p(z))) .$$

The splitting of the algebraic K-theory localization exact sequence is the special case $A = F$ of 10.14 (ii).
(ii)+(iii) Immediate from (i). □

Proposition 19.3 (i) *The unique factorization of every non-zero polynomial* $p(z) \in F[z]$

$$p(z) = a\prod_{j=1}^{k} p_j(z)^{m_j} \in F[z] \quad (m_j \geq 0)$$

as a product of a unit $a \in F^\bullet$ *and powers of distinct irreducible monic polynomials* $p_j(z) \in \mathcal{M}(F)$ *determines an isomorphism*

$$K_1(F(z)) = F(z)^\bullet \xrightarrow{\simeq} F^\bullet \oplus \mathbb{Z}[\mathcal{M}(F)] \; ; \; \tau(p(z)) \longrightarrow (a, \sum_{j=1}^{k} m_j[p_j(z)]) .$$

(ii) *For any endomorphism* $f : V \longrightarrow V$ *of a finite dimensional* F*-vector space* V *the unique factorization in* $F[z]$ *of the characteristic polynomial*

$$\mathrm{ch}_z(V, f) = \prod_{j=1}^{k} p_j(z)^{m_j} \in F[z] \quad (m_j \geq 0)$$

determines an isomorphism

$$\mathrm{End}_0(F) \xrightarrow{\simeq} \mathbb{Z}[\mathcal{M}(F)]\ ;\ [V,f] \longrightarrow \sum_{j=1}^{k} m_j[p_j(z)]\ ,$$

which is compatible with the isomorphism in (i) *via the injection*

$$\mathrm{End}_0(F) \longrightarrow K_1(F(z))\ ;\ [V,f] \longrightarrow \tau(z-f : V(z) \longrightarrow V(z))\ .$$

(iii) *For any automorphism $f : V \longrightarrow V$ of a finite dimensional F-vector space V the unique factorization in $F[z,z^{-1}]$ of $\mathrm{ch}_z(V,f)$ determines an isomorphism*

$$\mathrm{Aut}_0(F) \xrightarrow{\simeq} \mathbb{Z}[\mathcal{M}(F)\backslash\{z\}]\ ;\ [V,f] \longrightarrow \sum_{j=1}^{k} m_j[p_j(z)]$$

as in (ii).

Proof These are special cases of 19.2. □

Example 19.4 (i) For an algebraically closed field F (e.g. $\mathbb{C}$)

$$\mathcal{M}(F) = \{z-a\,|\,a\in F\} \cong F$$

so that the isomorphism of 19.3 (ii) can be written as

$$\mathrm{End}_0(F) \xrightarrow{\simeq} \mathbb{Z}[F]\ ;\ [V,f] \longrightarrow \sum_{j=1}^{k} m_j\lambda_j\ ,$$

sending an endomorphism $f : V \longrightarrow V$ of a finite dimensional F-vector space V to the linear combination of the distinct eigenvalues $\lambda_1, \lambda_2, \ldots, \lambda_k \in F$ counted with their multiplicities $m_1, m_2, \ldots, m_k$, such that

$$\mathrm{ch}_z(V,f) = \prod_{j=1}^{k}(z-\lambda_j)^{m_j} \in F[z]\ .$$

This can also be viewed as the devissage isomorphism

$$K_1(F[z],S) = \mathrm{End}_0(F) \xrightarrow{\simeq} \sum_{a\in F} K_0(F[z]/(z-a)) = \mathbb{Z}[F]\ ;$$

$$[V,f] \longrightarrow \sum_{j=1}^{k} m_j\lambda_j$$

with $S = F[z]\backslash\{0\} \subset F[z]$ and $S^{-1}F[z] = F(z)$. The near-projections (12.22)

$$p_j(f) = \prod_{i\neq j}(f-\lambda_i)^{m_i}/(\lambda_j-\lambda_i)^{m_i}\ :\ V \longrightarrow V \quad (1 \leq j \leq k)$$

are such that

$$(p_j(f)(1-p_j(f)))^{m_j} = 0\ :\ V \longrightarrow V\ .$$

The projection

$$p_j(f)_\omega = (p_j(f)^{m_j} + (1-p_j(f))^{m_j})^{-1}p_j(f)^{m_j} : V \longrightarrow V$$

has image the $(z-\lambda_j)$-primary component of the $(F[z],S)$-torsion module V

$$\begin{aligned} V_j &= \operatorname{im}(p_j(f)_\omega) = \ker((f-\lambda_j)^{m_j} : V \longrightarrow V) \\ &= F[z]_{(z-\lambda_j)} \otimes_{F[z]} V , \end{aligned}$$

with z acting by $f : V \longrightarrow V$ on V and by the restriction $f_j = f| : V_j \longrightarrow V_j$ on V_j. As in Lam, Ranicki and Smith [142] the Jordan normal form decomposition is given by

$$(V,f) = \bigoplus_{j=1}^{k} (V_j, f_j)$$

with $f_j - \lambda_i : V_j \longrightarrow V_j$ nilpotent for $i=j$ and an automorphism for $i \neq j$.
(ii) For $F = \mathbb{R}$ there is one degree 1 irreducible monic polynomial $z-\alpha \in \mathbb{R}[z]$ for each $\alpha \in \mathbb{R}$, and one degree 2 irreducible monic polynomial $(z-\omega)(z-\overline{\omega}) \in \mathbb{R}[z]$ for each unordered pair $\{\omega,\bar{\omega}\}$ of distinct complex conjugates in $\mathbb{C}$, so that

$$\max(\mathbb{R}[z]) = \mathbb{C}/\sim \quad (\sim\, = \text{ complex conjugation})$$

(cf. 14.9). The isomorphism of 19.3 (i) is given by

$$\operatorname{End}_0(\mathbb{R}) \xrightarrow{\simeq} \mathbb{Z}[\mathbb{C}/\sim] \; ; \; [V,f] \longrightarrow \sum_{j=1}^{p} m_j[\alpha_j] + \sum_{k=1}^{q} n_k[\omega_k]$$

with $\alpha_j \in \mathbb{R}$, $\omega_k \in \mathbb{C}\backslash\mathbb{R}$, $m_j, n_k \geq 0$ such that

$$\operatorname{ch}_z(V,f) = \prod_{j=1}^{p}(z-\alpha_j)^{m_j} \prod_{k=1}^{q} \big((z-\omega_k)(z-\overline{\omega}_k)\big)^{n_k} \in \mathbb{R}[z] .$$

(iii) If F is a finite field with q elements

$$\mathcal{M}(F) = \coprod_{n=1}^{\infty} \mathcal{M}(q,n)$$

with $\mathcal{M}(q,n)$ the finite set of irreducible monic polynomials $p_{j,n}(z) \in F[z]$ of degree n. Let $M(q,n)$ be the number of elements in $\mathcal{M}(q,n)$. The cyclotomic identity

$$\frac{1}{1-qz} = \prod_{n=1}^{\infty} \left(\frac{1}{1-z^n}\right)^{M(q,n)}$$

and the formula

$$M(q,n) = \frac{1}{n} \sum_{d|n} \mu(d) q^{\frac{n}{d}}$$

are due to Gauss and MacMahon, with μ the Möbius function

$$\mu(d) = \begin{cases} 1 & \text{if } d = 1 \\ (-1)^r & \text{if } d \text{ is a product of } r \text{ distinct primes} \\ 0 & \text{otherwise} . \end{cases}$$

The isomorphism of 19.3 (ii) is given by

$$\mathrm{End}_0(F) \xrightarrow{\simeq} \mathbb{Z}[\mathcal{M}(F)] = \sum_{n=1}^{\infty} \mathbb{Z}^{M(q,n)} ; \ [V,f] \longrightarrow \sum_{n=1}^{\infty} \sum_{j=1}^{M(q,n)} m_{j,n}[p_{j,n}(z)]$$

with

$$\mathrm{ch}_z(V,f) = \prod_{n=1}^{\infty} \prod_{j=1}^{M(q,n)} p_{j,n}(z)^{m_{j,n}} \quad (m_{j,n} \geq 0) .$$

See Dress and Siebeneicher [61], [62] for the connections between the numbers $M(q,n)$ and various combinatorial facts. □

The inclusion $F[z] \longrightarrow F(z)$ has the property that a finite f.g. free $F[z]$-module chain complex E^+ is F-finitely dominated if and only if E^+ is $F(z)$-contractible (8.5 (i)), if and only if $\dim_F H_*(E^+) < \infty$. Similarly, the inclusion $F[z,z^{-1}] \longrightarrow F(z)$ has the property that a finite f.g. free $F[z,z^{-1}]$-module chain complex E is F-finitely dominated if and only if E is $F(z)$-contractible (8.5 (ii)), if and only if $\dim_F H_*(E) < \infty$.

The general definition of Reidemeister torsion (16.1) is now specialized to function fields.

Definition 19.5 (i) The *Reidemeister torsion* of an F-finitely dominated finite f.g. free $F[z]$-module chain complex E^+ is the torsion

$$\begin{aligned} \Delta^+(E^+) &= \tau(F(z) \otimes_{F[z]} E^+) = \sum_{r=0}^{\infty} (-)^r [H_r(E^+), \zeta] \\ &\in \mathrm{coker}\big(K_1(F[z]) \longrightarrow K_1(F(z))\big) = \mathrm{End}_0(F) = \mathbb{Z}[\mathcal{M}(F)] , \end{aligned}$$

which is independent of the choice of basis for E^+ used in the definition.
(ii) The *Reidemeister torsion* of an F-finitely dominated finite f.g. free $F[z,z^{-1}]$-module chain complex E is the torsion

$$\begin{aligned} \Delta(E) &= \tau(F(z) \otimes_{F[z,z^{-1}]} E) = \sum_{r=0}^{\infty} (-)^r [H_r(E), \zeta] \\ &\in \mathrm{coker}\big(K_1(F[z,z^{-1}]) \longrightarrow K_1(F(z))\big) = \mathrm{Aut}_0(F) = \mathbb{Z}[\mathcal{M}(F) \backslash \{z\}] , \end{aligned}$$

which is independent of the choice of basis for E used in the definition. □

The projection

$$\operatorname{coker}(K_1(F[z])\longrightarrow K_1(F(z))) \longrightarrow \operatorname{coker}(K_1(F[z,z^{-1}])\longrightarrow K_1(F(z)))$$

sends $\Delta^+(E^+)$ to $\Delta(E)$ with $E = F[z,z^{-1}] \otimes_{F[z]} E^+$.

Example 19.6 Given a polynomial $p(z) \in F[z]$, define

$$\begin{aligned} E^+ &= \mathcal{C}(p(z) : F[z]\longrightarrow F[z]) , \\ E &= \mathcal{C}(p(z) : F[z,z^{-1}]\longrightarrow F[z,z^{-1}]) . \end{aligned}$$

If $p(z) \neq 0$ then E^+ and E are F-finitely dominated, with Reidemeister torsions

$$\begin{aligned} \Delta^+(E^+) &= [H_0(E^+),\zeta] = [F[z]/(p(z)),\zeta] \in \operatorname{End}_0(F) , \\ \Delta(E) &= [H_0(E),\zeta] = [F[z,z^{-1}]/(p(z)),\zeta] \in \operatorname{Aut}_0(F) . \end{aligned}$$

See 19.13 for further details. □

Example 19.7 Let X be a finite CW complex with an infinite cyclic cover $\overline{X}$ such that the F-vector space $H_*(\overline{X};F)$ is finite dimensional, so that the F-coefficient cellular chain complex $C(\overline{X};F)$ is an F-finitely dominated finite f.g. free $F[z,z^{-1}]$-module chain complex. The F-coefficient Reidemeister torsion of X considered by Milnor [194], [195] is

$$\Delta(X;F) = \Delta(C(\overline{X};F)) \in \operatorname{coker}(K_1(F[z,z^{-1}])\longrightarrow K_1(F(z))) . \quad \square$$

The K-theoretic properties of the Reidemeister torsion are now reformulated in terms of the computation of $K_1(S^{-1}F[z])$ for the localizations $S^{-1}F[z] \subseteq F(z)$ inverting various multiplicative subsets $S \subset F[z]$.

Terminology 19.8 Given a subset $J \subseteq \mathcal{M}(F)$ let $\{p_j(z) \in F[z] \,|\, j \in J\}$ be the corresponding collection of irreducible monic polynomials, and let $S_J \subset F[z]$ be the multiplicative subset of all the finite products

$$p(z) = p_{j_1}(z)^{m_1} p_{j_2}(z)^{m_2} \dots p_{j_k}(z)^{m_k} \in F[z]$$

for $j_1, j_2, \dots, j_k \in J, m_1, m_2, \dots, m_k \geq 0$. For each $j \in J$ write $S_{\{j\}}$ as S_j, and let

$$F_j = F[z]/(p_j(z))$$

be the residue class field of $F[z]$. Also, let

$$T_J = S_{\mathcal{M}(F)\backslash J} \subset F[z] ,$$

so that $S_J, T_J \subset F[z]$ are coprime multiplicative subsets with

$$(S_J T_J)^{-1} F[z] = F(z)$$

and there is defined a cartesian square of rings

$$\begin{array}{ccc} F[z] & \longrightarrow & S_J^{-1}F[z] \\ \downarrow & & \downarrow \\ T_J^{-1}F[z] & \longrightarrow & F(z) \end{array}$$

□

Proposition 19.9 (i) *For any subset $J \subseteq \mathcal{M}(F)$ the algebraic K-theory localization exact sequence for $F[z] \longrightarrow S_J^{-1}F[z]$*

$$\dots \longrightarrow K_n(F[z]) \longrightarrow K_n(S_J^{-1}F[z]) \xrightarrow{\partial_+} K_n(F[z], S_J) \longrightarrow K_{n-1}(F[z]) \longrightarrow \dots$$

breaks up into split exact sequences

$$0 \longrightarrow K_n(F[z]) \longrightarrow K_n(S_J^{-1}F[z]) \xrightarrow{\partial_+} K_n(F[z], S_J) \longrightarrow 0 \ (n \in \mathbb{Z}),$$

with

$$\begin{aligned} K_n(F[z], S_J) &= \mathrm{End}_{n-1}^{S_J}(F) \\ &= \sum_{j \in J} \mathrm{End}_{n-1}^{S_j}(F) = \sum_{j \in J} K_{n-1}(F_j) . \end{aligned}$$

(ii) *The S_J-primary endomorphism class group of F is such that*

$$\mathrm{End}_0^{S_J}(F) = \mathbb{Z}[\max(T_J^{-1}F[z])] = \mathbb{Z}[J] .$$

Proof (i) Every finite dimensional F_j-vector space V_j is an $(F[z], S_j)$-module, and the functors

$$\begin{aligned} &\mathbb{P}(F_j) \longrightarrow \mathbb{H}(F[z], S_j) \ ; \ V_j \longrightarrow V_j , \\ &\coprod_{j \in J} \mathbb{P}(F_j) \longrightarrow \mathbb{H}(F[z], S_J) \ ; \ \{V_j \,|\, j \in J\} \longrightarrow \sum_{j \in J} V_j \end{aligned}$$

induce isomorphisms in algebraic K-theory, by $\mathcal{P}$-primary decomposition and devissage (as in 19.4), so that

$$K_*(F[z], S_J) = \sum_{j \in J} K_*(F[z], S_j) = \sum_{j \in J} K_{*-1}(F_j).$$

There is defined an isomorphism of exact categories

$$\mathbb{H}(F[z], S_J) \xrightarrow{\simeq} \mathrm{End}^{S_J}(F) \ ; \ V_J \longrightarrow (V_J, \zeta) ,$$

so that

$$\begin{aligned} K_*(F[z], S_J) &= K_{*-1}(\mathbb{H}(F[z], S_J)) \\ &= K_{*-1}(\mathrm{End}^{S_J}(F)) = \mathrm{End}_{*-1}^{S_J}(F) . \end{aligned}$$

(ii) By (i)

$$\mathrm{End}_0^{S_J}(F) \;=\; \sum_{j\in J} K_0(F_j) \;=\; \mathbb{Z}[J]\;,$$

with the generator $[F_j] = 1 \in K_0(F_j) = \mathbb{Z}$ corresponding to the endomorphism

$$\zeta \;:\; F_j \longrightarrow F_j \;;\; x \longrightarrow zx$$

of the finite dimensional F-vector space F_j with characteristic polynomial

$$\mathrm{ch}_z(F_j,\zeta) \;=\; p_j(z) \in S_j \subset F[z]\;.$$

An endomorphism $f : V \longrightarrow V$ of a finite dimensional F-vector space is S_J-primary if and only if $\mathrm{ch}_z(V,f) \in S_J$, in which case the factorization

$$\mathrm{ch}_z(V,f) \;=\; \prod_{j=1}^{k} p_j(z)^{m_j} \in S_J$$

determines the endomorphism class

$$[V,f] \;=\; \sum_{j=1}^{k} m_j[p_j(z)] \in \mathrm{End}_0^{S_J}(F) \;=\; \mathbb{Z}[J]\;. \qquad \square$$

Example 19.10 (i) For any irreducible monic polynomial $p(z) \in F[z]$ let $J = \{(p(z))\} \in \mathcal{M}(F)$. The multiplicative subsets

$$\begin{aligned} S_J &= p(z)^\infty \;=\; \{p(z)^k \mid k \geq 0\}\;, \\ T_J &= \{q(z) \mid q(z) \text{ monic and coprime to } p(z)\} \subset F[z] \end{aligned}$$

are such that

$$\begin{aligned} S_J^{-1}F[z] &= F[z,p(z)^{-1}] \;\;,\;\; T_J^{-1}F[z] \;=\; F[z]_{(p(z))}\;, \\ K_n(F[z,p(z)^{-1}]) &= K_n(F[z]) \oplus K_n(F[z],p(z)^\infty) \\ &= K_n(F) \oplus \mathrm{End}_{n-1}^{p(z)^\infty}(F) \\ &= K_n(F) \oplus K_{n-1}(F[z]/(p(z)))\;, \\ \mathrm{End}_*^{p(z)^\infty}(F) &= K_*(F[z]/(p(z)))\;. \end{aligned}$$

The generator of

$$K_1(F[z],p(z)^\infty) \;=\; K_0(F[z]/(p(z))) \;=\; \mathbb{Z}$$

is represented by the $(p(z))$-primary endomorphism $(F[z]/(p(z)),\zeta)$, and

$$\mathrm{End}_0^{p(z)^\infty}(F) \;=\; \mathbb{Z}[\max(F[z]_{(p(z))})] \;=\; \mathbb{Z}[\mathcal{M}(F)\backslash(p(z))]\;.$$

If $p(z) \neq z$ then

$$\begin{aligned} \mathrm{End}^{p(z)^\infty}(F) &= \mathrm{Aut}^{p(z)^\infty}(F) , \\ \mathrm{End}_*^{p(z)^\infty}(F) &= \mathrm{Aut}_*^{p(z)^\infty}(F) = K_*(F[z]/(p(z))) , \end{aligned}$$

If $p(z) = z$ then

$$\begin{aligned} \mathrm{End}^{(z)^\infty}(F) &= \mathrm{Nil}(F) , \\ \mathrm{End}_*^{(z)^\infty}(F) &= \mathrm{Nil}_*(F) = K_*(F) . \end{aligned}$$

(ii) For $J = \mathcal{M}(F)$ the set $\{p_j(z) \mid j \in J\}$ consists of all the irreducible monic polynomials, and $S_J \subset F[z]$ is the multiplicative subset of all the monic polynomials. The localization is the quotient field

$$S_J^{-1}F[z] = F(z)$$

and 19.9 gives

$$\begin{aligned} K_*(F[z], S_J) &= \mathrm{End}_*(F) = \sum_{j\in J} K_{*-1}(F_j) , \\ K_*(F(z)) &= K_*(F) \oplus \sum_{j\in J} K_{*-1}(F_j) . \end{aligned}$$

In particular, for $* = 1$

$$\begin{aligned} K_1(F[z], S_J) &= \mathrm{End}_0(F) = \mathbb{Z} \oplus W(F) = \mathbb{Z}[\mathcal{M}(F)] , \\ K_1(F(z)) &= F^\bullet \oplus \mathbb{Z}[\mathcal{M}(F)] = F(z)^\bullet . \end{aligned}$$

For the completion $F[[z]]$

$$\begin{aligned} \max(F[[z]]) &= \{(z)\} ,\ K_1(F[[z]]) = K_1(F) \oplus \widehat{W}(F) , \\ K_1(F[[z]], Z) &= K_1(F[z], Z) = \mathrm{Nil}_0(F) = K_0(F) , \end{aligned}$$

so that

$$\begin{aligned} K_1\big(F((z))\big) &= K_1(F[[z]]) \oplus K_1(F[[z]], Z) \\ &= K_1(F) \oplus K_0(F) \oplus \widehat{W}(F) = F((z))^\bullet , \end{aligned}$$

with

$$K_1(F(z)) = F(z)^\bullet \longrightarrow K_1\big(F((z))\big) = F((z))^\bullet$$

the evident inclusion.

(iii) For $n = 1$ and any subset $J \subseteq \mathcal{M}(F)$ 19.9 gives:

$$\begin{aligned} K_1(S_J^{-1}F[z]) &= \{p(z)/q(z) \in F(z)^\bullet \mid q(z) \in S_J\} \\ &= K_1(F[z]) \oplus K_1(F[z], S_J) , \\ K_1(F[z], S_J) &= \mathrm{End}_0^{S_J}(F) \\ &= \sum_{j\in J} K_1(F[z], S_j) = \sum_{j\in J} K_0(F_j) = \mathbb{Z}[J] , \end{aligned}$$

with an isomorphism

$$\mathbb{Z}[J] \xrightarrow{\simeq} \mathrm{End}_0^{S_J}(F) \; ; \; \sum_{j\in J} m_j j \longrightarrow \sum_{j\in J} m_j [F_j, \zeta] \; .$$

The boundary map $\partial_+ : K_1(S_J^{-1}F[z]) \longrightarrow K_1(F[z], S_J)$ is split by

$$\begin{aligned} K_1(F[z], S_J) \; &= \; \mathbb{Z}[J] \longrightarrow K_1(S_J^{-1}F[z]) \; ; \\ \sum_{j\in J} m_j j &\longrightarrow \tau(\prod_{j\in J} p_j(z)^{m_j} : S_J^{-1}F[z] \longrightarrow S_J^{-1}F[z]) \; , \end{aligned}$$

with

$$\begin{aligned} \partial_+ \; : K_1(S_J^{-1}F[z]) &\longrightarrow K_1(F[z], S_J) \; = \; \mathbb{Z}[J] \; ; \\ \tau(a \prod_{j\in J} p_j(z)^{m_j} : S_J^{-1}F[z] \longrightarrow S_J^{-1}F[z]) &\longrightarrow \sum_{j\in J} m_j j \; . \end{aligned}$$

If $(z) \in J$ then the reverse characteristic polynomial defines an isomorphism

$$\begin{aligned} \widetilde{\mathrm{ch}}_z \; : \; \mathrm{End}_0^{S_J}(F) \; = \; \mathbb{Z}[J] &\xrightarrow{\simeq} \mathbb{Z} \oplus W^{S_J}(F) \; ; \\ \sum_{j\in J} m_j j &\longrightarrow (m_{(z)}, \prod_{j\in J\backslash\{(z)\}} \widetilde{p}_j(z)^{m_j}) \; , \end{aligned}$$

with $W^{S_J}(F)$ the S_J-primary Witt vector group (14.21). If $(z) \notin J$ the reverse characteristic polynomial defines an isomorphism

$$\begin{aligned} \widetilde{\mathrm{ch}}_z \; : \; \mathrm{Aut}_0^{S_J}(F) \; = \; \mathrm{End}_0^{S_J}(F) \; = \; \mathbb{Z}[J] &\xrightarrow{\simeq} W^{(z)^\infty S_J}(F) \; ; \\ \sum_{j\in J} m_j j &\longrightarrow \prod_{j\in J} \widetilde{p}_j(z)^{m_j} \; . \end{aligned}$$

□

Proposition 19.11 (i) *The Reidemeister torsion of an F-finitely dominated finite f.g. free $F[z]$-module chain complex E^+ is given by*

$$\begin{aligned} \Delta^+(E^+) \; &= \; (m_0, \prod_{j=1}^{k} \widetilde{p}_j(z)^{m_j}) \\ &\in \mathrm{coker}(K_1(F[z]) \longrightarrow K_1(F(z))) \; = \; \mathbb{Z} \oplus W(F) \; , \end{aligned}$$

with $p_j(z)$ the irreducible monic polynomials $\neq z$ appearing in the torsion

$$\begin{aligned} \tau(F(z) \otimes_{F[z]} E^+) \; &= \; a z^{m_0} \prod_{j=1}^{k} p_j(z)^{m_j} \\ &\in K_1(F(z)) \; = \; F(z)^\bullet \quad (a \in F^\bullet, m_j \in \mathbb{Z}) \; . \end{aligned}$$

(ii) *The Reidemeister torsion of an F-finitely dominated finite f.g. free $F[z,z^{-1}]$-module chain complex E is given by*

$$\Delta(E) = \prod_{j=1}^{k} \widetilde{p}_j(z)^{m_j} \in \mathrm{coker}\big(K_1(F[z,z^{-1}]) \longrightarrow K_1(F(z))\big) = W(F) ,$$

with $p_j(z)$ the irreducible monic polynomials appearing in the torsion

$$\tau(F(z) \otimes_{F[z,z^{-1}]} E) = a \prod_{j=0}^{k} p_j(z)^{m_j} \in K_1(F(z)) = F(z)^{\bullet}$$

$$(a \in F^{\bullet}, m_j \in \mathbb{Z}, p_0(z) = z, \widetilde{p}_0(z) = 1) .$$

Proof Let

$$J = \mathcal{M}(F)\backslash\{(z)\} \subset J' = \mathcal{M}(F) ,$$

so that J' indexes all the irreducible monic polynomials, and $S = S_{J'} \subset F[z]$ is such that

$$S^{-1}F[z] = F(z) .$$

In this case 19.10 gives

$$\text{(i) } \mathrm{coker}\big(K_1(F[z]) \longrightarrow K_1(F(z))\big) = \mathrm{End}_0(F) = K_0(F) \oplus W(F) = \mathbb{Z} \oplus \mathbb{Z}[J] ,$$

$$\text{(ii) } \mathrm{coker}\big(K_1(F[z,z^{-1}]) \longrightarrow K_1(F(z))\big) = \mathrm{Aut}_0(F) = W(F) = \mathbb{Z}[J] .$$

□

Proposition 19.12 (i) *The Reidemeister torsion of an F-finitely dominated finite f.g. free $F[z]$-module chain complex E^+ is given by*

$$\Delta^+(E^+) = \Big(\sum_{r=0}^{\infty}(-)^r n_r, \prod_{r=0}^{\infty} \widetilde{\mathrm{ch}}_z(H_r(E^+),\zeta)^{(-)^r}\Big) \in \mathrm{coker}\big(K_1(F[z]) \longrightarrow K_1(F(z))\big) = \mathbb{Z} \oplus W(F) ,$$

with

$$n_r = \dim_F\big(\ker(\zeta : H_r(E^+) \longrightarrow H_r(E^+))\big) = \dim_F H_r(E^+) - \dim_F H_r(F[z,z^{-1}] \otimes_{F[z]} E^+) \in \mathbb{Z} .$$

(ii) *The Reidemeister torsion of an F-finitely dominated finite f.g. free $F[z,z^{-1}]$-module chain complex E is given by*

$$\Delta(E) = \prod_{r=0}^{\infty} \widetilde{\mathrm{ch}}_z(H_r(E),\zeta)^{(-)^r} \in \mathrm{coker}\big(K_1(F[z,z^{-1}]) \longrightarrow K_1(F(z))\big) = W(F) .$$

Proof This is just an invariant formulation of 19.11. □

The expression in 19.12 (ii) of the Reidemeister torsion $\Delta(E)$ as an alternating product of characteristic polynomials was first obtained in Assertion 7 of Milnor [195].

Example 19.13 (i) Let E^+ be the 1-dimensional $F(z)$-contractible f.g. free $F[z]$-module chain complex defined by

$$E^+ = \mathcal{C}(p(z) : F[z] \longrightarrow F[z])$$

for a polynomial

$$p(z) = z^n \sum_{j=0}^{m} a_j z^j \in F[z] \quad (a_0, a_m \neq 0 \in F, n \geq 0) .$$

The Reidemeister torsion of E^+ is given by

$$\begin{aligned} \Delta^+(E^+) &= (n, \widetilde{\mathrm{ch}}_z(F[z]/(p(z)), \zeta)) \\ &= (n, (a_0)^{-1}(\sum_{j=0}^{m} a_{m-j} z^j)) \in \mathbb{Z} \oplus W(F) . \end{aligned}$$

(ii) Let E be the 1-dimensional $F(z)$-contractible f.g. free $F[z, z^{-1}]$-module chain complex defined by

$$E = \mathcal{C}(p(z) : F[z, z^{-1}] \longrightarrow F[z, z^{-1}])$$

for a polynomial

$$p(z) = z^n \sum_{j=0}^{m} a_j z^j \in F[z, z^{-1}] \quad (a_0, a_m \neq 0 \in F, n \in \mathbb{Z}) .$$

The Reidemeister torsion of E is given by

$$\Delta(E) = \widetilde{\mathrm{ch}}_z(F[z, z^{-1}]/(p(z)), \zeta) = (a_0)^{-1}(\sum_{j=0}^{m} a_{m-j} z^j) \in W(F) . \quad □$$

Example 19.14 Let $J = \mathcal{M}(F)$, as in 19.10 (ii). Given an element $\omega \in F$ let

$$\begin{aligned} J_\omega &= \{p(z) \in \mathcal{M}(F) \,|\, p(\omega) \neq 0 \in F\} \subseteq J , \\ S_\omega &= S_{J_\omega} = \{q(z) \in S_J \,|\, q(\omega) \neq 0 \in F\} \subseteq S . \end{aligned}$$

(i) The local ring of $F[z]$ at $\omega \in F$ is the localization

$$S_\omega^{-1} F[z] = \{p(z)/q(z) \in F(z) \,|\, q(\omega) \neq 0\} .$$

The evaluation map

$$\tau_\omega \ : \ F[z] \longrightarrow F \ ; \ \sum_{j=0}^{\infty} a_j z^j \longrightarrow \sum_{j=0}^{\infty} a_j \omega^j$$

has a canonical factorization

$$\tau_\omega \ : \ F[z] \longrightarrow S_\omega^{-1}F[z] \xrightarrow{\tau_\omega} F \ ,$$

with

$$\tau_\omega \ : \ S_\omega^{-1}F[z] \longrightarrow F \ ; \ p(z)/q(z) \longrightarrow p(\omega)/q(\omega)$$

such that

$$\ker(S_f^{-1}F[z] \longrightarrow F) \ = \ \{p(z)/q(z) \in F(z) \,|\, q(\omega) \neq 0, p(\omega) = 0\}$$

is the unique maximal ideal of $S_\omega^{-1}F[z]$.
The following conditions on a finite f.g. free $F[z]$-module chain complex E^+ are equivalent:

(a) E^+ is F-contractible via τ_ω,
(b) E^+ is $S_\omega^{-1}F[z]$-contractible,
(c) the reverse characteristic polynomials

$$c_r(z) \ = \ \widetilde{\mathrm{ch}}_z(H_r(E^+), \zeta) \in F[z] \ \ (r \geq 0)$$

are such that $c_r(\omega) \neq 0 \in F$.

By 19.9 there is defined an isomorphism

$$K_1(S_\omega^{-1}F[z]) \xrightarrow{\simeq} F^\bullet \oplus \mathbb{Z}[J_\omega] \ ;$$
$$\tau(p(z) : S_\omega^{-1}F[z] \longrightarrow S_\omega^{-1}F[z]) \longrightarrow (a, \sum_{j \in J_\omega} m_j j)$$
$$(p(z) \ = \ a \prod_{j \in J_\omega} p_j(z)^{m_j} \in S_\omega \subset F[z] \ , a \in F^\bullet) \ .$$

The evaluation map

$$\tau_\omega \ : \ S_\omega^{-1}F[z] \longrightarrow F \ ; \ z \longrightarrow \omega$$

induces

$$\tau_\omega \ : \ K_1(S_\omega^{-1}F[z]) \longrightarrow K_1(F) \ = \ F^\bullet \ ; \ \tau(p(z)) \longrightarrow p(\omega) \ .$$

If E^+ is a based f.g. free $F[z]$-module chain complex which is F-contractible via τ_ω the torsion of the induced contractible based f.g. free $S_\omega^{-1}F[z]$-module chain complex $S_\omega^{-1}E^+$ is given by

$$\tau(S_\omega^{-1}E^+) \ = \ (\tau(r^+), \Delta^+(E^+))$$
$$\in K_1(S_\omega^{-1}F[z]) \ = \ K_1(F) \oplus \mathbb{Z}[J_\omega] \ ,$$

with

$$r^+ \;:\; \mathcal{C}(\zeta - z : E^+[z] \longrightarrow E^+[z]) \longrightarrow E^+$$

the $F[z]$-module chain equivalence of 5.15.
(ii) Let ω be as in (i), and assume also that $\omega \neq 0 \in F$, so that

$$z \in S_\omega \;,\; F[z, z^{-1}] \subseteq S_\omega^{-1}F[z] \;,$$

and the projection

$$\tau_\omega \;:\; F[z,z^{-1}] \longrightarrow F \;;\; \sum_{j=-\infty}^{\infty} a_j z^j \longrightarrow \sum_{j=-\infty}^{\infty} a_j \omega^j$$

has a canonical factorization

$$\tau_\omega \;:\; F[z,z^{-1}] \longrightarrow S_\omega^{-1}F[z] \longrightarrow F$$

with

$$S_\omega^{-1}F[z] \longrightarrow F \;;\; p(z)/q(z) \longrightarrow p(\omega)/q(\omega) \;.$$

The following conditions on a finite f.g. free $F[z,z^{-1}]$-module chain complex E are equivalent:

(a) E is F-contractible via τ_ω,
(b) E is $S_\omega^{-1}F[z]$-contractible,
(c) the characteristic polynomials

$$\widetilde{\mathrm{ch}}_z(H_r(E), \zeta_*) \;=\; c_r(z) \in F[z] \;\; (r \geq 0)$$

are such that $c_r(\omega) \neq 0 \in F$.

The inclusion $S_\omega^{-1}F[z] \longrightarrow F(z)$ induces the inclusion

$$K_1(S_\omega^{-1}F[z]) \;=\; K_1(F) \oplus \mathbb{Z}[J_\omega] \longrightarrow K_1(F(z)) \;=\; K_1(F) \oplus \mathbb{Z}[J] \;.$$

If E is a based f.g. free $F[z,z^{-1}]$-module chain complex which is F-contractible via τ_ω the torsion of the induced contractible based f.g. free $S_\omega^{-1}F[z]$-module chain complex $S_\omega^{-1}E$ is given by

$$\begin{aligned}\tau(S_\omega^{-1}E) \;&=\; (\Phi^+(E), \Delta(E)) \\ &\in K_1(S_\omega^{-1}F[z]) \;=\; K_1(F[z,z^{-1}]) \oplus \mathbb{Z}[J_\omega \backslash \{(z)\}] \;,\end{aligned}$$

with

$$\begin{aligned}\Phi^+(E) \;&=\; \tau(q^+ : \mathcal{C}(1 - z\zeta^{-1} : E[z,z^{-1}] \longrightarrow E[z,z^{-1}]) \longrightarrow E) \\ &\in K_1(F[z,z^{-1}])\end{aligned}$$

the absolute version of the fibering obstruction (3.10). The evaluation map

$$\tau_\omega \;:\; S_\omega^{-1}F[z] \longrightarrow F \;;\; z \longrightarrow \omega$$

induces

$$K_1(S_\omega^{-1}F[z]) = K_1(F)\oplus \mathbb{Z}[J_\omega] \longrightarrow K_1(F) = F^\bullet\ ;$$
$$(a, \sum_{j\in J_\omega} m_j j) \longrightarrow a \prod_{j\in J_\omega} p_j(\omega)^{m_j}\ ,$$

sending

$$\Delta(E) = \sum_{j\in J_\omega\backslash\{(z)\}} m_j j \in \mathbb{Z}[J_\omega\backslash\{(z)\}]$$

to the Reidemeister torsion of E with respect to ω

$$\Delta_\omega(E) = \prod_{j\in J_\omega\backslash\{(z)\}} p_j(\omega)^{m_j} \in F^\bullet$$

as defined by Milnor [194, p. 387] in the case $F=\mathbb{R}$. □

Proposition 19.15 *The algebraic mapping torus of a chain map $f:C\longrightarrow C$ of a finite dimensional F-vector space chain complex C is an F-finitely dominated finite f.g. free $F[z,z^{-1}]$-module chain complex*

$$T(f) = \mathcal{C}(z-f: C[z,z^{-1}]\longrightarrow C[z,z^{-1}])$$

with Reidemeister torsion

$$\Delta(T(f)) = \prod_{r=0}^{\infty} \widetilde{\mathrm{ch}}_z(H_r(C),f_*)^{(-)^r} \in W(F)\ .$$

Proof The finite f.g. free $F[z]$-module chain complex

$$T(f)^+ = \mathcal{C}(z-f: C[z]\longrightarrow C[z])$$

is such that

$$(H_*(T(f)^+),\zeta) = (H_*(C),f_*)\ ,\ \ T(f) = F[z,z^{-1}]\otimes_{F[z]} T(f)^+\ .$$

Now 19.12 (i) gives the Reidemeister torsion of $T(f)^+$ to be

$$\begin{aligned}\Delta^+(T(f)^+) &= (n, \prod_{r=0}^{\infty} \widetilde{\mathrm{ch}}_z(H_r(T(f)^+),\zeta)^{(-)^r})\\ &= (n, \prod_{r=0}^{\infty} \widetilde{\mathrm{ch}}_z(H_r(C),f_*)^{(-)^r}) \in \mathbb{Z}\oplus W(F)\ ,\end{aligned}$$

with first component

$$n = \sum_{r=0}^{\infty} (-)^r \dim_F\big(\ker(f_*: H_r(C)\longrightarrow H_r(C))\big) \in \mathbb{Z}$$

and second component $\Delta(T(f))\in W(F)$. □

Example 19.16 Let $f : X \longrightarrow X$ be a map, with X a finitely dominated CW complex. The cellular $F[z,z^{-1}]$-module chain complex of the infinite cyclic cover $\overline{T}(f)$ of the mapping torus $T(f)$ is the algebraic mapping torus of $f : C(X;F) \longrightarrow C(X;F)$

$$\begin{aligned} C(\overline{T}(f);F) &= T(f : C(X;F) \longrightarrow C(X;F)) \\ &= \mathcal{C}(z - f : C(X;F)[z,z^{-1}] \longrightarrow C(X;F)[z,z^{-1}]) \end{aligned}$$

which is F-finitely dominated with Reidemeister torsion

$$\Delta(\overline{T}(f);F) = \prod_{r=0}^{\infty} \widetilde{\mathrm{ch}}_z(H_r(X;F), f_*)^{(-)^r} \in W(F) .$$

If F has characteristic 0 this is the Lefschetz ζ-function of f

$$\Delta(\overline{T}(f);F) = \exp\Big(\sum_{n=1}^{\infty} L(f^n;F)z^n/n\Big)^{-1} \in W(F) ,$$

with

$$L(f^n;F) = \sum_{r=0}^{\infty}(-)^r \mathrm{tr}(f^n_* : H_r(X;F) \longrightarrow H_r(X;F)) \in F \quad (n \geq 1)$$

the Lefschetz numbers of the iterates $f^n : X \longrightarrow X$. The Lefschetz zeta function was used by Weil in the study of the fixed points of the iterates of the Frobenius map on the algebraic variety over the algebraic closure of a finite field, and by Smale in dynamical systems. □

Part Two

Algebraic *L*-theory

20. Algebraic Poincaré complexes

The algebraic K-theory treatment of chain complexes will be extended in Part Two to algebraic Poincaré complexes, i.e. to algebraic L-theory, and this will be applied to high-dimensional knots.

An n-dimensional algebraic Poincaré complex over a ring with involution A is an n-dimensional A-module chain complex E with a chain equivalence to the n-dual E^{n-*}

$$E^{n-*} \simeq E$$

with either a symmetric or a quadratic structure. See Ranicki [235], [237] for the algebraic theory of Poincaré complexes. This chapter reviews the definitions, and the two basic types of cobordism groups of algebraic Poincaré complexes which occur in applications:

(i) the L-groups $L_*(A)$ of quadratic Poincaré complexes over A,
(ii) the Γ-groups $\Gamma_*(A \longrightarrow B)$ of quadratic complexes which are defined over A and become Poincaré over B, for a morphism of rings with involution $A \longrightarrow B$.

The applications of the L-groups to knot theory require many variations of these types, as well as the symmetric analogues. See Chap. 37 for examples of actual computations of the L-groups. The computation of the high-dimensional knot cobordism groups C_* is related in Chap. 42 to the L-groups of algebraic number fields and their rings of integers.

20A. $\dot{L}$-groups

Given a ring A let A^{op} be the opposite ring, the ring with one element a^{op} for each element $a \in A$ and

$$a^{op} + b^{op} = (a+b)^{op} \quad , \quad a^{op}b^{op} = (ba)^{op} \in A^{op} \; .$$

An involution $^- : A \longrightarrow A; a \longrightarrow \bar{a}$ is an isomorphism of rings

$$A \longrightarrow A^{op} \; ; \; a \longrightarrow \bar{a}^{op}$$

such that

$$\bar{\bar{a}} = a \in A \quad (a \in A) \; .$$

Use the involution on A to define a duality functor

$$* \ : \ \{A\text{-modules}\} \longrightarrow \{A\text{-modules}\} \ ; \ M \longrightarrow M^* \ = \ \mathrm{Hom}_A(M,A)$$

with

$$A \times M^* \longrightarrow M^* \ ; \ (a,f) \longrightarrow (x \longrightarrow f(x).\bar{a}) \ .$$

The dual of an A-module morphism $f : M \longrightarrow N$ is the A-module morphism

$$f^* \ : \ N^* \longrightarrow M^* \ ; \ x \longrightarrow (g \longrightarrow g(f(x))) \ .$$

If M is a f.g. projective A-module the natural A-module morphism

$$M \longrightarrow M^{**} \ ; \ x \longrightarrow (f \longrightarrow \overline{f(x)})$$

is an isomorphism.

Definition 20.1 (i) An *ϵ-symmetric form* (M,ϕ) is a f.g. projective A-module M together with an A-module morphism

$$\phi \ : \ M \longrightarrow M^* \ ; \ x \longrightarrow (y \longrightarrow \phi(x)(y))$$

such that $\epsilon\phi^* = \phi$. Equivalently, regard ϕ as a bilinear pairing

$$\phi \ : \ M \times M \longrightarrow A \ ; \ (x,y) \longrightarrow \phi(x,y) \ = \ \phi(x)(y)$$

such that

$$\phi(ax,by) \ = \ b\phi(x,y)\bar{a} \ , \ \overline{\epsilon\phi(y,x)} \ = \ \phi(x,y) \in A$$
$$(x,y \in M \ , \ a,b \in A) \ .$$

The form (M,ϕ) is *nonsingular* if $\phi : M \longrightarrow M^*$ is an isomorphism.
(ii) An *ϵ-quadratic form* (M,ψ) is a f.g. projective A-module M together with an equivalence class of A-module morphisms $\psi : M \longrightarrow M^*$, subject to the equivalence relation

$$\psi \sim \psi + \chi - \epsilon\chi^* \quad (\chi \in \mathrm{Hom}_A(M,M^*)) \ .$$

The *ϵ-symmetrization* of (M,ψ) is the ϵ-symmetric form $(M,\psi+\epsilon\psi^*)$.
(iii) A *sublagrangian* of an ϵ-symmetric form (M,ϕ) is a direct summand $L \subset M$ such that $L \subseteq L^{\perp}$, with

$$L^{\perp} \ = \ \{x \in M \,|\, \phi(x)(L) = 0\} \ .$$

A *lagrangian* is a sublagrangian L such that

$$L \ = \ L^{\perp} \ .$$

Similarly for ϵ-quadratic forms. □

For an A-module chain complex C write the dual A-modules as

$$C^r = (C_r)^* \quad (r \in \mathbb{Z}) .$$

The n-dual of an A-module chain complex C is the A-module chain complex C^{n-*} defined by

$$d_{C^{n-*}} = (d_C)^* : (C^{n-*})_r = C^{n-r} \longrightarrow (C^{n-*})_{r-1} = C^{n-r+1} .$$

Given an A-module chain complex C and a central unit $\epsilon \in A$ such that

$$\overline{\epsilon} = \epsilon^{-1} \in A$$

define the ϵ-transposition involution on $C \otimes_A C$ by

$$T_\epsilon : C_p \otimes_A C_q \longrightarrow C_q \otimes_A C_p ; \ x \otimes y \longrightarrow (-)^{pq}\epsilon y \otimes x ,$$

and let

$$\begin{cases} W^{\%}C = \mathrm{Hom}_{\mathbb{Z}[\mathbb{Z}_2]}(W, C \otimes_A C) \\ W_{\%}C = W \otimes_{\mathbb{Z}[\mathbb{Z}_2]} (C \otimes_A C) \end{cases}$$

with

$$W : \ldots \longrightarrow \mathbb{Z}[\mathbb{Z}_2] \xrightarrow{1-T} \mathbb{Z}[\mathbb{Z}_2] \xrightarrow{1+T} \mathbb{Z}[\mathbb{Z}_2] \xrightarrow{1-T} \mathbb{Z}[\mathbb{Z}_2]$$

the standard free $\mathbb{Z}[\mathbb{Z}_2]$-module resolution of $\mathbb{Z}$.

For chain complexes C which are f.g. projective (as will usually be the case) there are natural identifications

$$C \otimes_A C = \mathrm{Hom}_A(C^*, C) ,$$

$$T_\epsilon : \mathrm{Hom}_A(C^p, C_q) \longrightarrow \mathrm{Hom}_A(C^q, C_p) ; \ \phi \longrightarrow (-)^{pq}\phi^* .$$

A chain $\begin{cases} \phi \in W^{\%}C_n \\ \psi \in W_{\%}C_n \end{cases}$ is a collection of A-module morphisms

$$\begin{cases} \phi_s : C^{n-r+s} \longrightarrow C_r \\ \psi_s : C^{n-r-s} \longrightarrow C_r \end{cases} \quad (r, s \geq 0) ,$$

with the boundary $\begin{cases} d\phi \in W^{\%}C_{n-1} \\ d\psi \in W_{\%}C_{n-1} \end{cases}$ given by

$$\begin{cases} (d\phi)_s = d\phi_s + (-)^r\phi_s d^* + (-)^{n+s-1}(\phi_{s-1} + (-)^s T_\epsilon \phi_{s-1}) \\ \qquad\qquad : C^{n-r+s-1} \longrightarrow C_r \quad (\phi_{-1} = 0) \\ (d\psi)_s = d\psi_s + (-)^r\psi_s d^* + (-)^{n-s-1}(\psi_{s+1} + (-)^{s+1} T_\epsilon \psi_{s+1}) \\ \qquad\qquad : C^{n-r-s-1} \longrightarrow C_r . \end{cases}$$

Definition 20.2 (i) The $\begin{cases} \epsilon\text{-}symmetric \\ \epsilon\text{-}quadratic \end{cases}$ *Q-groups* of an A-module chain complex C are the $\begin{cases} \mathbb{Z}_2\text{-cohomology} \\ \mathbb{Z}_2\text{-homology} \end{cases}$ groups of $C \otimes_A C$

$$\begin{cases} Q^n(C, \epsilon) = H^n(\mathbb{Z}_2; C \otimes_A C) = H_n(W^{\%}C) \\ Q_n(C, \epsilon) = H_n(\mathbb{Z}_2; C \otimes_A C) = H_n(W_{\%}C) . \end{cases}$$

(ii) An *n-dimensional* $\begin{cases} \epsilon\text{-}symmetric \\ \epsilon\text{-}quadratic \end{cases}$ *complex* over A $\begin{cases} (C,\phi) \\ (C,\psi) \end{cases}$ is a finite f.g. projective A-module chain complex C together with a cycle $\begin{cases} \phi \in (W^{\%}C)_n \\ \psi \in (W_{\%}C)_n \end{cases}$ representing an element $\begin{cases} \phi \in Q^n(C,\epsilon) \\ \psi \in Q_n(C,\epsilon) \ . \end{cases}$ Such a complex is *Poincaré* if the A-module chain map

$$\begin{cases} \phi_0 \ : \ C^{n-*} \longrightarrow C \\ (1+T_\epsilon)\psi_0 \ : \ C^{n-*} \longrightarrow C \end{cases}$$

is a chain equivalence. □

In the applications, the chain complex C is usually n-dimensional.

Example 20.3 (i) A 0-dimensional ϵ-symmetric (Poincaré) complex (C,ϕ) with C 0-dimensional

$$C \ : \ \dots \longrightarrow 0 \longrightarrow C_0 \longrightarrow 0 \longrightarrow \dots$$

is an ϵ-symmetric form (C^0,ϕ_0) (with $\phi_0 : C^0 \longrightarrow C_0$ an isomorphism).
(ii) A 1-dimensional ϵ-symmetric (Poincaré) complex (C,ϕ) with C 1-dimensional

$$C \ : \ \dots \longrightarrow 0 \longrightarrow C_1 \xrightarrow{d} C_0 \longrightarrow 0 \longrightarrow \dots$$

is an *ϵ-symmetric formation*

$$(C_1 \oplus C^1, \begin{pmatrix} 0 & 1 \\ \epsilon & \phi_1 \end{pmatrix}; C_1, \mathrm{im}(\begin{pmatrix} \phi_0 \\ d^* \end{pmatrix} : C^0 \longrightarrow C_1 \oplus C^1))$$

that is an ϵ-symmetric form with a lagrangian and a sublagrangian (with the sublagrangian $L = \mathrm{im}(\begin{pmatrix} \phi_0 \\ d^* \end{pmatrix})$ a lagrangian in the nonsingular case).
Similarly in the ϵ-quadratic case.
See Ranicki [230] for more on forms and formations. □

A chain map $f : C \longrightarrow D$ of f.g. projective A-module chain complexes induces a $\mathbb{Z}$-module chain map

$$\begin{cases} f^{\%} \ : \ W^{\%}C \longrightarrow W^{\%}D \ ; \ \phi \ = \ \{\phi_s\} \longrightarrow f^{\%}\phi \ = \ \{f\phi_s f^*\} \\ f_{\%} \ : \ W_{\%}C \longrightarrow W_{\%}D \ ; \ \psi \ = \ \{\psi_s\} \longrightarrow f_{\%}\psi \ = \ \{f\psi_s f^*\} \end{cases}$$

with induced morphisms

$$\begin{cases} f^{\%} \ : \ Q^n(C,\epsilon) \longrightarrow Q^n(D,\epsilon) \\ f_{\%} \ : \ Q_n(C,\epsilon) \longrightarrow Q_n(D,\epsilon) \ . \end{cases}$$

An A-module chain homotopy $g : f \simeq f' : C \longrightarrow D$ induces a $\mathbb{Z}$-module chain homotopy

$$g^{\%} \ : \ f^{\%} \simeq f'^{\%} \ : \ W^{\%}C \longrightarrow W^{\%}D$$

with

$$\begin{aligned} g^{\%}(\phi)_s &= g\phi_s f^* + (-)^q f'\phi_s g^* + (-)^{q+s-1} gT_\epsilon\phi_{s-1}g^* \\ &\in W^{\%}D_{n-s+1} = \sum_{q=0}^{\infty} \mathrm{Hom}_A(D^{n-q-s+1}, D_q) \ \ (s \geq 0, \phi \in W^{\%}C_n) \ . \end{aligned}$$

Definition 20.4 (i) A *morphism* of n-dimensional ϵ-symmetric complexes

$$(f,\chi) \ : \ (C,\phi) \longrightarrow (D,\theta)$$

is a chain map $f : C \longrightarrow D$ together with a chain $\chi \in W^{\%}D_{n+1}$ such that

$$\theta - f^{\%}(\phi) \ = \ d(\chi) \in W^{\%}D_n \ .$$

(ii) A *homotopy* of morphisms of n-dimensional ϵ-symmetric complexes

$$(g,\nu) \ : \ (f,\chi) \ \simeq \ (f',\chi') \ : \ (C,\phi) \longrightarrow (D,\theta)$$

is a chain homotopy $g : f \simeq f' : C \longrightarrow D$ together with a chain $\nu \in W^{\%}D_{n+2}$ such that

$$\chi' - \chi \ = \ g^{\%}(\phi) + d(\nu) \in W^{\%}D_{n+1} \ .$$

Similarly in the ϵ-quadratic case. □

A morphism of ϵ-symmetric complexes $(f,\chi) : (C,\phi) \longrightarrow (D,\theta)$ is a homotopy equivalence if and only if $f : C \longrightarrow D$ is a chain equivalence. The homotopy type of an ϵ-symmetric complex (C,ϕ) over A depends only on the chain homotopy class of C and the homology class

$$\phi \in Q^n(C,\epsilon) \ = \ H_n(W^{\%}C) \ .$$

Similarly in the ϵ-quadratic case.

Example 20.5 (i) Given a CW complex X with universal cover $\widetilde{X}$ let $C(\widetilde{X})$ of the cellular chain complex of free $\mathbb{Z}[\pi_1(X)]$-modules. The *symmetric construction* of Ranicki [236] is the natural transformation

$$\Delta \ : \ H_n(X) \longrightarrow Q^n(C(\widetilde{X}))$$

which is obtained by applying $H_n(\mathbb{Z} \otimes_{\mathbb{Z}[\pi_1(X)]} -)$ to the Alexander-Whitney-Steenrod diagonal chain approximation

$$\Delta_{\widetilde{X}} \ : \ C(\widetilde{X}) \longrightarrow W^{\%}C(\widetilde{X}) \ .$$

For finite X and any homology class $[X] \in H_n(X)$ there is thus defined an n-dimensional symmetric complex $(C(\widetilde{X}),\phi)$ over $\mathbb{Z}[\pi_1(X)]$ with

$$\phi \ = \ \Delta[X] \in Q^n(C(\widetilde{X}))$$

such that

$$\phi_0 = [X] \cap - \ : \ C(\widetilde{X})^{n-*} \longrightarrow C(\widetilde{X}) \ .$$

By definition, X is an n-dimensional *geometric Poincaré complex* with fundamental class $[X] \in H_n(X)$ if and only if $(C(\widetilde{X}), \phi)$ is an n-dimensional symmetric Poincaré complex over $\mathbb{Z}[\pi_1(X)]$, with $\phi_0 : C(\widetilde{X})^{n-*} \longrightarrow C(\widetilde{X})$ a chain equivalence.

(ii) An n-dimensional *normal space* (X, ν_X, ρ_X) is a finite CW complex together with a spherical fibration $\nu_X : X \longrightarrow BSG(k)$ and a map $\rho_X : S^{n+k} \longrightarrow T(\nu_X)$ (Quinn [226]). Geometric Poincaré complexes are normal spaces, with the Spivak normal structure. See Ranicki [236], [237] for the *quadratic construction* and its application to the *quadratic kernel* of a normal map $(f, b) : M \longrightarrow X$ from an n-dimensional geometric Poincaré complex M to an n-dimensional normal space X. The quadratic kernel is an n-dimensional quadratic complex $(\mathcal{C}(f^!), \psi)$ over $\mathbb{Z}[\pi_1(X)]$ with $f^!$ the *Umkehr* $\mathbb{Z}[\pi_1(X)]$-module chain map

$$f^! \; : \; C(\widetilde{X})^{n-*} \xrightarrow{\widetilde{f}^*} C(\widetilde{M})^{n-*} \simeq C(\widetilde{M})$$

with $\mathcal{C}(f^!) \simeq \mathcal{C}(\widetilde{f})^{n-*+1}$. If X is an n-dimensional geometric Poincaré complex with fundamental class $[X] = f_*[M] \in H_n(X)$ (so that f has degree 1) then $f^!$ can be regarded as a chain map

$$f^! \; : \; C(\widetilde{X}) \simeq C(\widetilde{X})^{n-*} \xrightarrow{\widetilde{f}^*} C(\widetilde{M})^{n-*} \simeq C(\widetilde{M})$$

and $(\mathcal{C}(f^!), \psi)$ is an n-dimensional quadratic Poincaré complex over $\mathbb{Z}[\pi_1(X)]$, such that there is defined a homotopy equivalence of n-dimensional symmetric Poincaré complexes over $\mathbb{Z}[\pi_1(X)]$)

$$(C(\widetilde{M}), \Delta[M]) \simeq (\mathcal{C}(f^!), (1+T)\psi) \oplus (C(\widetilde{X}), \Delta[X]) \, .$$

See Ranicki [235], [236], [237] for further details. □

Definition 20.6 ([235, Chap. 3], [237, Chap. 1.5])

(i) The *relative* $\left\{\begin{array}{l}\textit{ε-symmetric}\\ \textit{ε-quadratic}\end{array}\right.$ *Q-groups* $\left\{\begin{array}{l}Q^n(f,\epsilon)\\ Q_n(f,\epsilon)\end{array}\right.$ of an A-module chain map $f : C \longrightarrow D$ are the $\left\{\begin{array}{l}\mathbb{Z}_2\text{-cohomology}\\ \mathbb{Z}_2\text{-homology}\end{array}\right.$ groups of the $\mathbb{Z}[\mathbb{Z}_2]$-module chain map

$$f \otimes f \; : \; C \otimes_A C \longrightarrow D \otimes_A D$$

with a long exact sequence

$$\begin{cases} \dots \longrightarrow Q^n(C,\epsilon) \xrightarrow{f^{\%}} Q^n(D,\epsilon) \longrightarrow Q^n(f,\epsilon) \longrightarrow Q^{n-1}(C,\epsilon) \longrightarrow \dots \\ \dots \longrightarrow Q_n(C,\epsilon) \xrightarrow{f_{\%}} Q_n(D,\epsilon) \longrightarrow Q_n(f,\epsilon) \longrightarrow Q_{n-1}(C,\epsilon) \longrightarrow \dots \, . \end{cases}$$

(ii) An *n-dimensional ε-symmetric pair* over A $(f, (\delta\phi, \phi))$ is a chain map $f : C \longrightarrow D$ of f.g. projective A-module chain complexes together with a cycle representing an element $(\delta\phi, \phi) \in Q^n(f, \epsilon)$. Such a pair is *Poincaré* if the A-module chain map

$$(\delta\phi, \phi)_0 = \begin{pmatrix} \delta\phi_0 \\ \phi_0 f^* \end{pmatrix} : D^{n-*} \longrightarrow \mathcal{C}(f)$$

is a chain equivalence. Similarly in the ϵ-quadratic case. □

Example 20.7 The constructions of algebraic Poincaré complexes from geometric Poincaré complexes in 20.4 generalize to pairs. Geometric Poincaré pairs determine symmetric Poincaré pairs. Likewise, normal maps of geometric Poincaré pairs determine quadratic Poincaré pairs. □

The terminology of 1.2 is extended to:

Terminology 20.8 The *Whitehead group* of a ring A is

$$Wh_1(A) = \begin{cases} \widetilde{K}_1(A) & \text{for arbitrary } A \\ \widetilde{K}_1(B[z, z^{-1}])/\{\tau(z)\} & \text{if } A = B[z, z^{-1}] \\ Wh(\pi) & \text{for a group ring } A = \mathbb{Z}[\pi]. \end{cases}$$

It should be clear from the context which of the three (inconsistent) definitions is being used. □

Proposition 20.9 Let $\begin{cases} (C, \phi) \\ (C, \psi) \end{cases}$ be an n-dimensional $\begin{cases} \epsilon\text{-symmetric} \\ \epsilon\text{-quadratic} \end{cases}$ Poincaré complex over A with C based f.g. free.

(i) The *torsion* of the complex is given by

$$\begin{cases} \tau(C, \phi) = \tau(\phi_0 : C^{n-*} \longrightarrow C) \in Wh_1(A)/\{\tau(\epsilon)\} \\ \tau(C, \psi) = \tau((1 + T_\epsilon)\psi_0 : C^{n-*} \longrightarrow C) \in Wh_1(A)/\{\tau(\epsilon)\} . \end{cases}$$

(ii) The complex is *simple* if

$$\begin{cases} \tau(C, \phi) = 0 \\ \tau(C, \psi) = 0 . \end{cases}$$ □

The involution on A determines the duality involutions

$$* : \widetilde{K}_0(A) \longrightarrow \widetilde{K}_0(A) \; ; \; [P] \longrightarrow [P^*] ,$$

$$* : Wh_1(A) \longrightarrow Wh_1(A) \; ; \; \tau(f : M \longrightarrow M) \longrightarrow \tau(f^* : M^* \longrightarrow M^*) .$$

Definition 20.10 Let $U \subseteq \widetilde{K}_0(A)$ (resp. $Wh_1(A)/\{\tau(\epsilon)\}$) be a $*$-invariant subgroup. The *n-dimensional U-intermediate* $\begin{cases} \epsilon\textit{-symmetric} \\ \epsilon\textit{-quadratic} \end{cases}$ *L-group of A* $\begin{cases} L^n_U(A, \epsilon) \\ L_n^U(A, \epsilon) \end{cases}$ is the cobordism group of n-dimensional $\begin{cases} \epsilon\text{-symmetric} \\ \epsilon\text{-quadratic} \end{cases}$ Poincaré complexes (C, θ) over A $(n \geq 0)$ which are f.g. projective (resp. based f.g. free), with

$$[C] \in U \subseteq \widetilde{K}_0(A) \;\; (\text{resp. } \tau(C, \theta) \in U \subseteq Wh_1(A)/\{\tau(\epsilon)\}) .$$ □

Example 20.11 The 0-dimensional L-group $\begin{cases} L^0_U(A,\epsilon) \\ L_0^U(A,\epsilon) \end{cases}$ is the *Witt group* of nonsingular $\begin{cases} \epsilon\text{-symmetric} \\ \epsilon\text{-quadratic} \end{cases}$ forms $\begin{cases} (M,\phi : M \longrightarrow M^*) \\ (M,\psi : M \longrightarrow M^*) \end{cases}$ over A, with

$$\begin{cases} \phi \;=\; \epsilon\phi^* \;:\; M \longrightarrow M^* \\ \psi + \epsilon\psi^* \;:\; M \longrightarrow M^* \end{cases}$$

an isomorphism. Such a form is $\begin{cases} \textit{metabolic} \\ \textit{hyperbolic} \end{cases}$ if there exists a lagrangian (20.1), in which case

$$\begin{cases} (M,\phi) \;=\; 0 \in L^0_U(A,\epsilon) \\ (M,\psi) \;=\; 0 \in L_0^U(A,\epsilon)\ . \end{cases}$$

See Ranicki [235] for a more detailed account of the correspondence between the 0-dimensional algebraic Poincaré cobordism groups and the Witt groups of forms. □

Proposition 20.12 *For $U \subseteq V$ there is a Rothenberg-type exact sequence*

$$\begin{cases} \dots \longrightarrow \widehat{H}^n(\mathbb{Z}_2;V/U) \longrightarrow L^n_U(A,\epsilon) \longrightarrow L^n_V(A,\epsilon) \\ \qquad\qquad\qquad\qquad\qquad\qquad \longrightarrow \widehat{H}^n(\mathbb{Z}_2;V/U) \longrightarrow \dots \\ \dots \longrightarrow \widehat{H}^{n+1}(\mathbb{Z}_2;V/U) \longrightarrow L_n^U(A,\epsilon) \longrightarrow L_n^V(A,\epsilon) \\ \qquad\qquad\qquad\qquad\qquad\qquad \longrightarrow \widehat{H}^n(\mathbb{Z}_2;V/U) \longrightarrow \dots \end{cases}$$

with the Tate $\mathbb{Z}_2$-cohomology groups defined by

$$\widehat{H}^n(\mathbb{Z}_2;V/U) \;=\; \{a \in V/U \,|\, a^* = (-)^n a\}/\{b + (-)^n b^* \,|\, b \in V/U\}\ . \quad \square$$

In the extreme cases $U = \{0\}, \widetilde{K}_0(A), Wh_1(A)/\{\tau(\epsilon)\}$ the ϵ-symmetric L-groups $L^n_U(A,\epsilon)$ are denoted by

$$\begin{aligned} L^n_{\widetilde{K}_0(A)}(A,\epsilon) \;&=\; L^n_p(A,\epsilon) \\ L^n_{\{0\}\subseteq\widetilde{K}_0(A)}(A,\epsilon) \;&=\; L^n_{Wh_1(A)/\{\tau(\epsilon)\}}(A,\epsilon) \;=\; L^n_h(A,\epsilon) \\ L^n_{\{0\}\subseteq Wh_1(A)/\{\tau(\epsilon)\}}(A,\epsilon) \;&=\; L^n_s(A,\epsilon) \end{aligned}$$

with exact sequences

$$\begin{aligned} \dots \longrightarrow L^n_h(A,\epsilon) \longrightarrow L^n_p(A,\epsilon) \longrightarrow \widehat{H}^n(\mathbb{Z}_2;\widetilde{K}_0(A)) \\ \longrightarrow L^{n-1}_h(A,\epsilon) \longrightarrow \dots \\ \dots \longrightarrow L^n_s(A,\epsilon) \longrightarrow L^n_h(A,\epsilon) \longrightarrow \widehat{H}^n(\mathbb{Z}_2;Wh_1(A)/\{\tau(\epsilon)\}) \\ \longrightarrow L^{n-1}_s(A,\epsilon) \longrightarrow \dots\ , \end{aligned}$$

and similarly in the ϵ-quadratic case. The ϵ-symmetrization maps

$$1+T_\epsilon \;:\; L_n^U(A,\epsilon) \longrightarrow L_U^n(A,\epsilon)\;;\;(C,\psi)\longrightarrow(C,\phi)$$

are defined by

$$\phi_s = \begin{cases}(1+T_\epsilon)\psi_0 & \text{if } s=0\\ 0 & \text{if } s\geq 1\,.\end{cases}$$

Remark 20.13 The ϵ-symmetrization maps $1+T_\epsilon : L_*^U(A,\epsilon)\longrightarrow L_U^*(A,\epsilon)$ are isomorphisms modulo 8-torsion, and are actual isomorphisms if there exists a central element $s\in A$ with

$$s+\overline{s}=1\in A\,,$$

e.g. $s=1/2\in A$. □

In the case $\epsilon=1$ the terminology is abbreviated

1-symmetric = symmetric , 1-quadratic = quadratic

$$L_U^*(A,1) \;=\; L_U^*(A)\;\;,\;\; L_*^U(A,1) \;=\; L_U^*(A)\,.$$

The free L-groups are written

$$L_h^*(A) \;=\; L^*(A)\;\;,\;\; L_*^h(A) \;=\; L_*(A)\,.$$

The *skew-suspension* maps of Ranicki [235, p. 105]

$$\overline{S}\;:\;L_n^U(A,\epsilon)\longrightarrow L_{n+2}^U(A,-\epsilon)\;;\;(C,\psi)\longrightarrow(SC,\overline{S}\psi)\;\;(\overline{S}\psi_s=\psi_s)\;,$$
$$\overline{S}\;:\;L_U^n(A,\epsilon)\longrightarrow L_U^{n+2}(A,-\epsilon)\;;\;(C,\phi)\longrightarrow(SC,\overline{S}\phi)\;\;(\overline{S}\phi_s=\phi_s)$$

are isomorphisms for the ϵ-quadratic L-groups L_* for all A, but are isomorphisms for the ϵ-symmetric L-groups L^* only for certain A, e.g. if A has homological dimension 1, such as a Dedekind ring, or if $1/2\in A$. The ϵ-quadratic L-groups of any ring with involution A are thus 4-periodic

$$L_n^U(A,\epsilon) \;=\; L_{n+2}^U(A,-\epsilon) \;=\; L_{n+4}^U(A,\epsilon)\,,$$

with

$$L_{2i}^U(A,\epsilon)=L_0^U(A,(-)^i\epsilon)\;\;,\;\;L_{2i+1}^U(A,\epsilon)=L_1^U(A,(-)^i\epsilon)\,,$$

(as in the original treatment by Wall [302]), while the ϵ-symmetric L-groups $L_U^*(A,\epsilon)$ are not 4-periodic in general.

The algebraic L-groups $L_*(A), L^*(A)$ of a ring with involution A are algebraic analogues of the manifold and geometric Poincaré bordism groups.

Example 20.14 (i) The *symmetric signature* of an n-dimensional geometric Poincaré complex X is the symmetric Poincaré cobordism class introduced by Mishchenko [201]

$$\sigma^*(X) \;=\; (C(\widetilde{X}),\Delta[X])\in L^n(\mathbb{Z}[\pi_1(X)])\,,$$

with $\widetilde{X}$ the universal cover of X.

(ii) The *quadratic signature* of a (degree 1) normal map $(f,b): M \longrightarrow X$ of n-dimensional geometric Poincaré complexes is the quadratic Poincaré cobordism class of the kernel n-dimensional quadratic Poincaré complex $(\mathcal{C}(f^!),\psi)$ over $\mathbb{Z}[\pi_1(X)]$ (20.4).

$$\sigma_*(f,b) = (\mathcal{C}(f^!),\psi) \in L_n(\mathbb{Z}[\pi_1(X)]) .$$

See Ranicki [236] for a detailed exposition of these signatures, including the identification of the quadratic signature of a normal map $(f,b): M \longrightarrow X$ from a manifold to a geometric Poincaré complex with the *surgery obstruction* of Wall [302]. □

Recall from Chap. 1 that a finite chain complex C of f.g. projective A-modules is round if $[C] = 0 \in K_0(A)$. It is convenient to have available the 'round L-groups' $L^*_r(A)$ which are related to the projective L-groups $L^*_p(A)$ in much the way that the absolute K-groups $K_*(A)$ are related to the reduced K-groups $\widetilde{K}_*(A)$.

Definition 20.15 (Hambleton, Ranicki and Taylor [104])
(i) The *round ϵ-symmetric L-groups* $L^n_r(A,\epsilon)$ $(n \geq 0)$ are the cobordism groups of n-dimensional ϵ-symmetric Poincaré complexes (C,ϕ) over A with C round, that is finite f.g. projective with

$$[C] = 0 \in K_0(A) .$$

(ii) The *round simple ϵ-symmetric L-groups* $L^n_{rs}(A,\epsilon)$ $(n \geq 0)$ are the cobordism groups of based f.g. free n-dimensional ϵ-symmetric Poincaré complexes (C,ϕ) over A with C round and

$$\tau(\phi_0 : C^{n-*} \longrightarrow C) = 0 \in K_1(A) .$$

Similarly for the *round ϵ-quadratic L-groups* $L^r_*(A,\epsilon)$ and the *round simple ϵ-quadratic L-groups* $L^{rs}_*(A,\epsilon)$. □

Proposition 20.16 [104]
(i) *The projective and round ϵ-symmetric L-groups are related by a Rothenberg-type exact sequence*

$$\dots \longrightarrow \widehat{H}^n(\mathbb{Z}_2;K_0(A)) \longrightarrow L^n_r(A,\epsilon) \longrightarrow L^n_p(A,\epsilon) \longrightarrow \widehat{H}^n(\mathbb{Z}_2;K_0(A)) \longrightarrow \dots .$$

(ii) *The round and round simple ϵ-symmetric L-groups are related by a Rothenberg-type exact sequence*

$$\dots \longrightarrow \widehat{H}^n(\mathbb{Z}_2;K_1(A)) \longrightarrow L^n_{rs}(A,\epsilon) \longrightarrow L^n_r(A,\epsilon) \longrightarrow \widehat{H}^n(\mathbb{Z}_2;K_1(A)) \longrightarrow \dots .$$

Similarly in the ϵ-quadratic case. □

20B. Γ-groups

The algebraic L-groups $L_*(A)$ of Wall [302] are the obstruction groups for surgery up to homotopy equivalence. The algebraic Γ-groups $\Gamma_*(A\longrightarrow B)$ of Cappell and Shaneson [40] are the obstruction groups for surgery up to homology equivalence. This section just deals with the algebraic definitions of the Γ-groups – see Chap. 22 for the applications to codimension 2 surgery theory.

Definition 20.17 Let $\mathcal{F} : A\longrightarrow B$ be a morphism of rings with involution, and let $U \subseteq \widetilde{K}_0(B)$ (resp. $Wh_1(B)/\{\tau(\epsilon)\}$) be a $*$-invariant subgroup. The n*-dimensional* U*-intermediate* $\left\{\begin{array}{l}\epsilon\textit{-symmetric}\\ \epsilon\textit{-quadratic}\end{array}\right.$ Γ*-group* $\left\{\begin{array}{l}\Gamma^n_U(\mathcal{F},\epsilon)\\ \Gamma_n^U(\mathcal{F},\epsilon)\end{array}\right.$ $(n \geq 0)$ is the cobordism group of n-dimensional $\left\{\begin{array}{l}\epsilon\text{-symmetric}\\ \epsilon\text{-quadratic}\end{array}\right.$ B-Poincaré complexes (C,θ) over A which are f.g. projective (resp. based f.g. free) for $i = 0$ (resp. 1) with

$$[B\otimes_A C] \in U \subseteq \widetilde{K}_0(B) \quad (\text{resp. } \tau(B\otimes_A (C,\theta)) \in U \subseteq Wh_1(B)/\{\tau(\epsilon)\}) \ . \ \square$$

By analogy with 20.12:

Proposition 20.18 *For* $U \subseteq V$ *there is a Rothenberg-type exact sequence*

$$\left\{\begin{array}{l}\dots \longrightarrow \widehat{H}^{n+1}(\mathbb{Z}_2;V/U) \longrightarrow \Gamma^n_U(\mathcal{F},\epsilon) \longrightarrow \Gamma^n_V(\mathcal{F},\epsilon)\\ \qquad\qquad\qquad\qquad\qquad\qquad\qquad \longrightarrow \widehat{H}^n(\mathbb{Z}_2;V/U) \longrightarrow \dots\\ \dots \longrightarrow \widehat{H}^{n+1}(\mathbb{Z}_2;V/U) \longrightarrow \Gamma_n^U(\mathcal{F},\epsilon) \longrightarrow \Gamma_n^V(\mathcal{F},\epsilon)\\ \qquad\qquad\qquad\qquad\qquad\qquad\qquad \longrightarrow \widehat{H}^n(\mathbb{Z}_2;V/U) \longrightarrow \dots \ .\end{array}\right.$$

$\square$

Remark 20.19 (i) See Cappell and Shaneson [40] for the original definition of the Γ-groups and B-homology surgery theory, and Ranicki [237, Chap. 7.7] for the chain complex treatment.
(ii) For a morphism of rings with involution $\mathcal{F} : A\longrightarrow B$ which is locally epic (9.10) the forgetful map from the free ϵ-quadratic Γ-groups of $\mathcal{F}$ to the free ϵ-quadratic L-groups of B

$$\Gamma^h_n(\mathcal{F},\epsilon) \longrightarrow L^h_n(B,\mathcal{F}(\epsilon)) \ ; \ (C,\psi) \longrightarrow B\otimes_A (C,\psi)$$

is surjective for even n and injective for odd n. The Γ-groups for $\mathcal{F} = 1 : A\longrightarrow B = A$ are just the L-groups of A

$$\Gamma_*(1 : A\longrightarrow A,\epsilon) \ = \ L_*(A,\epsilon) \ .$$

(iii) Let $(X,\partial X)$ be an n-dimensional geometric Λ-coefficient Poincaré pair, for some morphism of rings with involution $\mathcal{F} : \mathbb{Z}[\pi_1(X)]\longrightarrow\Lambda$, and let $U \subseteq Wh_1(\Lambda)$ be a $*$-invariant subgroup such that $\tau(X,\partial X;\Lambda) \in U$. The quadratic kernel of a degree 1 normal map $(f,b):(M,\partial M)\longrightarrow(X,\partial X)$ with $\partial f:\partial M\longrightarrow\partial X$ a Λ-homology equivalence with Λ-coefficient torsion $\tau(\partial f;\Lambda)\in U$ is an n-dimensional quadratic Λ-Poincaré complex $(\mathcal{C}(f^!),\psi)$ over $\mathbb{Z}[\pi_1(X)]$ with torsion $\tau(\mathcal{C}(f^!),\psi)\in U$. The *$\Lambda$-coefficient quadratic signature*

$$\sigma_*^\Lambda(f,b) \;=\; (\mathcal{C}(f^!),\psi)\in \Gamma_n^U(\mathcal{F})$$

is the cobordism class of the kernel n-dimensional quadratic Λ-Poincaré over $\mathbb{Z}[\pi_1(X)]$. This is the Λ-homology surgery obstruction, such that $\sigma_*^\Lambda(f,b)=0$ if (and for $\mathcal{F}$ locally epic and $n\geq 5$ only if) (f,b) is normal bordant rel ∂ to a Λ-homology equivalence of pairs with torsion in U.

(iv) Let $(X,\partial X)$ be an n-dimensional geometric Poincaré pair such that $H_*(\partial X;\Lambda)=0$ for some morphism of rings with involution $\mathcal{F}:\mathbb{Z}[\pi_1(X)]\longrightarrow\Lambda$, and let $U\subseteq Wh_1(\Lambda)$ be a $*$-invariant subgroup such that $\tau(X,\partial X;\Lambda)\in U$. The *$\Lambda$-coefficient symmetric signature*

$$\sigma_\Lambda^*(X,\partial X) \;=\; (C(\widetilde{X},\widetilde{\partial X}),\phi)\in \Gamma_U^n(\mathcal{F})$$

is the cobordism class of the n-dimensional symmetric Λ-Poincaré complex $(C(\widetilde{X},\widetilde{\partial X}),\phi)$ over $\mathbb{Z}[\pi_1(X)]$ with $\phi=\Delta[X]$ the usual symmetric structure. □

The relative ϵ-quadratic Γ-groups $\Gamma_*(\Phi,\epsilon)$ are defined for any commutative square of rings with involution

$$\begin{array}{ccc} A & \longrightarrow & A' \\ \big\downarrow & \Phi & \big\downarrow \\ B & \longrightarrow & B' \end{array}$$

to fit into the exact sequence

$$\begin{aligned} \dots \longrightarrow \Gamma_n(A\longrightarrow B,\epsilon) \longrightarrow \Gamma_n(A'\longrightarrow B',\epsilon) \longrightarrow \Gamma_n(\Phi,\epsilon) \\ \longrightarrow \Gamma_{n-1}(A\longrightarrow B,\epsilon) \longrightarrow \dots . \end{aligned}$$

Similarly for the ϵ-symmetric Γ-groups Γ^*.

Definition 20.20 For any morphism of rings with involution $\mathcal{F}:A\longrightarrow B$ let $\Phi_{\mathcal{F}}$ be the commutative square

$$\begin{array}{ccc} A & \longrightarrow & A \\ \big\downarrow & \Phi_{\mathcal{F}} & \big\downarrow \\ A & \longrightarrow & B \end{array}$$

so that the relative ϵ-quadratic Γ-groups $\Gamma_*(\Phi_{\mathcal{F}},\epsilon)$ fit into an exact sequence

$$\dots \longrightarrow L_n(A,\epsilon) \longrightarrow \Gamma_n(\mathcal{F},\epsilon) \xrightarrow{\partial} \Gamma_n(\Phi_{\mathcal{F}},\epsilon) \longrightarrow L_{n-1}(A,\epsilon) \longrightarrow \dots .$$

Similarly for the relative ϵ-symmetric Γ-groups $\Gamma^*(\Phi_{\mathcal{F}},\epsilon)$. □

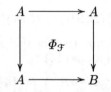

Proposition 20.21 (Ranicki [237, 2.5.7])
For a locally epic morphism of rings with involution $\mathcal{F} : A \longrightarrow B$ the relative ϵ-quadratic Γ-group $\Gamma_n(\Phi_{\mathcal{F}}, \epsilon)$ is (isomorphic to) the cobordism group of $(n-1)$-dimensional ϵ-quadratic B-contractible Poincaré complexes over A. The map ∂ in the exact sequence of 20.20 given by

$$\partial \ : \ \Gamma_n(\mathcal{F}, \epsilon) \longrightarrow \Gamma_n(\Phi_{\mathcal{F}}, \epsilon) \ ; \ (C, \psi) \longrightarrow \partial(C, \psi)$$

with (C, ψ) an n-dimensional ϵ-quadratic B-Poincaré complex over A, and $\partial(C, \psi)$ the boundary $(n-1)$-dimensional ϵ-quadratic B-contractible Poincaré complex over A. Similarly for the relative ϵ-symmetric Γ-groups $\Gamma^*(\Phi_{\mathcal{F}}, \epsilon)$. □

20C. Thickenings, unions and triads

This section sets out various algebraic Poincaré analogues of standard topological constructions, which will be useful further on.

For any n-dimensional geometric Poincaré pair $(W, \partial W)$ the n-dimensional symmetric Poincaré pair $(C(\partial W) \longrightarrow C(W), \Delta[W])$ is determined algebraically by the n-dimensional symmetric complex $(C(W, \partial W), \Delta[W]/\Delta[\partial W])$, as follows.

Definition 20.22 (Ranicki [235, Chap. 3])
(i) The *thickening* of an n-dimensional ϵ-symmetric complex (C, ϕ) is the n-dimensional ϵ-symmetric Poincaré pair $(f : \partial C \longrightarrow C^{n-*}, (0, \partial\phi))$ with

$$\begin{aligned}
\partial C \ &= \ S^{-1}\mathcal{C}(\phi_0 : C^{n-*} \longrightarrow C) \ , \\
f \ &= \ (0\ 1) \ : \ \partial C_r \ = \ C_{r+1} \oplus C^{n-r} \longrightarrow C^{n-r} \ , \\
d_{\partial C} \ &= \ \begin{pmatrix} d_C & (-)^r \phi_0 \\ 0 & (-)^r d_C^* \end{pmatrix} : \\
&\qquad \partial C_r \ = \ C_{r+1} \oplus C^{n-r} \longrightarrow \partial C_{r-1} \ = \ C_r \oplus C^{n-r+1} \ , \\
\partial\phi_0 \ &= \ \begin{pmatrix} (-)^{n-r-1} T_\epsilon \phi_1 & (-)^{r(n-r-1)}\epsilon \\ 1 & 0 \end{pmatrix} : \\
&\qquad \partial C^{n-r-1} \ = \ C^{n-r} \oplus C_{r+1} \longrightarrow \partial C_r \ = \ C_{r+1} \oplus C^{n-r} \ , \\
\partial\phi_s \ &= \ \begin{pmatrix} (-)^{n-r+s-1} T_\epsilon \phi_{s+1} & 0 \\ 0 & 0 \end{pmatrix} : \\
&\qquad \partial C^{n-r+s-1} = C^{n-r+s} \oplus C_{r-s+1} \\
&\qquad\qquad \longrightarrow \partial C_r = C_{r+1} \oplus C^{n-r} \quad (s \geq 1) \ .
\end{aligned}$$

The $(n-1)$-dimensional ϵ-symmetric Poincaré complex

$$\partial(C,\phi) \;=\; (\partial C, \partial\phi)$$

is the *boundary* of (C,ϕ).
(ii) The *Thom complex* of an n-dimensional ϵ-symmetric Poincaré pair $(f : C \longrightarrow D, (\delta\phi, \phi))$ over A is the n-dimensional ϵ-symmetric complex $(D/C, \delta\phi/\phi)$ over A with

$$D/C = \mathcal{C}(f : C \longrightarrow D)$$

and $\delta\phi/\phi \in Q^n(D/C,\epsilon)$ the image of $(\delta\phi,\phi) \in Q^n(f,\epsilon)$ under the canonical map.
Similarly in the ϵ-quadratic case. □

Proposition 20.23 ([235, Chap. 3])
(i) *The thickening and the Thom complex define inverse bijections between the homotopy equivalence classes of n-dimensional ϵ-symmetric complexes over A and $(n+1)$-dimensional ϵ-symmetric Poincaré pairs over A.*
(ii) *An n-dimensional ϵ-symmetric complex (C,ϕ) over A if and only if the boundary $(n-1)$-dimensional ϵ-symmetric Poincaré complex $\partial(C,\phi)$ is contractible.*
Similarly in the ϵ-quadratic case. □

The union of n-dimensional geometric Poincaré pairs $(X_+, \partial X_+)$, $(X_-, \partial X_-)$ with $\partial X_+ = -\partial X_-$ is an n-dimensional geometric Poincaré complex $X_+ \cup_\partial X_-$.

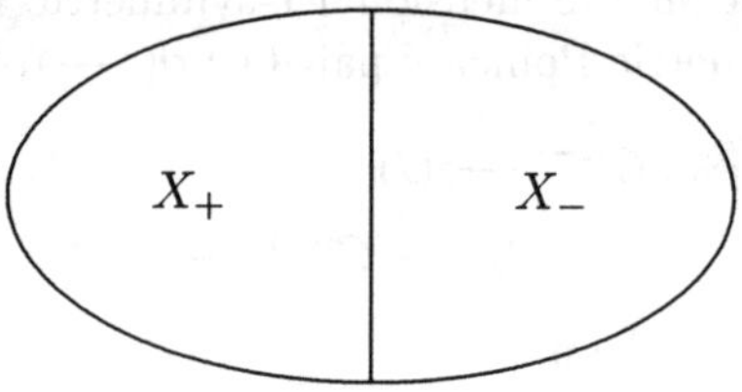

$X_+ \cup_\partial X_-$

Similarly for algebraic Poincaré pairs:

Definition 20.24 ([235, p. 135], [237, 1.7])
The *union* of n-dimensional ϵ-symmetric Poincaré pairs over A

$$c_+ \;=\; (f_+ : C \longrightarrow D_+, (\delta_+\phi, \phi)) \;\;,\;\; c_- \;=\; (f_- : C \longrightarrow D_-, (\delta_-\phi, -\phi))$$

is the n-dimensional ϵ-symmetric Poincaré complex over A

$$c_+ \cup c_- \;=\; (D_+ \cup_C D_-, \delta_+\phi \cup_\phi \delta_-\phi)$$

with

$$(D_+ \cup_C D_-) = \mathcal{C}(\begin{pmatrix} f_+ \\ f_- \end{pmatrix} : C \longrightarrow D_+ \oplus D_-) ,$$

$$(\delta_+\phi \cup_\phi \delta_-\phi)_s = \begin{pmatrix} \delta_+\phi_s & 0 & 0 \\ (-)^{n-r-1}\phi_s f_+^* & (-)^{n-r+s}T_\epsilon\phi_{s-1} & 0 \\ 0 & (-)^s f_-\phi_s & \delta_-\phi_s \end{pmatrix} :$$

$$(D_+ \cup_C D_-)^{n-r+s} = (D_+)^{n-r+s} \oplus C^{n-r+s-1} \oplus (D_-)^{n-r+s}$$

$$\longrightarrow (D_+ \cup_C D_-)_r = (D_+)_r \oplus C_{r-1} \oplus (D_-)_r \quad (s \geq 0, \phi_{-1} = 0) .$$

Similarly in the ϵ-quadratic case. □

The glueing of algebraic Poincaré pairs is a special case of the glueing of algebraic Poincaré cobordisms. In the applications of algebraic Poincaré complexes to topology in general (and Chap. 30 in particular) it is convenient to have available the following splitting construction, which characterizes the algebraic Poincaré complexes which are unions.

Proposition 20.25 *Let (C, ϕ) be an n-dimensional ϵ-symmetric Poincaré complex over A. A morphism of n-dimensional ϵ-symmetric complexes*

$$(f, \chi) : (C, \phi) \longrightarrow (D, \theta)$$

determines a homotopy equivalence

$$(C, \phi) \simeq c_+ \cup c_-$$

(regarding (C, ϕ) as the n-dimensional ϵ-symmetric Poincaré pair $(0 \longrightarrow C, (\phi, 0))$) with c_+, c_- the n-dimensional ϵ-symmetric Poincaré pairs

$$c_+ = (g : \partial D \longrightarrow D^{n-*}, (0, \partial\theta)) \ , \ c_- = (g_- : \partial D \longrightarrow D_-, (\theta', \partial\theta))$$

with c_+ the thickening of (D, θ), c_- homotopy equivalent to the thickening of $(\mathcal{C}(\phi_0 f^* : D^{n-*} \longrightarrow C), e^{\%}(\theta))$ and

$$D_- = \mathcal{C}(f : C \longrightarrow D)_{*+1} \ ,$$

$$e = inclusion : C \longrightarrow \mathcal{C}(\phi_0 f^*) ,$$

$$g_- = \begin{pmatrix} 1 & 0 \\ 0 & \phi_0 f^* \end{pmatrix} :$$

$$\partial D_r = D_{r+1} \oplus D^{n-r} \longrightarrow (D_-)_r = D_{r+1} \oplus C_r . \quad □$$

A manifold triad $(M; \partial_0 M, \partial_1 M)$ is a cobordism of manifolds with boundary, or equivalently a manifold M with the boundary expressed as a union of codimension 0 submanifolds $\partial_0 M, \partial_1 M \subset \partial M$

$$\partial M = \partial_0 M \cup \partial_1 M$$

such that

$$\partial_0 M \cap \partial_1 M = \partial\partial_0 M = \partial\partial_1 M .$$

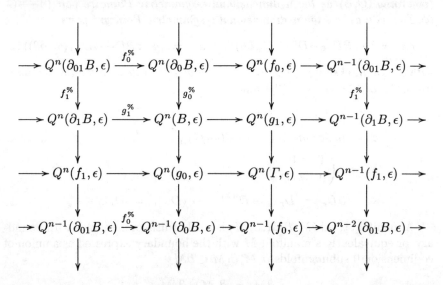

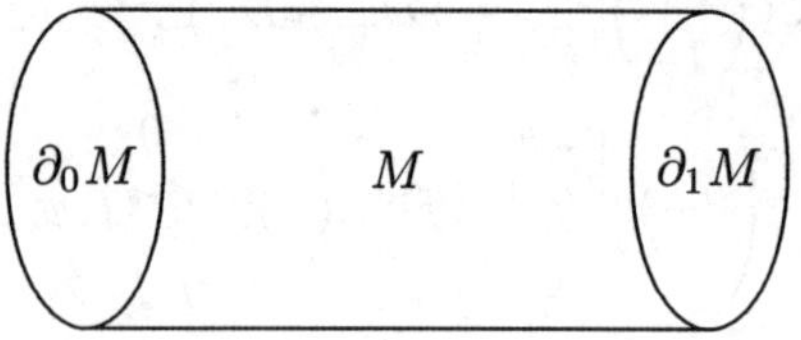

Algebraic and geometric Poincaré triads are the evident analogues of manifold triads, and have already been studied in Ranicki [237]. The standard constructions of algebraic complexes (resp. pairs) from geometric Poincaré complexes (resp. pairs) extend to triads. The algebraic properties of triads will be used in the treatment of codimension q surgery in Chap. 21 and beyond.

Definition 20.26 (i) An *n-dimensional ϵ-symmetric triad* over A

$$(\Gamma,\Theta) = \left(\begin{array}{ccc} \partial_{01}B & \xrightarrow{f_0} & \partial_0 B \\ {\scriptstyle f_1}\big\downarrow & & \big\downarrow{\scriptstyle g_0} \\ \partial_1 B & \xrightarrow{g_1} & B \end{array}\ ,\ \begin{array}{ccc} \partial_{01}\theta & \longrightarrow & \partial_0\theta \\ \big\downarrow & & \big\downarrow \\ \partial_1\theta & \longrightarrow & \theta \end{array}\right)$$

is a commutative square Γ of f.g. projective A-module chain complexes with an element $\Theta \in Q^n(\Gamma,\epsilon)$ in the triad Q-group which fits into the commutative diagram of exact sequences

$$\begin{array}{ccccccccc}
 & \big\downarrow & & \big\downarrow & & \big\downarrow & & \big\downarrow & \\
\longrightarrow & Q^n(\partial_{01}B,\epsilon) & \xrightarrow{f_0^{\%}} & Q^n(\partial_0 B,\epsilon) & \longrightarrow & Q^n(f_0,\epsilon) & \longrightarrow & Q^{n-1}(\partial_{01}B,\epsilon) & \longrightarrow \\
 & {\scriptstyle f_1^{\%}}\big\downarrow & & \big\downarrow{\scriptstyle g_0^{\%}} & & \big\downarrow & & {\scriptstyle f_1^{\%}}\big\downarrow & \\
\longrightarrow & Q^n(\partial_1 B,\epsilon) & \xrightarrow{g_1^{\%}} & Q^n(B,\epsilon) & \longrightarrow & Q^n(g_1,\epsilon) & \longrightarrow & Q^{n-1}(\partial_1 B,\epsilon) & \longrightarrow \\
 & \big\downarrow & & \big\downarrow & & \big\downarrow & & \big\downarrow & \\
\longrightarrow & Q^n(f_1,\epsilon) & \longrightarrow & Q^n(g_0,\epsilon) & \longrightarrow & Q^n(\Gamma,\epsilon) & \longrightarrow & Q^{n-1}(f_1,\epsilon) & \longrightarrow \\
 & \big\downarrow & & \big\downarrow & & \big\downarrow & & \big\downarrow & \\
\longrightarrow & Q^{n-1}(\partial_{01}B,\epsilon) & \xrightarrow{f_0^{\%}} & Q^{n-1}(\partial_0 B,\epsilon) & \longrightarrow & Q^{n-1}(f_0,\epsilon) & \longrightarrow & Q^{n-2}(\partial_{01}B,\epsilon) & \longrightarrow \\
 & \big\downarrow & & \big\downarrow & & \big\downarrow & & \big\downarrow &
\end{array}$$

(ii) An n-dimensional ϵ-symmetric triad (Γ,Θ) is *Poincaré* if the $(n-1)$-dimensional ϵ-symmetric pairs $(f_0,\partial_0\Theta)$, $(f_1,\partial_1\Theta)$ are Poincaré, and the A-module chain map

$$\Theta_0 \;:\; \mathcal{C}(g_0)^{n-*} \longrightarrow \mathcal{C}(g_1)$$

is a chain equivalence, in which case there is defined an n-dimensional ϵ-symmetric Poincaré pair

$$(g_0 \cup g_1 : \partial_0 B \cup_{\partial_{01} B} \partial_1 B \longrightarrow B, (\theta, \partial_0\theta \cup_{\partial_{01}\theta} -\partial_1\theta))$$

with boundary the union

$$(g_0 : \partial_{01} B \longrightarrow \partial_0 B, (\partial_0\theta, \partial_{01}\theta)) \cup (g_1 : \partial_{01} B \longrightarrow \partial_1 B, (-\partial_1\theta, -\partial_{01}\theta)) \;.$$

Similarly for ϵ-quadratic (Poincaré) pairs and triads. □

The following triad version of the one-one correspondence between the homotopy equivalence classes of algebraic Poincaré pairs and complexes (20.23) will be used in Chaps. 22, 26:

Proposition 20.27 *The homotopy equivalence classes of n-dimensional ϵ-symmetric Poincaré triads (Γ, Θ) over A are in one-one correspondence with the homotopy equivalence classes of n-dimensional ϵ-symmetric complexes (C, ϕ) over A together with a factorization*

$$\phi_0 \;=\; gf \;:\; C^{n-*} \xrightarrow{\;f\;} D \xrightarrow{\;g\;} C$$

up to chain homotopy, for some finite f.g. projective A-module chain complex D and chain maps $f : C^{n-*} \longrightarrow D$, $g : D \longrightarrow C$. The triad (Γ, Θ) (as in 20.22) corresponding to $(C, \phi), f, g$ has

$$\begin{aligned}
B \;&=\; C^{n-*} \;,\\
\partial_0 B \;&=\; \mathcal{C}(f : C^{n-*} \longrightarrow D)_{*+1} \;,\quad \partial_1 B \;=\; \mathcal{C}(g^* : C^{n-*} \longrightarrow D^{n-*})_{*+1} \;,\\
\partial_{01} B \;&=\; \mathcal{C}(\mathcal{C}(f^*)^{n-1-*} \longrightarrow \mathcal{C}(f^*))_{*+1} \;=\; \mathcal{C}(\mathcal{C}(g)^{n-1-*} \longrightarrow \mathcal{C}(g))_{*+1} \;.
\end{aligned}$$

Similarly for the ϵ-quadratic case, with an ϵ-quadratic complex (C, ψ) and a factorization

$$(1 + T_\epsilon)\psi_0 \;=\; gf \;:\; C^{n-*} \xrightarrow{\;f\;} D \xrightarrow{\;g\;} C \;.$$ □

Remark 20.28 Here is a useful interpretation of 20.27, which was already implicit in the odd-dimensional surgery obstruction of Wall [302, Chap. 6] (and the subsequent reformulations of Novikov [218] and Ranicki [230] using formations – see 20.29 (ii) below), and in the construction of the algebraic surgery transfer map in Lück and Ranicki [175]. Let $\mathcal{D}_n(A)$ be the *n-dimensional derived category*, the additive category of n-dimensional f.g. free A-module chain complexes and chain homotopy classes and chain maps. Regard $\mathcal{D}_n(A)$ as an additive category with involution $C \longrightarrow C^{n-*}$, so that the L-theory of $\mathcal{D}_n(A)$ is defined as in Ranicki [243]. Given $(C, \phi), f, g$ as in 20.8 note that the morphisms defined in $\mathcal{D}_n(A)$ by

$$\gamma = g^* : L = C^{n-*} \longrightarrow D^{n-*} ,$$
$$\mu = f : L = C^{n-*} \longrightarrow D$$

are the components of a morphism of symmetric forms in $\mathcal{D}_n(A)$

$$\begin{pmatrix} \gamma \\ \mu \end{pmatrix} : (L, 0) \longrightarrow (D \oplus D^{n-*}, \begin{pmatrix} 0 & 1 \\ 1 & 0 \end{pmatrix})$$

which is the inclusion of a sublagrangian with *hessian form*

$$\begin{pmatrix} \gamma \\ \mu \end{pmatrix}^* \begin{pmatrix} 0 & 1 \\ 0 & 0 \end{pmatrix} \begin{pmatrix} \gamma \\ \mu \end{pmatrix} = \gamma^*\mu = gf = \phi_0 : C^{n-*} \longrightarrow C$$

and boundary

$$L^{\perp}/L = \partial_{01}B .$$

(See [230] for more on the hessian form). □

Example 20.29 (i) An n-dimensional geometric Poincaré triad

$$(X; \partial_0 X, \partial_1 X; \partial_{01} X)$$

determines an n-dimensional symmetric Poincaré triad over $\mathbb{Z}[\pi_1(X)]$

$$(\Gamma, \Theta) = \left(\begin{array}{ccc} C(\widetilde{\partial_{01}X}) & \longrightarrow & C(\widetilde{\partial_0 X}) \\ \big\downarrow & & \big\downarrow \\ C(\widetilde{\partial_1 X}) & \longrightarrow & C(\widetilde{X}) \end{array} , \begin{array}{ccc} \Delta[\partial_{01}X] & \longrightarrow & \Delta[\partial_0 X] \\ \big\downarrow & & \big\downarrow \\ \Delta[\partial_1 X] & \longrightarrow & \Delta[X] \end{array} \right)$$

with

$$\partial_0 X \cap \partial_1 X = \partial_{01} X \ , \ \partial_0 X \cup \partial_1 X = \partial X$$

and $\widetilde{X}$ the universal cover of X. The corresponding n-dimensional symmetric complex over $\mathbb{Z}[\pi_1(X)]$

$$(C, \phi) = (C(\widetilde{X}, \widetilde{\partial X}), \Delta[X])$$

is such that

$$\begin{aligned} \phi_0 = [X] \cap - = gf : C^{n-*} &= C(\widetilde{X}, \widetilde{\partial X})^{n-*} \simeq C(\widetilde{X}) \\ &\xrightarrow{f} D = C(\widetilde{X}, \widetilde{\partial_1 X})^{n-*} \simeq C(\widetilde{X}, \widetilde{\partial_0 X}) \\ &\xrightarrow{g} C(\widetilde{X}, \widetilde{\partial X}) = C \end{aligned}$$

with f, g induced by the inclusions and

$$C(\widetilde{\partial_0 X}) \simeq \mathcal{C}(f)_{*+1} \ , \ C(\widetilde{\partial_1 X}) \simeq \mathcal{C}(g^* : C^{n-*} \longrightarrow D^{n-*})_{*+1} .$$

The $(n-1)$-dimensional symmetric Poincaré pairs $(\partial_0 X, \partial_{01} X), (\partial_0 X, \partial_{01} X)$ correspond to the $(n-1)$-dimensional symmetric complexes

$$(C(\widetilde{\partial_0 X},\widetilde{\partial_{01} X}),(e_0)^{\%}(\partial\phi))\ ,\ (C(\widetilde{\partial_1 X},\widetilde{\partial_{01} X}),(e_1)^{\%}(\partial\phi))$$

with

$$\begin{aligned} e_0\ :\ \partial C\ \simeq\ C(\widetilde{\partial X})\ &\xrightarrow{1\oplus g^*} \\ &\mathcal{C}(f)^{n-*}\ \simeq\ C(\widetilde{\partial_0 X})^{n-1-*}\ \simeq\ C(\widetilde{\partial X},\widetilde{\partial_1 X})\ \simeq\ C(\widetilde{\partial_0 X},\widetilde{\partial_{01} X})\ , \\ e_1\ :\ \partial C\ \simeq\ C(\widetilde{\partial X})\ &\xrightarrow{1\oplus f} \\ &\mathcal{C}(g^*)^{n-*}\ \simeq\ C(\widetilde{\partial_1 X})^{n-1-*}\ \simeq\ C(\widetilde{\partial X},\widetilde{\partial_0 X})\ \simeq\ C(\widetilde{\partial_1 X},\widetilde{\partial_{01} X})\ . \end{aligned}$$

Similarly for kernels of normal maps and quadratic Poincaré triads.
(ii) Let

$$(f,b)\ :\ (M;\partial_0 M,\partial_1 M;\partial_{01} M)\longrightarrow (X;\partial_0 X,\partial_1 X;\partial_{01} X)$$

be a normal map of n-dimensional geometric Poincaré triads, so that the quadratic kernel is an n-dimensional quadratic Poincaré triad (Γ,Ψ) over $\mathbb{Z}[\pi_1(X)]$. Suppose that $n=2i$, that f is $(i-1)$-connected, $\partial_0 f,\partial_1 f$ are $(i-2)$-connected and that $\partial_{01} f$ is a homotopy equivalence. In this case, the $\mathbb{Z}[\pi_1(X)]$-module morphisms of 20.24

$$\begin{aligned} \gamma\ =\ f\ :\ L\ =\ C^{2i-*}\ &=\ K_i(M)\longrightarrow D\ =\ K_i(M,\partial_1 M)\ , \\ \mu\ =\ g^*\ :\ L\ =\ C^{2i-*}\ &=\ K_i(M)\longrightarrow D^{2i-*}\ =\ K_i(M,\partial_0 M) \end{aligned}$$

are the components of the inclusion of a lagrangian in a hyperbolic $(-)^{i-1}$-quadratic form

$$\begin{pmatrix}\gamma\\ \mu\end{pmatrix}\ :\ (L,0)\longrightarrow (D\oplus D^{n-*},\begin{pmatrix}0&1\\0&0\end{pmatrix})$$

with hessian the kernel $(-)^i$-quadratic intersection form $(K_i(M),\psi)$

$$\gamma^*\mu\ =\ \psi+(-)^i\psi^*\ :\ K_i(M)\longrightarrow K_i(M)^*\ =\ K^i(M)\ .$$

The quadratic kernel determines (and is determined by) the nonsingular $(-)^{i-1}$-quadratic formation

$$(D\oplus D^{2i-*},\begin{pmatrix}0&1\\0&0\end{pmatrix};D,\mathrm{im}\begin{pmatrix}\gamma\\ \mu\end{pmatrix})\ ,$$

corresponding to the automorphism of the hyperbolic $(-)^{i-1}$-quadratic form sending D to $\mathrm{im}\begin{pmatrix}\gamma\\ \mu\end{pmatrix}$ used by Wall [302, Chap. 6] to construct the $(2i-1)$-dimensional surgery obstructions of (f_0,b_0) and (f_1,b_1)

$$\sigma_*(\partial_0 f,\partial_0 b)\ =\ \sigma_*(\partial_1 f,\partial_1 b)\in L_{2i-1}(\mathbb{Z}[\pi_1(X)])$$

assuming $\pi_1(\partial_0 X) = \pi_1(\partial_1 X) = \pi_1(X)$. Furthermore, if $\partial_0 f, \partial_1 f$ are homotopy equivalences then $(K_i(M), \psi)$ is a nonsingular $(-)^i$-quadratic form over $\mathbb{Z}[\pi_1(X)]$, and the $2i$-dimensional surgery obstruction of (f, b) ([302, Chap. 5]) is given by

$$\sigma_*(f,b) \;=\; (K_i(M), \psi) \in L_{2i}(\mathbb{Z}[\pi_1(X)]) \; . \qquad \square$$

It is also possible to glue together algebraic Poincaré triads. The following construction decomposing algebraic Poincaré pairs as unions of algebraic Poincaré triads is a relative version of 20.25, which will be used in Chap. 30:

Proposition 20.30 *Let $(i : \partial C \longrightarrow C, (\phi, \partial\phi))$ be an n-dimensional ϵ-symmetric Poincaré pair over A. A morphism of n-dimensional ϵ-symmetric pairs*

$$(f, \partial f; \chi, \partial\chi) \;:\; (i : \partial C \longrightarrow C, (\phi, \partial\phi)) \longrightarrow (j : \partial D \longrightarrow D, (\theta, \partial\theta))$$

determines n-dimensional ϵ-symmetric Poincaré triads

$$(\Gamma_+, \Theta_+) \;=\; \left(\begin{array}{ccc} \partial\partial D & \longrightarrow & \mathcal{C}(\partial f)_{*+1} \\ \big\downarrow & & \big\downarrow \\ \partial_+ D & \longrightarrow & \mathcal{C}(f)_{*+1} \end{array} \;,\; \begin{array}{ccc} \partial\partial\theta & \longrightarrow & \partial_+\theta_+ \\ \big\downarrow & & \big\downarrow \\ \partial_+\theta & \longrightarrow & \theta_+ \end{array} \right) ,$$

$$(\Gamma_-, \Theta_-) \;=\; \left(\begin{array}{ccc} \partial\partial D & \longrightarrow & \partial D^{n-*-1} \\ \big\downarrow & & \big\downarrow \\ \partial_+ D & \longrightarrow & \mathcal{C}(j)^{n-*} \end{array} \;,\; \begin{array}{ccc} -\partial\partial\theta & \longrightarrow & \partial_+\theta_- \\ \big\downarrow & & \big\downarrow \\ -\partial_+\theta & \longrightarrow & \theta_- \end{array} \right)$$

such that

$$(i : \partial C \longrightarrow C, (\phi, \partial\phi)) \;\simeq\; (\Gamma_+, \Theta_+) \cup (\Gamma_-, \Theta_-) \; ,$$

with

$$\begin{aligned} \partial\partial D \;&=\; \mathcal{C}(\partial\theta_0 : \partial D^{n-*-1} \longrightarrow \partial D)_{*+1} \; , \\ \partial_+ D \;&=\; \mathcal{C}((\theta_0 \;\; j\partial\theta_0) : \mathcal{C}(j)^{n-*} \longrightarrow D)_{*+1} \; . \end{aligned}$$

Similarly in the ϵ-quadratic case. $\square$

21. Codimension q surgery

This chapter reviews codimension q surgery theory for $q \geq 1$. Refer to Chap. 7 of Ranicki [237] for a previous account of the algebraic theory of codimension q surgery. For $q \geq 3$ this is the same as the ordinary surgery obstruction theory on the submanifold, while for $q = 1, 2$ the situation is considerably more complicated. It may appear that case $q = 2$ is the one of most direct application to knot theory, although in fact the case $q = 1$ (applied to the Seifert surfaces of knots) is just as relevant.

21A. Surgery on submanifolds

Here is the basic operation of codimension q surgery, for any $q \geq 1$:

Definition 21.1 Let M be an n-dimensional manifold, and let $N^{n-q} \subset M^n$ be a submanifold of codimension $q \geq 1$. An *ambient* (or *codimension* q) *surgery on* N *inside* M is the operation $(M, N) \longrightarrow (M, N')$ determined by an embedding

$$(D^{r+1}, S^r) \times D^{n-q-r} \longrightarrow (M, N)$$

with

$$N' = \mathrm{cl.}(N \backslash (S^r \times D^{n-q-r})) \cup (D^{r+1} \times S^{n-q-r-1}) \subset M. \qquad \square$$

Proposition 21.2 *The effect of an ambient surgery on a codimension q submanifold $N \subset M$ is a codimension q submanifold $N' \subset M$ which is related to N by the trace cobordism*

$$(W; N, N') = (N \times I \cup D^{r+1} \times D^{n-q-r}; N \times \{0\}, N')$$

with a codimension q embedding

$$(W; N, N') \subset M \times (I; \{0\}, \{1\}) .$$

Conversely, every ambient cobordism $(W; N, N') \subset M \times (I; \{0\}, \{1\})$ is the union of traces of a sequence of ambient surgeries.

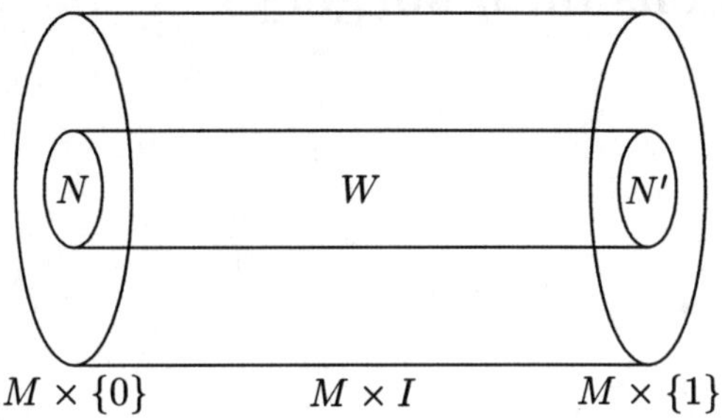

□

Definition 21.3 (i) A *codimension q CW pair* (X, Y) is a CW complex with a decomposition

$$X = E(\xi) \cup_{S(\xi)} Z$$

for some $(q-1)$-spherical fibration $\xi : Y \longrightarrow BG(q)$ over a subcomplex $Y \subset X$, with $Z \subset X$ a disjoint subcomplex and

$$(D^q, S^{q-1}) \longrightarrow (E(\xi), S(\xi)) \longrightarrow Y$$

the associated (D^q, S^{q-1})-fibration.

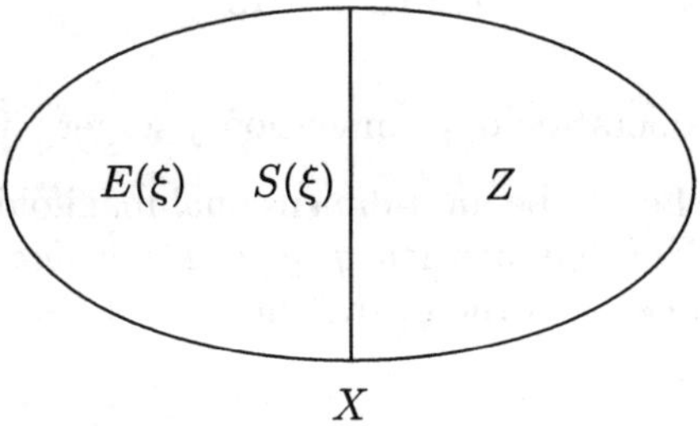

(ii) A *topological normal structure* on a a codimension q CW pair (X, Y) is a topological q-block bundle

$$\widetilde{\xi} : Y \longrightarrow B\widetilde{TOP}(q)$$

such that

$$\xi = J\widetilde{\xi} : Y \longrightarrow BG(q) .$$

(iii) An $(n, n-q)$*-dimensional geometric Poincaré pair* (X, Y) is a codimension q CW pair such that:

(a) X is an n-dimensional Poincaré complex,
(b) Y is an $(n-q)$-dimensional Poincaré complex,
(c) $(Z, S(\xi))$ is an n-dimensional Poincaré pair.

In this case, the normal $(q-1)$-spherical fibration is written

$$\xi = \nu_{Y \subset X} : Y \longrightarrow BG(q) .$$

The pair (X,Y) is *simple* if the Poincaré complexes and pairs in (a),(b),(c) are simple, i.e. have torsion $\tau = 0 \in Wh$. □

Example 21.4 For $q = 1,2$

$$\widetilde{BTOP}(q) = BO(q) = BG(q)$$

so that every codimension q CW pair (X,Y) has a canonical topological normal structure. □

21B. The splitting obstruction

Proposition 21.5 *Let (X,Y) be a codimension q CW pair with a topological normal structure.*

(i) *Every map $f: M \longrightarrow X$ from an n-dimensional manifold M can be made topologically transverse at $Y \subset X$, with*

$$N = f^{-1}(Y) \subset M$$

a codimension q submanifold, defining a simple $(n, n-q)$-dimensional geometric Poincaré pair (M,N) with

$$\nu_{N \subset M} : N \xrightarrow{f|} Y \xrightarrow{\widetilde{\xi}} \widetilde{BTOP}(q) .$$

(ii) *If $(f,b): M \longrightarrow X$ is a normal map (i.e. if f is covered by a map of stable topological block bundles $b: \nu_M \longrightarrow \eta$) then the restrictions define normal maps*

$$(g,c) = (f,b)| : N \longrightarrow Y ,$$
$$(h,d) = (f,b)| : (P, \partial P) = f^{-1}(Z, \partial Z) \longrightarrow (Z, \partial Z)$$

with $\partial Z = S(\xi)$, $\partial P = S(\nu_{N \subset M})$. □

Definition 21.6 Let (X,Y) be a codimension q CW pair with a topological normal structure.

(i) A homotopy equivalence $h: M \longrightarrow M'$ from an n-dimensional manifold M to an n-dimensional geometric Poincaré complex M' *splits along* $Y \subset X$ if there is given a reference map $r: M' \longrightarrow X$ which is Poincaré transverse at $Y \subset X$, with (M',N') an $(n,n-q)$-dimensional geometric Poincaré pair, such that the restrictions

$$(f,b) = h| : N = (rh)^{-1}(Y) \longrightarrow N' = r^{-1}(Y) ,$$
$$(g,c) = h| : P = (rh)^{-1}(Z) \longrightarrow P' = r^{-1}(Z)$$

are homotopy equivalences, i.e. if the normal maps (f,b) and (g,c) obtained by transversality can be improved by ambient surgeries on the codimension

q submanifold $N \subset M$ to be homotopy equivalences.
(ii) The homotopy equivalence h *h-splits along* $Y \subset X$ if h is h-cobordant to a split homotopy equivalence.
(iii) If (M', N') is a simple codimension q Poincaré pair and h is a simple homotopy equivalence then h *s-splits along* $Y \subset X$ if h is s-cobordant to a split homotopy equivalence such that the restrictions f, g are simple homotopy equivalences.
(iv) There are analogous notions of splitting for a homotopy equivalence from a manifold with boundary to a geometric Poincaré pair. □

Note that for $n \geq 5$ it is possible to replace 's-cobordant' by 'homotopic to' in the definition of s-splitting, by the s-cobordism theorem.

The codimension q splitting obstruction theory of Wall [302] was originally formulated for the problem of s-splitting homotopy equivalences, with obstruction groups $LS_*(\Phi)$. These will be denoted by $LS_*^s(\Phi)$. There is also an h-splitting version with obstruction groups $LS_*^h(\Phi)$, and it is these which will be denoted by $LS_*(\Phi)$. In the original treatment the LS-groups were defined geometrically, but in Ranicki [237] the LS-groups were expressed algebraically as the cobordism groups of algebraic Poincaré triads.

The fundamental groups (or groupoids) of the spaces in a codimension q CW pair (X, Y) fit into the pushout square

$$\begin{array}{ccc} \pi_1(S(\xi)) & \longrightarrow & \pi_1(Z) \\ \big\downarrow & \Phi & \big\downarrow \\ \pi_1(E(\xi)) & \longrightarrow & \pi_1(X) \end{array}$$

given by the Seifert–van Kampen theorem

$$\pi_1(X) \ = \ \pi_1(E(\xi)) *_{\pi_1(S(\xi))} \pi_1(Z)$$

(working with groupoids in the disconnected case).

Definition 21.7 (Wall [302, pp. 127, 138], Ranicki [237, p. 565])
Let (X, Y) be a codimension q CW pair with topological normal structure.
(i) The *quadratic LS-groups* $LS_*(\Phi)$ of the pushout square of groups Φ associated to (X, Y) are the relative groups appearing in the exact sequence

$$\dots \longrightarrow L_{n+1}(\mathbb{Z}[\pi_1(Z)] \longrightarrow \mathbb{Z}[\pi_1(X)]) \longrightarrow LS_{n-q}(\Phi) \longrightarrow L_{n-q}(\mathbb{Z}[\pi_1(Y)])$$
$$\xrightarrow{p_\xi^!} L_n(\mathbb{Z}[\pi_1(Z)] \longrightarrow \mathbb{Z}[\pi_1(X)]) \longrightarrow \dots .$$

Every element of $L_{n-q}(\mathbb{Z}[\pi_1(Y)])$ is the rel ∂ surgery obstruction $\sigma_*(f, b)$ of a normal map $(f, b) : (N, \partial N) \longrightarrow (N', \partial N')$ from an $(n-q)$-dimensional manifold with boundary to an $(n-q)$-dimensional geometric Poincaré pair, such that $\partial f : \partial N \longrightarrow \partial N'$ is a homotopy equivalence, and with a π_1-isomorphism reference map $N' \longrightarrow Y$. The map $p\xi^!$ is the composite of the transfer map

$$\xi^! \ : \ L_{n-q}(\mathbb{Z}[\pi_1(Y)]) \longrightarrow L_n(\mathbb{Z}[\pi_1(S(\xi))] \longrightarrow \mathbb{Z}[\pi_1(E(\xi))]) \ ;$$
$$\sigma_*(f,b) \longrightarrow \sigma_*(E(f),E(b))$$

sending the rel ∂ surgery obstruction $\sigma_*(f,b)$ to the rel ∂ surgery obstruction $\sigma_*(E(f),E(b))$ of the induced normal map of (D^k,S^{k-1})-bundles from an n-dimensional manifold with boundary to an n-dimensional geometric Poincaré pair

$$(E(f),E(b)) \ : \ (E(\widetilde{\xi}_N),\partial E(\widetilde{\xi}_N)) \longrightarrow (E(\widetilde{\xi}_{N'}),\partial E(\widetilde{\xi}_{N'}))$$

with

$$\widetilde{\xi}_N \ : \ N \longrightarrow B\widetilde{TOP}(q) \ \ , \ \ \widetilde{\xi}_{N'} \ : \ N' \longrightarrow B\widetilde{TOP}(q)$$

the pullbacks of $\widetilde{\xi} : Y \longrightarrow B\widetilde{TOP}(q)$, and

$$p \ : \ L_n(\mathbb{Z}[\pi_1(S(\xi))] \longrightarrow \mathbb{Z}[\pi_1(E(\xi))]) \longrightarrow L_n(\mathbb{Z}[\pi_1(Z)] \longrightarrow \mathbb{Z}[\pi_1(X)])$$

is the morphism of relative L-groups induced functorially by Φ.

(ii) Let $h : (M,\partial M) \longrightarrow (M',\partial M')$ be a homotopy equivalence from an n-dimensional manifold with boundary to an n-dimensional geometric Poincaré pair, with a π_1-isomorphism reference map $r : M' \longrightarrow X$ which is Poincaré transverse at $Y \subset X$, such that $\partial h : \partial M \longrightarrow \partial M'$ is a homotopy equivalence which is h-split along $Y \subset X$. The *codimension q splitting obstruction* of h

$$s_Y(h) \in LS_{n-q}(\Phi)$$

has image the surgery obstruction $\sigma_*(f,b) \in L_{n-q}(\mathbb{Z}[\pi_1(Y)])$ of the codimension q normal map obtained by transversality

$$(f,b) \ = \ h| \ : \ (N,\partial N) \ = \ (rh,\partial r \partial h)^{-1}(Y) \longrightarrow (N',\partial N') \ = \ (r,\partial r)^{-1}(Y)$$

which is determined by the bordism

$$h \times 1 \ : \ (M \times I; M \times \{0\}, E(\nu_{N \subset M}) \times \{1\}; P \times \{1\})$$
$$\longrightarrow (M' \times I; M' \times \{0\}, E(\nu_{N' \subset M'}) \times \{1\}; P' \times \{1\})$$
$$(P = (rh)^{-1}(Z) = \text{cl.}(M \backslash E(\nu_{N \subset M})))$$

of

$$(f,b)^! \ : \ (E(\nu_{N \subset M}),S(\nu_{N \subset M})) \longrightarrow (E(\nu_{N' \subset M'}),S(\nu_{N' \subset M'}))$$

to the homotopy equivalence $h : (M,\partial M) \longrightarrow (M',\partial M')$, as in the diagram

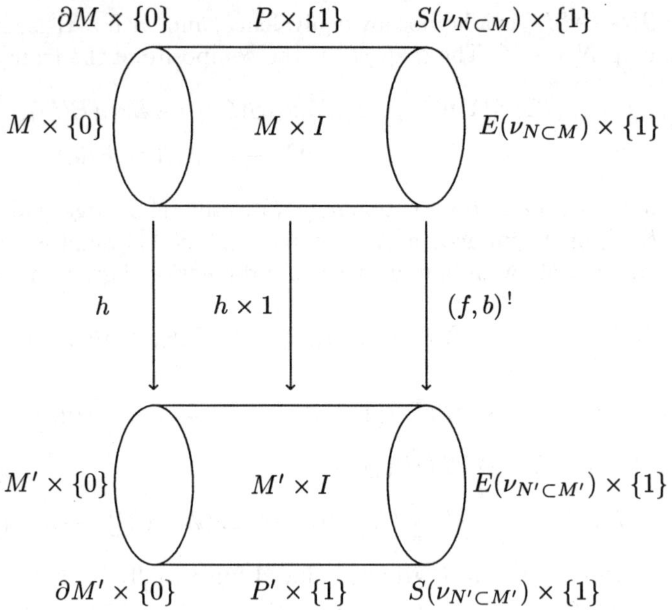

(iii) The *quadratic LN-groups LN_** are the LS-groups for a codimension q CW pair (X,Y) with

$$\pi_1(X) = \pi_1(Y) \ , \ \pi_1(S(\xi)) = \pi_1(Z)$$

(e.g. if $X = E(\xi)$ with $Z = S(\xi)$), that is

$$LS_*(\Phi) = LN_*(\mathbb{Z}[\pi_1(S(\xi))] \longrightarrow \mathbb{Z}[\pi_1(Y)]) \ .$$

(iv) The *simple quadratic LS-groups $LS^s_*(\Phi)$* and the *simple quadratic LN-groups LN^s_** are defined similarly, using the simple quadratic L-groups L^s_*. □

Proposition 21.8 (Wall [302, p. 126], Ranicki [237, 7.2.5])
(i) *The codimension q splitting obstruction of a homotopy equivalence from an n-dimensional manifold to an n-dimensional geometric Poincaré pair $h : (M, \partial M) \longrightarrow (M', \partial M')$ with $\partial h : \partial M \longrightarrow \partial M'$ split along $Y \subset X$ is such that $s_Y(h) = 0 \in LS_{n-q}(\Phi)$ if (and for $n - q \geq 5$ only if) h h-splits rel ∂ along $Y \subset X$. Similarly for s-splitting.*
(ii) *Every element $x \in LS_{n-q}(\Phi)$ is the codimension q splitting obstruction $x = s_Y(h)$ with h as in* (i).
Proof The original proof in [302] was in terms of the smoothing theory of Poincaré embeddings. The basic idea goes back to Browder: first try to perform (non-ambient) surgery on the submanifold $N \subset M$ to obtain a normal bordism from $h : M \longrightarrow M'$ to a split homotopy equivalence $h' : M'' \longrightarrow M'$, and then try to perform rel ∂ surgery on the normal bordism. □

Example 21.9 (i) For a codimension q CW pair (X, Y) with $q \geq 3$

$$\pi_1(S(\xi)) \;=\; \pi_1(E(\xi)) \;=\; \pi_1(Y) \;=\; \pi_1(Z) \;=\; \pi_1(X)$$

and the codimension q splitting obstruction is just the ordinary surgery obstruction

$$s_Y(h) \;=\; \sigma_*(f, b) \in LS_{n-q}(\Phi) \;=\; L_{n-q}(\mathbb{Z}[\pi_1(Y)]) \;,$$

by the Browder–Casson–Sullivan–Wall embedding theorem ([302, 11.3.1]).
(ii) For $q = 1$ there are three cases:

(A) ξ is trivial, and Y separates X (e.g. $Y = \{\text{pt.}\} \subset X = S^1 \vee S^1$),
(B) ξ is trivial, but Y does not separate X (e.g. $Y = \{\text{pt.}\} \subset X = S^1$),
(C) ξ is non-trivial (e.g. $Y = S^1 \subset X = \mathbb{RP}^2$).

See [302, Chaps. 12A, 12B, 12C] and Ranicki [237, 7.6] for accounts of codimension 1 splitting obstruction theory. □

22. Codimension 2 surgery

Lopez de Medrano [170] formulated "the general philosophy for dealing with surgery problems in codimension 2: do not insist on obtaining homotopy equivalences when you are doing surgery on the complement of a submanifold, be happy if you can obtain the correct homology conditions." Cappell and Shaneson [40], [46] developed the appropriate homology surgery theory, with the Γ-groups (20.17) as obstruction groups, and applied the theory to codimension 2 surgery. Freedman [84] and Matsumoto [184] worked out a somewhat different approach to codimension 2 surgery, using $\pm z$-quadratic forms over $\mathbb{Z}[\pi_1(S(\xi))]$ to formulate the obstructions to individual ambient surgeries. The relationship between these approaches has already been studied in Ranicki [237, 7.6], using both algebra and geometry. This chapter starts by recalling and extending the chain complex reformulation in [237] of the theory of [84] and [184].

22A. Characteristic submanifolds

In order to avoid twisted coefficients only oriented 2-plane bundles $\xi : M \longrightarrow BSO(2)$ are considered. Manifolds and submanifolds will be all assumed to be oriented.

Proposition 22.1 *If $(M, \partial M)$ is an n-dimensional manifold with boundary the following sets are in natural one-one correspondence:*

(i) *the cohomology group $H^2(M)$,*
(ii) *the homology group $H_{n-2}(M, \partial M)$,*
(iii) *the set $[M, BSO(2)]$ of isomorphism classes of 2-plane bundles ξ over M,*
(iv) *the ambient cobordism classes of codimension 2 submanifolds $(N, \partial N) \subset (M, \partial M)$.*

Proof Immediate from Poincaré duality, manifold transversality and the identifications

$$BSO(2) = MSO(2) = K(\mathbb{Z}, 2) = \mathbb{CP}^\infty . \qquad \square$$

For a 2-plane bundle $\xi : M \longrightarrow BSO(2)$ over a connected space M the S^1-fibration

$$S^1 \longrightarrow S(\xi) \xrightarrow{p} M$$

induces an exact sequence of groups

$$\pi_1(S^1) \longrightarrow \pi_1(S(\xi)) \xrightarrow{p_*} \pi_1(M) \longrightarrow \{1\}$$

such that the image of $1 \in \pi_1(S^1)$ is central in $\pi_1(S(\xi))$. The image of 1 will be denoted by $z \in \pi_1(S(\xi))$. Note that there is defined an exact sequence of group rings

$$\mathbb{Z}[\pi_1(S(\xi))] \xrightarrow{1-z} \mathbb{Z}[\pi_1(S(\xi))] \xrightarrow{p_*} \mathbb{Z}[\pi_1(M)] \longrightarrow 0 .$$

Definition 22.2 Let $(M, \partial M)$ be an n-dimensional manifold with boundary, and let $\xi : M \longrightarrow BSO(2)$ be a 2-plane bundle.
(i) The *exterior* of a codimension 2 submanifold with boundary

$$(N^{n-2}, \partial N) \subset (M^n, \partial M)$$

is the pair

$$(P^n, \partial_+ P) = (\text{cl.}(M \backslash E(\nu_{N \subset M})), \text{cl.}(\partial M \backslash E(\nu_{\partial N \subset \partial M}))) ,$$

such that $(P; \partial_+ P, S(\nu_{N \subset M}))$ is an n-dimensional manifold triad.
(ii) A *ξ-characteristic codimension 2 submanifold* of $(M, \partial M)$ is a codimension 2 submanifold with boundary $(N, \partial N) \subset (M, \partial M)$ representing the Poincaré dual of the Euler class of ξ

$$[N] = e(\xi) \in H_{n-2}(M, \partial M) = H^2(M) ,$$

with an identification

$$\xi|_N = \nu_{N \subset M} : N \longrightarrow BSO(2)$$

and an extension of the trivialization

$$\xi|_{S(\nu_{N \subset M})} = \epsilon^2 : S(\nu_{N \subset M}) \longrightarrow BSO(2)$$

to a trivialization

$$\xi|_P = \epsilon^2 : P \longrightarrow BSO(2) .$$

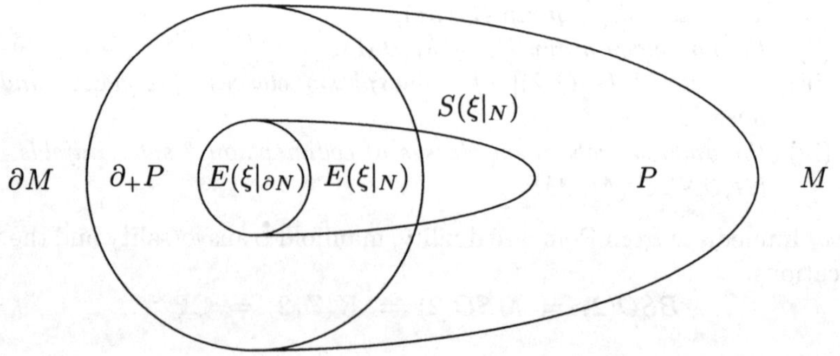

□

Proposition 22.3 (i) *For a ξ-characteristic codimension 2 submanifold $(N,\partial N)\subset(M,\partial M)$ the total spaces of S^1-fibrations define an $(n+1)$-dimensional manifold with boundary*

$$(M',\partial M') = (S(\xi),S(\xi|_{\partial M}))$$

such that

$$(N',\partial N') = (S(\nu_{N\subset M}),S(\nu_{\partial N\subset\partial M}))\subset(M',\partial M')$$

is an ϵ^2-characteristic codimension 2 submanifold with exterior

$$\begin{aligned}(P',\partial_+P') &= (\mathrm{cl.}(M'\backslash(N'\times D^2)),\mathrm{cl.}(\partial M'\backslash(\partial N'\times D^2)))\\ &= (P\times S^1,\partial_+P\times S^1)\ .\end{aligned}$$

(ii) *Every codimension 2 submanifold $(N,\partial N)\subset(M,\partial M)$ is ξ-characteristic, with $\xi: M\longrightarrow BSO(2)$ the 2-plane bundle with Euler class the Poincaré dual of the homology class represented by $(N,\partial N)$*

$$e(\xi) = [N]\in H^2(M) = H_{n-2}(M,\partial M)\ .$$

(iii) *For any 2-plane bundle $\xi: M\longrightarrow BSO(2)$ over an n-dimensional manifold with boundary $(M,\partial M)$ there exist ξ-characteristic codimension 2 submanifolds $(N,\partial N)\subset(M,\partial M)$.*

Proof (i)+(ii) Immediate from 22.1.
(iii) The pullback of $\xi: M\longrightarrow BSO(2)$ along the projection $p: S(\xi)\longrightarrow M$ is a trivial 2-plane bundle, since $p^*\xi\simeq\{*\}: M\longrightarrow BSO(2)$. □

22B. The antiquadratic construction

The algebraic theory of pseudoquadratic structures on chain complexes was developed in Ranicki [237, pp. 781–810] in order to describe the L-theory transfer maps

$$\xi^!\ :\ L_n(\mathbb{Z}[\pi_1(M)])\longrightarrow L_{n+2}(\mathbb{Z}[\pi_1(S(\xi))]\longrightarrow\mathbb{Z}[\pi_1(E(\xi))])$$

induced on the chain level by a 2-plane bundle $\xi: M\longrightarrow BSO(2)$. This theory will now be used to associate an 'antiquadratic complex' to a codimension 2 submanifold, such that geometric surgery on the submanifold corresponds to algebraic surgery on the complex.

Definition 22.4 ([237, p. 807]) The *antiquadratic complex* of a ξ-characteristic codimension 2 submanifold $(N,\partial N)\subset(M,\partial M)$ of an n-dimensional manifold with boundary is the n-dimensional $(-z)$-quadratic complex

$$\sigma_*(M,N,\xi) = (C,\psi)$$

with $C = C(\widetilde{P},\widetilde{\partial_+ P})$ the $\mathbb{Z}[\pi_1(S(\xi))]$-module chain complex of the cover $(\widetilde{P},\widetilde{\partial_+ P})$ of the exterior $(P,\partial_+ P)$ induced from the universal cover $\widetilde{S(\xi)}$ of $S(\xi)$, and $\psi \in Q_n(C,-z)$ the evaluation of the pseudoquadratic construction. □

Remark 22.5 (i) The 'anti'-terminology refers to the antistructure in the sense of Wall [303] which is determined on $\mathbb{Z}[\pi_1(S(\xi))]$ by an arbitrary 2-plane bundle $\xi : M \longrightarrow BO(2)$ using the 'pseudosymmetric' and 'pseudoquadratic' constructions of [237, pp. 799–802], the S^1-fibration analogues of the symmetric and quadratic constructions of Ranicki [236]. As indicated in [237, 7.8.9] it is possible to obtain these analogues by induction on the cells in the base CW complex (assumed finite). However, it would be more satisfactory to use an equivariant cellular Eilenberg-Zilber theorem, developing further the methods of Smith [271].
(ii) Since it is assumed that ξ is oriented, there is no actual need to consider antistructures here.
(iii) See 27.5 below for an explicit formula for the $(-z)$-quadratic structure $\psi \in Q_n(C,-z)$ in the framed case $\xi = \epsilon^2$. □

Proposition 22.6 *The antiquadratic complex $\sigma_*(M,N,\xi) = (C,\psi)$ has the following properties:*
(i) *The boundary $(n-1)$-dimensional $(-z)$-quadratic Poincaré complex over $\mathbb{Z}[\pi_1(S(\xi))]$*

$$\partial(C,\psi) = (\partial C, \partial\psi)$$

has

$$\partial C = \mathcal{C}((1+T_{-z})\psi_0 : C^{n-*} \longrightarrow C)_{*+1} \simeq C(\widetilde{S(\xi)},\widetilde{\partial_+ P})_{*+1} .$$

The antiquadratic complex of $\partial N \subset \partial M$ is the $(n-1)$-dimensional $(-z)$-quadratic complex over $\mathbb{Z}[\pi_1(S(\xi))]$

$$\sigma_*(\partial M, \partial N, \xi|_{\partial N}) = (C(\widetilde{\partial_+ P}), e_{\%}\psi)$$

with

$$e : \partial C \simeq C(\widetilde{S(\xi)},\widetilde{\partial_+ P})_{*+1} \xrightarrow{\text{proj.}} C(\widetilde{\partial_+ P}) , \quad \partial C(\widetilde{\partial_+ P}) \simeq C(S(\widetilde{\xi|_{\partial N}}))_{*+1} .$$

(ii) *Let*

$$i : C = C(\widetilde{P},\widetilde{\partial_+ P}) \longrightarrow C(\widetilde{P},\widetilde{\partial P})$$

be the inclusion. The induced morphism $i_{\%}$ of $(-z)$-quadratic Q-groups sends $\psi \in Q_n(C,-z)$ to

$$i_{\%}\psi = \phi_0 \in Q_n(C(\widetilde{P},\widetilde{\partial P}),-z)$$

with $\phi = \Delta[P] \in Q^n(C(\widetilde{P},\widetilde{\partial P}))$ the symmetric structure of $(P,\partial P)$, and

$$\phi_0 = [P] \cap - : C(\widetilde{P},\widetilde{\partial P})^{n-*} \longrightarrow C(\widetilde{P},\widetilde{\partial P}) .$$

(iii) *Inclusion defines a normal map*

$$(f,b)\ :\ (N,\partial N) \longrightarrow (M,\partial N)$$

from an $(n-2)$-dimensional manifold with boundary $(N,\partial N)$ to an $(n-2)$-dimensional normal pair $(M,\partial N)$, with fundamental class

$$f_*[N] \in H_{n-2}(M,\partial N)$$

and normal fibration $\nu_M \oplus \xi$. The quadratic kernel of (f,b) is the $(n-2)$-dimensional quadratic complex over $\mathbb{Z}[\pi_1(M)]$

$$\sigma_*(f,b)\ =\ \mathbb{Z}[\pi_1(M)]\otimes_{\mathbb{Z}[\pi_1(S(\xi))]}(S^{-1}C,\overline{S}^{-1}\psi)$$

with $\overline{S} : Q_{n-2}(S^{-1}C,z) \cong Q_n(C,-z)$ the skew-suspension isomorphism, and

$$\mathbb{Z}[\pi_1(M)]\otimes_{\mathbb{Z}[\pi_1(S(\xi))]}(S^{-1}C)\ \simeq\ \mathcal{C}(f^!)$$

the algebraic mapping cone of the Umkehr $\mathbb{Z}[\pi_1(M)]$-module chain map

$$f^!\ :\ C(\widetilde{M},\widetilde{\partial N})^{n-2-*} \xrightarrow{\widetilde{f}^*} C(\widetilde{N},\widetilde{\partial N})^{n-2-*}\ \simeq\ C(\widetilde{N})$$

where $\widetilde{M}$ is the universal cover of M and $\widetilde{N},\widetilde{\partial N}$ are the induced covers of $N,\partial N$. The quadratic kernel of the induced normal map of S^1-fibrations from an $(n-1)$-dimensional manifold with boundary to a normal pair

$$(S(f),S(b))\ :\ (S(\xi|_N),S(\xi|_{\partial N}))\longrightarrow(S(\xi),S(\xi|_{\partial N}))$$

is the $(n-1)$-dimensional quadratic complex over $\mathbb{Z}[\pi_1(S(\xi))]$

$$\sigma_*(S(f),S(b))\ =\ (\mathcal{C}(1-z:C{\longrightarrow}C)_{*+1},\theta)$$

with

$$\begin{aligned}\mathcal{C}(S(f)^!:C(\widetilde{S(\xi)},\widetilde{S(\xi|_{\partial N})})^{n-1-*}{\longrightarrow}C(\widetilde{S(\xi|_N)}))\ &\simeq\ C(\widetilde{S(\xi)},\widetilde{S(\xi|_N)})^{n-*}\\ \simeq\ C(\widetilde{S(\xi|_P)},\widetilde{S(\xi|_{\partial_+P})})_{*+1}\ &\simeq\ \mathcal{C}(1-z:C{\longrightarrow}C)_{*+1}\ ,\end{aligned}$$

$$\begin{aligned}\theta_s\ =\ \begin{pmatrix}0 & \psi_s\\ z^{-1}\psi_s & 0\end{pmatrix}\ :\ \mathcal{C}(S(f)^!)^{n-1-r+s}\ &=\ C^{n-r+s}\oplus C^{n-r+s-1}\\ \longrightarrow \mathcal{C}(S(f)^!)_r\ &=\ C_{r+1}\oplus C_r\quad(s\geq 0)\ .\end{aligned}$$

(iv) *The zero section inclusion $M{\longrightarrow}E(\xi)$ deforms to a normal map from an n-dimensional manifold triad to a normal space triad*

$$(h,d)\ :\ (M;E(\xi|_{\partial N}),\partial_+P)\longrightarrow(E(\xi);E(\xi|_{\partial N}),S(\xi))$$

such that

$$\begin{aligned}h\ &=\ E(f)\cup_{S(f)}g\ :\\ M\ &=\ E(\xi|_N)\cup_{S(\xi|_N)}P\longrightarrow E(\xi)\ =\ E(\xi)\cup_{S(\xi)\times\{1\}}S(\xi)\times I\ ,\end{aligned}$$

with

$$\begin{aligned}(g,c)\ &: (P;S(\xi|_N),\partial_+P;S(\xi|_{\partial N}))\\ &\longrightarrow (S(\xi)\times I;S(\xi)\times\{0\},S(\xi)\times\{1\};S(\xi|_{\partial N})\times I)\end{aligned}$$

a normal map from an n-dimensional manifold triad to a normal space triad, such that

$$(g,c)|\ =\ (S(f),S(b))\ :\ (S(\xi|_N),S(\xi|_{\partial N}))\longrightarrow (S(\xi),S(\xi|_{\partial N}))$$

is the induced normal map from an $(n-1)$-dimensional manifold with boundary (with (f,b) as in (iii)) which is the identity on the boundaries. Let

$$\sigma_*(g,c)\ =\ \left(\begin{array}{ccc} 0 & \longrightarrow & \mathcal{C}(\partial_+g^!) \\ \Big\downarrow & \Gamma & \Big\downarrow \\ \mathcal{C}(S(f)^!) & \longrightarrow & \mathcal{C}(g^!) \end{array}\ ,\ \Psi\right)$$

be the kernel n-dimensional quadratic triad over $\mathbb{Z}[\pi_1(S(\xi))]$, with the Umkehr $\mathbb{Z}[\pi_1(S(\xi))]$-module chain maps given by

$$\begin{aligned} g^!\ &:\ C(\widetilde{S(\xi)},\widetilde{S(\xi|_{\partial N})})^{n-1-*}\longrightarrow C(\widetilde{P})\ ,\\ \partial_+g^!\ &:\ C(\widetilde{S(\xi)},\widetilde{S(\xi|_{\partial N})})^{n-1-*}\longrightarrow C(\widetilde{\partial_+P})\ ,\\ S(f)^!\ &:\ C(\widetilde{S(\xi)},\widetilde{S(\xi|_{\partial N})})^{n-1-*}\longrightarrow C(\widetilde{S(\xi|_N)})\ .\end{aligned}$$

Let (Γ',Ψ') the n-dimensional quadratic Poincaré triad over $\mathbb{Z}[\pi_1(S(\xi))]$ associated by 20.27 to the n-dimensional quadratic complex

$$(C',\psi')\ =\ (C,(1-z^{-1})\psi)$$

and the factorization

$$(1+T)\psi'_0\ :\ C^{n-*}\xrightarrow{1-z^{-1}}C^{n-*}\xrightarrow{(1+T_{-z})\psi_0}C\ ,$$

with

$$\begin{array}{ccc} B & \longrightarrow & \mathcal{C}((1+T_{-z})\psi_0)_{*+1} \\ \Big\downarrow & \Gamma' & \Big\downarrow \\ \mathcal{C}(1-z^{-1}:C^{n-*}\longrightarrow C^{n-*})_{*+1} & \longrightarrow & C^{n-*} \end{array}$$

$$\simeq\quad \begin{array}{ccc} B & \longrightarrow & C(\widetilde{S(\xi)},\widetilde{\partial_+P})_{*+1} \\ \Big\downarrow & & \Big\downarrow \\ C(\widetilde{S(\xi)},\widetilde{S(\xi|_N)})_{*+1} & \longrightarrow & C(\widetilde{S(\xi)},\widetilde{P})_{*+1} \end{array}$$

and

$$\begin{aligned} B &= \mathcal{C}(1-z^{-1} : \mathcal{C}((1+T_{-z})\psi_0) \longrightarrow \mathcal{C}((1+T_{-z})\psi_0))_{*+2} \\ &\simeq \mathcal{C}(C(\widetilde{S(\xi)}, \widetilde{S(\xi|_{\partial N})})^{n-1-*} \longrightarrow C(\widetilde{S(\xi)}))_{*+1} . \end{aligned}$$

The kernel quadratic triad $\sigma_*(g,c) = (\Gamma, \Psi)$ is obtained from (Γ', Ψ') by collapsing B, with a homotopy equivalence

$$(\Gamma, \Psi) \simeq (\Gamma', \Psi')/B$$

$$= \left(\begin{array}{ccc} 0 & \longrightarrow & \mathcal{C}((1+T_{-z})\psi_0)_{*+1} \\ \big\downarrow & & \big\downarrow \\ \mathcal{C}(1-z : C \longrightarrow C)_{*+1} & \longrightarrow & \mathcal{C}(B \longrightarrow C^{n-*}) \end{array} , \ \Psi'/B \right)$$

□

Example 22.7 (i) Let $\xi = \epsilon^2 : M \longrightarrow BSO(2)$ be the trivial 2-plane bundle over an n-dimensional manifold with boundary $(M, \partial M)$, so that

$$(E(\xi), S(\xi)) = (M \times D^2, M \times S^1) .$$

The empty submanifold $N = \emptyset \subset M$ is ξ-characteristic, representing the Euler class $e(\xi) = 0 \in H^2(M)$, with exterior $(P, \partial_+ P) = (M, \partial M)$. The universal cover of $S(\xi) = M \times S^1$ is $\widetilde{S(\xi)} = \widetilde{M} \times \mathbb{R}$ with $\widetilde{M}$ the universal cover M, and the induced cover of $(P, \partial_+ P)$ is

$$(\widetilde{P}, \widetilde{\partial_+ P}) = (\widetilde{M} \times \mathbb{Z}, \widetilde{\partial M} \times \mathbb{Z}) .$$

The antiquadratic complex of 22.4

$$\sigma_*(M, \emptyset, \epsilon^2) = (C(\widetilde{P}, \widetilde{\partial_+ P}), \psi)$$

is given by the cellular chain complex

$$C(\widetilde{P}, \widetilde{\partial_+ P}) = C(\widetilde{M}, \widetilde{\partial M})[z, z^{-1}] ,$$

and the n-dimensional $(-z)$-quadratic structure $\psi \in Q_n(C(\widetilde{P}, \widetilde{\partial_+ P}), -z)$ over $\mathbb{Z}[\pi_1(M)][z, z^{-1}]$ is determined by the n-dimensional symmetric structure $\phi = \Delta[M] \in Q^n(C(\widetilde{M}, \widetilde{\partial M}))$ over $\mathbb{Z}[\pi_1(M)]$, with

$$\psi_s = \begin{cases} \phi_0 & \text{if } s = 0 \\ 0 & \text{if } s \geq 1 \end{cases} : C(\widetilde{P}, \widetilde{\partial_+ P})^{n-r-s} \longrightarrow C(\widetilde{P}, \widetilde{\partial_+ P})_r .$$

(ii) Let $\xi_k : M = \mathbb{CP}^k \longrightarrow BSO(2)$ be the Hopf 2-plane bundle over the k-dimensional complex projective space $\mathbb{CP}^k$ ($k \geq 1$), so that $n = 2k$ and

$$\begin{aligned} (E(\xi_k), S(\xi_k)) &= (\mathrm{cl.}(\mathbb{CP}^{k+1} \backslash D^{2k+2}), S^{2k+1}) , \\ z &= 1 \in \pi_1(S(\xi_k)) = \{1\} . \end{aligned}$$

The $(k-1)$-dimensional complex projective space is a ξ_k-characteristic submanifold

$$N^{2k-2} = \mathbb{CP}^{k-1} \subset M^{2k} = \mathbb{CP}^k$$

representing the Euler class $e(\xi_k) = 1 \in H^2(M) = \mathbb{Z}$, with exterior $P = D^{2k}$ and $C(P) = \mathbb{Z}$ (up to chain equivalence). The antiquadratic complex of 22.4

$$\sigma_*(\mathbb{CP}^k, \mathbb{CP}^{k-1}, \xi_k) = (C(P), \psi)$$

has the (-1)-quadratic structure $\psi \in Q_{2k}(C(P), -1)$ with

$$\psi_{2k} = 1 : C(P)^0 = \mathbb{Z} \longrightarrow C(P)_0 = \mathbb{Z} . \qquad \square$$

Theorem 22.8 (Freedman [84], Matsumoto [184])
Let $(M, \partial M)$ be an n-dimensional manifold with boundary. Given $\xi : M \longrightarrow BSO(2)$, let $(N, \partial N) \subset (M, \partial M)$ be a ξ-characteristic codimension 2 submanifold with normal bundle

$$\nu_{N \subset M} = \xi|_N : N \longrightarrow BSO(2)$$

and exterior $(P, \partial_+ P)$. Let $\sigma_*(M, N, \xi) = (C, \psi)$ be the corresponding antiquadratic complex over $\mathbb{Z}[\pi_1(S(\xi))]$, with $C = C(\widetilde{P}, \widetilde{\partial_+ P})$ and $\psi \in Q_n(C, -z)$. If $(P, \partial_+ P)$ is $(r-1)$-connected with $2r \leq n$ it is possible to kill an element $x \in \pi_r(P, S(\xi|_N)) = H^{n-r}(C)$ by algebraic surgery on (C, ψ) if (and for $n \geq 7$ only if) it is possible to kill x by an ambient surgery on (M, N) rel ∂, and the effect of the ambient surgery corresponds to the effect of the algebraic surgery.

Proof The algebraic theory of surgery on ϵ-quadratic complexes was developed in Ranicki [235], [236]. The exterior $(P, \partial_+ P)$ is $(r-1)$-connected if and only if the antiquadratic complex (C, ψ) is $(r-1)$-connected, and there are no obstructions to either geometric or algebraic surgery on any element $x \in \pi_r(P, S(\xi|_N)) = H^{n-r}(C)$ if $2r < n$ or $n = 2r+1$ (although some new r-dimensional homology may be created if $n = 2r+1$).
For $n = 2r$ geometric intersection and self intersection numbers define a $(-)^{r+1}z$-symmetric pairing

$$\lambda : \pi_r(P, S(\xi|_N)) \times \pi_r(P, S(\xi|_N)) \longrightarrow \mathbb{Z}[\pi_1(S(\xi))]$$

with a $(-)^{r+1}z$-quadratic refinement

$$\mu : \pi_r(P, S(\xi|_N)) \longrightarrow \mathbb{Z}[\pi_1(S(\xi))]/\{a + (-)^r z\overline{a} \mid a \in \mathbb{Z}[\pi_1(S(\xi))]\}$$

such that $\mu(x) = 0$ if (and for $r \geq 4$ only if) is possible to kill $x \in \pi_r(P, S(\xi|_N))$ by an ambient surgery, as follows.
Every element $x \in \pi_r(P, S(\xi|_N))$ is represented by an immersion

$$(f, \partial f) : (D^r, S^{r-1}) \longrightarrow (P, S(\xi|_N))$$

with a choice of path in P from the base point in $S(\xi|_N)$ to ∂f(base point in S^{r-1}), such that $\partial f : S^{r-1} \longrightarrow S(\xi|_N)$ is an embedding with the composite

$$p\partial f \ : \ S^{r-1} \xrightarrow{g} S(\xi|_N) \xrightarrow{p} N^{2r-2}$$

an immersion. Any two elements $x, x' \in \pi_r(P, S(\xi|_N))$ have representatives $(f, \partial f)$, $(f', \partial f')$ which intersect transversely at a finite number of points, and the evaluation of λ is defined by

$$\lambda(x,x') \ = \ \alpha(x,x') + (1-z)\beta(x,x') \in \mathbb{Z}[\pi_1(S(\xi))]$$

with

$$\alpha(x,x') \ = \sum_{u \in S^{r-1}, p\,\partial f(u) = p\,\partial f'(u)} \pm g(u) \in \mathbb{Z}[\pi_1(S(\xi))] \ ,$$

$$\beta(x,x') \ = \sum_{v \in D^r \backslash S^{r-1}, f(v) = f'(v)} \pm h(v) \in \mathrm{im}(\mathbb{Z}[\pi_1(P)] \longrightarrow \mathbb{Z}[\pi_1(S(\xi))])$$

with $g(u) \in \pi_1(S(\xi))$ (resp. $h(v) \in \mathrm{im}(\pi_1(P) \longrightarrow \pi_1(S(\xi)))$) represented by the loop obtained by starting at the base point in $S(\xi|_N)$, using the prescribed paths, and switching from $f(D^r)$ (resp. $\partial f(S^{r-1})$) to $f'(D^r)$ (resp. $\partial f'(D^r)$) at u (resp. v), joining $\partial f(u)$ to $\partial f'(u)$ by the path in the fibre $p^{-1}(p\,\partial f(u)) = S^1$ determined by the orientations. Similarly for the self intersection $\mu(x)$.
The $(-z)$-quadratic complex (C, ψ) is compatible with the geometrically defined $(-)^{r+1}z$-quadratic form $(\pi_r(P, S(\xi|_N)), \lambda, \mu)$ via the Hurewicz map

$$\begin{aligned}\pi_r(P, S(\xi|_N)) \ &= \ \pi_r(\widetilde{P}, \widetilde{S(\xi|_N)}) \\ &\longrightarrow H_r(\widetilde{P}, \widetilde{S(\xi|_N)}) \ = \ H^{n-r}(\widetilde{P}, \widetilde{\partial_+ P}) \ = \ H^{n-r}(C) \ .\end{aligned}$$

If $(P, S(\xi|_N))$ is $(r-1)$-connected then the Hurewicz map is an isomorphism, and $\mu(x) = 0$ is just the condition $\psi(x)(x) = 0$ for killing $x \in H^{n-r}(C)$ by algebraic surgery on (C, ψ). □

Remark 22.9 A codimension 2 submanifold $N^{2i} \subset M^{2i+2}$ is *taut* if the exterior $(P, \partial P)$ is i-connected

$$\pi_r(P, \partial P) \ = \ 0 \ \ (r \leq i) \ .$$

Lefschetz proved that nonsingular algebraic hypersurfaces in complex projective space are taut. Many authors have studied the representation of codimension 2 homology classes $\xi \in H_{2i}(M) = H^2(M)$ by taut submanifolds $N \subset M$ for $i \geq 1$, using a variety of methods: Thom, Rochlin, Massey, Hsiang-Sczarba,.... In particular, see Thomas and Wood [288] for results on taut embeddings in the smooth category, obtained by means of the signatures of the cyclic branched covers of manifolds M branched over codimension 2 submanifolds $N \subset M$ (cf. 27.10) and the Atiyah-Singer G-signature theorem. See Kato and Matsumoto [119] and Freedman [84] for results on taut embeddings obtained by means of codimension 2 surgery methods. □

22C. Spines

Definition 22.10 Let $(M, \partial M)$ be an n-dimensional manifold with boundary, with a 2-plane bundle $\xi : M \longrightarrow BSO(2)$, and let $\mathcal{F} : \mathbb{Z}[\pi_1(S(\xi))] \longrightarrow \Lambda$ be a morphism of rings with involution.
(i) A *Λ-homology boundary spine* of $(M, \partial M)$ is a $\xi|_{\partial M}$-characteristic codimension 2 submanifold $K \subset \partial M$ with exterior

$$L = \text{cl.}(\partial M \backslash E(\nu_{K \subset \partial M})) \quad (\nu_{K \subset \partial M} = \xi|_K)$$

such that the inclusion $L \longrightarrow S(\xi)$ is a Λ-homology equivalence. The *torsion* of K is

$$\tau(L \longrightarrow S(\xi); \Lambda) \in Wh_1(\Lambda) = K_1(\Lambda)/\{\tau(\pm z^j) \,|\, j \in \mathbb{Z}\} .$$

(ii) A *Λ-homology spine* of $(M, \partial M)$ is a ξ-characteristic codimension 2 submanifold with boundary $(N, \partial N) \subset (M, \partial M)$ with exterior

$$(P, \partial_+ P) = (\text{cl.}(M \backslash E(\xi|_N)), \text{cl.}(\partial M \backslash E(\xi|_{\partial N})))$$

such that the inclusions $\partial_+ P \longrightarrow P$, $P \longrightarrow S(\xi)$ are Λ-homology equivalences. The *torsions* of $(N, \partial N)$ are

$$\tau(\partial_+ P \longrightarrow P; \Lambda) \ , \ \tau(P \longrightarrow S(\xi); \Lambda) \in Wh_1(\Lambda) .$$

Note that $K = \partial N$ is a Λ-homology boundary spine of $(M, \partial M)$ with exterior $L = \partial_+ P$ and torsion

$$\tau(L \longrightarrow S(\xi); \Lambda) = \tau(\partial_+ P \longrightarrow P; \Lambda) + \tau(P \longrightarrow S(\xi); \Lambda) \in Wh_1(\Lambda) ,$$

and that $(P; \partial_+ P, S(\xi|_N))$ is a relative Λ-coefficient H-cobordism with torsion $\tau(\partial_+ P \longrightarrow P; \Lambda)$.
(iii) A *codimension 2 Seifert surface* of a $\xi|_{\partial M}$-characteristic codimension 2 submanifold $K \subset \partial M$ with exterior L is a ξ-characteristic codimension 2 submanifold $(N, \partial N) \subset (M, \partial M)$ with exterior $(P, \partial_+ P)$, such that

$$\partial N = K \ , \ \partial_+ P = L . \qquad \square$$

Remark 22.11 Codimension 2 Seifert surfaces are evident generalizations of the codimension 1 Seifert surfaces which arise in knot theory. See Chap. 27 for the connection between the two types of Seifert surface in the framed case $\xi = \epsilon^2$. $\square$

Example 22.12 Given an $(n-2)$-dimensional manifold with boundary $(N, \partial N)$ and a 2-plane bundle $\eta : N \longrightarrow BSO(2)$ define the n-dimensional manifold with boundary

$$(M, \partial M) = (E(\eta), \partial E(\eta)) ,$$

where

$$\partial E(\eta) = E(\eta|_{\partial N}) \cup_{S(\eta|_{\partial N})} S(\eta) .$$

Let $\xi : M \longrightarrow BSO(2)$ be the 2-plane bundle with

$$e(\xi) = e(\eta) \in H^2(M) = H^2(N) .$$

Then $(N, \partial N) \subset (M, \partial M)$ is a ξ-characteristic codimension 2 submanifold with normal bundle

$$\nu_{N \subset M} = \xi|_N = \eta : N \longrightarrow BSO(2) .$$

The exterior is given up to homeomorphism by

$$(P, \partial_+ P) = (S(\eta) \times I, S(\eta) \times \{0\}) ,$$

so that $(N, \partial N) \subset (M, \partial M)$ is a Λ-homology spine for any ring morphism $\mathcal{F} : \mathbb{Z}[\pi_1(S(\eta))] \longrightarrow \Lambda$. □

Proposition 22.13 *Let $(M, \partial M)$ be an n-dimensional manifold with boundary, and let $\xi : M \longrightarrow BSO(2)$ be a 2-plane bundle.*
Let $K \subset \partial M$ be a $\xi|_{\partial M}$-characteristic codimension 2 submanifold, with exterior L.
(i) *K has codimension 2 Seifert surfaces, and any two such are cobordant inside M rel K.*
(ii) *The antiquadratic complex $\sigma_*(M, N, \xi) = (C, \psi)$ associated by 22.4 to any codimension 2 Seifert surface N of K is an n-dimensional $(-z)$-quadratic complex over $\mathbb{Z}[\pi_1(S(\xi))]$ with*

$$C = C(\widetilde{P}, \widetilde{L}) \ , \ \partial C = \mathcal{C}((1+T_{-z})\psi_0 : C^{n-*} \longrightarrow C)_{*+1} \simeq C(\widetilde{S(\xi)}, \widetilde{L})_{*+1} .$$

(iii) *K is a Λ-homology boundary spine of $(M, \partial M)$ if and only if (C, ψ) is Λ-Poincaré, for any codimension 2 Seifert surface N.*
(iv) *A codimension 2 Seifert surface N of K is a Λ-homology spine (N, K) of $(M, \partial M)$ if and only if (C, ψ) is Λ-Poincaré and Λ-contractible.*
Proof The classifying map $\xi : M \longrightarrow BSO(2) = \mathbb{CP}^\infty$ can be made transverse regular at $\mathbb{CP}^{\infty-1} \subset \mathbb{CP}^\infty$ with

$$(\xi|_{\partial M})^{-1}(\mathbb{CP}^{\infty-1}) = K \subset \partial M$$

and

$$N = \xi^{-1}(\mathbb{CP}^{\infty-1}) \subset M$$

a codimension 2 Seifert surface for K. □

Definition 22.14 Let $(M, \partial M)$ be an n-dimensional manifold with boundary, and with a Λ-homology boundary spine K, for some $\xi : M \longrightarrow BSO(2)$, $\mathcal{F} : \mathbb{Z}[\pi_1(S(\xi))] \longrightarrow \Lambda$. The *$\Lambda$-homology spine obstruction* is the cobordism class

$$\sigma_*^\Lambda(M, K) = \sigma_*(M, N, \xi) \in \Gamma_n(\mathcal{F}, -z)$$

of the n-dimensional $(-z)$-quadratic Λ-Poincaré complex $\sigma_*(M,N,\xi) = (C,\psi)$ over $\mathbb{Z}[\pi_1(S(\xi))]$ associated by 22.13 to any codimension 2 Seifert surface N. □

Proposition 22.15 *Let $(M,\partial M)$ be an n-dimensional manifold with boundary, with $\xi : M\longrightarrow BSO(2)$, $\mathcal{F} : \mathbb{Z}[\pi_1(S(\xi))]\longrightarrow\Lambda$ as before.*
(i) *The Λ-homology spine obstruction of $(M,\partial M)$ with respect to a Λ-homology boundary spine $K \subset \partial M$ is such that*

$$\sigma_*^\Lambda(M,K) \;=\; 0 \in \Gamma_n(\mathcal{F},-z)$$

if (and for $n \geq 7$, locally epic $\mathcal{F}$ only if) there exists a codimension 2 Seifert surface N such that (N,K) is a Λ-homology spine of $(M,\partial M)$.
(ii) *Let $\Phi_\mathcal{F}$ be the commutative square of rings with involutions*

$$\begin{array}{ccc}\mathbb{Z}[\pi_1(S(\xi))] & \longrightarrow & \mathbb{Z}[\pi_1(S(\xi))] \\ \Big\downarrow & \Phi_\mathcal{F} & \Big\downarrow \\ \mathbb{Z}[\pi_1(S(\xi))] & \longrightarrow & \Lambda\end{array}$$

so that the relative Γ-groups $\Gamma_*(\Phi_\mathcal{F},-z)$ fit into an exact sequence

$$\begin{aligned}\dots \longrightarrow L_n(\mathbb{Z}[\pi_1(S(\xi))],-z) \longrightarrow \Gamma_n(\mathcal{F},-z) \longrightarrow \Gamma_n(\Phi_\mathcal{F},-z) \\ \longrightarrow L_{n-1}(\mathbb{Z}[\pi_1(S(\xi))],-z) \longrightarrow \dots\end{aligned}$$

with $\Gamma_n(\Phi_\mathcal{F},-z)$ the cobordism group of $(n-1)$-dimensional $(-z)$-quadratic Λ-contractible Poincaré complexes over $\mathbb{Z}[\pi_1(S(\xi))]$ (20.21). If $(N,\partial N) \subset (M,\partial M)$ is a ξ-characteristic codimension 2 submanifold such that $K = \partial N$ is a Λ-homology boundary spine the image of the Λ-homology spine obstruction $\sigma_*^\Lambda(M,\partial N) = (C,\psi) \in \Gamma_n(\mathcal{F},-z)$ is the cobordism class

$$\partial(C,\psi) \;=\; (\partial C,\partial\psi) \in \Gamma_n(\Phi_\mathcal{F},-z)$$

of the boundary $(n-1)$-dimensional $(-z)$-quadratic Λ-contractible Poincaré complex over $\mathbb{Z}[\pi_1(S(\xi))]$ with $\partial C \simeq C(\widetilde{S(\xi)},\widetilde{\partial_+ P})_{*+1}$. Moreover, the normal maps $(S(f),S(b))$, (g,c) of 22.6 are such that

$$(S(f),S(b)) \;:\; (S(\xi_N),S(\xi|_{\partial N})) \longrightarrow (S(\xi),S(\xi|_{\partial N}))$$

is a normal map from an $(n-1)$-dimensional manifold with boundary to an $(n-1)$-dimensional geometric Λ-coefficient Poincaré pair, and

$$\begin{aligned}(g,c) \;:\; (P;S(\xi|_N),\partial_+P;S(\xi|_{\partial N})) \\ \longrightarrow (S(\xi)\times I;S(\xi)\times\{0\},S(\xi)\times\{1\};S(\xi|_{\partial N})\times I)\end{aligned}$$

is a normal map from an n-dimensional manifold triad to an n-dimensional geometric Λ-coefficient Poincaré triad, such that

$$(\partial_+ g, \partial_+ c) \ : \ (\partial_+ P, S(\xi|_{\partial N})) \longrightarrow (S(\xi), S(\xi|_{\partial N}))$$

is a Λ-homology equivalence. The quadratic kernel of (g,c) is the n-dimensional quadratic Λ-Poincaré triad $\sigma_*(g,c)$ described in 22.6 (iv), corresponding to the factorization

$$(1+T)((1-z^{-1})\psi_0) \ : \ C^{n-*} \xrightarrow{1-z^{-1}} C^{n-*} \xrightarrow{(1+T_{-z})\psi_0} C$$

with $(1+T_{-z})\psi_0 : C^{n-*} \longrightarrow C$ a Λ-equivalence.
(iii) *Let Σ be the set of Λ-invertible square matrices in $\mathbb{Z}[\pi_1(S(\xi))]$. If the natural map $\mathbb{Z}[\pi_1(S(\xi))] \longrightarrow \Sigma^{-1}\mathbb{Z}[\pi_1(S(\xi))]$ is injective and $1-z \in \Lambda^\bullet$ (i.e. $1-z \in \Sigma$) then*

$$\begin{aligned} \Gamma_n(\mathcal{F}, -z) &= L_n(\Sigma^{-1}\mathbb{Z}[\pi_1(S(\xi))], -z) \\ &= \Gamma_n(\mathcal{F}) \ = \ L_n(\Sigma^{-1}\mathbb{Z}[\pi_1(S(\xi))]) \\ &= \Gamma^n(\mathcal{F}, -z) \ = \ L^n(\Sigma^{-1}\mathbb{Z}[\pi_1(S(\xi))], -z) \\ &= \Gamma^n(\mathcal{F}) \ = \ L^n(\Sigma^{-1}\mathbb{Z}[\pi_1(S(\xi))]) \end{aligned}$$

and the Λ-homology spine obstruction is the Λ-coefficient symmetric signature of the n-dimensional geometric Λ-coefficient Poincaré pair $(P, \partial P)$

$$\sigma_*^\Lambda(M,K) \ = \ \sigma_\Lambda^*(P, \partial P) \in \Gamma_n(\mathcal{F}, -z) \ = \ \Gamma^n(\mathcal{F}) \ .$$

Proof (i) Immediate from 22.8.
(ii) Immediate from 20.21 and 22.6.
(iii) The central unit

$$s \ = \ (1-z)^{-1} \in \Sigma^{-1}\mathbb{Z}[\pi_1(S(\xi))]^\bullet$$

is such that $s + \overline{s} = 1$, so that the ϵ-symmetrization maps

$$1 + T_\epsilon \ : \ L_*(\Sigma^{-1}\mathbb{Z}[\pi_1(S(\xi))], \epsilon) \longrightarrow L^*(\Sigma^{-1}\mathbb{Z}[\pi_1(S(\xi))], \epsilon)$$

are isomorphisms, for any central unit $\epsilon \in \Sigma^{-1}\mathbb{Z}[\pi_1(S(\xi))]$ such that $\overline{\epsilon} = \epsilon^{-1}$. See Chap. 25 below for the identifications

$$\begin{aligned} \Gamma_*(\mathcal{F}, \epsilon) &= L_*(\Sigma^{-1}\mathbb{Z}[\pi_1(S(\xi))], \epsilon) \ , \\ \Gamma^*(\mathcal{F}, \epsilon) &= L^*(\Sigma^{-1}\mathbb{Z}[\pi_1(S(\xi))], \epsilon) \ . \end{aligned}$$ □

Definition 22.16 Let X be a space with a 2-plane bundle $\xi : X \longrightarrow BSO(2)$, let $\mathcal{F} : \mathbb{Z}[\pi_1(S(\xi)] \longrightarrow \Lambda$ be a morphism of rings with involution, and let $U \subseteq Wh_1(\Lambda)$ be a $*$-invariant subgroup.
(i) The *U-intermediate bounded spine bordism group* $BB_n^U(X, \xi, \mathcal{F})$ is the group of bordism classes of objects (M, K, f), where $(M, \partial M)$ is an $(n+2)$-dimensional manifold with boundary, together with a map $f : M \longrightarrow X$ and a $(\partial f)^*\xi$-characteristic codimension 2 submanifold $K^{n-1} \subset \partial M$ with the exterior $\partial_+ P = \mathrm{cl.}(\partial M \backslash E((\partial f)^*\xi|_K))$ such that

$$\begin{aligned}(\partial_+ f)^* e(\xi) &= 0 \in H^2(\partial_+ P) ,\\ H_*(\partial_+ P; \Lambda) &= H_*(S(f^*\xi); \Lambda) ,\\ \tau(\partial_+ P \longrightarrow S(f^*\xi); \Lambda) &\in U \subseteq Wh_1(\Lambda) ,\end{aligned}$$

i.e. such that K is a Λ-homology boundary spine for $(M, \partial M)$ with torsion in U.

(ii) The *U-intermediate empty spine bordism group* $AB_n^U(X, \xi, \mathcal{F})$ is the group of bordism classes of objects (M, f), where $(M, \partial M)$ is an $(n+2)$-dimensional manifold with boundary with a map $f : M \longrightarrow X$ such that

$$\begin{aligned}(\partial f)^* e(\xi) &= 0 \in H^2(\partial M) ,\\ H_*(\partial M; \Lambda) &= H_*(S(f^*\xi); \Lambda) ,\\ \tau(\partial M \longrightarrow S(f^*\xi); \Lambda) &\in U \subseteq Wh_1(\Lambda) ,\end{aligned}$$

i.e. such that $\emptyset$ is a Λ-homology boundary spine for $(M, \partial M)$ with torsion in U.

(iii) Suppose that $1 - z \in \Lambda^\bullet$ and $\tau(1 - z) \in U$. The *U-intermediate closed spine bordism group* $\Delta_n^U(X, \xi, \mathcal{F})$ is the group of bordism classes of objects (M, f), where M is a closed $(n+1)$-dimensional manifold with a map $f : M \longrightarrow S(\xi)$ such that

$$H_*(M; \Lambda) = 0 \ , \ \ \tau(M; \Lambda) \in U \subseteq Wh_1(\Lambda) ,$$

i.e. such that $\emptyset$ is a Λ-homology spine for $(M, \emptyset)$ with torsion in U. □

For $U = Wh_1(\Lambda)$ the groups defined in 22.16 are denoted by

$$\begin{aligned}BB_n^{Wh_1(\Lambda)}(X, \mathcal{F}, \xi) &= BB_n(X, \mathcal{F}, \xi) ,\\ AB_n^{Wh_1(\Lambda)}(X, \mathcal{F}, \xi) &= AB_n(X, \mathcal{F}, \xi) ,\\ \Delta_n^{Wh_1(\Lambda)}(X, \mathcal{F}, \xi) &= \Delta_n(X, \mathcal{F}, \xi) .\end{aligned}$$

Proposition 22.17 *Let X be a space with a 2-plane bundle $\xi : X \longrightarrow BSO(2)$, and let*

$$\mathcal{F} : A = \mathbb{Z}[\pi_1(S(\xi)] \longrightarrow \Lambda$$

be a morphism of rings with involution. Let $U \subseteq Wh_1(\Lambda)$ be a $*$-invariant subgroup, denoting the preimage in $Wh_1(A)$ by U also.

(i) *The bounded and empty spine bordism groups are isomorphic*

$$AB_*^U(X, \xi, \mathcal{F}) = BB_*^U(X, \xi, \mathcal{F}) .$$

(ii) *The spine obstruction defines morphisms*

$$BB_n^U(X, \xi, \mathcal{F}) \longrightarrow \Gamma_{n+2}^U(\mathcal{F}, -z) \ ; \ (M, K, f) \longrightarrow \sigma_*^\Lambda(M, K)$$

which are isomorphisms for locally epic $\mathfrak{F}$ and $n \geq 5$ with $\pi_1(X)$ finitely presented (e.g. if X has a finite 2-skeleton), in which case the spine obstruction defines a natural isomorphism of exact sequences

$$\begin{array}{ccccccccc}
\to & BB_n^U(X,\xi,A) & \to & BB_n^U(X,\xi,\mathfrak{F}) & \to & \Gamma_{n+2}^U(\Phi_{\mathfrak{F}},-z) & \to & BB_{n-1}^U(X,\xi,A) & \to \\
& \sigma_* \big\downarrow \cong & & \sigma_* \big\downarrow \cong & & \big\| & & \sigma_* \big\downarrow \cong & \\
\to & L_{n+2}^U(A,-z) & \longrightarrow & \Gamma_{n+2}^U(\mathfrak{F},-z) & \to & \Gamma_{n+2}^U(\Phi_{\mathfrak{F}},-z) & \longrightarrow & L_{n+1}^U(A,-z) & \longrightarrow
\end{array}$$

with $\Phi_{\mathfrak{F}}$ as in 22.15, and $\Gamma_{n+2}^U(\Phi_{\mathfrak{F}},-z)$ the cobordism group of based f.g. free $(n+1)$-dimensional $(-z)$-quadratic Poincaré complexes (C,ψ) over A which are Λ-contractible with $\tau(C;\Lambda) \in U$.

(iv) *If $1-z \in \Lambda^\bullet$ and $\tau(1-z) \in U$ the empty and closed spine bordism groups are related by an exact sequence*

$$\begin{aligned}
\ldots \longrightarrow \Omega_{n+2}(S(\xi)) \longrightarrow AB_n^U(X,\xi,\mathfrak{F}) \longrightarrow \Delta_n^U(X,\xi,\mathfrak{F}) \\
\longrightarrow \Omega_{n+1}(S(\xi)) \longrightarrow \ldots
\end{aligned}$$

with $\Omega_*(S(\xi))$ the ordinary bordism groups of $S(\xi)$, and

$$\begin{aligned}
&\Omega_{n+2}(S(\xi)) \longrightarrow AB_n^U(X,\xi,\mathfrak{F}) \ ; \ (M,f) \longrightarrow (M,pf) \ , \\
&AB_n^U(X,\xi,\mathfrak{F}) \longrightarrow \Delta_n^U(X,\xi,\mathfrak{F}) \ ; \ (M,f) \longrightarrow (\partial M,\overline{\partial f}) \ , \\
&\Delta_n^U(X,\xi,\mathfrak{F}) \longrightarrow \Omega_{n+1}(S(\xi)) \ ; \ (M,f) \longrightarrow (M,f)
\end{aligned}$$

with $\overline{\partial f} : \partial M \longrightarrow S(\xi)$ the lift of $\partial f : \partial M \longrightarrow X$ determined by the trivialization $(\partial f)^*\xi = \epsilon^2 : \partial M \longrightarrow BSO(2)$.

(v) *If $1-z \in \Lambda^\bullet$ and $\mathfrak{F}$ is locally epic with factorization through an injective localization map i*

$$\mathfrak{F} : A \xrightarrow{\ i\ } \Sigma^{-1}A \longrightarrow \Lambda$$

(with Σ as in 22.15), then for $n \geq 5$, finitely presented $\pi_1(X)$ and locally epic $\mathfrak{F}$

$$\begin{aligned}
AB_n^U(X,\xi,\mathfrak{F}) &= BB_n^U(X,\xi,\mathfrak{F}) \\
&= \Gamma_{n+2}^U(\mathfrak{F},-z) = \Gamma_{n+2}^U(\mathfrak{F}) \\
&= L_{n+2}^U(\Sigma^{-1}A,-z) = L_{n+2}^U(\Sigma^{-1}A) \\
&= \Gamma_U^{n+2}(\mathfrak{F},-z) = \Gamma_U^{n+2}(\mathfrak{F}) \\
&= L_U^{n+2}(\Sigma^{-1}A,-z) = L_U^{n+2}(\Sigma^{-1}A) \ ,
\end{aligned}$$

the symmetric signature defines a natural transformation of exact sequences

$$\begin{array}{ccccccc} \cdots \to \Omega_{n+2}(S(\xi)) & \to & AB^U_n(X,\xi,\mathcal{F}) & \to & \Delta^U_n(X,\xi,\mathcal{F}) & \to & \Omega_{n+1}(S(\xi)) \to \cdots \\ \downarrow \sigma^* & & \cong \downarrow \sigma^* & & \downarrow \sigma^* & & \downarrow \sigma^* \\ \cdots \longrightarrow L^{n+2}_U(A) & \longrightarrow & \Gamma^{n+2}_U(\mathcal{F}) & \longrightarrow & \Gamma^{n+2}_U(\Phi_{\mathcal{F}}) & \longrightarrow & L^{n+1}_U(A) \longrightarrow \cdots \end{array}$$

and there is defined an exact sequence

$$\begin{aligned} \ldots \longrightarrow L^{n+2}_U(A) \longrightarrow \Delta^U_n(X,\xi,\mathcal{F}) \longrightarrow \Gamma^{n+2}_U(\Phi_{\mathcal{F}}) \oplus \Omega_{n+1}(S(\xi)) \\ \longrightarrow L^{n+1}_U(A) \longrightarrow \ldots . \end{aligned}$$

Proof (i) Given an $(n+2)$-dimensional manifold with boundary $(M,\partial M)$ with a map $f : M \longrightarrow X$ and a Λ-homology boundary spine $K \subset \partial M$ with exterior L let $N \subset M$ be a codimension 2 Seifert surface (22.13 (ii)) with $\partial N = K$, and define an $(n+2)$-dimensional manifold with boundary

$$(M',\partial M') = (M \cup_{E(\nu_{K\subset \partial M})} E(\nu_{N\subset M}), L \cup_{S(\nu_{K\subset\partial M})} S(\nu_{N\subset M}))$$

with empty Λ-homology boundary spine $\emptyset \subset \partial M'$. The construction defines an isomorphism

$$BB^U_n(X,\xi,\mathcal{F}) \longrightarrow AB^U_n(X,\xi,\mathcal{F}) \ ; \ (M,K,f) \longrightarrow (M',f')$$

inverse to the evident forgetful map

$$AB^U_n(X,\xi,\mathcal{F}) \longrightarrow BB^U_n(X,\xi,\mathcal{F}) \ ; \ (M,f) \longrightarrow (M,\emptyset,f) \ .$$

(ii)+(iii) Immediate from 22.15, realizing elements of $\Gamma^U_{n+2}(\mathcal{F},-z)$ by elements of $BB^U_n(X,\mathcal{F})$ as in Matsumoto [184, 5.2]. Here are the details of the construction for $n = 2i \geq 6$, realizing a $(-)^i z$-quadratic form (A^ℓ,ψ) over $A = \mathbb{Z}[\pi_1(S(\xi))]$. Choose a $(2i-1)$-dimensional manifold with boundary $(K_0,\partial K_0)$ with a π_1-isomorphism map $K_0 \longrightarrow X$, and use the method of Wall [302, Chap. 5] to construct a $2i$-dimensional relative cobordism $(N;K_0,K_1)$ with

$$N^{2i} = K_0 \times I \cup_{1\otimes\psi+(-)^i 1\otimes\psi^*} \bigcup_\ell D^i \times D^i$$

the trace of surgeries on ℓ disjoint trivial embeddings $S^{i-1} \times D^i \subset \mathrm{int}(K_0)$ with self intersection form $(\mathbb{Z}[\pi_1(X)]^\ell, 1\otimes\psi)$. By construction, there is an $(i-1)$-connected normal map of $2i$-dimensional manifold triads

$$(f,b) \ : \ (N;K_0,K_1) \longrightarrow K_0 \times (I;\{0\},\{1\})$$

with

$$(f,b)| = \mathrm{id.} \ : \ K_0 \longrightarrow K_0 \ \ , \ \ K_i(N) = \mathbb{Z}[\pi_1(X)]^\ell \ ,$$

and quadratic kernel the $2i$-dimensional quadratic complex $(S^i\mathbb{Z}[\pi_1(X)]^\ell, 1\otimes\psi)$ over $\mathbb{Z}[\pi_1(X)]$. Define the $(2i+2)$-dimensional manifold with boundary

$$(M,\partial M) = (E(\xi|_{K_0\times I}), \partial E(\xi|_{K_0\times I})) \ ,$$

regarding

$$\partial(K_0 \times I) \times \{0\} \subset \partial M \;=\; E(\xi|_{\partial(K_0\times I)}) \cup_{S(\xi|_{\partial(K_0\times I)})} S(\xi|_{K_0\times I})$$

as a ξ-characteristic codimension 2 submanifold. Use the method of [184, 5.2] (cf. 22.8) to perform ambient surgeries on ℓ disjoint trivial embeddings

$$S^{i-1} \times D^i \subset \text{int}(K_0) \times \{0\} \subset \partial M$$

with self intersections ψ, and trace a ξ-characteristic codimension 2 submanifold triad

$$(N \cup_{\partial K_0 \times I} K_0 \times I; \partial(K_0 \times I), \partial N) \subset \partial M \times (I; \{0\}, \{1\}) .$$

Let

$$P^{2i+2} \;=\; S(\xi|_N) \times I \cup \bigcup_{\ell} D^{i+1} \times D^{i+1} ,$$

the trace of surgeries on ℓ disjoint embeddings $S^i \times D^{i+1} \subset \text{int}(S(\xi|_N))$ representing generators of the kernel A-module

$$K_i(S(\xi|_N)) \;=\; A^\ell$$

of the $(i-1)$-connected normal map of $(2i+1)$-dimensional manifolds with boundary

$$(S(f), S(b)) \;:\; (S(\xi|_N), S(\xi|_{\partial N})) \longrightarrow (S(\xi|_{K_0\times I}), S(\xi|_{\partial(K_0\times I)})) ,$$

and let $\partial_+ P$ be the effect of these surgeries. The ξ-characteristic codimension 2 submanifold $(N, \partial N) \subset (M, \partial M)$ has exterior $(P, \partial_+ P)$, so that up to homeomorphism

$$(M, \partial M) \;=\; (E(\xi|_N) \cup_{S(\xi|_N)} P, E(\xi|_{\partial N}) \cup_{S(\xi|_{\partial N})} \partial_+ P)$$

and there is defined a normal map of $(2i+2)$-dimensional manifold triads

$$\begin{aligned}&(g, c) \;:\; (P; \partial_+ P, S(\xi|_N); S(\xi|_{\partial N}))\\ &\longrightarrow (S(\xi|_{K_0\times I}) \times I; S(\xi|_{K_0\times I}) \times \{0\}, S(\xi|_{K_0\times I}) \times \{1\}; S(\xi|_{\partial(K_0\times I)}) \times I) .\end{aligned}$$

The kernel $(2i+2)$-dimensional quadratic Poincaré triad over A is the one associated by 20.27 to the $(2i+2)$-dimensional quadratic complex $(S^{i+1}A^\ell, (1 - z^{-1})\psi)$ and the factorization

$$\begin{aligned}&(1 - z^{-1})\psi + (-)^{i+1}((1 - z^{-1})\psi)^* \;=\; (\psi + (-)^i z\psi^*)(1 - z^{-1}) \;:\\ &\mathcal{C}(g^!)^{2i+2-*} \;=\; S^{i+1}A^\ell \xrightarrow{1-z^{-1}} S^{i+1}A^\ell \xrightarrow{\psi+(-)^i z\psi^*} \mathcal{C}(g^!) \;=\; S^{i+1}A^\ell .\end{aligned}$$

Now

$$\mathcal{C}(\partial_+ g^! : C(\widetilde{S(\xi|_{K_0\times I})}) \longrightarrow C(\widetilde{\partial_+ P})) \;\simeq\; S^i \mathcal{C}(\psi + (-)^i z\psi^* : A^\ell \longrightarrow A^\ell) ,$$

so that ∂N is a Λ-homology boundary spine of $(M,\partial M)$ if and only if the $(-)^i z$-quadratic form (A^ℓ,ψ) is Λ-nonsingular, in which case the Λ-homology spine obstruction of $(N,\partial N)\subset(M,\partial M)$ is

$$\sigma_*^\Lambda(M,\partial N) \;=\; (A^\ell,\psi)\in\Gamma_{2i+2}(\mathcal{F},-z) \;=\; \Gamma_0(\mathcal{F},(-)^i z)\ .$$

(iv) If M is a closed $(n+2)$-dimensional manifold with a map $f:M\longrightarrow S(\xi)$ then $\emptyset$ is a Λ-homology boundary spine of $(M,\emptyset)$, with $H_*(S(f^*\xi);\Lambda)=0$ and

$$\tau(S(f^*\xi);\Lambda) \;=\; \chi(M)\tau(1-z)\in U\subseteq Wh_1(\Lambda)\ .$$

(v) Combine 22.6, 22.15 and (i)–(iv). □

Remark 22.18 If $1-z\in\Lambda^\bullet$, $\tau(1-z)\in U$ the isomorphism $BB_n^U(X,\xi,\mathcal{F})\cong AB_n^U(X,\xi,\mathcal{F})$ of 22.17 (i) can also be expressed as

$$BB_n^U(X,\xi,\mathcal{F}) \xrightarrow{\;\simeq\;} AB_n^U(X,\xi,\mathcal{F})\ ;\ (M,K,f)\longrightarrow(P,f|)$$

with $P=\mathrm{cl.}(M\backslash E(\nu_{N\subset M}))$ the exterior of any codimension 2 Seifert surface $(N,\partial N)\subset(M,\partial M)$ of $\partial N=K\subset\partial M$. □

For the remainder of Chap. 22 only the case $\Lambda=\mathbb{Z}[\pi_1(M)]$ will be considered.

Example 22.19 Let $(M,\partial M)$ be an n-dimensional manifold with boundary, with a 2-plane bundle $\xi:M\longrightarrow BSO(2)$ and let

$$\mathcal{F} \;=\; p_* \;:\; A \;=\; \mathbb{Z}[\pi_1(S(\xi))] \longrightarrow \Lambda \;=\; \mathbb{Z}[\pi_1(M)]\ .$$

A $\xi|_{\partial M}$-characteristic codimension 2 submanifold $K\subset\partial M$ with exterior L is a Λ-homology boundary spine of $(M,\partial M)$ if and only if (M,K) is an $(n-2)$-dimensional geometric Poincaré pair, with

$$\begin{aligned} H_*(M,K;\Lambda) \;&\cong\; H_*(M,E(\xi|_K);\Lambda) \;\cong\; H^{n-*}(M,L;\Lambda)\\ &\cong\; H^{n-*}(E(\xi),S(\xi);\Lambda) \;\cong\; H^{n-2-*}(M;\Lambda)\ ,\\ \tau(M,K;\Lambda) \;&=\; \tau(L\longrightarrow S(\xi);\Lambda)\in Wh_1(\Lambda)\ .\end{aligned}$$

If $K\subset\partial M$ is a Λ-homology boundary spine for $(M,\partial M)$ and $N\subset M$ is a codimension 2 Seifert surface for K the inclusion $N\subset M$ defines a normal map $(g,c):(N,K)\longrightarrow(M,K)$ and the antiquadratic complex $\sigma_*(M,N,\xi)=(C,\psi)$ of 22.4 is an n-dimensional $(-z)$-quadratic complex over A which is Λ-Poincaré, with

$$\begin{aligned} &C \;=\; C(\widetilde{P},\widetilde{L})\ \ ,\ \ \Lambda\otimes_A C \;\simeq\; C(\widetilde{M},\widetilde{N})\ ,\\ &\partial C \;\simeq\; C(\widetilde{S(\xi)},\widetilde{L})_{*+1}\ \ ,\ \ \tau(\Lambda\otimes_A C) \;=\; \tau(M,K;\Lambda)\ .\end{aligned}$$

A codimension 2 Seifert surface of K is a Λ-homology spine (N,K) of $(M,\partial M)$ if and only if (g,c) is a homotopy equivalence, if and only if (C,ψ) is Λ-contractible. The Λ-homology spine obstruction

$$\sigma_*^\Lambda(M,K) \;=\; (C,\psi)\in BB_n(X,\mathcal{F},\xi) \;=\; \Gamma_n(\mathcal{F},-z)$$

is the obstruction of Matsumoto [184], with image the surgery obstruction of (g,c)

$$\Lambda \otimes_A \sigma_*^{\Lambda}(M,K) \;=\; \sigma_*(g,c) \in L_n(\Lambda,-1) \;=\; L_{n-2}(\Lambda)\ .$$

Cappell and Shaneson [45] showed that there exists an immersed Λ-homology spine, i.e. a homotopy equivalence $N \longrightarrow M$ which is a codimension 2 immersion with boundary $\partial N = K$. □

22D. The homology splitting obstruction

Definition 22.20 (i) A *weak* $(n, n-2)$*-dimensional geometric Poincaré pair* (X, Y) is a codimension 2 CW pair such that:

(a) X is an n-dimensional Poincaré complex,
(b) Y is an $(n-2)$-dimensional Poincaré complex,
(c) $(Z, S(\xi))$ is an n-dimensional $\mathbb{Z}[\pi_1(X)]$-coefficient Poincaré pair.

(ii) Let (X, Y) be a codimension 2 CW pair with a topological normal structure. A homotopy equivalence $h : M \longrightarrow X'$ from an n-dimensional manifold M to an n-dimensional geometric Poincaré complex X' *weakly splits along* $Y \subset X$ if there is given a reference map $r : X' \longrightarrow X$ which is weakly Poincaré transverse at $Y \subset X$, with (X', Y') a weak $(n, n-2)$-dimensional geometric Poincaré pair, such that the restriction

$$(f, b) \;=\; h| \;:\; N \;=\; (rh)^{-1}(Y) \longrightarrow Y' \;=\; r^{-1}(Y)$$

is a homotopy equivalence, and the restriction

$$(g, c) \;=\; f| \;:\; P \;=\; (rh)^{-1}(Z) \longrightarrow Z' \;=\; r^{-1}(Z)$$

is a $\mathbb{Z}[\pi_1(X)]$-coefficient homology equivalence.
(iii) There are relative versions of (i) and (ii). In particular, a *weak* $(n, n-2)$*-dimensional geometric Poincaré triad* $(X, Y; \partial X, \partial Y)$ is a codimension 2 CW pair (X, Y) such that $(X, \partial X)$ is an n-dimensional geometric Poincaré pair for some subcomplex $\partial X \subset X$, and such that $(Y, \partial Y)$ is an $(n-2)$-dimensional geometric Poincaré pair with boundary $\partial Y = \partial X \cap Y$, and such that $(Z; \partial_+ Z, S(\xi), S(\xi|_{\partial Y}))$ is an n-dimensional geometric $\mathbb{Z}[\pi_1(X)]$-coefficient Poincaré triad with $\partial_+ Z = \partial X \cap Z$.

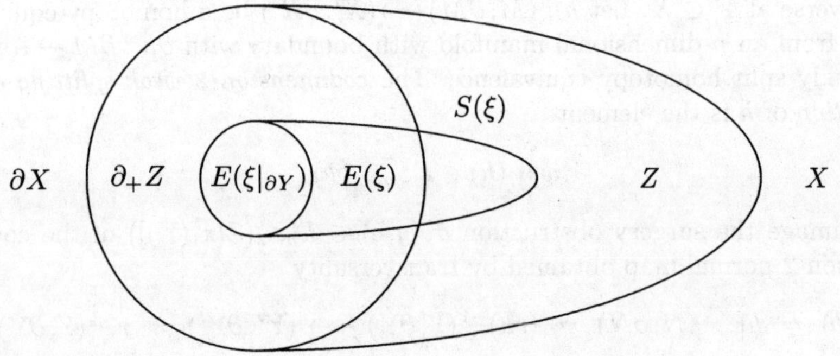

□

The ΓS- and ΓN-groups of Ranicki [237] are the analogues of the LS- and LN-groups in the context of the homology surgery theory of Cappell and Shaneson [40].

Definition 22.21 (Ranicki [237, p. 770])
(i) The *quadratic ΓS-groups* $\Gamma S_*(\Phi)$ of the pushout square of groups associated to a codimension 2 CW pair (X, Y)

$$\begin{array}{ccc} \pi_1(S(\xi)) & \longrightarrow & \pi_1(Z) \\ \Big\downarrow & \Phi & \Big\downarrow \\ \pi_1(E(\xi)) & \longrightarrow & \pi_1(X) \end{array}$$

are the relative groups appearing in the exact sequence

$$\dots \longrightarrow \Gamma_{n+1}(\Phi_X) \longrightarrow \Gamma S_{n-2}(\Phi) \longrightarrow L_{n-2}(\mathbb{Z}[\pi_1(Y)]) \xrightarrow{p\xi^!} \Gamma_n(\Phi_X) \longrightarrow \dots$$

with Φ_X the commutative square of rings with involution

$$\begin{array}{ccc} \mathbb{Z}[\pi_1(Z)] & \longrightarrow & \mathbb{Z}[\pi_1(X)] \\ \Big\downarrow & \Phi_X & \Big\downarrow \\ \mathbb{Z}[\pi_1(X)] & \longrightarrow & \mathbb{Z}[\pi_1(X)] \end{array}$$

and $p\xi^!$ the composite of the transfer map

$$\xi^! \, : \, L_{n-2}(\mathbb{Z}[\pi_1(X)]) \longrightarrow L_n(\mathbb{Z}[\pi_1(S(\xi))] \longrightarrow \mathbb{Z}[\pi_1(E(\xi))])$$

and the natural map

$$p \, : \, L_n(\mathbb{Z}[\pi_1(S(\xi))] \longrightarrow \mathbb{Z}[\pi_1(E(\xi))]) \longrightarrow \Gamma_n(\Phi_X) \, .$$

(iii) Let $(X', Y'; \partial X', \partial Y')$ be a weak $(n, n-2)$-dimensional geometric Poincaré triad, with a π_1-isomorphism reference map $r : X' \longrightarrow X$ weakly Poincaré transverse at $Y \subset X$. Let $h : (M, \partial M) \longrightarrow (X', \partial X')$ be a homotopy equivalence from an n-dimensional manifold with boundary with $\partial h : \partial M \longrightarrow \partial X'$ a weakly split homotopy equivalence. The *codimension 2 weak splitting obstruction* of h is the element

$$ws_Y(h) \in \Gamma S_{n-2}(\Phi)$$

with image the surgery obstruction $\sigma_*(g, c) \in L_{n-2}(\mathbb{Z}[\pi_1(Y)])$ of the codimension 2 normal map obtained by transversality

$$(f, b) \, = \, h| \, : \, (N, \partial N) \, = \, (rh)^{-1}(Y, \partial Y) \longrightarrow (Y', \partial Y') \, = \, r^{-1}(Y, \partial Y)$$

which is determined by the bordism

$$\begin{aligned} h\times 1\ :\ &(M\times I; M\times\{0\}, E(\nu_{N\subset M})\times\{1\}; P\times\{1\}) \\ &\qquad\longrightarrow (X'\times I; X'\times\{0\}, E(\nu_{Y'\subset X'})\times\{1\}; Z'\times\{1\}) \\ &\qquad(P=(rh)^{-1}(Z)=\mathrm{cl.}(M\backslash E(\nu_{N\subset M}))) \end{aligned}$$

of

$$(f,b)^!\ :\ (E(\nu_{N\subset M}), S(\nu_{N\subset M}))\longrightarrow (E(\nu_{Y'\subset X'}), S(\nu_{Y'\subset X'}))$$

to the homotopy equivalence $h:(M,\partial M)\longrightarrow(X',\partial X')$.

(iv) The *quadratic ΓN-groups* $\Gamma N_*(\mathcal{F})$ are the ΓS-groups for a codimension 2 *CW* pair (X,Y) with

$$\pi_1(X)\ =\ \pi_1(Y)\quad,\quad \pi_1(S(\xi))\ =\ \pi_1(Z)$$

(e.g. if $X=E(\xi)$ with $Z=S(\xi)$) and $\mathcal{F}:\mathbb{Z}[\pi_1(S(\xi))]\longrightarrow\mathbb{Z}[\pi_1(Y)]$ the induced map, that is

$$\Gamma N_*(\mathcal{F})\ =\ \Gamma S_*\left(\begin{array}{ccc} \pi_1(S(\xi)) & \longrightarrow & \pi_1(S(\xi)) \\ {\scriptstyle\mathcal{F}}\downarrow & & \downarrow{\scriptstyle\mathcal{F}} \\ \pi_1(Y) & \longrightarrow & \pi_1(Y) \end{array}\right).$$

□

Proposition 22.22 (Ranicki [237, 7.8.1, 7.8.2])

(i) *The LS- and ΓS-groups of a codimension 2 CW pair (X,Y) are related by the commutative braid of exact sequences*

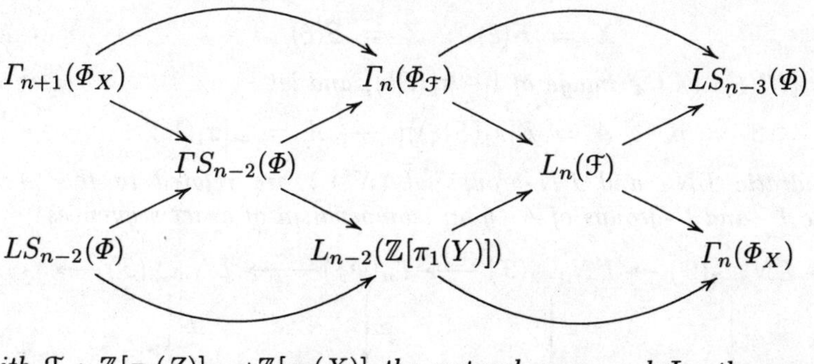

with $\mathcal{F}:\mathbb{Z}[\pi_1(Z)]\longrightarrow\mathbb{Z}[\pi_1(X)]$ the natural map, and $\Phi_{\mathcal{F}}$ the commutative square of rings with involution

$$\begin{array}{ccc} \mathbb{Z}[\pi_1(Z)] & \longrightarrow & \mathbb{Z}[\pi_1(Z)] \\ \Big\downarrow & \Phi_{\mathcal{F}} & \Big\downarrow \\ \mathbb{Z}[\pi_1(Z)] & \longrightarrow & \mathbb{Z}[\pi_1(X)] \end{array}$$

(ii) *The codimension 2 weak splitting obstruction of a homotopy equivalence from an n-dimensional manifold with boundary to an n-dimensional geometric Poincaré triad $(h,\partial h):(M,\partial M)\longrightarrow(X',\partial X')$ with ∂h weakly split along $Y\subset X$ is such that*

$$ws_Y(h) \;=\; 0 \in \Gamma S_{n-2}(\Phi)$$

if (and for $n\geq 7$ only if) h can be weakly split rel ∂ along $Y\subset X$. Similarly for s-splitting.
(iii) *Every element $x\in\Gamma S_{n-2}(\Phi)$ is the codimension 2 weak splitting obstruction $x=ws_Y(h)$ of a homotopy equivalence h as in* (ii), *with images*

$$\sigma_*(f,b)\in L_{n-2}(\mathbb{Z}[\pi_1(Y)]) \;\;,\;\; \sigma_*^{\mathbb{Z}[\pi_1(X)]}(g,c)\in\Gamma_n(\Phi_{\mathcal{F}})$$

the surgery obstructions of the restrictions

$$\begin{array}{l}(f,b) \;=\; h| \;:\; (N,\partial N)\longrightarrow(Y',\partial Y')\;,\\ (g,c) \;=\; h| \;:\; (P;\partial_+P,S(\xi|_N);S(\xi|_{\partial N}))\longrightarrow(Z';\partial_+Z',S(\xi|_{Y'});S(\xi|_{\partial Y'}))\\ \qquad\qquad (P \;=\; \mathrm{cl.}(M\backslash E(\xi|_N))\;,\;\partial_+P \;=\; \mathrm{cl.}(\partial M\backslash E(\xi|_{\partial N})))\end{array}$$

with (g,c) a normal bordism from the $(n-1)$-dimensional normal map

$$(g,c)| \;=\; (S(f),S(b)) \;:\; (S(\xi|_N),S(\xi|_{\partial N}))\longrightarrow(S(\xi|_{Y'}),S(\xi|_{\partial Y'}))$$

to the $\mathbb{Z}[\pi_1(X)]$-homology equivalence

$$(g,c)| \;:\; (\partial_+P,S(\xi|_{\partial N}))\longrightarrow(\partial_+Z',S(\xi|_{\partial Y'}))\;.$$ □

Proposition 22.23 (Ranicki [237, 7.8.12])
Given a 2-plane bundle $\xi:Y\longrightarrow BSO(2)$ over a CW complex Y let (X,Y) be the codimension 2 CW pair defined by

$$X \;=\; E(\xi)\;\;,\;\; Z \;=\; S(\xi)\;.$$

Let $z\in\pi_1(S(\xi))$ be the image of $1\in\pi_1(S^1)$, and let

$$\mathcal{F} \;=\; p_* \;:\; A \;=\; \mathbb{Z}[\pi_1(S(\xi))]\longrightarrow\Lambda \;=\; \mathbb{Z}[\pi_1(Y)]\;.$$

The quadratic LN- and ΓN-groups of (X,Y) are related to the $(-z)$-quadratic L- and Γ-groups of A by an isomorphism of exact sequences

$$\begin{array}{ccccccccc}\dots\to & LN_{n-2}(\mathcal{F}) & \to & \Gamma N_{n-2}(\mathcal{F}) & \to & \Gamma_n(\Phi_{\mathcal{F}}) & \to & LN_{n-3}(\mathcal{F}) & \to\dots\\ & \downarrow\cong & & \downarrow\cong & & \downarrow\cong & & \downarrow\cong & \\ \dots\to & L_n(A,-z) & \to & \Gamma_n(\mathcal{F},-z) & \to & \Gamma_n(\Phi_{\mathcal{F}},-z) & \to & L_{n-1}(A,-z) & \to\dots\end{array}$$

with $\Phi_{\mathcal{F}}$ the commutative square of rings with involution

$$\begin{array}{ccc}A & \longrightarrow & A\\ \big\downarrow & \Phi_{\mathcal{F}} & \big\downarrow\\ A & \longrightarrow & \Lambda\end{array}$$

Proof Every element $x \in LN_{n-2}(\mathcal{F})$ (resp. $\Gamma N_{n-2}(\mathcal{F})$) is the codimension 2 splitting obstruction $x = s_Y(h)$ (resp. weak splitting obstruction $x = ws_Y(h)$) of a homotopy equivalence $h : (M, \partial M) \longrightarrow (X', \partial X')$ from an n-dimensional manifold with boundary to an n-dimensional geometric Poincaré pair $(X', \partial X')$ with a weakly Poincaré transverse reference map $(r, \partial r) : (X', \partial X') \longrightarrow X$ such that

$$\begin{aligned}
&\nu_{N\subset M} \ : \ N \ = \ (rh)^{-1}(Y) \longrightarrow Y' = r^{-1}(Y) \xrightarrow{\nu_{Y'\subset X'}} BSO(2) \ , \\
&\nu_{Y'\subset X'} \ : \ Y' \longrightarrow X' \longrightarrow Y \xrightarrow{\xi} BSO(2) \ , \\
&\pi_1(M) \ = \ \pi_1(\partial M) \ = \ \pi_1(X') \ = \ \pi_1(\partial X') \ = \ \pi_1(Y) \ , \\
&\pi_1(P) \ = \ \pi_1(\partial_+ P) \ = \ \pi_1(Z') \ = \ \pi_1(\partial_+ Z') \ = \ \pi_1(S(\xi)) \\
&(P \ = \ \mathrm{cl.}(M \backslash E(\nu_{N\subset M})) \ , \ \partial_+ P \ = \ \mathrm{cl.}(\partial M \backslash E(\nu_{\partial N\subset \partial M})))
\end{aligned}$$

and such that $\partial h : \partial M \longrightarrow \partial X'$ is split (resp. weakly split) along $Y \subset X$. The kernel version of the antiquadratic complex of 22.4 is an n-dimensional $(-z)$-quadratic complex (C, ψ) over $\mathbb{Z}[\pi_1(S(\xi))]$ with

$$C \ = \ \mathcal{C}(h^! : C(\widetilde{Z}', \widetilde{\partial_+ Z}') \longrightarrow C(\widetilde{P}, \widetilde{\partial_+ P}))$$

which is Poincaré (resp. $\mathbb{Z}[\pi_1(Y)]$-Poincaré), with $\widetilde{P}, \widetilde{Z}'$ the universal covers of P, Z' and

$$\begin{aligned}
h^! \ : \ C(\widetilde{Z}', \widetilde{\partial_+ Z}') \ &\simeq \ C(\widetilde{Z}', \widetilde{S(\nu_{Y'\subset X'})})^{n-*} \\
&\longrightarrow C(\widetilde{P}, \widetilde{S(\nu_{N\subset M})})^{n-*} \ \simeq \ C(\widetilde{P}, \widetilde{\partial_+ P})
\end{aligned}$$

the Umkehr $\mathbb{Z}[\pi_1(S(\xi))]$-module chain map. The morphisms

$$\begin{aligned}
&LN_{n-2}(\mathcal{F}) \longrightarrow L_n(\mathbb{Z}[\pi_1(S(\xi))], -z) \ ; \ s_Y(h) \longrightarrow (C, \psi) \ , \\
&\Gamma N_{n-2}(\mathcal{F}) \longrightarrow \Gamma_n(\mathcal{F}, -z) \ ; \ ws_Y(h) \longrightarrow (C, \psi)
\end{aligned}$$

are isomorphisms.
A finite f.g. free $\mathbb{Z}[\pi_1(S(\xi))]$-module chain complex C is $\mathbb{Z}[\pi_1(Y)]$-contractible if and only if the $\mathbb{Z}[\pi_1(S(\xi))]$-module chain map $1-z : C \longrightarrow C$ is a $\mathbb{Z}[\pi_1(Y)]$-equivalence, so that there is defined an isomorphism

$$\Gamma_n(\Phi_{\mathcal{F}}) \xrightarrow{\simeq} \Gamma_n(\Phi_{\mathcal{F}}, -z) \ ; \ (C, \psi) \longrightarrow (C, (1-z)\psi) \ . \qquad \square$$

Example 22.24 Let $(M, \partial M)$ be an n-dimensional manifold with boundary, and let $\xi : M \longrightarrow BSO(2)$, $(N, \partial N) \subset (M, \partial M)$, $(P, \partial_+ P)$ as in 22.2, and $\mathcal{F} = p_* : \mathbb{Z}[\pi_1(S(\xi))] \longrightarrow \Lambda = \mathbb{Z}[\pi_1(M)]$ as in 22.24. Define the codimension 2 *CW* pair

$$(X, Y) \ = \ (M, N) \ .$$

As in 22.19 assume that $K = \partial N$ is a Λ-homology boundary spine of $(M, \partial M)$, so that (M, K) is an $(n-2)$-dimensional geometric Poincaré pair. The weak $(n, n-2)$-dimensional geometric Poincaré triad defined by

$$(X', Y'; \partial X', \partial Y') = (E(\xi), M; S(\xi) \cup_{S(\xi|_K)} E(\xi|_K), K)$$

is equipped with a reference map $r : X' \longrightarrow X$ which is weakly Poincaré transverse at $Y \subset X$. The homotopy equivalence defined by the inclusion $M \subset E(\xi)$ as the zero section

$$(h, \partial h) : (M, \partial M) \longrightarrow (X', \partial X')$$

has $\partial h : \partial M \longrightarrow \partial X'$ a weakly split homotopy equivalence, and the anti-quadratic complex (C, ψ) of 22.4 (with $C = C(\widetilde{P}, \widetilde{\partial_+ P})$) is an n-dimensional $(-z)$-quadratic Λ-Poincaré complex over $\mathbb{Z}[\pi_1(S(\xi))]$. By 22.22 and 22.23 the Λ-homology spine obstruction of 22.14 is the weak codimension 2 splitting obstruction of 22.21

$$\sigma_*^\Lambda(M, K) = ws_Y(h) = (C, \psi) \in BB_n(X, \xi, \mathcal{F}) = \Gamma N_{n-2}(\mathcal{F}) = \Gamma_n(\mathcal{F}, -z) .$$

Inclusion defines a normal map

$$(f, b) : (N, K) \longrightarrow (M, K)$$

with quadratic kernel $\sigma_*(f, b)$ an $(n-2)$-dimensional quadratic Poincaré complex over Λ. Use 22.6 (iii) to identify the n-dimensional (-1)-quadratic Poincaré complexes over Λ

$$\Lambda \otimes_{\mathbb{Z}[\pi_1(S(\xi))]} (C, \psi) = \overline{S}\sigma_*(f, b) ,$$

showing that the Λ-homology spine obstruction has image the surgery obstruction of (f, b)

$$\Lambda \otimes_{\mathbb{Z}[\pi_1(S(\xi))]} \sigma_*^\Lambda(M, K) = \sigma_*(f, b) \in L_n(\Lambda, -1) = L_{n-2}(\Lambda) .$$

The restriction of $(h, \partial h)$ defines a normal map from an n-dimensional manifold triad to an n-dimensional geometric Poincaré triad

$$\begin{aligned}(h, \partial h)| = (g, c) &: (P; \partial_+ P, S(\xi|_N); S(\xi|_K)) \\ &\longrightarrow (S(\xi) \times I; S(\xi) \times \{0\}, S(\xi) \times \{1\}; S(\xi|_K) \times I)\end{aligned}$$

with $\partial_+ P \longrightarrow S(\xi)$ a Λ-homology equivalence. The boundary map

$$\partial : \Gamma_n(\mathcal{F}, -z) = \Gamma N_{n-2}(\mathcal{F}) \longrightarrow \Gamma_n(\Phi_{\mathcal{F}}, -z) = \Gamma_n(\Phi_{\mathcal{F}}) ; (C, \psi) \longrightarrow \partial(C, \psi)$$

(with $\Phi_{\mathcal{F}}$ as in 22.23) sends the Λ-homology spine obstruction $\sigma_*^\Lambda(M, K) = (C, \psi)$ to the relative Λ-coefficient homology surgery obstruction

$$\sigma_*^\Lambda(g, c) = \partial(C, \psi) \in \Gamma_n(\Phi_{\mathcal{F}}) ,$$

which is the obstruction of Cappell and Shaneson [44] to the existence of a Λ-homology spine. □

23. Manifold and geometric Poincaré bordism of $X \times S^1$

The free L-groups of the Laurent polynomial extension $A[z,z^{-1}]$ of a ring with involution A split as

$$\begin{aligned} L_h^{n+1}(A[z,z^{-1}]) &= L_h^{n+1}(A) \oplus L_p^n(A) , \\ L^h_{n+1}(A[z,z^{-1}]) &= L^h_{n+1}(A) \oplus L^p_n(A) \end{aligned}$$

(Shaneson [263], Wall [302, 12.6], Novikov [218], Ranicki [231], Milgram and Ranicki [189]). The L-theory splittings are algebraic analogues of the splittings of the manifold bordism groups of $X \times S^1$

$$\Omega_{n+1}(X \times S^1) = \Omega_{n+1}(X) \oplus \Omega_n(X)$$

and also of the geometric Poincaré bordism groups

$$\Omega^h_{n+1}(X \times S^1) = \Omega^h_{n+1}(X) \oplus \Omega^p_n(X) .$$

These geometric splittings will now be described in detail.

In Chap. 23 manifolds are understood to be oriented manifolds in one of the standard three categories: smooth, combinatorial or topological.

Definition 23.1 For any space X let $\Omega_n(X)$ $(n \geq 0)$ be the group of equivalence classes of pairs (M,f) with M a closed n-dimensional manifold and $f : M \longrightarrow X$ a map, subject to the equivalence relation:

$(M,f) \simeq (M',f')$ if there exists an $(n+1)$-dimensional cobordism $(L;M,M')$ with a map $e : L \longrightarrow X$ such that $e| = f : M \longrightarrow X$, $e| = f' : M' \longrightarrow X$.

□

Choose a base point $* \in S^1$, and let

$$\begin{aligned} i &: X \longrightarrow X \times S^1 ~;~ x \longrightarrow (x,*) , \\ j &: X \times S^1 \longrightarrow X ~;~ (x,s) \longrightarrow x . \end{aligned}$$

Proposition 23.2 *The manifold bordism group $\Omega_{n+1}(X \times S^1)$ fits into a direct sum system*

$$\Omega_{n+1}(X) \underset{j}{\overset{i}{\rightleftarrows}} \Omega_{n+1}(X \times S^1) \underset{C}{\overset{B}{\rightleftarrows}} \Omega_n(X)$$

with

$$\begin{aligned} i \ &: \ \Omega_{n+1}(X) \longrightarrow \Omega_{n+1}(X \times S^1) \ ; \ (M,e) \longrightarrow (M,ie) \ , \\ j \ &: \ \Omega_{n+1}(X \times S^1) \longrightarrow \Omega_{n+1}(X) \ ; \ (M,f) \longrightarrow (M,jf) \ , \\ B \ &: \ \Omega_{n+1}(X \times S^1) \longrightarrow \Omega_n(X) \ ; \ (M,f) \longrightarrow (f^{-1}(X \times \{*\}), f|) \\ &\qquad (\textit{assuming } f : M \longrightarrow X \times S^1 \textit{ transverse at } X \times \{*\} \subset X \times S^1 \ , \\ C \ &: \ \Omega_n(X) \longrightarrow \Omega_{n+1}(X \times S^1) \ ; \ (N,g) \longrightarrow (N \times S^1, g \times 1) \ . \end{aligned}$$

Proof Given an $(n+1)$-dimensional manifold M and a map $f : M \longrightarrow X \times S^1$ which is transverse at $X \times \{*\} \subset X \times S^1$ let

$$g \ = \ f| \ : \ N \ = \ f^{-1}(X \times \{*\}) \longrightarrow X \times \{*\} \ = \ X \ ,$$

with $N \subset M$ a codimension 1 framed submanifold. Cutting M along N there is obtained an $(n+1)$-dimensional manifold

$$M_N \ = \ \mathrm{cl.}(M \backslash (N \times [0,1]))$$

with boundary $\partial M_N = N \sqcup -zN$ two copies of N. The bordism $(M_N; N, zN)$ is a fundamental domain for the infinite cyclic cover of M

$$\overline{M} \ = \ f^*(X \times \mathbb{R}) \ = \ \bigcup_{j=-\infty}^{\infty} z^j(M_N; N, zN)$$

with $z : \overline{M} \longrightarrow \overline{M}$ the generating covering translation. Regard $M \times [0,1]$ as a relative bordism $(M \times [0,1]; M_N, M \sqcup N \times [0,1])$, defining a fundamental domain for a bordism $(\delta M, \delta f : \delta M \longrightarrow X \times S^1)$ between (M,f) and $(M, ijf) \sqcup (N \times S^1, g \times S^1)$ realizing

$$(M,f) \ = \ (M,ijf) + (N \times S^1, g \times S^1) \in \Omega_{n+1}(X \times S^1) \ ,$$

and verifying the identity

$$1 \ = \ ij + CB \ : \ \Omega_{n+1}(X \times S^1) \longrightarrow \Omega_{n+1}(X \times S^1) \ . \qquad \square$$

Definition 23.3 For any space X let $\Omega^h_n(X)$ (resp. $\Omega^p_n(X)$) be the group of bordism classes of pairs (P,f) with P a finite (resp. finitely dominated) n-dimensional geometric Poincaré complex and $f : P \longrightarrow X$ a map. $\square$

Proposition 23.4 (Pedersen and Ranicki [224])
(i) *For $n \geq 5$ the finite and finitely dominated geometric Poincaré bordism groups are related by a Rothenberg-type exact sequence*

$$\ldots \longrightarrow \Omega_n^h(X) \longrightarrow \Omega_n^p(X) \longrightarrow \widehat{H}^n(\mathbb{Z}_2\,;\widetilde{K}_0(\mathbb{Z}[\pi_1(X)])) \longrightarrow \Omega_{n-1}^h(X) \longrightarrow \ldots\,.$$

(ii) *The geometric Poincaré bordism class of a (finitely dominated, finite) n-dimensional geometric Poincaré pair $(P, \partial P)$ relative to a map $P \longrightarrow X$ is the Tate $\mathbb{Z}_2$-cohomology class of the Wall finiteness obstruction*

$$[P] \in \widehat{H}^n(\mathbb{Z}_2\,;\widetilde{K}_0(\mathbb{Z}[\pi_1(X)]))\,. \qquad \square$$

The projective surgery theory of [224] also gives:

Proposition 23.5 *For $n \geq 5$ and any space X a map $f : M \longrightarrow X \times S^1$ from a finite $(n+1)$-dimensional geometric Poincaré complex M is finitely dominated Poincaré transverse at $X \times \{*\} \subset X \times S^1$, in that (up to finite Poincaré bordism)*

$$f = f_N \cup g \times 1 \;:\; M = M_N \cup N \times [0,1] \longrightarrow X \times S^1 = X \times [0,1] \cup X \times [0,1]$$

with $N = f^{-1}(X \times \{*\})$ and

$$(f_N; g, zg) \;:\; (M_N; N, zN) \longrightarrow X \times ([0,1]; \{0\}, \{1\})$$

a finitely dominated $(n+1)$-dimensional geometric Poincaré bordism such that

$$[M_N] = [N] = B\tau(M) \in \widetilde{K}_0(\mathbb{Z}[\pi_1(X)])\,. \qquad \square$$

Proposition 23.6 *The finite geometric Poincaré bordism group $\Omega_{n+1}^h(X \times S^1)$ fits into a direct sum system*

$$\Omega_{n+1}^h(X) \underset{j'}{\overset{i}{\rightleftarrows}} \Omega_{n+1}^h(X \times S^1) \underset{C}{\overset{B}{\rightleftarrows}} \Omega_n^p(X)$$

with

$$i \;:\; \Omega_{n+1}^h(X) \longrightarrow \Omega_{n+1}^h(X \times S^1)\;;\; (M, e) \longrightarrow (M, ie)\,,$$

$$j' \;:\; \Omega_{n+1}^h(X \times S^1) \longrightarrow \Omega_{n+1}^h(X)\;;\; (M, f) \longrightarrow (M, jf) + (\partial P, \partial h)$$

(assuming $f : M \longrightarrow X \times S^1$ Poincaré transverse at $X \times \{*\} \subset X \times S^1$,

for any (finitely dominated, finite) $(n+2)$-dimensional geometric

Poincaré pair $(P, \partial P)$ with a map $h : P \longrightarrow X$ such that

$[P] = [N] \in \widetilde{K}_0(\mathbb{Z}[\pi_1(X)])$, with $N = f^{-1}(X \times \{*\}) \subset M$) ,

$$B \;:\; \Omega_{n+1}^h(X \times S^1) \longrightarrow \Omega_n^p(X)\;;\; (M, f) \longrightarrow (f^{-1}(X \times \{*\}), f|) = (N, g)\,,$$

$$C \;:\; \Omega_n^p(X) \longrightarrow \Omega_{n+1}^h(X \times S^1)\;;\; (N, g) \longrightarrow (N \times S^1, g \times 1)\,.$$

Proof Suppose given a finite $(n+1)$-dimensional geometric Poincaré complex M and a map $f : M \longrightarrow X \times S^1$ which is finitely dominated Poincaré transverse at $X \times \{*\} \subset X \times S^1$, as above. Regard $M \times [0,1]$ as a relative finitely dominated geometric Poincaré bordism $(M \times [0,1]; M_N, M \sqcup N \times [0,1])$, defining a fundamental domain for a finitely dominated geometric Poincaré bordism $(\delta M, \delta f : \delta M \longrightarrow X \times S^1)$ between (M, f) and $(M, ijf) \sqcup (N \times S^1, g \times S^1)$ realizing

$$(M, f) \;=\; (M, ijf) + (N \times S^1, g \times S^1) \in \Omega^p_{n+1}(X \times S^1) \;,$$

with finiteness obstruction

$$[\delta M] \;=\; -i[N] \in \widetilde{K}_0(\mathbb{Z}[\pi_1(X)][z, z^{-1}]) \;.$$

For any (finitely dominated, finite) $(n+2)$-dimensional geometric Poincaré pair $(P, \partial P)$ with a map $h : P \longrightarrow X$ such that

$$[P] \;=\; [N] \in \widetilde{K}_0(\mathbb{Z}[\pi_1(X)])$$

there is defined a (homotopy) finite geometric Poincaré bordism $(\delta M, \delta f) \sqcup (P, h)$ between (M, f) and the disjoint union

$$(M, ijf) \sqcup (N \times S^1, g \times S^1) \sqcup (\partial P, i\partial h)$$

realizing the identity

$$1 \;=\; ij' + CB \;:\; \Omega^h_{n+1}(X \times S^1) \longrightarrow \Omega^h_{n+1}(X \times S^1) \;. \qquad \square$$

24. L-theory of Laurent extensions

The algebraic mapping torus of a self homotopy equivalence of an algebraic Poincaré complex over a ring with involution A is an A-finitely algebraic Poincaré complex over the Laurent polynomial extension $A[z,z^{-1}]$ ($\overline{z}=z^{-1}$). Following a recollection of the algebraic L-theory splitting theorems

$$\begin{aligned} L^n_h(A[z,z^{-1}],\epsilon) &= L^n_h(A,\epsilon)\oplus L^{n-1}_p(A,\epsilon)\ , \\ L_n^h(A[z,z^{-1}],\epsilon) &= L_n^h(A,\epsilon)\oplus L^p_{n-1}(A,\epsilon) \end{aligned}$$

it will now be proved that an algebraic Poincaré complex over $A[z,z^{-1}]$ is A-finitely dominated if and only if it is homotopy equivalent to the algebraic mapping torus of a self homotopy equivalence of an algebraic Poincaré complex over A. See Chap. 34 for the L-theory of the polynomial extensions $A[x]$, $A[x,x^{-1}]$ with involution $\overline{x}=x$.

Fundamental domains for infinite cyclic covers of compact manifolds may be constructed by geometric transversality. Likewise, algebraic transversality provides finitely dominated Poincaré fundamental domains over A for finite f.g. free algebraic Poincaré complexes over the Laurent polynomial extension $A[z,z^{-1}]$ with the involution

$$\bar{}\ :\ A[z,z^{-1}]\longrightarrow A[z,z^{-1}]\ ;\ \sum_{j=0}^{\infty}a_jz^j\longrightarrow \sum_{j=0}^{\infty}\overline{a}_jz^{-j}\ .$$

Definition 24.1 (i) Given A-module chain complexes C,D and chain maps $f,g:C\longrightarrow D$ define the *double relative* $\left\{\begin{array}{l}\epsilon\text{-}symmetric\\ \epsilon\text{-}quadratic\end{array}\right.$ *Q-groups* of (C,f,g)

$$\left\{\begin{aligned} Q^{n+1}(f,g,\epsilon) &= H_{n+1}(f^{\%}-g^{\%}:W^{\%}C\longrightarrow W^{\%}D) \\ Q_{n+1}(f,g,\epsilon) &= H_{n+1}(f_{\%}-g_{\%}:W_{\%}C\longrightarrow W_{\%}D) \end{aligned}\right.$$

to fit into an exact sequence

$$\left\{\begin{array}{l} \dots\longrightarrow Q^{n+1}(D,\epsilon)\longrightarrow Q^{n+1}(f,g,\epsilon) \\ \qquad\qquad\longrightarrow Q^n(C,\epsilon)\xrightarrow{f^{\%}-g^{\%}}Q^n(D,\epsilon)\longrightarrow\dots \\ \\ \dots\longrightarrow Q_{n+1}(D,\epsilon)\longrightarrow Q_{n+1}(f,g,\epsilon) \\ \qquad\qquad\longrightarrow Q_n(C,\epsilon)\xrightarrow{f_{\%}-g_{\%}}Q_n(D,\epsilon)\longrightarrow\dots\ . \end{array}\right.$$

(ii) An n-dimensional $\begin{cases} \epsilon\text{-symmetric} \\ \epsilon\text{-quadratic} \end{cases}$ pair over A

$$\Gamma = \left((f\ g) : C \oplus C' \longrightarrow D\,, \begin{cases} (\delta\phi, \phi \oplus -\phi') \\ (\delta\psi, \psi \oplus -\psi') \end{cases} \right)$$

is *fundamental* if $\begin{cases} (C', \phi') = (C, \phi) \\ (C', \psi') = (C, \psi)\,, \end{cases}$ so that $\begin{cases} (\delta\phi, \phi) \in Q^{n+1}(f, g, \epsilon) \\ (\delta\psi, \psi) \in Q_{n+1}(f, g, \epsilon). \end{cases}$

(iii) The *union* of a fundamental n-dimensional $\begin{cases} \epsilon\text{-symmetric} \\ \epsilon\text{-quadratic} \end{cases}$ (Poincaré) pair Γ over A is the n-dimensional $\begin{cases} \epsilon\text{-symmetric} \\ \epsilon\text{-quadratic} \end{cases}$ (Poincaré) complex over $A[z, z^{-1}]$

$$U(\Gamma) = \begin{cases} (E, \Phi) \\ (E, \Psi) \end{cases}$$

with

$$\begin{aligned} d_E &= \begin{pmatrix} d_D & (-)^{r-1}(f - gz) \\ 0 & d_C \end{pmatrix} : \\ E_r &= D_r[z, z^{-1}] \oplus C_{r-1}[z, z^{-1}] \\ &\qquad \longrightarrow E_{r-1} = D_{r-1}[z, z^{-1}] \oplus C_{r-2}[z, z^{-1}]\,, \\ \Phi_s &= \begin{pmatrix} \delta\phi_s & (-)^s g\phi_s z \\ (-)^{n-r-1}\phi_s f^* & (-)^{n-r+s} T_\epsilon \phi_{s-1} \end{pmatrix} : \\ E^{n-r+s} &= D^{n-r+s}[z, z^{-1}] \oplus C^{n-r+s-1}[z, z^{-1}] \\ &\qquad \longrightarrow E_r = D_r[z, z^{-1}] \oplus C_{r-1}[z, z^{-1}]\,, \\ \Psi_s &= \begin{pmatrix} \delta\psi_s & (-)^s g\psi_s z \\ (-)^{n-r-1}\psi_s f^* & (-)^{n-r-s} T_\epsilon \psi_{s+1} \end{pmatrix} : \\ E^{n-r-s} &= D^{n-r-s}[z, z^{-1}] \oplus C^{n-r-s-1}[z, z^{-1}] \\ &\qquad \longrightarrow E_r = D_r[z, z^{-1}] \oplus C_{r-1}[z, z^{-1}]\,. \end{aligned}$$

(iii) A fundamental pair Γ over A is *finitely balanced* if C and D are finitely dominated and

$$[C] = [D] \in \widetilde{K}_0(A)\,,$$

in which case

$$[U(\Gamma)] = [D[z, z^{-1}]] - [C[z, z^{-1}]] = 0 \in \widetilde{K}_0(A[z, z^{-1}])$$

and the union $U(\Gamma)$ is homotopy finite. A choice of chain equivalence $C \simeq D$ (if any) determines a round finite structure on $U(\Gamma)$. □

Proposition 24.2 (Ranicki [244, Chap. 16])

(i) *Every f.g. free n-dimensional* $\begin{cases} \epsilon\text{-symmetric} \\ \epsilon\text{-quadratic} \end{cases}$ *(Poincaré) complex* (E,θ) *over* $A[z,z^{-1}]$ *is homotopy equivalent to the union* $U(\Gamma)$ *of a finitely balanced n-dimensional* $\begin{cases} \epsilon\text{-symmetric} \\ \epsilon\text{-quadratic} \end{cases}$ *(Poincaré) cobordism* Γ. *If* (E,θ) *is based and simple then* Γ *can be chosen to be based f.g. free, with* (E,θ) *simple homotopy equivalent to* $U(\Gamma)$.

(ii) *The simple and free* $\begin{cases} \epsilon\text{-symmetric} \\ \epsilon\text{-quadratic} \end{cases}$ *L-groups of* $A[z,z^{-1}]$ *fit into geometrically significant direct sum systems*

$$L^n_t(A,\epsilon) \underset{j'}{\overset{i_!}{\rightleftarrows}} L^n_t(A[z,z^{-1}],\epsilon) \underset{C}{\overset{B}{\rightleftarrows}} L^{n-1}_u(A,\epsilon)\ ,$$

$$L_n^t(A,\epsilon) \underset{j'}{\overset{i_!}{\rightleftarrows}} L_n^t(A[z,z^{-1}],\epsilon) \underset{C}{\overset{B}{\rightleftarrows}} L_{n-1}^u(A,\epsilon)$$

for $(t,u)=(s,h)$ *or* (h,p), *with* $i_!$ *induced by the inclusion* $i: A \longrightarrow A[z,z^{-1}]$ *and* C *defined by product with the symmetric Poincaré complex* $\sigma^*(S^1)$ *over* $\mathbb{Z}[z,z^{-1}]$

$$\begin{aligned} C &= \sigma^*(S^1)\otimes - \ :\ L^{n-1}_u(A,\epsilon) \longrightarrow L^n_t(A[z,z^{-1}],\epsilon)\ , \\ C &= \sigma^*(S^1)\otimes - \ :\ L^r_{n-1}(A,\epsilon) \longrightarrow L^t_n(A[z,z^{-1}],\epsilon)\ . \end{aligned}$$ □

Let

$$j\ :\ A[z,z^{-1}] \longrightarrow A\ ;\ z \longrightarrow 1\ .$$

Note that in general $j' \neq j_!$ – see 24.4 below for a description of $j'-j_!$.

The symmetric signature defines a natural transformation of direct sum systems

$$\begin{array}{ccccc} \Omega^h_{n+1}(X) & \underset{j'}{\overset{i}{\rightleftarrows}} & \Omega^h_n(X\times S^1) & \underset{C}{\overset{B}{\rightleftarrows}} & \Omega^p_{n-1}(X) \\ \downarrow \sigma^* & & \downarrow \sigma^* & & \downarrow \sigma^* \\ L^n_h(\mathbb{Z}[\pi_1(X)]) & \underset{j'}{\overset{i_!}{\rightleftarrows}} & L^n_h(\mathbb{Z}[\pi_1(X)][z,z^{-1}]) & \underset{C}{\overset{B}{\rightleftarrows}} & L^{n-1}_p(\mathbb{Z}[\pi_1(X)]) \end{array}$$

from geometric Poincaré bordism to symmetric Poincaré bordism.

Definition 24.3 Let

$$(h,\chi)\ :\ (C,\phi) \longrightarrow (C,\phi)$$

be a self map of a finitely dominated n-dimensional ϵ-symmetric complex (C,ϕ) over A.

(i) The *algebraic mapping torus* of (h,χ) is the homotopy finite $(n+1)$-dimensional ϵ-symmetric complex over $A[z,z^{-1}]$

$$T(h,\chi) \;=\; U(\Gamma) \;=\; (E,\theta)$$

defined by the union of the finitely balanced $(n+1)$-dimensional ϵ-symmetric cobordism

$$\Gamma \;=\; ((h\ 1) : C\oplus C \longrightarrow C, (\chi, \phi\oplus -\phi))\ ,$$

with

$$\begin{aligned} E \;&=\; \mathcal{C}(h-z : C[z,z^{-1}] \longrightarrow C[z,z^{-1}])\ ,\\ \theta_s \;&=\; \begin{pmatrix} \chi_s & (-)^s\phi_s z \\ (-)^{n-r}\phi_s h^* & (-)^{n-r+s+1}T_\epsilon\phi_{s-1}\end{pmatrix} :\\ &E^{n-r+s+1} \;=\; C^{n-r+s+1}[z,z^{-1}]\oplus C^{n-r+s}[z,z^{-1}]\\ &\qquad\longrightarrow E_r \;=\; C_r[z,z^{-1}]\oplus C_{r-1}[z,z^{-1}]\quad (s\geq 0)\ . \end{aligned}$$

(ii) The *A-coefficient algebraic mapping torus* of (h,χ) is the homotopy finite $(n+1)$-dimensional ϵ-symmetric complex over A

$$T_A(h,\chi) \;=\; A\otimes_{A[z,z^{-1}]} T(h,\chi) \;=\; (E_A,\theta_A)$$

$$\begin{aligned} E_A \;&=\; \mathcal{C}(h-1 : C \longrightarrow C)\ ,\\ (\theta_A)_s \;&=\; \begin{pmatrix} \chi_s & (-)^s\phi_s \\ (-)^{n-r}\phi_s h^* & (-)^{n-r+s+1}T_\epsilon\phi_{s-1}\end{pmatrix} :\\ (E_A)^{n-r+s+1} \;&=\; C^{n-r+s+1}\oplus C^{n-r+s} \longrightarrow (E_A)_r \;=\; C_r\oplus C_{r-1}\quad (s\geq 0)\ . \end{aligned}$$

Similarly in the ϵ-quadratic case. □

Proposition 24.4 *The algebraic mapping torus $T(h,\chi)$ of a self homotopy equivalence $(h,\chi) : (C,\phi) \longrightarrow (C,\phi)$ of a finitely dominated n-dimensional ϵ-symmetric Poincaré complex (C,ϕ) over A is a homotopy finite $(n+1)$-dimensional ϵ-symmetric Poincaré complex over $A[z,z^{-1}]$ with a canonical (round) finite structure, with respect to which it has torsion*

$$\begin{aligned} \tau(T(h,\chi)) \;&=\; \tau(-zh : C[z,z^{-1}] \longrightarrow C[z,z^{-1}]) \;=\; (\tau(h),[C],0,0)\\ &\in Wh_1(A[z,z^{-1}]) \;=\; Wh_1(A)\oplus \widetilde{K}_0(A)\oplus \widetilde{\mathrm{Nil}}_0(A)\oplus \widetilde{\mathrm{Nil}}_0(A)\ . \end{aligned}$$

The cobordism class $T(h,\chi)\in L^{n+1}_h(A[z,z^{-1}],\epsilon)$ is such that

$$BT(h,\chi) \;=\; (C,\phi)\in L^n_p(A,\epsilon)\ ,$$

and

$$\begin{aligned} j'T(h,\chi) - T_A(h,\chi) \;&=\; [\tau(-1 : C \longrightarrow C)]\\ &\in \mathrm{im}(\widehat{H}^{n+2}(\mathbb{Z}_2;Wh_1(A)) \longrightarrow L^{n+1}_h(A,\epsilon))\ . \end{aligned}$$

If C is f.g. free then $\tau(-1 : C \longrightarrow C) = 0 \in Wh_1(A)$ and

$$\begin{aligned} j'T(h,\chi) \;&=\; T_A(h,\chi) \;=\; j_!T(h,\chi)\in L^{n+1}_h(A,\epsilon)\ ,\\ T(h,\chi) \;&=\; i_!j_!T(h,\chi)\oplus(\sigma^*(S^1)\otimes(C,\phi))\in L^{n+1}_h(A[z,z^{-1}],\epsilon)\ , \end{aligned}$$

with $j : A[z, z^{-1}] \longrightarrow A; z \longrightarrow 1$ *the projection.* □

An infinite cyclic cover $\overline{X}$ of a finite $(n+1)$-dimensional geometric Poincaré complex X is finitely dominated if and only if X is homotopy equivalent to the mapping torus $T(h)$ of a self homotopy equivalence $h : Y \longrightarrow Y$ of a finitely dominated n-dimensional geometric Poincaré complex Y, with $(Y, h) \simeq (\overline{X}, \zeta)$. Similarly for algebraic Poincaré complexes:

Proposition 24.5 *An* $(n+1)$*-dimensional f.g. free* ϵ*-symmetric Poincaré complex* (E, θ) *over* $A[z, z^{-1}]$ *is A-finitely dominated if and only if it is homotopy equivalent to the algebraic mapping torus* $T(h, \chi)$ *of a self homotopy equivalence* $(h, \chi) : (C, \phi) \longrightarrow (C, \phi)$ *of a finitely dominated n-dimensional* ϵ*-symmetric Poincaré complex* (C, ϕ) *over A. Similarly in the* ϵ*-quadratic case.*

Proof The $A[z, z^{-1}]$-module chain complex

$$T(h) \;=\; \mathcal{C}(h - z : C[z, z^{-1}] \longrightarrow C[z, z^{-1}])$$

is A-module chain equivalent to C, for any self chain equivalence $h : C \longrightarrow C$ of a finitely dominated A-module chain complex C. In particular, it follows that the algebraic mapping torus $T(h, \chi)$ of a self homotopy equivalence $(h, \chi) : (C, \phi) \longrightarrow (C, \phi)$ of a finitely dominated n-dimensional ϵ-symmetric Poincaré complex (C, ϕ) over A is A-finitely dominated.

Conversely, let (E, θ) be an A-finitely dominated $(n+1)$-dimensional f.g. free ϵ-symmetric Poincaré complex over $A[z, z^{-1}]$. Let $i^! E$ denote the A-module chain complex defined by E, with A acting by the restriction of the $A[z, z^{-1}]$-module action on E to the inclusion $i : A \longrightarrow A[z, z^{-1}]$. Since $i^! E$ is A-finitely dominated there is defined an $A[z, z^{-1}]$-module chain equivalence

$$q \;:\; T(\zeta) \;=\; \mathcal{C}(\zeta - z : i^! E[z, z^{-1}] \longrightarrow i^! E[z, z^{-1}]) \xrightarrow{\;\simeq\;} E$$

as in 3.10. Regard the ϵ-symmetric structure on E as an element of the $\mathbb{Z}_2$-hyperhomology group $\theta \in Q^{n+1}(E, \epsilon) = H_{n+1}(W^{\%} E)$ with

$$\begin{aligned} W^{\%} E \;&=\; \mathrm{Hom}_{\mathbb{Z}[\mathbb{Z}_2]}(W, \mathrm{Hom}_{A[z,z^{-1}]}(E^*, E)) \\ &=\; \mathrm{Hom}_{\mathbb{Z}[\mathbb{Z}_2]}(W, E \otimes_{A[z,z^{-1}]} E)\;. \end{aligned}$$

Applying $E \otimes_{A[z,z^{-1}]} -$ to q there is obtained a $\mathbb{Z}[\mathbb{Z}_2]$-module chain equivalence

$$\begin{aligned} 1 \otimes q^+ \;:\; E \otimes_{A[z,z^{-1}]} T(\zeta) \;&=\; \mathcal{C}(1 - \zeta \otimes \zeta : i^! E \otimes_A i^! E \longrightarrow i^! E \otimes_A i^! E) \\ &\longrightarrow E \otimes_{A[z,z^{-1}]} E\;, \end{aligned}$$

The relative ϵ-symmetric structure groups $Q^*(1, \zeta, \epsilon)$ in the exact sequence

$$\begin{aligned} \dots \longrightarrow Q^{n+1}(i^! E, \epsilon) &\xrightarrow{1-\zeta^{\%}} Q^{n+1}(i^! E, \epsilon) \\ &\longrightarrow Q^{n+1}(1, \zeta, \epsilon) \longrightarrow Q^n(i^! E, \epsilon) \longrightarrow \dots \end{aligned}$$

are thus such that

$$Q^*(1,\zeta,\epsilon) = Q^*(E,\epsilon) .$$

The absolute ϵ-symmetric Poincaré structure $\theta \in Q^{n+1}(E,\epsilon)$ corresponds to a relative ϵ-symmetric Poincaré structure $(\chi, i^!\theta) \in Q^{n+1}(1,\zeta,\epsilon)$, defining a finitely dominated n-dimensional ϵ-symmetric Poincaré complex $(i^!E, i^!\theta)$ over A with a homotopy equivalence $(\zeta,\chi) : (i^!E, i^!\theta) \longrightarrow (i^!E, i^!\theta)$ such that (E,θ) is homotopy equivalent to $T(\zeta,\chi)$. □

For any self chain equivalence $h : P \longrightarrow P$ of a finitely dominated A-module chain complex P and any $n \geq 0$ the $(n+1)$-dual $T(h)^{n+1-*}$ of the algebraic mapping torus $T(h) = \mathcal{C}(h - z : P[z,z^{-1}] \longrightarrow P[z,z^{-1}])$ is $A[z,z^{-1}]$-module chain equivalent to the algebraic mapping torus of $(h^*)^{-1} : P^{n-*} \longrightarrow P^{n-*}$. Thus if (E,θ) is an A-finitely dominated finite f.g. free $(n+1)$-dimensional ϵ-symmetric Poincaré complex over $A[z,z^{-1}]$ there is an $A[z,z^{-1}]$-module chain equivalence

$$\begin{aligned}(i^!E)^{n-*} &= \mathrm{Hom}_A(E,A)_{*-n} \xrightarrow{\simeq} \\ &E^{n+1-*} = \mathrm{Hom}_{A[z,z^{-1}]}(E, A[z,z^{-1}]))_{*-n-1}\end{aligned}$$

and the n-dimensional ϵ-symmetric Poincaré structure $i^!\theta \in Q^n(i^!E,\epsilon)$ defined in the proof of 24.5 is such that

$$i^!\theta_0 : (i^!E)^{n-*} \longrightarrow E^{n+1-*} \xrightarrow{\theta_0} E = i^!E .$$

Example 24.6 Let (E,θ) be a 1-dimensional ϵ-symmetric Poincaré complex over $A[z,z^{-1}]$ with

$$d = \sum_{j=m}^{n} a_j z^j : E_1 = A[z,z^{-1}] \longrightarrow E_0 = A[z,z^{-1}]$$

for some Laurent polynomial

$$p(z) = \sum_{j=m}^{n} a_j z^j \in A[z,z^{-1}] .$$

If E is A-finitely dominated (e.g. if A is commutative and the extreme coefficients $a_m, a_n \in A$ are units) then (E,θ) is homotopy equivalent to the algebraic mapping torus $T(h,\chi)$ of the self homotopy equivalence

$$(h,\chi) : (P,\phi) \longrightarrow (P,\phi)$$

of the f.g. free 0-dimensional ϵ-symmetric Poincaré complex (= nonsingular ϵ-symmetric form) (P,ϕ) over A defined by

$$\begin{aligned} h &= \zeta^{-1} : P_0 = H_0(E) = A[z,z^{-1}]/(p(z)) \longrightarrow P_0 , \\ \phi &= \theta_0 : P^0 = H^1(E) \longrightarrow H_0(E) = P_0 , \ \chi = 0 . \end{aligned}$$

□

25. Localization and completion in L-theory

Localization and completion techniques are a standard feature of the theory of quadratic forms, as evidenced by the Hasse-Minkowski local-global principle. Refer to Chap. 3 of Ranicki [237] for an account of the relevant L-theory. The main novelty here is that the localization exact sequences of [237] in ϵ-symmetric and ϵ-quadratic L-theory

$$\ldots \longrightarrow L^n(A,\epsilon) \longrightarrow L^n(S^{-1}A,\epsilon) \longrightarrow L^n(A,S,\epsilon) \longrightarrow L^{n-1}(A,\epsilon) \longrightarrow \ldots\ ,$$

$$\ldots \longrightarrow L_n(A,\epsilon) \longrightarrow L_n(S^{-1}A,\epsilon) \longrightarrow L_n(A,S,\epsilon) \longrightarrow L_{n-1}(A,\epsilon) \longrightarrow \ldots$$

are extended to noncommutative localizations $\Sigma^{-1}A$ of a ring with involution A.

As in Chap. 9 let Σ be a set of square matrices with entries in a ring A, so that the localization $\Sigma^{-1}A$ is defined. An involution $^- : A \longrightarrow A$ is extended to the ring $M_n(A)$ of $n \times n$ matrices with entries in A by

$$^- : M_n(A) \longrightarrow M_n(A)\ ;\ M = (a_{ij}) \longrightarrow \overline{M} = (\bar{a}_{ji})\ .$$

A set Σ of square matrices in A is involution-invariant if $\overline{M} \in \Sigma$ for all $M \in \Sigma$, in which case the localization $\Sigma^{-1}A$ is a ring with involution, and the natural map

$$i : A \longrightarrow \Sigma^{-1}A\ ;\ a \longrightarrow a/1$$

is a morphism of rings with involution.

As in Part One only localizations in the case when $i : A \longrightarrow \Sigma^{-1}A$ is injective will be considered, so that there is a localization exact sequence in algebraic K-theory (9.8)

$$\begin{aligned} \ldots \longrightarrow K_1(A) \xrightarrow{i} K_1(\Sigma^{-1}A) &\xrightarrow{\partial} K_1(A,\Sigma) \\ &\xrightarrow{j} K_0(A) \xrightarrow{i} K_0(\Sigma^{-1}A) \longrightarrow \ldots \end{aligned}$$

with $K_1(A,\Sigma) = K_0(\mathbb{H}(A,\Sigma))$ the class group of (A,Σ)-modules. It will be assumed that the natural maps $\mathbb{Z} \longrightarrow A$, $\mathbb{Z} \longrightarrow \Sigma^{-1}A$ induce split injections $K_*(\mathbb{Z}) \longrightarrow K_*(A)$, $K_*(\mathbb{Z}) \longrightarrow K_*(\Sigma^{-1}A)$ $(* = 0,1)$ so there is a reduced version of the localization exact sequence

$$\ldots \longrightarrow Wh_1(A) \xrightarrow{i} Wh_1(\Sigma^{-1}A) \xrightarrow{\partial} Wh_1(A,\Sigma) \xrightarrow{j} \widetilde{K}_0(A) \xrightarrow{i} \widetilde{K}_0(\Sigma^{-1}A) \longrightarrow \ldots$$

with

$$Wh_1(A,\Sigma) = K_1(A,\Sigma) ,$$

$$Wh_1(A) = \begin{cases} \widetilde{K}_1(A) & \text{for arbitrary } A \\ Wh(\pi) & \text{for a group ring } A = \mathbb{Z}[\pi] , \end{cases}$$

$$Wh_1(\Sigma^{-1}A) = \begin{cases} \widetilde{K}_1(\Sigma^{-1}A) & \text{for arbitrary } A \\ K_1(\Sigma^{-1}\mathbb{Z}[\pi])/\{\pm g \,|\, g \in \mathbb{Z}[\pi]\} & \text{for a group ring } A = \mathbb{Z}[\pi] . \end{cases}$$

For a ring with involution A and an involution-invariant set Σ of square matrices in A the dual of a $\Sigma^{-1}A$-isomorphism $d \in \mathrm{Hom}_A(P_1,P_0)$ is a $\Sigma^{-1}A$-isomorphism $d^* \in \mathrm{Hom}_A(P_0^*,P_1^*)$.

Definition 25.1 The *Σ-dual* of an (A,Σ)-module M with f.g. projective A-module resolution

$$0 \longrightarrow P_1 \xrightarrow{d} P_0 \longrightarrow M \longrightarrow 0$$

is the (A,Σ)-module $M\hat{}$ with f.g. projective A-module resolution

$$0 \longrightarrow P_0^* \xrightarrow{d^*} P_1^* \longrightarrow M\hat{} \longrightarrow 0 . \qquad \square$$

The Σ-duality involution

$$\mathbb{H}(A,\Sigma) \longrightarrow \mathbb{H}(A,\Sigma) \ ; \ M \longrightarrow M\hat{}$$

induces an involution

$$* : Wh_1(A,\Sigma) \longrightarrow Wh_1(A,\Sigma) \ ; \ \tau_\Sigma(M) \longrightarrow \tau_\Sigma(M\hat{}) .$$

Remark 25.2 For a multiplicative subset $\Sigma = S \subset A$ the S-dual of an (A,S)-module M can be expressed as

$$M\hat{} = \mathrm{Hom}_A(M, S^{-1}A/A)$$

with A acting by

$$A \times M\hat{} \longrightarrow M\hat{} \ ; \ (a,f) \longrightarrow (x \longrightarrow f(x).\bar{a}) ,$$

and

$$P_1^* \longrightarrow M\hat{} \ ; \ f \longrightarrow ([x] \longrightarrow f(d^{-1}(x))) \ \ (x \in P_0 \ , \ [x] \in M) ,$$

with $d^{-1} \in \mathrm{Hom}_{S^{-1}A}(S^{-1}P_0, S^{-1}P_1)$. $\square$

The algebraic K-theory localization exact sequence has the following L-theory analogues, involving the torsion L-groups:

Definition 25.3 Let A be a ring with involution and let Σ be an involution-invariant set of square matrices with entries in A such that $i : A \longrightarrow \Sigma^{-1}A$ is injective. For any $*$-invariant subgroup $U \subseteq Wh_1(A,\Sigma)$ the *ϵ-symmetric (A,Σ)-torsion L-group* $L^n_U(A,\Sigma,\epsilon)$ $(n \geq 0)$ is the group of cobordism classes of $\Sigma^{-1}A$-contractible f.g. projective $(n-1)$-dimensional ϵ-symmetric Poincaré complexes (C,ϕ) over A with

$$\tau_\Sigma(C) \in U \subseteq Wh_1(A,\Sigma) .$$

Similarly for the *ϵ-quadratic (A,Σ)-torsion L-group* $L_n^U(A,\Sigma,\epsilon)$. □

See Ranicki [237, Chap. 1.12] for the definition of (-1)-dimensional algebraic Poincaré complexes required for the definition of $L^0_U(A,\Sigma,\epsilon)$ and $L_0^U(A,\Sigma,\epsilon)$.

Proposition 25.4 (i) *The ϵ-symmetric L-group $L^n_{\partial^{-1}U}(\Sigma^{-1}A,\epsilon)$ is (isomorphic to) the cobordism group of n-dimensional f.g. projective ϵ-symmetric $\Sigma^{-1}A$-Poincaré complexes (C,ϕ) over A with $[\partial C] \in U \subseteq Wh_1(A,\Sigma)$.*
(ii) *The ϵ-symmetric (A,Σ)-torsion L-groups are the relative groups in the localization exact sequence*

$$\begin{aligned} \dots \longrightarrow L^n_{jU}(A,\epsilon) &\xrightarrow{\;i\;} L^n_{\partial^{-1}U}(\Sigma^{-1}A,\epsilon) \\ &\xrightarrow{\;\partial\;} L^n_U(A,\Sigma,\epsilon) \xrightarrow{\;j\;} L^{n-1}_{jU}(A,\epsilon) \longrightarrow \dots , \end{aligned}$$

with

$$\begin{aligned} \partial \;:\; L^n_{\partial^{-1}U}(\Sigma^{-1}A,\epsilon) \;=\; \Gamma^n_{\partial^{-1}U}(A \longrightarrow \Sigma^{-1}A,\epsilon) &\longrightarrow L^n_U(A,\Sigma,\epsilon) \;; \\ \Sigma^{-1}(C,\phi) &\longrightarrow \partial(C,\phi) . \end{aligned}$$

(iii) *The ϵ-symmetric L-groups associated to a pair $(U_2, U_1 \subseteq U_2)$ of $*$-invariant subgroups $U_1 \subseteq U_2 \subseteq Wh_1(A,\Sigma)$ are related by a Rothenberg-type exact sequence*

$$\begin{aligned} \dots \longrightarrow \widehat{H}^{n+1}(\mathbb{Z}_2;U_2/U_1) &\longrightarrow L^n_{U_1}(A,\Sigma,\epsilon) \\ \longrightarrow L^n_{U_2}(A,\Sigma,\epsilon) &\longrightarrow \widehat{H}^n(\mathbb{Z}_2;U_2/U_1) \longrightarrow \dots . \end{aligned}$$

Similarly for the ϵ-quadratic case.
Proof As for the localization $S^{-1}A$ inverting an involution-invariant multiplicative subset $S \subset A$ in Ranicki [237, Chap. 3]. The injection $i : A \longrightarrow \Sigma^{-1}A$ is locally epic, and the relative L-groups are given by 20.21 to be

$$L_*(i,\epsilon) \;=\; \Gamma_*(\Phi_i,\epsilon) \;=\; L_*(A,\Sigma,\epsilon)$$

with

$$\begin{array}{ccc} A & \longrightarrow & A \\ \big\downarrow & \Phi_i & \big\downarrow \\ A & \longrightarrow & \Sigma^{-1}A \end{array}$$

□

Proposition 25.5 (Cappell and Shaneson [40], Vogel [297])
Let π be a finitely presented group, and let $\mathcal{F} : \mathbb{Z}[\pi]\longrightarrow\Lambda$ be a locally epic morphism of rings with involution. Assume that the localization of $\Sigma^{-1}\mathbb{Z}[\pi]$ inverting the set Σ of Λ-invertible matrices in $\mathbb{Z}[\pi]$ is such that the natural map $i : \mathbb{Z}[\pi]\longrightarrow\Sigma^{-1}\mathbb{Z}[\pi]$ is injective. Let $U \subseteq K_1(\Lambda)$ be a $*$-invariant subgroup containing the image of $\pm\pi$.
(i) *A normal map $(f,b) : (M,\partial M)\longrightarrow(X,\partial X)$ from an n-dimensional manifold with boundary to an n-dimensional Λ-coefficient n-dimensional geometric Poincaré pair with $\pi_1(X) = \pi$, $\tau(X,\partial X;\Lambda) \in U$ and with $\partial f : \partial M\longrightarrow\partial X$ a Λ-coefficient homology equivalence with $\tau(\partial f;\Lambda) \in U$ has a* Λ-homology surgery obstruction

$$\sigma_*^\Lambda(f,b) \in \Gamma_n^U(\mathcal{F}) \;=\; L_n^U(\Sigma^{-1}\mathbb{Z}[\pi])$$

such that $\sigma_*^\Lambda(f,b) = 0$ if (f,b) is normal bordant to a Λ-coefficient homology equivalence with torsion in U.
(ii) *Let $n \geq 5$. If (f,b) (as in* (i)*) is such that $\sigma_*^\Lambda(f,b) = 0$ then (f,b) is normal bordant to a Λ-coefficient homology equivalence with torsion in U. Moreover, every element in $\Gamma_n^U(\mathcal{F})$ is realized as $\sigma_*^\Lambda(f,b)$ for some (f,b).* □

Terminology 25.6 In the special cases

$$U \;=\; \{0\}\ ,\ \mathrm{im}(Wh_1(\Sigma^{-1}A)\longrightarrow Wh_1(A,\Sigma))\ ,\ Wh_1(A,\Sigma)$$

write the ϵ-symmetric L-groups $L_U^*(A,\Sigma,\epsilon)$ of 25.4 as

$$\begin{aligned} L^n_{\{0\}\subseteq Wh_1(A,\Sigma)}(A,\Sigma,\epsilon) \;&=\; L^n_s(A,\Sigma,\epsilon)\ , \\ L^n_{\mathrm{im}(Wh_1(\Sigma^{-1}A))}(A,\Sigma,\epsilon) \;&=\; L^n_h(A,\Sigma,\epsilon)\ , \\ L^n_{Wh_1(A,\Sigma)}(A,\Sigma,\epsilon) \;&=\; L^n_p(A,\Sigma,\epsilon)\ . \end{aligned}$$

Similarly in the ϵ-quadratic case. □

Proposition 25.7 *The torsion ϵ-symmetric L-groups of 25.6 fit into the commutative braids of exact sequences*

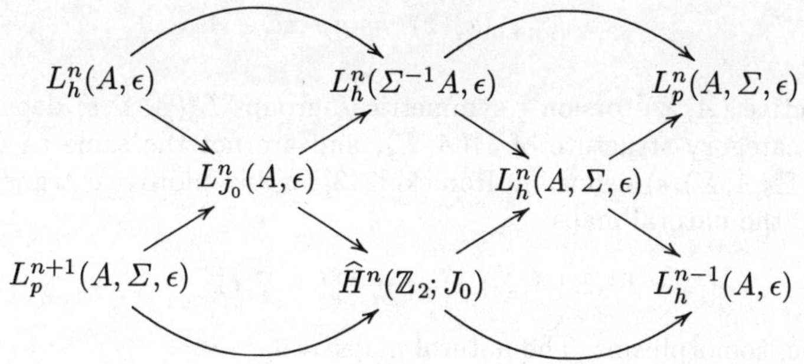

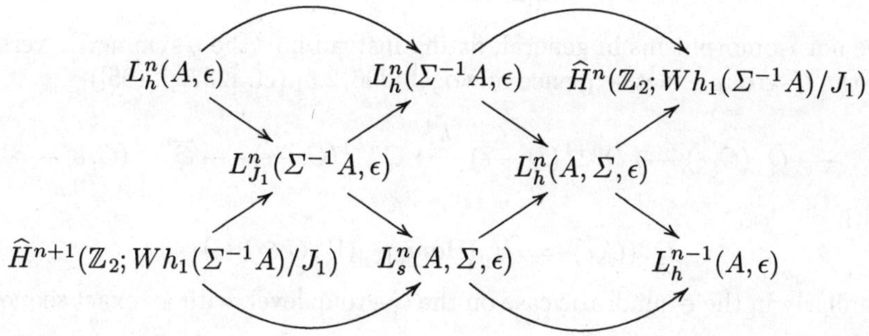

with

$$\begin{aligned} J_0 &= \mathrm{im}(Wh_1(A,\Sigma)\longrightarrow\widetilde{K}_0(A)) \;=\; \ker(\widetilde{K}_0(A)\longrightarrow\widetilde{K}_0(\Sigma^{-1}A))\ , \\ J_1 &= \mathrm{im}(Wh_1(A)\longrightarrow Wh_1(\Sigma^{-1}A)) \;=\; \ker(Wh_1(\Sigma^{-1}A)\longrightarrow Wh_1(A,\Sigma))\ . \end{aligned}$$

Similarly in the ϵ-quadratic case. □

Remark 25.8 As in the case of a multiplicative subset $S\subset A$ considered in Ranicki [237, 3.2] it is possible to view $L_U^n(A,\Sigma,\epsilon)$ as the cobordism group of n-dimensional ϵ-symmetric Poincaré complexes (C,ϕ) over (A,Σ), where:

(i) C is an n-dimensional chain complex of (A,Σ)-modules such that

$$\tau_\Sigma(C) \;=\; \sum_{r=0}^{n}(-)^r\tau_\Sigma(C_r)\in U\subseteq Wh_1(A,\Sigma)\ ,$$

(ii) $\phi\in Q^{n+1}(D,-\epsilon)$ for some $(n+1)$-dimensional $\Sigma^{-1}A$-contractible f.g. projective A-module chain complex D with an A-module chain map $h: D\longrightarrow C$ inducing homology isomorphisms $h_*: H_*(D)\cong H_*(C)$,

(iii) the A-module chain map

$$\phi_0\ :\ D^{n+1-*}\;\simeq\; C^{n-\wedge}\;=\;\mathrm{Hom}_A(C,\Sigma^{-1}A/A)_{*-n}\longrightarrow D\simeq C$$

is a homology equivalence,

(iv) for each $x\in D^{n+1}$ there exists $a\in A$ such that

$$\phi_{n+1}(x)(x) \;=\; a - \epsilon\bar{a} \in A \ .$$

The projective (A,Σ)-torsion ϵ-symmetric L-groups $L_p^*(A,\Sigma,\epsilon)$ depend on the exact category structure of $\mathbb{H}(A,\Sigma)$, and are not the same as the L-groups $L^*(\mathbb{H}(A,\Sigma),\epsilon)$ defined in Ranicki [243] for any additive category with involution: the natural maps

$$L^*(\mathbb{H}(A,\Sigma),\epsilon) \longrightarrow L_p^*(A,\Sigma,\epsilon)$$

may not be isomorphisms. The natural maps

$$\overline{Q}^n(C,\epsilon) \;=\; H_n(\mathrm{Hom}_{\mathbb{Z}[\mathbb{Z}_2]}(W,\mathrm{Hom}_A(C^\frown,C))) \longrightarrow Q^{n+1}(D,-\epsilon)$$

are not isomorphisms in general, fitting instead into the ϵ-symmetric version of the Q-group exact sequence of Vogel [296, 2.5] (cf. [237, p. 195])

$$\dots \longrightarrow \overline{Q}^n(C,\epsilon) \longrightarrow Q^{n+1}(D,-\epsilon) \xrightarrow{h^\%} Q^{n+1}(C,-\epsilon) \longrightarrow \overline{Q}^{n-1}(C,\epsilon) \longrightarrow \dots$$

with

$$Q^n(C,\epsilon) \;=\; H_n(\mathrm{Hom}_{\mathbb{Z}[\mathbb{Z}_2]}(W,C\otimes_A C)) \ .$$

Similarly in the ϵ-quadratic case on the Q-group level, with an exact sequence

$$\dots \longrightarrow \overline{Q}_n(C,\epsilon) \longrightarrow Q_{n+1}(D,-\epsilon) \xrightarrow{h_\%} Q_{n+1}(C,-\epsilon) \longrightarrow \overline{Q}_{n-1}(C,\epsilon) \longrightarrow \dots \ .$$

However, in this case it is proved in [296] that on the L-group level

$$L_*(\mathbb{H}(A,\Sigma),\epsilon) \;=\; L_*^p(A,\Sigma,\epsilon) \ . \qquad \square$$

Definition 25.9 (i) An involution $T : H \longrightarrow H$ on an abelian group H is *hyperbolic* if there exists a direct sum decomposition $H = H^+ \oplus H^-$ and inverse isomorphisms

$$T^+ : H^+ \xrightarrow{\simeq} H^- \ , \ T^- : H^- \xrightarrow{\simeq} H^+$$

such that

$$T \;=\; \begin{pmatrix} 0 & T^- \\ T^+ & 0 \end{pmatrix} : H \;=\; H^+ \oplus H^- \longrightarrow H \;=\; H^+ \oplus H^- \ .$$

(ii) An involution $T : \mathbb{H} \longrightarrow \mathbb{H}$ on an exact category $\mathbb{H}$ is *hyperbolic* if there exists a product decomposition $\mathbb{H} = \mathbb{H}^+ \times \mathbb{H}^-$ and inverse isomorphisms

$$T^+ : \mathbb{H}^+ \xrightarrow{\simeq} \mathbb{H}^- \ , \ T^- : \mathbb{H}^- \xrightarrow{\simeq} \mathbb{H}^+$$

such that

$$T \;=\; \begin{pmatrix} 0 & T^- \\ T^+ & 0 \end{pmatrix} : \mathbb{H} \;=\; \mathbb{H}^+ \times \mathbb{H}^- \longrightarrow \mathbb{H} \;=\; \mathbb{H}^+ \times \mathbb{H}^- \ . \qquad \square$$

Proposition 25.10 *If the Σ-duality involution on $\mathbb{H}(A,\Sigma)$ is hyperbolic then*

$$\widehat{H}^*(\mathbb{Z}_2;Wh_1(A,\Sigma)) = 0$$

and for any $*$-invariant subgroup $U \subseteq Wh_1(A,\Sigma)$

$$\begin{aligned} L^*_U(A,\Sigma,\epsilon) &= L^U_*(A,\Sigma,\epsilon) = 0 , \\ L^{jU}_*(A,\epsilon) &= L^*_{jU}(A,\epsilon) \\ &= L^{\partial^{-1}U}_*(\Sigma^{-1}A,\epsilon) = L^*_{\partial^{-1}U}(\Sigma^{-1}A,\epsilon) . \end{aligned}$$

Proof Let $\mathbb{H}(A,\Sigma) = \mathbb{H}^+ \times \mathbb{H}^-$, with $\mathbb{H}^+$, $\mathbb{H}^-$ interchanged by the Σ-duality involution $T : \mathbb{H}(A,\Sigma) \longrightarrow \mathbb{H}(A,\Sigma)$. Every $\Sigma^{-1}A$-contractible f.g. projective A-module chain complex C is chain equivalent to $C^+ \oplus C^-$ with $C^{\pm}$ homology equivalent to a chain complex in $\mathbb{H}^{\pm}$, such that

$$\begin{aligned} Q_n(C,\epsilon) &= Q^n(C,\epsilon) = H_n(\mathrm{Hom}_A((C^+)^*,C^-)) , \\ Q_n(C^{\pm},\epsilon) &= Q^n(C^{\pm},\epsilon) = 0 . \end{aligned}$$

In particular, the ϵ-symmetric and ϵ-quadratic Q-groups coincide, and similarly for the L-groups. Every $\Sigma^{-1}A$-contractible ϵ-symmetric Poincaré complex (C,ϕ) admits a $\Sigma^{-1}A$-contractible ϵ-symmetric Poincaré null-cobordism $(C \longrightarrow C^+, (j^+\phi,\phi))$. □

Proposition 25.11 *Let A be a ring with involution such that there exists a central element $s \in A$ with*

$$s + \overline{s} = 1 \in A .$$

(i) *For any f.g. projective A-module chain complex C the ϵ-duality involution $T_\epsilon : H_n \longrightarrow H_n$ on $H_n = H_n(\mathrm{Hom}_A(C^*,C))$ is hyperbolic, and*

$$\begin{aligned} Q^n(C,\epsilon) &= \ker(1 - T_\epsilon : H_n \longrightarrow H_n) \\ &= \mathrm{coker}(1 - T_\epsilon : H_n \longrightarrow H_n) = Q_n(C,\epsilon) . \end{aligned}$$

The L-groups of A are such that for any $*$-invariant subgroup $U \subseteq \widetilde{K}_i(A)$ $(i = 0,1)$

$$L^*_U(A,\epsilon) = L^U_*(A,\epsilon) .$$

(ii) *If $s \in A$ is a non-zero divisor then*

$$S = \{s^j(1-s)^k \,|\, j,k \geq 0\} \subset A$$

is a multiplicative subset such that for any $*$-invariant subgroup $U \subseteq K_1(A,S)$

$$L^U_*(A,S,\epsilon) = L^*_U(A,S,\epsilon) = 0 ,$$

and

$$\begin{aligned} L_*^{jU}(A,\epsilon) &= L^*_{jU}(A,\epsilon) \\ &= L_*^{\partial^{-1}U}(S^{-1}A,\epsilon) \ = \ L^*_{\partial^{-1}U}(S^{-1}A,\epsilon) \ . \end{aligned}$$

Proof (i) The forgetful map

$$Q^n(C,\epsilon) \longrightarrow \ker(1-T_\epsilon : H_n \longrightarrow H_n) \ ; \ \phi \longrightarrow \phi_0$$

is an isomorphism with inverse

$$\ker(1-T_\epsilon : H_n \longrightarrow H_n) \longrightarrow Q^n(C,\epsilon) \ ; \ \theta \longrightarrow \phi \ , \ \phi_j \ = \ \begin{cases} s\theta & \text{if } j=0 \\ 0 & \text{if } j \geq 1 \ . \end{cases}$$

The forgetful map

$$Q_n(C,\epsilon) \longrightarrow \operatorname{coker}(1-T_\epsilon : H_n \longrightarrow H_n) \ ; \ \psi \longrightarrow s(1+T_\epsilon)\psi_0$$

is an isomorphism with inverse

$$\operatorname{coker}(1-T_\epsilon : H_n \longrightarrow H_n) \longrightarrow Q_n(C,\epsilon) \ ; \ \theta \longrightarrow \psi \ , \ \psi_j \ = \ \begin{cases} s\theta & \text{if } j=0 \\ 0 & \text{if } j \geq 1 \ . \end{cases}$$

The ϵ-symmetrization map

$$1+T_\epsilon \ : \ Q_n(C,\epsilon) \longrightarrow Q^n(C,\epsilon) \ ; \ \psi \longrightarrow \phi \ , \ \phi_j \ = \ \begin{cases} (1+T_\epsilon)\psi_0 & \text{if } j=0 \\ 0 & \text{if } j \geq 1 \end{cases}$$

is an isomorphism with inverse

$$(1+T_\epsilon)^{-1} \ : \ Q^n(C,\epsilon) \longrightarrow Q_n(C,\epsilon) \ ; \ \phi \longrightarrow \psi \ , \ \psi_j \ = \ \begin{cases} s\phi_0 & \text{if } j=0 \\ 0 & \text{if } j \geq 1. \end{cases}$$

(ii) The multiplicative subsets

$$S_0 \ = \ \{s^j \,|\, j \geq 0\} \ , \ S_1 \ = \ \{(1-s)^k \,|\, k \geq 0\} \subset A$$

are such that

$$S \ = \ S_0S_1 \ , \ \overline{S}_0 \ = \ S_1 \ , \ \overline{S}_1 \ = \ S_0 \ .$$

An A-module M is S-torsion if and only if $s : M \longrightarrow M$ is a near-projection (12.23) with

$$(s(1-s))^N \ = \ 0 \ : \ M \longrightarrow M$$

for some $N \geq 0$, in which case the projection defined by

$$s_\omega \ = \ (s^N + (1-s)^N)^{-1}s^N \ : \ M \longrightarrow M$$

is such that $M = M_0 \oplus M_1$ with

$$\begin{aligned} M_0 &= \{x \in M \,|\, s^jx = 0 \text{ for some } j \geq 0\} \ = \ (1-s_\omega)(M) \ , \\ M_1 &= \{x \in M \,|\, (1-s)^kx = 0 \text{ for some } k \geq 0\} \ = \ s_\omega(M) \ , \\ (M\hat{}\,)_0 &= (M_1)\hat{} \ , \ (M\hat{}\,)_1 \ = \ (M_0)\hat{} \ . \end{aligned}$$

The S-duality involution

$$\mathbb{H}(A,S) \longrightarrow \mathbb{H}(A,S) \; ; \; M \longrightarrow M\hat{}$$

is hyperbolic, corresponding to

$$\mathbb{H}(A,S_0)\times\mathbb{H}(A,S_1) \longrightarrow \mathbb{H}(A,S_0)\times\mathbb{H}(A,S_1) \; ; \; (M_0,M_1) \longrightarrow (M_1\hat{}, M_0\hat{})$$

under the isomorphism of exact categories

$$\mathbb{H}(A,S) \xrightarrow{\simeq} \mathbb{H}(A,S_0)\times\mathbb{H}(A,S_1) \; ; \; M \longrightarrow (M_0,M_1) \; ,$$

so that 25.10 applies. □

Example 25.12 If $2 \in A$ is invertible then $s = 1/2 \in A$ is such that $s+\overline{s} = 1 \in A$, and 25.11 gives

$$\begin{aligned} L^U_*(A,\epsilon) \;&=\; L^*_U(A,\epsilon) \;=\; L^U_*(A[1/2],\epsilon) \;=\; L^*_U(A[1/2],\epsilon) \; , \\ L^U_*(A,(2)^\infty,\epsilon) \;&=\; L^*_U(A,(2)^\infty,\epsilon) \; . \end{aligned}$$

□

Definition 25.13 Given a ring with involution A let $A[s]$ be the polynomial extension ring of A, with the involution extended by

$$\overline{s} \;=\; 1-s \; .$$

□

See Chap. 36 for a detailed account of the L-theory of $A[s]$.

Proposition 25.14 (i) *The ϵ-symmetrization maps for $A[s]$ are isomorphisms*

$$1+T_\epsilon \;:\; L^U_n(A,\epsilon) \xrightarrow{\simeq} L^n_U(A[s],\epsilon)$$

for any $*$-invariant subgroup $U \subseteq Wh_1(A)$.
(ii) *The morphism of rings with involution*

$$A[s] \longrightarrow A[z,z^{-1},(1-z)^{-1}] \; ; \; s \longrightarrow (1-z)^{-1}$$

induces isomorphisms in Tate $\mathbb{Z}_2$-cohomology

$$\widehat{H}^n(\mathbb{Z}_2;K_1(A[s])) \xrightarrow{\simeq} \widehat{H}^n(\mathbb{Z}_2;K_1(A[z,z^{-1},(1-z)^{-1}]))$$

and also in L-theory

$$L^n_U(A[s],\epsilon) \xrightarrow{\simeq} L^n_U(A[z,z^{-1},(1-z)^{-1}],\epsilon)$$

for any $*$-invariant subgroup $U \subseteq Wh_1(A[s])$.
Proof (i) Immediate from 25.11 (i).
(ii) As in the proof of 25.11 (ii) define the multiplicative subsets

$$\begin{aligned} S \;&=\; \{s^j(1-s)^k \,|\, j,k \geq 0\} \; , \\ S_0 \;&=\; \{s^j \,|\, j \geq 0\} \; , \; S_1 \;=\; \{(1-s)^k \,|\, k \geq 0\} \subset A[s] \; , \end{aligned}$$

such that $S = S_0S_1$ is involution-invariant and $\overline{S}_0 = S_1$, $\overline{S}_1 = S_0$. An $(A[s], S)$-module M is a f.g. projective A-module with a near-projection $s : M \longrightarrow M$, such that $(s(1-s))^N = 0 : M \longrightarrow M$ for some $N \geq 0$. As in 12.23 there is defined a projection

$$s_\omega = (s^N + (1-s)^N)^{-1}s^N = s_\omega^2 : M \longrightarrow M ,$$

so that $M = M_0 \oplus M_1$ with

$$M_0 = (1-s_\omega)(M) , M_1 = s_\omega(M) .$$

The endomorphisms

$$\nu_0 = s| : M_0 \longrightarrow M_0 , \nu_1 = (1-s)| : M_1 \longrightarrow M_1$$

are nilpotent, and the functor

$$\mathbb{H}(A[s], S) \longrightarrow \mathrm{Nil}(A) \times \mathrm{Nil}(A) ; M \longrightarrow ((M_0, \nu_0), (M_1, \nu_1))$$

is an isomorphism of exact categories. For $i = 0$ (resp. 1) an $(A[s], S_i)$-module M_i is a f.g. projective A-module with a nilpotent endomorphism $s : M_0 \longrightarrow M_0$ (resp. $1 - s : M_1 \longrightarrow M_1$), and the functors

$$\begin{aligned}\mathbb{H}(A[s], S_0) &\longrightarrow \mathrm{Nil}(A) ; M_0 \longrightarrow [M_0, s] ,\\ \mathbb{H}(A[s], S_1) &\longrightarrow \mathrm{Nil}(A) ; M_1 \longrightarrow [M_1, 1-s]\end{aligned}$$

are also isomorphisms of exact categories. The S-duality involution on $\mathbb{H}(A[s], S)$ is hyperbolic, corresponding to

$$\begin{aligned}&\mathrm{Nil}(A) \times \mathrm{Nil}(A) \longrightarrow \mathrm{Nil}(A) \times \mathrm{Nil}(A) ;\\ &\qquad ((M_0, \nu_0), (M_1, \nu_1)) \longrightarrow ((M_1^*, \nu_1^*), (M_0^*, \nu_0^*)) .\end{aligned}$$

Use $s = (1-z)^{-1}$ to identify

$$S^{-1}A[s] = A[z, z^{-1}, (1-z)^{-1}] ,$$

with

$$\overline{z} = 1 - (1-s)^{-1} = (1-s^{-1})^{-1} = z^{-1} \in S^{-1}A[s] .$$

The short exact sequence of $\mathbb{Z}[\mathbb{Z}_2]$-modules

$$0 \longrightarrow K_1(A[s]) \longrightarrow K_1(A[s, s^{-1}, (1-s)^{-1}]) \longrightarrow K_1(A[s], S) \longrightarrow 0$$

has

$$\begin{aligned}&K_1(A[s]) = K_1(A) \oplus \mathrm{Nil}_0(A) ,\\ &K_1(A[s, s^{-1}, (1-s)^{-1}]) = K_1(A) \oplus K_0(A) \oplus \mathrm{Nil}_0(A) \oplus \mathrm{Nil}_0(A) \oplus \mathrm{Nil}_0(A) ,\\ &K_1(A[s], S) = K_1(A[s], S_0) \oplus K_1(A[s], S_1) = \mathrm{Nil}_0(A) \oplus \mathrm{Nil}_0(A)\end{aligned}$$

with the duality involution determined by $\overline{s} = 1 - s$ interchanging the two summands in $K_1(A[s], S)$, so that

$$\widehat{H}^*(\mathbb{Z}_2; K_1(A[s], S)) = 0$$

and

$$\widehat{H}^*(\mathbb{Z}_2; K_1(A[s])) = \widehat{H}^*(\mathbb{Z}_2; K_1(A[s, s^{-1}, (1-s)^{-1}])) .$$

It is immediate from 25.11 (ii) that

$$L_*^U(A[s], S, \epsilon) = L_U^*(A[s], S, \epsilon) = 0$$

for any $*$-invariant subgroup $U \subseteq Wh_1(A[s])$, and hence that

$$\begin{aligned} L_U^*(A[s], \epsilon) &= L_U^*(A[z, z^{-1}, (1-z)^{-1}], \epsilon) \\ &= L_*^U(A[s], \epsilon) = L_*^U(A[z, z^{-1}, (1-z)^{-1}], \epsilon) . \end{aligned}$$ □

Definition 25.15 A central element $t \in A$ is *coprime* to a multiplicative subset $S \subset A$ if for every $s \in S$ there exist $a, b \in A$ such that

$$as + bt = 1 \in A ,$$

or equivalently if S and $T = \{t^k \mid k \geq 0\} \subset A$ are coprime multiplicative subsets in the sense of 4.13. □

If $t \in A$ is coprime to $S \subset A$ there is defined a cartesian morphism

$$(A, S) \longrightarrow (T^{-1}A, S) ,$$

and hence a cartesian square of rings

$$\begin{array}{ccc} A & \longrightarrow & T^{-1}A \\ \downarrow & & \downarrow \\ S^{-1}A & \longrightarrow & (ST)^{-1}A \end{array}$$

with an isomorphism of exact categories

$$\mathbb{H}(A, S) \xrightarrow{\simeq} \mathbb{H}(T^{-1}A, S) \; ; \; M \longrightarrow T^{-1}M$$

inducing excision isomorphisms in the relative L-groups. See Ranicki [237, Chap. 3] for the corresponding L-theory Mayer–Vietoris exact sequences.

Proposition 25.16 *Let $S \subset A$ be an involution-invariant multiplicative subset in a ring with involution, and such that there exists a central element $t \in A$ coprime to S and with*

$$t + \overline{t} = t\overline{t} \in A .$$

(i) *For any $S^{-1}A$-contractible f.g. free A-module chain complex C the ϵ-duality involution $T_\epsilon : H_n \longrightarrow H_n$ on $H_n = H_n(\mathrm{Hom}_A(C^*, C))$ is hyperbolic, and*

$$\begin{aligned} Q^n(C,\epsilon) &= \ker(1 - T_\epsilon : H_n \longrightarrow H_n) \\ &= \mathrm{coker}(1 - T_\epsilon : H_n \longrightarrow H_n) = Q_n(C,\epsilon) . \end{aligned}$$

(ii) *The L-groups of (A,S) are such that*

$$L^*_U(A,S,\epsilon) = L^U_*(A,S,\epsilon)$$

for any $*$-invariant subgroup $U \subseteq Wh_1(A,S)$.

Proof (i) The multiplicative subset

$$T = \{t^j \bar{t}^k \,|\, j,k \geq 0\} \subset A$$

is involution-invariant and coprime to S. The cartesian morphism

$$(A,S) \longrightarrow (T^{-1}A, S)$$

determines a cartesian square of rings with involution

$$\begin{array}{ccc} A & \longrightarrow & T^{-1}A \\ \big\downarrow & & \big\downarrow \\ S^{-1}A & \longrightarrow & (ST)^{-1}A \end{array}$$

and an isomorphism of exact categories with duality

$$\mathbb{H}(A,S) \longrightarrow \mathbb{H}(T^{-1}A,S) \; ; \; M \longrightarrow T^{-1}M .$$

The element

$$a = t^{-1} \in T^{-1}A$$

is such that

$$a + \bar{a} = t\bar{t}/(t+\bar{t}) = 1 \in T^{-1}A .$$

The inclusion $C \longrightarrow T^{-1}C$ is a homology equivalence for any $S^{-1}A$-contractible finite f.g. free A-module chain complex C, so that the localization maps are isomorphisms

$$\begin{array}{l} H_*(C \otimes_A C) \xrightarrow{\simeq} H_*(T^{-1}C \otimes_{T^{-1}A} T^{-1}C) , \\ Q^*(C,\epsilon) \xrightarrow{\simeq} Q^*(T^{-1}C,\epsilon) \; , \; Q_*(C,\epsilon) \xrightarrow{\simeq} Q_*(T^{-1}C,\epsilon) \end{array}$$

and 25.11 applies to $T^{-1}C$.

(ii) Immediate from (i). □

Example 25.17 (i) If $S = \{\text{odd integers}\} \subset \mathbb{Z} \subseteq A$ then $t = 2 \in A$ is coprime to S with $t + \bar{t} = t\bar{t} = 4 \in A$, so that

$$L_*^U(A,S,\epsilon) \;=\; L_U^*(A,S,\epsilon) \;.$$

(ii) For any ring with involution A the set Π of A-invertible square matrices in $A[z,z^{-1}]$ is coprime to

$$t \;=\; 1-z \in A[z,z^{-1}]$$

with

$$t+\bar{t} \;=\; (1-z)(1-z^{-1}) \;=\; t\bar{t} \in A[z,z^{-1}] \;,$$

so that

$$L_*^U(A[z,z^{-1}],\Pi,\epsilon) \;=\; L_U^*(A[z,z^{-1}],\Pi,\epsilon) \;. \qquad \square$$

Definition 25.18 (i) An *ϵ-symmetric linking form (M,λ) over (A,Σ)* is an (A,Σ)-module M with a pairing

$$\lambda \;:\; M\times M \longrightarrow \Sigma^{-1}A/A \;;\; (x,y) \longrightarrow \lambda(x,y)$$

such that

$$\lambda(y,x) \;=\; \epsilon\overline{\lambda(x,y)} \in \Sigma^{-1}A/A \;.$$

(ii) A linking form (M,λ) is *nonsingular* if the adjoint A-module morphism

$$M \longrightarrow M\widehat{\ } \;;\; x \longrightarrow (y \longrightarrow \lambda(x,y))$$

is an isomorphism.
(iii) A linking form (M,λ) is *metabolic* if there exists an (A,S)-submodule $L\subseteq M$ such that $L=L^\perp$, with

$$L^\perp \;=\; \{y\in M \,|\, \lambda(x,y)=0\in\Sigma^{-1}A/A \text{ for all } x\in L\} \;.$$

(iv) The *ϵ-symmetric linking Witt group $W^\epsilon(A,\Sigma)$* is the group of equivalence classes of nonsingular ϵ-symmetric linking forms over (A,Σ) with

$(M,\lambda)\sim(M',\lambda')$ if there exists an isomorphism

$$(M,\lambda)\oplus(N,\mu)\cong(M',\lambda')\oplus(N',\mu')$$

for some metabolic $(N,\mu),(N',\mu')$. $\square$

Proposition 25.19 (i) *The isomorphism classes of (nonsingular) ϵ-symmetric linking forms over (A,Σ) are in natural one-one correspondence with the homotopy equivalence classes of $\Sigma^{-1}A$-contractible 1-dimensional $(-\epsilon)$-symmetric (Poincaré) complexes (C,ϕ) over A.*
(ii) *For $i\geq 1$ there is a natural map from the ϵ-symmetric linking form Witt group of (A,Σ) to the $2i$-dimensional (A,Σ)-torsion $(-)^i\epsilon$-symmetric L-group*

$$W^\epsilon(A,\Sigma) \longrightarrow L^{2i}(A,\Sigma,(-)^i\epsilon) \;;\; (M,\lambda) \longrightarrow S^{i-1}(C,\phi) \;.$$

Proof Given a linking form (M,λ) let

$$0 \longrightarrow C_1 \longrightarrow C_0 \longrightarrow M^\frown \longrightarrow 0$$

be a f.g. projective A-module resolution of $M^\frown$. Then C is an $\Sigma^{-1}A$-contractible A-module chain complex such that

$$Q^1(C,-\epsilon) \;=\; \ker(1-T_\epsilon : \mathrm{Hom}_A(M,M^\frown) \longrightarrow \mathrm{Hom}_A(M,M^\frown)) \;. \qquad \square$$

Example 25.20 Let A be a ring with involution, with an involution-invariant multiplicative subset

$$S \;=\; \{u^j s^k \,|\, j \in \mathbb{Z}, k \geq 0\} \subset A$$

for some central non-zero divisor $s \in A$ and unit $u \in A^\bullet$ such that

$$u\overline{s} \;=\; s \;,\; \overline{u} \;=\; u^{-1} \in A \;.$$

The localization of A inverting S is a ring with involution

$$S^{-1}A \;=\; A[1/s] \;.$$

An (A,S)-module M is an A-module of homological dimension 1 such that $s^k M = 0$ for some $k \geq 0$. The A-module morphism

$$\mathrm{Hom}_{A/s^kA}(M, A/s^kA) \longrightarrow \mathrm{Hom}_A(M, S^{-1}A/A) \;;\; f \longrightarrow (x \longrightarrow f(x)/s^k)$$

is an isomorphism if M is a f.g. projective A/s^kA-module. Thus for each $k \geq 1$ there is defined a one-one correspondence

$$\begin{aligned} &\{u^k\epsilon\text{-symmetric forms over } A/s^kA\} \longrightarrow \\ &\{\epsilon\text{-symmetric linking forms } (M,\mu) \text{ over } (A,S) \\ &\qquad\qquad \text{with } M \text{ a f.g. projective } A/s^kA\text{-module}\} \;; \\ &\qquad\qquad (M,\lambda) \longrightarrow (M,\mu) \;\; (\mu(x,y) = \lambda(x,y)/s^k) \;. \end{aligned}$$

The morphism of Witt groups

$$L^0(A/sA, u\epsilon) \longrightarrow W^\epsilon(A,S) \;;\; (M,\lambda) \longrightarrow (M,\mu)$$

will be shown in Chap. 38 to be an isomorphism if A is a Dedekind ring and A/sA is a field, with

$$L^0(A/sA, u\epsilon) \;=\; W^\epsilon(A,S) \;=\; L^{2i}(A,S,(-)^i\epsilon) \;\; (i \geq 1) \;. \qquad \square$$

Example 25.21 Let X be an n-dimensional geometric Poincaré complex, and let $\widetilde{X}$ be a regular covering with group of covering translations π (e.g. the universal cover with $\pi_1(X) = \pi$). If Σ is an involution-invariant set of square matrices in $A = \mathbb{Z}[\pi]$ such that

$$\Sigma^{-1}H_*(\widetilde{X}) \;=\; 0$$

there is defined an n-dimensional $\Sigma^{-1}A$-contractible symmetric Poincaré complex $(C(\widetilde{X}),\phi)$ over A, with $\phi = \Delta[X]$, and hence a symmetric signature

$$\sigma^*(X) \;=\; (C(\widetilde{X}),\phi) \in L^{n+1}(A,\Sigma)\ .$$

The symmetric Poincaré structure ϕ determines a linking pairing

$$\begin{aligned}\mu\ :\ H_r(\widetilde{X}) \times H_{n-r-1}(\widetilde{X}) &\longrightarrow \Sigma^{-1}A/A\ ;\\ (x,y) &\longrightarrow \phi_0(w,y)/\sigma\ \ (\sigma\in\Sigma, w\in C_{r+1}, d(w)=\sigma x)\end{aligned}$$

such that

$$\begin{aligned}\mu(ax,by) \;&=\; b\mu(x,y)\overline{a}\ \ (a,b\in A)\ ,\\ \mu(y,x) \;&=\; (-)^{r(n-r-1)}\overline{\mu(x,y)} \in \Sigma^{-1}A/A\ .\end{aligned}$$

(i) The construction of μ generalizes the classical linking pairing of Seifert [261]

$$T_r(X)\times T_{n-r-1}(X) \longrightarrow \mathbb{Q}/\mathbb{Z}$$

on the torsion subgroups $T_r(X)\subseteq H_r(X)$.
(ii) If X is the exterior of a knot $k: S^{n-2}\subset S^n$ and $\overline{X}$ is the canonical infinite cyclic cover then

$$P^{-1}H_*(\overline{X}) \;=\; 0\ \ (*\neq 0)$$

with $P=\{p(z)\,|\,p(1)=1\}\subset\mathbb{Z}[z,z^{-1}]$, and

$$\mu\ :\ H_r(\overline{X})\times H_{n-r-1}(\overline{X}) \longrightarrow P^{-1}\mathbb{Z}[z,z^{-1}]/\mathbb{Z}[z,z^{-1}]$$

is the linking pairing of Blanchfield [23]. See Chap. 32 for a more detailed exposition. □

See Ranicki [237, Chap. 3.5] for a detailed account of linking forms, including the precise relationship between $L^{2i}(A,\Sigma,\epsilon)$ and the Witt group of nonsingular $(-)^i\epsilon$-symmetric linking forms over (A,Σ), as well as the ϵ-quadratic versions. The ϵ-quadratic L-groups $L^U_*(A,S,\epsilon)$ are 4-periodic

$$L^U_*(A,S,\epsilon) \;=\; L^U_{*+2}(A,S,-\epsilon) \;=\; L^U_{*+4}(A,S,\epsilon)\ ,$$

with $L^U_{2i}(A,S,\epsilon)=L^U_0(A,S,(-)^i\epsilon)$ the Witt group of nonsingular split $(-)^i\epsilon$-quadratic linking forms (M,λ,ν) over (A,S) with

$$[M]\in U\subseteq Wh_1(A,S)\ .$$

The ϵ-symmetric L-groups $L^*_U(A,S,\epsilon)$ are not 4-periodic in general, but there are defined skew-suspension maps

$$L^n_U(A,S,\epsilon) \longrightarrow L^{n+2}_U(A,S,-\epsilon) \longrightarrow L^{n+4}_U(A,S,\epsilon)\ .$$

Proposition 25.22 *Let A be a ring with involution and let $S \subset A$ be an involution-invariant multiplicative subset such that there exists a central element $t \in A$ coprime to $S \subset A$ with $t + \bar{t} = t\bar{t} \in A$ the ϵ-symmetric L-groups $L_U^*(A, S, \epsilon)$ coincide with the ϵ-quadratic L-groups $L_*^U(A, S, \epsilon)$, and are 4-periodic for $* \geq 0$. In particular, $L_U^{2i}(A, S, \epsilon)$ $(i \geq 0)$ is the Witt group of nonsingular $(-)^i\epsilon$-symmetric linking forms (M, λ) over (A, S) with $[M] \in U \subseteq Wh_1(A, S)$.*

Proof Immediate from 25.16 and the 4-periodicity of the ϵ-quadratic L-groups. □

The following results on the L-theory localization exact sequence will be used later on:

Proposition 25.23 *Let Σ_1, Σ_2 be involution-invariant coprime sets of square matrices in a ring with involution A (9.16) such that the natural maps*

$$i_1 : A \longrightarrow (\Sigma_1)^{-1}A \ , \ i_2 : A \longrightarrow (\Sigma_2)^{-1}A$$

are injective. The localizations fit into a cartesian square of rings with involution

$$\begin{array}{ccc} A & \longrightarrow & (\Sigma_1)^{-1}A \\ \downarrow & & \downarrow \\ (\Sigma_2)^{-1}A & \longrightarrow & (\Sigma_1 \cup \Sigma_2)^{-1}A \end{array}$$

and the ϵ-symmetric torsion L-groups are such that

$$\begin{aligned} L_p^*(A, \Sigma_1 \cup \Sigma_2, \epsilon) &= L_p^*(A, \Sigma_1, \epsilon) \oplus L_p^*(A, \Sigma_2, \epsilon) \ , \\ L_h^*(A, \Sigma_1 \cup \Sigma_2, \epsilon) &= L_h^*(A, \Sigma_1, \epsilon) \oplus L_h^*(A, \Sigma_2, \epsilon) \end{aligned}$$

with commutative braids of exact sequences

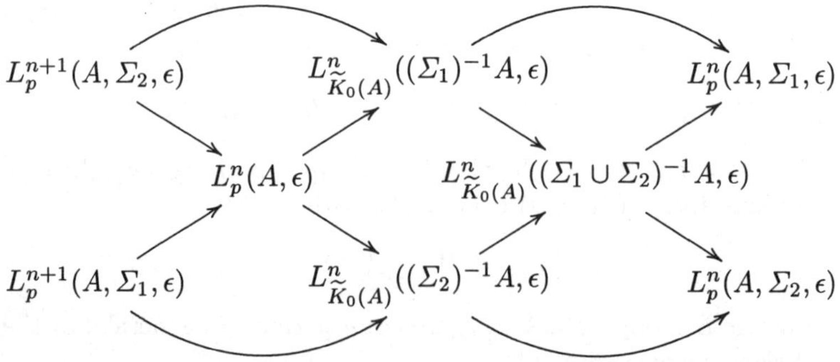

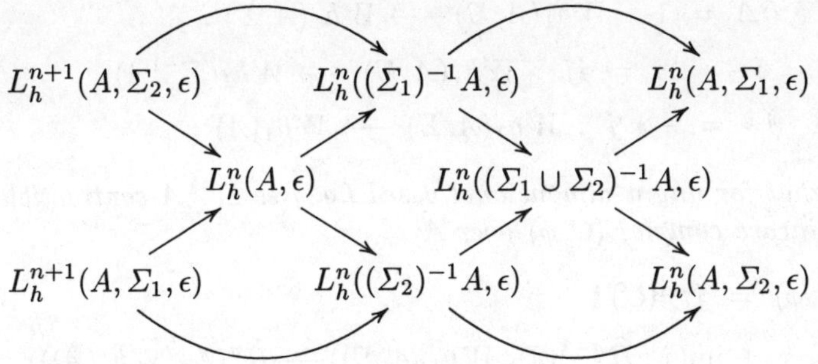

with $L^* \equiv L_h^*$. *Similarly for* ϵ*-quadratic L-theory.* □

Proposition 25.24 *Let* $\Sigma, \widetilde{\Sigma}$ *be involution-invariant sets of square matrices with entries in a ring with involution* A*, such that the localizations* $\Sigma^{-1}A$*,* $\widetilde{\Sigma}^{-1}A$ *are defined, with the natural maps* $A\longrightarrow\Sigma^{-1}A$*,* $A\longrightarrow\widetilde{\Sigma}^{-1}A$ *injections and* $\widetilde{\Sigma}\subseteq\Sigma$*. The symmetric L-theory localization exact sequences for* (A,Σ)*,* $(A,\widetilde{\Sigma})$ *are related by a commutative braid of exact sequences*

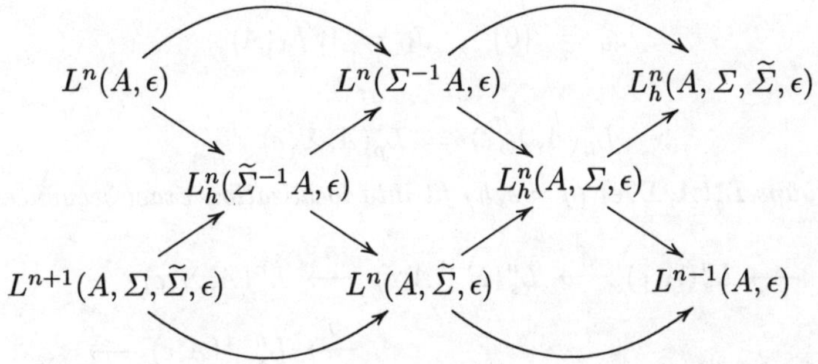

with $L^* \equiv L_h^*$ *and* $L^n(A,\Sigma,\widetilde{\Sigma},\epsilon)$ *the cobordism group of f.g. free* $(n-1)$*-dimensional* ϵ*-symmetric complexes over* A *which are* $\widetilde{\Sigma}^{-1}A$*-Poincaré and* $\Sigma^{-1}A$*-contractible. Similarly for* ϵ*-quadratic L-theory.* □

Proposition 25.25 *Let* A *be a ring with involution and let* Σ *be an involution-invariant set of square matrices with entries in* A *such that the natural map* $\mathcal{F}: A\longrightarrow\Sigma^{-1}A$ *is injective (as in 25.4), and also such that the algebraic K-theory exact sequence*

$$0 \longrightarrow Wh_1(A) \xrightarrow{\ i\ } Wh_1(\Sigma^{-1}A) \xrightarrow{\ \partial\ } Wh_1(A,\Sigma) \longrightarrow 0$$

is split exact, with morphisms

$$\Delta : Wh_1(A,\Sigma) \longrightarrow Wh_1(\Sigma^{-1}A)\ ,\ j : Wh_1(A,\Sigma) \longrightarrow Wh_1(A)$$

such that

$$\begin{aligned} \partial\Delta &= 1 : Wh_1(A,\Sigma) \longrightarrow Wh_1(A,\Sigma) , \\ \Delta - *\Delta* &= ij : Wh_1(A,\Sigma) \longrightarrow Wh_1(\Sigma^{-1}A) , \\ j* &= -*j : Wh_1(A,\Sigma) \longrightarrow Wh_1(A) , \end{aligned}$$

and such that for any n-dimensional based f.g. free $\Sigma^{-1}A$-contractible symmetric Poincaré complex (C,ϕ) over A

$$\begin{aligned} \tau(C,\phi) &= j\tau_\Sigma(C) \\ &\in \mathrm{im}(j : \widehat{H}^{n-1}(\mathbb{Z}_2;Wh_1(A,\Sigma)) \longrightarrow \widehat{H}^n(\mathbb{Z}_2;Wh_1(A))) . \end{aligned}$$

(i) *The L-groups $L^*_U(A,\Sigma,\epsilon)$ associated to any $*$-invariant subgroup $U \subseteq Wh_1(A,\Sigma)$ fit into the localization exact sequence*

$$\begin{aligned} \dots \longrightarrow L^n_{jU}(A,\epsilon) \xrightarrow{i} L^n_{ijU\oplus\Delta U}(\Sigma^{-1}A,\epsilon) \xrightarrow{\partial} L^n_U(A,\Sigma,\epsilon) \\ \xrightarrow{j} L^{n-1}_{jU}(A,\epsilon) \longrightarrow \dots . \end{aligned}$$

(ii) *The J-groups of 25.7 are given by*

$$J_0 = \{0\} \ , \ J_1 = Wh_1(A)$$

and

$$L^*_h(A,\Sigma,\epsilon) = L^*_p(A,\Sigma,\epsilon) .$$

The L-groups $L^*_q(A,\Sigma,\epsilon)$ $(q = s,h)$ fit into localization exact sequences

$$\begin{aligned} \dots \longrightarrow L^n_s(A,\epsilon) \xrightarrow{i} L^n_s(\Sigma^{-1}A,\epsilon) \xrightarrow{\partial} L^n_s(A,\Sigma,\epsilon) \\ \xrightarrow{j} L^{n-1}_s(A,\epsilon) \longrightarrow \dots , \\ \dots \longrightarrow L^n_h(A,\epsilon) \xrightarrow{i} L^n_h(\Sigma^{-1}A,\epsilon) \xrightarrow{\partial} L^n_h(A,\Sigma,\epsilon) \\ \xrightarrow{j} L^{n-1}_h(A,\epsilon) \longrightarrow \dots \end{aligned}$$

and are such that there is defined a commutative diagram

$$\begin{array}{ccccccccc}
 & & \downarrow & & \downarrow & & \downarrow & & \downarrow & \\
\longrightarrow & L_s^n(A,\epsilon) & \longrightarrow & L_h^n(A,\epsilon) & \longrightarrow & \widehat{H}^n(\mathbb{Z}_2;Wh_1(A)) & \longrightarrow & L_s^{n-1}(A,\epsilon) & \longrightarrow \\
 & \downarrow{\scriptstyle i} & & \downarrow{\scriptstyle i} & & \downarrow{\scriptstyle i} & & \downarrow{\scriptstyle i} & \\
\longrightarrow & L_s^n(\Sigma^{-1}A,\epsilon) & \longrightarrow & L_h^n(\Sigma^{-1}A,\epsilon) & \longrightarrow & \widehat{H}^n(\mathbb{Z}_2;Wh_1(\Sigma^{-1}A)) & \longrightarrow & L_s^{n-1}(\Sigma^{-1}A,\epsilon) & \longrightarrow \\
 & \downarrow{\scriptstyle \partial} & & \downarrow{\scriptstyle \partial} & & \downarrow{\scriptstyle \partial} & & \downarrow{\scriptstyle \partial} & \\
\longrightarrow & L_s^n(A,\Sigma,\epsilon) & \longrightarrow & L_h^n(A,\Sigma,\epsilon) & \longrightarrow & \widehat{H}^n(\mathbb{Z}_2;Wh_1(A,\Sigma)) & \longrightarrow & L_s^{n-1}(A,\Sigma,\epsilon) & \longrightarrow \\
 & \downarrow{\scriptstyle j} & & \downarrow{\scriptstyle j} & & \downarrow{\scriptstyle j} & & \downarrow{\scriptstyle j} & \\
\longrightarrow & L_s^{n-1}(A,\epsilon) & \longrightarrow & L_h^{n-1}(A,\epsilon) & \longrightarrow & \widehat{H}^{n-1}(\mathbb{Z}_2;Wh_1(A)) & \longrightarrow & L_s^{n-2}(A,\epsilon) & \longrightarrow \\
 & \downarrow & & \downarrow & & \downarrow & & \downarrow &
\end{array}$$

with exact rows and columns.
Similarly for the ϵ-quadratic L-groups. □

By analogy with the localization-completion excision property of K-theory:

Proposition 25.26 (Ranicki [237, 3.7.3])
Let A be a ring with involution, and let $S \subset A$ be an involution-invariant multiplicative subset. The localization $S^{-1}A$ and completion $\widehat{A}_S$ fit into a cartesian square of rings with involution

$$\begin{array}{ccc}
A & \longrightarrow & S^{-1}A \\
\downarrow & & \downarrow \\
\widehat{A}_S & \longrightarrow & \widehat{S}^{-1}\widehat{A}_S
\end{array}$$

The inclusion $A \longrightarrow \widehat{A}$ induces excision isomorphisms in the relative ϵ-symmetric L-groups

$$L_p^*(A,S,\epsilon) \;\cong\; L_p^*(\widehat{A},\widehat{S},\epsilon)$$

and a Mayer–Vietoris exact sequence in the absolute ϵ-symmetric L-groups

$$\begin{aligned}
\dots \longrightarrow L_I^n(A,\epsilon) \longrightarrow L_h^n(S^{-1}A,\epsilon) \oplus L_{\widehat{I}}^n(\widehat{A}_S,\epsilon) \longrightarrow L_h^n(\widehat{S}^{-1}\widehat{A}_S,\epsilon) \\
\longrightarrow L_I^{n-1}(A,\epsilon) \longrightarrow \dots
\end{aligned}$$

with

$$\begin{aligned}
I &= \ker(\widetilde{K}_0(A) \longrightarrow \widetilde{K}_0(S^{-1}A)) \subseteq \widetilde{K}_0(A) , \\
\widehat{I} &= \ker(\widetilde{K}_0(\widehat{A}_S) \longrightarrow \widetilde{K}_0(\widehat{S}^{-1}\widehat{A}_S)) \subseteq \widetilde{K}_0(\widehat{A}_S) .
\end{aligned}$$

Similarly for ϵ-quadratic L-theory.

Proof As for the K-theory case in 4.17, noting that $A \longrightarrow \widehat{A}_S$ induces an isomorphism of exact categories with involution

$$\mathbb{H}(A,S) \cong \mathbb{H}(\widehat{A}_S, \widehat{S}) .$$

The condition

$$\delta = 0 : \widehat{H}^0(\mathbb{Z}_2; \widehat{S}^{-1}\widehat{A}_S/\widehat{A}_S, \epsilon) \longrightarrow \widehat{H}^1(\mathbb{Z}_2; A, \epsilon)$$

required for the ϵ-symmetric case in [228, 3.7.3] is actually redundant, since $Q^*(D, -\epsilon) \cong Q^*(\widehat{D}, -\epsilon)$ for any $S^{-1}A$-contractible finite f.g. projective A-module chain complex D by the exact sequence of Vogel [296, 2.5] (quoted in 25.8 above). □

26. Asymmetric L-theory

An asymmetric complex (C, λ) is a chain complex with a chain map $\lambda : C^{n-*} \longrightarrow C$. The complex is Poincaré if λ is a chain equivalence. In Chap. 26 will develop asymmetric L-theory, in preparation for the applications to topology in subsequent chapters. See Chap. 39 for the computation of the asymmetric L-groups of a field F.

There are two distinct situations in topology where asymmetric complexes arise:

(a) A knot in a boundary. Given an n-dimensional manifold with boundary $(M, \partial M)$ and a codimension 2 submanifold $K^{n-3} \subset \partial M$ and a Seifert surface $J^{n-2} \subset \partial M$, there is defined in Chap. 27 an n-dimensional asymmetric complex $(C(M,J), \lambda)$. This construction generalizes the Seifert form of an $(n-3)$-knot $k : S^{n-3} \subset S^{n-1}$ – the special case $(M, \partial M) = (D^n, S^{n-1})$, $K = S^{n-3}$.

(b) A twisted double (e.g. an open book) boundary. Given an n-dimensional manifold with boundary $(M, \partial M)$ and a codimension 1 submanifold $P^{n-2} \subset \partial M$ such that $\partial M = Q \cup_h Q$ is a twisted double with $h : P = \partial Q \longrightarrow \partial Q$, there is defined in Chap. 28 an n-dimensional asymmetric Poincaré complex

$$(\mathcal{C}(i_0 - i_1 : C(Q) \longrightarrow C(M,P)), \lambda) \ ,$$

with $i_0, i_1 : Q \longrightarrow M$ the two inclusions. This construction generalizes the nonsingular Seifert form of a fibred $(n-3)$-knot $k : S^{n-3} \subset S^{n-1}$ – the special case $(M, \partial M) = (D^n, S^{n-1})$, $Q = J \times I$, $h = g \cup 1 : \partial Q = J \cup_{\partial J} J \longrightarrow \partial Q$ with $g : J \longrightarrow J$ the monodromy automorphism of a Seifert surface $J^{n-2} \subset S^{n-3}$.

The asymmetric complex construction (a) will be used in Chap. 27 to express the obstruction groups of Chap. 22 for surgery on submanifolds of codimension 2 in the framed case $\xi = \epsilon^2$ as asymmetric L-groups. The applications to knot theory will be considered in Chap. 33.

The asymmetric Poincaré complex construction (b) will be used in Chaps. 28, 29, 30 to express the twisted double and open book bordism groups as asymmetric L-groups, and to describe the relationship with the automorphism bordism groups.

Terminology 26.1 Given a ring with involution A let

$$\mathcal{F} \ : \ A[z,z^{-1}] \longrightarrow \Lambda \ \ (\overline{z} = z^{-1})$$

be a locally epic morphism of rings with involution, and let Σ be the set of Λ-invertible matrices in A. □

As in 20.20 let $\Phi_{\mathcal{F}}$ be the commutative square of rings with involution

$$\begin{array}{ccc} A[z,z^{-1}] & \longrightarrow & A[z,z^{-1}] \\ \big\downarrow & \Phi_{\mathcal{F}} & \big\downarrow \\ A[z,z^{-1}] & \longrightarrow & \Lambda \end{array}$$

so that the relative quadratic Γ-groups $\Gamma_*(\Phi_{\mathcal{F}},\epsilon)$ fit into an exact sequence

$$\begin{aligned} \dots \longrightarrow L_n(A[z,z^{-1}],\epsilon) \longrightarrow \Gamma_n(\mathcal{F},\epsilon) \longrightarrow \Gamma_n(\Phi_{\mathcal{F}},\epsilon) \\ \longrightarrow L_{n-1}(A[z,z^{-1}],\epsilon) \longrightarrow \dots \end{aligned}$$

with $\epsilon \in A[z,z^{-1}]$ a central unit such that $\overline{\epsilon} = \epsilon^{-1}$. By 20.21 $\Gamma_n(\Phi_{\mathcal{F}},\epsilon)$ is the cobordism group of $(n-1)$-dimensional ϵ-quadratic Λ-contractible Poincaré complexes over $A[z,z^{-1}]$.

The framed codimension 2 surgery obstruction groups are the algebraic Γ-groups $\Gamma_*(\mathcal{F},\epsilon)$, $\Gamma_*(\Phi_{\mathcal{F}},\epsilon)$, for appropriate $\mathcal{F},\epsilon$. The factorization

$$\mathcal{F} \ : \ A[z,z^{-1}] \longrightarrow \Sigma^{-1}A[z,z^{-1}] \longrightarrow \Lambda$$

has the universal property that a finite f.g. free $A[z,z^{-1}]$-module chain complex C is Λ-contractible if and only if it is $\Sigma^{-1}A[z,z^{-1}]$-contractible (9.15). It will be the case in the applications that

$$A[z,z^{-1}] \longrightarrow \Sigma^{-1}A[z,z^{-1}]$$

is an injection, so that the algebraic Γ-groups are the L-groups of the localization

$$\begin{aligned} \Gamma_*(\mathcal{F},\epsilon) &= \Gamma_*(A[z,z^{-1}] \longrightarrow \Sigma^{-1}A[z,z^{-1}],\epsilon) \\ &= L_*(\Sigma^{-1}A[z,z^{-1}],\epsilon) \end{aligned}$$

(where $L_* = L^h_*$) and the L-groups of the localization fit into the L-theory exact sequence of Chap. 25

$$\begin{aligned} \dots \longrightarrow L_n(A[z,z^{-1}],\epsilon) \longrightarrow L_n(\Sigma^{-1}A[z,z^{-1}],\epsilon) \\ \longrightarrow L_n(A[z,z^{-1}],\Sigma,\epsilon) \longrightarrow L_{n-1}(A[z,z^{-1}],\epsilon) \longrightarrow \dots \end{aligned}$$

and

$$\Gamma_*(\Phi_{\mathcal{F}},\epsilon) \ = \ L_*(A[z,z^{-1}],\Sigma,\epsilon) \ .$$

In the original application of the ϵ-quadratic Γ-groups to codimension 2 surgery due to Cappell and Shaneson [40] $\epsilon = 1$, as in the original application of the algebraic L-groups to ordinary surgery due to Wall [302]. As recalled in Chap. 22, the codimension 2 surgery groups LS_* of [302] and their Γ-analogues were expressed in Ranicki [237, 7.8] in terms of the $(-z)$-quadratic Γ-groups $\Gamma_*(\mathcal{F}, -z), \Gamma_*(\Phi_{\mathcal{F}}, -z)$, making use of the work of Matsumoto [184] and Freedman [84]. The main results of Chap. 26 express $\Gamma_*(\mathcal{F}, -z), \Gamma_*(\Phi_{\mathcal{F}}, -z)$ in terms of asymmetric complexes over A, using an algebraic transversality analogue of the relationship between knots and Seifert surfaces.

Definition 26.2 Let $\epsilon \in A[z, z^{-1}]$ be a central unit such that $\overline{\epsilon} = \epsilon^{-1}$.
(i) An *n-dimensional asymmetric complex over A (C, λ)* is an n-dimensional f.g. projective A-module chain complex

$$C : \ldots \longrightarrow 0 \longrightarrow C_n \longrightarrow C_{n-1} \longrightarrow \ldots \longrightarrow C \longrightarrow C_0$$

together with a chain map $\lambda : C^{n-*} \longrightarrow C$. The asymmetric complex (C, λ) is *(Λ, ϵ)-Poincaré* if the $A[z, z^{-1}]$-module chain map

$$(1 + T_\epsilon)\lambda \ : \ C^{n-*}[z, z^{-1}] \longrightarrow C[z, z^{-1}]$$

is a Λ-equivalence, and for based f.g. free C the *torsion* is

$$\tau(C, \lambda) \ = \ \tau((1 + T_\epsilon)\lambda : \Lambda \otimes_{A[z,z^{-1}]} C^{n-*} \longrightarrow \Lambda \otimes_{A[z,z^{-1}]} C) \in Wh_1(\Lambda) \ .$$

(ii) An *$(n+1)$-dimensional asymmetric pair over A $(f : C \longrightarrow D, (\delta\lambda, \lambda))$* is a chain map $f : C \longrightarrow D$ of f.g. projective A-module chain complexes with C n-dimensional and D $(n+1)$-dimensional, together with a chain homotopy

$$\delta\lambda \ : \ f\lambda f^* \ \simeq \ 0 \ : \ D^{n-*} \longrightarrow D \ .$$

The asymmetric pair is *(Λ, ϵ)-Poincaré* if the $A[z, z^{-1}]$-module chain maps

$$\begin{aligned} &(1 + T_\epsilon)\begin{pmatrix} \delta\lambda \\ \lambda f^* \end{pmatrix} \ : \ D^{n+1-*}[z, z^{-1}] \longrightarrow \mathcal{C}(f)[z, z^{-1}] \ , \\ &(1 + T_\epsilon)\,(\delta\lambda \quad f\lambda) \ : \ \mathcal{C}(f)^{n+1-*}[z, z^{-1}] \longrightarrow D[z, z^{-1}] \end{aligned}$$

are Λ-equivalences, in which case (C, λ) is an n-dimensional asymmetric (Λ, ϵ)-Poincaré complex called the *boundary* of the pair.
(iii) (Λ, ϵ)-asymmetric Poincaré complexes (C, λ), (C', λ') over A are *cobordant* if $(C, \lambda) \oplus (C', -\lambda')$ is the boundary of a (Λ, ϵ)-asymmetric Poincaré pair.
(iv) The $\left\{\begin{array}{l} \textit{projective} \\ \textit{free} \\ \textit{simple} \end{array}\right.$ *(Λ, ϵ)-asymmetric L-group* $\left\{\begin{array}{l} LAsy_p^n(A, \Lambda, \epsilon) \\ LAsy_h^n(A, \Lambda, \epsilon) \\ LAsy_s^n(A, \Lambda, \epsilon) \end{array}\right.$ is the cobordism group of n-dimensional (Λ, ϵ)-asymmetric Poincaré complexes

$$(C,\lambda) \text{ over } A \text{ with } C \begin{cases} \text{f.g. projective} \\ \text{f.g. free} \\ \text{based f.g. free and } \tau(C,\lambda)=0 \ . \end{cases}$$

□

It will be proved in 26.11 that

$$\begin{aligned} L_n(A[z,z^{-1}],z) &= \widehat{H}^{n+1}(\mathbb{Z}_2;\widetilde{K}_0(A)) \ , \\ \Gamma_n^q(\mathcal{F},z) &= L_n^q(\Sigma^{-1}A[z,z^{-1}],z) = LAsy_q^n(A,\Lambda,z) \quad (q=s,h) \ , \\ \Gamma_n^h(\Phi_{\mathcal{F}},z) &= L_n^h(A[z,z^{-1}],\Sigma,z) = LAsy_p^n(A,\Lambda,z) \ . \end{aligned}$$

The 0-dimensional asymmetric complexes are asymmetric forms:

Definition 26.3 (i) An *asymmetric form over* A (L,λ) is a f.g. projective A-module L together with an A-module morphism $\lambda : L \longrightarrow L^*$, corresponding to a sesquilinear pairing

$$\lambda \ : \ L \times L \longrightarrow A \ ; \ (x,y) \longrightarrow \lambda(x,y) \ = \ \lambda(x)(y)$$

which is additive in each variable, and such that

$$\lambda(ax,by) \ = \ b\lambda(x,y)\overline{a} \in A \quad (a,b \in A, x,y \in L) \ .$$

(ii) An asymmetric form (L,λ) over A is (Λ,ϵ)-*nonsingular* if the $A[z,z^{-1}]$-module morphism

$$\lambda + \epsilon\lambda^* \ : \ L[z,z^{-1}] \longrightarrow L^*[z,z^{-1}]$$

is a Λ-isomorphism.

(iii) A *lagrangian* of a (Λ,ϵ)-nonsingular asymmetric form (L,λ) over A is a submodule $K \subset L$ such that:

(a) $\lambda(x,y) = 0 \in A$ for all $x \in K$, $y \in L$,
(b) $\lambda(x,y) = 0 \in A$ for all $x \in L$, $y \in K$,
(c) $\Lambda \otimes_{A[z,z^{-1}]} K[z,z^{-1}]$ is a f.g. projective lagrangian of the nonsingular ϵ-symmetric form $(\Lambda \otimes_{A[z,z^{-1}]} L[z,z^{-1}], \lambda + \epsilon\lambda^*)$ over Λ.

A form is *metabolic* if it has a lagrangian.

(iv) The (Λ,ϵ)-*asymmetric form Witt group of* A is the group of equivalence classes of (Λ,ϵ)-nonsingular asymmetric form over A, with $(L,\lambda) \sim (L',\lambda')$ if there exists an isomorphism

$$(L,\lambda) \oplus (M,\mu) \cong (L',\lambda') \oplus (M',\mu')$$

for some metabolic forms $(M,\mu),(M',\mu')$. □

Asymmetric forms (L,λ) which are nonsingular in the sense that $\lambda : L \longrightarrow L^*$ is an A-module isomorphism will appear in Chap. 28.

Proposition 26.4 *The 0-dimensional (Λ,ϵ)-asymmetric L-group of A $LAsy^0_q(A,\Lambda,\epsilon)$ $(q = p,h,s)$ is isomorphic to the Witt group of (Λ,ϵ)-nonsingular asymmetric forms (L,λ) over A with L f.g. projective (resp. f.g. free, resp. based f.g. free with $\tau = 0$) for $q = p$ (resp. h, resp. s).* □

Example 26.5 (i) Let Ω be the set of Fredholm matrices in $A[z,z^{-1}]$ (13.3), and let

$$\mathcal{F} \,:\, A[z,z^{-1}] \longrightarrow \Lambda \,=\, \Omega^{-1}A[z,z^{-1}]$$

be the natural injection. An n-dimensional asymmetric complex (C,λ) over A is (Λ,ϵ)-Poincaré if and only if the $A[z,z^{-1}]$-module chain complex

$$\mathcal{C}((1+T_\epsilon)\lambda : C^{n-*}[z,z^{-1}] \longrightarrow C[z,z^{-1}])$$

is A-finitely dominated. See Chap. 28 below for a more detailed account of the (Ω,ϵ)-asymmetric L-groups $LAsy^*(A,\Omega,\epsilon)$, which will be applied to the bordism groups of automorphisms of manifolds and to open book decompositions in Chap. 29 below.
(ii) Let

$$\mathcal{F} \,:\, A[z,z^{-1}] \longrightarrow \Lambda \,=\, A$$

be the natural projection, so that the localization $\Pi^{-1}A[z,z^{-1}]$ inverting the set Π of A-invertible matrices in $A[z,z^{-1}]$ is defined as in 10.18. An n-dimensional asymmetric complex (C,λ) over A is (Π,z)-Poincaré if and only if the A-module chain map $(1+T)\lambda : C^{n-*} \longrightarrow C$ is a chain equivalence. See Chap. 33 for a more detailed account of the (Π,ϵ)-asymmetric L-groups $LAsy^*(A,\Pi,\epsilon)$, and the applications to knot theory. □

Proposition 26.6 (i) *The projective and free (Λ,ϵ)-asymmetric L-groups are related by an exact sequence*

$$\begin{aligned}\dots \longrightarrow \widehat{H}^{n+1}(\mathbb{Z}_2;\widetilde{K}_0(A)) \longrightarrow LAsy^n_h(A,\Lambda,\epsilon) \longrightarrow LAsy^n_p(A,\Lambda,\epsilon)\\ \longrightarrow \widehat{H}^n(\mathbb{Z}_2;\widetilde{K}_0(A)) \longrightarrow LAsy^{n-1}_h(A,\Lambda,\epsilon) \longrightarrow \dots \,.\end{aligned}$$

(ii) *The free and simple (Λ,ϵ)-asymmetric L-groups are related by an exact sequence*

$$\begin{aligned}\dots \longrightarrow \widehat{H}^{n+1}(\mathbb{Z}_2;Wh_1(\Lambda)) \longrightarrow LAsy^n_s(A,\Lambda,\epsilon) \longrightarrow LAsy^n_h(A,\Lambda,\epsilon)\\ \longrightarrow \widehat{H}^n(\mathbb{Z}_2;Wh_1(\Lambda)) \longrightarrow LAsy^{n-1}_s(A,\Lambda,\epsilon) \longrightarrow \dots \,.\end{aligned}$$

Proof As for the symmetric L-groups in Ranicki [235, Chap. 10]. □

Definition 26.7 (i) For any A-modules M,N there is defined a natural injection of $\mathbb{Z}[z,z^{-1}]$-modules

$$\begin{aligned}&\mathrm{Hom}_A(M,N)[z,z^{-1}] \longrightarrow \mathrm{Hom}_{A[z,z^{-1}]}(M[z,z^{-1}],N[z,z^{-1}])\,;\\ &\sum_{j=-\infty}^{\infty} z^j f_j \longrightarrow \left(\sum_{k=-\infty}^{\infty} z^k x_k \longrightarrow \sum_{j=-\infty}^{\infty}\sum_{k=-\infty}^{\infty} z^{j+k} f_j(x_k)\right).\end{aligned}$$

(ii) An $A[z,z^{-1}]$-module morphism $f : M[z,z^{-1}] \longrightarrow N[z,z^{-1}]$ is *positive* if it is in the image of the natural $\mathbb{Z}[z]$-module morphism

$$\mathrm{Hom}_A(M,N)[z] \longrightarrow \mathrm{Hom}_{A[z,z^{-1}]}(M[z,z^{-1}],N[z,z^{-1}]) ;$$
$$\sum_{j=0}^{\infty} z^j f_j \longrightarrow \Big(\sum_{k=-\infty}^{\infty} z^k x_k \longrightarrow \sum_{j=0}^{\infty}\sum_{k=-\infty}^{\infty} z^{j+k} f_j(x_k) \Big) .$$

Let

$$\begin{aligned}&\mathrm{Hom}^+_{A[z,z^{-1}]}(M[z,z^{-1}],N[z,z^{-1}]) \\ &= \mathrm{im}(\mathrm{Hom}_A(M,N)[z]) \subseteq \mathrm{Hom}_{A[z,z^{-1}]}(M[z,z^{-1}],N[z,z^{-1}])\end{aligned}$$

be the $\mathbb{Z}[z]$-module of positive $A[z,z^{-1}]$-module morphisms $f : M[z,z^{-1}] \longrightarrow N[z,z^{-1}]$.

(iii) An $A[z,z^{-1}]$-module morphism $f : M[z,z^{-1}] \longrightarrow N[z,z^{-1}]$ is *negative* if it is in the image of the natural $\mathbb{Z}[z^{-1}]$-module injection

$$z^{-1}\mathrm{Hom}_A(M,N)[z^{-1}] \longrightarrow \mathrm{Hom}_{A[z,z^{-1}]}(M[z,z^{-1}],N[z,z^{-1}]) ;$$
$$\sum_{j=-\infty}^{-1} z^j f_j \longrightarrow \Big(\sum_{k=-\infty}^{\infty} z^k x_k \longrightarrow \sum_{j=-\infty}^{-1}\sum_{k=-\infty}^{\infty} z^{j+k} f_j(x_k) \Big) .$$

Let

$$\begin{aligned}&\mathrm{Hom}^-_{A[z,z^{-1}]}(M[z,z^{-1}],N[z,z^{-1}]) \\ &= \mathrm{im}(z^{-1}\mathrm{Hom}_A(M,N)[z^{-1}]) \subseteq \mathrm{Hom}_{A[z,z^{-1}]}(M[z,z^{-1}],N[z,z^{-1}])\end{aligned}$$

be the $\mathbb{Z}[z^{-1}]$-module of negative $A[z,z^{-1}]$-module morphisms $f : M[z,z^{-1}] \longrightarrow N[z,z^{-1}]$. □

Proposition 26.8 *If M is a f.g. projective A-module then the morphism in 26.7 (i) is an isomorphism, allowing the identifications*

$$\begin{aligned}\mathrm{Hom}_{A[z,z^{-1}]}(M[z,z^{-1}],N[z,z^{-1}]) &= \mathrm{Hom}_A(M,N)[z,z^{-1}] ,\\ \mathrm{Hom}^+_{A[z,z^{-1}]}(M[z,z^{-1}],N[z,z^{-1}]) &= \mathrm{Hom}_A(M,N)[z] ,\\ \mathrm{Hom}^-_{A[z,z^{-1}]}(M[z,z^{-1}],N[z,z^{-1}]) &= z^{-1}\mathrm{Hom}_A(M,N)[z^{-1}] ,\end{aligned}$$
$$\begin{aligned}&\mathrm{Hom}_{A[z,z^{-1}]}(M[z,z^{-1}],N[z,z^{-1}]) \\ &= \mathrm{Hom}^+_{A[z,z^{-1}]}(M[z,z^{-1}],N[z,z^{-1}]) \oplus \mathrm{Hom}^-_{A[z,z^{-1}]}(M[z,z^{-1}],N[z,z^{-1}]) .\end{aligned}$$

Every $A[z,z^{-1}]$-module morphism $f : M[z,z^{-1}] \longrightarrow N[z,z^{-1}]$ has a unique expression as a Laurent polynomial

$$f = \sum_{j=-\infty}^{\infty} z^j f_j \quad (f_j \in \mathrm{Hom}_A(M,N))$$

and hence a unique decomposition as a sum of a positive and a negative morphism

$$f = f^+ + f^- : M[z,z^{-1}] \longrightarrow N[z,z^{-1}]$$

with

$$f^+ = \sum_{j=0}^{\infty} z^j f_j \ , \ f^- = \sum_{j=-\infty}^{-1} z^j f_j \ . \qquad \square$$

Definition 26.9 The *reduced asymmetric L-groups* $\widetilde{LAsy}_h^*(A,\Lambda,\epsilon)$ are the relative groups in the exact sequence

$$\dots \longrightarrow L_h^n(A,\epsilon) \longrightarrow LAsy_h^n(A,\Lambda,\epsilon) \longrightarrow \widetilde{LAsy}_h^n(A,\Lambda,\epsilon) \longrightarrow L_h^{n-1}(A,\epsilon) \longrightarrow \dots \ . \qquad \square$$

It is convenient (although potentially confusing) to have two terminologies for the asymmetric L-groups:

Terminology 26.10 For any involution-invariant set Σ_0 of square matrices in $A[z,z^{-1}]$ such that the natural morphism of rings with involution

$$A[z,z^{-1}] \longrightarrow \Lambda_0 = (\Sigma_0)^{-1}A[z,z^{-1}]$$

is injective write the asymmetric L-groups of (A,Λ_0,ϵ) as

$$LAsy_q^*(A,\Lambda_0,\epsilon) = LAsy_q^*(A,\Sigma_0,\epsilon) \ \ (q = p,h,s) \ ,$$

calling (Λ_0,ϵ)-Poincaré asymmetric complexes (resp. L-groups) the (Σ_0,ϵ)-*Poincaré asymmetric* complexes (resp. L-groups). Similarly in the reduced case

$$\widetilde{LAsy}_q^*(A,\Lambda_0,\epsilon) = \widetilde{LAsy}_q^*(A,\Sigma_0,\epsilon) \ \ (q = p,h,s) \ . \qquad \square$$

In particular,

$$LAsy_q^*(A,\Lambda,\epsilon) = LAsy_q^*(A,\Sigma,\epsilon) \ ,$$
$$\widetilde{LAsy}_q^*(A,\Lambda,\epsilon) = \widetilde{LAsy}_q^*(A,\Sigma,\epsilon) \ .$$

Proposition 26.11 *Let $\mathcal{F} : A[z,z^{-1}] \longrightarrow \Lambda$ be a locally epic morphism of rings with involution, such that the localization map $A[z,z^{-1}] \longrightarrow \Sigma^{-1}A[z,z^{-1}]$ inverting the set Σ of Λ-invertible matrices in $A[z,z^{-1}]$ is injective.*
Given a central unit $\epsilon \in A$ such that $\bar{\epsilon} = \epsilon^{-1} \in A$ write

$$\epsilon_1 = \epsilon z^{\pm 1} \in A_1 = A[z,z^{-1}] \ .$$

(i) *There are defined isomorphisms of exact sequences*

$$\begin{array}{ccccccccccc}
\cdots \longrightarrow & \widehat{H}^{n+1} & \longrightarrow & LAsy^n_h(A,\Lambda,\epsilon_1) & \longrightarrow & LAsy^n_p(A,\Lambda,\epsilon_1) & \longrightarrow & \widehat{H}^n & \longrightarrow \cdots \\
& \cong\big\downarrow & & \cong\big\downarrow & & \cong\big\downarrow & & \cong\big\downarrow & \\
\cdots \longrightarrow & L^h_n(A_1,\epsilon_1) & \longrightarrow & \Gamma^h_n(\mathcal{F},\epsilon_1) & \longrightarrow & \Gamma^h_n(\Phi_{\mathcal{F}},\epsilon_1) & \longrightarrow & L^h_{n-1}(A_1,\epsilon_1) & \longrightarrow \cdots \\
& \big\| & & \cong\big\downarrow & & \cong\big\downarrow & & \big\| & \\
\cdots \longrightarrow & L^h_n(A_1,\epsilon_1) & \longrightarrow & L^h_n(\Sigma^{-1}A_1,\epsilon_1) & \xrightarrow{\partial} & L^h_n(A_1,\Sigma,\epsilon_1) & \longrightarrow & L^h_{n-1}(A_1,\epsilon_1) & \longrightarrow \cdots \\
& \cong\big\downarrow 1+T_{\epsilon_1} & & \cong\big\downarrow 1+T_{\epsilon_1} & & \cong\big\downarrow 1+T_{\epsilon_1} & & \cong\big\downarrow 1+T_{\epsilon_1} & \\
\cdots \longrightarrow & L_h^n(A_1,\epsilon_1) & \longrightarrow & L_h^n(\Sigma^{-1}A_1,\epsilon_1) & \xrightarrow{\partial} & L_h^n(A_1,\Sigma,\epsilon_1) & \longrightarrow & L_h^{n-1}(A_1,\epsilon_1) & \longrightarrow \cdots
\end{array}$$

with $\widehat{H}^n = \widehat{H}^n(\mathbb{Z}_2;\widetilde{K}_0(A))$*, including the isomorphisms*

$$\begin{array}{l}
LAsy^n_h(A,\Lambda,\epsilon_1) \longrightarrow \Gamma^h_n(\mathcal{F},\epsilon_1) \; ; \; (C,\lambda) \longrightarrow (C[z,z^{-1}],\lambda) \; , \\
\Gamma^h_n(\mathcal{F},\epsilon_1) \longrightarrow L^n_h(\Sigma^{-1}A_1,\epsilon_1) \; ; \; (C,\psi) \longrightarrow (\Sigma^{-1}C,(1+T_{\epsilon_1})\psi) \; , \\
LAsy^n_p(A,\Lambda,\epsilon_1) \longrightarrow L^n_h(A_1,\Lambda,\epsilon_1) \; ; \; (C,\lambda) \longrightarrow \partial(C[z,z^{-1}],(1+T_{\epsilon_1})\lambda) \; .
\end{array}$$

Thus up to isomorphism

$$\begin{array}{lllllll}
L^h_n(A_1,\epsilon_1) & = & L^n_h(A_1,\epsilon_1) & = & \widehat{H}^{n+1}(\mathbb{Z}_2;\widetilde{K}_0(A)) \; , & & \\
\Gamma^h_n(\mathcal{F},\epsilon_1) & = & L^h_n(\Sigma^{-1}A_1,\epsilon_1) & = & L^n_h(\Sigma^{-1}A_1,\epsilon_1) & = & LAsy^n_h(A,\Lambda,\epsilon_1) \; , \\
\Gamma^h_n(\Phi_{\mathcal{F}},\epsilon_1) & = & L^h_n(A_1,\Lambda,\epsilon_1) & = & L^n_h(A_1,\Lambda,\epsilon_1) & = & LAsy^n_p(A,\Lambda,\epsilon_1) \; .
\end{array}$$

The simple asymmetric L-groups are such that there are defined isomorphisms

$$LAsy^n_s(A,\Lambda,\epsilon_1) \longrightarrow \Gamma^s_n(\mathcal{F},\epsilon_1) \; ; \; (C,\lambda) \longrightarrow (C[z,z^{-1}],\lambda) \; .$$

(ii) *The various asymmetric L-groups are 4-periodic*

$$LAsy^n_q(A,\Lambda,\epsilon_1) \;=\; LAsy^{n+2}_q(A,\Lambda,-\epsilon_1) \;=\; LAsy^{n+4}_q(A,\Lambda,\epsilon_1) \;\; (q=p,h,s) \; .$$

(iii) *If $\Sigma = \Sigma_1 \cup \Sigma_2$ for involution-invariant sets of square matrices Σ_1, Σ_2 in A which are coprime (9.16) then for any central unit $\eta \in A[z,z^{-1}]$ such that $\overline{\eta} = \eta^{-1}$ the projective asymmetric L-groups split as*

$$\begin{array}{lll}
LAsy^n_p(A,\Lambda,\eta) & = & LAsy^n_p(A,\Sigma,\eta) \\
& = & LAsy^n_p(A,\Sigma_1,\eta) \oplus LAsy^n_p(A,\Sigma_2,\eta) \; .
\end{array}$$

For $\eta = \epsilon_1$ the free asymmetric L-groups fit into a commutative braid of exact sequences

$$\begin{array}{ccccc}
LAsy_p^{n+1}(A,\Sigma_2,\epsilon_1) & & LAsy_h^n(A,\Sigma_1,\epsilon_1) & & LAsy_p^n(A,\Sigma_1,\epsilon_1) \\
& \searrow \quad \nearrow & & \searrow \quad \nearrow & \\
& \widehat{H}^{n+1}(\mathbb{Z}_2;\widetilde{K}_0(A)) & & LAsy_h^n(A,\Sigma,\epsilon_1) & \\
& \nearrow \quad \searrow & & \nearrow \quad \searrow & \\
LAsy_p^{n+1}(A,\Sigma_1,\epsilon_1) & & LAsy_h^n(A,\Sigma_2,\epsilon_1) & & LAsy_p^n(A,\Sigma_2,\epsilon_1)
\end{array}$$

(iv) *If $1-z\in\Sigma$ there is defined an isomorphism*

$$L_h^n(\Sigma^{-1}A_1,\epsilon) \xrightarrow{\simeq} L_h^n(\Sigma^{-1}A_1,-\epsilon_1)\ ;\ \Sigma^{-1}(E,\theta)\longrightarrow \Sigma^{-1}(E,(1-z^{\mp 1})\theta)\ ,$$

and the localization exact sequence for the ϵ-symmetric L-theory of $\Sigma^{-1}A_1$ is such that there is defined an isomorphism

$$\begin{array}{ccccc}
L_h^n(A_1,\epsilon) & \longrightarrow & L_h^n(\Sigma^{-1}A_1,\epsilon) & \xrightarrow{\partial} & L_h^n(A_1,\Sigma,\epsilon) \\
\downarrow{\cong} & & \downarrow{\cong} & & \downarrow{\cong} \\
L_h^n(A,\epsilon)\oplus L_p^{n-1}(A,\epsilon) & \rightarrow & LAsy_h^n(A,\Sigma,\epsilon) & \rightarrow & \widetilde{LAsy}_h^n(A,\Sigma,\epsilon)\oplus L_p^{n-2}(A,\epsilon)
\end{array}$$

with

$$L_h^n(A,\epsilon)\oplus L_p^{n-1}(A,\epsilon)\longrightarrow LAsy_h^n(A,\Sigma,\epsilon)\ ;\ ((C,\phi),(C',\phi'))\longrightarrow (C,\phi_0)\ ,$$

so that

$$\begin{aligned}
L_h^n(\Sigma^{-1}A_1,\epsilon) &= L_h^n(\Sigma^{-1}A_1,-\epsilon_1) = LAsy_h^n(A,\Lambda,-\epsilon_1) \\
&= LAsy_h^n(A,\Lambda,\epsilon) = LAsy_h^n(A,\Sigma,\epsilon)\ .
\end{aligned}$$

Similarly for ϵ-quadratic L-theory, with an isomorphism

$$\begin{array}{ccccc}
L_n^h(A_1,\epsilon) & \longrightarrow & L_n^h(\Sigma^{-1}A_1,\epsilon) & \xrightarrow{\partial} & L_n^h(A_1,\Sigma,\epsilon) \\
\downarrow{\cong} & & \downarrow{\cong} & & \downarrow{\cong} \\
L_n^h(A,\epsilon)\oplus L_{n-1}^p(A,\epsilon) & \rightarrow & LAsy_h^n(A,\Lambda,-\epsilon_1) & \rightarrow & \widetilde{LAsy}_h^n(A,\Lambda,-\epsilon_1)\oplus L_{n-2}^p(A,\epsilon)
\end{array}$$

(v) *If Σ is coprime to $1-z$ there are defined isomorphisms*

$$L_h^n(A_1,\Sigma,\epsilon_1) \xrightarrow{\simeq} L_h^n(A_1,\Sigma,-\epsilon)\ ;\ (E,\theta)\longrightarrow (E,(1-z^{\mp 1})\theta)$$

and

$$\begin{aligned}
L_h^n(A_1,\Sigma,-\epsilon) &= L_h^n(A_1,\Sigma,\epsilon_1) \\
&= LAsy_p^n(A,\Lambda,\epsilon_1) = LAsy_p^n(A,\Sigma,\epsilon_1)\ .
\end{aligned}$$

Let

$$A_2 \;=\; ((1-z)^\infty)^{-1}A_1 \;=\; A[z,z^{-1},(1-z)^{-1}]\ .$$

The commutative braid of exact sequences given by 25.24

$$\begin{array}{ccccc}
L_h^{n+1}(A_1,(1-z)^\infty,\epsilon_1) & & L_h^n(\Sigma^{-1}A_1,\epsilon_1) & & L_h^n(A_1,\Sigma,\epsilon_1) \\
& \searrow \quad \nearrow & & \searrow \quad \nearrow & \\
& L_h^n(A_1,\epsilon_1) & & L_h^n(\Sigma^{-1}A_2,\epsilon_1) & \\
& \nearrow \quad \searrow & & \nearrow \quad \searrow & \\
L_h^{n+1}(A_1,\Sigma,\epsilon_1) & & L_h^n(A_2,\epsilon_1) & & L_h^n(A_1,(1-z)^\infty,\epsilon_1)
\end{array}$$

is isomorphic to

$$\begin{array}{ccccc}
LAsy_p^{n+1}(A_1,(1-z)^\infty,\epsilon_1) & & LAsy_h^n(A,\Sigma,\epsilon_1) & & LAsy_p^n(A_1,\Sigma,\epsilon_1) \\
& \searrow \quad \nearrow & & \searrow \quad \nearrow & \\
& \widehat{H}^{n+1}(\mathbb{Z}_2;\widetilde{K}_0(A)) & & LAsy_h^n(A,\Sigma(1-z)^\infty,\epsilon_1) & \\
& \nearrow \quad \searrow & & \nearrow \quad \searrow & \\
LAsy_p^{n+1}(A_1,\Sigma,\epsilon_1) & & LAsy_h^{n+1}(A,(1-z)^\infty,\epsilon_1) & & LAsy_p^n(A,(1-z)^\infty,\epsilon_1)
\end{array}$$

If A is such that $\widetilde{K}_0(A)=0$ then

$$\begin{aligned}
L_h^n(\Sigma^{-1}A_2,\epsilon_1) &= L_h^n(\Sigma^{-1}A_2,-\epsilon) \\
&= L_h^n(A_2,-\epsilon)\oplus L_h^n(A_1,\Sigma,-\epsilon) \\
&= LAsy_h^n(A,(1-z)^\infty,\epsilon_1)\oplus LAsy_p^n(A,\Sigma,\epsilon_1)\ .
\end{aligned}$$

Proof For definiteness, take $\epsilon_1 = \epsilon z^{-1}$.
(i)+(ii) Let E be a finite based f.g. free $A[z,z^{-1}]$-module chain complex. By 7.3 there exists a Mayer–Vietoris presentation (E^+,E^-). In particular, there exists a finite based f.g. free $A[z]$-module subcomplex $E^+\subset E$ such that

$$A[z,z^{-1}]\otimes_{A[z]}E^+ \;=\; E\ .$$

Let $F_r\subset E_r^+$ be the A-submodule generated by the $A[z]$-module basis, so that

$$E_r^+ \;=\; F_r[z]\ ,\quad E_r \;=\; F_r[z,z^{-1}]\ .$$

Every element

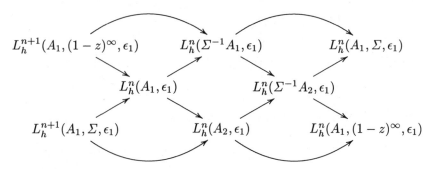

$$\theta \in (E \otimes_{A[z,z^{-1}]} E)_n \;=\; \mathrm{Hom}_{A[z,z^{-1}]}(E^*, E)_n$$

is represented by $A[z,z^{-1}]$-module morphisms

$$\theta \;:\; E^r \;=\; F^r[z,z^{-1}] \longrightarrow E_{n-r} \;=\; F_{n-r}[z,z^{-1}] \;\; (r \in \mathbb{Z})\ .$$

Let $(E \otimes_{A[z,z^{-1}]} E)^+$ (resp. $(E \otimes_{A[z,z^{-1}]} E)^-$) be the $\mathbb{Z}$-module subcomplex of $E \otimes_{A[z,z^{-1}]} E$ consisting of the elements θ for which these $A[z,z^{-1}]$-module morphisms are positive (resp. negative), so that by 26.8

$$E \otimes_{A[z,z^{-1}]} E \;=\; (E \otimes_{A[z,z^{-1}]} E)^+ \oplus (E \otimes_{A[z,z^{-1}]} E)^-\ .$$

In particular, for $E = A[z,z^{-1}]$

$$A[z,z^{-1}] \;=\; A[z,z^{-1}]^+ \oplus A[z,z^{-1}]^-$$

with $A[z,z^{-1}]^+ = A[z]$, $A[z,z^{-1}]^- = z^{-1}A[z^{-1}]$. Since $\epsilon_1 = \epsilon z^{-1}$ the ϵ_1-transposition involution

$$T_{\epsilon_1} \;:\; A[z,z^{-1}] \longrightarrow A[z,z^{-1}]\ ;\ a \;=\; \sum_{j=-\infty}^{\infty} a_j z^j \longrightarrow \epsilon_1 \overline{a} \;=\; \sum_{j=-\infty}^{\infty} \epsilon \overline{a}_{-j-1} z^j$$

is such that

$$T_{\epsilon_1}(A[z,z^{-1}]^{\pm}) \;=\; A[z,z^{-1}]^{\mp}$$

and $(A[z,z^{-1}], T_{\epsilon_1})$ is a free $\mathbb{Z}[\mathbb{Z}_2]$-module. Similarly, the ϵ_1-transposition involution

$$T_{\epsilon_1} \;:\; E \otimes_{A[z,z^{-1}]} E \longrightarrow E \otimes_{A[z,z^{-1}]} E$$

is such that

$$T_{\epsilon_1}(E \otimes_{A[z,z^{-1}]} E)^{\pm} \;=\; (E \otimes_{A[z,z^{-1}]} E)^{\mp}\ .$$

Thus $(E \otimes_{A[z,z^{-1}]} E, T_{\epsilon_1})$ is a free $\mathbb{Z}[\mathbb{Z}_2]$-module chain complex and

$$Q_n(E,\epsilon_1) \;=\; Q^n(E,\epsilon_1) \;=\; H_n((E \otimes_{A[z,z^{-1}]} E)^+)\ .$$

More precisely, a positive chain map $\theta : E^{n-*} \longrightarrow E$ determines an n-dimensional ϵ_1-quadratic structure $\psi \in (W_\% E)_n$ with

$$\psi_s \;=\; \begin{cases} \theta & \text{if } s = 0 \\ 0 & \text{if } s \geq 1 \end{cases}$$

and the $\mathbb{Z}$-module morphisms

$$H_n((E \otimes_{A[z,z^{-1}]} E)^+) \longrightarrow Q_n(E,\epsilon_1)\ ;\ \theta \longrightarrow \psi\ ,$$
$$1 + T_{\epsilon_1} \;:\; Q_n(E,\epsilon_1) \longrightarrow Q^n(E,\epsilon_1)$$

are isomorphisms. Similarly for L-groups.

Every element of $L^n_h(\Sigma^{-1}A_1, \epsilon_1)$ is represented by $\Sigma^{-1}(E, (1+T_{\epsilon_1})\theta)$ for some $\theta \in H_n((E \otimes_{A[z,z^{-1}} E)^+)$ such that the $A[z,z^{-1}]$-module chain map

$1+T_{\epsilon_1} : E^{n-*}\longrightarrow E$ is a $\Sigma^{-1}A[z,z^{-1}]$-equivalence. It will be proved that the morphism

$$LAsy_h^n(A,\Lambda,\epsilon_1) \longrightarrow L_h^n(\Sigma^{-1}A_1,\epsilon_1) ~;~ (B,\lambda) \longrightarrow (\Sigma^{-1}B[z,z^{-1}],(1+T_{\epsilon_1})\lambda)$$

is an isomorphism by constructing an explicit inverse, using the algebraic transversality technique of Ranicki [244, Chap. 16], as follows. Given finite based f.g. free A-module chain complexes C,D and A-module chain maps $f,g: C\longrightarrow D$ define the A-module chain map

$$\Gamma_{f,g} ~=~ f\otimes f - g\otimes g ~:~ C\otimes_A C \longrightarrow D\otimes_A D~,$$

the finite based f.g. free $A[z,z^{-1}]$-module chain complex

$$E ~=~ \mathcal{C}(f-zg: C[z,z^{-1}]\longrightarrow D[z,z^{-1}])~,$$

and the *union* $\mathbb{Z}$-module chain map

$$U ~:~ \mathcal{C}(\Gamma_{f,g}) \longrightarrow (E\otimes_{A[z,z^{-1}]} E)^+ ~;~ (\delta\mu,\mu) \longrightarrow \theta$$

with

$$\begin{aligned}\theta ~=~ \begin{pmatrix} \delta\mu & zg\mu \\ \mu f^* & 0 \end{pmatrix} ~:~ \\ E^{n-r} ~=~ (D^{n-r}\oplus C^{n-r-1})[z,z^{-1}] \longrightarrow E_r ~=~ (D_r\oplus C_{r-1})[z,z^{-1}] \\ ((\delta\mu,\mu)\in \mathcal{C}(\Gamma_{f,g})_n)~.\end{aligned}$$

Working as in [244, Chap. 16] it can be shown that every n-dimensional ϵ_1-symmetric complex over $A[z,z^{-1}]$ is homotopy equivalent to the union (E,θ) of some n-dimensional asymmetric pair over A

$$((f~g): C\oplus C\longrightarrow D, (\delta\mu,\mu\oplus -\mu))$$

with

$$\theta ~=~ U(\delta\mu,\mu)\in Q^n(E,\epsilon_1) ~=~ H_n((E\otimes_{A[z,z^{-1}]} E)^+)~.$$

Define the n-dimensional asymmetric complex over A

$$(D',\delta\mu') ~=~ (\mathcal{C}(f\mu^* - g\mu : C^{n-*-1}\longrightarrow D), \begin{pmatrix} \delta\mu & f \\ g^* & 0 \end{pmatrix})~.$$

Use the $(n+1)$-dimensional ϵ_1-symmetric pair $(p: E\longrightarrow SC[z,z^{-1}],(0,\theta))$ (p = projection) over $A[z,z^{-1}]$ to perform algebraic surgery on (E,θ), with trace an $(n+1)$-dimensional ϵ_1-symmetric $\Sigma^{-1}A[z,z^{-1}]$-Poincaré pair over $A[z,z^{-1}]$ $(E\oplus E'\longrightarrow \delta E,(0,\theta\oplus -\theta'))$ with $\delta E = \mathcal{C}(C^{n-*}[z,z^{-1}]\longrightarrow E)$ and

$$(E',\theta') ~=~ (D'[z,z^{-1}],(1+T_{\epsilon_1})\delta\mu')~.$$

The function

$$L_h^n(\Sigma^{-1}A_1,\epsilon_1) \longrightarrow \mathrm{LAsy}_h^n(A,\Sigma,\epsilon_1)\ ;$$
$$\Sigma^{-1}(E,\theta) = \Sigma^{-1}(E',\theta') \longrightarrow (D',\delta\mu')$$

is the inverse isomorphism.

In particular, taking $\Sigma = \{1\}$ there are defined isomorphisms

$$\mathrm{LAsy}_h^n(A,\{1\},\epsilon_1) \longrightarrow L_h^n(A_1,\epsilon_1)\ ;\ (B,\lambda) \longrightarrow (B[z,z^{-1}],(1+T_{\epsilon_1})\lambda)\ .$$

For any n-dimensional f.g. free asymmetric $(\{1\},\epsilon_1)$-Poincaré complex (B,λ) over A the $A[z,z^{-1}]$-module chain map

$$(1+T_{\epsilon_1})\lambda\ :\ B^{n-*}[z,z^{-1}] \longrightarrow B[z,z^{-1}]$$

is a chain equivalence, and the A-module chain complexes defined by

$$P = \mathcal{C}((1+T_{\epsilon_1})\lambda : zB^{n-*}[z] \longrightarrow B[z])\ ,$$
$$Q = \mathcal{C}((1+T_{\epsilon_1})\lambda : B^{n-*}[z^{-1}] \longrightarrow B[z^{-1}])$$

are such that

$$\lambda \simeq \begin{pmatrix} 0 & 0 \\ \mu & 0 \end{pmatrix}\ :\ B^{n-*} \simeq P^{n-*}\oplus Q^{n-*} \longrightarrow B \simeq P\oplus Q$$

for a chain equivalence $\mu : P^{n-*} \longrightarrow Q$. The morphisms

$$\mathrm{LAsy}_h^n(A,\{1\},\epsilon_1) \longrightarrow \widehat{H}^{n+1}(\mathbb{Z}_2;\widetilde{K}_0(A))\ ;\ (B,\lambda) \longrightarrow [P]\ ,$$
$$\widehat{H}^{n+1}(\mathbb{Z}_2;\widetilde{K}_0(A)) \longrightarrow \mathrm{LAsy}_h^n(A,\{1\},\epsilon_1)\ ;\ [P] \longrightarrow (P\oplus P^{n-*},\begin{pmatrix} 0 & 0 \\ 1 & 0 \end{pmatrix})$$

are thus inverse isomorphisms, allowing the identification

$$\mathrm{LAsy}_h^n(A,\{1\},\epsilon_1) = L_h^n(A_1,\epsilon_1) = \widehat{H}^{n+1}(\mathbb{Z}_2;\widetilde{K}_0(A))\ .$$

The morphisms

$$\mathrm{LAsy}_p^n(A,\Sigma,\epsilon_1) \longrightarrow L_h^n(A_1,\Sigma,\epsilon_1)\ ;\ (B,\lambda) \longrightarrow \partial(B[z,z^{-1}],(1+T_{\epsilon_1})\lambda)$$

fit into a morphism of exact sequences

$$\begin{array}{ccccccccc}
\longrightarrow & \widehat{H}^{n+1} & \longrightarrow & \mathrm{LAsy}_h^n(A,\Sigma,\epsilon_1) & \longrightarrow & \mathrm{LAsy}_p^n(A,\Sigma,\epsilon_1) & \longrightarrow & \widehat{H}^n & \longrightarrow \\
& \downarrow\cong & & \downarrow\cong & & \downarrow & & \downarrow\cong & \\
\longrightarrow & L_h^n(A_1,\epsilon_1) & \longrightarrow & L_h^n(\Sigma^{-1}A_1,\epsilon_1) & \xrightarrow{\partial} & L_h^n(A_1,\Sigma,\epsilon_1) & \longrightarrow & L_h^{n-1}(A_1,\epsilon_1) & \longrightarrow
\end{array}$$

It now follows from the 5-lemma that there is also an identification

$$\mathrm{LAsy}_p^n(A,\Sigma,\epsilon_1) = L_h^n(A_1,\Sigma,\epsilon_1)\ .$$

(iii) If C is a finite f.g. projective A-module chain complex which is $\Sigma^{-1}A_1$-contractible then the induced $(\Sigma_i)^{-1}A_1$-module chain complexes

$$C(i) = (\Sigma_i)^{-1}C[z,z^{-1}] \quad (i=1,2)$$

are such that the natural A_1-module chain map

$$C[z,z^{-1}] \longrightarrow C(1) \oplus C(2)$$

is a chain equivalence. Thus $C(1)$ (resp. $C(2)$) is A_1-module chain equivalent to a finite f.g. projective A_1-module chain complex which is $(\Sigma_2)^{-1}A_1$- (resp. $(\Sigma_1)^{-1}A_1$)-contractible, and

$$\begin{aligned} Q^n(C[z,z^{-1}],\eta) &= Q^n(C(1),\eta) \oplus Q^n(C(2),\eta) \oplus H_n(C(1) \otimes_{A_1} C(2)) \\ &= Q^n(C(1),\eta) \oplus Q^n(C(2),\eta) \ . \end{aligned}$$

The commutative exact braid for $\eta = \epsilon_1$ is obtained by using (i) to express the commutative exact braid of 25.24

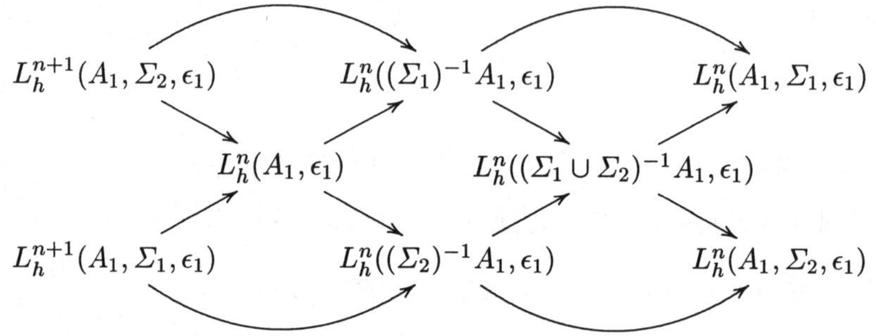

in terms of the asymmetric L-groups.
(iv) If $1-z \in \Sigma$ then

$$u = 1-z \in \Sigma^{-1}A_1$$

is a central unit such that

$$\overline{u} = -z^{-1}u \ , \quad -\epsilon_1 = \frac{\overline{u}}{u}.\epsilon \in A_1 \ .$$

(v) Σ is coprime to $1-z$ if and only if the augmentation $A[z,z^{-1}]\longrightarrow A; z\longrightarrow 1$ sends every matrix in Σ to an invertible matrix in A, in which case $1-z$: $E\longrightarrow E$ is an $A[z,z^{-1}]$-module chain equivalence for every $\Sigma^{-1}A[z,z^{-1}]$-contractible finite f.g. free $A[z,z^{-1}]$-module chain complex E and the morphisms

$$1-z \ : \ Q^n(E,\epsilon_1) \longrightarrow Q^n(E,-\epsilon) \ ; \ \theta \longrightarrow (1-z)\theta$$

are isomorphisms. □

Remark 26.12 (i) The identifications of 26.11

$$L_n^h(A[z,z^{-1}],z) = \widehat{H}^{n+1}(\mathbb{Z}_2;\widetilde{K}_0(A))$$

are analogues of the identifications of the codimension 2 surgery obstruction LN_*

$$LN_n = L_n^s(A[z,z^{-1}],z) = \widehat{H}^{n+1}(\mathbb{Z}_2;\widetilde{K}_1(A))$$

obtained geometrically for a group ring $A = \mathbb{Z}[\pi]$ by Wall [302, 13A.10] and algebraically by Ranicki [237, Prop. 7.8.14] for any ring with involution A. The other identifications of 26.11 are also motivated by codimension 2 surgery obstruction theory, as described in Chap. 27 below.
(ii) Let $\Sigma = \Omega$ be the set of Fredholm matrices in $A[z,z^{-1}]$. In particular, $1-z \in \Omega$, so that by 26.11 (iii)

$$L_h^n(\Omega^{-1}A[z,z^{-1}]) = L_h^n(\Omega^{-1}A[z,z^{-1}],-z) = LAsy_h^n(A,\Omega,-z) .$$

In Chap. 28 the asymmetric L-groups $LAsy_h^n(A,\Omega,-z)$ will be shown to be 2-periodic in n, being the Witt group of nonsingular asymmetric forms over A if n is even, and 0 if n is odd.
(iii) Let $\Sigma = (1-z)^\infty$, so that

$$\Sigma^{-1}A[z,z^{-1}] = A[z,z^{-1},(1-z)^{-1}]$$

and by 26.11 (iii)

$$L_h^n(A[z,z^{-1},(1-z)^{-1}],\epsilon) = LAsy_h^n(A,\Sigma,-\epsilon z)$$

for any central unit $\epsilon \in A$ with $\overline{\epsilon} = \epsilon^{-1} \in A$. The algebraic L-theory of $A[z,z^{-1},(1-z)^{-1}]$ will be studied in detail in Chap. 36, and the asymmetric $(\Sigma,-\epsilon z)$-Poincaré L-groups $LAsy_h^*(A,\Sigma,-\epsilon z)$ will be identified there with the 'almost ϵ-symmetric' L-groups of A. □

27. Framed codimension 2 surgery

The algebraic results on asymmetric L-theory of Chap. 26 will now be applied to the codimension 2 surgery theory of Chap. 22 with trivial normal bundle $\xi = \epsilon^2$. This will be used describe the bordism of automorphisms of manifolds in Chap. 28, the algebraic theory of open books in Chap. 29, and high-dimensional knot cobordism in Chap. 33.

27A. Codimension 1 Seifert surfaces

In general, a knot $k : N^n \subset M^{n+2}$ need not be spanned by a codimension 1 Seifert surface. The following condition ensures the existence of such a spanning surface.

Definition 27.1 Let $(M, \partial M)$ be an n-dimensional manifold with boundary. A codimension 2 submanifold $(N, \partial N) \subset (M, \partial M)$ is *homology framed* if it is ϵ^2-characteristic in the sense of 22.2, that is if

$$[N] \;=\; 0 \in H_{n-2}(M, \partial M)$$

and there is given a particular identification

$$\nu_{N\subset M} \;=\; \epsilon^2 \;:\; N \longrightarrow BSO(2)$$

with an extension of the projection $S(\nu_{N\subset M}) = N \times S^1 \longrightarrow S^1$ to the *canonical projection* on the exterior

$$(p, \partial_+ p) \;:\; (P, \partial_+ P) \;=\; (\mathrm{cl.}(M \backslash N \times D^2), \mathrm{cl.}(\partial M \backslash \partial N \times D^2)) \longrightarrow S^1 \;,$$

corresponding to a lift of

$$\begin{aligned}[N] &\in \ker(H_{n-2}(N, \partial N) \longrightarrow H_{n-2}(M, \partial M)) \\ &= \mathrm{im}(H_{n-1}(M, N \times D^2 \cup \partial_+ P) \longrightarrow H_{n-2}(N, \partial N))\end{aligned}$$

to an element

$$p \in H_{n-1}(M, N \times D^2 \cup \partial_+ P) \;=\; H_{n-1}(P, \partial P) \;=\; H^1(P) \;=\; [P, S^1] \;. \;\square$$

Remark 27.2 The homology framing condition $[N] = 0 \in H_{n-2}(M, \partial M)$ is equivalent to the condition $\nu_{N \subset M} = \epsilon^2$ if $H^2(M) \longrightarrow H^2(N)$ is injective. However, in general the two conditions are not equivalent. For example,

$$N = \{\text{pt.}\} \subset M = S^2$$

has $\nu_{N \subset M} = \epsilon^2$, but it is not homology framed (and does not admit a Seifert surface) – in fact, $N \subset M$ is η-characteristic, with $\eta : M \longrightarrow BSO(2)$ the Hopf bundle, such that

$$[N] = e(\eta) = 1 \in H_0(M) = H^2(M) = \mathbb{Z} . \qquad \square$$

Definition 27.3 Let $(M, \partial M)$ be an n-dimensional manifold with boundary, and let $(N, \partial N) \subset (M, \partial M)$ be a homology framed codimension 2 submanifold with exterior $(P, \partial_+ P)$, so that

$$M = N \times D^2 \cup_{N \times S^1} P \ , \ P = \text{cl.}(M \backslash N \times D^2) \ ,$$
$$\partial M = \partial N \times D^2 \cup_{\partial N \times S^1} \partial_+ P \ , \ \partial_+ P = \text{cl.}(\partial M \backslash \partial N \times D^2) \ .$$

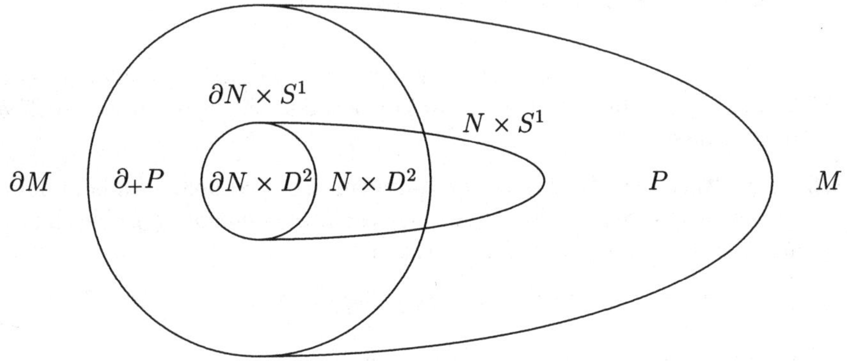

(i) The *canonical infinite cyclic cover* of the exterior $(P, \partial_+ P)$ is the infinite cyclic cover obtained from the universal cover $\mathbb{R}$ of S^1 by pullback along the canonical projection $(p, \partial_+ p) : (P, \partial_+ P) \longrightarrow S^1$

$$(\overline{P}, \overline{\partial_+ P}) = (p, \partial_+ p)^* \mathbb{R} ,$$

which restricts to $(N, \partial_+ N) \times \mathbb{R}$ over $(N, \partial N) \times S^1 \subseteq (P, \partial_+ P)$.

(ii) A *codimension 1 Seifert surface* $(F, \partial_+ F)$ for $(N, \partial N)$ is a codimension 1 submanifold $F \subset P$ such that

$$\partial F = N \cup_{\partial N} \partial_+ F \text{ with } \partial_+ F = F \cap \partial_+ P ,$$
$$\partial(\partial_+ F) = \partial N \ , \ \nu_{F \subset P} = \epsilon : F \longrightarrow BO(1) ,$$
$$[F] = p \in H_{n-1}(M, N \times D^2 \cup \partial_+ P) ,$$

with $\partial_+ F$ a codimension 1 Seifert surface for $\partial N \subset \partial M$.

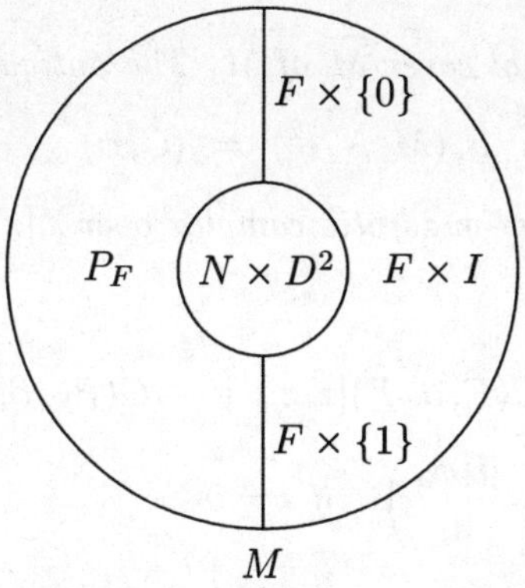

(iii) A *Seifert fundamental domain* for the canonical infinite cyclic cover $\overline{P}$ of P is the relative cobordism $(P_F; i_0F, i_1F)$ obtained by cutting P along a codimension 1 Seifert surface $F \subset P$, with

$$\begin{aligned} i_k \ &: \ F \longrightarrow F \times \{0,1\} \subset P_F \ ; \ x \longrightarrow (x,k) \ \ (k=0,1) \ , \\ P &= P_F \cup_{F\times\{0,1\}} F \times I \ . \end{aligned}$$

The relative cobordism $(\partial_+ P_{\partial_+ F}; i_0\partial_+ F, i_1\partial_+ F)$ is then a Seifert fundamental domain for the canonical infinite cyclic cover $\overline{\partial_+ P}$ of $\partial_+ P$. □

Remark 27.4 The special case $N = S^{n-2} \subset M = S^n$ of 27.3 for $n = 3$ is the classical knot theory situation in which codimension 1 Seifert surfaces first appeared (Frankl and Pontrjagin [83], Seifert [262]). □

The antiquadratic complex (22.4) of a homology framed codimension 2 submanifold has the following expression in terms of a codimension 1 Seifert surface:

Proposition 27.5 *Let $(M, \partial M)$ be an n-dimensional manifold with boundary, and let $(N, \partial N) \subset (M, \partial M)$ be a homology framed codimension 2 submanifold with exterior $(P, \partial_+ P)$.*
(i) *There exist codimension 1 Seifert surfaces $(F, \partial_+ F) \subset (M, \partial M)$ for $(N, \partial N)$, with*

$$\partial F \ = \ N \cup_{\partial N} \partial_+ F \subset \partial P \ = \ N \times S^1 \cup_{\partial N \times S^1} \partial_+ P \ .$$

(ii) *A codimension 1 Seifert surface $(F, \partial_+ F)$ for $(N, \partial N)$ determines an n-dimensional manifold triad $(P_F; i_0F, i_1F, N)$ and hence an n-dimensional symmetric Poincaré triad over $\mathbb{Z}[\pi_1(M)]$*

$$(\Gamma, \Phi) \ = \ \left(\begin{array}{ccc} C(\widetilde{N}, \widetilde{\partial N}) & \longrightarrow & C(\widetilde{F}, \widetilde{\partial_+ F}) \\ \Big\downarrow & & \Big\downarrow {\scriptstyle i_0} \\ C(\widetilde{F}, \widetilde{\partial_+ F}) & \xrightarrow{\ i_1\ } & C(\widetilde{P}_F, \widetilde{\partial_+ P}_{\partial_+ F}) \end{array} \ , \ \begin{array}{ccc} \partial\phi & \longrightarrow & \phi \\ \Big\downarrow & & \Big\downarrow \\ \phi & \longrightarrow & \delta\phi \end{array} \right)$$

with $\widetilde{N}, \widetilde{\partial N}, \widetilde{F}, \widetilde{\partial_+ F}, \widetilde{P}_F, \widetilde{\partial_+ P}_{\partial_+ F}$ the covers of $N, \partial N, F, \partial_+ F, P_F, \partial_+ P_{\partial_+ F}$

induced from the universal cover $\widetilde{M}$ *of* M. *The antiquadratic complex*

$$\sigma_*(M,N,\epsilon^2) \;=\; (C,\psi)$$

is the n*-dimensional* $(-z)$*-quadratic complex over* $\mathbb{Z}[\pi_1(M)][z,z^{-1}]$ *given by*

$$\begin{aligned} C \;&=\; C(\widetilde{P},\widetilde{\partial_+ P}) \\ &=\; \mathcal{C}(i_0 - z i_1 : C(\widetilde{F},\widetilde{\partial_+ F})[z,z^{-1}] \longrightarrow C(\widetilde{P}_F,\widetilde{\partial_+ P}_{\partial_+ F})[z,z^{-1}])\ , \end{aligned}$$

$$\psi_s \;=\; \begin{cases} \begin{pmatrix} \delta\phi_0 & z i_1 \phi_0 \\ \pm\phi_0 i_0^* & 0 \end{pmatrix} & \text{if } s = 0 \\ 0 & \text{if } s \geq 1 \end{cases}$$

(as in 24.1) and such that

$$\mathcal{C}((1+T_{-z})\psi : C^{n-*} \longrightarrow C) \;\simeq\; C(\widetilde{M}\times\mathbb{R},\widetilde{\partial_+ P})$$

with $\widetilde{P},\widetilde{\partial_+ P}$ *the pullbacks to* $P,\partial_+ P$ *of the universal cover* $\widetilde{M}\times\mathbb{R}$ *of* $M\times S^1$.
Proof Make $(p,\partial_+ p) : (P,\partial_+ P)\longrightarrow S^1$ transverse regular at $\{\text{pt.}\}\subset S^1$, with

$$(F,\partial_+ F) \;=\; (p,\partial_+ p)^{-1}(\{\text{pt.}\}) \subset (P,\partial_+ P)\ . \qquad \square$$

Here is how asymmetric complexes arise in homology framed codimension 2 surgery :

Definition 27.6 Let $(M,\partial M)$ be an n-dimensional manifold with boundary, and let $K\subset\partial M$ be a homology framed codimension 2 submanifold, with exterior L and a codimension 1 Seifert surface $J\subset\partial M$, so that

$$\partial J \;=\; K \subset \partial M \;=\; K\times D^2 \cup_{K\times S^1} L\ .$$

Let $\widetilde{M}$ be the universal cover of M, and let $\widetilde{J},\widetilde{L}_J$ be the induced covers of J, L_J, so that $(M;L_J,J\times I)$ is a relative cobordism with

$$L \;=\; L_J \cup_{J\times\{0,1\}} J\times I$$

and there is defined a $\mathbb{Z}[\pi_1(M)]$-module chain equivalence

$$[M]\cap - \;:\; C(\widetilde{M},\widetilde{L}_J)^{n-*} \;\simeq\; C(\widetilde{M},\widetilde{J}\times I)\ .$$

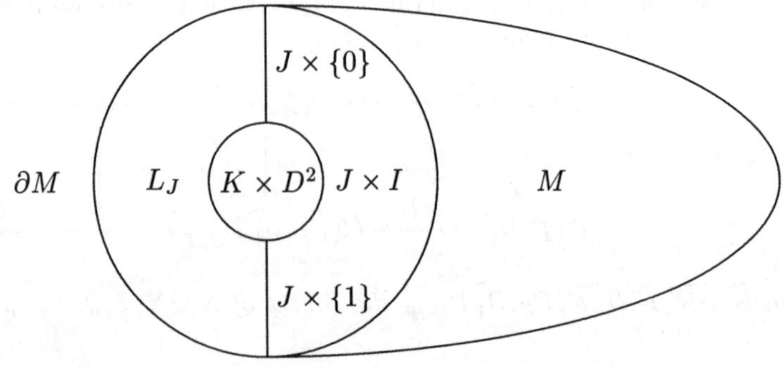

(i) The *asymmetric complex* of (M, J) is the n-dimensional asymmetric complex (C, λ) over $\mathbb{Z}[\pi_1(M)]$ with

$$\begin{aligned} C &= C(\widetilde{M}, \widetilde{L}_J) , \\ \lambda : C^{n-*} &= C(\widetilde{M}, \widetilde{L}_J)^{n-*} \xrightarrow[\simeq]{[M]\cap -} C(\widetilde{M}, \widetilde{J} \times I) \simeq C(\widetilde{M}, \widetilde{J}) \\ &\xrightarrow{\widetilde{i}_0} C(\widetilde{M}, \widetilde{L}_J) = C , \end{aligned}$$

such that up to $\mathbb{Z}[\pi_1(M)]$-module chain equivalence

$$\begin{aligned} \mathcal{C}(\lambda : C^{n-*} \longrightarrow C) &\simeq \mathcal{C}(\widetilde{i}_0 : C(\widetilde{J}) \longrightarrow C(\widetilde{L}_J))_{*-1} , \\ \mathcal{C}(T\lambda : C^{n-*} \longrightarrow C) &\simeq \mathcal{C}(\widetilde{i}_1 : C(\widetilde{J}) \longrightarrow C(\widetilde{L}_J))_{*-1} \end{aligned}$$

and up to $\mathbb{Z}[\pi_1(M)][z, z^{-1}]$-module chain equivalence

$$\begin{aligned} &\mathcal{C}((1 + T_{-z})\lambda : C^{n-*}[z, z^{-1}] \longrightarrow C[z, z^{-1}]) \\ &\simeq \mathcal{C}(\widetilde{i}_0 - z\widetilde{i}_1 : C(\widetilde{M}, \widetilde{J})[z, z^{-1}] \longrightarrow C(\widetilde{M}, \widetilde{L}_J)[z, z^{-1}]) \\ &\simeq C(\widetilde{M \times \mathbb{R}}, \widetilde{L}) \end{aligned}$$

with $\widetilde{L}$ the pullback to L of the universal cover $\widetilde{M \times \mathbb{R}}$ of $M \times S^1$.
(ii) If K is a Λ-homology boundary spine of $(M, \partial M)$ with respect to some morphism of rings with involution $\mathcal{F} : \mathbb{Z}[\pi_1(M)][z, z^{-1}] \longrightarrow \Lambda$ then (C, λ) is an n-dimensional asymmetric complex which is $(\Lambda, -z)$-Poincaré, and the *asymmetric signature* of (M, J) is the cobordism class

$$\sigma^*(M, J) = (C, \lambda) \in LAsy^n_h(\mathbb{Z}[\pi_1(M)], \Lambda, -z) . \qquad \square$$

Remark 27.7 As in Remark 20.28 and Example 20.29 the $\mathbb{Z}[\pi_1(M)]$-module chain maps induced by the relative cobordism $(M; J \times I, L_J)$

$$\begin{aligned} \gamma : G &= C(\widetilde{M}) \longrightarrow F = C(\widetilde{M}, \widetilde{L}_J) , \\ \mu : G &= C(\widetilde{M}) \longrightarrow C(\widetilde{M}, \widetilde{J} \times I) \simeq F^{n-*} \end{aligned}$$

are the components of the inclusion of a sublagrangian in a metabolic (-1)-symmetric form in the derived category with involution $\mathcal{D}_n(\mathbb{Z}[\pi_1(M)])$ of n-dimensional f.g. free $\mathbb{Z}[\pi_1(M)]$-module chain complexes and chain homotopy classes of chain maps

$$\begin{pmatrix} \gamma \\ \mu \end{pmatrix} : (G, 0) \longrightarrow (F \oplus F^{n-*}, \begin{pmatrix} 0 & 1 \\ -1 & 0 \end{pmatrix})$$

such that

$$\mathcal{C}(\gamma) \simeq C(\widetilde{L}_J)_{*-1} , \quad \mathcal{C}(\mu) \simeq C(\widetilde{J} \times I)_{*-1} ,$$

with boundary

$$G^{\perp}/G \simeq C(\partial(\widetilde{J} \times I))_{*-1}$$

and hessian form the intersection pairing

$$\begin{aligned} \gamma^* \mu &= [M] \cap - : \\ G &= C(\widetilde{M}) \simeq C(\widetilde{M}, \partial\widetilde{M})^{n-*} \longrightarrow C(\widetilde{M}, \partial\widetilde{M}) \simeq G^{n-*} . \end{aligned}$$

The special feature of the situation in 27.6 is the existence of the asymmetric pairing (i.e. chain map) $\lambda : F^{n-*} \longrightarrow F$ such that

$$\gamma \simeq \lambda\mu \simeq \lambda^*\mu : G \longrightarrow F ,$$

with

$$\begin{aligned} \lambda \oplus 1 &\simeq \widetilde{i}_0 : \mathcal{C}(\mu) \simeq C(\widetilde{J})_{*-1} \longrightarrow \mathcal{C}(\gamma) \simeq C(\widetilde{L}_J)_{*-1} , \\ \lambda^* \oplus 1 &\simeq \widetilde{i}_1 : \mathcal{C}(\mu) \simeq C(\widetilde{J})_{*-1} \longrightarrow \mathcal{C}(\gamma) \simeq C(\widetilde{L}_J)_{*-1} . \end{aligned}$$ □

27B. Codimension 2 Seifert surfaces

Proposition 27.8 *Let $(M, \partial M)$ be an n-dimensional manifold with boundary, and let $K \subset \partial M$ be a homology framed codimension 2 submanifold, with exterior L.*

(i) *Let $N \subset M$ be a codimension 2 Seifert surface for K, so that $(N, K) \subset (M, \partial M)$ is a homology framed codimension 2 submanifold, and let (P, L) be the exterior. Inclusion defines a normal map from an $(n-2)$-dimensional manifold with boundary to an $(n-2)$-dimensional normal pair*

$$(f, b) : (N, K) \longrightarrow (M, K)$$

with $(f, b)| = 1 : \partial N = K \longrightarrow K$. Let $p : P \longrightarrow S^1$ be the canonical projection, and let $q : P \longrightarrow I$ a Morse function on the relative cobordism $(P; N \times S^1, L)$. Then

$$g : P \longrightarrow M \times S^1 \times I \; ; \; x \longrightarrow (x, p(x), q(x))$$

defines a normal map from an n-dimensional manifold triad to an n-dimensional normal triad

$$\begin{aligned} (g, c) : (P; N \times S^1, L; K \times S^1) &\longrightarrow \\ &(M \times S^1 \times I; M \times S^1 \times \{0\}, M \times S^1 \times \{1\}; K \times S^1) \end{aligned}$$

such that

$$(g, c)| = (f, b) \times 1 : (N, K) \times S^1 \longrightarrow (M, K) \times S^1$$

and

$$\begin{aligned} (g, c) \cup \mathrm{id.} : M &= P \cup_{N \times S^1} (N \times D^2) \\ &\longrightarrow M \times S^1 \times I \cup_{N \times S^1 \times \{1\}} (N \times D^2) \end{aligned}$$

is a deformation of the inclusion $M \longrightarrow M \times D^2$ (as in 22.6 (iv)).

(ii) *A codimension 1 Seifert surface $J \subset \partial M$ for K determines a codimension 2 Seifert surface $N \subset M$ for K, with the exterior (P, L) such that*

$$C(\widetilde{P},\widetilde{L}) \;=\; C(\widetilde{M},\widetilde{L}_J)[z,z^{-1}] \;=\; C[z,z^{-1}]\ ,$$

with $\widetilde{M}$ the universal cover of M, $\widetilde{L}_J$ the pullback cover of L_J, and with $\widetilde{P},\widetilde{L}$ the pullbacks to P,L of the universal cover $\widetilde{M}\times\mathbb{R}$ of $M\times S^1$. The n-dimensional asymmetric complex (C,λ) over $\mathbb{Z}[\pi_1(M)]$ (27.6) determines the antiquadratic complex (22.4)

$$\sigma_*(M,N,\epsilon^2) \;=\; (C[z,z^{-1}],\psi)\ ,$$

the n-dimensional $(-z)$-quadratic complex over $\mathbb{Z}[\pi_1(M)][z,z^{-1}]$ with

$$\psi_s \;=\; \begin{cases}\lambda & \textit{if } s=0\\ 0 & \textit{if } s\geq 1\ ,\end{cases}$$

$$\begin{aligned}&\mathcal{C}((1+T_{-z})\psi_0 : C^{n-*}[z,z^{-1}]\longrightarrow C[z,z^{-1}])\\ &\qquad =\; \mathcal{C}((1+T_{-z})\lambda : C^{n-*}[z,z^{-1}]\longrightarrow C[z,z^{-1}])\\ &\qquad \simeq\; C(\widetilde{M}\times\mathbb{R},\widetilde{L}) \;\simeq\; \mathcal{C}(\widetilde{f} : C(\widetilde{L})\longrightarrow C(\widetilde{M}\times\mathbb{R}))\ .\end{aligned}$$

The quadratic kernel of (f,b) is the $(n-2)$-dimensional quadratic complex over $\mathbb{Z}[\pi_1(M)]$

$$\sigma_*(f,b) \;=\; (S^{-1}C,\lambda)$$

with $S^{-1}C\simeq\mathcal{C}(f^!)$ the algebraic mapping cone of the Umkehr chain map

$$f^! \;:\; C(\widetilde{M},\widetilde{K})^{n-2-*} \xrightarrow{\ \widetilde{f}^*\ } C(\widetilde{N},\widetilde{K})^{n-2-*} \;\simeq\; C(\widetilde{N})\ .$$

Let

$$\sigma_*(g,c) \;=\; \left(\begin{array}{ccc} 0 & \longrightarrow & \mathcal{C}(\partial_+g^!)\\ \big\downarrow & \Gamma & \big\downarrow\\ \mathcal{C}(S(f)^!) & \longrightarrow & \mathcal{C}(g^!)\end{array}\ ,\ \Psi\right)$$

be the kernel n-dimensional quadratic triad over $\mathbb{Z}[\pi_1(M)][z,z^{-1}]$, with the Umkehr $\mathbb{Z}[\pi_1(M)][z,z^{-1}]$-module chain maps given by

$$\begin{aligned} g^! \;&:\; C(\widetilde{M}\times\mathbb{R},\widetilde{K}\times\mathbb{R})^{n-1-*} \longrightarrow C(\widetilde{P})\ ,\\ \partial_+g^! \;&:\; C(\widetilde{M}\times\mathbb{R},\widetilde{K}\times\mathbb{R})^{n-1-*} \longrightarrow C(\widetilde{L})\ ,\\ S(f)^! \;&:\; C(\widetilde{M}\times\mathbb{R},\widetilde{K}\times\mathbb{R})^{n-1-*} \longrightarrow C(\widetilde{M}\times\mathbb{R})\ .\end{aligned}$$

Let (Γ',Ψ') be the n-dimensional quadratic Poincaré triad over $\mathbb{Z}[\pi_1(M)]$ $[z,z^{-1}]$ associated by 20.27 to the n-dimensional quadratic complex

$$(C',\psi') \;=\; (C[z,z^{-1}],(1-z^{-1})\psi)$$

and the factorization

$$(1+T)\psi'_0 \;:\; C^{n-*}[z,z^{-1}] \xrightarrow{\ 1-z^{-1}\ } C^{n-*}[z,z^{-1}] \xrightarrow{\ (1+T_{-z})\psi_0\ } C[z,z^{-1}]\ ,$$

corresponding to the inclusion of the lagrangian in a hyperbolic (-1)-quadratic form in the derived category with involution $\mathcal{D}_n(\mathbb{Z}[\pi_1(M)][z,z^{-1}])$ (20.28)

$$\begin{pmatrix}(1+T_{-z})\psi_0\\ 1-z^{-1}\end{pmatrix} : (C^{n-*}[z,z^{-1}],0) \longrightarrow (C[z,z^{-1}]\oplus C^{n-*}[z,z^{-1}], \begin{pmatrix}0&1\\0&0\end{pmatrix})$$

with hessian form $(C^{n-*}[z,z^{-1}],\psi'_0)$ and

$$\Gamma' \qquad \begin{array}{ccc} B & \longrightarrow & \mathcal{C}((1+T_{-z})\psi_0)_{*+1} \\ \downarrow & & \downarrow \\ \mathcal{C}(1-z^{-1}:C^{n-*}[z,z^{-1}]\longrightarrow C^{n-*}[z,z^{-1}])_{*+1} & \longrightarrow & C^{n-*}[z,z^{-1}] \end{array}$$

$$\simeq \qquad \begin{array}{ccc} B & \longrightarrow & C(\widetilde{M}\times\mathbb{R},\widetilde{L})_{*+1} \\ \downarrow & & \downarrow \\ C(\widetilde{M}\times\mathbb{R},\widetilde{K}\times\mathbb{R})_{*+1} & \longrightarrow & C(\widetilde{M}\times\mathbb{R},\widetilde{P})_{*+1} \end{array}$$

and

$$\begin{aligned} B &= \mathcal{C}(1-z^{-1}:\mathcal{C}((1+T_{-z})\psi_0)\longrightarrow\mathcal{C}((1+T_{-z})\psi_0))_{*+2} \\ &\simeq \mathcal{C}(C(\widetilde{M}\times\mathbb{R},\widetilde{K}\times\mathbb{R})^{n-1-*}\longrightarrow C(\widetilde{M}\times\mathbb{R}))_{*+1} . \end{aligned}$$

The kernel quadratic triad $\sigma_*(g,c)=(\Gamma,\Psi)$ is obtained from (Γ',Ψ') by collapsing B, with a homotopy equivalence

$$(\Gamma,\Psi) \simeq (\Gamma',\Psi')/B$$

$$= \left(\begin{array}{ccc} 0 & \longrightarrow & \mathcal{C}((1+T_{-z})\psi_0)_{*+1} \\ \downarrow & & \downarrow \\ \mathcal{C}(1-z:C[z,z^{-1}]\longrightarrow C[z,z^{-1}])_{*+1} & \longrightarrow & \mathcal{C}(B\longrightarrow C^{n-*}[z,z^{-1}]) \end{array} , \Psi'/B \right)$$

In particular, the kernel $(n-1)$-dimensional quadratic complex

$$\sigma_*(\partial_+g,\partial_+c) = (\partial_+C,\partial_+\psi)$$

is given by

$$\begin{aligned} \partial_+C &= \mathcal{C}(\partial_+g^! : C(\widetilde{M}\times\mathbb{R},\widetilde{K}\times\mathbb{R})^{n-1-*}\longrightarrow C(\widetilde{L})) \\ &\simeq C(\widetilde{M}\times\mathbb{R},\widetilde{L})^{n-*} \simeq C(\widetilde{M}\times\mathbb{R},\widetilde{L})_{*+1} \\ &\simeq \mathcal{C}((1+T_{-z})\lambda : C^{n-*}[z,z^{-1}]\longrightarrow C[z,z^{-1}])_{*+1} , \end{aligned}$$

$$\partial_+\psi_s = \begin{cases} \begin{pmatrix}0&1\\z&0\end{pmatrix} & \text{if } s=0 \\ 0 & \text{if } s\geq 1 . \end{cases}$$

(iii) *For any morphism of rings with involution $\mathcal{F}:\mathbb{Z}[\pi_1(M)][z,z^{-1}]\longrightarrow\Lambda$ the following conditions are equivalent:*

(a) (C,λ) *is an asymmetric* $(\Lambda,-z)$*-Poincaré complex (26.2),*
(b) $\partial_+ g : L \longrightarrow M \times S^1$ *is a* Λ*-homology equivalence,*
(c) $(M \times S^1, K \times S^1)$ *is an* $(n-1)$*-dimensional* Λ*-coefficient geometric Poincaré pair with fundamental class* $[L] \in H_{n-1}(M \times S^1, K \times S^1)$,
(d) K *is a* Λ*-homology boundary spine for* $(M, \partial M)$ *(22.10),*
(e) $(C[z,z^{-1}],\psi)$ *is a* Λ*-Poincaré* $(-z)$*-quadratic complex,*
(f) $(\partial_+ C, \partial_+ \psi)$ *is a* Λ*-contractible quadratic complex.*

(iv) *If* $\mathcal{F}$ *is locally epic and the conditions of* (iii) *are satisfied, the isomorphism of 26.11*

$$LAsy_h^n(\mathbb{Z}[\pi_1(M)],\Lambda,-z) \cong \Gamma_n^h(\mathcal{F},-z)$$

sends the cobordism class (C,λ) *to the* Λ*-homology spine obstruction (22.14)*

$$\sigma_*^\Lambda(M,K) \;=\; (C[z,z^{-1}],\psi) \in \Gamma_n^h(\mathcal{F},-z)$$

allowing the identification

$$\sigma_*^\Lambda(M,K) \;=\; (C,\lambda) \in LAsy_h^n(\mathbb{Z}[\pi_1(M)],\Lambda,-z) \;=\; \Gamma_n^h(\mathcal{F},-z)\ .$$

Proof (i) Formal.
(ii) Define a *push-in* of the codimension 1 Seifert surface $J \subset \partial M$ to be the codimension 2 Seifert surface $N \subset M$ for $K \subset \partial M$ obtained by pushing J rel ∂ into the interior of N, with

$$N \cong J\ ,\ \partial N \;=\; \partial J \;=\; K \subset \partial M\ .$$

As in the proof of 27.5 the exterior of the homology framed codimension 2 submanifold $(N,\partial N) \subset (M,\partial M)$

$$(P,\partial_+ P) \;=\; (\mathrm{cl.}(M\backslash(N \times D^2)), \mathrm{cl.}(\partial M\backslash(\partial N \times D^2)))$$

is equipped with a map $(p,\partial_+ p)\ :\ (P,\partial_+ P)\longrightarrow S^1$ transverse regular at $\{\mathrm{pt.}\} \subset S^1$. In this case it is possible to arrange the inverse image to be a codimension 1 Seifert surface for $(N,\partial N) \subset (M,\partial M)$

$$(F,\partial_+ F) \;=\; (p,\partial_+ p)^{-1}(\{\mathrm{pt.}\}) \subset (P,\partial_+ P)$$

such that

$$(F,\partial_+ F) \;\cong\; (J \times I, J \times \{0\})\ \ ,\ \ (P_F, \partial_+ P_{\partial_+ F}) \;\cong\; (M, L_J)\ .$$

The antiquadratic complex $(C(\widetilde{P},\widetilde{\partial_+ P}),\psi)$ is given by 27.5, with

$$\begin{aligned} C(\widetilde{P},\widetilde{\partial_+ P}) \;&=\; \mathcal{C}(i_0 - z i_1 : C(\widetilde{F},\widetilde{\partial_+ F})[z,z^{-1}]\longrightarrow C(\widetilde{P}_F,\widetilde{\partial_+ P}_{\partial_+ F})[z,z^{-1}]) \\ &\simeq\; C(\widetilde{P}_F,\widetilde{\partial_+ P}_{\partial_+ F})[z,z^{-1}] \\ &\simeq\; C(\widetilde{M},\widetilde{L}_J)[z,z^{-1}] \;=\; C[z,z^{-1}]\ . \end{aligned}$$

(iii) The structures of the quadratic kernel complex $\sigma_*(f,b)$ and triad $\sigma_*(g,c)$ are given by 22.6.
(iv) Immediate from (iii). □

Example 27.9 (i) Let π be a finitely presented group, and let $(\mathbb{Z}[\pi]^\ell, \lambda)$ be an asymmetric form on a f.g. free $\mathbb{Z}[\pi]$-module of rank $\ell \geq 1$. For any integer $i \geq 3$ it is possible to realize the $(2i+2)$-dimensional asymmetric complex $(S^{i+1}\mathbb{Z}[\pi]^\ell, \lambda)$ as the asymmetric complex of a $(2i+2)$-dimensional manifold with boundary $(M, \partial M)$ and a homology framed codimension 2 submanifold $K \subset \partial M$ with a codimension 1 Seifert surface $J \subset \partial M$, as follows.
As in the proof of 22.17 choose a $(2i-1)$-dimensional manifold with boundary $(K_0, \partial K_0)$ such that $\pi_1(K_0) = \pi$, and use the method of Wall [302, Chap. 5] to construct a $2i$-dimensional relative cobordism $(N; K_0, K_1)$ with

$$N^{2i} = K_0 \times I \cup_{\lambda + (-)^i \lambda^*} \bigcup_{\ell} D^i \times D^i$$

the trace of surgeries on ℓ disjoint trivial embeddings $S^{i-1} \times D^i \subset \operatorname{int}(K_0)$ with self intersections λ, writing the boundary of N as

$$K = \partial N = K_0 \cup_\partial -K_1 .$$

By construction, there is an $(i-1)$-connected normal map of $2i$-dimensional manifold triads

$$(f, b) : (N; K_0, K_1) \longrightarrow K_0 \times (I; \{0\}, \{1\})$$

with

$$(f, b)| = \text{id.} : K_0 \longrightarrow K_0 \ , \ K_i(N) = \mathbb{Z}[\pi]^\ell ,$$

and quadratic kernel the $2i$-dimensional quadratic complex $(S^i\mathbb{Z}[\pi]^\ell, \lambda)$ over $\mathbb{Z}[\pi]$. The quadratic kernel of the normal map of closed $(2i-1)$-dimensional manifolds

$$(\partial f, \partial b) = (f, b)| : K \longrightarrow \partial(K_0 \times I) = K_0 \cup_\partial -K_0$$

is the boundary $(2i-1)$-dimensional quadratic Poincaré complex $\partial(S^i\mathbb{Z}[\pi]^\ell, \lambda)$ over $\mathbb{Z}[\pi]$. Define the $(2i+2)$-dimensional manifold with boundary

$$(M, \partial M) = (K_0 \times I \times D^2, \partial(K_0 \times I \times D^2))$$

and regard

$$\partial(K_0 \times I) \times \{0\} \subset \partial M = \partial(K_0 \times I) \times D^2 \cup_{\partial(K_0 \times I) \times S^1} (K_0 \times I \times S^1)$$

as a homology framed codimension 2 submanifold. Use the method of Matsumoto [184, 5.2] (cf. 22.8) to perform ambient surgeries on ℓ disjoint trivial embeddings

$$S^{i-1} \times D^i \subset \operatorname{int}(K_0) \times \{0\} \subset \partial(K_0 \times I) \times D^2$$

with self intersections λ, and trace a homology framed codimension 2 submanifold triad

$$(N \cup_{\partial K_0 \times I} K_0 \times I; \partial(K_0 \times I), K) \subset \partial(K_0 \times I) \times D^2 \times (I; \{0\}, \{1\}) .$$

The homology framed codimension 2 submanifold $(N, K) \subset (M, \partial M)$ is such that

$$(M, \partial M) \cong (N \times D^2 \cup_{N \times S^1} P, K \times D^2 \cup_{K \times S^1} L)$$

with the exterior (P, L) given by

$$P^{2i+2} = N \times S^1 \times I \cup \bigcup_{\ell} D^{i+1} \times D^{i+1}$$

the trace of surgeries on ℓ disjoint embeddings $S^i \times D^{i+1} \subset \text{int}(N \times S^1)$ representing generators of the kernel $\mathbb{Z}[\pi][z, z^{-1}]$-module

$$K_i(N \times S^1) = \mathbb{Z}[\pi][z, z^{-1}]^\ell$$

of the $(i-1)$-connected normal map $e \times \text{id}_{S^1} : N \times S^1 \longrightarrow K_0 \times I \times S^1$, and L the effect of these surgeries. The kernel $(2i+2)$-dimensional quadratic Poincaré triad over $\mathbb{Z}[\pi][z, z^{-1}]$ of the corresponding i-connected normal map of $(2i+2)$-dimensional manifold triads

$$(g, c) : (P; N \times S^1, L) \longrightarrow K_0 \times I \times S^1 \times (I; \{0\}, \{1\})$$

is the one associated by 20.27 to the factorization of the duality chain map of the $(2i+2)$-dimensional quadratic complex over $\mathbb{Z}[\pi][z, z^{-1}]$

$$(\mathcal{C}(g^!), \psi) = (S^{i+1}\mathbb{Z}[\pi][z, z^{-1}]^\ell, (1 - z^{-1})\lambda)$$

given by

$$\begin{aligned}(1+T)\psi_0 &= (1-z^{-1})\lambda + (-)^{i+1}(1-z)\lambda^* = (1-z^{-1})(\lambda + (-)^i z\lambda^*)\\ &: \mathcal{C}(g^!)^{2i+2-*} = S^{i+1}\mathbb{Z}[\pi][z, z^{-1}]^\ell \xrightarrow{1-z^{-1}} S^{i+1}\mathbb{Z}[\pi][z, z^{-1}]^\ell\\ &\qquad \xrightarrow{(1+T_{-z})\lambda} \mathcal{C}(g^!) = S^{i+1}\mathbb{Z}[\pi][z, z^{-1}]^\ell .\end{aligned}$$

The homology framed codimension 2 submanifold $K \subset \partial M$ has a codimension 1 Seifert surface $J \subset \partial M$ homeomorphic to N (so that N is a push-in of J), such that the canonical infinite cyclic cover $\overline{L}$ of L admits a Seifert fundamental domain $(L_J; i_0 J, i_1 J)$ with a normal map

$$(g, c) : (L_J; i_0 J, i_1 J) \longrightarrow K_0 \times I \times (I; \{0\}, \{1\})$$

and

$$\begin{aligned}i_0 &= \lambda : \mathcal{C}(\partial_0 g^!) = S^i\mathbb{Z}[\pi]^\ell \longrightarrow \mathcal{C}(g^!) = S^i\mathbb{Z}[\pi]^\ell ,\\ i_1 &= (-)^{i+1}\lambda^* : \mathcal{C}(\partial_1 g^!) = S^i\mathbb{Z}[\pi]^\ell \longrightarrow \mathcal{C}(g^!) = S^i\mathbb{Z}[\pi]^\ell ,\\ C(\widetilde{L}, \widetilde{K}_0 \times I \times \mathbb{R}) &= \mathcal{C}(i_0 - zi_1 : \mathcal{C}(\partial_0 g^!)[z, z^{-1}] \longrightarrow \mathcal{C}(\partial_1 g^!)[z, z^{-1}])\\ &= \mathcal{C}(\lambda + (-)^i z\lambda^* : S^i\mathbb{Z}[\pi][z, z^{-1}]^\ell \longrightarrow S^i(\mathbb{Z}[\pi][z, z^{-1}]^\ell)^*) .\end{aligned}$$

The cellular chain complexes of the universal covers $\widetilde{M},\widetilde{N},\widetilde{L}_J$ of M,N,L_J are such that

$$C(\widetilde{M},\widetilde{L}_J) \simeq C(\widetilde{M},\widetilde{N})^{2i+2-*} \simeq S^{i+1}\mathbb{Z}[\pi]^{\ell} ,$$

and the asymmetric complex given by 27.6 is $(S^{i+1}\mathbb{Z}[\pi]^{\ell},\lambda)$.
(ii) Suppose that $\pi=\{1\}$, $(K_0,\partial K_0)=(D^{2i-1},S^{2i-2})$ in (i), and that $(\mathbb{Z}^k,\lambda)$ is an asymmetric form over $\mathbb{Z}$ such that

$$\lambda+(-)^i\lambda^* : \mathbb{Z}^{\ell} \longrightarrow (\mathbb{Z}^{\ell})^*$$

is a $\mathbb{Z}$-module isomorphism, i.e. a $(-)^i$-symmetric Seifert form over $\mathbb{Z}$ in the terminology of 32.1. In this case $(M,\partial M)=(D^{2i+2},S^{2i+1})$, and the homology framed codimension 2 submanifold $K\subset S^{2i+1}$ is homeomorphic to S^{2i-1} by the generalized Poincaré conjecture – in effect, this is the surgery construction of high-dimensional knots due to Kervaire [132] and Levine [156]. The knot $k : K\cong S^{2i-1}\subset S^{2i+1}$ has a codimension 1 Seifert surface $J^{2i}\subset S^{2i+1}$ with $\dot{H}_*(J)=0$ $(*\neq i)$, $H_i(J)=\mathbb{Z}^{\ell}$ and Seifert form $(\mathbb{Z}^{\ell},\lambda)$. □

27C. Branched cyclic covers

Cyclic branched cyclic covers are manifolds which are cyclic covers off a submanifold, usually codimension 2. See Burde and Zieschang [34, Chap. 8E] for an account of the traditional applications of branched covers in knot theory. This section relates cyclic branched covers and asymmetric complexes, generalizing the relationship between a Seifert matrix for a 1-knot $k : S^1\subset S^3$ and the homology of the cyclic covers of S^3 branched over k.

Definition 27.10 Let $(M,\partial M)$ be an n-dimensional manifold with boundary, and let $(N,\partial N)\subset(M,\partial M)$ be a homology framed codimension 2 submanifold with exterior (P,∂_+P), so that

$$(M,\partial M) = (N\times D^2\cup_{N\times S^1}P,\partial N\times D^2\cup_{\partial N\times S^1}\partial_+P) .$$

For any integer $a\geq 2$ the *a-fold branched cyclic cover* of $(M,\partial M)$ branched over $(N,\partial N)$ is the n-dimensional manifold with boundary

$$(M',\partial M') = (N\times D^2\cup_{N\times S^1}P',\partial N\times D^2\cup_{\partial N\times S^1}\partial_+P')$$

with homology framed codimension 2 submanifold $(N,\partial N)\subset(M',\partial M')$ and exterior

$$(P',\partial_+P') = (\overline{P},\overline{\partial_+P})/z^a$$

the a-fold (unbranched) cyclic cover of (P,∂_+P), with $(\overline{P},\overline{\partial_+P})$ the canonical infinite cyclic cover of (P,∂_+P). □

What is the asymmetric complex of a branched cyclic cover?

Proposition 27.11 *Given an n-dimensional manifold with boundary* $(M, \partial M)$ *and a homology framed codimension 2 submanifold* $K \subset \partial M$ *with a codimension 1 Seifert surface* $J \subset \partial M$, *let* (C, λ) *be the asymmetric complex of* (M, J) *(27.6), the n-dimensional asymmetric complex* $\mathbb{Z}[\pi_1(M)]$ *with*

$$C = C(\widetilde{M}, \widetilde{\partial_+ M_J}) \simeq C(\widetilde{M}, \widetilde{J})^{n-*} \simeq C(\widetilde{M}, \widetilde{N})^{n-*} .$$

Let $N \subset M$ *be a push-in codimension 2 Seifert surface for* K *obtained by pushing* J *rel*∂ *into the interior of* M, *with* $\partial N = K$, *and let* $(M', \partial M')$ *be the a-fold cyclic cover of* $(M, \partial M)$ *branched over* $(N, \partial N)$, *as in 27.10. Then* $N \subset M'$ *is a push-in of the codimension 1 Seifert surface* $J \subset \partial M'$ *for* $\partial N \subset \partial M'$ *into* M'. *Let* $\widetilde{M}'$ *be the pullback to* M' *of the universal cover* $\widetilde{M}$ *of* M *along the branched covering projection* $M' \longrightarrow M$. *The asymmetric complex of* (M', J) *over* $\mathbb{Z}[\pi_1(M)]$ *is given by*

$$(C', \lambda') = \left(\bigoplus_a C, \begin{pmatrix} \lambda & -T\lambda & 0 & \dots & 0 \\ -\lambda & \lambda + T\lambda & -T\lambda & \dots & 0 \\ 0 & -\lambda & \lambda + T\lambda & \dots & 0 \\ \vdots & \vdots & \vdots & \ddots & \vdots \\ 0 & 0 & 0 & \dots & \lambda + T\lambda \end{pmatrix} \right) ,$$

with

$$C' = C(\widetilde{M}', \widetilde{\partial_+ M'_J}) \simeq C(\widetilde{M}', \widetilde{N})^{n-*} .$$

Proof Regarding M as a relative cobordism

$$(M; J \times I, J \times I; \partial_+ M_J \cup (J \times I))$$

gives a fundamental domain for the canonical infinite cyclic cover $(\overline{P}, \overline{\partial_+ P})$, so that $(P', \partial_+ P')$ is the cyclic concatenation of a copies of M, and M' is the linear concatenation of a copies of M

$$(M'; J \times I, J \times I; \partial_+ M'_J \cup (J \times I)) = \bigcup_a (M; J \times I, J \times I; \partial_+ M_J \cup (J \times I))$$

as in the diagram

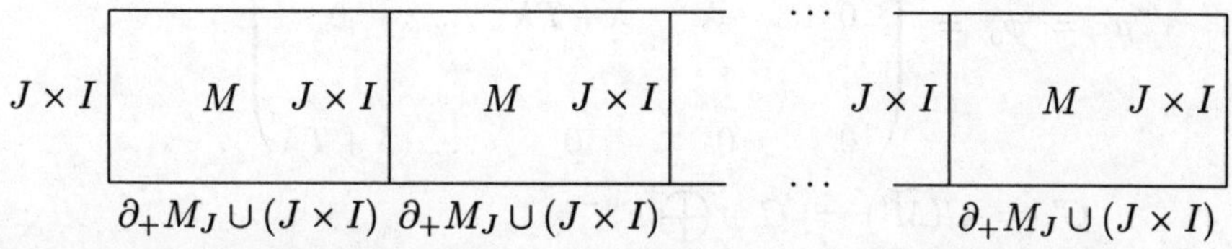

As in 27.7 the chain maps

$$\gamma : G = C(\widetilde{M}) \longrightarrow F = C(\widetilde{M}, \widetilde{\partial_+ M_J}) ,$$
$$\mu : G = C(\widetilde{M}) \longrightarrow C(\widetilde{M}, \widetilde{J} \times I) \simeq F^{n-*}$$

are the components of the inclusion of a sublagrangian in a metabolic (-1)-symmetric form in the derived category with involution $\mathcal{D}_n(\mathbb{Z}[\pi_1(M)])$

$$\begin{pmatrix}\gamma\\ \mu\end{pmatrix} : (G,0) \longrightarrow (F\oplus F^{n-*}, \begin{pmatrix}0 & 1\\ -1 & 0\end{pmatrix}),$$

with $\lambda : C^{n-*} = F^{n-*} \longrightarrow C = F$ such that

$$\gamma = \lambda\mu = \lambda^*\mu : G \longrightarrow F ,$$

and hessian the intersection pairing of $(M, \partial M)$

$$\gamma^*\mu = \phi_0 : G = C(\widetilde{M}) \longrightarrow G^{n-*} = C(\widetilde{M}, \widetilde{\partial M}) .$$

The components of the inclusion of the sublagrangian corresponding to the branched cover are given by

$$\gamma' = \begin{pmatrix}\gamma & -T\lambda & 0 & \dots & 0\\ -\gamma & \lambda+T\lambda & -T\lambda & \dots & 0\\ 0 & -\lambda & \lambda+T\lambda & \dots & 0\\ \vdots & \vdots & \vdots & \ddots & \vdots\\ 0 & 0 & 0 & \dots & \lambda+T\lambda\end{pmatrix} :$$

$$G' = C(\widetilde{M}') = G\oplus\bigoplus_{a-1} F^{n-*} \longrightarrow F' = C(\widetilde{M}', \widetilde{\partial_+ M'_J}) = \bigoplus_a F ,$$

$$\mu' = \begin{pmatrix}\mu & 0 & 0 & \dots & 0\\ 0 & 1 & 0 & \dots & 0\\ 0 & 0 & 1 & \dots & 0\\ \vdots & \vdots & \vdots & \ddots & \vdots\\ 0 & 0 & 0 & \dots & 1\end{pmatrix} : G' = G\oplus\bigoplus_{a-1} F^{n-*} \longrightarrow F'^{n-*} = \bigoplus_a F^{n-*} ,$$

with hessian the intersection pairing of $(M', \partial M')$

$$\gamma'^*\mu' = \phi'_0 = \begin{pmatrix}\phi_0 & -\gamma^* & 0 & \dots & 0\\ -\gamma & \lambda+T\lambda & -T\lambda & \dots & 0\\ 0 & -\lambda & \lambda+T\lambda & \dots & 0\\ \vdots & \vdots & \vdots & \ddots & \vdots\\ 0 & 0 & 0 & \dots & \lambda+T\lambda\end{pmatrix} :$$

$$G' = C(\widetilde{M}') = G\oplus\bigoplus_{a-1} F^{n-*}$$

$$\longrightarrow G'^{n-*} = C(\widetilde{M}', \widetilde{\partial M}') = G^{n-*}\oplus\bigoplus_{a-1} F . \qquad \square$$

Example 27.12 Let

$$(M, \partial M) = (D^{2i+2}, S^{2i+1}) \ (i \geq 1) ,$$

and let

$$K = S^{2i-1} \subset S^{2i+1}$$

be a knot with Seifert surface J. The statement and proof of 27.11 give the homological invariants of the corresponding a-fold branched cover $(M', \partial M')$ of $(M, \partial M)$. Only the middle-dimensional homology modulo torsion will be considered here. The relevant part of the asymmetric complex (C, λ) of (M, J) is the $(-)^i$-symmetric Seifert form (L, λ) over $\mathbb{Z}$, with

$$\lambda + (-)^i\lambda^* : H^{i+1}(C) = L = H_i(J) \longrightarrow L^*$$

an isomorphism. In this case

$$\begin{aligned} H_{i+1}(G) &= H_{i+1}(D^{2i+2}) = 0 , \\ H_{i+1}(F) &= H^{i+1}(D^{2i+2}, J) = H^i(J) \end{aligned}$$

and the inclusion of the sublagrangian in a metabolic $(-)^i$-symmetric form over $\mathbb{Z}$

$$\begin{pmatrix} \gamma' \\ \mu' \end{pmatrix} : (H_{i+1}(G'), 0) \longrightarrow (H_{i+1}(F') \oplus H_{i+1}(F')^*, \begin{pmatrix} 0 & 1 \\ (-)^i & 0 \end{pmatrix})$$

has components

$$\gamma' = \begin{pmatrix} (-)^i\lambda^* & 0 & 0 & \dots & 0 \\ \lambda + (-)^{i+1}\lambda^* & (-)^i\lambda^* & 0 & \dots & 0 \\ -\lambda & \lambda + (-)^{i+1}\lambda^* & (-)^i\lambda^* & \dots & 0 \\ \vdots & \vdots & \vdots & \ddots & \vdots \\ 0 & 0 & 0 & \dots & \lambda + (-)^{i+1}\lambda^* \end{pmatrix} :$$

$$\begin{aligned} H_{i+1}(G') &= H_{i+1}(M') = \bigoplus_{a-1} L \\ &\longrightarrow H_{i+1}(F') = H_{i+1}(M', \partial_+ M'_J) = \bigoplus_{a} L^* , \end{aligned}$$

$$\mu' = \begin{pmatrix} 0 & 0 & 0 & \dots & 0 \\ 1 & 0 & 0 & \dots & 0 \\ 0 & 1 & 0 & \dots & 0 \\ \vdots & \vdots & \vdots & \ddots & \vdots \\ 0 & 0 & 0 & \dots & 1 \end{pmatrix} : H_{i+1}(G') = \bigoplus_{a-1} L \longrightarrow H_{i+1}(F')^* = \bigoplus_{a} L ,$$

with hessian the $(-)^{i+1}$-symmetric intersection pairing of $(M', \partial M')$

$$\gamma'^*\mu' = \phi'_0$$

$$= \begin{pmatrix} \lambda+(-)^{i+1}\lambda^* & (-)^i\lambda^* & 0 & \dots & 0 \\ -\lambda & \lambda+(-)^{i+1}\lambda^* & (-)^i\lambda^* & \dots & 0 \\ 0 & -\lambda & \lambda+(-)^{i+1}\lambda^* & \dots & 0 \\ \vdots & \vdots & \vdots & \ddots & \vdots \\ 0 & 0 & 0 & \dots & \lambda+(-)^{i+1}\lambda^* \end{pmatrix}$$

$$: H_{i+1}(G') = \bigoplus_{a-1} L \longrightarrow H_{i+1}(G')^* = \bigoplus_{a-1} L^*$$

and

$$\mathrm{coker}(\phi'_0 : \bigoplus_{a-1} L \longrightarrow \bigoplus_{a-1} L^*) = H_i(\partial M') .$$

The construction and homology computation of $(M', \partial M')$ go back to Seifert [262] (cf. Burde and Zieschang [34, Chap. 8E]), and the formula for the intersection pairing was obtained by Cappell and Shaneson [41] and Kauffman [121, 5.6], [122, 12.2]. See 40.15 for the signature of this pairing. □

27D. Framed spines

Definition 27.13 Given a space X let $\xi = \epsilon^2 : X \longrightarrow BSO(2)$ be the trivial 2-plane bundle, and let

$$\mathcal{F} : \mathbb{Z}[\pi_1(S(\xi))] = \mathbb{Z}[\pi_1(X)][z, z^{-1}] \longrightarrow \Lambda$$

be a morphism of rings with involution.
The *framed spine bordism groups* $BB_*(X, \mathcal{F})$, $AB_*(X, \mathcal{F})$, $\Delta_*(X, \mathcal{F})$ are the spine bordism groups of 22.16 with

$$BB_n(X, \mathcal{F}) = BB_n(X, \epsilon^2, \mathcal{F}) ,$$
$$AB_n(X, \mathcal{F}) = AB_n(X, \epsilon^2, \mathcal{F}) ,$$
$$\Delta_n(X, \mathcal{F}) = \Delta_n(X, \epsilon^2, \mathcal{F}) .$$

More explicitly:
(i) The *bounded framed spine bordism group* $BB_n(X, \mathcal{F})$ is the bordism group of $(n+2)$-dimensional manifolds with boundary $(M, \partial M)$ with a map $(M, \partial M) \longrightarrow X$ and a homology framed codimension 2 submanifold $K \subset \partial M$ with exterior L such that K is a Λ-homology boundary spine, i.e.

$$H_*(L; \Lambda) = H_*(M \times S^1; \Lambda) .$$

(ii) The *framed empty spine bordism group* $AB_n(X, \mathcal{F})$ is the bordism group of $(n+2)$-dimensional manifolds with boundary $(M, \partial M)$ with a map $(M, \partial M) \longrightarrow X \times S^1$ such that $K = \emptyset$ in (i), i.e.

$$H_*(\partial M; \Lambda) = H_*(M \times S^1; \Lambda) .$$

(iii) Assuming $1-z \in \Lambda^\bullet$, the *framed closed spine bordism group* $\Delta_n(X,\mathcal{F})$ is the bordism group of closed $(n+1)$-dimensional manifolds N with a map $N \longrightarrow X \times S^1$ such that $H_*(N;\Lambda)=0$.
Similarly for the U-intermediate groups. □

Proposition 27.14 (i) *The bounded framed and empty spine cobordism groups are isomorphic*

$$BB_n(X,\mathcal{F}) \;=\; AB_n(X,\mathcal{F})\ .$$

(ii) *The asymmetric complex construction of 27.6 defines a morphism*

$$\sigma^* \;:\; BB_n(X,\mathcal{F}) \longrightarrow LAsy_h^{n+2}(\mathbb{Z}[\pi_1(X)],\Lambda,-z)\,;\,(M,K)\longrightarrow (C(\widetilde{M},\widetilde{L}_J),\lambda)\,,$$

which is an isomorphism if $n \geq 5$, $\pi_1(X)$ *is finitely presented and* $\mathcal{F}$ *is locally epic. Moreover, in this case the asymmetric signature defines a natural isomorphism of exact sequences*

$$\begin{array}{ccccccccc}
\cdots \longrightarrow & AB_n(X,1) & \longrightarrow & AB_n(X,\mathcal{F}) & \longrightarrow & \Gamma^h_{n+2}(\Phi_{\mathcal{F}},-z) & \longrightarrow \cdots \\
& \cong\big\downarrow\sigma^* & & \cong\big\downarrow\sigma^* & & \cong\big\downarrow\sigma^* & \\
\cdots \longrightarrow & \widehat{H}^{n+3} & \longrightarrow & LAsy_h^{n+2}(A,\Lambda,-z) & \longrightarrow & LAsy_p^{n+2}(A,\Lambda,-z) & \longrightarrow \cdots
\end{array}$$

with $A=\mathbb{Z}[\pi_1(X)]$, $\Gamma^h_{n+2}(\Phi_{\mathcal{F}},-z)$ *the cobordism group of* $(n+1)$*-dimensional* $(-z)$*-quadratic* Λ*-contractible Poincaré complexes over* $A[z,z^{-1}]$, *and*

$$\widehat{H}^{n+3} \;=\; L^h_{n+2}(A[z,z^{-1}],-z) \;=\; \widehat{H}^{n+3}(\mathbb{Z}_2;\widetilde{K}_0(A))\ .$$

(iv) *If* $\mathcal{F}$ *is locally epic and* $1-z\in\Lambda^\bullet$ *there is defined a natural transformation of exact sequences*

$$\begin{array}{ccccccc}
\cdots \longrightarrow \Omega_{n+2}(X\times S^1) & \longrightarrow & AB_n(X,\mathcal{F}) & \longrightarrow & \Delta_n(X,\mathcal{F}) & \longrightarrow \cdots \\
\big\downarrow\sigma^* & & \big\downarrow\sigma^* & & \big\downarrow\sigma^* & \\
\cdots \longrightarrow L_h^{n+2}(A[z,z^{-1}]) & \longrightarrow & LAsy_h^{n+2}(A,\Lambda,-z) & \longrightarrow & \Gamma_h^{n+2}(\Phi_{\mathcal{F}}) & \longrightarrow \cdots
\end{array}$$

with $\Gamma_h^{n+2}(\Phi_{\mathcal{F}})$ *the cobordism group of* $(n+1)$*-dimensional symmetric* Λ*-contractible Poincaré complexes over* $A[z,z^{-1}]$, *and*

$$\sigma^* \;:\; \Delta_n(X,\mathcal{F}) \longrightarrow \Gamma_h^{n+2}(\Phi_{\mathcal{F}})\ ;\ M \longrightarrow (C(\widetilde{M}),\Delta[M])$$

the Λ*-contractible symmetric signature map. If* $n\geq 5$, $\pi_1(X)$ *is finitely presented and* $\mathcal{F}$ *is locally epic*

$$\sigma^* \;:\; AB_n(X,\mathcal{F}) \cong LAsy_h^{n+2}(A,\Lambda,-z)\ ,$$

and there is defined an exact sequence

$$\begin{aligned}
\cdots \longrightarrow L_h^{n+2}(A[z,z^{-1}]) \longrightarrow \Delta_n(X,\mathcal{F}) \longrightarrow \Gamma_h^{n+2}(\Phi_{\mathcal{F}}) \oplus \Omega_{n+1}(X\times S^1) \\
\longrightarrow L_h^{n+1}(A[z,z^{-1}]) \longrightarrow \ \cdots\ .
\end{aligned}$$

Proof This is the special case $\xi=\epsilon^2$ of 22.18. □

Example 27.15 For

$$\mathcal{F} \,:\, \mathbb{Z}[\pi_1(X)][z,z^{-1}] \longrightarrow \Lambda \;=\; \mathbb{Z}[\pi_1(X)]\ ;\ z \longrightarrow 1$$

there are identifications

$$\begin{aligned} AB_n(X,\mathcal{F}) \;&=\; BB_n(X,\mathcal{F}) \;=\; LAsy_h^{n+2}(\mathbb{Z}[\pi_1(X)],\Lambda,-z)\ , \\ \Gamma_{n+2}(\Phi_{\mathcal{F}}) \;&=\; \Gamma_{n+2}(\Phi_{\mathcal{F}},-z) \;=\; LAsy_p^{n+2}(\mathbb{Z}[\pi_1(X)],\Lambda,-z)\ . \end{aligned}$$

In particular, for $X = \{\text{pt.}\}$ and $\Lambda = \mathbb{Z}$

$$\begin{aligned} AB_{n+1}(\{\text{pt.}\},\mathcal{F}) \;&=\; BB_{n+1}(\{\text{pt.}\},\mathcal{F}) \;=\; LAsy_h^{n+3}(\mathbb{Z},\mathbb{Z},-z) \\ &=\; \Gamma_{n+3}(\Phi_{\mathcal{F}}) \;=\; \Gamma_{n+3}(\Phi_{\mathcal{F}},-z) \;=\; LAsy^{n+3}(\mathbb{Z},\mathbb{Z},-z)\ , \end{aligned}$$

and these are the high-dimensional knot cobordism groups C_n for knots $k : S^n \subset S^{n+2}$ – see Chap. 33 for a further discussion. □

Remark 27.16 (i) For any rings with involution A, A' the product of an n-dimensional asymmetric complex (C,λ) over A and an n'-dimensional asymmetric complex (C',λ') over A' is an $(n+n')$-dimensional asymmetric complex over $A \otimes_{\mathbb{Z}} A'$

$$(C,\lambda)\otimes(C',\lambda') \;=\; (C\otimes C',\lambda\otimes\lambda')\ .$$

(ii) Let $(M,\partial M)$ be an n-dimensional manifold with boundary, with a homology framed codimension 2 submanifold $(N,\partial N) \subset (M,\partial M)$ such that N is a push-in of a codimension 1 Seifert surface $J \subset \partial M$ for ∂N. As in 27.6 there is defined an n-dimensional asymmetric complex (C,λ) over $\mathbb{Z}[\pi_1(M)]$, with $C = C(\widetilde{M},\widetilde{N})^{n-*}$. Likewise, let $(M',\partial M')$ be an n'-dimensional manifold with boundary, with a homology framed codimension 2 submanifold $(N',\partial N') \subset (M',\partial M')$ such that N' is a push-in, and let (C',λ') be the corresponding n'-dimensional asymmetric complex over $\mathbb{Z}[\pi_1(M')]$. Then

$$N'' \;=\; N\times M'\cup M\times N' \subset M'' \;=\; M\times M'$$

is a homology framed codimension 2 submanifold of an $(n+n')$-dimensional manifold with boundary, such that the corresponding $(n+n')$-dimensional asymmetric complex over

$$\mathbb{Z}[\pi_1(M'')] \;=\; \mathbb{Z}[\pi_1(M)\times\pi_1(M')] \;=\; \mathbb{Z}[\pi_1(M)]\otimes_{\mathbb{Z}}\mathbb{Z}[\pi_1(M')]$$

is the product

$$(C'',\lambda'') \;=\; (C,\lambda)\otimes(C',\lambda')\ ,$$

with

$$\begin{aligned} C'' \;&=\; C(\widetilde{M}'',\widetilde{N}'')^{n+n'-*} \\ &=\; C(\widetilde{M},\widetilde{N})^{n-*}\otimes_{\mathbb{Z}} C(\widetilde{M}',\widetilde{N}')^{n'-*} \;=\; C\otimes_{\mathbb{Z}} C'\ . \end{aligned}$$

For $M = D^n$, $M' = D^{n'}$ this is the knot product operation of Kauffman [120] and Kauffman and Neumann [123] (cf. 29.22 below). □

28. Automorphism L-theory

This chapter deals with the L-theory of automorphisms of algebraic Poincaré complexes, using the localization $\Omega^{-1}A[z,z^{-1}]$ inverting the set Ω (13.3) of Fredholm matrices in the Laurent polynomial extension $A[z,z^{-1}]$ ($\overline{z} = z^{-1}$) of a ring with involution A. In the first instance, the duality involution on $Wh_1(\Omega^{-1}A[z,z^{-1}])$ and the fibering obstruction are used to prove that every manifold band is cobordant to a fibre bundle over S^1, allowing the identification of the closed framed spine bordism groups $\Delta_*(X,\mathfrak{F})$ of Chap. 27 for

$$\mathfrak{F} = \text{inclusion} : \mathbb{Z}[\pi_1(X)][z,z^{-1}] \longrightarrow \Omega^{-1}\mathbb{Z}[\pi_1(X)][z,z^{-1}]$$

with the bordism groups $\Delta_*(X)$ of automorphisms of manifolds over a space X.

The automorphism L-groups $L\text{Aut}^*_q(A)$ ($q = s,h,p$) are the cobordism groups of self homotopy equivalences of algebraic Poincaré complexes over a ring with involution A, the algebraic analogues of $\Delta_*(X)$. The main algebraic results of Chap. 28 are the identifications (28.17, 28.33)

$$L^*_h(A[z,z^{-1}],\Omega) = L\text{Aut}^{*-2}_p(A) \ , \ L^*_h(\Omega^{-1}A[z,z^{-1}]) = L\text{Asy}^*_h(A)$$

with $L\text{Asy}^*_h(A)$ the asymmetric L-groups of Chap. 26. The automorphism L-groups thus fit into the exact sequence

$$\begin{aligned}\dots \longrightarrow L^n_h(A[z,z^{-1}]) \longrightarrow L\text{Asy}^n_h(A) \longrightarrow L\text{Aut}^{n-2}_p(A) \\ \longrightarrow L^{n-1}_h(A[z,z^{-1}]) \longrightarrow \dots .\end{aligned}$$

The connection between the automorphism and asymmetric L-groups will be considered further in Chap. 29 below, in connection with the obstruction theory for open book decompositions. See Chap. 39 for the computation of the automorphism and asymmetric L-groups of a field with involution F.

The algebraic K-theory localization exact sequence

$$\begin{aligned}\dots \longrightarrow Wh_1(A[z,z^{-1}]) \xrightarrow{\ i\ } Wh_1(\Omega^{-1}A[z,z^{-1}]) \xrightarrow{\ \partial\ } Wh_1(A[z,z^{-1}],\Omega) \\ \xrightarrow{\ j\ } \widetilde{K}_0(A[z,z^{-1}]) \longrightarrow \dots\end{aligned}$$

splits for any ring A (by 13.19), with $j = 0$ and ∂ split by

$$\begin{aligned} \Delta : Wh_1(A[z,z^{-1}],\Omega) &= \mathrm{Aut}_0(A) \longrightarrow Wh_1(\Omega^{-1}A[z,z^{-1}]) ; \\ [P,h] &\longrightarrow \tau(z-h : \Omega^{-1}P[z,z^{-1}] \longrightarrow \Omega^{-1}P[z,z^{-1}]) . \end{aligned}$$

A based f.g. free $A[z,z^{-1}]$-module chain complex E is $\Omega^{-1}A[z,z^{-1}]$-contractible if and only if it is A-finitely dominated (i.e. a band), in which case

$$\begin{aligned} \tau(\Omega^{-1}E) &= (\Phi^+(E),[E,\zeta]) \\ &\in Wh_1(\Omega^{-1}A[z,z^{-1}]) = Wh_1(A[z,z^{-1}]) \oplus \mathrm{Aut}_0(A) . \end{aligned}$$

For a ring with involution A the involution

$$A[z,z^{-1}] \longrightarrow A[z,z^{-1}] ; \ z \longrightarrow z^{-1}$$

extends to isomorphisms of rings

$$A((z)) \longrightarrow A((z^{-1}))^{op} ; \ \sum_{j=-\infty}^{\infty} a_j z^j \longrightarrow \sum_{j=-\infty}^{\infty} \overline{a}_j z^{-j} ,$$

$$A((z^{-1})) \longrightarrow A((z))^{op} ; \ \sum_{j=-\infty}^{\infty} a_j z^j \longrightarrow \sum_{j=-\infty}^{\infty} \overline{a}_j z^{-j} .$$

28A. Algebraic Poincaré bands

Definition 28.1 An n-dimensional *algebraic Poincaré band* (C,ϕ) is an n-dimensional algebraic (= ϵ-symmetric or ϵ-quadratic) Poincaré complex over $A[z,z^{-1}]$ such that C is a chain complex band in the sense of 3.2, that is based f.g. free and A-finitely dominated. □

Example 28.2 The symmetric Poincaré complex of a geometric Poincaré band (15.13) is an algebraic Poincaré band. □

Proposition 28.3 (i) *The set Ω of Fredholm matrices in $A[z,z^{-1}]$ is involution-invariant, such that the Ω-duality involution on $Wh_1(A[z,z^{-1}],\Omega)$ corresponds to the involution*

$$* : \mathrm{Aut}_0(A) \longrightarrow \mathrm{Aut}_0(A) ; \ (P,h) \longrightarrow (P^*,(h^*)^{-1})$$

on the automorphism class group $\mathrm{Aut}_0(A)$.
(ii) *The algebraic K-theory localization exact sequence of 13.19*

$$0 \longrightarrow Wh_1(A[z,z^{-1}]) \xrightarrow{i} Wh_1(\Omega^{-1}A[z,z^{-1}]) \xrightarrow{\partial} \mathrm{Aut}_0(A) \longrightarrow 0$$

is a short exact sequence of $\mathbb{Z}[\mathbb{Z}_2]$-modules, which splits as a $\mathbb{Z}$-module exact sequence. The splitting map for ∂

$$\begin{aligned}\Delta\ :\ \mathrm{Aut}_0(A) &\longrightarrow Wh_1(\Omega^{-1}A[z,z^{-1}])\ ;\\ [P,h] &\longrightarrow \tau(z-h:\Omega^{-1}P[z,z^{-1}]\longrightarrow\Omega^{-1}P[z,z^{-1}])\end{aligned}$$

is such that

$$\Delta - *\Delta* \ =\ i\delta\ :\ \mathrm{Aut}_0(A)\longrightarrow Wh_1(\Omega^{-1}A[z,z^{-1}])\ ,$$

with

$$\begin{aligned}\delta\ :\ \mathrm{Aut}_0(A) &\longrightarrow Wh_1(A[z,z^{-1}])\ ;\\ [P,h] &\longrightarrow \tau(-zh:P[z,z^{-1}]\longrightarrow P[z,z^{-1}])\end{aligned}$$

such that

$$\delta * \ =\ -*\delta\ :\ \mathrm{Aut}_0(A)\longrightarrow Wh_1(A[z,z^{-1}])\ .$$

(iii) *A based f.g. free n-dimensional ϵ-symmetric Poincaré complex (C,ϕ) over $A[z,z^{-1}]$ is $\Omega^{-1}A[z,z^{-1}]$-contractible if and only if it is an algebraic Poincaré band (i.e. C is A-finitely dominated), in which case*

$$\begin{aligned}\tau_\Omega(C) &= [C,\zeta]\in Wh_1(A[z,z^{-1}],\Omega)\ =\ \mathrm{Aut}_0(A)\ ,\\ \tau(C,\phi) &= \Phi^+(C)+(-)^n\Phi^+(C)^*+\delta[C,\zeta]\in Wh_1(A[z,z^{-1}])\end{aligned}$$

and

$$\begin{aligned}\delta\ :\ \widehat{H}^{n-1}(\mathbb{Z}_2;\mathrm{Aut}_0(A)) &\longrightarrow \widehat{H}^n(\mathbb{Z}_2;Wh_1(A[z,z^{-1}]))\ ;\\ [C,\zeta] &\longrightarrow \delta[C,\zeta]\ =\ \tau(C,\phi)\ .\end{aligned}$$

(iv) *A based f.g. free n-dimensional ϵ-symmetric complex (C,ϕ) over $A[z,z^{-1}]$ is $\Omega^{-1}A[z,z^{-1}]$-Poincaré if and only if ∂C is A-finitely dominated, in which case*

$$\begin{aligned}\tau(\Omega^{-1}(C,\phi)) &= \tau(\Omega^{-1}\partial C)\ =\ (\Phi^+(\partial C),[\partial C,\zeta])\\ &\quad\in Wh_1(\Omega^{-1}A[z,z^{-1}])\ =\ Wh_1(A[z,z^{-1}])\oplus\mathrm{Aut}_0(A)\ ,\\ \tau(\partial C,\partial\phi) &= \Phi^+(\partial C)+(-)^{n+1}\Phi^+(\partial C)^*+\delta[\partial C,\zeta]\in Wh_1(A[z,z^{-1}])\ .\end{aligned}$$

Proof (i) Given a $k\times k$ Fredholm matrix $\omega=(\omega_{ij})$ in $A[z,z^{-1}]$ there is defined a f.g. projective A-module

$$P\ =\ \mathrm{coker}(z-\omega:A[z,z^{-1}]^k\longrightarrow A[z,z^{-1}]^k)$$

with an automorphism

$$h\ :\ P\longrightarrow P\ ;\ x\longrightarrow zx$$

and an $A[z,z^{-1}]$-module exact sequence

$$0\longrightarrow A[z,z^{-1}]^k\xrightarrow{z-\omega}A[z,z^{-1}]^k\longrightarrow P\longrightarrow 0\ .$$

The dual $k \times k$ matrix $\omega^* = (\overline{\omega}_{ji})$ is Fredholm, with an $A[z,z^{-1}]$-module exact sequence

$$0 \longrightarrow A[z,z^{-1}]^k \xrightarrow{z^{-1}-\omega^*} A[z,z^{-1}]^k \longrightarrow P^* \longrightarrow 0$$

where

$$z = (h^*)^{-1} : P^* \longrightarrow P^* .$$

(ii) Immediate from the definitions.
(iii) The algebraic mapping torus of a self homotopy equivalence $(h,\chi) : (E,\theta)\longrightarrow(E,\theta)$ of a finitely dominated $(n-1)$-dimensional ϵ-symmetric Poincaré complex (E,θ) over A is a homotopy finite n-dimensional ϵ-symmetric Poincaré complex $T(h,\chi)$ over $A[z,z^{-1}]$. The torsion of $T(h,\chi)$ with respect to the canonical finite structure is given by

$$\begin{aligned}\tau(T(h,\chi)) &= \tau(-zh : C[z,z^{-1}]\longrightarrow C[z,z^{-1}]) \\ &= \delta[C,h] = \delta[E,\zeta] \in Wh_1(A[z,z^{-1}]) .\end{aligned}$$

By 23.5 every $\Omega^{-1}A[z,z^{-1}]$-contractible based f.g. free n-dimensional ϵ-symmetric Poincaré complex (C,ϕ) over $A[z,z^{-1}]$ is homotopy equivalent to the algebraic mapping torus $T(h,\chi)$ of a homotopy equivalence $(h,\chi) : (E,\theta)\longrightarrow(E,\theta)$ of a f.g. projective $(n-1)$-dimensional ϵ-symmetric Poincaré complex (E,θ) over A with

$$[C,\zeta] = [E,h] \in \mathrm{Aut}_0(A) .$$

(iv) Apply (iii) to the boundary $(n-1)$-dimensional ϵ-symmetric Poincaré complex $\partial(C,\phi) = (\partial C, \partial\phi)$ over $A[z,z^{-1}]$. □

28B. Duality in automorphism K-theory

Let

$$\begin{aligned}\delta &: \mathrm{Aut}_0(A) \longrightarrow Wh_1(A[z,z^{-1}]) ; \\ &\qquad [P,h] \longrightarrow \tau(-zh : P[z,z^{-1}]\longrightarrow P[z,z^{-1}]) , \\ \delta_0 &: \mathrm{Aut}_0(A) \longrightarrow \widetilde{K}_0(A) ; \ [P,h] \longrightarrow [P]\end{aligned}$$

(with δ as in 28.3).

Definition 28.4 (i) The *simple automorphism class group* of A is

$$\begin{aligned}\mathrm{Aut}_0^s(A) &= \ker(\delta : \mathrm{Aut}_0(A)\longrightarrow Wh_1(A[z,z^{-1}])) \\ &= \{[P,h] \in \mathrm{Aut}_0(A) \,|\, P \text{ f.g. free and } \tau(h) = 0 \in Wh_1(A)\} \\ &= \{[P,f] + [P,g] - [P,gf] - [P,1]\} + \{[A,\pm 1]\} \subseteq \mathrm{Aut}_0(A) .\end{aligned}$$

(ii) The *free automorphism class group* of A is

$$\begin{aligned}\mathrm{Aut}_0^h(A) &= \ker(\delta_0 : \mathrm{Aut}_0(A) \longrightarrow \widetilde{K}_0(A)) \\ &= \{[P,h] \in \mathrm{Aut}_0(A) \,|\, P \text{ f.g. free}\} .\end{aligned}$$
□

Proposition 28.5 (i) *The $*$-invariant subgroups*

$$\mathrm{Aut}_0^s(A) \subseteq \mathrm{Aut}_0^h(A) \subseteq \mathrm{Aut}_0(A)$$

are such that there is defined a commutative braid of exact sequences

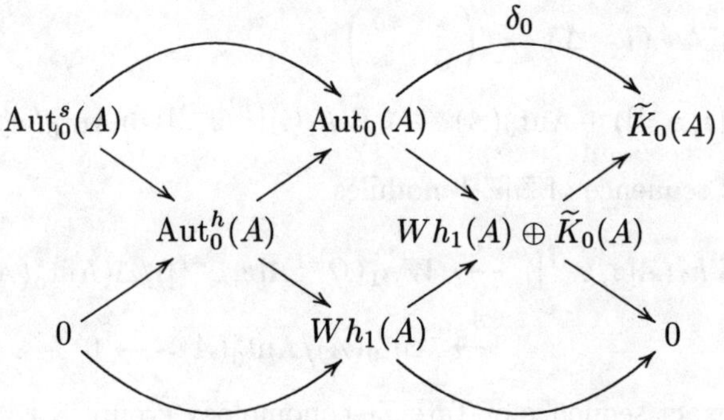

with $\delta_0 : \mathrm{Aut}_0(A) \longrightarrow \widetilde{K}_0(A)$ *split by*

$$\Delta_0 \;:\; \widetilde{K}_0(A) \longrightarrow \mathrm{Aut}_0(A) \;;\; [P] \longrightarrow [P,1] .$$

(ii) *The $*$-invariant subgroup*

$$\Delta(\mathrm{Aut}_0^s(A)) \subseteq Wh_1(\Omega^{-1}A[z,z^{-1}])$$

is such that

$$\begin{aligned}\widehat{H}^*(\mathbb{Z}_2;\Delta(\mathrm{Aut}_0^s(A))) &= \widehat{H}^*(\mathbb{Z}_2;Wh_1(\Omega^{-1}A[z,z^{-1}])) , \\ L^*_{\Delta(\mathrm{Aut}_0^s(A))}(\Omega^{-1}A[z,z^{-1}],\epsilon) &= L^*_h(\Omega^{-1}A[z,z^{-1}],\epsilon) .\end{aligned}$$

Proof (i) Immediate from the definitions.
(ii) The (geometrically significant) decomposition

$$Wh_1(A[z,z^{-1}]) = Wh_1(A) \oplus \widetilde{K}_0(A) \oplus \widetilde{\mathrm{Nil}}_0(A) \oplus \widetilde{\mathrm{Nil}}_0(A)$$

is such that

$$\begin{aligned}\mathrm{Aut}_0(A)/\mathrm{Aut}_0^s(A) &= \mathrm{im}(\delta : \mathrm{Aut}_0(A) \longrightarrow Wh_1(A[z,z^{-1}])) \\ &= Wh_1(A) \oplus \widetilde{K}_0(A) .\end{aligned}$$

The duality involution on $Wh_1(A[z,z^{-1}])$ interchanges the two $\widetilde{\mathrm{Nil}}$-components, so that

$$\begin{aligned}\widehat{H}^*(\mathbb{Z}_2;\mathrm{Aut}_0(A)/\mathrm{Aut}_0^s(A)) &= \widehat{H}^{*+1}(\mathbb{Z}_2;Wh_1(A[z,z^{-1}]))\\ &= \widehat{H}^{*+1}(\mathbb{Z}_2;Wh_1(A))\oplus \widehat{H}^*(\mathbb{Z}_2;\widetilde{K}_0(A))\ .\end{aligned}$$

The isomorphism

$$(i\ \ \Delta)\ :\ Wh_1(A[z,z^{-1}])\oplus \mathrm{Aut}_0(A) \xrightarrow{\simeq} Wh_1(\Omega^{-1}A[z,z^{-1}])$$

is such that

$$(i\ \ \Delta)^{-1}*(i\ \ \Delta)\ =\ \begin{pmatrix} * & *\delta \\ 0 & * \end{pmatrix}\ :$$

$$Wh_1(A[z,z^{-1}])\oplus \mathrm{Aut}_0(A)\longrightarrow Wh_1(A[z,z^{-1}])\oplus \mathrm{Aut}_0(A)\ .$$

The short exact sequence of $\mathbb{Z}[\mathbb{Z}_2]$-modules

$$\begin{aligned}0\longrightarrow Wh_1(A[z,z^{-1}]) &\xrightarrow{i} Wh_1(\Omega^{-1}A[z,z^{-1}])/\Delta(\mathrm{Aut}_0^s(A))\\ &\xrightarrow{\partial} \mathrm{Aut}_0(A)/\mathrm{Aut}_0^s(A)\longrightarrow 0\end{aligned}$$

induces a long exact sequence of Tate $\mathbb{Z}_2$-cohomology groups

$$\begin{aligned}\dots\longrightarrow\ &\widehat{H}^n(\mathbb{Z}_2;Wh_1(A[z,z^{-1}]))\\ \xrightarrow{i}\ &\widehat{H}^n(\mathbb{Z}_2;Wh_1(\Omega^{-1}A[z,z^{-1}])/\Delta(\mathrm{Aut}_0^s(A)))\\ \xrightarrow{\partial}\ &\widehat{H}^n(\mathbb{Z}_2;\mathrm{Aut}_0(A)/\mathrm{Aut}_0^s(A))\\ \xrightarrow{\delta}\ &\widehat{H}^{n-1}(\mathbb{Z}_2;Wh_1(A[z,z^{-1}]))\longrightarrow\dots\ .\end{aligned}$$

The connecting maps δ are isomorphisms, so that

$$\begin{aligned}\widehat{H}^*(\mathbb{Z}_2;Wh_1(\Omega^{-1}A[z,z^{-1}])/\Delta(\mathrm{Aut}_0^s(A))) &= 0\ ,\\ \widehat{H}^*(\mathbb{Z}_2;\Delta(\mathrm{Aut}_0^s(A))) &= \widehat{H}^*(\mathbb{Z}_2;Wh_1(\Omega^{-1}A[z,z^{-1}]))\ ,\\ L^*_{\Delta(\mathrm{Aut}_0^s(A))}(\Omega^{-1}A[z,z^{-1}],\epsilon) &= L^*_h(\Omega^{-1}A[z,z^{-1}],\epsilon)\ .\end{aligned}$$ □

Proposition 28.6 *For $n\geq 6$ every n-dimensional manifold band M is cobordant to a fibre bundle M' over S^1 by a cobordism $(W;M,M')$ which is a band.*
Proof Let $\pi_1(\overline{M})=\pi$, so that $\pi_1(M)=\pi\times\mathbb{Z}$, and write

$$\Lambda\ =\ \Omega^{-1}\mathbb{Z}[\pi][z,z^{-1}]\ .$$

The cellular $\mathbb{Z}[\pi][z,z^{-1}]$-module chain complex $C(\widetilde{M})$ of the universal cover $\widetilde{M}$ of M is Λ-contractible, with the Λ-coefficient torsion

$$\tau(M;\Lambda)\ =\ (\Phi^+(M),[\overline{M},\zeta])\in Wh_1(\Lambda)\ =\ Wh_1(\pi\times\mathbb{Z})\oplus \mathrm{Aut}_0(\mathbb{Z}[\pi])$$

such that

$$\begin{aligned}\tau(M;\Lambda)+(-)^{n+1}\tau(M;\Lambda)^* &= \Lambda\otimes\tau([M]\cap - : C(\widetilde{M})^{n-*}\longrightarrow C(\widetilde{M}))\\ &= 0\in Wh_1(\Lambda)\ .\end{aligned}$$

Given a $k\times k$ Fredholm matrix ω in $\mathbb{Z}[\pi][z,z^{-1}]$ let

$$\tau(\omega) \;=\; (\alpha,\beta)\in Wh_1(\Lambda) \;=\; Wh_1(\pi\times\mathbb{Z})\oplus \mathrm{Aut}_0(\mathbb{Z}[\pi])\ ,$$

so that

$$\tau(\omega)^* \;=\; \tau(\omega^*) \;=\; (\alpha^*+\delta\beta^*,\beta^*)\in Wh_1(\Lambda) \;=\; Wh_1(\pi\times\mathbb{Z})\oplus\mathrm{Aut}_0(\mathbb{Z}[\pi])\ .$$

For any integer $r\geq 2$ such that $2r<n$ let $(W;M,M')$ be the Λ-coefficient H-cobordism obtained by trivially attaching k r-handles to $M\times I$, and then using ω to attach k $(r+1)$-handles

$$W \;=\; (M\times I\cup\bigcup_k D^r\times D^{n-r+1})\cup_\omega\bigcup_k D^{r+1}\times D^{n-r}\ .$$

Then

$$\pi_1(M) \;=\; \pi_1(W) \;=\; \pi_1(M') \;=\; \pi\times\mathbb{Z}$$

and the universal covers $\widetilde{W},\widetilde{M}'$ of W,M' are such that the relative cellular chain complexes of $(\widetilde{W},\widetilde{M})$, $(\widetilde{W},\widetilde{M}')$ are given by

$$\begin{aligned}C(\widetilde{W},\widetilde{M}) &= S^r\mathcal{C}(\omega:\mathbb{Z}[\pi][z,z^{-1}]^k\longrightarrow\mathbb{Z}[\pi][z,z^{-1}]^k)\ ,\\ C(\widetilde{W},\widetilde{M}') &= S^{n-r+1}\mathcal{C}(\omega^*:\mathbb{Z}[\pi][z,z^{-1}]^k\longrightarrow\mathbb{Z}[\pi][z,z^{-1}]^k)\ .\end{aligned}$$

Thus W,M' are Λ-acyclic, and hence bands (13.9), with

$$\begin{aligned}\tau(W;\Lambda) &= \tau(M;\Lambda)+(-)^r\tau(\omega)\\ &= \tau(M';\Lambda)+(-)^{n-r+1}\tau(\omega)^*\in Wh_1(\Lambda)\ ,\end{aligned}$$

and

$$\begin{aligned}\Phi^+(W) &= \Phi^+(M)+(-)^r\alpha \;=\; \Phi^+(M')+(-)^{n-r+1}(\alpha^*+\delta\beta^*)\ ,\\ \Phi^+(M') &= \Phi^+(M)+(-)^r(\alpha+(-)^n\alpha^*+(-)^n\delta\beta^*)\in Wh_1(\pi\times\mathbb{Z})\ ,\\ [\overline{M'},\zeta'] &= [\overline{M},\zeta]+(-)^r(\beta+(-)^n\beta^*)\in\mathrm{Aut}_0(\mathbb{Z}[\pi])\ .\end{aligned}$$

By 28.5 the inclusion $\Delta(\mathrm{Aut}_0^s(\mathbb{Z}[\pi]))\longrightarrow Wh_1(\Lambda)$ induces an isomorphism

$$\widehat{H}^n(\mathbb{Z}_2;\Delta(\mathrm{Aut}_0^s(\mathbb{Z}[\pi])))\xrightarrow{\;\simeq\;}\widehat{H}^n(\mathbb{Z}_2;Wh_1(\Lambda))\ ,$$

so that for some $k\times k$ Fredholm matrix ω in $\mathbb{Z}[\pi][z,z^{-1}]$

$$\tau(M;\Lambda)+(-)^r(\tau(\omega)+(-)^n\tau(\omega)^*)\in\Delta(\mathrm{Aut}_0^s(\mathbb{Z}[\pi]))\subseteq Wh_1(\Lambda)\ .$$

In this case the fibering obstruction of M' is

$$\begin{aligned}\Phi^+(M') &= \Phi^+(M) + (-)^r(\alpha + (-)^n\alpha^* + (-)^n\delta\beta^*) \\ &= 0 \in Wh_1(\pi \times \mathbb{Z}) ,\end{aligned}$$

so that M' is a fibre bundle over S^1. □

Remark 28.7 (i) Given an n-dimensional manifold band M and an integer r such that $2r < n$ as in 28.6 let $\nu : \mathbb{Z}[\pi]^k \longrightarrow \mathbb{Z}[\pi]^k$ be a nilpotent endomorphism such that $(-)^{r+1}(\mathbb{Z}[\pi]^k, \nu) \in \widetilde{\mathrm{Nil}}_0(\mathbb{Z}[\pi])$ is one of the $\widetilde{\mathrm{Nil}}$-components of the fibering obstruction $\Phi^+(M) \in Wh_1(\pi \times \mathbb{Z})$, with

$$\begin{aligned}&\Phi^+(M) + (-)^r(\tau(1 - z\nu : \mathbb{Z}[\pi][z,z^{-1}]^k \longrightarrow \mathbb{Z}[\pi][z,z^{-1}]^k) \\ &\qquad + (-)^{n+1}\tau(1 - z^{-1}\nu^* : \mathbb{Z}[\pi][z,z^{-1}]^k \longrightarrow \mathbb{Z}[\pi][z,z^{-1}]^k)) \\ &\qquad \in \mathrm{im}(Wh_1(\pi) \oplus \widetilde{K}_0(\mathbb{Z}[\pi]) \longrightarrow Wh_1(\pi \times \mathbb{Z})) .\end{aligned}$$

Taking $\omega = 1 - z\nu$ in the construction of the proof of 28.6 gives the 'relaxation' h-cobordism $(W; M, M')$ from M to a 'relaxed' band M' with fibering obstruction

$$\begin{aligned}\Phi^+(M') &= \Phi^+(M) + (-)^r(\tau(\omega) + (-)^{n+1}\tau(\omega)^*) \\ &\in \mathrm{im}(Wh_1(\pi) \oplus \widetilde{K}_0(\mathbb{Z}[\pi]) \longrightarrow Wh_1(\pi \times \mathbb{Z}))\end{aligned}$$

such that the $\widetilde{\mathrm{Nil}}$-components are 0. See Siebenmann [266] and Hughes and Ranicki [113, Chaps. 18, 24] for the various relaxing properties of bands.
(ii) There is also a purely algebraic version of 28.6 : every based f.g. free simple n-dimensional algebraic Poincaré band over $A[z,z^{-1}]$ is band cobordant to the mapping torus of a simple self homotopy equivalence of a based f.g. free simple $(n-1)$-dimensional algebraic Poincaré complex over A. □

28C. Bordism of automorphisms of manifolds

The framed spine bordism groups $AB_*(X, \mathfrak{F})$, $\Delta_*(X, \mathfrak{F})$ of Chap. 27 for the Fredholm localization

$$\mathfrak{F} = \text{inclusion} : \mathbb{Z}[\pi_1(X)][z,z^{-1}] \longrightarrow \Omega^{-1}\mathbb{Z}[\pi_1(X)][z,z^{-1}]$$

will now be identified with the bordism groups $AB_*(X)$ of manifolds with boundary a fibre bundle over S^1 and the bordism groups $\Delta_*(X)$ of automorphisms of manifolds. The bordism groups $AB_*(X)$, $\Delta_*(X)$ will be related to automorphism L-theory in 28D, and to asymmetric L-theory in Chaps. 29, 30.

Definition 28.8 (i) The *automorphism bordism group* $\Delta_n(X)$ is the group of bordism classes of triples (F, g, h) with F a closed n-dimensional manifold, $g : F \longrightarrow X$ a map, and $h : F \longrightarrow F$ an automorphism together with a homotopy $g \simeq gh : F \longrightarrow X$, so that there is induced a map

$$T(g) \;:\; T(h) \longrightarrow T(1 : X \longrightarrow X) \;=\; X \times S^1 \;.$$

(ii) The *bounded fibre bundle bordism group* $AB_n(X)$ is the group of bordism classes of quintuples $(M, f; F, g, h)$ with (F, g, h) as in (i), M an $(n+2)$-dimensional manifold with boundary the fibre bundle $\partial M = T(h)$ over S^1, and $f : M \longrightarrow X \times S^1$ a map such that

$$f| \;=\; T(g) \;:\; \partial M \;=\; T(h) \longrightarrow X \times S^1 \;. \qquad \square$$

In other words, $\Delta_n(X)$ is the bordism group of closed $(n+1)$-dimensional manifolds with the structure of a fibre bundle over S^1 and a map to X.

Remark 28.9 Browder [29, 2.29] initiated the study of the bordism of diffeomorphisms of manifolds, using the mapping torus to pass from the bordism class of a diffeomorphism $h : F \longrightarrow F$ of a closed n-dimensional manifold F to the bordism class of the closed $(n+1)$-dimensional manifold

$$T(h) \;=\; F \times [0,1]/\{(x,0) = (h(x),1) \,|\, x \in F\} \;. \qquad \square$$

Proposition 28.10 (i) *Given a space X let*

$$\mathcal{F} \;:\; \mathbb{Z}[\pi_1(X)][z, z^{-1}] \longrightarrow \Omega^{-1}\mathbb{Z}[\pi_1(X)][z, z^{-1}]$$

be the inclusion in the Fredholm localization. The forgetful maps from the automorphism and fibre bundle groups to the framed spine bordism groups of Chap. 27

$$AB_n(X) \longrightarrow AB_n(X, \mathcal{F}) \;\;,\;\; \Delta_n(X) \longrightarrow \Delta_n(X, \mathcal{F})$$

are isomorphisms for $n \geq 5$, *so that for* $* \geq 5$ *there are identifications*

$$AB_*(X) \;=\; AB_*(X, \mathcal{F}) \;\;,\;\; \Delta_*(X) \;=\; \Delta_*(X, \mathcal{F}) \;.$$

(ii) *The automorphism and bounded fibre bundle bordism groups fit into an exact sequence*

$$\dots \longrightarrow \Omega_{n+2}(X \times S^1) \longrightarrow AB_n(X) \longrightarrow \Delta_n(X) \xrightarrow{T} \Omega_{n+1}(X \times S^1) \longrightarrow \dots$$

with

$$\Omega_{n+2}(X \times S^1) \longrightarrow AB_n(X) \;;\; (M, f) \longrightarrow (M, f; \emptyset, \emptyset, \emptyset) \;,$$

$$AB_n(X) \longrightarrow \Delta_n(X) \;;\; (M, f; F, g, h) \longrightarrow (F, g, h) \;,$$

$$T \;:\; \Delta_n(X) \longrightarrow \Omega_{n+1}(X \times S^1) \;;\; (F, g, h) \longrightarrow (T(h), T(g)) \;.$$

Proof (i) Immediate from 28.6.
(ii) Formal. $\square$

28D. The automorphism signature

The $(A[z,z^{-1}],\Omega)$-torsion L-groups $L^*(A[z,z^{-1}],\Omega,\epsilon)$ will now be identified with the automorphism L-groups $LAut^{*-2}(A,\epsilon)$. The automorphism signature of a self homotopy equivalence of a Poincaré complex is identified with signature of the mapping torus, in both algebra and topology.

Definition 28.11 (i) The $\begin{cases} \textit{$\epsilon$-symmetric} \\ \textit{$\epsilon$-quadratic} \end{cases}$ *autometric Q-groups* of a chain map $h: C\longrightarrow C$ are the Q-groups

$$\begin{cases} Q^*_{aut}(C,h,\epsilon) &= Q^*(C,h,1,\epsilon) \\ Q_*^{aut}(C,h,\epsilon) &= Q_*(C,h,1,\epsilon) \end{cases}$$

which fit into the exact sequence

$$\begin{cases} \dots \longrightarrow Q^{n+1}(C,\epsilon) \longrightarrow Q^{n+1}_{aut}(C,h,\epsilon) \\ \qquad\qquad \longrightarrow Q^n(C,\epsilon) \xrightarrow{h^{\%}-1^{\%}} Q^n(C,\epsilon) \longrightarrow \dots \\ \dots \longrightarrow Q_{n+1}(C,\epsilon) \longrightarrow Q^{aut}_{n+1}(C,h,\epsilon) \\ \qquad\qquad \longrightarrow Q_n(C,\epsilon) \xrightarrow{h_{\%}-1_{\%}} Q_n(C,\epsilon) \longrightarrow \dots \; . \end{cases}$$

(ii) An *autometric structure* $(h,\delta\phi)$ for an n-dimensional ϵ-symmetric complex (C,ϕ) over A is a self chain map $h: C\longrightarrow C$ together with a chain $\delta\phi\in W^{\%}C_{n+1}$ such that

$$\begin{aligned} h\phi_s h^* - \phi_s &= d\delta\phi_s + (-)^r\delta\phi_s d^* + (-)^{n+s-1}(\delta\phi_{s-1} + (-)^s T_\epsilon\delta\phi_{s-1}) \\ &: C^{n-r+s} \longrightarrow C_r \quad (r,s\geq 0, \delta\phi_{-1}=0) \; , \end{aligned}$$

representing an element $(\delta\phi,\phi)\in Q^{n+1}_{aut}(C,h,\epsilon)$, or equivalently a morphism $(h,\delta\phi):(C,\phi)\longrightarrow(C,\phi)$.
Similarly in the ϵ-quadratic case. □

Example 28.12 (i) An *autometric structure* for an ϵ-symmetric form (M,ϕ) over A is an endomorphism

$$h : (M,\phi) \longrightarrow (M,\phi) \; ,$$

that is $h\in \mathrm{Hom}_A(M,M)$ such that

$$h^*\phi h = \phi : M \longrightarrow M^* \; .$$

This is just an autometric structure $(h^*,0)$ on the 0-dimensional ϵ-symmetric complex (C,ϕ) with $C_0 = M^*$, $C_r = 0$ for $r\neq 0$.
(ii) An automorphism $h:(M,\phi)\longrightarrow(M,\phi)$ of an ϵ-symmetric form over A is an autometric structure such that $h: M\longrightarrow M$ is an automorphism. □

Proposition 28.13 (i) *The ϵ-symmetric Q-groups of an A-finitely dominated ($= \Omega^{-1}A[z,z^{-1}]$-contractible) finite f.g. free $A[z,z^{-1}]$-module chain complex D are the autometric ϵ-symmetric Q-groups of $(D^!,\zeta)$*

$$Q^*(D,\epsilon) \;=\; Q^*_{aut}(D^!,\zeta,\epsilon)\ .$$

(ii) *The homotopy equivalence classes of f.g. free n-dimensional $\Omega^{-1}A[z,z^{-1}]$ -contractible ϵ-symmetric Poincaré complexes (C,ϕ) over $A[z,z^{-1}]$ are in one-one correspondence with the homotopy equivalence classes of self homotopy equivalences of finitely dominated $(n-1)$-dimensional ϵ-symmetric Poincaré complexes over A.*

(iii) *The L-group $L^n_h(\Omega^{-1}A[z,z^{-1}],\epsilon)$ is (isomorphic to) the cobordism group of f.g. free n-dimensional ϵ-symmetric Poincaré pairs $(f: C\longrightarrow D,(\delta\phi,\phi))$ over $A[z,z^{-1}]$ such that (C,ϕ) is A-finitely dominated.*

Proof (i) Let (C,h) be a finite chain complex of $(A[z,z^{-1}],\Omega)$-modules, i.e. a finite f.g. projective A-module chain complex C together an automorphism $h: C\longrightarrow C$. The f.g. projective $A[z,z^{-1}]$-module chain complex

$$D \;=\; \mathcal{C}(z-h: C[z,z^{-1}]\longrightarrow C[z,z^{-1}])$$

is homology equivalent to (C,h), and every A-finitely dominated f.g. free $A[z,z^{-1}]$-module chain complex is chain equivalent to one of this type. The $\mathbb{Z}[\mathbb{Z}_2]$-module chain map

$$\mathcal{C}(h\otimes h - 1\otimes 1: C\otimes_A C\longrightarrow C\otimes_A C) \longrightarrow D\otimes_{A[z,z^{-1}]} D$$

defined by

$$\begin{aligned}\mathcal{C}(h\otimes h-1\otimes 1)_r \;&=\; (C\otimes_A C)_r\oplus(C\otimes_A C)_{r-1}\\ \longrightarrow (D\otimes_{A[z,z^{-1}]}D)_r \;&=\; (C[z,z^{-1}]\otimes_{A[z,z^{-1}]}C[z,z^{-1}])_r\\ &\qquad\oplus(C[z,z^{-1}]\otimes_{A[z,z^{-1}]}C[z,z^{-1}])_{r-1}\\ &\qquad\oplus(C[z,z^{-1}]\otimes_{A[z,z^{-1}]}C[z,z^{-1}])_{r-1}\\ &\qquad\oplus(C[z,z^{-1}]\otimes_{A[z,z^{-1}]}C[z,z^{-1}])_{r-2}\ ;\\ (u,v)\longrightarrow &((h^{-1}\otimes 1)u,(zh^{-1}\otimes 1)v,(1\otimes 1)v,0)\end{aligned}$$

is a homology equivalence inducing isomorphisms

$$Q^*_{aut}(C,h,\epsilon)\;\cong\; Q^*(D,\epsilon)\ ,$$

so that there is defined an exact sequence

$$\dots\longrightarrow Q^{n+1}(D,\epsilon)\longrightarrow Q^n(C,\epsilon)\xrightarrow{h^{\%}-1} Q^n(C,\epsilon)\longrightarrow Q^n(D,\epsilon)\longrightarrow\dots\ .$$

An $(n+1)$-dimensional ϵ-symmetric structure $\theta\in Q^{n+1}(D,\epsilon)$ on D is thus the same as an n-dimensional ϵ-symmetric structure $\phi\in Q^n(C,\epsilon)$ with a refinement to an autometric structure $(\delta\phi,\phi)\in Q^{n+1}_{aut}(C,h,\epsilon)$. Moreover

$$\theta_0 \; : \; H^{n+1-*}(D) \; \cong \; H^{n-*}(C) \xrightarrow{\phi_0} H_*(C) \; \cong \; H_*(D) \; ,$$

so that (D,θ) is a Poincaré complex over $A[z,z^{-1}]$ if and only if (C,ϕ) is a Poincaré complex over A.
(ii) The mapping torus defines a one-one correspondence between the homotopy equivalence classes of self homotopy equivalences and autometric structures of finitely dominated $(n-1)$-dimensional ϵ-symmetric Poincaré complexes over A. Self homotopy equivalences can thus be taken to be given by autometric structures, and (i) applies.
(iii) Every f.g. free n-dimensional ϵ-symmetric Poincaré complex over $A[z,z^{-1}]$ is (homotopy equivalent to) $\Omega^{-1}(E,\theta)$ for an n-dimensional ϵ-symmetric $\Omega^{-1}A[z,z^{-1}]$-Poincaré complex (E,θ) over $A[z,z^{-1}]$ (by 24.4 (i)), and the thickening of (E,θ) (20.22) is an n-dimensional ϵ-symmetric Poincaré pair $(f : C \longrightarrow D, (\delta\phi,\phi))$ over $A[z,z^{-1}]$ such that $(C,\phi) = \partial(E,\theta)$ is A-finitely dominated. □

Proposition 28.14 *The ϵ-symmetrization maps*

$$1+T_\epsilon \; : \; L_n^U(\Omega^{-1}A[z,z^{-1}],\epsilon) \longrightarrow L_U^n(\Omega^{-1}A[z,z^{-1}],\epsilon)$$

are isomorphisms, for any $*$-invariant subgroup $U \subseteq Wh_1(\Omega^{-1}A[z,z^{-1}])$.
Proof The central unit

$$s \; = \; (1-z)^{-1} \in \Omega^{-1}A[z,z^{-1}]$$

is such that

$$s+\overline{s} \; = \; 1 \in \Omega^{-1}A[z,z^{-1}] \; ,$$

so that 25.11 (i) applies. □

Remark 28.15 (i) A nonsingular ϵ-symmetric form (M,ϕ) over A determines a ring $\mathrm{Hom}_A(M,M)$ with involution $f \longrightarrow \phi^{-1}f^*\phi$. The surgery transfer group $L^0(R,A,\epsilon)$ of Lück and Ranicki [177] is defined for any rings with involution R,A to be the Witt group of triples (M,ϕ,U) with (M,ϕ) a nonsingular ϵ-symmetric form over A and $U : R \longrightarrow \mathrm{Hom}_A(M,M)^{op}$ a morphism of rings with involution. (See 28.24 below for the connection with surgery transfer). For a group ring $R = \mathbb{Z}[\pi]$ $L^0(\mathbb{Z}[\pi],A,\epsilon)$ is the π-equivariant Witt group studied by Dress [60] for finite π and by Neumann [212] for infinite π.
(ii) If A is a Dedekind ring with quotient field $F = S^{-1}A$ $(S = A\backslash\{0\})$ the L-groups $L^n(R,A,\epsilon)$ are defined in [177] for all $n \geq 0$. The even-dimensional group

$$L^{2i}(R,A,\epsilon) \; = \; L^0(R,A,(-)^i\epsilon)$$

is just the $(-)^i\epsilon$-symmetric L-group of (R,A), as in (i). A nonsingular ϵ-symmetric linking form (L,λ) over A determines a ring $\mathrm{Hom}_A(L,L)$ with involution $f \longrightarrow \lambda^{-1}f^{\wedge}\lambda$. The odd-dimensional group

$$L^{2i+1}(R,A,\epsilon) \; = \; L^1(R,A,(-)^i\epsilon)$$

is the quotient of the Witt group of triples (L,λ,T) with (L,λ) a nonsingular $(-)^{i+1}\epsilon$-symmetric linking form over (A,S) and $T: R\longrightarrow \mathrm{Hom}_A(L,L)^{op})$ a morphism of rings with involution by the boundaries $\partial(N,\theta,V)$ of F-nonsingular $(-)^{i+1}\epsilon$-symmetric forms (N,θ) over A with a morphism of rings $V: R\longrightarrow \mathrm{Hom}_A(N,N)^{op}$ such that the composite

$$R \xrightarrow{V} \mathrm{Hom}_A(N,N)^{op} \longrightarrow \mathrm{Hom}_F(F\otimes_A N, F\otimes_A N)^{op}$$

is a morphism of rings with involution. In particular, for $R=\mathbb{Z}$

$$L^n(\mathbb{Z},A,\epsilon) \;=\; L^n(A,\epsilon)\ .$$

(iii) An n-dimensional ϵ-symmetric Poincaré complex (C,ϕ) over A determines a ring $H_0(\mathrm{Hom}_A(C,C))^{op}$ with involution $f\longrightarrow \phi^{-1}f^*\phi$. An autometric structure $(h,\delta\phi)$ on an n-dimensional ϵ-symmetric Poincaré complex (C,ϕ) over A determines an element $h\in H_0(\mathrm{Hom}_A(C,C))^{op}$ such that $\overline{h}=h^{-1}$, so that there is defined a morphism of rings with involution

$$\rho\;:\;\mathbb{Z}[z,z^{-1}]\longrightarrow H_0(\mathrm{Hom}_A(C,C))^{op}\ ;\ z\longrightarrow h\ .$$

If $n=0$ or if A is a Dedekind ring it is possible to identify

$$L\mathrm{Aut}^n(A,\epsilon)\;=\;L^n(\mathbb{Z}[z,z^{-1}],A,\epsilon)\ .$$ □

Definition 28.16 Let $U\subseteq \mathrm{Aut}_0(A)$ be a $*$-invariant subgroup. The *ϵ-symmetric automorphism L-groups* $L\mathrm{Aut}^n_U(A,\epsilon)$ are the cobordism groups of self homotopy equivalences $(h,\chi):(E,\theta)\longrightarrow(E,\theta)$ of finitely dominated n-dimensional ϵ-symmetric Poincaré complexes (E,θ) over A with automorphism class

$$[E,h]\in U\subseteq \mathrm{Aut}_0(A)\ .$$

Similarly for the *ϵ-quadratic automorphism L-groups* $L\mathrm{Aut}^U_n(A,\epsilon)$. □

Proposition 28.17 (i) *For any $*$-invariant subgroup $U\subseteq \mathrm{Aut}_0(A)$ there are natural identifications of the $(A[z,z^{-1}],\Omega)$-torsion L-groups of $A[z,z^{-1}]$ and the automorphism L-groups of A*

$$L^n_U(A[z,z^{-1}],\Omega,\epsilon)\;=\;L\mathrm{Aut}^{n-2}_U(A,\epsilon)\ ,$$

and there is a localization exact sequence

$$\begin{aligned}\dots\longrightarrow L^n_{\delta(U)}(A[z,z^{-1}],\epsilon)&\xrightarrow{i} L^n_{\delta(U)\oplus\Delta(U)}(\Omega^{-1}A[z,z^{-1}],\epsilon)\\ &\xrightarrow{\partial} L\mathrm{Aut}^{n-2}_U(A,\epsilon)\xrightarrow{T} L^{n-1}_{\delta(U)}(A[z,z^{-1}],\epsilon)\longrightarrow\dots\ ,\end{aligned}$$

with T defined by the algebraic mapping torus.
(ii) *The ϵ-symmetric automorphism L-groups associated to a pair $(U_2,U_1\subseteq U_2)$ of $*$-invariant subgroups $U_1\subseteq U_2\subseteq \mathrm{Aut}_0(A)$ are related by a Rothenberg-type exact sequence*

$$\ldots \longrightarrow \widehat{H}^{n+1}(\mathbb{Z}_2;U_2/U_1) \longrightarrow LAut^n_{U_1}(A,\epsilon) \\ \longrightarrow LAut^n_{U_2}(A,\epsilon) \longrightarrow \widehat{H}^n(\mathbb{Z}_2;U_2/U_1) \longrightarrow \ldots$$

with

$$LAut^n_{U_2}(A,\epsilon) \longrightarrow \widehat{H}^n(\mathbb{Z}_2;U_2/U_1) \; ; \; ((h,\chi),(E,\theta)) \longrightarrow [E,h] \; .$$

(iii) *If* $U \subseteq \mathrm{Aut}_0(A)$ *is a* $*$*-invariant subgroup such that* $\mathrm{Aut}^s_0(A) \subseteq U$ *then*

$$L^*_{\delta(U)\oplus\Delta(U)}(\Omega^{-1}A[z,z^{-1}],\epsilon) \;=\; L^*_h(\Omega^{-1}A[z,z^{-1}],\epsilon) \; .$$

Similarly in the ϵ*-quadratic case.*

Proof (i) Apply 25.25, with (A,Σ) replaced by $(A[z,z^{-1}],\Omega)$.
(ii) As for the ordinary L-groups in Ranicki [235, Chap. 10].
(iii) As in 28.5 (ii) (the special case $U = \mathrm{Aut}^s_0(A)$) the inclusion

$$\delta(U) \oplus \Delta(U) \longrightarrow Wh_1(\Omega^{-1}A[z,z^{-1}])$$

induces isomorphisms in the Tate $\mathbb{Z}_2$-cohomology $\widehat{H}^*(\mathbb{Z}_2;\ )$. □

Terminology 28.18 (i) Write

$$\begin{cases} LAut^n_p(A,\epsilon) \;=\; LAut^n_{\mathrm{Aut}_0(A)}(A,\epsilon) \\ LAut^n_h(A,\epsilon) \;=\; LAut^n_{\mathrm{Aut}^h_0(A)}(A,\epsilon) \\ LAut^n_s(A,\epsilon) \;=\; LAut^n_{\mathrm{Aut}^s_0(A)}(A,\epsilon) \; . \end{cases}$$

Thus $\begin{cases} LAut^n_p(A,\epsilon) \\ LAut^n_h(A,\epsilon) \\ LAut^n_s(A,\epsilon) \end{cases}$ is the cobordism group of $\begin{cases} - \\ - \\ \text{simple} \end{cases}$ self homotopy equivalences $(h,\chi) : (E,\theta) \longrightarrow (E,\theta)$ of $\begin{cases} \text{f.g. projective} \\ \text{f.g. free} \\ \text{f.g. free} \end{cases}$ n-dimensional ϵ-symmetric Poincaré complexes (E,θ) over A.
(ii) In the case $\epsilon = 1$ the terminology is abbreviated

$$LAut^*_U(A,1) \;=\; LAut^*_U(A) \; ,$$
$$LAut^*_q(A,1) \;=\; LAut^*_q(A) \; .$$

□

Proposition 28.19 *The* ϵ*-symmetric automorphism* L*-groups* $LAut^*_q(A,\epsilon)$ *for* $q = s,h,p$ *are related to each other and to the* ϵ*-symmetric* L*-groups of* $A[z,z^{-1}]$ *and* $\Omega^{-1}A[z,z^{-1}]$ *by the commutative braids of exact sequences*

$$\begin{array}{ccccccccc}
L_h^{n+2}(\Omega^{-1}A[z,z^{-1}],\epsilon) & & & & LAut_p^n(A,\epsilon) & & & & \widehat{H}^{n+1}(\mathbb{Z}_2;Wh_1(A[z,z^{-1}])) \\
& \searrow & & \nearrow & & \searrow & & \nearrow & \\
& & LAut_s^n(A,\epsilon) & & & & L_h^{n+1}(A[z,z^{-1}],\epsilon) & & \\
& \nearrow & & \searrow & & \nearrow & & \searrow & \\
\widehat{H}^{n+2}(\mathbb{Z}_2;Wh_1(A[z,z^{-1}])) & & & & L_s^{n+1}(A[z,z^{-1}],\epsilon) & & & & L_h^{n+1}(\Omega^{-1}A[z,z^{-1}],\epsilon)
\end{array}$$

$$\begin{array}{ccccccccc}
L_h^{n+2}(\Omega^{-1}A[z,z^{-1}],\epsilon) & & & & LAut_p^n(A,\epsilon) & & & & \widehat{H}^{n}(\mathbb{Z}_2;\widetilde{K}_0(A)) \\
& \searrow & & \nearrow & & \searrow & & \nearrow & \\
& & LAut_h^n(A,\epsilon) & & & & L_h^{n+1}(A[z,z^{-1}],\epsilon) & & \\
& \nearrow & & \searrow & & \nearrow & & \searrow & \\
\widehat{H}^{n+1}(\mathbb{Z}_2;\widetilde{K}_0(A)) & & & & L_{Wh_1(A)}^{n+1}(A[z,z^{-1}],\epsilon) & & & & L_h^{n+1}(\Omega^{-1}A[z,z^{-1}],\epsilon)
\end{array}$$

$$\begin{array}{ccccccccc}
L_h^{n+2}(\Omega^{-1}A[z,z^{-1}],\epsilon) & & & & LAut_h^n(A,\epsilon) & & & & \widehat{H}^{n+1}(\mathbb{Z}_2;Wh_1(A)) \\
& \searrow & & \nearrow & & \searrow & & \nearrow & \\
& & LAut_s^n(A,\epsilon) & & & & L_{Wh_1(A)}^{n+1}(A[z,z^{-1}],\epsilon) & & \\
& \nearrow & & \searrow & & \nearrow & & \searrow & \\
\widehat{H}^{n+2}(\mathbb{Z}_2;Wh_1(A)) & & & & L_s^{n+1}(A[z,z^{-1}],\epsilon) & & & & L_h^{n+1}(\Omega^{-1}A[z,z^{-1}],\epsilon)
\end{array}$$

$$\begin{array}{ccccccccc}
LAut_s^n(A,\epsilon) & & & & LAut_p^n(A,\epsilon) & & & & \widehat{H}^{n-1}(\mathbb{Z}_2;\widetilde{K}_0(A)) \\
& \searrow & & \nearrow & & \searrow & & \nearrow & \\
& & LAut_h^n(A,\epsilon) & & & & \widehat{H}^{n}(\mathbb{Z}_2;Wh_1(A[z,z^{-1}])) & & \\
& \nearrow & & \searrow & & \nearrow & & \searrow & \\
\widehat{H}^{n}(\mathbb{Z}_2;\widetilde{K}_0(A)) & & & & \widehat{H}^{n}(\mathbb{Z}_2;Wh_1(A)) & & & & LAut_s^{n-1}(A,\epsilon)
\end{array}$$

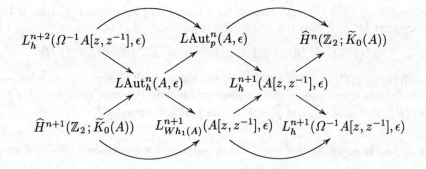

with

$$\begin{aligned}
&LAut^n_p(A,\epsilon) \longrightarrow \widehat{H}^{n+1}(\mathbb{Z}_2;Wh_1(A[z,z^{-1}])) \ ; \\
&\qquad ((h,\chi):(E,\theta)\longrightarrow(E,\theta)) \longrightarrow \tau(-zh:E[z,z^{-1}]\longrightarrow E[z,z^{-1}]) \ , \\
&LAut^n_h(A,\epsilon) \longrightarrow \widehat{H}^{n}(\mathbb{Z}_2;Wh_1(A)) \ ; \ ((h,\chi):(E,\theta)\longrightarrow(E,\theta)) \longrightarrow \tau(h) \ .
\end{aligned}$$

Similarly in the ϵ-quadratic case. □

Definition 28.20 Let $h: F\longrightarrow F$ be a self homotopy equivalence of a finitely dominated n-dimensional geometric Poincaré complex F with a map $g: F\longrightarrow X$ and a homotopy $gh \simeq g: F\longrightarrow X$, for some space X with universal cover $\widetilde{X}$ and fundamental group $\pi_1(X)=\pi$. The *automorphism symmetric signature* of (F,h,f) with respect to a $*$-invariant subgroup $U\subseteq Aut_0(\mathbb{Z}[\pi])$ such that $[F,h]\in U$ is the cobordism class

$$\sigma^*(F,h) \ = \ (\widetilde{h},\chi)\in LAut^n_U(\mathbb{Z}[\pi]) \ = \ L^{n+2}_U(\mathbb{Z}[\pi][z,z^{-1}],\Omega)$$

with $(\widetilde{h},\chi):(C(\widetilde{F}),\phi)\longrightarrow(C(\widetilde{F}),\phi)$ the induced self homotopy equivalence of the finitely dominated n-dimensional symmetric Poincaré complex $\sigma^*(F)=(C(\widetilde{F}),\phi)$ over $\mathbb{Z}[\pi]$, and $\widetilde{F}=g^*\widetilde{X}$ the pullback cover of F. □

Proposition 28.21 *A* $\begin{cases} - \\ - \\ \textit{simple}\end{cases}$ *self homotopy equivalence $h: F\longrightarrow F$ of a* $\begin{cases}\textit{finitely dominated} \\ \textit{finite} \\ \textit{finite}\end{cases}$ *n-dimensional geometric Poincaré complex F with a map $g: F\longrightarrow X$ and a homotopy $gh\simeq g: F\longrightarrow X$ for some space X with universal cover $\widetilde{X}$ and fundamental group $\pi_1(X)=\pi$ has an invariant*

$$\begin{cases}\sigma^*(F,h)\in LAut^n_p(\mathbb{Z}[\pi]) \\ \sigma^*(F,h)\in LAut^n_h(\mathbb{Z}[\pi]) \\ \sigma^*(F,h)\in LAut^n_s(\mathbb{Z}[\pi]) \ .\end{cases}$$

The mapping torus $T(h)$ of a self homotopy equivalence $h: F\longrightarrow F$ of a $\begin{cases}\textit{finitely dominated} \\ \textit{finite}\end{cases}$ *n-dimensional geometric Poincaré complex F is a* $\begin{cases}\textit{homotopy finite} \\ \textit{simple}\end{cases}$ *$(n+1)$-dimensional geometric Poincaré complex, with the symmetric signature given by*

$$\begin{cases}
\sigma^*(T(h)) \ = \ [\sigma^*(F,h)] \\
\qquad \in \mathrm{im}(T: LAut^n_p(\mathbb{Z}[\pi])\longrightarrow L^{n+1}_h(\mathbb{Z}[\pi][z,z^{-1}])) \\
\qquad\quad = \ \ker(i: L^{n+1}_h(\mathbb{Z}[\pi][z,z^{-1}])\longrightarrow L^{n+1}_h(\Omega^{-1}\mathbb{Z}[\pi][z,z^{-1}])) \ , \\
\sigma^*(T(h)) \ = \ [\sigma^*(F,h)] \\
\qquad \in \mathrm{im}(T: LAut^n_h(\mathbb{Z}[\pi])\longrightarrow L^{n+1}_s(\mathbb{Z}[\pi][z,z^{-1}])) \\
\qquad = \ \ker(i: L^{n+1}_s(\mathbb{Z}[\pi][z,z^{-1}])\longrightarrow L^{n+1}_h(\Omega^{-1}\mathbb{Z}[\pi][z,z^{-1}])) \ .
\end{cases}$$

Proof The symmetric Poincaré complex of a mapping torus of a self homotopy equivalence of a geometric Poincaré complex is the algebraic mapping torus of the induced self chain equivalence of a symmetric Poincaré complex. □

Proposition 28.22 (Ranicki [248])
Let X be a connected space, and let

$$A = \mathbb{Z}[\pi_1(X)] \ , \ \Lambda = \Omega^{-1}A[z,z^{-1}] \ .$$

(i) *The symmetric signatures define a natural transformation of exact sequences*

$$\begin{array}{ccccccc} \to \Omega_{n+2}(X\times S^1) & \to & AB_n(X) & \longrightarrow & \Delta_n(X) & \xrightarrow{T} & \Omega_{n+1}(X\times S^1) \to \\ \downarrow \sigma^* & & \downarrow \sigma^* & & \downarrow \sigma^* & & \downarrow \sigma^* \\ \to L^{n+2}_t(A[z,z^{-1}]) & \xrightarrow{i} & L^{n+2}_h(\Lambda) & \xrightarrow{\partial} & L^{n+2}_t(A[z,z^{-1}],\Omega) & \xrightarrow{j} & L^{n+1}_t(A[z,z^{-1}]) \to \end{array}$$

with

$$L^{n+2}_t(A[z,z^{-1}],\Omega) = L\mathrm{Aut}^n_u(A) \quad (28.19)$$

for $(t,u)=(h,p)$, $(Wh_1(A),h)$ or (s,s).
(ii) *For finitely presented $\pi_1(X)$ and $n\geq 5$ the symmetric signature maps are isomorphisms*

$$\sigma^* \ : \ AB_n(X) \xrightarrow{\simeq} L^{n+2}_h(\Lambda) \ ,$$

and the automorphism bordism groups $\Delta_*(X)$ fit into an exact sequence

$$\begin{aligned} \dots \longrightarrow L^{n+2}_t(A[z,z^{-1}]) \longrightarrow \Delta_n(X) \longrightarrow L\mathrm{Aut}^n_u(A)\oplus \Omega_{n+1}(X\times S^1) \\ \longrightarrow L^{n+1}_t(A[z,z^{-1}]) \longrightarrow \dots \end{aligned}$$

with (t,u) as in (i).
Proof (i) Use 28.20, taking into account that an automorphism $h: F\longrightarrow F$ of a compact manifold is a simple self homotopy equivalence of a simple geometric Poincaré complex, and that $T(h)$ is Ω-contractible with torsion

$$\tau(\Omega^{-1}T(h)) = \Delta[F,h] \in \Delta(\mathrm{Aut}^s_0(A)) \subseteq Wh_1(A[z,z^{-1}]) \ .$$

The braids of 28.19 include the L-theory localization exact sequences for each of the pairs of categories (t,u).
(ii) This is a special case of 27.14. □

Example 28.23 (i) Let $h: F\longrightarrow F$ be an automorphism of a $2k$-dimensional manifold F, with a map $g: F\longrightarrow X$ to a space X and a homotopy $gh\simeq g: F\longrightarrow X$. For any morphism $\mathbb{Z}[\pi_1(X)]\longrightarrow A$ to a Dedekind ring with involution A there is defined an automorphism symmetric signature

$$\begin{aligned} A\otimes_{\mathbb{Z}[\pi_1(X)]}\sigma^*(F,h) &= (H_k(F;A),1\otimes\phi_0,h_*) \\ &\in L\mathrm{Aut}^{2k}(A) = L\mathrm{Aut}^0(A,(-)^k) \ . \end{aligned}$$

This is the isometric invariant of López de Medrano [172], the Witt class of the automorphism

$$h_* \ : \ (H_k(F;A)/\text{torsion}, 1 \otimes \phi_0) \longrightarrow (H_k(F;A)/\text{torsion}, 1 \otimes \phi_0)$$

of the nonsingular $(-)^k$-symmetric intersection form over A. See Neumann [212] for the computation of the $\mathbb{Z}$-equivariant $(-)^k$-symmetric Witt ring $LAut^{2k}(\mathbb{Z}) = LAut^0(\mathbb{Z}, (-)^k)$. The exact sequence of 28.22 for $X = \{\text{pt.}\}$

$$\begin{aligned} \dots \longrightarrow L^{n+2}(\mathbb{Z}[z,z^{-1}]) \longrightarrow \Delta_n(\text{pt.}) \\ \longrightarrow LAut^n(\mathbb{Z}) \oplus \Omega_{n+1}(S^1) \longrightarrow L^{n+1}(\mathbb{Z}[z,z^{-1}]) \longrightarrow \dots \end{aligned}$$

recovers the computation of $\Delta_n(\text{pt.})$ $(n \geq 3)$ due to Kreck [141, 5.7].
(ii) See Bonahon [24] and Edmonds and Ewing [62] for the computation of $\Delta_2(\text{pt.})$, the bordism group of automorphisms of surfaces. □

Example 28.24 The algebraic surgery transfer map of Lück and Ranicki [175], [177]

$$p^! \ : \ L_m(\mathbb{Z}[\pi_1(B)]) \longrightarrow L_{m+n}(\mathbb{Z}[\pi_1(E)])$$

is defined for any fibration $F \longrightarrow E \xrightarrow{p} B$ with the fibre F an n-dimensional geometric Poincaré complex. The surgery transfer group $L^n(\mathbb{Z}[\pi_1(B)], \mathbb{Z})$ is defined in [177] (cf. 28.15), with products

$$L^n(\mathbb{Z}[\pi_1(B)], \mathbb{Z}) \otimes L_m(\mathbb{Z}[\pi_1(B)]) \longrightarrow L_{m+n}(\mathbb{Z}[\pi_1(B)]) \ .$$

The $\pi_1(B)$-equivariant symmetric signature

$$\sigma^*(F,p) \in L^n(\mathbb{Z}[\pi_1(B)], \mathbb{Z})$$

is such that

$$p_! p^! \ = \ \sigma^*(F,p) \otimes - \ : \ L_m(\mathbb{Z}[\pi_1(B)]) \longrightarrow L_{m+n}(\mathbb{Z}[\pi_1(B)]) \ .$$

In particular, for any element $(F, g, h) \in \Delta_n(X)$ there is defined a fibre bundle

$$F \longrightarrow E \ = \ T(h) \xrightarrow{p} B \ = \ S^1$$

and the $\mathbb{Z}$-equivariant symmetric signature is just the $\mathbb{Z}$-coefficient automorphism symmetric signature

$$\sigma^*(F,p) \ = \ \mathbb{Z} \otimes_{\mathbb{Z}[\pi_1(X)]} \sigma^*(F,h) \in L^n(\mathbb{Z}, \mathbb{Z}) \ = \ LAut^n(\mathbb{Z}) \ . \qquad \square$$

28E. The trace map χ_z

The identification given by 28.17

$$L^0(A[z,z^{-1}],\Omega,\epsilon) \;=\; LAut^0(A,-\epsilon)$$

is somewhat indirect: this will now be remedied by obtaining a direct one-one correspondence between ϵ-symmetric linking forms over $(A[z,z^{-1}],\Omega)$ and automorphisms of $(-\epsilon)$-symmetric forms over A. This extends to the L-theory of $A[z,z^{-1}]$ ($\overline{z}=z^{-1}$) the identification of exact categories given by 13.16

$$\mathbb{H}(A[z,z^{-1}],\Omega) \;=\; Aut(A)\;.$$

Let $A[[z,z^{-1}]]$ be the $A[z,z^{-1}]$-module consisting of all the formal Laurent polynomials $\sum\limits_{j=-\infty}^{\infty} a_j z^j$, without any finiteness conditions on the coefficients $a_j \in A$. The inclusions

$$\begin{array}{l} i_+ \;:\; A[z,z^{-1}] \longrightarrow A((z))\;,\; i_- \;:\; A[z,z^{-1}] \longrightarrow A((z^{-1}))\;,\\ j_+ \;:\; A((z)) \longrightarrow A[[z,z^{-1}]]\;,\; j_- \;:\; A((z^{-1})) \longrightarrow A[[z,z^{-1}]] \end{array}$$

fit into a short exact sequence of $A[z,z^{-1}]$-modules

$$0 \longrightarrow A[z,z^{-1}] \xrightarrow{\begin{pmatrix} i_+ \\ i_- \end{pmatrix}} A((z))\oplus A((z^{-1})) \xrightarrow{(j_+ \;\; -j_-)} A[[z,z^{-1}]] \longrightarrow 0\;.$$

Every Fredholm matrix $\omega \in \Omega$ is invertible over $A((z))$ and $A((z^{-1}))$ (13.1), so that there are defined ring morphisms

$$k_+ \;:\; \Omega^{-1}A[z,z^{-1}] \longrightarrow A((z))\;,\;\; k_- \;:\; \Omega^{-1}A[z,z^{-1}] \longrightarrow A((z^{-1}))$$

such that

$$\begin{array}{l} i_+ \;:\; A[z,z^{-1}] \longrightarrow \Omega^{-1}A[z,z^{-1}] \xrightarrow{k_+} A((z))\;,\\ i_- \;:\; A[z,z^{-1}] \longrightarrow \Omega^{-1}A[z,z^{-1}] \xrightarrow{k_-} A((z^{-1}))\;,\\ j_+k_+ \;=\; j_-k_- \;:\; \Omega^{-1}A[z,z^{-1}] \longrightarrow A[[z,z^{-1}]]\;. \end{array}$$

Definition 28.25 The *universal trace map* is the A-module morphism

$$\chi_z \;:\; \Omega^{-1}A[z,z^{-1}]/A[z,z^{-1}] \longrightarrow A\;;\; [\omega] \longrightarrow a_0^+ - a_0^-$$

with $a_0^+, a_0^- \in A$ the constant coefficients in

$$k_+(\omega) \;=\; \sum_{j=-\infty}^{\infty} a_j^+ z^j \in A((z))\;,\; k_-(\omega) \;=\; \sum_{j=-\infty}^{\infty} a_j^- z^j \in A((z^{-1}))\;. \quad \square$$

Remark 28.26 For a field $A = F$ the universal trace map

$$\chi_z \ : \ \Omega^{-1}A[z,z^{-1}]/A[z,z^{-1}] \ = \ F(z)/F[z,z^{-1}] \longrightarrow F$$

is the version due to Litherland [168, A3] of the trace function of Trotter [291], [292]. □

Proposition 28.27 (i) *If*

$$p(z) \ = \ \sum_{j=0}^{d} a_j z^j \ , \ q(z) = \sum_{k=0}^{d-1} b_k z^k \in A[z,z^{-1}]$$

are such that the coefficients $a_j \in A$ are central and $a_0, a_d \in A^\bullet$ then

$$\chi_z\left(\frac{q(z)}{p(z)}\right) \ = \ \frac{b_0}{a_0} \in A \ .$$

(ii) *For any $\omega \in \Omega^{-1}A[z,z^{-1}]/A[z,z^{-1}]$*

$$\overline{\chi_z(\omega)} \ = \ -\chi_z(\overline{\omega}) \in A \ .$$

(iii) *For any $(A[z,z^{-1}],\Omega)$-module L (= f.g. projective A-module L with an automorphism $z : L \longrightarrow L$) the universal trace map χ_z induces an $A[z,z^{-1}]$-module isomorphism*

$$L^\frown \ = \ \mathrm{Hom}_{A[z,z^{-1}]}(L, \Omega^{-1}A[z,z^{-1}]/A[z,z^{-1}]) \xrightarrow{\ \simeq\ } L^* \ = \ \mathrm{Hom}_A(L,A) \ ;$$
$$f \longrightarrow \chi_z f \ .$$

The identification of 28.15

$$L^0(A[z,z^{-1}],\Omega,\epsilon) \ = \ LAut^0(A,-\epsilon)$$

is induced by the one-one correspondence

$$\{\epsilon\text{-symmetric linking forms } (L,\lambda) \text{ over } (A[z,z^{-1}],\Omega)\} \rightleftarrows$$
$$\{(-\epsilon)\text{-symmetric forms } (L,\phi) \text{ over } A$$
$$\text{with an automorphism } h : (L,\phi) \longrightarrow (L,\phi)\}$$

with

$$h \ = \ z \ : \ L \longrightarrow L \ ,$$
$$\phi \ = \ \chi_z\lambda \ : \ L \times L \xrightarrow{\ \lambda\ } \Omega^{-1}A[z,z^{-1}]/A[z,z^{-1}] \xrightarrow{\ \chi_z\ } A \ .$$

Proof (i) The polynomial $p(z) \in A[z,z^{-1}]$ is a 1×1 Fredholm matrix, with

$$k_+\left(\frac{q(z)}{p(z)}\right)_0 \ = \ \frac{b_0}{a_0} \ , \ k_-\left(\frac{q(z)}{p(z)}\right)_0 \ = \ 0 \in A \ .$$

(ii) The identity $\overline{\chi_z(\omega)} = -\chi_z(\overline{\omega})$ is immediate from the identities

$$\begin{aligned}\overline{k_+(\omega)} &= k_-(\overline{\omega}) \in A((z^{-1}))\ ,\\ \overline{k_-(\omega)} &= k_+(\overline{\omega}) \in A((z))\ .\end{aligned}$$

(iii) The $(A[z,z^{-1}],\Omega)$-dual $L\hat{}$ of L has the universal property that for any 1-dimensional f.g. projective $A[z,z^{-1}]$-module resolution of L

$$0 \longrightarrow P_1 \xrightarrow{\ d\ } P_0 \longrightarrow L \longrightarrow 0$$

there is a dual resolution

$$0 \longrightarrow P_0^* \xrightarrow{\ d^*\ } P_1^* \longrightarrow L\hat{} \longrightarrow 0\ .$$

For the standard resolution of L

$$d = z - h\ :\ P_1 = L[z,z^{-1}] \longrightarrow P_0 = L[z,z^{-1}]$$

the universal trace map χ_z identifies the dual resolution with the standard resolution of L^*

$$0 \longrightarrow L^*[z,z^{-1}] \xrightarrow{\ z^{-1}-h^*\ } L^*[z,z^{-1}] \longrightarrow L^* \longrightarrow 0\ . \qquad \square$$

Example 28.28 Let A be commutative ring with involution, so that by 13.12 (ii)

$$\Omega^{-1}A[z,z^{-1}] = S^{-1}A[z,z^{-1}]$$

with $S \subset A[z,z^{-1}]$ the multiplicative subset of the bionic polynomials

$$p(z) = \sum_{j=0}^{d} a_j z^j \in A[z,z^{-1}]\ \ (a_d = 1, a_0 \in A^\bullet)\ .$$

An $(A[z,z^{-1}],S)$-module L is a f.g. projective A-module (also denoted by L) together with an automorphism $h : L \longrightarrow L$. For any such L there exists $p(z) \in S$ with

$$p(h) = 0 : L \longrightarrow L\ ,\ \ a_0 z^d \overline{p(z)} = p(z) \in A[z,z^{-1}]$$

(e.g. $p(z) = \mathrm{ch}_z(L,h)$ if L is a f.g. free A-module), and there is defined an $A[z,z^{-1}]$-module isomorphism

$$\begin{aligned}&\mathrm{Hom}_{A[z,z^{-1}]}(L, A[z,z^{-1}]/(p(z)))\\ &\xrightarrow{\ \simeq\ } \mathrm{Hom}_{A[z,z^{-1}]}(L, S^{-1}A[z,z^{-1}]/A[z,z^{-1}])\ ;\ f \longrightarrow (x \longrightarrow f(x)/p(x))\ .\end{aligned}$$

Every $f \in \mathrm{Hom}_{A[z,z^{-1}]}(L, A[z,z^{-1}]/(p(z)))$ can be expressed as

$$f = \sum_{j=0}^{d-1} z^j f_j\ :\ L \longrightarrow A[z,z^{-1}]/(p(z)) = \sum_{j=0}^{d-1} z^j A\ ,$$

with $f_j \in L^* = \mathrm{Hom}_A(L,A)$ and

$$\chi_z\left(\frac{f(x)}{p(z)}\right) = f_0(x) \in A ,$$

$$f_j(x) = (a_0)^{-1} f_0(\sum_{i=0}^{j} a_i h^{i-j} x) \in A ,$$

$$(x \in L, 0 \leq j \leq d-1) .$$

The isomorphism of 28.27 is given in this case by

$$L\hat{} = \mathrm{Hom}_{A[z,z^{-1}]}(L, A[z,z^{-1}]/(p(z))) \xrightarrow{\simeq} L^* = \mathrm{Hom}_A(L,A) ; f \longrightarrow f_0 .$$

□

Example 28.29 Let X be an n-dimensional geometric Poincaré complex, and let $\widetilde{X}$ be a regular cover of X with group of covering translations $\pi \times \mathbb{Z}$. Let $A = R[\pi]$ for some commutative ring R, so that $A[z,z^{-1}] = R[\pi \times \mathbb{Z}]$. As before, let Ω be the involution-invariant set of Fredholm matrices in $A[z,z^{-1}]$. The finite f.g. free $A[z,z^{-1}]$-module chain complex $C(\widetilde{X};R)$ is A-finitely dominated if and only if

$$\Omega^{-1} H_*(\widetilde{X};R) = 0 ,$$

in which case the infinite cyclic covering $\overline{X}$ is an $(n-1)$-dimensional geometric R-coefficient Poincaré complex and as in 25.21 there is defined a linking pairing

$$\mu : H_r(\widetilde{X};R) \times H_{n-r-1}(\widetilde{X};R) \longrightarrow \Omega^{-1}A[z,z^{-1}]/A[z,z^{-1}] .$$

The composite

$$\chi_z\mu : H_r(\widetilde{X};R) \times H_{n-r-1}(\widetilde{X};R) \xrightarrow{\mu} \Omega^{-1}A[z,z^{-1}]/A[z,z^{-1}] \xrightarrow{\chi_z} A$$

is the nonsingular intersection pairing of $\widetilde{X}$ regarded as a cover of $\overline{X}$. Two cases are particularly interesting:

(i) If X is an n-dimensional geometric Poincaré band take $R = \mathbb{Z}$, and let $\widetilde{X}$ be the universal cover of X, with the infinite cyclic cover $\overline{X}$ a finitely dominated $(n-1)$-dimensional geometric Poincaré band. The exterior X of a fibred $(n-2)$-knot $k : S^{n-2} \subset S^n$ is a special case.

(ii) Given a map $X \longrightarrow S^1$ with pullback infinite cyclic cover $\overline{X}$ take $\pi = \{1\}$, and let $R = F$ be a field, so that

$$A = F , \ \Omega^{-1}A[z,z^{-1}] = F(z) .$$

If $H_*(X;F) \cong H_*(S^1;F)$ then

$$\Omega^{-1}H_*(\overline{X};F) = 0 \ , \ \dim_F H_*(\overline{X};F) < \infty \ ,$$
$$\mu \ : \ H_r(\overline{X};F) \times H_{n-r-1}(\overline{X};F) \longrightarrow F(z)/F[z,z^{-1}]$$

with

$$\chi_z\mu \ : \ H_r(\overline{X};F) \times H_{n-r-1}(\overline{X};F) \longrightarrow F$$

the pairing of Milnor [195]. The exterior X of any $(n-2)$-knot $k : S^{n-2} \subset S^n$ is a special case – $H_*(\overline{X};F)$ has the homological properties properties of a fibred $(n-2)$-knot. □

28F. Automorphism and asymmetric L-theory

The ϵ-symmetric L-groups of the Fredholm localization $\Omega^{-1}A[z,z^{-1}]$ were identified in Chap. 26 with the $(\Omega,-z\epsilon)$-asymmetric L-groups of A, with

$$\begin{aligned} L^n_h(\Omega^{-1}A[z,z^{-1}],\epsilon) &= L^n_h(\Omega^{-1}A[z,z^{-1}],-z\epsilon) \\ &= LAsy^n_h(A,\Omega,-z\epsilon) \ . \end{aligned}$$

The $(\Omega,-z\epsilon)$-asymmetric L-groups $LAsy^*(A,\Omega,-z\epsilon)$ will now be identified with the L-groups $LAsy^*(A)$ of asymmetric complexes (C,λ) over A which are Poincaré in the sense that $\lambda : C^{n-*}\longrightarrow C$ is an A-module chain equivalence. Such asymmetric complexes arise in the obstruction theory for open book decompositions, as will be described in Chap. 29. In 28.33 it will be shown that

$$\begin{aligned} LAsy^n_p(A) &\longrightarrow L^n_h(\Omega^{-1}A[z,z^{-1}],\epsilon) \ ; \\ (C,\lambda) &\longrightarrow (\Omega^{-1}C[z,z^{-1}],(1-z^{-1})\lambda+(1-z)T_\epsilon\lambda) \end{aligned}$$

is an isomorphism. This is an abstract version of the expression in 27.8 of the symmetric complex of the exterior of a framed codimension 2 submanifold $N \subset M$ in terms of an asymmetric complex.

Definition 28.30 (i) An n-dimensional asymmetric complex over A (C,λ) is *Poincaré* if the chain map $\lambda : C^{n-*}\longrightarrow C$ is a chain equivalence. The *torsion* of a based f.g. free n-dimensional asymmetric Poincaré complex (C,λ) over A is

$$\tau(C,\lambda) = \tau(\lambda : C^{n-*}\longrightarrow C) \in Wh_1(A) \ .$$

(ii) An $(n+1)$-dimensional asymmetric pair over A $(f : C\longrightarrow D,(\delta\lambda,\lambda))$ is *Poincaré* if the chain maps

$$\begin{pmatrix} \delta\lambda \\ \lambda f^* \end{pmatrix} : D^{n+1-*} \longrightarrow \mathcal{C}(f) \ , \ (\delta\lambda \quad f\lambda) \ : \ \mathcal{C}(f)^{n+1-*} \longrightarrow D$$

are chain equivalences, in which case (C,λ) is an asymmetric Poincaré complex. The n-dimensional asymmetric complex (C,λ) is the *boundary* of the pair.

(iii) Asymmetric Poincaré complexes (C, λ), (C', λ') are *cobordant* if $(C, \lambda) \oplus (C', -\lambda')$ is the boundary of an asymmetric Poincaré pair.
(iv) The *projective (resp. free, simple) asymmetric L-group* $LAsy^n_p(A)$ (resp. $LAsy^n_h(A)$, $LAsy^n_s(A)$) is the cobordism group of n-dimensional asymmetric Poincaré complexes (C, λ) over A with C f.g. projective (resp. f.g. free, based f.g. free with $\tau(\lambda : C^{n-*} \longrightarrow C) = 0 \in Wh_1(A)$). □

Example 28.31 The 0-dimensional asymmetric L-group $LAsy^0_q(A)$ ($q = s, h, p$) is the *Witt group* of nonsingular asymmetric forms (L, λ) over A, with $\lambda : L \longrightarrow L^*$ an isomorphism. Such a form is *metabolic* if there exists a *lagrangian*, i.e. a direct summand $K \subset L$ such that $K = K^\perp$, with

$$K^\perp = \{x \in L \,|\, \lambda(x)(K) = 0\} ,$$

in which case

$$(L, \lambda) = 0 \in LAsy^0_q(A) . \qquad \square$$

Proposition 28.32 (i) *The projective and free asymmetric L-groups are related by a Rothenberg-type exact sequence*

$$\dots \longrightarrow LAsy^n_h(A) \longrightarrow LAsy^n_p(A) \longrightarrow \widehat{H}^n(\mathbb{Z}_2; \widetilde{K}_0(A)) \longrightarrow LAsy^{n-1}_h(A) \longrightarrow \dots .$$

(ii) *The forgetful maps $LAsy^n_s(A) \longrightarrow LAsy^n_h(A)$ are isomorphisms, so the free and simple asymmetric L-groups coincide*

$$LAsy^*_s(A) = LAsy^*_h(A) .$$

Proof (i) As for the symmetric L-groups in Ranicki [235, Chap. 10].
(ii) Given an n-dimensional based f.g. free asymmetric Poincaré complex (B, λ) over A let $\alpha : C \longrightarrow C$ be a self chain equivalence of an n-dimensional based f.g. free A-module chain complex C such that

$$\tau(\alpha) = -\tau(\lambda : B^{n-*} \longrightarrow B) \in Wh_1(A) .$$

The morphism

$$LAsy^n_h(A) \longrightarrow LAsy^n_s(A) \; ; \; (B, \lambda) \longrightarrow (B, \lambda) \oplus (C \oplus C^{n-*}, \begin{pmatrix} 0 & 1 \\ \alpha & 0 \end{pmatrix})$$

is an isomorphism inverse to the forgetful map $LAsy^n_s(A) \longrightarrow LAsy^n_h(A)$. □

As in Chap. 13 let Ω be the involution-invariant set of Fredholm matrices M in $A[z, z^{-1}]$, i.e. the square matrices such that $\mathrm{coker}(M)$ is a f.g. projective A-module.

Proposition 28.33 *Let $\epsilon \in A$ be a central unit such that $\bar{\epsilon} = \epsilon^{-1}$, and let $(t,u) = (h,s)$ or (h,h) or $(\widetilde{K}_0(A), p)$.*

(i) *The asymmetric L-groups of 28.30 are such that up to natural isomorphism*

$$\begin{aligned} LAsy^*_u(A) &= LAsy^*_u(A,\Omega,-\epsilon z) \\ &= L^*_t(\Omega^{-1}A[z,z^{-1}],-\epsilon z) \\ &= L^*_t(\Omega^{-1}A[z,z^{-1}],\epsilon) \ . \end{aligned}$$

In particular, there are defined natural isomorphisms

$$\begin{aligned} LAsy^n_u(A) &\xrightarrow{\simeq} L^n_t(\Omega^{-1}A[z,z^{-1}],\epsilon) \ ; \\ &(C,\lambda) \longrightarrow (\Omega^{-1}C[z,z^{-1}],(1-z^{-1})\lambda + (1-z)T_\epsilon\lambda) \ . \end{aligned}$$

(ii) *The asymmetric and automorphism L-groups are related by an exact sequence*

$$\begin{aligned} \dots \longrightarrow L^n_t(A[z,z^{-1}],\epsilon) \longrightarrow LAsy^n_u(A) \longrightarrow LAut^{n-2}_p(A,\epsilon) \\ \longrightarrow L^{n-1}_t(A[z,z^{-1}],\epsilon) \longrightarrow \dots \ . \end{aligned}$$

Proof (i) An n-dimensional asymmetric Poincaré complex (C,λ) over A is an n-dimensional asymmetric $(\Omega,\epsilon z)$-Poincaré complex over A, so that there is defined a morphism

$$LAsy^n_u(A) \longrightarrow LAsy^n_u(A,\Omega,-\epsilon z) \ ; \ (C,\lambda) \longrightarrow (C,\lambda) \ .$$

Given an n-dimensional asymmetric $(\Omega,-\epsilon z)$-Poincaré complex (D,μ) over A define an n-dimensional asymmetric Poincaré complex (C,λ) over A by

$$\begin{aligned} C &= \mathcal{C}((1+T_{-\epsilon z})\mu : D^{n-*}[z,z^{-1}]\longrightarrow D[z,z^{-1}])_{*+1} \ , \\ \lambda &= \begin{pmatrix} 0 & -\epsilon z \\ 1 & 0 \end{pmatrix} : C^{n-r} = D_{r+1}[z,z^{-1}] \oplus D^{n-r}[z,z^{-1}] \\ &\qquad\qquad \longrightarrow C_r = D^{n-r}[z,z^{-1}] \oplus D_{r+1}[z,z^{-1}] \ . \end{aligned}$$

The construction defines a morphism

$$LAsy^n_u(A,\Omega,-\epsilon z) \longrightarrow LAsy^n_u(A) \ ; \ (D,\mu) \longrightarrow (C,\lambda) \ .$$

The composite

$$LAsy^n_u(A) \longrightarrow LAsy^n_u(A,\Omega,-\epsilon z) \longrightarrow LAsy^n_u(A)$$

is the identity on the level of objects. The composite

$$LAsy^n_u(A,\Omega,-\epsilon z) \longrightarrow LAsy^n_u(A) \longrightarrow LAsy^n_u(A,\Omega,-\epsilon z)$$

sends an n-dimensional asymmetric $(\Omega, -\epsilon z)$-Poincaré complex (D, μ) over A to the n-dimensional asymmetric $(\Omega, -\epsilon z)$-Poincaré complex (C, λ) over A constructed as above. The A-module chain complex defined by

$$E = \mathcal{C}((1 + T_{-\epsilon z})\mu : D^{n-*}[z] \longrightarrow D[z])_{*+1}$$

is such that there is a natural identification

$$E^{n-*} = \mathcal{C}((1 + T_{-\epsilon z})\mu : D^{n-*}[z^{-1}] \longrightarrow z^{-1}D[z^{-1}])_{*+1}$$

with a homotopy equivalence of n-dimensional asymmetric $(\Omega, -\epsilon z)$-Poincaré complexes over A

$$(D, \mu) \simeq (C, \lambda) \oplus (E \oplus E^{n-*}, \begin{pmatrix} 0 & 1 \\ 0 & 0 \end{pmatrix}) ,$$

so that

$$(D, \mu) = (C, \lambda) \in LAsy^n_u(A, \Omega, -\epsilon z)$$

and the composite

$$LAsy^n_u(A, \Omega, -\epsilon z) \longrightarrow LAsy^n_u(A) \longrightarrow LAsy^n_u(A, \Omega, \epsilon z)$$

is also the identity. The identifications

$$LAsy^*_u(A, \Omega, -\epsilon z) = L^*_t(\Omega^{-1}A[z, z^{-1}], -\epsilon z) = L^*_t(\Omega^{-1}A[z, z^{-1}], \epsilon)$$

were already obtained in 25.11.
(ii) Combine (i) and 28.19. □

Proposition 28.34 (i) *The asymmetric L-groups are 2-periodic*

$$LAsy^*_q(A) = LAsy^{*+2}_q(A) \quad (q = s, h, p) .$$

(ii) *The even-dimensional asymmetric L-groups of asymmetric Poincaré complexes are the L-groups of nonsingular asymmetric forms*

$$LAsy^{2*}_q(A) = LAsy^0_q(A) \quad (q = s, h, p) .$$

(iii) *The free odd-dimensional asymmetric L-groups vanish*

$$LAsy^{2*+1}_h(A) = LAsy^1_h(A) = 0 ,$$

and the projective odd-dimensional asymmetric L-groups

$$LAsy^{2*+1}_p(A) = LAsy^1_p(A)$$

fit into an exact sequence

$$0 \longrightarrow LAsy^1_p(A) \longrightarrow \widehat{H}^1(\mathbb{Z}_2; \widetilde{K}_0(A)) \longrightarrow LAsy^0_h(A) \longrightarrow LAsy^0_p(A) \longrightarrow \widehat{H}^0(\mathbb{Z}_2; \widetilde{K}_0(A)) \longrightarrow 0 .$$

Proof It will be proved that for $n = 2i$ or $2i + 1$ the i-fold suspension map

$$S^i : LAsy^{n-2i}_q(A) \longrightarrow LAsy^n_q(A) ; \ (C, \lambda) \longrightarrow (S^iC, \lambda)$$

is an isomorphism, by constructing an explicit inverse. (The construction is the asymmetric L-theory version of the instant quadratic Poincaré surgery obstruction of Ranicki [235].)

Given an n-dimensional asymmetric Poincaré complex (C,λ) let (C',λ') be the n-dimensional asymmetric Poincaré complex defined by

$$C'_r = \begin{cases} C^{n-r} \\ \ker\left(\begin{pmatrix} d_C & (-)^{i-1}\lambda \\ 0 & d_C^* \end{pmatrix} : C_i \oplus C^{n-i+1} \longrightarrow C_{i-1} \oplus C^{n-i+2}\right) \\ C_r \end{cases}$$

$$\text{if } \begin{cases} r \le i-1 \\ r = i \\ r \ge i+1 , \end{cases}$$

$$d_{C'} = \begin{cases} d_C^* : C'_r = C^{n-r} \longrightarrow C'_{r-1} = C^{n-r+1} \\ [0\ 1] : C'_r = C'_i \longrightarrow C'_{r-1} = C^{i+1} \\ \begin{bmatrix} d_C \\ 0 \end{bmatrix} : C'_r = C_{i+1} \longrightarrow C'_{r-1} = C'_i \\ d_C : C'_r = C_r \longrightarrow C'_{r-1} = C_{r-1} \end{cases} \quad \text{if } \begin{cases} r \le i-1 \\ r = i \\ r = i+1 \\ r \ge i+2 , \end{cases}$$

$$\lambda' = \begin{cases} 1 : C'^{n-r} = C^{n-r} \longrightarrow C'_r = C^{n-r} \\ \begin{bmatrix} \lambda & d_C \\ (-)^i d_C^* & 0 \end{bmatrix} : C'^{n-r} = C'^i \longrightarrow C'_r = C'_i \\ 1 : C'^{n-r} = C_r \longrightarrow C'_r = C_r \\ \begin{bmatrix} \lambda \\ (-)^i d_C^* \end{bmatrix} : C'^{n-r} = C'^{i+1} \longrightarrow C'_r = C'_i \\ [\lambda\ d_C] : C'^{n-r} = C'^i \longrightarrow C'_r = C'_i \\ 1 : C'^{n-r} = C_r \longrightarrow C'_r = C_r \end{cases}$$

$$\text{if } \begin{cases} r \le i-1 \\ r = i \text{ and } n = 2i \\ r \ge i+1 \text{ and } n = 2i \\ r = i \text{ and } n = 2i+1 \\ r = i+1 \text{ and } n = 2i+1 \\ r \ge i+2 \text{ and } n = 2i+1 . \end{cases}$$

The chain equivalence $h : C' \longrightarrow C$ defined by

$$h = \begin{cases} \lambda : C'_r = C^{n-r} \longrightarrow C_r = C_r & \text{if } r \le i-1 \\ [1\ 0] : C'_i \longrightarrow C_i & \text{if } r = i \\ 1 : C'_r = C_r \longrightarrow C_r & \text{if } r \ge i+1 \end{cases}$$

is a homotopy equivalence $h : (C',\lambda') \longrightarrow (C,\lambda)$, so that

$$(C,\lambda) = (C',\lambda') \in L\mathrm{Asy}^n_q(A)$$

with $\lambda' : C'^{n-r} \longrightarrow C'_r$ an isomorphism $\begin{cases} \text{for all } r \text{ if } n = 2i \\ \text{for } r \ne i, i+1 \text{ if } n = 2i+1 . \end{cases}$

The $(i-1)$-connected n-dimensional asymmetric Poincaré complex (C'', λ'') defined by

$$\lambda'' = \begin{cases} \lambda' : C''^{n-r} = C'^{n-r} \longrightarrow C''_r = C'_r \\ \qquad \text{if } \begin{cases} n = 2i \text{ and } r = i \\ n = 2i+1 \text{ and } r = i, i+1, \end{cases} \\ 0 : C''^{n-r} = 0 \longrightarrow C''_r = 0 \text{ otherwise} . \end{cases}$$

is cobordant to (C', λ'), with a cobordism

$$((f' \ f'') : C' \oplus C'' \longrightarrow D, (0, \lambda' \oplus -\lambda''))$$

defined by

$$D_r = \begin{cases} C'_r & \text{if } r \geq i \\ 0 & \text{if } r \leq i-1 \end{cases}$$

$$d_D = d_{C'} : D_r = C_r \longrightarrow D_{r-1} = C_{r-1} \text{ if } r \geq i+2 ,$$

$$f' = \text{projection} : C' \longrightarrow D ,$$

$$f'' = \text{projection} : C'' \longrightarrow D .$$

Thus

$$(C, \lambda) = (C', \lambda') = (C'', \lambda'') \in LAsy^n_q(A) ,$$

and the maps

$$LAsy^n_q(A) \longrightarrow LAsy^{n-2i}_q(A) \ ; \ (C, \lambda) \longrightarrow (S^{-i}C''^i, \lambda'') ,$$

$$LAsy^{n-2i}_q(A) \longrightarrow LAsy^n_q(A) \ ; \ (C, \lambda) \longrightarrow (S^i C, \lambda)$$

are inverse isomorphisms.

It remains to prove that $LAsy^1_h(A) = 0$. Note first that for any 1-dimensional f.g. free asymmetric Poincaré complex (C, λ) over A such that $\lambda : C^{1-*} \longrightarrow C$ is an isomorphism there is defined a null-cobordism $(f : C \longrightarrow D, (0, \lambda))$ with

$$f = 1 : C_1 \longrightarrow D_1 = C_1 \ , \ D_r = 0 \text{ for } r \neq 1 ,$$

so that

$$(C, \lambda) = 0 \in LAsy^1_h(A) .$$

It therefore suffices to prove that every 1-dimensional f.g. free asymmetric Poincaré complex (C, λ) over A is homotopy equivalent to one with $\lambda : C^{1-*} \longrightarrow C$ an isomorphism (which is the case if and only if $\lambda : C^0 \longrightarrow C_1$ is an isomorphism). This is done using the following adaptation to asymmetric L-theory of Neumann's stabilization lemma (cf. Winkelnkemper [308], Quinn [227, §8]).

Let (C', λ') be the 1-dimensional f.g. free asymmetric Poincaré complex homotopy equivalent to (C, λ) defined by

$$d' = d \oplus 1 : C'_1 = C_1 \oplus C^0 \longrightarrow C'_0 = C_0 \oplus C^0 ,$$
$$\lambda' = \lambda \oplus 0 : \begin{cases} C'^0 = C^0 \oplus C_0 \longrightarrow C'_1 = C_1 \oplus C^0 \\ C'^1 = C^1 \oplus C_0 \longrightarrow C'_0 = C_0 \oplus C^0 . \end{cases}$$

The short exact sequence

$$0 \longrightarrow C^0 \xrightarrow{\begin{pmatrix} \lambda \\ d^* \end{pmatrix}} C_1 \oplus C^1 \xrightarrow{(d \;\; -\lambda)} C_0 \longrightarrow 0$$

splits, so there exist A-module morphisms

$$\alpha : C_0 \longrightarrow C_1 \;,\; \beta : C_0 \longrightarrow C^1$$

such that

$$d\alpha - \lambda\beta = 1 : C_0 \longrightarrow C_0 ,$$

and the A-module morphism

$$\begin{pmatrix} \lambda & \alpha \\ d^* & \beta \end{pmatrix} : C^0 \oplus C_0 \longrightarrow C_1 \oplus C^1$$

is an isomorphism. Choose an A-module isomorphism

$$h : C^1 \xrightarrow{\simeq} C^0 .$$

(Recall that A is assumed to be such that the rank of f.g. free A-modules is well-defined, so that C^0, C^1 are isomorphic A-modules). Use the A-module morphism

$$\gamma = \begin{pmatrix} 0 & \alpha \\ h & h\beta \end{pmatrix} : C'^1 = C^1 \oplus C_0 \longrightarrow C'_1 = C_1 \oplus C^0$$

to define an isomorphism

$$\lambda'' = \lambda' + d'\gamma + \gamma d'^* : C'^{1-*} \xrightarrow{\simeq} C'$$

in the chain homotopy class of $\lambda' : C'^{1-*} \longrightarrow C'$, with

$$\begin{aligned} \lambda'' &= \begin{pmatrix} \lambda & 0 \\ 0 & 0 \end{pmatrix} + \begin{pmatrix} 0 & \alpha \\ h & h\beta \end{pmatrix} \begin{pmatrix} d^* & 0 \\ 0 & 1 \end{pmatrix} \\ &= \begin{pmatrix} 1 & 0 \\ 0 & h \end{pmatrix} \begin{pmatrix} \lambda & \alpha \\ d^* & \beta \end{pmatrix} : \\ & C'^0 = C^0 \oplus C_0 \longrightarrow C_1 \oplus C^1 \longrightarrow C'_1 = C_1 \oplus C^0 . \end{aligned}$$

□

Definition 28.35 (i) A self chain equivalence $h : E \longrightarrow E$ is *fibred* if

$$h - 1 \;:\; E \longrightarrow E$$

is a chain equivalence, generalizing the notion of fibred automorphism (13.2).
(ii) The *projective (resp. free) ϵ-symmetric fibred automorphism L-groups* $LAut^n_{q,fib}(A,\epsilon)$ for $q = p$ (resp. h) are the cobordism groups of fibred self homotopy equivalences $(h,\chi) : (E,\theta) \longrightarrow (E,\theta)$ of finitely dominated (resp. f.g. free) n-dimensional ϵ-symmetric Poincaré complexes (E,θ) over A. □

The fibred automorphism L-groups will be related in Chapter 31 to the 'isometric' L-groups.

By analogy with the round L-groups (20.15, 20.16):

Definition 28.36 The *round ϵ-symmetric automorphism L-group* $LAut^n_r(A,\epsilon)$ ($n \geq 0$) is the cobordism group of self homotopy equivalences (h,χ) : $(E,\theta) \longrightarrow (E,\theta)$ of finitely dominated n-dimensional ϵ-symmetric Poincaré complexes (E,θ) over A such that $[E,h] = 0 \in \mathrm{Aut}_0(A)$.
Similarly for the *round ϵ-quadratic L-groups* $LAut^r_*(A,\epsilon)$. □

Proposition 28.37 *The projective and round ϵ-symmetric automorphism L-groups are related by a Rothenberg-type exact sequence*

$$\begin{aligned} \dots \longrightarrow \widehat{H}^{n+1}(\mathbb{Z}_2;\mathrm{Aut}_0(A)) \longrightarrow LAut^n_r(A,\epsilon) \longrightarrow LAut^n_p(A,\epsilon) \\ \longrightarrow \widehat{H}^n(\mathbb{Z}_2;\mathrm{Aut}_0(A)) \longrightarrow \dots \end{aligned}$$

with

$$LAut^n_p(A,\epsilon) \longrightarrow \widehat{H}^n(\mathbb{Z}_2;\mathrm{Aut}_0(A)) \;;\; (E,h,\theta,\chi) \longrightarrow [E,h] \;.$$

Similarly in the ϵ-quadratic case. □

In 39.20 it will be shown that $LAut^0(F)$ maps onto $\mathbb{Z}_2[H^0(\mathbb{Z}_2;\mathcal{M}_z(F))]$, for any field with involution F.

29. Open books

An n-dimensional manifold M is an 'open book' if it is the relative mapping torus of a rel ∂ automorphism of an $(n-1)$-dimensional manifold with boundary, or equivalently if there exists a codimension 2 homology framed submanifold $N^{n-2} \subset M^n$ such that the exterior fibres over S^1.

A closed n-dimensional manifold M is a 'twisted double' if it is obtained from two copies of an n-dimensional manifold with boundary $(Q, \partial Q)$ by glueing along a homeomorphism $h : \partial Q \longrightarrow \partial Q$, with $M = Q \cup_h -Q$. Open books are particular examples of twisted doubles. The handlebody methods of Smale's h-cobordism theorem were applied in the 1960's to study twisted double decompositions of manifolds, initially in the differentiable case. In the 1970's Winkelnkemper [307], [308] and Quinn [227] combined handlebody theory, Stallings engulfing and high-dimensional surgery theory to obtain an obstruction theory for the existence and uniqueness of twisted double and open book decompositions. This obstruction theory for open book decompositions will now be developed from the chain complex point of view. Twisted doubles will be considered in Chap. 30 – in fact, the high-dimensional obstruction theory for twisted doubles is the same as for open books.

The geometric definitions of open books are given in 29A. The asymmetric signature of an n-dimensional manifold M is defined in 29B

$$\sigma^*(M) \in LAsy^n(\mathbb{Z}[\pi_1(M)]) ,$$

such that $\sigma^*(M) = 0$ if (and for $n \geq 5$ only if) M has an open book decomposition. The asymmetric signature is just the image of the symmetric signature $\sigma^*(M) \in L^n(\mathbb{Z}[\pi_1(M)])$. The asymmetric L-groups vanish for odd n, so odd-dimensional manifolds have (non-unique) open book decompositions.

See the Appendix by Winkelnkemper for a detailed account of the history and applications of open books.

29A. Geometric open books

Definition 29.1 Let $(F, \partial F \subset F)$ be a pair of spaces. The *relative mapping torus* of a rel ∂ self map $(h, 1) : (F, \partial F) \longrightarrow (F, \partial F)$ is the space

$$t(h) = T(h) \cup_{\partial F \times S^1} \partial F \times D^2 . \qquad \square$$

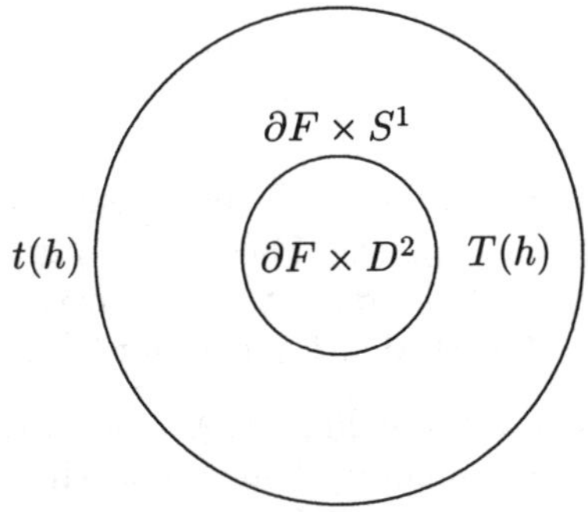

Proposition 29.2 *Let $(h, \partial h) : (F, \partial F) \longrightarrow (F, \partial F)$ be an automorphism of an n-dimensional manifold with boundary $(F, \partial F)$.*
(i) *The mapping torus pair $(T(h), T(\partial h))$ is an $(n+1)$-dimensional manifold with boundary which fibres over S^1*

$$(F, \partial F) \longrightarrow (T(h), T(\partial h)) \longrightarrow S^1 .$$

(ii) *If $\partial h = 1 : \partial F \longrightarrow \partial F$ the relative mapping torus $t(h)$ is a closed $(n+1)$-dimensional manifold with a homology framed codimension 2 submanifold $\partial F \subset t(h)$ such that the exterior*

$$\text{cl.}(t(h) \backslash \partial F \times D^2) = T(h)$$

fibres over S^1. $\square$

Definition 29.3 An *open book decomposition* of a closed $(n+1)$-dimensional manifold M is a homeomorphism $M \cong t(h)$ for some rel ∂ automorphism $(h, 1) : (F, \partial F) \longrightarrow (F, \partial F)$ of an n-dimensional manifold with boundary $(F, \partial F)$. This is abbreviated to M is an *open book*

$$M = t(h) .$$

The codimension 1 submanifold $F \subset M$ is the *page* and the codimension 2 submanifold $\partial F \subset M$ is the *binding* of the open book. $\square$

Example 29.4 The mapping torus of an automorphism $h : F \longrightarrow F$ of a closed n-dimensional manifold F is a closed $(n+1)$-dimensional manifold with an open book decomposition

$$M = T(h) = t(h)$$

with the fibre F as page and empty binding $\partial F = \emptyset$. □

Example 29.5 The only closed (connected, orientable) surfaces with open book decompositions are S^2 and $S^1 \times S^1$. For if $M^2 = t(h : F \longrightarrow F)$ has genus g then the Euler characteristic of the 0-dimensional manifold ∂F is

$$\chi(\partial F) \;=\; \chi(M) \;=\; 2 - 2g \geq 0$$

and $g = 0$ or 1. However, for every surface M the 6-dimensional manifold $M \times \mathbb{CP}^2$ does have an open book decomposition (30.14). □

Proposition 29.6 *Let M be a closed $(n+2)$-dimensional manifold with a codimension 2 homology framed submanifold $N^n \subset M^{n+2}$ with exterior P.*
(i) *If the projection $\partial P = N \times S^1 \longrightarrow S^1$ extends to the projection $P \longrightarrow S^1$ of a fibre bundle $P = T(h : F \longrightarrow F)$, with $F^{n+1} \subset M$ a codimension 1 Seifert surface for N, then M has an open book decomposition with page F and binding ∂F.*
(ii) *The manifold M has an open book decomposition with page F and binding N if and only if there exists a codimension 1 Seifert surface $F^{n+1} \subset M$ for N such that the Seifert fundamental domain $(P_F; i_0F, i_1F)$ (27.3) for the canonical infinite cyclic cover $\overline{P}$ of P is homeomorphic to $F \times (I; \{0\}, \{1\})$. For $n \geq 4$ this is the case if and only if the inclusion $i_0F \longrightarrow P_F$ is a simple homotopy equivalence, by the s-cobordism theorem.*

Proof (i) This is just the restatement of the definition of an open book in the language of codimension 2 manifolds.
(ii) Immediate from (i). □

There are also relative versions:

Proposition 29.7 *If $(F; \partial_+F, \partial_-F)$ is an $(n+1)$-dimensional relative cobordism with an automorphism*

$$(h; 1, \partial_-h) \;:\; (F; \partial_+F, \partial_-F) \longrightarrow (F; \partial_+F, \partial_-F)\ ,$$

the relative mapping tori of h, ∂_-h constitute an $(n+2)$-dimensional manifold with boundary

$$(t_+(h), t(\partial_-h)) \;=\; (T(h) \cup \partial_+F \times D^2, T(\partial_-h) \cup \partial\partial_-F \times D^2)\ . \qquad \square$$

Thus $(\partial_+F, \partial\partial_+F) \subset (t_+(h), t(\partial_-h))$ is a homology framed codimension 2 submanifold such that the exterior is a fibre bundle over S^1, with fibre the codimension 1 Seifert surface $(F, \partial_-F) \subset (t_+(h), t(\partial_-h))$.

Definition 29.8 (i) An *open book decomposition* of an $(n+2)$-dimensional manifold with boundary $(M, \partial M)$ is a homeomorphism

$$(M, \partial M) \;\cong\; (t_+(h), t(\partial_-h))$$

for some automorphism of a relative $(n+1)$-dimensional cobordism

$$(h; 1, \partial_-h) \;:\; (F; \partial_+F, \partial_-F) \longrightarrow (F; \partial_+F, \partial_-F)\ ,$$

with *page* (F, ∂_-F) and *binding* $(\partial_+F, \partial\partial_+F)$.

(ii) An *open book cobordism* $(t_+(h); t(\partial_- h), t(\partial'_- h))$ is an open book decomposition of a cobordism.

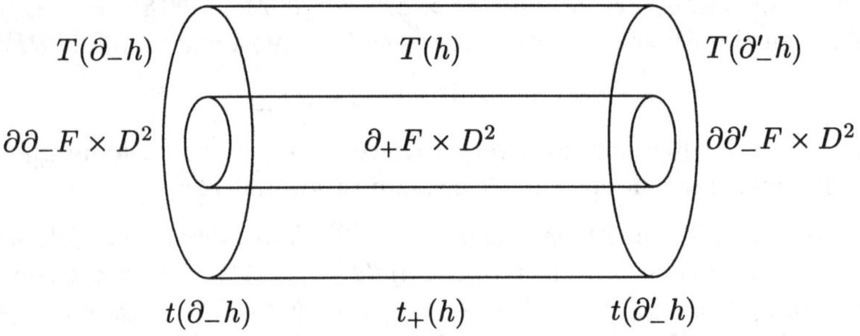

(iii) The *open book bordism group* $OB_n(X)$ is defined for any space X and any $n \geq 0$ to be the group of bordism classes of triples (F, g, h) consisting of an n-dimensional manifold with boundary $(F, \partial F)$, a map $g : F \longrightarrow X$ and a rel ∂ automorphism $(h, 1) : (F, \partial F) \longrightarrow (F, \partial F)$ with a homotopy $g \simeq gh : F \longrightarrow X$ which is constant on ∂F.
(iv) The *bounded book bordism group* $BB_n(X)$ is the group of bordism classes of quintuples $(M, f; F, g, h)$ with $X, (F, g, h)$ as in (iii), together with an $(n+2)$-dimensional manifold M with boundary the open book $\partial M = t(h)$, and a map $f : M \longrightarrow X$ such that $f| = t(g) : \partial M = t(h) \longrightarrow X$. (Here, $t(g)$ is the map induced by f, rather than the relative mapping torus of g). □

In other words, $OB_n(X)$ is the bordism group of closed $(n+1)$-dimensional manifolds with an open book structure and a compatible map to X.

Proposition 29.9 *The book bordism groups are related by the exact sequence*

$$\dots \longrightarrow \Omega_{n+2}(X) \longrightarrow BB_n(X) \longrightarrow OB_n(X) \xrightarrow{t} \Omega_{n+1}(X) \longrightarrow \dots$$

with

$$t : OB_n(X) \longrightarrow \Omega_{n+1}(X) \; ; \; (F, g, h) \longrightarrow (t(h), t(g)) \; .$$

Proof Formal. □

Proposition 29.10 *Given an $(n+2)$-dimensional manifold M with an open book boundary $\partial M = t(h : F \longrightarrow F)$, let $N^n \subset M^{n+2}$ be the codimension 2 Seifert surface for $\partial N = \partial F \subset \partial M$ obtained by pushing F into the interior of M, and let*

$$h' = h \cup 1 : F' = F \cup_\partial -F \longrightarrow F' \; ,$$

so that the exterior

$$M' = \text{cl.}(M \backslash N \times D^2)$$

has boundary the fibre bundle

$$\partial M' = T(h' : F' \longrightarrow F')$$

and $(M; F \times I, F \times I)$ is a fundamental domain for the canonical infinite cyclic cover $\overline{M}'$ of M'. For $n \geq 4$ the following conditions are equivalent:

(i) *the open book decomposition on ∂M extends to an open book decomposition of M with page $F \times I$ and binding F',*

(ii) *the fibre bundle decomposition on $\partial M'$ extends to a fibre bundle decomposition of M' with fibre $F \times I$,*

(iii) *the inclusion $i : F \longrightarrow M$ is a simple homotopy equivalence.*

Proof By the relative s-cobordism theorem. From the algebraic point of view, note that the $\mathbb{Z}[\pi][z, z^{-1}]$-module chain equivalence

$$\mathcal{C}((1-z)i : C(\widetilde{F})[z, z^{-1}] \longrightarrow C(\widetilde{M})[z, z^{-1}]) \simeq C(\widetilde{M}')$$

determines an $\Omega^{-1}\mathbb{Z}[\pi][z, z^{-1}]$-module chain equivalence

$$\Omega^{-1}C(\widetilde{M}, \widetilde{F})[z, z^{-1}] \simeq \Omega^{-1}C(\widetilde{M}')$$

(since $1 - z \in \Omega^{-1}\mathbb{Z}[\pi][z, z^{-1}]$ is a unit), so that $\overline{M}'$ is finitely dominated if and only if $i : F \longrightarrow M$ is a homotopy equivalence. Here, it is assumed that

$$\pi_1(F) = \pi_1(M) = \pi \ , \ \pi_1(M') = \pi \times \mathbb{Z} \ ,$$

$\widetilde{F}, \widetilde{M}, \widetilde{M}'$ are the universal covers of F, M, M' respectively, and $\overline{M}' = \widetilde{M}'/\mathbb{Z}$ is the corresponding infinite cyclic cover of M'. □

Every open book is open book cobordant to a fibre bundle over S^1:

Proposition 29.11 *Let $(h, 1) : (F, \partial F) \longrightarrow (F, \partial F)$ be a rel ∂ automorphism of an n-dimensional manifold with boundary $(F, \partial F)$. For any $(n+1)$-dimensional manifold N with boundary $\partial N = \partial F$ (e.g. $N = F$) the mapping torus of the automorphism*

$$h' = h \cup 1 : F' = F \cup_\partial N \longrightarrow F \cup_\partial N$$

is a fibre bundle $T(h') = T(h) \cup_\partial N \times S^1$ over S^1, such that there is defined an open book cobordism $(t_+(g); t(h), T(h'))$ with

$$g = h' \times 1 : F' \times I \longrightarrow F' \times I \ , \ \partial F' = \emptyset \ , \ \partial_+ G = N \ . \quad \square$$

Proposition 29.12 *Let M be an $(n+2)$-dimensional manifold with open book boundary $\partial M = t(h : F \longrightarrow F)$.*

(i) *If $(N, \partial N) \subset (M, \partial M)$ is a codimension 2 Seifert surface for $\partial N = \partial F \subset \partial M$ then the exterior of $(N, \partial N) \subset (M, \partial M)$ is an $(n+2)$-dimensional manifold with boundary a fibre bundle over S^1*

$$(M', \partial M') = (\mathrm{cl.}(M \backslash N \times D^2), T(h'))$$

where

$$h' = h \cup 1 : F' = F \cup_\partial N \longrightarrow F' \ .$$

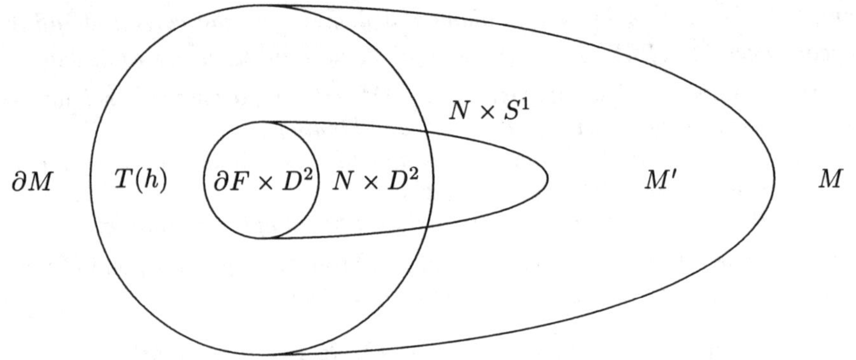

The manifolds with boundary $(M, \partial M)$, $(M', \partial M')$ are connected by a relative bordism $(L; M, M'; \partial_+ L)$ such that

$$(\partial_+ L; \partial M, \partial M') \;=\; (t_+(g); t(h), T(h'))$$

is the open book cobordism of 29.11.

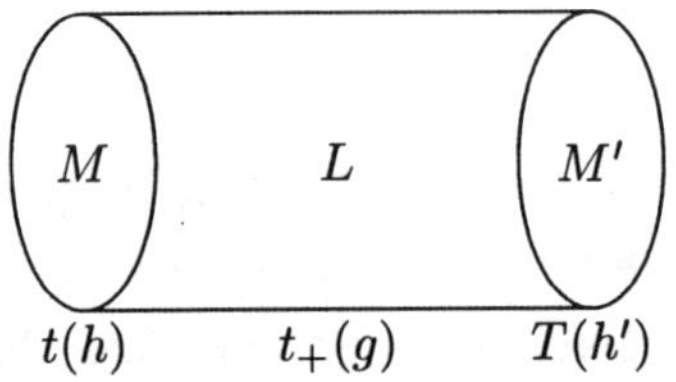

(ii) *It is possible to extend the open book decomposition on ∂M to an open book decomposition of M with page F' and binding N if and only if there exists a codimension 2 Seifert surface $N \subset M$ for $\partial M \subset \partial N$ such that*

$$(M'; F' \times I, F' \times I) \;\cong\; F' \times I \times (I; \{0\}, \{1\}) \;.$$

For $n \geq 5$ this is the case if and only if $F' \longrightarrow M'$ is a simple homotopy equivalence, by the relative s-cobordism theorem.
Proof (i) Identify

$$M \;=\; M' \cup_{T(h')} t_+(g)$$

and set

$$L \;=\; M \times I \;\;,\;\; g \;=\; h \times 1 \;:\; G \;=\; F \times I \longrightarrow G \;.$$

(ii) Apply 29.10. □

Corollary 29.13 (i) *The bounded book bordism groups are isomorphic to the bounded fibre bundle bordism groups of 28.8*

$$BB_n(X) \;\cong\; AB_n(X) \;\;(n \geq 0) \;.$$

(ii) *The bounded book bordism groups are related to the automorphism bordism groups by the exact sequence*

$$\ldots \longrightarrow \Omega_{n+2}(X\times S^1) \longrightarrow BB_n(X) \longrightarrow \Delta_n(X) \xrightarrow{T} \Omega_{n+1}(X\times S^1) \longrightarrow \ldots$$

with

$$\begin{aligned} &\Omega_{n+2}(X \times S^1) \longrightarrow BB_n(X) \ ; \ M \longrightarrow (M, \emptyset) \ , \\ &BB_n(X) \longrightarrow \Delta_n(X) \ ; \ (M, f; F, g, h) \longrightarrow (F \cup F, h \cup 1, f \cup f) \ , \\ &T \ : \ \Delta_n(X) \longrightarrow \Omega_{n+1}(X \times S^1) \ ; \ (F, g, h) \longrightarrow (T(h), T(g)) \ . \end{aligned}$$

□

Proof (i) There is an evident forgetful map

$$AB_n(X) \longrightarrow BB_n(X) \ ; \ (M, f; F, g, h) \longrightarrow (M, jf; F, g, h) \ ,$$

with $j : X \times S^1 \longrightarrow X$ the projection. Use the construction of 29.11 with $N = F$ to define a morphism

$$BB_n(X) \longrightarrow AB_n(X) \ ; \ (M, f; F, g, h) \longrightarrow (M', f'; F', g', h')$$

with

$$\begin{aligned} M' &= M \cup_{F\times I\times\partial I} (F \times I \times I) \ , \\ F' &= \partial(F \times I) = F \cup_{\partial F} F \ , \quad h' = h \cup 1 \ \ \text{(etc.)} \end{aligned}$$

In order to define $f' : M' \longrightarrow X \times S^1$ identify $t(h)$ with the h'-twisted double of $F \times I$

$$t(h) = (F \times I) \cup_{h'} -(F \times I) \ ,$$

regard M as a relative cobordism $(M; F \times I, F \times I)$ with a Morse function

$$e \ : \ (M; F \times I, F \times I) \longrightarrow (I; \{0\}, \{1\})$$

and set

$$\begin{aligned} f' = (F \times e) \cup (h \times pr_2) \ : \ M' &= M \cup_{F\times I\times\partial I} (F \times I \times I) \\ &\longrightarrow (X \times I) \cup_{X\times\partial I} X \times I = X \times S^1 \ . \end{aligned}$$

The composite

$$AB_n(X) \longrightarrow BB_n(X) \longrightarrow AB_n(X)$$

is the identity, by construction. The composite

$$BB_n(X) \longrightarrow AB_n(X) \longrightarrow BB_n(X)$$

is the identity, by the relative open book bordism between $(M, t(h))$ and $(M', T(h'))$ of 29.12 (i). (The method of proof is essentially the same as for 22.17 (i)).

(ii) Immediate from (i). □

Open book decompositions are the same as framed $\Omega^{-1}\mathbb{Z}[\pi][z, z^{-1}]$-homology spines:

Proposition 29.14 *Let $(M, \partial M)$ be an $(n+2)$-dimensional manifold with boundary, and let*

$$\pi_1(M) = \pi \ , \ \Lambda = \Omega^{-1}\mathbb{Z}[\pi][z, z^{-1}] \ .$$

(i) *If $\partial M = t(h : F \longrightarrow F)$ is an open book with page F then the binding $K = \partial F \subset \partial M$ is a framed Λ-homology boundary spine for $(M, \partial M)$ with exterior*

$$L = \mathrm{cl.}(\partial M \backslash K \times D^2) = T(h : F \longrightarrow F)$$

and Λ-coefficient torsion

$$\begin{aligned} \tau(L \longrightarrow M \times S^1; \Lambda) &= (0, [M, 1] - [F, h]) \\ &\in \Delta(\mathrm{Aut}_0^s(\mathbb{Z}[\pi])) \subseteq Wh_1(\Lambda) = Wh_1(\pi \times \mathbb{Z}) \oplus \mathrm{Aut}_0(\mathbb{Z}[\pi]) \ . \end{aligned}$$

(ii) *If $n \geq 5$ and $K \subset \partial M$ is a Λ-homology boundary spine with exterior L and Λ-coefficient torsion*

$$\tau(L \longrightarrow M \times S^1; \Lambda) \in \Delta(\mathrm{Aut}_0^s(\mathbb{Z}[\pi])) \subseteq Wh_1(\Lambda)$$

then ∂M has an open book decomposition with binding K and page a codimension 1 Seifert surface $J \subset \partial M$ such that $(L_J; i_0 J, i_1 J)$ is an s-cobordism.
(iii) *Let $(M, \partial M) = (t(h), t(\partial_+ h))$ be an open book with page $(F, \partial_+ F)$ and binding $(N, \partial N)$, so that*

$$(h, \partial_+ h) : (F, \partial_+ F) \longrightarrow (F, \partial_+ F) \ , \ \partial F = \partial_+ F \cup N$$

as in 29.2 (iii). Then $(N, \partial N) \subset (M, \partial M)$ is a framed Λ-homology spine for $(M, \partial M)$ with exterior

$$\begin{aligned} (P, \partial_+ P) &= (\mathrm{cl.}(M \backslash N \times D^2), \mathrm{cl.}(\partial M \backslash \partial N \times D^2)) \\ &= (T(h : F \longrightarrow F), T(\partial_+ h : \partial_+ F \longrightarrow \partial_+ F)) \end{aligned}$$

and Λ-coefficient torsions

$$\begin{aligned} \tau(\partial_+ P \longrightarrow P; \Lambda) &= (0, [F, h] - [\partial_+ F, \partial_+ h]) \ , \\ \tau(P \longrightarrow M \times S^1; \Lambda) &= (0, [M, 1] - [F, h]) \\ &\in \Delta(\mathrm{Aut}_0^s(\mathbb{Z}[\pi])) \subseteq Wh_1(\Lambda) = Wh_1(\pi \times \mathbb{Z}) \oplus \mathrm{Aut}_0(\mathbb{Z}[\pi]) \ . \end{aligned}$$

(iv) *If $n \geq 6$ and $(N, \partial N) \subset (M, \partial M)$ is a Λ-homology boundary spine with exterior $(P, \partial_+ P)$ and Λ-coefficient torsions*

$$\tau(\partial_+ P \longrightarrow P; \Lambda) \ , \ \tau(P \longrightarrow M \times S^1; \Lambda) \in \Delta(\mathrm{Aut}_0^s(\mathbb{Z}[\pi])) \subseteq Wh_1(\Lambda)$$

then $(M, \partial M)$ has an open book decomposition with binding $(N, \partial N)$. The dimension restriction $n \geq 6$ can be reduced to $n \geq 5$ if ∂M already has an open book decomposition (e.g. if $\partial M = \emptyset$).
Proof Translate the Λ-homology spine terminology of Chap. 27 into open book terminology. □

29B. The asymmetric signature

The asymmetric signature will now be defined for a manifold with open book boundary. The asymmetric signature vanishes if and only if the open book decomposition extends to the interior. In particular, the asymmetric signature is defined for a closed manifold, in which case it is the obstruction to the existence of an open book decomposition.

Definition 29.15 Let $(M,\partial M)$ be an $(n+2)$-dimensional manifold with open book boundary

$$\partial M \;=\; t(h:F\longrightarrow F)$$

with $(h,1):(F,\partial F)\longrightarrow(F,\partial F)$ a rel ∂ automorphism of an n-dimensional manifold with boundary $(F,\partial F)$. The exterior of $\partial F\subset\partial M$ is

$$L \;=\; \mathrm{cl.}(t(h)\backslash\partial F\times D^2) \;=\; T(h:F\longrightarrow F)$$

with codimension 1 Seifert surface $F\subset L$ and

$$L_F \;=\; \mathrm{cl.}(L\backslash F\times I) \;\cong\; F\times I\ .$$

(i) The *asymmetric complex* of (M,F,h) is the $(n+2)$-dimensional asymmetric Poincaré complex (C,λ) over $\mathbb{Z}[\pi_1(M)]$ given by the construction of 27.6, with $\widetilde{M}$ the universal cover of M, $\widetilde{F}$ the pullback cover of F and $\lambda:C^{n-*}\longrightarrow C$ the $\mathbb{Z}[\pi_1(M)]$-module chain equivalence given by

$$\begin{aligned} C \;&=\; C(\widetilde{M},\widetilde{L}_F)\;\simeq\; C(\widetilde{M},\widetilde{F})\ ,\\ \lambda\;:\;C^{n-*}\;&=\;C(\widetilde{M},\widetilde{L}_F)^{n-*}\xrightarrow[\simeq]{[M]\cap-} C(\widetilde{M},\widetilde{F}\times I)\;\simeq\;C(\widetilde{M},\widetilde{F})\\ &\qquad\xrightarrow[\simeq]{\widetilde{i}_0} C(\widetilde{M},\widetilde{L}_F)\;=\;C\ ,\end{aligned}$$

with $\widetilde{M}$ the universal cover of M and $\widetilde{F}$ the pullback cover of F.
(ii) The *asymmetric signature* of (M,F,h) is the cobordism class

$$\sigma^*(M,F,h)\;=\;(C,\lambda)\in LAsy_h^{n+2}(\mathbb{Z}[\pi_1(M)])\ . \qquad\square$$

Proposition 29.16 *The asymmetric signature is a bounded open book bordism invariant, defining morphisms*

$$\sigma^*\;:\;BB_n(X)\longrightarrow LAsy_h^{n+2}(\mathbb{Z}[\pi_1(X)])\ ;\ (M,f;F,g,h)\longrightarrow\sigma^*(M,F,h)$$

Proof By the relative version of 29.15 a relative $(n+2)$-dimensional cobordism $(M;\partial_+M,\partial_-M)$ with

$$(\partial_+M,\partial\partial_+M)\;=\;(t_+(h),t(\partial_-h))$$

an $(n+1)$-dimensional open book with boundary (29.8) determines an $(n+2)$-dimensional asymmetric Poincaré pair $(C(\widetilde{\partial_- M},\widetilde{\partial_- F}) \longrightarrow C(\widetilde{M},\widetilde{F}),(\lambda,\partial_-\lambda))$ over $\mathbb{Z}[\pi_1(M)]$. □

Alexander [3] proved that every 3-manifold has an open book decomposition. More generally:

Theorem 29.17 (Winkelnkemper [308] for M closed, $\pi_1(M) = \{1\}$ and $n \geq 7$, T. Lawson [152] for M closed and odd $n \geq 7$, Quinn [227] for $n \geq 3$)
(i) *If $(M,\partial M)$ is an $(n+2)$-dimensional manifold with open book boundary $\partial M = t(h : F \longrightarrow F)$ the asymmetric signature is*

$$\sigma^*(M,F,h) \;=\; 0 \in LAsy_h^{n+2}(\mathbb{Z}[\pi_1(M)])$$

if (and for $n \geq 3$ only if) the open book decomposition of ∂M extends to M. The obstruction group is 0 for odd n, so in particular every odd-dimensional closed manifold of dimension ≥ 3 is an open book.
(ii) *The asymmetric signature map is an isomorphism for $n \geq 4$ and finitely presented $\pi_1(X)$*

$$\sigma^* \;:\; BB_n(X) \xrightarrow{\simeq} LAsy_h^{n+2}(\mathbb{Z}[\pi_1(X)]) \;.$$

Proof (i) It is clear from 29.16 that if the open book decomposition ∂M extends to M then $\sigma^*(M,F,h) = 0$, so only the converse need be considered. Three methods are presented in the high-dimensional case $n \geq 5$. See [227] for the special low-dimensional arguments required for $n = 3, 4$.

Method 1. An open book structure is the same as a Λ-homology spine (29.14), with $\Lambda = \Omega^{-1}\mathbb{Z}[\pi_1(M)][z,z^{-1}]$. By 27.8 the asymmetric signature is the Λ-homology spine obstruction of 22.15

$$\sigma^*(M,F,h) \in \Gamma_{n+2}^h(\mathcal{F},-z) \;=\; LAsy_h^{n+2}(\mathbb{Z}[\pi_1(M)]) \;.$$

The asymmetric signature is the obstruction to modifying the pushed in codimension 2 Seifert surface $F^n \subset M^{n+2}$ for $\partial F \subset \partial M$ by ambient surgery to a submanifold $N^n \subset M^{n+2}$ such that the exterior is a fibre bundle over S^1.

Method 2 (Chain complex version of [227, §11]).
A codimension 1 Seifert surface $(G,\partial_+G) \subset M$ for $(F,\partial F) \subset \partial M$ is a page for an open book decomposition of M extending the open book decomposition of ∂M if and only if the relative cobordism obtained by cutting M along $G \subset M$

$$(M_G;G_0,G_1) \;=\; (\mathrm{cl.}(M \backslash G \times I);G \times \{0\},G \times \{1\})$$

is homeomorphic to $G \times (I;\{0\},\{1\})$. Conversely, given a codimension 1 Seifert surface $(G,\partial_+G) \subset M$ such that there is defined an $(n+3)$-dimensional asymmetric pair over $\mathbb{Z}[\pi_1(M)]$

$$d_G \;=\; (f_G \;: C(\widetilde{M},\widetilde{F}) \longrightarrow C(\widetilde{M},\widetilde{G}),(\delta\lambda,\lambda))$$

then

$$\begin{aligned}\mathcal{C}(f_G)^{n+3-*} &\simeq C(\widetilde{G},\widetilde{F})^{n+2-*}\\ &\simeq C(\widetilde{G}\times I,\widetilde{G}\times\{0\}\cup\widetilde{\partial_+G}\times I\cup\widetilde{G}\times\{1\})\ ,\\ C(\widetilde{M},\widetilde{G}) &\simeq C(\widetilde{M}_G,\widetilde{G}\times\{0\}\cup\widetilde{\partial_+G}\times I\cup\widetilde{G}\times\{1\})\ ,\\ \mathcal{C}((\delta\lambda,\lambda):\mathcal{C}(f_G)^{n+3-*}\longrightarrow C(\widetilde{M},\widetilde{G})) &\simeq C(\widetilde{M}_G,\widetilde{G}_0)\ ,\end{aligned}$$

and d_G is a simple Poincaré pair if and only if $(M_G;G_0,G_1)$ is a relative s-cobordism. Thus for $n\geq 5$ d_G is a simple Poincaré pair if and only if G is a page of an open book decomposition of M extending the open book decomposition of ∂M. The asymmetric signature

$$\sigma^*(M,F,h)\in LAsy_h^{n+2}(\mathbb{Z}[\pi_1(M)])\ =\ LAsy_s^{n+2}(\mathbb{Z}[\pi_1(M)])$$

is 0 if and only if there exists a simple asymmetric null-cobordism

$$d\ =\ (f:C(\widetilde{M},\widetilde{F})\longrightarrow D,(\delta\lambda,\lambda))\ ,$$

which can be arranged (as in 28.34) to be such that $D_r=0$ for $2r<n+2$. If d can be realized as $d=d_G$ for some codimension 1 Seifert surface $(G,\partial_+G)\subset M$ the open book decomposition of ∂M extends to M. A *ribbon handle* of index $r+1$ for a codimension 1 Seifert surface $(G,\partial_+G)\subset M$ is an embedding

$$(D^{r+1}\times D^{n-r},S^r\times D^{n-r})\times I\subset(M,\partial_+G\times I)\ .$$

Attaching a ribbon handle to (G,∂_+G) results in a new codimension 1 Seifert surface

$$(G',\partial_+G')\ =\ (G\cup D^{r+1}\times D^{n-r},\mathrm{cl.}(\partial_+G\backslash S^r\times D^{n-r})\cup D^{r+1}\times S^{n-r-1})\ .$$

The method is to start with an arbitrary codimension 1 Seifert surface (G,∂_+G) (e.g. a collar $(F\times I,F\times\{1\})\subset M$) and then realize d by attaching ribbon handles of index $r+1\leq[(n+2)/2]$.

Method 3 (Chain complex version of [227, §§4-10]).
For any codimension 2 Seifert surface $\partial_+G\subset M$ for $\partial F\subset\partial M$ the exterior is an $(n+2)$-dimensional manifold with fibre bundle boundary

$$(M',\partial M')\ =\ (\mathrm{cl.}(M\backslash\partial_+G\times D^2),T(h'))$$

where

$$h'\ =\ h\cup 1\ :\ F'\ =\ F\cup_\partial\partial_+G\longrightarrow F'\ .$$

The relative mapping torus

$$t_+(h'\times 1:F'\times I\longrightarrow F'\times I)\ =\ T(h')\times I\cup\partial_+G\times D^2\times\{0\}$$

defines an open book cobordism $(t_+(h'\times 1);t(h),T(h'))$ such that

$$(M \times I; M \times \{0\}, M' \times \{1\}; t_+(h' \times 1))$$

is a relative bordism, so that as in 29.16 there is induced an asymmetric Poincaré cobordism over $\mathbb{Z}[\pi_1(M)]$

$$(C(\widetilde{M}, \widetilde{F}) \oplus C(\widetilde{M}', \widetilde{F}') \longrightarrow C(\widetilde{M}, \widetilde{F}'), (0, \lambda \oplus -\lambda')) .$$

Attach handles to $\partial F \times I \subset M$ of index $\leq [n/2]$, to obtain a codimension 2 Seifert surface $\partial_+ G \subset M$ for $\partial F \subset \partial M$ such that

$$H_r(\partial_+ \widetilde{G}, \partial \widetilde{F}) = H_r(\widetilde{M}, \widetilde{F}) \text{ for } r \neq \begin{cases} i \\ i, i+1 \end{cases} \text{ if } n+2 = \begin{cases} 2i \\ 2i+1 , \end{cases}$$

in which case

$$H_r(\widetilde{M}', \widetilde{F}') = 0 \text{ for } r \neq \begin{cases} i \\ i, i+1 \end{cases} \text{ if } n+2 = \begin{cases} 2i \\ 2i+1 . \end{cases}$$

In the even-dimensional case $n+2 = 2i$ the asymmetric signature

$$\begin{aligned} \sigma^*(M, F, h) &= (H_i(\widetilde{M}', \widetilde{F}'), \lambda') \\ &\in LAsy_h^{2i}(\mathbb{Z}[\pi_1(M)]) = LAsy_s^{2i}(\mathbb{Z}[\pi_1(M)]) \end{aligned}$$

is 0 if and only if the simple nonsingular asymmetric form $(H_i(\widetilde{M}', \widetilde{F}'), \lambda')$ admits a simple based f.g. free lagrangian L. If there exists such a lagrangian the basis elements $x_1, x_2, \ldots, x_\ell \in L$ can be represented by ℓ disjoint ribbon handles of index i

$$(D^i \times D^{i-1}, S^{i-1} \times D^{i-1}) \times I \subset (M'^{2i}, F'^{2i-2} \times I) .$$

The trace of the corresponding ℓ codimension 2 surgeries on $F' = F' \times \{0\} \subset M'$ is a codimension 1 Seifert surface for $(F, \partial F) \subset \partial M$

$$(G, \partial_+ G) = (F' \times I \cup \bigcup_\ell D^i \times D^{i-1}, \partial_+ G \times \{0\}) \subset M^{2i}$$

with $(M_G; G_0, G_1)$ a relative s-cobordism.

In the odd-dimensional case $n+2 = 2i+1$ it is possible to apply Neumann's stabilization lemma as in the proof of

$$LAsy_h^{2i+1}(\mathbb{Z}[\pi_1(M)]) = LAsy_s^{2i+1}(\mathbb{Z}[\pi_1(M)]) = 0$$

in 28.34, arranging for the chain equivalence in the $(i-1)$-connected $(2i+1)$-dimensional asymmetric Poincaré complex $(C(\widetilde{M}', \widetilde{F}'), \lambda')$

$$\lambda' : C(\widetilde{M}', \widetilde{F}')^{2i+1-*} \xrightarrow{\simeq} C(\widetilde{M}', \widetilde{F}')$$

to be a simple isomorphism. The basis elements of $C(\widetilde{M}', \widetilde{F}')_i$ can then be represented by disjoint ribbon handles of index i

$$(D^i \times D^i, S^{i-1} \times D^i) \times I \subset (M'^{2i+1}, F'^{2i-1} \times I)$$

such that the trace of the corresponding codimension 2 surgeries on $F' \subset M'$ is a codimension 1 Seifert surface for $(F, \partial F) \subset \partial M$

$$(G, \partial_+ G) = (F' \times I \cup \bigcup_{\ell} D^i \times D^i, \partial_+ G \times \{0\}) \subset M^{2i+1}$$

with $(M_G; G_0, G_1)$ a relative s-cobordism.

(ii) Immediate from (i), noting that the proof of (i) includes the realization of every nonsingular asymmetric form as the asymmetric complex of an even-dimensional manifold with open book boundary – see 29.20 below for a more detailed account. □

Remark 29.18 The actual result of Winkelnkemper [308] is that for $n \geq 7$ a simply-connected closed n-dimensional manifold M is an open book if $n \not\equiv 0(\text{mod}4)$, and that for $n = 4k \geq 8$ M is an open book if and only if

$$\text{signature}(M) = 0 \in L^{4k}(\mathbb{Z}) = \mathbb{Z} .$$

The asymmetric L-theory formulation is due to Quinn [227]. The natural map

$$L^{4k}(\mathbb{Z}) = \mathbb{Z} \longrightarrow LAsy^{4k}(\mathbb{Z}) = \mathbb{Z}^\infty \oplus \mathbb{Z}_2^\infty \oplus \mathbb{Z}_4^\infty$$

is a split injection (see Chaps. 37–42 below for this and related results), so that the signature is subsumed in the simply-connected asymmetric signature invariant of 29.17. See 30.12 below for the chain complex version of the method of [308]. □

Corollary 29.19 (i) *The bounded book bordism groups $BB_*(X)$ are related to the framed spine bordism groups $BB_*(X, \mathfrak{F})$ (27.13) for*

$$\mathfrak{F} : \mathbb{Z}[\pi_1(X)][z, z^{-1}] \longrightarrow \Lambda = \Omega^{-1}\mathbb{Z}[\pi_1(X)][z, z^{-1}]$$

by forgetful maps

$$BB_n(X) \longrightarrow BB_n(X, \mathfrak{F}) .$$

(ii) *For $n \geq 5$ and finitely presented $\pi_1(X)$ the forgetful maps of (i) are isomorphisms, so that there are identifications*

$$\begin{aligned} BB_n(X) &= BB_n(X, \mathfrak{F}) \\ &= AB_n(X) = L_h^{n+2}(\Omega^{-1}\mathbb{Z}[\pi_1(X)][z, z^{-1}]) \\ &= LAsy_h^{n+2}(\mathbb{Z}[\pi_1(X)]) \\ &= \begin{cases} LAsy_h^0(\mathbb{Z}[\pi_1(X)]) & \textit{if } n \textit{ is even} \\ 0 & \textit{if } n \textit{ is odd} . \end{cases} \end{aligned}$$

Proof The identification $BB_n(X) = LAsy_h^{n+2}(\mathbb{Z}[\pi_1(X)])$ was already obtained in 29.17. The identification

$$L_h^{n+2}(\Omega^{-1}\mathbb{Z}[\pi_1(X)][z, z^{-1}]) = LAsy_h^{n+2}(\mathbb{Z}[\pi_1(X)])$$

is a special case of 28.33. □

Remark 29.20 The inverse of the asymmetric signature isomorphism

$$\sigma^* \;:\; BB_n(X) \xrightarrow{\simeq} LAsy_h^{n+2}(A) \;=\; L_{n+2}^h(\Lambda)$$
$$(n = 2i \geq 6 \;,\; A \;=\; \mathbb{Z}[\pi_1(X)] \;,\; \Lambda \;=\; \Omega^{-1}A[z,z^{-1}])$$

can be described by the realization of asymmetric forms as in 27.9 or Quinn [227, §7]. Alternatively, it is possible to directly apply the realization theorem of Wall [302, Chap. 5], as follows.

Given a nonsingular asymmetric form (L,λ) over A define an asymmetric form $(L[z,z^{-1}],\psi)$ over $A[z,z^{-1}]$ by

$$\psi \;=\; (1-z^{-1})\lambda \;:\; L[z,z^{-1}] \longrightarrow L^*[z,z^{-1}] \;.$$

The $(-)^{i+1}$-symmetrization of ψ defines a Λ-isomorphism

$$\omega \;=\; \psi + (-)^{i+1}\psi^* \;=\; (1-z^{-1})(\lambda + (-)^i z\lambda^*) \;:\; L[z,z^{-1}] \longrightarrow L^*[z,z^{-1}] \;,$$

with a short exact sequence

$$0 \longrightarrow L[z,z^{-1}] \xrightarrow{\omega} L^*[z,z^{-1}] \longrightarrow L \oplus L^* \longrightarrow 0 \;.$$

The action of z on $L \oplus L^*$ defines an automorphism of a hyperbolic $(-)^i$-quadratic form over A

$$h \;=\; \begin{pmatrix} 1+(-)^{i+1}(\lambda^*)^{-1}\lambda & (\lambda^*)^{-1} \\ (-)^i\lambda & 0 \end{pmatrix}$$
$$: (L \oplus L^*, \begin{pmatrix} 0 & 1 \\ 0 & 0 \end{pmatrix}) \longrightarrow (L \oplus L^*, \begin{pmatrix} 0 & 1 \\ 0 & 0 \end{pmatrix})$$

with torsion

$$\tau(h) \;=\; \tau(\lambda) - \tau(\lambda)^* \in Wh_1(A) \;.$$

Let N^{2i} be a closed $2i$-dimensional manifold with a π_1-isomorphism reference map $N \longrightarrow X$. Let $\ell = \dim_A(L)$, and attach ℓ $(i+1)$-handles to $N \times S^1 \times I$ with self intersections ψ, to obtain an i-connected $(2i+2)$-dimensional normal map

$$(f,b) \;:\; (M;N \times S^1, \partial_+ M) \longrightarrow N \times S^1 \times (I;\{0\},\{1\})$$

with quadratic kernel the based f.g. free $(2i+2)$-dimensional quadratic Λ-Poincaré complex over $A[z,z^{-1}]$

$$\sigma_*(f,b) \;=\; (S^{i+1}L[z,z^{-1}],\psi)$$

and such that

$$M \;=\; N \times S^1 \times I \cup_\omega \bigcup_\ell D^{i+1} \times D^{i+1} \;.$$

The normal map

$$(\partial_+ f, \partial_+ b) \;=\; (f,b)| \;:\; \partial_+ M \longrightarrow N \times S^1$$

is a Λ-coefficient homology equivalence, so that $\partial_+ M$ is a band. The quadratic kernel of $(\partial_+ f, \partial_+ b)$ is the $(2i+1)$-dimensional Λ-contractible quadratic Poincaré complex over $A[z,z^{-1}]$

$$\begin{aligned}\sigma_*(\partial_+ f, \partial_+ b) &= \partial(S^{i+1}L[z,z^{-1}],\psi) \\ &= (S^i\mathcal{C}(\omega : L[z,z^{-1}]\longrightarrow L^*[z,z^{-1}]),\partial\psi)\end{aligned}$$

and the Λ-coefficient Whitehead torsion of $\partial_+ f$ is

$$\begin{aligned}\tau(\partial_+ f;\Lambda) \;=\; \tau(\omega) &= (\Phi^+(\partial_+ M),[\overline{\partial_+ M},\zeta]) - (\Phi^+(N),[N,1]) \\ &= (\tau(\lambda),[L\oplus L^*,h]) \\ &\in Wh_1(\Lambda) \;=\; Wh_1(A[z,z^{-1}])\oplus \mathrm{Aut}_0(A)\ .\end{aligned}$$

The fibering obstruction of $\partial_+ M$ is

$$\Phi^+(\partial_+ M) \;=\; \tau(\lambda)\in Wh_1(A)\subseteq Wh_1(A[z,z^{-1}])$$

so that $\partial_+ M$ fibres over S^1 if and only if $\tau(\lambda)=0$. Thus for simple (L,λ)

$$\partial_+ M \;=\; T(h' : N'\longrightarrow N')\ ,\quad N'^{2i} \;=\; N^{2i}\#\mathop{\#}_{\ell} S^i\times S^i$$

with $h' : N'\longrightarrow N'$ a homeomorphism inducing the $\mathbb{Z}[\pi]$-module self chain equivalence

$$\begin{aligned}&\widetilde{h}' \;=\; 1\oplus h\ : \\ &C(\widetilde{N}') \;=\; C(\widetilde{N})\oplus(L\oplus L^*) \xrightarrow{\;\simeq\;} C(\widetilde{N}') \;=\; C(\widetilde{N})\oplus(L\oplus L^*)\ .\end{aligned}$$

The asymmetric Witt class of (L,λ) is the obstruction to extending the open book decomposition of the boundary

$$\partial M \;=\; T(1\cup h' : N\cup N'\longrightarrow N\cup N')$$

to M, i.e. the asymmetric signature is given by

$$\begin{aligned}\sigma^*(M,N\cup N',1\cup h') &= \sigma_*(f,b) \;=\; (L,\lambda) \\ &\in BB_{2i}(X) \;=\; LAsy^0_h(A) \;=\; L^{2i+2}_h(\Lambda)\ .\end{aligned}$$

□

Example 29.21 (i) Suppose given an open book decomposition of S^{n+2}, i.e. a rel ∂ automorphism $(h,1) : (F,\partial F)\longrightarrow(F,\partial F)$ of an $(n+1)$-dimensional manifold with boundary $(F,\partial F)$ with relative mapping torus

$$t(h) \;=\; S^{n+2}\ .$$

The asymmetric signature of $(D^{n+3},F,h)\in BB_{n+1}(\{\mathrm{pt.}\})$ is given by

$$\sigma^*(D^{n+2},F,h) \;=\; (S\dot{C}(F),\lambda)\in LAsy^{n+3}(\mathbb{Z})\ ,$$

with $\dot{C}(F)$ the reverse chain complex of F, and $\lambda : S\dot{C}(F)^{n+3-*}\longrightarrow S\dot{C}(F)$ a chain equivalence inducing the isomorphisms

$$\lambda_*\ :\ H^{n+1-*}(F) \xrightarrow[\simeq]{[F]\cap -} H_*(F,\partial F) \xrightarrow[\simeq]{V} H_*(F)\quad (*\neq 0,n+1)$$

with V the variation map such that

$$1 - h_* \ : \ H_*(F) \longrightarrow H_*(F,\partial F) \xrightarrow{V} H_*(F) \ .$$

(The *variation* map $V : H_*(F,\partial F) \longrightarrow H_*(F)$ is defined for any open book $t(h : F \longrightarrow F)$, with an exact sequence

$$\dots \longrightarrow H_r(F,\partial F) \xrightarrow{V} H_r(F) \longrightarrow H_r(t(h)) \longrightarrow H_{r-1}(F,\partial F) \longrightarrow \dots$$

– see Lamotke [144], Durfee and Kauffman [65], Kauffman [121, §2]).
(ii) A fibred knot $k : S^n \subset S^{n+2}$ with Seifert surface

$$(F^{n+1}, \partial F = k(S^n)) \subset S^{n+2}$$

and monodromy

$$(h,1) \ : \ (F,\partial F) \longrightarrow (F,\partial F)$$

determines an open book decomposition of $S^{n+2} = t(h)$, and so determines an element

$$(D^{n+3}, G; F, h, f) \in BB_{n+1}(\{\text{pt.}\})$$

with an asymmetric signature as in (i)

$$\sigma^*(D^{n+3}, F, h) \ = \ (S\dot{C}(F), \lambda) \in LAsy^{n+3}(\mathbb{Z}) \ .$$

In particular, the trivial knot $k_0 : S^n \subset S^{n+2}$ is fibred, with monodromy $h_0 = \text{id.} : F_0 = D^{n+1} \longrightarrow D^{n+1}$, and $\sigma^*(D^{n+3}, F_0, h_0) = 0$.
(iii) For any integer $a \geq 2$ the cyclic permutation of period a

$$h_a \ : \ F_a \ = \ \{1, 2, \dots, a\} \longrightarrow F_a \ ; \ j \longrightarrow j + 1 (\text{mod}\, a)$$

has (relative) mapping torus

$$t(h_a) \ = \ T(h_a) \ = \ S^1 \ ,$$

defining an open book decomposition of S^1 with page F_a and empty binding $\partial F_a = \emptyset$. The nonsingular asymmetric form is given by

$$(H_1(D^2, F_a), \lambda_a) \ = \ \left(\bigoplus_{a-1} \mathbb{Z}, \begin{pmatrix} 1 & -1 & 0 & \dots & 0 \\ 0 & 1 & -1 & \dots & 0 \\ 0 & 0 & 1 & \dots & 0 \\ \vdots & \vdots & \vdots & \ddots & \vdots \\ 0 & 0 & 0 & \dots & 1 \end{pmatrix} \right)$$

and the asymmetric signature is

$$\sigma^*(D^2, F_a, h_a) \ = \ (C(D^2, F_a), \lambda_a) \in LAsy^2(\mathbb{Z}) \ ,$$

with $C(D^2, F_a) \simeq \bigoplus_{a-1} S\mathbb{Z}$. In the terminology of Kauffman and Neumann [123, Chap. 6] $S^1 = t(h_a)$ is the open book decomposition of S^1 associated to the empty fibred knot $[a] : S^{-1} = \emptyset \subset S^1$, corresponding to the fibration

$$F_a \longrightarrow S^1 \xrightarrow{a} S^1$$

with $a : S^1 \longrightarrow S^1 ; z \longrightarrow z^a$. This is the fibred knot construction of Milnor [196] associated to the isolated singular point 0 of $\mathbb{C} \longrightarrow \mathbb{C} ; z \longrightarrow z^a$. See 29.22 (iii) below for the application to the Brieskorn fibred knots.
(iv) Set $a = 2$ in (iii). The isomorphism $BB_0(\{\text{pt.}\}) \cong AB_0(\{\text{pt.}\})$ of 29.20 sends the element

$$(W, G; F, h, f) \ = \ (D^2, G; F_2, h_2, f) \in BB_0(\{\text{pt.}\})$$

to

$$(W \cup F \times I \times I, G \cup f(pr_1); F \cup -F, h \cup 1, f \cup f) \in AB_0(\{\text{pt.}\}) \ .$$

The map $G \cup f(pr_1) : W \cup F \times I \times I \longrightarrow S^1$ defines a null bordism over S^1 of

$$T(h_2 \cup 1) \ = \ (S^1, 2 : S^1 \longrightarrow S^1) \sqcup -(S^1 \sqcup S^1, 1 \sqcup 1 : S^1 \sqcup S^1 \longrightarrow S^1) \ ,$$

which is a double cover of the annulus $S^1 \times I$ branched over a point. This is a special case of the construction of Dold (López de Medrano [171, p. 75]), which for any fixed-point free involution $T : N \longrightarrow N$ of a closed n-dimensional manifold N with characteristic codimension 1 submanifold $M \subset N$ uses the double cover $\mathfrak{D}$ of $N \times I$ branched at $M/T \times \{1/2\}$ to define a cobordism $(\mathfrak{D}; N, N \sqcup N)$ with an involution

$$(S; T, \text{transposition}) \ : \ (\mathfrak{D}; N, N \sqcup N) \longrightarrow (\mathfrak{D}; N, N \sqcup N)$$

such that S has fixed point set M/T. (Here, $(N, T) = (S^1, -1)$, $M = S^0 \subset S^1$ and $M/T = \{\text{pt.}\}$, with $\mathfrak{D} = W \cup F \times I \times I$ a connected surface with three boundary components.) □

Remark 29.22 (i) For any rings with involution A, A' the product of an n-dimensional asymmetric Poincaré complex (C, λ) over A and an n'-dimensional asymmetric Poincaré complex (C', λ') over A' is an $(n + n')$-dimensional asymmetric Poincaré complex over $A \otimes_{\mathbb{Z}} A'$

$$(C, \lambda) \otimes (C', \lambda') \ = \ (C \otimes C', \lambda \otimes \lambda')$$

with a corresponding product in the asymmetric L-groups

$$LAsy^n(A) \otimes_{\mathbb{Z}} LAsy^{n'}(A') \longrightarrow LAsy^{n+n'}(A \otimes_{\mathbb{Z}} A') \ .$$

As in 27.16, there is a geometric interpretation: if M is an n-dimensional manifold with open book boundary $\partial M = t(h : F \longrightarrow F)$ and M' is an n'-dimensional manifold with open book boundary $\partial M' = t(h' : F' \longrightarrow F')$ then $M \times M'$ is an $(n + n')$-dimensional manifold with open book boundary

$$\begin{aligned} \partial(M \times M') \ &= \ M \times \partial M' \cup \partial M \times M' \\ &= \ t((1_M \times h') \cup (h \times 1_{M'}) : M \times F' \cup F \times M' \\ &\qquad\qquad \longrightarrow M \times F' \cup F \times M') \ , \end{aligned}$$

with the product asymmetric signature

$$\begin{aligned}&\sigma^*(M \times M', M \times F' \cup F \times M', (1_M \times h') \cup (h \times 1_{M'}))\\ &\quad = \sigma^*(M,F,h) \otimes \sigma^*(M',F',h') \in LAsy^{n+n'}(\mathbb{Z}[\pi_1(M \times M')]) .\end{aligned}$$

(ii) Suppose that $(M, \partial M)$ is an n-dimensional manifold with boundary and $(N, \partial N) \subset (M, \partial M)$ is a homology framed codimension 2 submanifold which is a push-in as in the proof of 27.8, with n-dimensional asymmetric complex (C, λ) over $\mathbb{Z}[\pi_1(M)]$ (27.6) where $C = C(\widetilde{M}, \widetilde{N})^{n-*}$. For any integer $a \geq 2$ let $\widehat{F}_a \subset D^2$ be a subset of a points in the interior of D^2, a push-in of the codimension 1 Seifert surface $F_a \subset S^1$ of 29.21 (iii) for the fibred knot $[a] : S^{-1} = \emptyset \subset S^1$. Define the *a-fold cyclic suspension* $(M \times D^2, M')$ of (M, N) to be the product $(n+2)$-dimensional manifold with boundary $(M \times D^2, \partial(M \times D^2))$ with the homology framed codimension 2 submanifold

$$M' = N \times D^2 \cup M \times \widehat{F}_a \subset M \times D^2$$

such that $(M', \partial M')$ is the a-fold cyclic branched cover of $(M, \partial M)$ branched over $(N, \partial N)$ (27.10). The $(n+2)$-dimensional asymmetric complex over $\mathbb{Z}[\pi_1(M)]$ of the cyclic suspension is the product

$$(C', \lambda') = (C, \lambda) \otimes (\bigoplus_{a-1} S\mathbb{Z}, \lambda_a) .$$

For $(M, \partial M) = (D^n, S^{n-1})$ this is just the a-fold cyclic suspension operation of Kauffman and Neumann [123].

(iii) The product formula for the Seifert form of the product of a knot and a fibred knot was first proved in [123, 6.5], including the a-fold cyclic suspension of a knot $K = \partial N \subset S^{n-1}$ as a special case. See Kauffman [120], [121] for the topological construction of the manifolds of Brieskorn [26] : the fibred knot $\Sigma(a_0, a_1, \ldots, a_i)^{2i-1} \subset S^{2i+1}$ at the singular point $0 \in \mathbb{C}^{i+1}$ of the function

$$f : \mathbb{C}^{i+1} \longrightarrow \mathbb{C} \; ; \; (z_0, z_1, \ldots, z_i) \longrightarrow z_0^{a_0} + z_1^{a_1} + \ldots + z_i^{a_i} \quad (a_0, \ldots, a_i \geq 2)$$

was obtained inductively as the a_i-fold cyclic suspension of the fibred knot $\Sigma(a_0, a_1, \ldots, a_{i-1})^{2i-3} \subset S^{2i-1}$, starting with $[a_0] : \Sigma(a_0) = S^{-1} = \emptyset \subset S^1$. The application of the product formula gave the Seifert form to be

$$\bigotimes_{j=0}^{i} (\bigoplus_{a_j - 1} \mathbb{Z}, \lambda_{a_j})$$

as originally obtained by Brieskorn and Pham (cf. Milnor [196, 9.1]).

(iv) The 2-fold periodicity isomorphism

$$LAsy^n(A) \xrightarrow{\simeq} LAsy^{n+2}(A) \; ; \; (C, \lambda) \longrightarrow (SC, \lambda)$$

is given by product with the element $(S\mathbb{Z}, \lambda_2) \in LAsy^2(\mathbb{Z})$ $(\lambda_2 = 1)$, corresponding to the 2-fold cyclic suspension operation of Bredon [25], Kauffman [120], Neumann [210] and Kauffman and Neumann [123]. □

30. Twisted doubles

A closed manifold is a twisted double if it is of the form

$$Q \cup_h -Q \;=\; Q \times \{0,1\}/\{(x,0) \simeq (h(x),1)\,|\, x \in P\}$$

for some manifold with boundary (Q,P) and a self homeomorphism

$$h : \partial Q = P \longrightarrow P .$$

Twisted doubles have played an important role in the history of manifolds, especially in the odd dimensions:

(i) every 3-dimensional manifold admits a Heegaard decomposition as a twisted double of solid tori

$$M^3 \;=\; (\#S^1 \times D^2) \cup_h -(\#S^1 \times D^2)$$

with $h : \#S^1 \times S^1 \longrightarrow \#S^1 \times S^1$,

(ii) the original exotic 7-spheres of Milnor [190] are twisted doubles

$$\Sigma^7 \;=\; S^3 \times D^4 \cup_{h_1} S^3 \times D^4 \;=\; D^7 \cup_{h_2} -D^7$$

with $h_1 : S^3 \times S^3 \longrightarrow S^3 \times S^3$, $h_2 : S^6 \longrightarrow S^6$ diffeomorphisms,

(iii) Smale [269] and Barden used the handlebody proof of the h-cobordism theorem to show that every simply-connected odd-dimensional manifold M^{2i+1} ($i \geq 2$) is a twisted double,

(iv) the original formulation in Wall [302] of the odd-dimensional surgery obstruction groups was in terms of automorphisms of hyperbolic quadratic forms, corresponding to twisted doubles in algebraic L-theory.

Although the existence and classification of twisted double structures appears at first to be a codimension 1 surgery problem, it is in fact susceptible to the methods of codimension 2 surgery.

An open book decomposition is a particular kind of twisted double. The asymmetric signature obstruction to an open book decomposition on a manifold M was identified in Chap. 29 with a framed 2 codimension surgery invariant, namely the obstruction to modifying the empty codimension 2 submanifold $\emptyset \subset M$ to a framed 2 codimension 2 submanifold $L \subset M$ such that the exterior fibres over S^1. In the original work of Winkelnkemper [307] and Quinn [227] the open book invariant arose via codimension 1 surgery, as the obstruction to the existence of a twisted double decomposition $M = Q \cup_h -Q$ such that $P = R \cup_k -R$ is a twisted double with the two inclusions $R \longrightarrow Q$ simple homotopy equivalences. This chapter will extend the open book theory of Chap. 29 to twisted doubles, reversing the historical order. The asymmetric signature invariant will be generalized to a manifold with twisted double boundary, and will be shown to be the obstruction to extending the twisted double decomposition to the interior.

The geometric theory of twisted doubles is developed in 30A. Twisted doubles arise as the boundaries of fundamental domains for infinite cyclic covers of manifolds with boundary a fibre bundle over S^1, as follows. If $(M, \partial M)$ is an $(n+2)$-dimensional manifold with boundary and a map $(p, \partial p) : (M, \partial M) \longrightarrow S^1$ such that $\partial p : \partial M \longrightarrow S^1$ is a fibre bundle with $\partial M = T(h : P \longrightarrow P)$ then a fundamental domain $(W; Q, \zeta Q)$ for the infinite cyclic cover $\overline{M} = p^*\mathbb{R}$ of M is an $(n+2)$-dimensional manifold W with boundary the twisted double $\partial W = Q \cup_h -Q$.

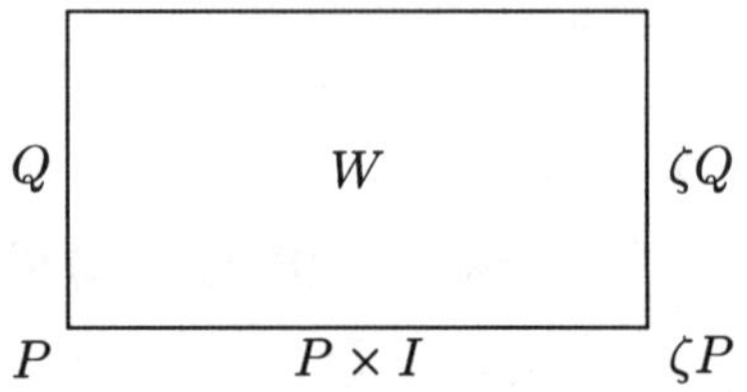

Conversely, a manifold with a twisted double boundary $(W, \partial W = Q \cup_h -Q)$ determines an $(n+2)$-dimensional manifold with fibre bundle boundary

$$(M, \partial M) = (W \cup Q \times I, T(h : P \longrightarrow P)) .$$

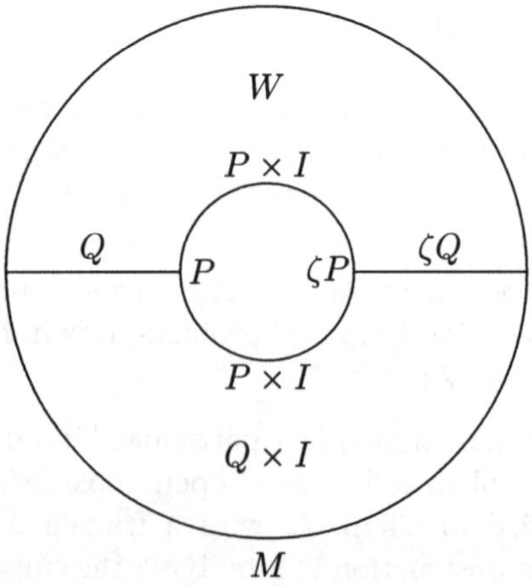

For any such W, M, P, Q, h the construction of Chap. 29 gives an $(n+2)$-dimensional asymmetric Poincaré complex (C, λ) over $\mathbb{Z}[\pi_1(W)]$ with

$$C = C(\widetilde{M}, \widetilde{P}) = \mathcal{C}((i \quad j_0 - j_1) : C(\widetilde{P}) \oplus C(\widetilde{Q}) \longrightarrow C(\widetilde{W}))$$
$$(i : P \longrightarrow W \ , \ j_0, j_1 : Q \longrightarrow W \text{ inclusions}) \ .$$

The algebraic theory of twisted doubles is developed in 30B, allowing the definition of the asymmetric signature

$$\sigma^*(W, Q, P, h) = (C, \lambda) \in LAsy_h^{n+2}(\mathbb{Z}[\pi_1(W)])$$

using only the algebraic Poincaré duality decomposition of (W, Q, P, h). The asymmetric signature is 0 if (and for $n \geq 4$ only if) the twisted double decomposition of the boundary ∂W extends to W. Thus a high-dimensional manifold is a twisted double if and only if it is an open book. The algebraic theory of twisted doubles is extended to open books in 30C, with the algebraic L-groups of twisted doubles defined in 30D. In particular, it is shown that the odd-dimensional L-groups are quotients of the odd-dimensional twisted double L-groups.

30A. Geometric twisted doubles

Definition 30.1 (i) The *double* of an n-dimensional geometric Poincaré pair (Q, P) is the n-dimensional geometric Poincaré complex

$$D(Q, P) = Q \cup_P -Q = \partial(Q \times I) \ .$$

(ii) The *twisted double* of an n-dimensional geometric Poincaré pair (Q, P) with respect to a self homotopy equivalence $h : P \longrightarrow P$ is the n-dimensional geometric Poincaré complex

$$D(Q, P, h) = Q \cup_h -Q = Q \times \{0,1\}/\{(x,0) = (h(x),1) \,|\, x \in P\} \ . \quad \square$$

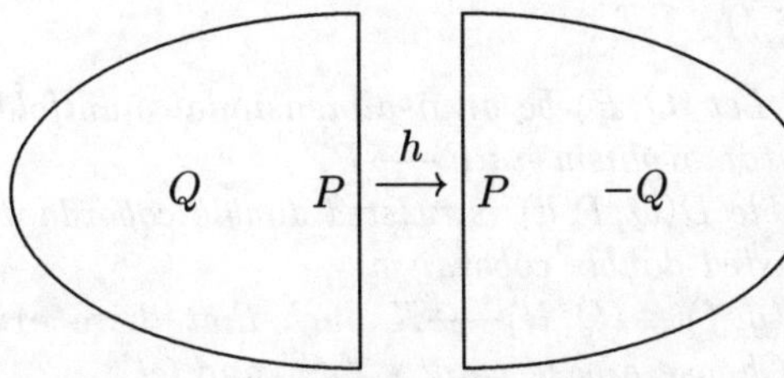

$D(Q, P, h)$

Remark 30.2 (i) The double of (Q, P) is just the (un)twisted double

$$D(Q, P) = D(Q, P, 1) \ .$$

(ii) The twisted double $D(Q,P,h)$ of an n-dimensional manifold with boundary (Q,P) with respect to an automorphism $h : P \longrightarrow P$ is a closed n-dimensional manifold.
(iii) If $h : P \longrightarrow P$ extends to a homotopy equivalence (resp. homeomorphism) $H : Q \longrightarrow Q$ there is defined a homotopy equivalence (resp. homeomorphism)

$$D(Q,P,h) \longrightarrow D(Q,P) \ ; \ (x,i) \longrightarrow \begin{cases} (H(x),0) & \text{if } i=0 \\ (x,1) & \text{if } i=1 \end{cases} \quad (x \in Q) \ .$$

(iv) The mapping torus of a homotopy equivalence of an n-dimensional geometric Poincaré complex $h : P \longrightarrow P$ is a twisted double

$$T(h : P \longrightarrow P) \ = \ D(P \times I, P \times \partial I, h \sqcup 1) \ ,$$

with

$$h \sqcup 1 \ : \ \partial(P \times I) \ = \ P \sqcup -P \longrightarrow P \sqcup -P \ .$$

(v) The relative mapping torus of a rel ∂ homeomorphism of an n-dimensional manifold with boundary $(h,1) : (P,\partial P) \longrightarrow (P,\partial P)$ is a twisted double

$$t(h : P \longrightarrow P) \ = \ D(P \times I, \partial(P \times I), h \cup 1) \ ,$$

with

$$h \cup 1 \ : \ \partial(P \times I) \ = \ D(P) \ = \ P \cup_{\partial P} -P \longrightarrow P \cup_{\partial P} -P \ . \qquad \square$$

By contrast with open book structures on surfaces (29.5):

Example 30.3 Every closed (connected, orientable) surface M^2 is the double

$$M \ = \ D(F, \partial F)$$

of the 2-dimensional manifold with boundary obtained by deleting the neighbourhoods $D_i \cong D^2 \subset D^2$ of g distinct points $x_i \in \text{int}(D^2)$ $(1 \leq i \leq g)$

$$(F, \partial F) \ = \ (\text{cl.}(D^2 \backslash \bigsqcup_{i=1}^{g} D_i), \bigsqcup_{g+1} S^1) \ ,$$

with g the genus of M. $\square$

Proposition 30.4 *Let (Q,P) be an n-dimensional manifold with boundary, together with an automorphism $h : P \longrightarrow P$.*
(i) *The twisted double $D(Q,P,h)$ is twisted double cobordant to the mapping torus $T(h)$ by a twisted double cobordism.*
(ii) *Given a map $(g,f) : (Q,P) \longrightarrow X$ such that there exists a homotopy $fh \simeq f : P \longrightarrow X$. Choose a base point $* \in S^1$ and let*

$$i \ : \ X \longrightarrow X \times S^1 \ ; \ x \longrightarrow (x, *) \ ,$$
$$j \ : \ X \times S^1 \longrightarrow X \ ; \ (x,s) \longrightarrow x \ .$$

Then g extends to a map

$$D(g,f) = g\cup g : D(Q,P,h) = Q\cup_h -Q \longrightarrow X$$

such that

$$\begin{aligned}(D(Q,P,h),iD(g,f)) &= (T(h),ijT(f)) = (T(h),T(f)) \\ &\in \mathrm{im}(i : \Omega_n(X)\longrightarrow\Omega_n(X\times S^1)) \subseteq \Omega_n(X\times S^1) .\end{aligned}$$

Proof (i) Use a collar $P\times I\subset Q$ to write the double as

$$D(Q,P,h) = (Q\times\{-1\}\cup_1 P\times[-1,0])\bigcup_h (P\times[0,1]\cup_1 Q\times\{1\}) .$$

The mapping torus $T(h)$ is obtained from $D(Q,P,h)$ by surgery on

$$Q\times\{-1,1\}\subset D(Q,P,h) ,$$

with the trace

$$\begin{aligned}R &= D(Q,P,h)\times I\cup_{Q\times\{-1,1\}\times\{1\}} Q\times[-1,1] \\ &= (Q\times[-1,0]\times I\cup_{Q\times\{-1\}\times\{1\}} Q\times[-1,0]) \\ &\qquad \bigcup_{1\cup h\cup 1} (Q\times[0,1]\times I\cup_{Q\times\{1\}\times\{1\}} Q\times[0,1])\end{aligned}$$

defining a twisted double cobordism $(R;D(Q,P,h),T(h))$.
(ii) By 23.2

$$\begin{aligned}(T(h),T(f)) &= (T(h),ijT(f))\oplus(P\times S^1,f\times 1) \\ &= (T(h),ijT(f))\oplus\partial(Q\times S^1,g\times 1) \\ &= (T(h),ijT(f))\in\Omega_n(X\times S^1) ,\end{aligned}$$

so it suffices to verify that

$$(D(Q,P,h),D(g,f)) = (T(h),jT(f))\in\Omega_n(X) .$$

The map $g:Q\longrightarrow X$ extends to a map on the cobordism defined in (i)

$$(G;D(g,f),jT(f)) : (R;D(Q,P,h),T(h)) \longrightarrow X ,$$

so that

$$(D(Q,P,h),D(g,f)) = (T(h),jT(f))\in\Omega_n(X) . \qquad \square$$

Definition 30.5 (i) The *twisted double bordism group* $DB_n(X)$ is the group of bordism classes of quintuples (Q,P,h,g,f) with (Q,P) an $(n+1)$-dimensional manifold with boundary, $h:P\longrightarrow P$ an automorphism and $(g,f):(Q,P)\longrightarrow X$ a map such that there exists a homotopy $fh\simeq f: P\longrightarrow X$. (ii) The *bounded twisted double bordism group* $CB_n(X)$ is the group of bordism classes of septuples $(W,G;Q,P,h,g,f)$ with $(W,\partial W)$ an $(n+2)$-dimensional manifold with boundary $\partial W = D(Q,P,h)$ the twisted double of an $(n+1)$-dimensional manifold with boundary (Q,P) with respect to an automorphism $h:P\longrightarrow P$, and $G:W\longrightarrow X$, $(g,f):(Q,P)\longrightarrow X$ maps such that there exists a homotopy $fh\simeq f:P\longrightarrow X$ with

$$G| \;=\; D(g,f)\;:\;\partial W \;=\; D(Q,P,h)\longrightarrow X\;. \qquad \square$$

In other words, $DB_n(X)$ is the bordism group of closed $(n+1)$-dimensional manifolds with a twisted double structure and a compatible map to X.

Proposition 30.6 (i) *There are natural identifications of the open book and twisted double bordism books*

$$OB_n(X) \;=\; DB_n(X) \;\;(n\geq 0)\;,$$

and also of the bounded automorphism, open book and twisted double bordism groups

$$AB_n(X) \;=\; BB_n(X) \;=\; CB_n(X) \;\;(n\geq 0)\;,$$

with a natural isomorphism of exact sequences

$$\begin{array}{ccccccccc}
\cdots\longrightarrow & \Omega_{n+2}(X) & \longrightarrow & BB_n(X) & \longrightarrow & OB_n(X) & \xrightarrow{\;t\;} & \Omega_{n+1}(X) & \longrightarrow\cdots\\
 & \| & & \downarrow\cong & & \downarrow\cong & & \| & \\
\cdots\longrightarrow & \Omega_{n+2}(X) & \longrightarrow & CB_n(X) & \longrightarrow & DB_n(X) & \xrightarrow{\;D\;} & \Omega_{n+1}(X) & \longrightarrow\cdots
\end{array}$$

with

$$\begin{aligned}
&BB_n(X)\longrightarrow CB_n(X)\;;\\
&\qquad (W,G;P,h,f)\longrightarrow (W,G;P\times I,D(P),h\cup 1,f(pr_P),f\cup f)\;,\\
&OB_n(X)\longrightarrow DB_n(X)\;;\;(P,h,f)\longrightarrow (P\times I,D(P),h\cup 1,f(pr_P))\;,\\
&t\;:\;OB_n(X)\longrightarrow \Omega_{n+1}(X)\;;\;(P,h,f)\longrightarrow (t(h),t(f))\;,\\
&D\;:\;DB_n(X)\longrightarrow \Omega_{n+1}(X)\;;\;(Q,P,h,g,f)\longrightarrow (D(Q,P,h),g\cup g)\;.
\end{aligned}$$

(ii) *The automorphism bordism groups split as*

$$\Delta_n(X) \;=\; OB_n(X)\oplus\Omega_n(X)\;.$$

The exact sequence

$$\dots\longrightarrow\Omega_{n+2}(X\times S^1)\longrightarrow BB_n(X)\longrightarrow\Delta_n(X)\xrightarrow{\;T\;}\Omega_{n+1}(X\times S^1)\longrightarrow\dots$$

is isomorphic to

$$\begin{aligned}
\dots\longrightarrow\;&\Omega_{n+2}(X)\oplus\Omega_{n+1}(X)\longrightarrow BB_n(X)\\
&\longrightarrow OB_n(X)\oplus\Omega_n(X)\xrightarrow{\;t\oplus 1\;}\Omega_{n+1}(X)\oplus\Omega_n(X)\longrightarrow\dots\;,
\end{aligned}$$

and contains the exact sequence

$$\ldots \longrightarrow \Omega_{n+2}(X) \longrightarrow BB_n(X) \longrightarrow OB_n(X) \xrightarrow{t} \Omega_{n+1}(X) \longrightarrow \ldots$$

as a direct summand.
(iii) *The asymmetric signature maps*

$$\sigma^* \ : \ AB_n(X) \ = \ BB_n(X) \ = \ CB_n(X) \longrightarrow LAsy_h^{n+2}(\mathbb{Z}[\pi_1(X)])$$

are isomorphisms for finitely presented $\pi_1(X)$ and $n \geq 5$. Moreover, using 28.33 to identify

$$LAsy_h^{n+2}(\mathbb{Z}[\pi_1(X)]) \ = \ L_h^{n+2}(\Omega^{-1}\mathbb{Z}[\pi_1(X)][z,z^{-1}])$$

the asymmetric signature is identified with the $\Omega^{-1}\mathbb{Z}[\pi_1(X)][z,z^{-1}]$-coefficient symmetric signature

$$\begin{aligned} CB_n(X) &\longrightarrow L_h^{n+2}(\Omega^{-1}\mathbb{Z}[\pi_1(X)][z,z^{-1}]) \ ; \\ &(W,G;Q,P,h,g,f) \longrightarrow \sigma^*(M,\partial M;\Omega^{-1}\mathbb{Z}[\pi_1(X)][z,z^{-1}]) \end{aligned}$$

with $(M,\partial M)$ the $(n+2)$-dimensional manifold with fibre bundle boundary given by

$$M \ = \ W \cup Q \times I \ , \ \partial M \ = \ T(h : P \longrightarrow P) \ .$$

(iv) *For finitely presented $\pi_1(X)$ and $n \geq 5$ the automorphism and open book bordism groups $\Delta_n(X)$, $OB_n(X)$ are related to the ordinary bordism groups $\Omega_*(X)$ by exact sequences*

$$\begin{aligned} 0 \longrightarrow \Delta_{2i+1}(X) \xrightarrow{T} \Omega_{2i+2}(X \times S^1) &\longrightarrow LAsy_h^0(\mathbb{Z}[\pi_1(X)]) \\ &\longrightarrow \Delta_{2i}(X) \xrightarrow{T} \Omega_{2i+1}(X \times S^1) \longrightarrow 0 \ , \\ 0 \longrightarrow OB_{2i+1}(X) \xrightarrow{t} \Omega_{2i+2}(X) &\longrightarrow LAsy_h^0(\mathbb{Z}[\pi_1(X)]) \\ &\longrightarrow OB_{2i}(X) \xrightarrow{t} \Omega_{2i+1}(X) \longrightarrow 0 \ . \end{aligned}$$

Proof (i) See 29.13 for the identifications $AB_n(X) = BB_n(X)$. The following argument shows that the natural maps

$$\begin{aligned} &CB_n(X) \longrightarrow AB_n(X) \ ; \\ &\quad (W,G;Q,P,h,g,f) \longrightarrow (M,G \cup g(pr_Q);P,h,f) \ , \ M \ = \ W \cup Q \times I \ , \\ &AB_n(X) \longrightarrow CB_n(X) \ ; \\ &\quad (M,F;P,h,f) \longrightarrow (M,jF;P \times I, P \times \partial I, h \cup 1, T(f)|, T(f)|) \end{aligned}$$

define inverse isomorphisms.
Given any representative of an element $(M,F;P,h,f) \in AB_n(X)$ it is possible to make the map

$$(F,T(f)) \ : \ (M,T(h)) \longrightarrow X \times S^1$$

transverse regular at $X \times \{\text{pt.}\} \subset X \times S^1$, with

$$(Q,P) = (F,T(f))^{-1}(X \times \{\text{pt.}\}) \subset M$$

a framed codimension 1 submanifold with boundary equipped with a map

$$(g,f) = (F,T(f))| : (Q,P) \longrightarrow X \times \{\text{pt.}\} = X .$$

Cutting M along Q defines an $(n+2)$-dimensional manifold with boundary

$$(N,\partial N) = (\text{cl.}(M\backslash \text{nbhd.}(Q)), D(Q,P,h))$$

with a map

$$(G,D(g,f)) : (N,D(Q,P,h)) \longrightarrow M \xrightarrow{jF} X$$

such that

$$F = G \cup g \times 1 : M = N \cup Q \times I \longrightarrow X \times S^1 .$$

The image of $(M,F;P,h,f) \in AB_n(X)$ under the map $AB_n(X) \longrightarrow CB_n(X)$ can thus be expressed as

$$\begin{aligned}(M,jF;P \times I, P \times \partial I, h \cup 1, T(f)|, T(f)|)\\ &= (N,G;Q,P,h,g,f) + (Q \times I, g(pr_1); Q,P,f,1)\\ &= (N,G;Q,P,h,g,f) \in CB_n(X) .\end{aligned}$$

Applying 30.4, it follows that both the composites

$$AB_n(X) \longrightarrow CB_n(X) \longrightarrow AB_n(X) ,$$
$$CB_n(X) \longrightarrow AB_n(X) \longrightarrow CB_n(X)$$

are the identity maps.
The maps

$$OB_n(X) \longrightarrow DB_n(X) ; (P,h,f) \longrightarrow (P \times I, D(P), h \cup 1, f(pr_P), f) ,$$
$$DB_n(X) \longrightarrow OB_n(X) ; (Q,P,h,g,f) \longrightarrow (P,h,f)$$

are inverse isomorphisms, noting that

$$(Q,P,h,g,f) = (P \times I, D(P), h \cup 1, f(pr_P), f) \in DB_n(X)$$

via the relative bordism $Q \times I$ between Q and $P \times I$.
(ii) By analogy with the direct sum system of 23.2

$$\Omega_{n+1}(X) \underset{j}{\overset{i}{\rightleftarrows}} \Omega_{n+1}(X \times S^1) \underset{C}{\overset{B}{\rightleftarrows}} \Omega_n(X)$$

define the direct sum system

$$OB_n(X) \underset{j}{\overset{i}{\rightleftarrows}} \varDelta_n(X) \underset{C}{\overset{B}{\rightleftarrows}} \varOmega_n(X)$$

with

$$\begin{aligned}
i &: OB_n(X) \longrightarrow \varDelta_n(X) \ ; \ (P,h,f) \longrightarrow (D(P,\partial P), h \cup 1, f \cup f) \ , \\
j &: \varDelta_n(X) \longrightarrow OB_n(X) \ ; \ (P,h,f) \longrightarrow (D(P,\emptyset), h \sqcup 1, f \sqcup f) \ , \\
B &: \varDelta_n(X) \longrightarrow \varOmega_n(X) \ ; \ (P,h,f) \longrightarrow (P,f) \ , \\
C &: \varOmega_n(X) \longrightarrow \varDelta_n(X) \ ; \ (P,f) \longrightarrow (P,1,f) \ .
\end{aligned}$$

The mapping tori define a natural transformation of direct sum systems

$$\begin{array}{ccccc}
OB_n(X) & \underset{j}{\overset{i}{\rightleftarrows}} & \varDelta_n(X) & \underset{C}{\overset{B}{\rightleftarrows}} & \varOmega_n(X) \\
\big\downarrow t & & \big\downarrow T & & \big\| \\
\varOmega_{n+1}(X) & \underset{j}{\overset{i}{\rightleftarrows}} & \varOmega_{n+1}(X \times S^1) & \underset{C}{\overset{B}{\rightleftarrows}} & \varOmega_n(X)
\end{array}$$

(iii) Combine (i) and 29.17 (ii).
(iv) Combine (ii) and (iii). □

It has already been remarked that an open book is a twisted double (30.2 (v)). The following result gives a sufficient homotopy theoretic condition for a manifold to be an open book, and *a fortiori* a twisted double.

Proposition 30.7 *Let W be a closed n-dimensional manifold. Suppose given a separating codimension 1 submanifold $V^{n-1} \subset W$, so that*

$$W = W_+ \cup_V W_- \ , \quad W_+ \cap W_- = \partial W_+ = \partial W_- = V \ ,$$

and also a separating codimension 1 submanifold $U^{n-2} \subset V$, so that

$$V = V_+ \cup_U V_- \ , \quad V_+ \cap V_- = \partial V_+ = \partial V_- = U \ .$$

If the inclusions $V_+ \longrightarrow W_+$, $V_+ \longrightarrow W_-$ are simple homotopy equivalences and $n \geq 6$ then W has an open book decomposition

$$W \cong t(h : P \longrightarrow P)$$

with $h : P \longrightarrow P$ a rel ∂ self homeomorphism of the $(n-1)$-dimensional manifold with boundary $(P, \partial P) = (V_+, U)$.

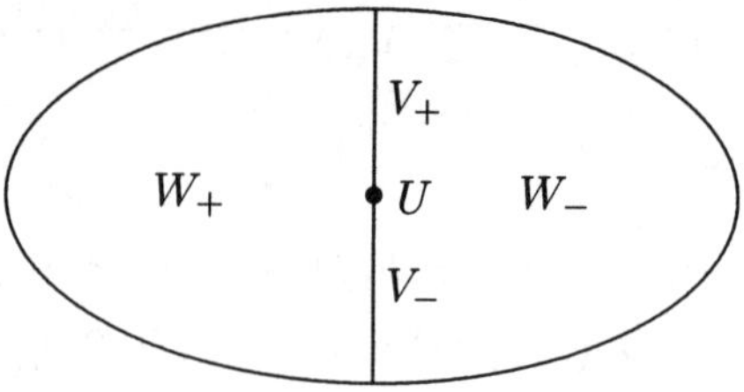

Proof Apply the relative s-cobordism theorem to the relative bordisms $(W_+; V_+, V_-; U)$ and $(W_-; V_+, V_-; U)$. □

30B. Algebraic twisted doubles

Algebraic twisted doubles are the algebraic Poincaré analogues of twisted doubles. A high-dimensional manifold has a twisted double structure if and only if its algebraic Poincaré complex has an algebraic twisted double structure, if and only if the asymmetric signature is 0.

Given an algebraic Poincaré pair over a ring with involution A with twisted double boundary there will now be constructed an asymmetric Poincaré complex over A, whose cobordism class is the obstruction to the extension (up to bordism) of the twisted double decomposition to the interior. The pair is a fundamental domain for an algebraic Poincaré complex over the Fredholm localization $\Omega^{-1}A[z, z^{-1}]$. The construction allows the inverse of the isomorphism of 28.33

$$\begin{aligned} LAsy_h^{n+2}(A) &\xrightarrow{\simeq} L_h^{n+2}(\Omega^{-1}A[z,z^{-1}],\epsilon) ; \\ &(C,\lambda) \longrightarrow (\Omega^{-1}C[z,z^{-1}], (1-z)\lambda + (1-z^{-1})T_\epsilon\lambda) \end{aligned}$$

to be interpreted as an asymmetric signature map

$$L_h^{n+2}(\Omega^{-1}A[z,z^{-1}],\epsilon) \xrightarrow{\simeq} LAsy_h^{n+2}(A) ,$$

an algebraic version of the asymmetric signature map of 30.7

$$\sigma^* : CB_n(X) = L_h^{n+2}(\Omega^{-1}\mathbb{Z}[\pi_1(X)][z,z^{-1}]) \longrightarrow LAsy_h^{n+2}(\mathbb{Z}[\pi_1(X)]) .$$

Definition 30.8 (i) The *double* of an n-dimensional ϵ-symmetric complex (E, θ) over A is the n-dimensional ϵ-symmetric Poincaré complex over A

$$\begin{aligned} D(E,\theta) &= (D(E), D(\theta)) \\ &= (\partial E \longrightarrow E^{n-*}, (0, \partial\theta)) \cup_{(E,\theta)} (\partial E \longrightarrow E^{n-*}, (0, -\partial\theta)) . \end{aligned}$$

(ii) The *twisted double* of an n-dimensional ϵ-symmetric Poincaré pair $c = (f : C \longrightarrow D, (\delta\phi, \phi))$ over A with respect to a self homotopy equivalence

$(h,\chi):(C,\phi)\longrightarrow(C,\phi)$ is the n-dimensional ϵ-symmetric Poincaré complex over A

$$\begin{aligned} c\cup_{(h,\chi)} -c &= (D\cup_h -D, \delta\phi\cup_\chi -\delta\phi) \\ &= (fh:C\longrightarrow D,(\delta\phi+f\chi f^*,\phi))\cup(f:C\longrightarrow D,(-\delta\phi,-\phi)) \end{aligned}$$

with

$$D\cup_h D = \mathcal{C}(\begin{pmatrix} fh \\ f \end{pmatrix}:C\longrightarrow D\oplus D)\ ,$$

$$(\delta\phi\cup_\chi -\delta\phi)_s = \begin{pmatrix} \delta\phi_s + f\chi_s f^* & 0 & 0 \\ (-)^{n-r}\phi_s h^* f^* & (-)^{n-r+s+1}T_\epsilon\phi_{s-1} & 0 \\ 0 & (-)^s f\phi_s & -\delta\phi_s \end{pmatrix} :$$

$$\begin{aligned} (D\cup_h D)^{n-r+s} &= D^{n-r+s}\oplus C^{n-r+s-1}\oplus D^{n-r+s} \\ &\longrightarrow (D\cup_h D)_r = D_r\oplus C_{r-1}\oplus D_r\ \ (s\geq 0)\ . \end{aligned}$$

The untwisted double of (i) is the twisted double of the n-dimensional ϵ-symmetric Poincaré pair $(\partial E\longrightarrow E^{n-*},(0,\partial\theta))$ with respect to the identity $(1,0):(\partial E,\partial\theta)\longrightarrow(\partial E,\partial\theta)$, which can be written as

$$(D\cup_h D,\delta\phi\cup_\chi -\delta\phi) = (D\cup_C D,\delta\phi\cup_\phi -\delta\phi)\ .$$

(iii) A *twisted double structure* on an n-dimensional ϵ-symmetric Poincaré complex (E,θ) is a homotopy equivalence $(E,\theta)\simeq c\cup_{(h,\chi)} -c$ to a twisted double. Similarly for pairs.

(iv) A *concordance* of twisted double structures on (E,θ) is a twisted double structure on the product cobordism $(E,\theta)\otimes\sigma^*(I;\{0\},\{1\})$ which agrees with the given structures on $(E,\theta)\otimes\sigma^*(\{0\})$ and $(E,\theta)\otimes\sigma^*(\{1\})$. □

Example 30.9 The twisted double of an n-dimensional geometric Poincaré pair (Q,P) with respect to a self homotopy equivalence $h:P\longrightarrow P$ is an n-dimensional geometric Poincaré complex

$$N = D(Q,P,h) = Q\cup_h -Q\ .$$

Let $(\widetilde{Q},\widetilde{P})$ be the pullback to (Q,P) of the universal cover $\widetilde{N}$ of N. The symmetric complex of N is the twisted symmetric Poincaré double over $\mathbb{Z}[\pi_1(N)]$

$$\sigma^*(N) = (C(\widetilde{Q})\cup_{\widetilde{h}} C(\widetilde{Q}),\delta\phi\cup_\chi -\delta\phi)$$

with

$$\sigma^*(Q,P) = (C(\widetilde{P})\longrightarrow C(\widetilde{Q}),(\delta\phi,\phi))$$

the n-dimensional symmetric Poincaré pair of (Q,P) over $\mathbb{Z}[\pi_1(N)]$ and

$$(\widetilde{h},\chi)\ :\ (C(\widetilde{P}),\phi)\longrightarrow(C(\widetilde{P}),\phi)$$

the induced self homotopy equivalence. □

Definition 30.10 Let $x = (g : \partial E \longrightarrow E, (\theta, \partial\theta))$ be an n-dimensional ϵ-symmetric Poincaré pair over A such that the boundary is the twisted double of an $(n-1)$-dimensional ϵ-symmetric Poincaré pair $(f : C \longrightarrow D, (\delta\phi, \phi))$ over A with respect to a self homotopy equivalence $(h, \chi) : (C, \phi) \longrightarrow (C, \phi)$

$$(\partial E, \partial\theta) \ = \ (D \cup_h D, \delta\phi \cup_\chi -\delta\phi)$$

with

$$g \ = \ (j_0 \ k \ j_1) \ : \ (D \cup_h D)_r \ = \ D_r \oplus C_{r-1} \oplus D_r \longrightarrow E_r \ .$$

(i) The *asymmetric complex* of x is the n-dimensional asymmetric Poincaré complex (B, λ) over A with

$$B \ = \ \mathcal{C}(j_0 - j_1 : D \longrightarrow \mathcal{C}(j_0 f : C \longrightarrow E))$$

and $\lambda : B^{n-*} \longrightarrow B$ the chain equivalence which fits into the chain homotopy commutative diagram

$$\begin{array}{ccccccccc}
0 & \longrightarrow & \mathcal{C}((j_0 \ j_1) : D \oplus D \longrightarrow E)^{n-*} & \longrightarrow & B^{n-*} & \longrightarrow & \mathcal{C}(f)^{n-*} & \longrightarrow & 0 \\
 & & \simeq \big\downarrow (\theta, \partial\theta)_0 & & \big\downarrow \lambda & & \simeq \big\downarrow (\delta\phi, \phi)_0 & & \\
0 & \longrightarrow & \mathcal{C}(j_0 f : C \longrightarrow E) & \longrightarrow & B & \longrightarrow & D_{*-1} & \longrightarrow & 0
\end{array}$$

In particular, $(B, \lambda) = (E, \theta_0)$ in the closed case $\partial E = 0$.
(ii) The *asymmetric signature* of x is the asymmetric Poincaré cobordism class

$$\sigma^*(x) \ = \ (B, \lambda) \in LAsy^n(A) \ . \qquad \square$$

Given an n-dimensional geometric Poincaré triad $(Q; \partial_0 Q, \partial_1 Q; \partial_{01} Q)$ and a self homotopy equivalence of the partial boundary $(n-1)$-dimensional geometric Poincaré pair

$$(h, \partial_0 h) \ : \ (\partial_1 Q, \partial_{01} Q) \longrightarrow (\partial_1 Q, \partial_{01} Q)$$

there is defined a twisted double n-dimensional geometric Poincaré pair

$$(Q \cup_h -Q, \partial_0 Q \cup_{\partial_0 h} -\partial_0 Q) \ .$$

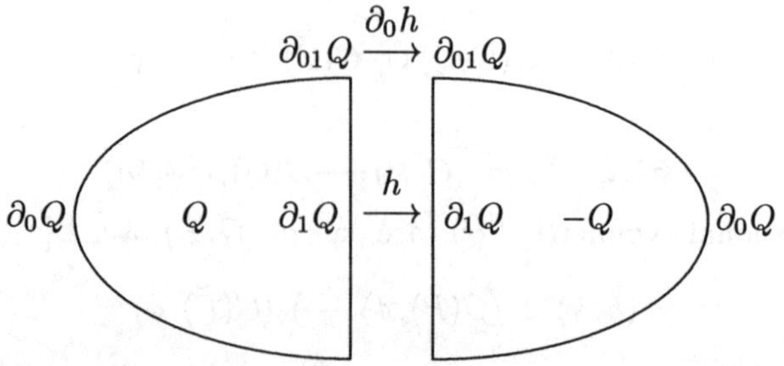

There is a corresponding construction of the twisted double of an n-dimensional ϵ-symmetric Poincaré triad with respect to a self homotopy equivalence of the partial boundary $(n-1)$-dimensional ϵ-symmetric Poincaré pair.

Proposition 30.11 (i) *The concordance classes of twisted double decompositions of an n-dimensional ϵ-symmetric Poincaré complex (E,θ) are in one-one correspondence with the rel ∂ asymmetric null-cobordisms of (E,θ_0). In particular, (E,θ) has a twisted double structure if and only if the asymmetric signature*

$$\sigma^*(x) \;=\; (E,\theta_0) \in LAsy^n(A)$$

is $\sigma^*(x)=0$.
(ii) *The asymmetric signature of an n-dimensional ϵ-symmetric Poincaré pair $x=(\partial E \longrightarrow E,(\theta,\partial\theta))$ with a twisted double structure on the boundary $\partial x = (\partial E,\partial\theta)$ is given by*

$$\sigma^*(x) \;=\; (E\cup_{\partial E} F, \theta_0 \cup_{\partial\theta_0} -\lambda) \in LAsy^n(A)$$

with $(\partial E \longrightarrow F,(\lambda,\partial\theta_0))$ the asymmetric null-cobordism corresponding to the twisted double structure of ∂x. It is possible to extend the twisted double structure from ∂x to x if and only if $\sigma^*(x)=0$.
(iii) *The cobordism group of f.g. free n-dimensional ϵ-symmetric Poincaré pairs x over A with a twisted double structure on the boundary is isomorphic to $L^n_h(\Omega^{-1}A[z,z^{-1}])$. The asymmetric signature isomorphism*

$$L^n_h(\Omega^{-1}A[z,z^{-1}],\epsilon) \xrightarrow{\;\simeq\;} LAsy^n_h(A)\ ;\ x \longrightarrow \sigma^*(x) \;=\; (B,\lambda)$$

is inverse to the isomorphism of 28.33

$$\begin{aligned} LAsy^n_h(A) &\xrightarrow{\;\simeq\;} L^n_h(\Omega^{-1}A[z,z^{-1}],\epsilon)\ ;\\ (C,\lambda) &\longrightarrow (\Omega^{-1}C[z,z^{-1}],(1-z)\lambda+(1-z^{-1})T_\epsilon\lambda)\ . \end{aligned}$$

Proof (i) Suppose that (E,θ) is homotopy equivalent to the twisted double

$$(D\cup_h D, \delta\phi\cup_\chi -\delta\phi)$$

of an n-dimensional ϵ-symmetric Poincaré pair $(f: C\longrightarrow D,(\delta\phi,\phi))$ with respect to a self homotopy equivalence $(h,\chi):(C,\phi)\longrightarrow(C,\phi)$. Define a chain complex

$$F \;=\; \mathcal{C}(f)\oplus C_{*-1}$$

and let $e: D\cup_h D\longrightarrow F$ be the chain map defined by

$$e \;=\; \begin{pmatrix} 1 & 0 & 1 \\ 0 & 1-h & 0 \\ 0 & 1 & 0 \end{pmatrix} :$$
$$(D\cup_h D)_r \;=\; D_r\oplus C_{r-1}\oplus D_r \longrightarrow F_r \;=\; D_r\oplus C_{r-1}\oplus C_{r-1}\ .$$

The composite chain map

$$g\ :\ E\ \simeq\ D\cup_h D\ \xrightarrow{e}\ F$$

and the A-module morphisms

$$\lambda\ =\ \begin{pmatrix} 0 & 0 & 0 \\ \chi_0 f^* & \phi_0(1-h^*) & \phi_0 \\ 0 & \phi_0 h^* & 0 \end{pmatrix}\ :$$

$$F^{n+1-r}\ =\ D^{n+1-r}\oplus C^{n-r}\oplus C^{n-r}\ \longrightarrow\ F_r\ =\ D_r\oplus C_{r-1}\oplus C_{r-1}$$

define an $(n+1)$-dimensional asymmetric Poincaré pair $(g : E\longrightarrow F, (\lambda,\theta_0))$, so that

$$(E,\theta_0)\ =\ 0\in LAsy^n(A)\ .$$

(Let $T(h,\chi)$ be the algebraic mapping torus of (h,χ) (24.3), and write

$$A\otimes_{A[z,z^{-1}]}T(h,\chi)\ =\ (B,\beta)\ ,$$

so that

$$d_B\ =\ \begin{pmatrix} d_C & (-)^r(1-h) \\ 0 & d_C \end{pmatrix}\ :$$

$$B_r\ =\ C_r\oplus C_{r-1}\longrightarrow B_{r-1}\ =\ C_{r-1}\oplus C_{r-2}\ ,$$

$$\beta_s\ =\ \begin{pmatrix} 0 & (-)^s\phi_s \\ (-)^{n-r}\phi_s & (-)^{n-r+s+1}T_\epsilon\phi_{s-1} \end{pmatrix}\ :$$

$$B^{n-r+s+1}\ =\ C^{n-r+s+1}\oplus C^{n-r+s}\ \longrightarrow\ B_r\ =\ C_r\oplus C_{r-1}\ .$$

The asymmetric Poincaré pair $(g,(\lambda,\beta_0))$ is the union of the asymmetrization of the twisted double symmetric Poincaré cobordism

$$(E\oplus B\longrightarrow\mathcal{C}(\Delta : D\longrightarrow E),(\delta\phi,\phi\oplus-\theta))$$

analogous to the geometric twisted double cobordism $(R;D(Q,P,h),T(h))$ used in the proof of 30.4 (i), and the asymmetric null-cobordism $(B\longrightarrow C_{*-1}, (0,\beta_0))$ of (B,β_0).)

Conversely, suppose that $(E,\theta_0)=0\in LAsy^n(A)$, so that there exists an $(n+1)$-dimensional asymmetric Poincaré pair $(g : E\longrightarrow F,(\lambda,\theta_0))$. Apply the construction of 20.25 to the morphism of n-dimensional ϵ-symmetric complexes

$$(g,0)\ :\ (E,\theta)\longrightarrow(F,g^{\%}(\theta))\ .$$

The construction gives a homotopy equivalence

$$(E,\theta)\ \simeq\ c_+\cup c_-$$

with c_+ the thickening of $(F,g^{\%}(\theta))$ (20.22) and c_- homotopy equivalent to the thickening of $(\mathcal{C}(\theta_0 g^* : F^{n-*}\longrightarrow E), j^{\%}(\theta))$, with $j : E\longrightarrow\mathcal{C}(\theta_0 g^*)$ the inclusion. This is a twisted double decomposition of (E,θ) since the A-module morphisms

$$h = (g\ \lambda) : \mathcal{C}(\theta_0 g^*)_r = E_r \oplus F^{n-r+1} \longrightarrow F_r$$

define a homotopy equivalence of n-dimensional ϵ-symmetric complexes

$$h : (\mathcal{C}(\theta_0 g^*), j^{\%}(\theta)) \simeq (F, g^{\%}(\theta))$$

inducing a homotopy equivalence of the thickenings $h : c_- \simeq c_+$.
(ii) This is just the relative version of (i), using the relative version of 20.25 given by 20.30.
(iii) By 24.2 every f.g. free n-dimensional ϵ-symmetric Poincaré complex over $\Omega^{-1}A[z,z^{-1}]$ is the union $U(\Gamma)$ (24.1) of the fundamental n-dimensional symmetric $\Omega^{-1}A[z,z^{-1}]$-Poincaré pair

$$\Gamma = (\mathcal{C}(f) \oplus \mathcal{C}(f) \longrightarrow \mathcal{C}(j_0 f), (\theta/\partial\theta, \delta\phi/\phi \oplus -\delta\phi/\phi))$$

obtained from a f.g. free n-dimensional ϵ-symmetric Poincaré pair $x = (g : \partial E \longrightarrow E, (\theta, \partial\theta))$ with boundary the twisted double of an $(n-1)$-dimensional ϵ-symmetric Poincaré pair $(f : C \longrightarrow D, (\delta\phi, \phi))$ over A with respect to a self homotopy equivalence $(h,\chi) : (C,\phi) \longrightarrow (C,\phi)$. □

Corollary 30.12 *Given an n-dimensional manifold with twisted double boundary $(W, \partial W = Q \cup_h -Q)$ and a map $G : W \longrightarrow X$ let (B,λ) be the n-dimensional asymmetric Poincaré complex constructed as in 30.10 from the n-dimensional symmetric Poincaré pair $\sigma^*(W,\partial W)$ over $\mathbb{Z}[\pi_1(X)]$ with twisted double boundary $\sigma^*(\partial W) = \sigma^*(Q) \cup_h -\sigma^*(Q)$. Up to chain equivalence*

$$B = \mathcal{C}(j_0 - j_1 : C(\widetilde{Q}) \longrightarrow C(\widetilde{W}, \widetilde{P})) = C(\widetilde{M}, \widetilde{P})$$

with $j_0, j_1 : Q \longrightarrow W$ the two inclusions and

$$(M, \partial M) = (W \cup Q \times I, T(h : P \longrightarrow P))$$

an n-dimensional manifold with fibre bundle boundary.
(i) (B,λ) is the asymmetric Poincaré complex associated to $(M,\partial M)$ in 29.15, with

$$\lambda = [M] \cap - : H^{n-*}(B) = H^{n-*}(\widetilde{M}, \widetilde{P}) \xrightarrow{\simeq} H_*(B) = H_*(\widetilde{M}, \widetilde{P}) ,$$

so that

$$\begin{aligned}\sigma^*(W; Q, P, h, g, f) &= \sigma^*(M, G \cup 1_{Q \times I}; P, h, f) \\ &= (B, \lambda) \in LAsy_h^n(\mathbb{Z}[\pi_1(X)]) .\end{aligned}$$

(ii) The asymmetric signature map on the twisted double bordism group is given by

$$\begin{aligned}\sigma^* &: CB_{n-2}(X) \longrightarrow LAsy_h^n(\mathbb{Z}[\pi_1(X)]) ; \\ &(W, G; Q, P, h, g, f) \longrightarrow \sigma^*(W, G; Q, P, h, g, f) = (B, \lambda) .\end{aligned}$$

(iii) *Take* $G = 1 : W \longrightarrow X = W$. *The asymmetric signature is such that* $\sigma^*(W) = 0 \in LAsy^n_h(\mathbb{Z}[\pi_1(W)])$ *if (and for* $n \geq 6$ *only if)* W *has a twisted double decomposition extending* $\partial W = Q \cup_h -Q$.

Proof (i) The construction of 30.11 is the algebraic analogue of the geometric construction of 29.15.

(ii) Immediate from (i).

(iii) It is clear from 30.11 that $\sigma^*(W) = 0$ for a twisted double W. For the converse, only the closed case $\partial W = \emptyset$ will be considered, for the sake of brevity. Let then W be a closed n-dimensional manifold with universal cover $\widetilde{W}$, with simple n-dimensional symmetric Poincaré complex and let $(C,\phi) = (C(\widetilde{W}), \Delta[W])$ over $\mathbb{Z}[\pi_1(W)]$, so that

$$\sigma^*(W) \; = \; (C,\phi_0) \in LAsy^n_h(\mathbb{Z}[\pi_1(W)]) \; = \; LAsy^n_s(\mathbb{Z}[\pi_1(W)]) \; .$$

If $\sigma^*(W) = 0$ there exists a simple asymmetric null-cobordism

$$(f : C(\widetilde{W}) \longrightarrow D, (\lambda, \phi_0))$$

with $D_r = 0$ for $2r < n$. The morphism of n-dimensional symmetric complexes

$$(f,0) \; : \; (C,\phi) \longrightarrow (D, f^{\%}(\phi))$$

determines by 20.25 an algebraic splitting

$$(C,\phi) \; \simeq \; c_+ \cup c_-$$

which can be realized geometrically (perhaps after some low-dimensional adjustments) as

$$W \; = \; W_+ \cup_V W_- \;\; , \;\; V^{n-1} \; = \; W_+ \cap W_- \; = \; \partial W_+ \; = \; \partial W_-$$

with

$$\begin{aligned} c_\pm \; &= \; (C(\widetilde{V}) \longrightarrow C(\widetilde{W}_\pm), \Delta[W_\pm]) \; , \\ f \; : \; C \; = \; C(\widetilde{W}) \; &\longrightarrow \; D \; = \; C(\widetilde{W}, \widetilde{W}_+) \; = \; C(\widetilde{W}_-, \widetilde{V}) \; . \end{aligned}$$

The chain map

$$g \; = \; (1 \;\; \lambda) \; : \; C(\widetilde{V}) \; = \; \partial D \longrightarrow D_{*+1}$$

determines a morphism of $(n-1)$-dimensional symmetric complexes

$$(g,0) \; : \; (C(\widetilde{V}), \Delta[V]) \; = \; (\partial D, \partial f^{\%}(\phi)) \longrightarrow (D_{*+1}, g^{\%}\partial f^{\%}(\phi)) \; .$$

Applying 20.25 again there is obtained an algebraic splitting

$$(\partial D, \partial f^{\%}(\phi)) \; \simeq \; d_+ \cup d_-$$

which can be realized geometrically (perhaps after some low-dimensional adjustments) as

$$V = V_+ \cup_U V_- \ , \ U^{n-2} = V_+ \cap V_- = \partial V_+ = \partial V_-$$

with

$$d_\pm = (C(\widetilde{U}) \longrightarrow C(\widetilde{V}_\pm), \Delta[V_\pm]) \ ,$$
$$g : \partial D = C(\widetilde{V}) \longrightarrow D_{*+1} = C(\widetilde{V}, \widetilde{V}_+) = C(\widetilde{V}_-, \widetilde{U}) \ .$$

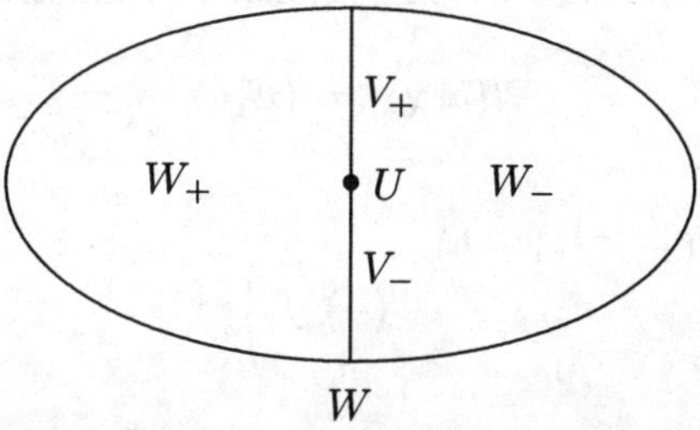

Now $(W_+; V_+, V_-; U)$ and $(W_-; V_+, V_-; U)$ are relative s-cobordisms, so that by 30.7 W is a twisted double, namely an open book with page V_+ and binding U. (This is a chain complex version of the original argument of Winkelnkemper [307], [308]). □

Corollary 30.13 *For $n \geq 6$ a closed n-dimensional manifold M has an open book decomposition if and only if it is a twisted double.*

Proof The vanishing of the asymmetric signature $\sigma^*(M) \in LAsy^n_h(\mathbb{Z}[\pi_1(M)])$ is the necessary and sufficient condition for both an open book decomposition and a twisted double decomposition. □

Remark 30.14 Every closed surface M^2 is a double (30.3), and hence so is $M \times \mathbb{CP}^2$. Therefore the 6-dimensional manifold $M \times \mathbb{CP}^2$ has an open book decomposition, whereas M itself only has an open book decomposition if it is S^2 or $S^1 \times S^1$ (29.5). □

30C. Algebraic open books

Algebraic open books are the algebraic Poincaré analogues of open books. A high-dimensional manifold has an open book structure if and only if its algebraic Poincaré complex has an algebraic open book structure, if and only if the asymmetric signature is 0.

Definition 30.15 (i) A *rel ∂ self homotopy equivalence*

$$(h, \chi) : (f : C \longrightarrow D, (\delta\phi, \phi)) \longrightarrow (f : C \longrightarrow D, (\delta\phi, \phi))$$

of an n-dimensional ϵ-symmetric Poincaré pair $(f : C \longrightarrow D, (\delta\phi, \phi))$ is a chain equivalence $h : D \longrightarrow D$ such that

$$hf = f : C \longrightarrow D$$

together with a chain $\chi \in (W^{\%}C)_{n+1}$ such that

$$h^{\%}(\delta\phi) - \delta\phi = d(\chi) \in (W^{\%}C)_n .$$

(ii) The *relative algebraic mapping torus* of a rel ∂ self homotopy equivalence (h,χ) (as in (i)) is the $(n+1)$-dimensional ϵ-symmetric Poincaré complex over A

$$t(h,\chi) = (E,\theta)$$

with

$$d_E = \begin{pmatrix} d_D & (-)^{r-1}(1-h) & 0 \\ 0 & d_D & (-)^r f \\ 0 & 0 & d_C \end{pmatrix} :$$
$$E_r = D_r \oplus D_{r-1} \oplus C_{r-2} \longrightarrow E_{r-1} = D_{r-1} \oplus D_{r-2} \oplus C_{r-3} ,$$

$$\theta_s = \begin{pmatrix} \chi_s & (-)^s\delta\phi_s & hf\phi_s \\ (-)^{n-r}\delta\phi_s h^* & (-)^{n-r+s+1}T_\epsilon\delta\phi_{s-1} & 0 \\ \phi_s f^* h^* & 0 & 0 \end{pmatrix} :$$
$$E^{n-r+s+1} = D^{n-r+s+1} \oplus D^{n-r+s} \oplus C^{n-r+s-1}$$
$$\longrightarrow E_r = D_r \oplus D_{r-1} \oplus C_{r-2} .$$

(iii) An *open book structure* on an n-dimensional ϵ-symmetric Poincaré complex (E,θ) is a homotopy equivalence $(E,\theta) \simeq t(h,\chi)$ to an open book. Similarly for pairs.
(iv) A *concordance* of open book structures on (E,θ) is an open book structure on the product cobordism $(E,\theta) \otimes \sigma^*(I;\{0\},\{1\})$ which agrees with the given structures on $(E,\theta) \otimes \sigma^*(\{0\})$ and $(E,\theta) \otimes \sigma^*(\{1\})$. □

Example 30.16 A rel ∂ self homotopy equivalence $h : (F,\partial F) \longrightarrow (F,\partial F)$ of an n-dimensional geometric Poincaré pair induces a rel ∂ self homotopy equivalence of an ϵ-symmetric Poincaré pair over $\mathbb{Z}[\pi_1(F)]$

$$(\widetilde{h},\widetilde{\chi}) : \sigma^*(F,\partial F) \longrightarrow \sigma^*(F,\partial F)$$

and the symmetric Poincaré complex of a geometric open book $t(h)$ is an algebraic open book

$$\sigma^*(t(h)) = t(\widetilde{h},\widetilde{\chi}) .$$ □

Open books are twisted doubles in algebra as well as in topology (30.4 (v)):

Proposition 30.17 *Let (h,χ) be a rel ∂ self homotopy equivalence of an n-dimensional ϵ-symmetric Poincaré pair $(f : C \longrightarrow D, (\delta\phi,\phi))$ over A.*
(i) The relative algebraic mapping torus is such that up to homotopy equivalence

$$t(h,\chi) \;=\; T_A(h,\chi)\cup_{T_A(C,\phi)}\delta T_A(C,\phi)$$

with

$$T_A(C,\phi) \;=\; (C,\phi)\otimes\sigma^*(S^1;\mathbb{Z}) \;=\; (C\oplus C_{*-1},\theta)$$

the A-coefficient algebraic mapping torus (24.3) of $1:(C,\phi)\longrightarrow(C,\phi)$ *and*

$$\begin{aligned} T_A(h,\chi) \;&=\; (f\oplus f : C\oplus C_{*-1}\longrightarrow \mathcal{C}(1-h:D\longrightarrow D),(\delta\theta,\theta))\ ,\\ \delta T_A(h,\chi) \;&=\; (1\oplus 0 : C\oplus C_{*-1}\longrightarrow C,(0,\theta)) \end{aligned}$$

the null-cobordisms defined by

$$\begin{aligned} \theta_s \;&=\; \begin{pmatrix} 0 & (-)^s\phi_s \\ (-)^{n-r-1}\phi_s & (-)^{n-r+s}T_\epsilon\phi_{s-1}\end{pmatrix} :\\ &\qquad C^{n-r+s}\oplus C^{n-r+s-1}\longrightarrow C_r\oplus C_{r-1}\ ,\\ \delta\theta_s \;&=\; \begin{pmatrix} \chi_s & (-)^s\delta\phi_s \\ (-)^{n-r}\delta\phi_s h^* & (-)^{n-r+s+1}T_\epsilon\delta\phi_{s-1}\end{pmatrix} :\\ &\qquad D^{n-r+s+1}\oplus D^{n-r+s}\longrightarrow D_r\oplus D_{r-1}\ \ (s\geq 0)\ . \end{aligned}$$ □

(ii) *There is defined an* $(n+2)$*-dimensional asymmetric Poincaré pair*

$$\delta t(h,\chi) \;=\; (g:E\longrightarrow \mathcal{C}(f)_{*-1},(0,\theta_0))$$

with

$$g=\begin{pmatrix} 0 & 1 & 0\\ 0 & 0 & 1\end{pmatrix} : E_r \;=\; D_r\oplus D_{r-1}\oplus C_{r-2}\longrightarrow \mathcal{C}(f)_{r-1} \;=\; D_{r-1}\oplus C_{r-2}\ ,$$

so that

$$t(h,\chi) \;=\; 0\in LAsy^{n+1}(A)\ .$$

(iii) *The relative algebraic mapping torus* $t(h,\chi)$ *is homotopy equivalent to the twisted double of the* $(n+1)$*-dimensional* ϵ*-symmetric Poincaré pair*

$$(1\cup 1 : D\cup_{1:C\to C}D\longrightarrow D,(0,\delta\phi\cup_\phi -\delta\phi))$$

over A with respect to the self homotopy equivalence

$$(h\cup 1,\chi\cup 0) : (D\cup_{1:C\to C}D,\delta\phi\cup_\phi-\delta\phi)\longrightarrow(D\cup_{1:C\to C}D,\delta\phi\cup_\phi-\delta\phi)\ .$$ □

All the results on algebraic twisted doubles of 30B have open book versions, such as the following version of 30.11:

Proposition 30.18 *The concordance classes of open book structures on an* n*-dimensional* ϵ*-symmetric Poincaré complex* (E,θ) *are in one-one correspondence with the rel* ∂ *cobordism classes of asymmetric null-cobordisms of* (E,θ_0)*. In particular,* (E,θ) *has an open book structure if and only if the asymmetric signature*

$$\sigma^*(x) \;=\; (E,\theta_0)\in LAsy^n(A)$$

is $\sigma^*(x)=0$*. Similarly for the relative case.* □

The concordance classes of open book structures are thus the same as the concordance classes of twisted double structures, and the asymmetric signature is the obstruction to the existence of an open book structure as well as the obstruction to the existence of a twisted double structure.

30D. Twisted double L-theory

The twisted double L-groups $DBL^*(A,\epsilon)$ are the algebraic analogues of the twisted double bordism groups $DB_*(X)$ of 30A.

Definition 30.19 The *projective ϵ-symmetric twisted double L-group* $DBL^n_p(A,\epsilon)$ is the cobordism group of f.g. projective $(n+1)$-dimensional ϵ-symmetric Poincaré complexes over A with a twisted double structure. □

An element of $DBL^n_p(A,\epsilon)$ is the cobordism class of a f.g. projective $(n+1)$-dimensional ϵ-symmetric Poincaré pair over A $(f: C\longrightarrow D,(\delta\phi,\phi))$ together with a self homotopy equivalence $(h,\chi):(C,\phi)\longrightarrow(C,\phi)$.

Proposition 30.20 (i) *The twisted double L-groups are such that*

$$\mathrm{LAut}^n_p(A,\epsilon) \;=\; L^n_p(A,\epsilon)\oplus DBL^n_p(A,\epsilon)\ .$$

(ii) *The twisted double L-group $DBL^n_p(A,\epsilon)$ is (isomorphic to) the cobordism group of $(n+1)$-dimensional f.g. projective ϵ-symmetric Poincaré complexes over A with an open book structure.*

Proof (i) Define a direct sum system

$$DBL^n_p(A,\epsilon) \;\underset{\longleftarrow}{\overset{\longrightarrow}{}}\; \mathrm{LAut}^n_p(A,\epsilon) \;\underset{\longleftarrow}{\overset{\longrightarrow}{}}\; L^n_p(A,\epsilon)$$

with

$$\begin{aligned}
&\mathrm{LAut}^n_p(A,\epsilon)\longrightarrow L^n_p(A,\epsilon)\ ;\ ((h,\chi):(C,\phi)\longrightarrow(C,\phi))\longrightarrow(C,\phi)\ ,\\
&L^n_p(A,\epsilon)\longrightarrow \mathrm{LAut}^n_p(A,\epsilon)\ ;\ (C,\phi)\longrightarrow((1,0):(C,\phi)\longrightarrow(C,\phi))\ ,\\
&DBL^n_p(A,\epsilon)\longrightarrow \mathrm{LAut}^n_p(A,\epsilon)\ ;\ (f:C\longrightarrow D,(\delta\phi,\phi),(h,\chi))\longrightarrow(h,\chi)\ ,\\
&\mathrm{LAut}^n_p(A,\epsilon)\longrightarrow DBL^n_p(A,\epsilon)\ ;\\
&((h,\chi):(C,\phi)\longrightarrow(C,\phi))\longrightarrow((1\ 1):C\oplus C\longrightarrow C,(0,\phi\oplus-\phi),h\oplus 1)\ .
\end{aligned}$$

The main ingredient is the following algebraic analogue of 30.4, which for any f.g. projective $(n+1)$-dimensional ϵ-symmetric Poincaré pair over A $(f: C\longrightarrow D,(\delta\phi,\phi))$ together with a self homotopy equivalence $(h,\chi):(C,\phi)\longrightarrow(C,\phi)$ gives a canonical twisted double cobordism between the twisted double $(D\cup_h D,\delta\phi\cup_\chi-\delta\phi)$ and the A-coefficient algebraic mapping torus $T_A(h,\chi)$. The $(n+1)$-dimensional ϵ-symmetric Poincaré pairs $(f: C\longrightarrow D,(\delta\phi,\phi))$, $((1\ 1): C\oplus C\longrightarrow C,(0,\phi\oplus-\phi))$ are cobordant via an $(n+2)$-dimensional ϵ-symmetric Poincaré triad (Γ,Φ) with

$$\begin{array}{ccc}
C \oplus (C \oplus C) & \xrightarrow{1 \oplus (f \oplus f)} & C \oplus D \\
\Big\downarrow{\scriptstyle f \oplus (1 \oplus 1)} & \Gamma & \Big\downarrow{\scriptstyle f \oplus 1} \\
D \oplus C & \xrightarrow{1 \oplus f} & D
\end{array}$$

Use the commutative diagram of chain complexes and chain maps

$$\begin{array}{ccccc}
D & \xrightarrow{1} & D & \xleftarrow{f} & C \\
\Big\uparrow{\scriptstyle f} & & \Big\uparrow{\scriptstyle f \oplus 1} & & \Big\uparrow{\scriptstyle 1 \oplus 1} \\
C & \xrightarrow{1 \oplus 0} & C \oplus D & \xleftarrow{1 \oplus f} & C \oplus C \\
\Big\downarrow{\scriptstyle h}\,\simeq & & \Big\downarrow{\scriptstyle h \oplus 1}\,\simeq & & \simeq\,\Big\downarrow{\scriptstyle h \oplus 1} \\
C & \xrightarrow{1 \oplus 0} & C \oplus D & \xleftarrow{1 \oplus f} & C \oplus C \\
\Big\downarrow{\scriptstyle f} & & \Big\downarrow{\scriptstyle f \oplus 1} & & \Big\downarrow{\scriptstyle 1 \oplus 1} \\
D & \xrightarrow{1} & D & \xleftarrow{f} & C
\end{array}$$

to define a twisted double cobordism

$$((u\ v) : (D \cup_h D) \oplus T_A(h) \longrightarrow E, (\delta\chi, (\delta\phi \cup_\chi -\delta\phi) \oplus T_A(\chi)))$$

between $(D \cup_h D, \delta\phi \cup_\chi -\delta\phi)$ and $T_A(h, \chi)$, with

$$E = \mathcal{C}(f(1-h) : C \longrightarrow D)$$

chain equivalent to the chain complex twisted double

$$D \cup_{h \oplus 1 : C \oplus D \longrightarrow C \oplus D} D = \mathcal{C}(\begin{pmatrix} f & 1 \\ fh & 1 \end{pmatrix} : C \oplus D \longrightarrow D \oplus D)$$

and

$$u : D \cup_h D \longrightarrow E \ , \ v : T_A(h) \longrightarrow E$$

the chain maps defined by

$$\begin{aligned} u = \begin{pmatrix} 1 & 0 & 1 \\ 0 & 1 & 0 \end{pmatrix} : (D \cup_h D)_r &= D_r \oplus C_{r-1} \oplus D_r \\ &\longrightarrow E_r = D_r \oplus C_{r-1} \ , \end{aligned}$$

$$v = \begin{pmatrix} f & 0 \\ 0 & 1 \end{pmatrix} : T_A(h)_r = C_r \oplus C_{r-1} \longrightarrow E_r = D_r \oplus C_{r-1} \ .$$

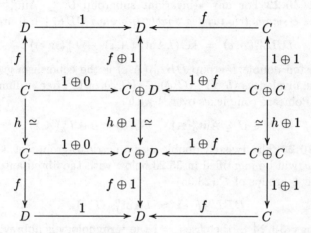

(ii) This algebraic analogue of the isomorphism $DB_n(X) \cong OB_n(X)$ of 30.6 (i) is immediate from 30.11 and 30.18. □

Remark 30.21 The relative version of the thickening construction of 20.22 can be used to identify $DBL^n(A,\epsilon)$ with the cobordism group of rel ∂ self homotopy equivalences $(h,\chi) : (C,\phi) \longrightarrow (C,\phi)$ of n-dimensional ϵ-symmetric complexes over A. □

Recall from 28.4 the definition of the free automorphism class group

$$\mathrm{Aut}_0^h(A) = \ker(\mathrm{Aut}_0(A) \longrightarrow \widetilde{K}_0(A)) .$$

Definition 30.22 For any $*$-invariant subgroup $U \subseteq \mathrm{Aut}_0^h(A)$ the *U-intermediate ϵ-symmetric twisted double L-groups* $DBL_U^*(A,\epsilon)$ are

$$DBL_U^n(A,\epsilon) = \ker(L\mathrm{Aut}_U^n(A,\epsilon) \longrightarrow L_h^n(A,\epsilon)) . \qquad \square$$

The twisted double L-group $DBL_U^n(A,\epsilon)$ is the cobordism group of self homotopy equivalences $(h,\chi) : (C,\phi) \longrightarrow (C,\phi)$ of f.g. free n-dimensional ϵ-symmetric Poincaré complexes over A with

$$[C,h] \in U \subseteq \mathrm{Aut}_0^h(A) \ , \quad (C,\phi) = 0 \in L_h^n(A,\epsilon) .$$

Remark 30.23 The twisted double ϵ-symmetric L-groups of a field with involution F will be identified in 35.20 below with the fibred automorphism ϵ-symmetric L-groups of F (28.35)

$$DBL^n(F,\epsilon) = L\mathrm{Aut}_{fib}^n(F,\epsilon) . \qquad \square$$

Terminology 30.24 In the case $\epsilon = 1$ the terminology is abbreviated

$$DBL_U^*(A,1) = DBL_U^*(A) .$$

Also, for $U = \mathrm{Aut}_0^h(A), \mathrm{Aut}_0^s(A)$ write

$$DBL_{\mathrm{Aut}_0^h(A)}^n(A,\epsilon) = DBL_h^n(A,\epsilon) = DBL^n(A,\epsilon) ,$$

$$DBL_{\mathrm{Aut}_0^s(A)}^n(A,\epsilon) = DBL_s^n(A,\epsilon) . \qquad \square$$

Proposition 30.25 *Let $(f : C \longrightarrow D, (\delta\phi,\phi))$ be a f.g. free n-dimensional ϵ-symmetric Poincaré pair over A together with a self homotopy equivalence $(h,\chi) : (C,\phi) \longrightarrow (C,\phi)$. The twisted double is such that*

$$(D \cup_h D, \delta\phi \cup_\chi -\delta\phi) = T_A(h,\chi) \in L_h^n(A,\epsilon) ,$$

$$i_!(D \cup_h D, \delta\phi \cup_\chi -\delta\phi) = T(h,\chi) \in L_h^n(A[z,z^{-1}],\epsilon) .$$

Proof By 24.4

$$T(h,\chi) = i_! T_A(h,\chi) \in \mathrm{im}(i_! : L_h^n(A,\epsilon) \longrightarrow L_h^n(A[z,z^{-1}],\epsilon)) ,$$

and by the cobordism constructed in the proof of 30.20

$$(D \cup_h D, \delta\phi \cup_\chi -\delta\phi) = T_A(h,\chi) \in L_h^n(A,\epsilon) . \qquad \square$$

Proposition 30.26 (i) *For any $*$-invariant subgroups $U \subseteq \mathrm{Aut}_0^h(A)$, $V \subseteq \widetilde{K}_0(A)$ there are natural direct sum splittings*

$$LAut^n_{U\oplus V}(A,\epsilon) \;=\; DBL^n_U(A,\epsilon) \oplus L^n_V(A,\epsilon) \;.$$

(ii) *For any $*$-invariant subgroup $U \subseteq \mathrm{Aut}_0^h(A)$ such that $\mathrm{Aut}_0^s(A) \subseteq U$ there is defined an exact sequence*

$$\begin{aligned}\dots \longrightarrow L^n_{\widetilde{\delta}(U)}(A,\epsilon) &\xrightarrow{\;i\;} L^n_h(\Omega^{-1}A[z,z^{-1}],\epsilon)\\ &\xrightarrow{\;\partial\;} DBL^{n-2}_U(A,\epsilon) \xrightarrow{\;D\;} L^{n-1}_{\widetilde{\delta}(U)}(A,\epsilon) \longrightarrow \dots \;,\end{aligned}$$

with

$$L^n_h(\Omega^{-1}A[z,z^{-1}],\epsilon) \;=\; LAsy^n_h(A) \quad (28.33)$$

and D given by the twisted double

$$\begin{aligned}D \;:\; DBL^{n-2}_U(A,\epsilon) &\longrightarrow L^{n-1}_{\widetilde{\delta}(U)}(A,\epsilon) \;;\\ ((h,\chi):(C,\phi)\longrightarrow(C,\phi)) &\longrightarrow (D\cup_h D, \delta\phi\cup_\chi -\delta\phi) \;.\end{aligned}$$

for any f.g. free null-cobordism $(f : C\longrightarrow D, (\delta\phi,\phi))$ of (C,ϕ).
(iii) *The cobordism class of the algebraic mapping torus $T(h,\chi)$ of a self homotopy equivalence $(h,\chi) : (E,\theta)\longrightarrow(E,\theta)$ of a f.g. free n-dimensional ϵ-symmetric Poincaré complex (E,θ) over A with $[E,h] \in U \subseteq \mathrm{Aut}_0^h(A)$ and $(E,\theta) = 0 \in L^n_h(A,\epsilon)$ is induced via the inclusion $i : A\longrightarrow A[z,z^{-1}]$ from the A-coefficient algebraic mapping torus $T_A(h,\chi)$*

$$T(h,\chi) \;=\; i_!T_A(h,\chi) \in \mathrm{im}(i_! : L^{n+1}_{\widetilde{\delta}(U)}(A,\epsilon)\longrightarrow L^{n+1}_{\widetilde{\delta}(U)}(A[z,z^{-1}],\epsilon)) \;.$$

(iv) *The twisted double map D in (ii) is given by the A-coefficient algebraic mapping torus*

$$D \;=\; T_A \;:\; DBL^n_U(A,\epsilon) \longrightarrow L^{n+1}_{\widetilde{\delta}(U)}(A,\epsilon) \;;\; (h,\chi) \longrightarrow T_A(h,\chi) \;.$$

Proof (i) Define a direct sum system

$$DBL^n_U(A,\epsilon) \;\underset{\longleftarrow}{\overset{\longrightarrow}{}}\; LAut^n_{U\oplus V}(A,\epsilon) \;\underset{\longleftarrow}{\overset{\longrightarrow}{}}\; L^n_V(A,\epsilon)$$

with

$$\begin{aligned}&LAut^n_{U\oplus V}(A,\epsilon)\longrightarrow L^n_V(A,\epsilon) \;;\; ((h,\chi):(C,\phi)\longrightarrow(C,\phi)) \longrightarrow (C,\phi) \;,\\ &L^n_V(A,\epsilon) \longrightarrow LAut^n_{U\oplus V}(A,\epsilon) \;;\; (C,\phi) \longrightarrow ((1,0):(C,\phi)\longrightarrow(C,\phi)) \;,\\ &LAut^n_{U\oplus V}(A,\epsilon) \longrightarrow DBL^n_U(A,\epsilon) \;;\\ &((h,\chi):(C,\phi)\longrightarrow(C,\phi)) \longrightarrow\\ &\qquad ((h,\chi)\oplus(1,0):(C,\phi)\oplus(C,-\phi)\longrightarrow(C,\phi)\oplus(C,-\phi)) \;,\\ &DBL^n_U(A,\epsilon) \longrightarrow LAut^n_{U\oplus V}(A,\epsilon) \;;\\ &\qquad ((h,\chi):(C,\phi)\longrightarrow(C,\phi)) \longrightarrow ((h,\chi):(C,\phi)\longrightarrow(C,\phi)) \;.\end{aligned}$$

(ii) The maps in the exact sequence of 28.17

$$\dots \longrightarrow L^n_{\widetilde{\delta}(U)\oplus V}(A[z,z^{-1}],\epsilon) \xrightarrow{i} L^n_h(\Omega^{-1}A[z,z^{-1}],\epsilon)$$
$$\xrightarrow{\partial} LAut^{n-2}_{U\oplus V}(A,\epsilon) \xrightarrow{T} L^{n-1}_{\widetilde{\delta}(U)\oplus V}(A[z,z^{-1}],\epsilon) \longrightarrow \dots$$

split as

$$\begin{aligned} i \ = \ \widetilde{i}\oplus 0 \ : \ & L^n_{\widetilde{\delta}(U)\oplus V}(A[z,z^{-1}],\epsilon) \ = \ L^n_{\widetilde{\delta}(U)}(A,\epsilon)\oplus L^{n-1}_V(A,\epsilon) \\ & \longrightarrow L^n_h(\Omega^{-1}A[z,z^{-1}],\epsilon) \\ \partial \ = \ \widetilde{\partial}\oplus 0 \ : \ & L^n_h(\Omega^{-1}A[z,z^{-1}],\epsilon) \\ & \longrightarrow LAut^{n-2}_{U\oplus V}(A,\epsilon) \ = \ DBL^{n-2}_U(A,\epsilon)\oplus L^{n-2}_V(A,\epsilon) \ , \\ T \ = \ D\oplus 1 \ : \ & LAut^{n-2}_{U\oplus V}(A,\epsilon) \ = \ DBL^{n-2}_U(A,\epsilon)\oplus L^{n-2}_V(A,\epsilon) \\ & \longrightarrow L^{n-1}_{\widetilde{\delta}(U)\oplus V}(A[z,z^{-1}],\epsilon) \ = \ L^{n-1}_{\widetilde{\delta}(U)}(A,\epsilon)\oplus L^{n-2}_V(A,\epsilon) \end{aligned}$$

for any $*$-invariant subgroup $V \subseteq \widetilde{K}_0(A)$, with $T = D\oplus 1$ by 30.25. Alternatively, use 30.11.
(iii)+(iv) Immediate from 24.4 and 30.24, since

$$\tau(-1 : C \longrightarrow C) \ = \ 0 \in Wh_1(A)$$

for a f.g. free A-module chain complex C. □

Proposition 30.27 (i) *The automorphism L-groups* $LAut^*_q$ *($q = s,h,p$) split as*

$$\begin{aligned} LAut^n_p(A,\epsilon) \ &= \ DBL^n_h(A,\epsilon)\oplus L^n_p(A,\epsilon) \ , \\ LAut^n_h(A,\epsilon) \ &= \ DBL^n_h(A,\epsilon)\oplus L^n_h(A,\epsilon) \ , \\ LAut^n_s(A,\epsilon) \ &= \ DBL^n_s(A,\epsilon)\oplus L^n_h(A,\epsilon) \ . \end{aligned}$$

(ii) *The twisted book L-groups* $DBL^*_q(A,\epsilon)$ *($q = s,h$) fit into a commutative braid of exact sequences*

$$\begin{array}{ccccc} L^{n+2}_h(\Omega^{-1}A[z,z^{-1}],\epsilon) & & DBL^n_h(A,\epsilon) & & \widehat{H}^{n+1}(\mathbb{Z}_2;Wh_1(A)) \\ & \searrow \quad \nearrow & & \searrow \quad \nearrow & \\ & DBL^n_s(A,\epsilon) & & L^{n+1}_h(A,\epsilon) & \\ & \nearrow \quad \searrow & & \nearrow \quad \searrow & \\ \widehat{H}^{n+2}(\mathbb{Z}_2;Wh_1(A)) & & L^{n+1}_s(A,\epsilon) & & L^{n+1}_h(\Omega^{-1}A[z,z^{-1}],\epsilon) \end{array}$$

Proof Immediate from 30.26 (i) and 28.19. □

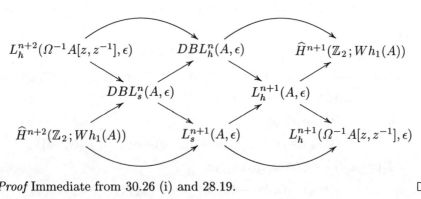

Proposition 30.28 (i) *The symmetric automorphism signature maps split as sums*

$$\begin{aligned}\sigma^* \;=\; \widetilde{\sigma}^* \oplus \sigma^* \;:\; \Delta_n(X) \;&=\; DB_n(X) \oplus \Omega_n(X) \\ &\longrightarrow \mathrm{LAut}^n_s(A) \;=\; DBL^n_s(A) \oplus L^n_h(A)\end{aligned}$$

for any space X, with $A = \mathbb{Z}[\pi_1(X)]$.
(ii) *The various signature maps define a natural transformation of exact sequences*

$$\begin{array}{ccccccccc} \dots \longrightarrow & \Omega_{n+2}(X) & \longrightarrow & AB_n(X) & \longrightarrow & DB_n(X) & \xrightarrow{D} & \Omega_{n+1}(X) & \longrightarrow \dots \\ & \big\downarrow \sigma^* & & \big\downarrow \sigma^* & & \big\downarrow \widetilde{\sigma}^* & & \big\downarrow \sigma^* & \\ \dots \longrightarrow & L^{n+2}_s(A) & \xrightarrow{i} & L^{n+2}_h(\Omega^{-1}A[z,z^{-1}]) & \xrightarrow{\partial} & DBL^n_s(A) & \xrightarrow{D} & L^{n+1}_s(A) & \longrightarrow \dots \end{array}$$

with $\sigma^* \;:\; AB_n(X) \xrightarrow{\simeq} L^{n+2}_h(\Omega^{-1}A[z,z^{-1}])$ the asymmetric signature of 28.22, and $\widetilde{\sigma}^*$ sending a rel ∂ automorphism $(h,1) : (P,\partial P) \longrightarrow (P,\partial P)$ of an n-dimensional manifold with boundary $(P,\partial P)$ to the cobordism class of the induced rel ∂ simple self homotopy equivalence $(\widetilde{h},\chi) : \sigma^*(P,\partial P) \longrightarrow \sigma^*(P,\partial P)$ of the f.g. free n-dimensional symmetric Poincaré pair $\sigma^*(P,\partial P)$ over A.
(iii) *For finitely presented $\pi_1(X)$ and $n \geq 5$ the asymmetric signature maps σ^* are isomorphisms, and there is defined an exact sequence*

$$\begin{aligned}\dots \longrightarrow L^{n+2}_s(A) \longrightarrow DB_n(X) \\ \xrightarrow{D \oplus \widetilde{\sigma}^*} \Omega_{n+1}(X) \oplus DBL^n_s(A) \xrightarrow{\sigma^* \oplus -D} L^{n+1}_s(A) \longrightarrow \dots \,.\end{aligned}$$ □

Remark 30.29 The identification (28.33)

$$L^{2i+2}_h(\Omega^{-1}A[z,z^{-1}],\epsilon) \;=\; \mathrm{LAsy}^0_h(A)$$

and the vanishing of the odd-dimensional asymmetric L-groups (28.34)

$$\mathrm{LAsy}^{2i+1}_h(A) \;=\; 0$$

give the exact sequences of 30.26 to be

$$\begin{aligned}0 \longrightarrow DBL^{2i+1}_q(A,\epsilon) \xrightarrow{t} L^{2i+2}_q(A,\epsilon) \longrightarrow \mathrm{LAsy}^0_h(A) \\ \longrightarrow DBL^{2i}_q(A,\epsilon) \xrightarrow{t} L^{2i+1}_q(A,\epsilon) \longrightarrow 0\end{aligned}$$

with $q = s, h$, and there are also quadratic L-theory versions

$$\begin{aligned}0 \longrightarrow DBL^q_{2i+1}(A,\epsilon) \xrightarrow{t} L^q_{2i+2}(A,\epsilon) \longrightarrow \mathrm{LAsy}^0_h(A) \\ \longrightarrow DBL^q_{2i}(A,\epsilon) \xrightarrow{t} L^q_{2i+1}(A,\epsilon) \longrightarrow 0\,.\end{aligned}$$

The quadratic L-groups $L^q_*(A,\epsilon)$, $DBL^q_*(A,\epsilon)$ are 4-periodic, and

$$t\ :\ DBL^q_{2i}(A,\epsilon)\ =\ DBL^q_0(A,(-)^i\epsilon)\longrightarrow L^q_{2i+1}(A,\epsilon)\ =\ L^q_1(A,(-)^i\epsilon)$$

is the surjection which sends an automorphism $\alpha : H_{(-)^i\epsilon}(L)\longrightarrow H_{(-)^i\epsilon}(L)$ of a hyperbolic $(-)^i\epsilon$-quadratic form to the Witt class of the nonsingular $(-)^i\epsilon$-quadratic formation $(H_{(-)^i\epsilon}(L);L,\alpha(L))$. The surjectivity of t in quadratic L-theory is thus essentially given by the original definition by Wall [302, Chap. 6] of the odd-dimensional surgery obstruction groups as stable unitary groups. The symmetric L-groups $L^*_q(A,\epsilon)$, $DBL^*_q(A,\epsilon)$ are not 4-periodic in general. The surjectivity of

$$t\ :\ DBL^{2i}_q(A,\epsilon)\longrightarrow L^{2i+1}_q(A,\epsilon)$$

is not obvious even in the low-dimensional case $i=0$, when t sends an automorphism $\alpha : (M,\phi)\longrightarrow(M,\phi)$ of a metabolic ϵ-symmetric form to the Witt class of the nonsingular ϵ-symmetric formation $(M,\phi;L,\alpha(L))$, for any lagrangian L. □

Remark 30.30 (i) The *Schneiden und Kleben* (= cutting and pasting) bordism groups $\overline{SK}_*(X)$ of Karras, Kreck, Neumann and Ossa [117] are the quotients of the ordinary bordism groups

$$\overline{SK}_n(X)\ =\ \Omega_n(X)/\!\sim$$

by the equivalence relation generated by

$$M\cup_f -N\ \sim\ M\cup_g -N \text{ for manifolds with boundary } (M,\partial M),\ (N,\partial N)$$
$$\text{and homeomorphisms } f,g:\partial M\longrightarrow\partial N\ ,$$

or equivalently

$$\overline{SK}_n(X)\ =\ \mathrm{coker}(D:DB_{n-1}(X)\longrightarrow\Omega_n(X))\ .$$

For finitely presented $\pi_1(X)$ and $n\geq 6$ the exact sequence of [117, 1.2]

$$0\longrightarrow F_n(X)\longrightarrow\Omega_n(X)\longrightarrow\overline{SK}_n(X)\longrightarrow 0$$

is isomorphic to the exact sequence given by 30.6

$$\begin{aligned}0\longrightarrow \mathrm{im}(D:DB_{n-1}(X)\longrightarrow\Omega_n(X))&\longrightarrow\Omega_n(X)\\ &\longrightarrow\mathrm{im}(\sigma^*:\Omega_n(X)\longrightarrow LAsy^n_h(\mathbb{Z}[\pi_1(X)]))\longrightarrow 0\end{aligned}$$

with $F_n(X)\subseteq\Omega_n(X)$ the subgroup of the bordism classes $M^n\longrightarrow X$ of closed manifolds which fibre over S^1. In particular

$$\begin{aligned}\overline{SK}_n(X)\ &=\ \mathrm{im}(\sigma^*:\Omega_n(X)\longrightarrow LAsy^n_h(\mathbb{Z}[\pi_1(X)]))\\ &=\ \ker(LAsy^n_h(\mathbb{Z}[\pi_1(X)])\longrightarrow DB_{n-2}(X))\end{aligned}$$

is the cobordism group of the n-dimensional asymmetric Poincaré complexes over $\mathbb{Z}[\pi_1(X)]$ which are realized by bordism classes $M^n \longrightarrow X$. The even-dimensional $\overline{SK}$-groups are non-zero in general, while the odd-dimensional $\overline{SK}$-groups vanish

$$\overline{SK}_{2*+1}(X) \;=\; 0 \;.$$

Thus for $n \geq 6$ the following conditions are equivalent for a closed n-dimensional manifold M:

(a) M is an open book,
(b) M is a twisted double,
(c) $\sigma^*(M) = 0 \in LAsy_h^n(\mathbb{Z}[\pi_1(M)])$,
(d) $(M,1) = 0 \in \overline{SK}_n(M)$.

Note that the asymmetric signature map factors through the symmetric L-theory assembly map A of Ranicki [245, Chap. 13]

$$\sigma^* \;:\; \Omega_n(X) \longrightarrow H_n(X;\mathbb{L}^\bullet(\mathbb{Z})) \xrightarrow{A} L_h^n(\mathbb{Z}[\pi_1(X)]) \longrightarrow LAsy_h^n(\mathbb{Z}[\pi_1(X)]) \;,$$

so that

$$\overline{SK}_n(X) \subseteq \mathrm{im}(A : H_n(X;\mathbb{L}^\bullet(\mathbb{Z})) \longrightarrow LAsy_h^n(\mathbb{Z}[\pi_1(X)])) \;.$$

The $\overline{SK}$-groups of a simply-connected space X are given by

$$\begin{aligned}\overline{SK}_n(X) \;&=\; \overline{SK}_n(\{*\}) \\ &=\; \mathrm{im}(L^n(\mathbb{Z}) \longrightarrow LAsy^n(\mathbb{Z})) \\ &=\; \begin{cases} \mathbb{Z} \text{ (signature)} & \text{if } n \equiv 0 (\mathrm{mod}\, 4) \\ 0 & \text{otherwise}\end{cases}\end{aligned}$$

(Neumann [211, Theorem 1]).

(ii) Define the *Schneiden und Kleben ϵ-symmetric L-groups* $SKL^n(A,\epsilon)$ to be the quotients of the ϵ-symmetric L-groups

$$SKL^n(A,\epsilon) \;=\; L^n(A,\epsilon)/\!\sim$$

by the equivalence relation generated by

$$C \cup_f -D \;\sim\; C \cup_g -D$$

for n-dimensional ϵ-symmetric Poincaré pairs $(C,\partial C)$, $(D,\partial D)$ and homotopy equivalences $f,g : \partial C \longrightarrow \partial D$.

The ϵ-symmetric SKL-groups are the images of the ϵ-symmetric L-groups in the asymmetric L-groups

$$\begin{aligned}SKL^n(A,\epsilon) \;&=\; \mathrm{coker}(D : DBL^{n-1}(A,\epsilon) \longrightarrow L^n(A,\epsilon)) \\ &=\; \mathrm{im}(L^n(A,\epsilon) \longrightarrow LAsy^n(A)) \;,\end{aligned}$$

with

$$SKL^{2*+1}(A,\epsilon) \;=\; 0 \;.$$

As usual, there is also an ϵ-quadratic version. □

Remark 30.31 A fibre bundle with fibre a d-dimensional manifold F

$$F^d \longrightarrow E \xrightarrow{p} B$$

induces geometric transfer maps in the ordinary bordism groups

$$p^! \,:\, \Omega_n(B) \longrightarrow \Omega_{n+d}(E) \;;\; (M, f : M \longrightarrow B) \longrightarrow (M^!, f^! : M^! \longrightarrow E)$$

with $M^! = f^*E$ the pullback fibre bundle over M. The geometric transfer maps are also defined for the twisted double bordism groups

$$p^! \,:\, DB_n(B) \longrightarrow DB_{n+d}(E) \;,$$

and the bordism SK-groups

$$p^! \,:\, \overline{SK}_n(B) \longrightarrow \overline{SK}_{n+d}(E) \;.$$

The transfer maps in the algebraic K- and quadratic L-groups

$$\begin{aligned} p^! \,&:\, K_i(\mathbb{Z}[\pi_1(B)]) \longrightarrow K_i(\mathbb{Z}[\pi_1(E)]) \;\; (i = 0, 1) \;, \\ p^! \,&:\, L_n(\mathbb{Z}[\pi_1(B)]) \longrightarrow L_{n+d}(\mathbb{Z}[\pi_1(E)]) \;\; (n \geq 0) \end{aligned}$$

can be defined both geometrically and algebraically: a morphism of based f.g. free $\mathbb{Z}[\pi_1(B)]$-modules

$$f \,:\, \mathbb{Z}[\pi_1(B)]^k \longrightarrow \mathbb{Z}[\pi_1(B)]^\ell$$

is sent to the chain homotopy class of the 'parallel transport' $\mathbb{Z}[\pi_1(E)]$-module chain map

$$f^! \,:\, \bigoplus_k C(\widetilde{F}) \longrightarrow \bigoplus_\ell C(\widetilde{F})$$

(Lück [173], Lück and Ranicki [175], [177]). The algebraic construction also gives transfer maps in the asymmetric L-groups

$$p^! \,:\, LAsy_h^n(\mathbb{Z}[\pi_1(B)]) \longrightarrow LAsy_h^{n+d}(\mathbb{Z}[\pi_1(E)]) \;\; (n \geq 0)$$

which fit into a natural transformation of exact sequences

$$\begin{array}{ccccccccc} \to & LAsy_h^{n+1}(\mathbb{Z}[\pi_1(B)]) & \to & DB_{n-1}(B) & \to & \Omega_n(B) & \to & LAsy_h^n(\mathbb{Z}[\pi_1(B)]) & \to \\ & \downarrow p^! & & \downarrow p^! & & \downarrow p^! & & \downarrow p^! & \\ \to & LAsy_h^{n+d+1}(\mathbb{Z}[\pi_1(E)]) & \to & DB_{n+d-1}(E) & \to & \Omega_{n+d}(E) & \to & LAsy_h^{n+d}(\mathbb{Z}[\pi_1(E)]) & \to \end{array}$$

For $n + d \equiv 0 (\bmod 4)$ the transfer map in asymmetric L-theory and the surgery transfer methods of [177] (cf. 28.24) can be used to provide an L-theoretic interpretation of the work of Karras, Kreck, Neumann and Ossa [117] and Neumann [211], [213] on the connection between open book decompositions, the SK-groups and the (non-)multiplicativity of signature in fibre bundles. □

31. Isometric L-theory

This chapter is concerned with the algebraic L-theory of the polynomial extension $A[s]$ with the involution $\overline{s} = 1 - s$. The connection with high-dimensional knot theory may appear remote at first, so here is a quote from the introduction to the memoir of Stoltzfus [277] in which the high-dimensional knot cobordism groups were finally computed:

We are, however, acutely aware that [the formalism used] is intimately related to the setting of Hermitian algebraic K-theory, particularly for the ring $\mathbb{Z}[X]$ and the involution induced by $X^* = 1 - X$. This ring is crucial in our study of the complement of a knot or any space which is a homology circle.

In fact, it is the L-theory of the Fredholm localization $\Omega_+^{-1}A[s]$ rather than $A[s]$ itself which is relevant to knot theory.

An isometric structure on an n-dimensional ϵ-symmetric complex (C, ϕ) is a chain map $f : C \longrightarrow C$ such that

$$f\phi_0 \simeq \phi_0(1 - f^*) : C^{n-*} \longrightarrow C$$

or equivalently a chain map $\widehat{\psi}_0 : C^{n-*} \longrightarrow C$ such that

$$(1 + T_\epsilon)\widehat{\psi}_0 \simeq \phi_0 : C^{n-*} \longrightarrow C .$$

The isometric L-group $L\mathrm{Iso}^n(A, \epsilon)$ is defined to be the cobordism group of n-dimensional ϵ-symmetric complexes over A with isometric structure. In Chap. 33 below the high-dimensional knot cobordism groups C_* are identified with the isometric L-groups $L\mathrm{Iso}^{*+1}(\mathbb{Z})$.

31A. Isometric structures

Terminology 31.1 Given a ring with involution A let $A[s]$ denote the polynomial extension of A with the involution extended by

$$\overline{s} = 1 - s . \qquad \square$$

The ϵ-symmetrization maps

$$1+T_\epsilon \ : \ L_n(A[s],\epsilon) \longrightarrow L^n(A[s],\epsilon)$$

are isomorphisms (25.14), so there is no difference between the ϵ-quadratic and ϵ-symmetric L-groups for the rings with involution and structures considered in this chapter.

As in Chap. 10 let Ω_+ be the set of Fredholm matrices in $A[s]$. By 10.3 a $k\times k$ matrix ω in $A[s]$ is Fredholm if and only if the $A[s]$-module morphism

$$s-\omega \ : \ A[s]^k \longrightarrow A[s]^k$$

is injective and the cokernel is a f.g. projective A-module. By 10.14 the Ω_+-torsion K-group of $A[s]$ is related to the endomorphism class group of A by the isomorphism

$$\begin{aligned}&\mathrm{End}_0(A) \xrightarrow{\simeq} K_1(A[s],\Omega_+) \ ;\\ &[P,f] \longrightarrow [\mathrm{coker}(s-f:P[s]{\longrightarrow}P[s])] \ = \ [P \text{ with } s \text{ acting by } f] \ .\end{aligned}$$

Let $A[z,z^{-1}]$ have the involution $\overline{z}=z^{-1}$, and let

$$j \ : \ A[z,z^{-1}] \longrightarrow A \ ; \ z \longrightarrow 1 \ .$$

The set Π (10.17) of square matrices ω in $A[z,z^{-1}]$ such that $j(\omega)$ is an invertible matrix in A is involution-invariant.

The main result of Chap. 31 is the identification of the ϵ-symmetric $\Omega_+^{-1}A[s]$-torsion L-groups of $A[s]$ with the $(-\epsilon)$-symmetric isometric L-groups

$$L^n(A[s],\Omega_+,\epsilon) \ = \ L\mathrm{Iso}^n(A,-\epsilon) \ .$$

More precisely, the localization exact sequence

$$\begin{aligned}\dots \longrightarrow L^n_q(A[s],\epsilon) \longrightarrow L^n_q(\Omega_+^{-1}A[s],\epsilon) \longrightarrow L^n_q(A[s],\Omega_+,\epsilon)\\ \longrightarrow L^{n-1}_q(A[s],\epsilon) \longrightarrow \dots \quad (q=s,h)\end{aligned}$$

will be shown to break up into split exact sequences

$$0 \longrightarrow L^n_q(A[s],\epsilon) \longrightarrow L^n_q(\Omega_+^{-1}A[s],\epsilon) \longrightarrow L^n_q(A[s],\Omega_+,\epsilon) \longrightarrow 0$$

with

$$\begin{aligned}L^n_t(A[s],\Omega_+,\epsilon) \ &= \ L\mathrm{Iso}^n_u(A,-\epsilon)\\ &= \ L^n_t(A[z,z^{-1}],\Pi,\epsilon) \ \ ((t,u)=(s,h),(h,p)) \ .\end{aligned}$$

Definition 31.2 The *isometric class group* $\mathrm{Iso}_0(A)$ is the endomorphism class group $\mathrm{End}_0(A)$ with the duality involution

$$\mathrm{Iso}_0(A) \longrightarrow \mathrm{Iso}_0(A) \ ; \ [P,f] \longrightarrow [P^*,1-f^*] \ . \qquad \square$$

Proposition 31.3 *The set Ω_+ of Fredholm matrices in $A[s]$ is involution-invariant, and the isomorphism of 10.14*

$$\mathrm{Iso}_0(A) \xrightarrow{\simeq} K_1(A[s],\Omega_+) \; ; \; [P,f] \longrightarrow [\mathrm{coker}(s-f : P[s]\longrightarrow P[s])]$$

preserves the involutions, and

$$K_1(A[s]) \underset{j_+}{\overset{i_+}{\rightleftarrows}} K_1(\Omega_+^{-1}A[s]) \underset{\Delta_+}{\overset{\partial_+}{\rightleftarrows}} \mathrm{Iso}_0(A)$$

is a direct sum system of $\mathbb{Z}[\mathbb{Z}_2]$-modules, with

$$\widehat{H}^*(\mathbb{Z}_2;K_1(\Omega_+^{-1}A[s])) = \widehat{H}^*(\mathbb{Z}_2;K_1(A[s])) \oplus \widehat{H}^*(\mathbb{Z}_2;\mathrm{Iso}_0(A)) .$$

Proof Given a $k \times k$ Fredholm matrix $\omega = (\omega_{ij})$ in $A[s]$ there is defined a f.g. projective A-module

$$P = \mathrm{coker}(s-\omega : A[s]^k \longrightarrow A[s]^k)$$

with an endomorphism

$$f : P \longrightarrow P \; ; \; x \longrightarrow sx$$

and an $A[s]$-module exact sequence

$$0 \longrightarrow A[s]^k \xrightarrow{s-\omega} A[s]^k \longrightarrow P \longrightarrow 0 .$$

The dual $k \times k$ matrix $\omega^* = (\overline{\omega}_{ji})$ is Fredholm, with an $A[s]$-module exact sequence

$$0 \longrightarrow A[s]^k \xrightarrow{1-s-\omega^*} A[s]^k \longrightarrow P^* \longrightarrow 0$$

where

$$s = 1-f^* : P^* \longrightarrow P^* . \qquad \square$$

Definition 31.4 (i) Given an A-module chain complex C and a chain map $f : C \longrightarrow C$ use the $\mathbb{Z}[\mathbb{Z}_2]$-module chain map

$$\Gamma_f^{iso} = f \otimes 1 - 1 \otimes (1-f) : (C \otimes_A C, T_\epsilon) \longrightarrow (C \otimes_A C, T_\epsilon)$$

to define the *ϵ-isometric Q-groups* of (C,f)

$$Q_{iso}^{n+1}(C,f,\epsilon) = H_{n+1}(\Gamma_f^{iso} : W^{\%}C \longrightarrow W^{\%}C)$$

to fit into an exact sequence

$$\dots \longrightarrow Q^{n+1}(C,\epsilon) \longrightarrow Q_{iso}^{n+1}(C,f,\epsilon)$$
$$\longrightarrow Q^n(C,\epsilon) \xrightarrow{\Gamma_f^{iso}} Q^n(C,\epsilon) \longrightarrow \dots .$$

(ii) An *isometric structure* $(f,\delta\phi)$ for an n-dimensional ϵ-symmetric complex (C,ϕ) over A is a chain map $f : C \longrightarrow C$ together with a chain $\delta\phi \in W^{\%}C_{n+1}$ such that

$$f\phi_q - \phi_q(1-f^*) = d\delta\phi_q + (-)^r\delta\phi_q d^* + (-)^{n+q-1}(\delta\phi_{q-1} + (-)^q T_\epsilon \delta\phi_{q-1})$$
$$: C^{n-r+q} \longrightarrow C_r \quad (q, r \geq 0, \delta\phi_{-1} = 0) ,$$

representing an element $(\delta\phi, \phi) \in Q^{n+1}_{iso}(C, f, \epsilon)$.
(iii) An isometric structure $(f, \delta\phi)$ on (C, ϕ) is *fibred* if $f : C \longrightarrow C$ is fibred (28.35), i.e. if $f, 1-f : C \longrightarrow C$ are chain equivalences. □

Proposition 31.5 (i) *The morphisms* $\Gamma^{iso}_f : Q^*(C, \epsilon) \longrightarrow Q^*(C, \epsilon)$ *are such that*

$$\Gamma^{iso}_f + 1 \;:\; Q^n(C,\epsilon) \longrightarrow H_n(C \otimes_A C) \xrightarrow{f\otimes 1} H_n(C \otimes_A C) \longrightarrow Q^n(C,\epsilon)$$

with

$$Q^n(C,\epsilon) \longrightarrow H_n(C \otimes_A C) \;;\; \phi \longrightarrow \phi_0 ,$$
$$H_n(C \otimes_A C) \longrightarrow Q^n(C,\epsilon) \;;$$
$$\psi \longrightarrow \phi \,,\; \phi_q = \begin{cases} (1+T_\epsilon)\psi & \text{if } q = 0 \\ 0 & \text{if } q \geq 1 . \end{cases}$$

(ii) *The isometric Q-groups are such that there is defined a commutative braid of exact sequences*

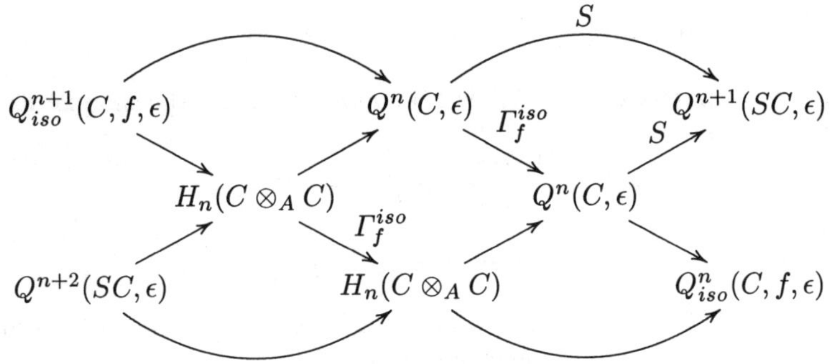

with

$$Q^{n+1}_{iso}(C, f, \epsilon) \longrightarrow H_n(C \otimes_A C) \;;\; \phi \longrightarrow (f \otimes 1)\phi_0 ,$$

and

$$S \,:\, Q^n(C,\epsilon) \longrightarrow Q^{n+1}(SC,\epsilon) \,;\, \{\phi_q \,|\, q \geq 0\} \longrightarrow \{\phi_{q-1} \,|\, q \geq 0\} \quad (\phi_{-1} = 0)$$

the suspension map of Ranicki [235]. □

Remark 31.6 The exact sequence in the braid of 31.5 (ii)

$$\ldots \longrightarrow H_{n+1}(C \otimes_A C) \longrightarrow Q^{n+1}_{iso}(C, f, \epsilon) \longrightarrow H_n(C \otimes_A C)$$
$$\xrightarrow{\Gamma^{iso}_f} H_n(C \otimes_A C) \longrightarrow \ldots$$

shows that an n-dimensional ϵ-symmetric (Poincaré) complex (C, ϕ) with an isometric structure $(\delta\phi, f)$ is essentially the same as an ϵ-ultraquadratic

(Poincaré) complex $(C, \widehat{\psi} \in H_n(C \otimes_A C))$ – see Ranicki [237, 7.8] and Chap. 32 below for ϵ-ultraquadratic L-theory. □

Example 31.7 (i) An *isometric structure* for an ϵ-symmetric form (M, ϕ) over A is an endomorphism $f : M \longrightarrow M$ such that

$$f^*\phi = \phi(1-f) : M \longrightarrow M^* .$$

An isometric structure on (M, ϕ) is just an isometric structure $(f^*, 0)$ on the 0-dimensional ϵ-symmetric complex (C, ϕ) with $C_0 = M^*$, $C_r = 0$ for $r \neq 0$.
(ii) Isometric structures on ϵ-symmetric forms were used by Levine [157], [158], Kervaire [133] and Stoltzfus [277] in the computation of the high-dimensional knot cobordism groups. See Chaps. 32, 33 for the connection between isometric structures and Seifert forms. □

By analogy with 28.13:

Proposition 31.8 (i) *The ϵ-symmetric Q-groups of an A-finitely dominated ($= \Omega_+^{-1}A[s]$-contractible) finite f.g. free $A[s]$-module chain complex D are the isometric ϵ-symmetric Q-groups of $(D^!, \sigma)$*

$$Q^*(D, \epsilon) = Q^*_{iso}(D^!, \sigma) ,$$

with $D^!$ the finitely dominated A-module chain complex defined by D and

$$\sigma : D^! \longrightarrow D^! ; w \longrightarrow sw .$$

(ii) *The homotopy equivalence classes of finitely dominated n-dimensional ϵ-symmetric Poincaré complexes (C, ϕ) over A with an isometric structure $(f, \delta\phi)$ are in one-one correspondence with the homotopy equivalence classes of f.g. free $(n+1)$-dimensional $\Omega_+^{-1}A[s]$-contractible ϵ-symmetric Poincaré complexes (D, θ) over $A[s]$.*
(iii) *A finitely dominated n-dimensional ϵ-symmetric Poincaré complex (C, ϕ) over A with an isometric structure $(f, \delta\phi)$ determines a f.g. projective n-dimensional $(-\epsilon)$-symmetric $\Omega_+^{-1}A[s]$-Poincaré complex $(C[s], \Phi)$ over $A[s]$ with*

$$\Phi_q = (s-f)\phi_q + \delta\phi_{q-1} : C[s]^{n-r+q} \longrightarrow C[s]_r \quad (q, r \geq 0) .$$

The skew-suspension $(SC[s], \Phi)$ is a f.g. projective $(n+2)$-dimensional ϵ-symmetric $\Omega_+^{-1}A[s]$-Poincaré complex over $A[s]$. The boundary

$$(D, \theta) = \partial(SC[s], \Phi)$$

is a (homotopy) f.g. free $(n+1)$-dimensional ϵ-symmetric Poincaré complex over $A[s]$ corresponding to (C, ϕ) and $(f, \delta\phi)$ in (i), such that

$$\begin{aligned}\tau_{\Omega_+}(D) &= [C, f] \in K_1(A[s], \Omega_+) = \mathrm{Iso}_0(A) ,\\ \tau(D, \theta) &= \Phi(D) + (-)^n\Phi(D)^* \in K_1(A[s])\end{aligned}$$

with respect to the canonical round finite structure on D.

Proof (i) Let (C,f) be a finite chain complex of $(A[s],\Omega_+)$-modules, i.e. a finite f.g. projective A-module chain complex C together a chain map $f : C \longrightarrow C$. The f.g. projective $A[s]$-module chain complex

$$D = \mathcal{C}(s - f : C[s] \longrightarrow C[s])$$

is homology equivalent to (C,f), and every A-finitely dominated ($=\Omega_+^{-1}A[s]$-contractible) f.g. free $A[s]$-module chain complex is chain equivalent to one of this type, with $(D^!,\sigma) \simeq (C,f)$. The $\mathbb{Z}[\mathbb{Z}_2]$-module chain map

$$\mathcal{C}(\Gamma_f^{iso} : C \otimes_A C \longrightarrow C \otimes_A C) \longrightarrow D \otimes_{A[s]} D$$

defined by

$$\begin{aligned}\mathcal{C}(\Gamma_f^{iso})_r &= (C \otimes_A C)_r \oplus (C \otimes_A C)_{r-1} \\ &\longrightarrow (D \otimes_{A[s]} D)_r = (C[s] \otimes_{A[s]} C[s])_r \oplus (C[s] \otimes_{A[s]} C[s])_{r-1} \\ &\oplus (C[s] \otimes_{A[s]} C[s])_{r-1} \oplus (C[s] \otimes_{A[s]} C[s])_{r-2} ~;~ (u,v) \longrightarrow (u,v,v,0)\end{aligned}$$

is a homology equivalence inducing isomorphisms

$$Q^*_{iso}(C,f,\epsilon) \cong Q^*(D,\epsilon) ,$$

so that there is defined an exact sequence

$$\dots \longrightarrow Q^{n+1}(D,\epsilon) \longrightarrow Q^n(C,\epsilon) \xrightarrow{\Gamma_f^{iso}} Q^n(C,\epsilon) \longrightarrow Q^n(D,\epsilon) \longrightarrow \dots .$$

An $(n+1)$-dimensional ϵ-symmetric structure $\theta \in Q^{n+1}(D,\epsilon)$ on D is thus the same as an n-dimensional ϵ-symmetric structure $\phi \in Q^n(C,\epsilon)$ with a refinement to an isometric structure $(\delta\phi,\phi) \in Q^{n+1}_{iso}(C,f,\epsilon)$. Moreover

$$\theta_0 ~:~ H^{n+1-*}(D) \cong H^{n-*}(C) \xrightarrow{\phi_0} H_*(C) \cong H_*(D) ,$$

so that (D,θ) is a Poincaré complex over $A[s]$ if and only if (C,ϕ) is a Poincaré complex over A.

(ii)+(iii) Immediate from (i). □

Definition 31.9 Let $U \subseteq \mathrm{Iso}_0(A)$ be a $*$-invariant subgroup.

(i) The *ϵ-isometric L-groups* $L\mathrm{Iso}^n_U(A,\epsilon)$ $(n \geq 0)$ are the cobordism groups of finitely dominated n-dimensional ϵ-symmetric Poincaré complexes (C,ϕ) over A with an isometric structure $(f,\delta\phi)$ such that

$$[C,f] \in U \subseteq \mathrm{End}_0(A) .$$

(ii) The *fibred ϵ-isometric L-groups* $L\mathrm{Iso}^n_{fib,U}(A,\epsilon)$ $(n \geq 0)$ are defined as in (i), using fibred isometric structures.

(iii) The *reduced ϵ-isometric L-groups* $\widetilde{L\mathrm{Iso}}^n_U(A,\epsilon)$ $(n \geq 0)$ are defined as in (i), but in addition requiring that (C, ϕ) be equipped with a finitely dominated null-cobordism with projective class in the image $U_0 \subseteq \widetilde{K}_0(A)$ of U. □

Proposition 31.10 *Let $U \subseteq \mathrm{Iso}_0(A)$ be a $*$-invariant subgroup.*
(i) *The isometric and reduced isometric L-groups are related by an exact sequence*

$$\ldots \longrightarrow \widetilde{LIso}^n_U(A,\epsilon) \longrightarrow LIso^n_U(A,\epsilon) \longrightarrow L^n_{U_0}(A,\epsilon) \longrightarrow \widetilde{LIso}^{n-1}_U(A,\epsilon) \longrightarrow \ldots .$$

(ii) *There are natural identifications of the $(A[s],\Omega_+)$-torsion ϵ-symmetric L-groups and the $(-\epsilon)$-isometric L-groups of A*

$$L^n_U(A[s],\Omega_+,\epsilon) = LIso^n_U(A,-\epsilon) .$$

(iii) *For any $*$-invariant subgroup $T \subseteq K_1(A[s])$ the localization exact sequence of 25.4*

$$\ldots \longrightarrow L^n_T(A[s],\epsilon) \xrightarrow{i_+} L^n_{i_+(T)\oplus\Delta_+(U)}(\Omega_+^{-1}A[s],\epsilon) \xrightarrow{\partial_+} LIso^n_U(A,-\epsilon) \xrightarrow{j_+} L^{n-1}_T(A[s],\epsilon) \longrightarrow \ldots$$

breaks up into split exact sequences, with $j_+ = 0$ and ∂_+ split by

$$\Delta_+ : LIso^n_U(A,-\epsilon) \longrightarrow L^n_{i_+(T)\oplus\Delta_+(U)}(\Omega_+^{-1}A[s],\epsilon) ;$$
$$(C,f,\delta\phi,\phi) \longrightarrow \Omega_+^{-1}(\mathcal{C}(s-f : C[s]\longrightarrow C[s]),\Phi)$$

(Φ as in 31.8 (iii)), so that there are defined direct sum systems

$$L^n_{i_+(T)}(A[s],\epsilon) \underset{j_+}{\overset{i_+}{\rightleftarrows}} L^n_{i_+(T)\oplus\Delta_+(U)}(\Omega_+^{-1}A[s],\epsilon) \underset{\Delta_+}{\overset{\partial_+}{\rightleftarrows}} LIso^n_U(A,-\epsilon) .$$

(iii) *The ϵ-symmetric endomorphism L-groups associated to a pair $(U_2, U_1 \subseteq U_2)$ of $*$-invariant subgroups $U_1 \subseteq U_2 \subseteq \mathrm{Iso}_0(A)$ are related by a Rothenberg-type exact sequence*

$$\ldots \longrightarrow \widehat{H}^{n+1}(\mathbb{Z}_2;U_2/U_1) \longrightarrow LIso^n_{U_1}(A,\epsilon) \longrightarrow LIso^n_{U_2}(A,\epsilon) \longrightarrow \widehat{H}^n(\mathbb{Z}_2;U_2/U_1) \longrightarrow \ldots$$

with

$$LIso^n_{U_2}(A,\epsilon) \longrightarrow \widehat{H}^n(\mathbb{Z}_2;U_2/U_1) ; \ (C,f,\delta\phi,\phi) \longrightarrow [C,f] .$$

Proof (i) By construction.
(ii) Immediate from 31.8 (ii).

(iii) By 31.8 (iii) the maps ∂_+, Δ_+ are such that $\partial_+\Delta_+ = 1$.
(iii) As for the ordinary L-groups in Ranicki [235, Chap. 10]. □

Remark 31.11 The algebraic surgery method of Ranicki [235] extends to prove that the ϵ-isometric L-groups of any ring with involution A are 4-periodic

$$LIso^n_U(A,\epsilon) \;=\; LIso^{n+2}_U(A,-\epsilon) \;=\; LIso^{n+4}_U(A,\epsilon) \quad (U \subseteq \mathrm{Iso}_0(A))$$

See Chap. 39B below for more on the isometric L-theory of fields. □

Terminology 31.12 Write the projective isometric L-groups as

$$LIso^n_{\mathrm{Iso}_0(A)}(A,\epsilon) \;=\; LIso^n_p(A,\epsilon)\ ,$$
$$\widetilde{LIso}^n_{\mathrm{Iso}_0(A)}(A,\epsilon) \;=\; \widetilde{LIso}^n_p(A,\epsilon)\ . \qquad \square$$

Proposition 31.13 (i) *The free ϵ-symmetric L-groups of $\Omega_+^{-1}A[s]$ are such that*

$$L^n_h(\Omega_+^{-1}A[s],\epsilon) \;=\; L^n_h(A[s],\epsilon) \oplus LIso^n_p(A,-\epsilon)\ ,$$
$$L^n_h(A[s],\Omega_+\epsilon) \;=\; LIso^n_p(A,-\epsilon)\ .$$

(ii) *The forgetful map*

$$LIso^n_p(A,\epsilon) \longrightarrow L^p_n(A,\epsilon)\ ;\ (C,f,\delta\phi,\phi) \longrightarrow (C,\psi)\ ,\ \psi_q \;=\; \begin{cases} f\phi_0 & \text{if } q=0 \\ 0 & \text{if } q \geq 1 \end{cases}$$

is onto for n even and one-one for n odd.
Proof (i) This is a special case of 31.10.
(ii) By the 4-periodicity of $LIso^*$ and L_* it is sufficient to consider the cases $n = 0, 1$.
For $n = 0$ every element $(M,\psi) \in L^p_0(A,\epsilon)$ is represented by a nonsingular ϵ-quadratic form (M,ψ) over A. A lift of

$$\psi \in Q_\epsilon(M) \;=\; \mathrm{coker}(1 - T_\epsilon : \mathrm{Hom}_A(M,M^*) \longrightarrow \mathrm{Hom}_A(M,M^*))$$

to an element $\widehat{\psi} \in \mathrm{Hom}_A(M,M^*)$ determines an isometric structure on (M,ψ)

$$f \;=\; (\psi + \epsilon\psi^*)^{-1}\psi \;:\; M \longrightarrow M$$

such that $(M,\psi + \epsilon\psi^*, f) \in LIso^0_p(A,\epsilon)$ has image $(M,\psi) \in L^p_0(A,\epsilon)$.
For $n = 1$ every element

$$(C,f,\delta\phi,\phi) \in \ker(LIso^1_p(A,\epsilon) \longrightarrow L^p_1(A,\epsilon))$$

is represented by a 1-dimensional ϵ-symmetric Poincaré complex (C,ϕ) with an isometric structure $(f,\delta\phi)$ such that the image 1-dimensional ϵ-quadratic Poincaré complex (C,ψ) admits a null-cobordism $(C \longrightarrow D, (\delta\psi,\psi))$ with $D_r = 0$ for $r \neq 1$. It is possible to lift $\delta\psi$ to a null-cobordism of the isometric structure, so that $(C,f,\delta\phi,\phi) = 0 \in LIso^1_p(A,\epsilon)$. □

By analogy with 28.17:

Remark 31.14 An isometric structure $(f, \delta\phi)$ on an m-dimensional ϵ-symmetric Poincaré complex (C, ϕ) over A determines an element

$$f \in R = H_0(\mathrm{Hom}_A(C,C))$$

such that $\overline{f} = 1 - f$, so that there is defined a morphism of rings with involution

$$\rho : \mathbb{Z}[s] \longrightarrow R \; ; \; s \longrightarrow f \; .$$

The morphisms

$$LIso^m(A,\epsilon) \longrightarrow L^m(\mathbb{Z}[s], A, \epsilon) \; ; \; (C, f, \delta\phi, \phi) \longrightarrow (C, \phi, \rho) \quad (m \geq 0)$$

are isomorphisms by 31.6, with $L^m(\mathbb{Z}[s], A, \epsilon)$ the surgery transfer group of Lück and Ranicki [177] (cf. 28.15, 28.24), such that either $m = 0$ or A is a Dedekind ring with involution. For any ring with involution B there are defined products

$$LIso^m(A,\epsilon) \otimes L_n(B[s],\eta) \longrightarrow L_{m+n}(A \otimes_{\mathbb{Z}} B, \epsilon\eta) \; . \qquad \square$$

Proposition 31.15 *Let*

$$\Lambda = A[z, z^{-1}, (1-z)^{-1}] \; .$$

(i) *Sending s to $(1-z)^{-1}$ defines isomorphisms of rings with involution*

$$A[s, s^{-1}, (1-s)^{-1}] \cong \Lambda \; , \quad \Omega_+^{-1}A[s] \cong \Pi^{-1}\Lambda \; .$$

(ii) *The inclusion*

$$A[s] \longrightarrow A[s, s^{-1}, (1-s)^{-1}] = \Lambda$$

induces an isomorphism of split exact sequences

$$\begin{array}{ccccccccc} 0 & \longrightarrow & L_h^n(A[s],\epsilon) & \longrightarrow & L_h^n(\Omega_+^{-1}A[s],\epsilon) & \longrightarrow & L_h^n(A[s],\Omega_+,\epsilon) & \longrightarrow & 0 \\ & & \downarrow \cong & & \downarrow \cong & & \downarrow \cong & & \\ 0 & \longrightarrow & L_h^n(\Lambda,\epsilon) & \longrightarrow & L_h^n(\Pi^{-1}\Lambda,\epsilon) & \longrightarrow & L_h^n(\Lambda,\Pi,\epsilon) & \longrightarrow & 0 \end{array}$$

and the inclusion $A[z, z^{-1}] \longrightarrow \Lambda$ induces an isomorphism

$$L_h^n(A[z, z^{-1}], \Pi, \epsilon) \cong L_h^n(\Lambda, \Pi, \epsilon) \; .$$

(iii) *The commutative braid of L-theory localization exact sequences*

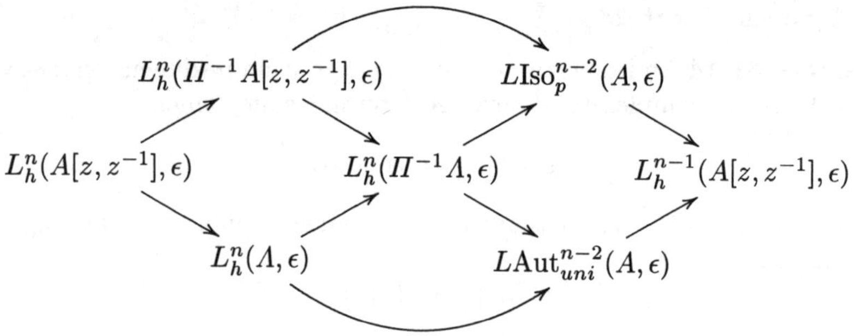

splits as a direct sum of the commutative braid of exact sequences

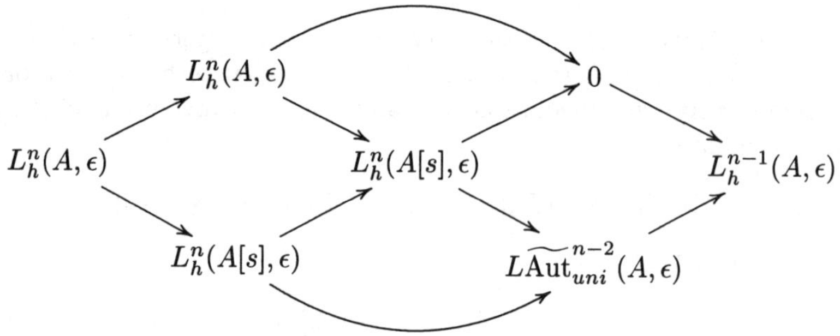

and the commutative braid of exact sequences

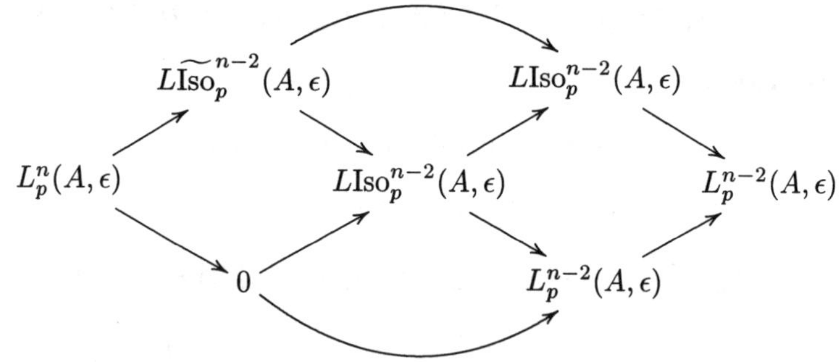

In particular,

$$\begin{aligned} L_h^n(A[z,z^{-1}],\Pi,\epsilon) &= L_h^n(A[s],\Omega_+,\epsilon) = LIso_p^{n-2}(A,\epsilon) , \\ L_h^n(\Pi^{-1}A[z,z^{-1}],\epsilon) &= L_h^n(A,\epsilon) \oplus \widetilde{LIso}_p^{n-2}(A,\epsilon) , \\ L_h^n(\Pi^{-1}\Lambda,\epsilon) &= L_h^n(\Omega_+^{-1}A[s],\epsilon) \\ &= L_h^n(A[s],\epsilon) \oplus LIso_p^{n-2}(A,\epsilon) . \end{aligned}$$

Proof (i) The morphism of rings with involution

$$A[s] \longrightarrow \Lambda \; ; \; s \longrightarrow (1-z)^{-1}$$

sends a monic Fredholm matrix in $A[s]$

$$\omega = \sum_{j=0}^{d} \omega_j z^j \in \Omega_{+,mon} \quad (\omega_d = 1)$$

to a matrix in Λ of the form $(1-z)^{-d}\tau$ with

$$\tau = \sum_{j=0}^{d} \omega_j (1-z)^{d-j}$$

A-invertible. The isomorphism of rings with involution

$$\Lambda \xrightarrow{\simeq} A[s, s^{-1}, (1-s)^{-1}] \; ; \; z \longrightarrow 1 - s^{-1}$$

sends an A-invertible matrix in $A[z, z^{-1}]$

$$\tau = \sum_{j=m}^{n} \tau_j z^j \in \Pi$$

with $\sum\limits_{j=m}^{n} \tau_j = 1$ to a matrix in $A[s]$ of the form $(s-1)^m s^{-n} \omega$ with

$$\omega = \sum_{k=0}^{n-m} \tau_{k+m} s^{n-m-k} (s-1)^k \in \Omega_{+,mon}$$

monic. By 10.7 (iii)

$$\Omega_{+,mon}^{-1} A[s] = \Omega_+^{-1} A[s] \; ,$$

so $s \longrightarrow (1-z)^{-1}$ defines an isomorphism of rings with involution

$$\Omega_+^{-1} A[s] \xrightarrow{\simeq} \Pi^{-1} \Lambda \; .$$

(ii) The isomorphism $L_h^n(A[s], \epsilon) \cong L_h^n(\Lambda, \epsilon)$ was obtained in 25.25 (i). Since Π is coprime to $(1-z)$ the inclusion $A[z, z^{-1}] \longrightarrow \Lambda$ induces an equivalence of exact categories with involution

$$\mathbb{H}(A[z, z^{-1}], \Pi) \cong \mathbb{H}(\Lambda, \Pi)$$

and hence an isomorphism of L-groups

$$L_h^n(A[z, z^{-1}], \Pi, \epsilon) \cong L_h^n(\Lambda, \Pi, \epsilon) \; .$$

(iii) For any A-module chain map $f : C \longrightarrow C$ the $\mathbb{Z}[\mathbb{Z}_2]$-module chain map of 31.4 is such that

$$\begin{aligned}\Gamma_f^{iso} &= f\otimes 1 - 1\otimes(1-f)\\ &= f\otimes f-(1-f)\otimes(1-f) \;=\; f^{\%}-(1-f)^{\%}\\ &: (C\otimes_A C,T_\epsilon)\longrightarrow(C\otimes_A C,T_\epsilon)\ .\end{aligned}$$

The ϵ-symmetric Q-groups of (C,f) thus fit into an exact sequence

$$\begin{aligned}\dots\longrightarrow Q^{n+1}(C,\epsilon)\longrightarrow Q^{n+1}_{iso}(C,f,\epsilon)&\\ \longrightarrow Q^n(C,\epsilon)\xrightarrow{f^{\%}-(1-f)^{\%}} Q^n(C,\epsilon)\longrightarrow\dots\ ,&\end{aligned}$$

and an isometric structure $(\delta\phi,\phi)\in Q^n_{iso}(C,f,\epsilon)$ is the same as a fundamental cobordism

$$\Gamma \;=\; ((f\ 1-f):C\oplus C\longrightarrow C,(\delta\phi,\phi\oplus-\phi)\in Q^{n+1}((f\ 1-f),\epsilon))\ .$$

Every A-contractible finite f.g. free $A[z,z^{-1}]$-module chain complex C is chain equivalent to $\mathcal{C}(1-f+zf:P[z,z^{-1}]\longrightarrow P[z,z^{-1}])$ for some endomorphism $f:P\longrightarrow P$ of a finite f.g. projective A-module chain complex P, so that the morphism

$$LIso_p^{n-2}(A,\epsilon)\longrightarrow L_h^n(A[z,z^{-1}],\Pi,\epsilon)\ ;\ (C,f,\delta\phi,\phi)\longrightarrow U(\Gamma)$$

is an isomorphism.
The isomorphism of (i) induces isomorphisms in L-theory which respect the direct sum splittings of 31.13 and

$$\begin{aligned}L_h^n(\Omega_+^{-1}A[s],\epsilon) &= L_h^n(\Pi^{-1}\Lambda,\epsilon)\\ &= L_h^n(\Lambda,\epsilon)\oplus L_h^n(A[z,z^{-1}],\Pi,\epsilon)\\ &= L_h^n(A[s],\epsilon)\oplus LIso_p^{n-2}(A,\epsilon)\ .\end{aligned}$$

□

Proposition 31.16 *The fibred automorphism L-groups of 28.35 are related to the fibred isometric L-groups by isomorphisms*

$$LAut^n_{fib,p}(A,\epsilon)\xrightarrow{\simeq} LIso^n_{fib,p}(A,\epsilon)\ ;\ (C,\phi,h,\chi)\longrightarrow(C,(1-h)^{-1},\delta\phi,\phi)$$

and a commutative braid of exact sequences

$$\begin{array}{ccccc} L_h^n(A[z,z^{-1}],\epsilon) & & L_h^n(\Pi^{-1}A[z,z^{-1}],\epsilon) & & L_h^n(A[z,z^{-1}],\Pi,\Omega_{fib},\epsilon)\\ & \searrow \quad \nearrow & & \searrow \quad \nearrow & \\ & L_h^n(\Omega_{fib}^{-1}A[z,z^{-1}],\epsilon) & & LIso_p^{n-2}(A,\epsilon) & \\ & \nearrow \quad \searrow & & \nearrow \quad \searrow & \\ L_h^{n+1}(A[z,z^{-1}],\Pi,\Omega_{fib},\epsilon) & & LAut^{n-2}_{fib,p}(A,\epsilon) & & L_h^{n-1}(A[z,z^{-1}],\epsilon)\end{array}$$

with

$$L\mathrm{Aut}^n_{fib,p}(A,\epsilon) \longrightarrow L\mathrm{Iso}^n_p(A,\epsilon)\ ;\ (C,\phi,h,\chi) \longrightarrow (C,(1-h)^{-1},\delta\phi,\phi)$$

and $L^{n+2}_h(A[z,z^{-1}],\Pi,\Omega_{fib},\epsilon)$ *the cobordism group of finitely dominated* n*-dimensional* ϵ*-symmetric complexes* (C,ϕ) *over* A *with an isometric structure* $(f,\delta\phi)$ *such that* $\partial f : \partial C \longrightarrow \partial C$ *is a fibred self chain equivalence, where*

$$\partial f = \begin{pmatrix} f & \delta\phi_0 \\ 0 & 1-f^* \end{pmatrix} : \partial C_r = C_{r+1} \oplus C^{n-r} \longrightarrow \partial C_r = C_{r+1} \oplus C^{n-r} .$$

The morphism

$$\begin{aligned} L^n_h(\Omega^{-1}_{fib}A[z,z^{-1}],\epsilon) &= L^n_h(A,\epsilon) \oplus \widetilde{L\mathrm{Aut}}^{n-2}_{fib,p}(A,\epsilon) \\ &\longrightarrow L^n_h(\Pi^{-1}A[z,z^{-1}],\epsilon) = L^n_h(A,\epsilon) \oplus \widetilde{L\mathrm{Iso}}^{n-2}_p(A,\epsilon) \end{aligned}$$

has components the identity $1 : L^n_h(A,\epsilon) \longrightarrow L^n_h(A,\epsilon)$ *and the morphism*

$$\widetilde{L\mathrm{Aut}}^{n-2}_{fib,p}(A,\epsilon) \longrightarrow \widetilde{L\mathrm{Iso}}^{n-2}_p(A,\epsilon)\ ;\ (C,\phi,h,\chi) \longrightarrow (C,\phi,(1-h)^{-1},\chi') .$$

Proof Combine the L-theory localization exact sequence of 31.15

$$\begin{aligned} \dots \longrightarrow L^n_h(A[z,z^{-1}],\epsilon) \longrightarrow L^n_h(\Pi^{-1}A[z,z^{-1}],\epsilon) \longrightarrow L\mathrm{Iso}^{n-2}_p(A,\epsilon) \\ \longrightarrow L^{n-1}_h(A[z,z^{-1}],\epsilon) \longrightarrow \dots \end{aligned}$$

and the L-theory localization exact sequence

$$\begin{aligned} \dots \longrightarrow L^n_h(A[z,z^{-1}],\epsilon) \longrightarrow L^n_h(\Omega^{-1}_{fib}A[z,z^{-1}],\epsilon) \longrightarrow L\mathrm{Aut}^{n-2}_{fib,p}(A,\epsilon) \\ \longrightarrow L^{n-1}_h(A[z,z^{-1}],\epsilon) \longrightarrow \dots . \end{aligned}$$

□

Example 31.17 If $A = F$ is a field with involution and

$$P = \{p(z) \in F[z,z^{-1}] \,|\, p(1) \in F^\bullet\}$$

then by 13.12

$$\begin{aligned} \Pi^{-1}F[z,z^{-1}] &= P^{-1}F[z,z^{-1}] \\ &= \Omega^{-1}_{fib}F[z,z^{-1}] = F[z,z^{-1}]_{(1-z)} , \end{aligned}$$

so that

$$L^*(F[z,z^{-1}],\Pi,\Omega_{fib},\epsilon) = 0$$

and the morphisms

$$L\mathrm{Aut}^n_{fib}(F,\epsilon) \longrightarrow L\mathrm{Iso}^n(F,\epsilon)\ ;\ (C,\phi,h,\chi) \longrightarrow (C,(1-h)^{-1},\delta\phi,\phi)$$

are isomorphisms. □

By analogy with the round L-groups (20.15, 20.16):

Definition 31.18 The *round ϵ-symmetric isometric L-groups* $LIso^n_r(A,\epsilon)$ $(n \geq 0)$ are the cobordism groups of finitely dominated n-dimensional ϵ-symmetric Poincaré complexes (C,ϕ) over A with an endometric structure $(f,\delta\phi) : C \longrightarrow C$ such that

$$[C,f] \;=\; 0 \in Iso_0(A) \ .$$

Similarly for the *round ϵ-quadratic isometric L-groups* $LIso^r_*(A,\epsilon)$. □

Proposition 31.19 *The projective and round ϵ-symmetric isometric L-groups are related by a Rothenberg-type exact sequence*

$$\dots \longrightarrow \widehat{H}^{n+1}(\mathbb{Z}_2; Iso_0(A)) \longrightarrow LIso^n_r(A,\epsilon) \longrightarrow LIso^n_p(A,\epsilon) \longrightarrow \widehat{H}^n(\mathbb{Z}_2; Iso_0(A)) \longrightarrow \dots$$

where

$$LIso^n_p(A,\epsilon) \longrightarrow \widehat{H}^n(\mathbb{Z}_2; Iso_0(A)) \ ; \ (C,f,\delta\phi,\phi) \longrightarrow [C,f] \ .$$

Similarly in the ϵ-quadratic case. □

In 39.12 it will be shown that $LIso^0(F)$ maps onto $\mathbb{Z}_2[H^0(\mathbb{Z}_2; \mathcal{M}_s(F))]$, for any field with involution F.

31B. The trace map χ_s

The identification given by 31.13

$$L^0_h(A[s],\Omega_+,\epsilon) \;=\; LIso^0_p(A,-\epsilon)$$

will now be made more explicit, extending to the L-theory of $A[s]$ $(\overline{s} = 1-s)$ the identification of exact categories given by 10.14

$$\mathbb{H}(A[s],\Omega_+) \;=\; End(A) \ .$$

A trace map $\chi_s : \Omega_+^{-1}A[s]/A[s] \longrightarrow A$ will be used to obtain a direct one-one correspondence between ϵ-symmetric linking forms over $(A[s],\Omega_+)$ and $(-\epsilon)$-symmetric forms over A with an isometric structure (by analogy with the automorphism case in Chap. 28E). In Chap. 32B a modified version of χ_s will be used to obtain a one-one correspondence between the S-equivalence classes of ϵ-symmetric Seifert forms over A and the isomorphism classes of $(-\epsilon)$-symmetric Blanchfield forms over $A[z,z^{-1}]$.

Every Fredholm matrix $\omega \in \Omega_+$ is invertible over $A((s^{-1}))$ (10.3), so that the inclusion $i : A[s] \longrightarrow A((s^{-1}))$ factors through a ring morphism

$$k : \Omega_+^{-1}A[s] \longrightarrow A((s^{-1}))$$

with

$$i : A[s] \longrightarrow \Omega_+^{-1}A[s] \xrightarrow{k} A((s^{-1})) .$$

By analogy with 28.25, 28.26, 28.27:

Definition 31.20 The *universal trace map* is the A-module morphism

$$\chi_s : \Omega_+^{-1}A[s]/A[s] \longrightarrow A ; [\omega] \longrightarrow a_{-1}$$

with a_{-1} the coefficient of s^{-1} in

$$k(\omega) = \sum_{j=-\infty}^{\infty} a_j s^j \in A((s^{-1})) . \qquad \square$$

Remark 31.21 For a field $A = F$ the universal trace map

$$\chi_s : \Omega_+^{-1}A[s]/A[s] = F(s) \longrightarrow F$$

is the trace function of Trotter [291], [292]. $\square$

Proposition 31.22 (i) *If*

$$p(s) = \sum_{j=0}^{d} a_j s^j , \quad q(s) = \sum_{k=0}^{d-1} b_k s^k \in A[s]$$

are such that the coefficients $a_j \in A$ are central and $a_d \in A^\bullet$ then

$$\chi_s\left(\frac{q(s)}{p(s)}\right) = \frac{b_{d-1}}{a_d} \in A .$$

(ii) *For any $\omega \in \Omega_+^{-1}A[s]/A[s]$*

$$\overline{\chi_s(\omega)} = -\chi_s(\overline{\omega}) \in \Omega_+^{-1}A[s]/A[s] .$$

(iii) *For any $(A[s], \Omega_+)$-module L (= f.g. projective A-module L with an endomorphism $s : L \longrightarrow L$) the universal trace map χ_s induces an $A[s]$-module isomorphism*

$$L\hat{} = \mathrm{Hom}_{A[s]}(L, \Omega_+^{-1}A[s]/A[s]) \xrightarrow{\simeq} L^* = \mathrm{Hom}_A(L, A) ;$$

$$f \longrightarrow \chi_s f .$$

The identification of 31.13

$$L^0(A[s], \Omega_+, \epsilon) = LIso^0(A, -\epsilon)$$

is induced by the one-one correspondence

$$\{\epsilon\text{-symmetric linking forms } (L,\lambda) \text{ over } (A[s],\Omega_+)\} \rightleftarrows$$
$$\{(-\epsilon)\text{-symmetric forms } (L,\phi) \text{ over } A$$
$$\text{with an isometric structure } f : L \longrightarrow L\}$$

with

$$f = s : L \longrightarrow L ,$$

$$\phi = \chi_s\lambda : L \times L \xrightarrow{\lambda} \Omega_+^{-1}A[s]/A[s] \xrightarrow{\chi_s} A .$$

Proof (i) The polynomial $p(s) \in A[s]$ is a 1×1 Fredholm matrix with

$$\frac{1}{p(s)} = (a_d)^{-1}s^{-d} + \text{lower powers of } s \in A((s^{-1})) ,$$

so that

$$k\left(\frac{q(s)}{p(s)}\right)_{-1} = \frac{b_{d-1}}{a_d} \in A .$$

(ii) A 1×1 Fredholm matrix in $A[s]$ is a polynomial

$$p(s) = \sum_{j=0}^{\infty} a_j s^j \in A[s]$$

with an inverse in $A((s^{-1}))$

$$\frac{1}{p(s)} = \sum_{j=-\infty}^{\infty} b_j s^j \in A((s^{-1})) .$$

The dual polynomial

$$\overline{p(s)} = \sum_{j=0}^{\infty} \bar{a}_j (1-s)^j \in A[s]$$

is again a 1×1 Fredholm matrix in $A[s]$, with inverse

$$\frac{1}{\overline{p(s)}} = \sum_{j=-\infty}^{\infty} \bar{b}_j (1-s)^j \in A((s^{-1}))$$

where

$$\begin{aligned}(1-s)^j &= (-s)^j(1-s^{-1})^j \\ &= (-)^j \sum_{i=-\infty}^{j} \binom{-1-i}{j-i} s^i \in A((s^{-1})) \quad (j \leq -1) .\end{aligned}$$

Thus for any element of the type

$$\omega = \frac{s^r}{p(s)} \in \Omega_+^{-1}A[s]/A[s] \quad (r \in \mathbb{Z})$$

it is the case that

$$\begin{aligned} \chi_s(\omega) &= b_{-r-1} \,, \\ \chi_s(\overline{\omega}) &= \chi_s\left(\frac{(1-s)^r}{\overline{p(s)}}\right) = -\overline{b}_{-r-1} = -\overline{\chi_s(\omega)} \in A \,. \end{aligned}$$

Similarly for more general numerators and denominators.
(iii) The $(A[s], \Omega_+)$-dual $L^\wedge$ of L has the universal property that for any 1-dimensional f.g. projective $A[s]$-module resolution of L

$$0 \longrightarrow P_1 \xrightarrow{d} P_0 \longrightarrow L \longrightarrow 0$$

there is a dual resolution

$$0 \longrightarrow P_0^* \xrightarrow{d^*} P_1^* \longrightarrow L^\wedge \longrightarrow 0 \,.$$

For the standard resolution of L

$$d = s - f : P_1 = L[s] \longrightarrow P_0 = L[s]$$

the universal trace map χ_s identifies the dual resolution with the standard resolution of L^*

$$0 \longrightarrow L^*[s] \xrightarrow{1-s-f^*} L^*[s] \longrightarrow L^* \longrightarrow 0 \,. \qquad \square$$

Example 31.23 Suppose that A is commutative, so that (as in 10.10 (iii))

$$\Omega_+^{-1}A[s] = S^{-1}A[s]$$

with $S \subset A[s]$ the multiplicative subset consisting of all the monic polynomials

$$p(s) = \sum_{j=0}^{d} a_j s^j \in A[s] \quad (a_d = 1) \,.$$

Let L be an $(A[s], S)$-module, i.e. a f.g. projective A-module (also denoted by L) together with an endomorphism $f : L \longrightarrow L$. If $p(s) \in S$ is such that $p(f) = 0$ (e.g. $p(s) = \mathrm{ch}_s(L, f)$ if L is a f.g. free A-module) there is defined an $A[s]$-module isomorphism

$$\begin{aligned} \mathrm{Hom}_{A[s]}(L, A[s]/(p(s))) &\xrightarrow{\simeq} \mathrm{Hom}_{A[s]}(L, S^{-1}A[s]/A[s]) \,; \\ g &\longrightarrow (y \longrightarrow \frac{g(y)}{p(s)}) \,. \end{aligned}$$

Every $g \in \mathrm{Hom}_{A[s]}(L, A[s]/(p(s)))$ can be expressed as

$$g = \sum_{j=0}^{d-1} z^j g_j \; : \; L \longrightarrow A[s]/(p(s)) = \sum_{j=0}^{d-1} s^j A \; ,$$

with $g_j \in L^* = \operatorname{Hom}_A(L, A)$ and

$$\chi_s\left(\frac{g(y)}{p(s)}\right) = g_{d-1}(y) \in A \; ,$$

$$g_j(y) = g_{d-1}\left(\sum_{i=j+1}^{d} a_i f^{i-j-1}(y)\right) \in A$$

$$(y \in L, 0 \leq j \leq d-1) \; .$$

The isomorphism of 31.22 is given in this case by

$$L\hat{} = \operatorname{Hom}_{A[s]}(L, A[s]/(p(s))) \xrightarrow{\simeq} L^* = \operatorname{Hom}_A(L, A) \; ; \; g \longrightarrow g_{d-1} \; . \; \square$$

32. Seifert and Blanchfield complexes

Seifert complexes are abstractions of the duality properties of a Seifert surface of an n-knot, generalizing the form of Seifert [262]. Blanchfield complexes are abstractions of the duality properties of the infinite cyclic cover of the exterior of an n-knot, generalizing the form of Blanchfield [23]. These types of algebraic Poincaré complexes were already introduced in Ranicki [237]. The object of this chapter is to recall the definitions, and to make precise the relationship between the Blanchfield and Seifert complexes, using the L-theory of the Fredholm localizations

$$\Omega_+^{-1}A[s]\ ,\ \Omega_+^{-1}A[s,s^{-1},(1-s)^{-1}]\ =\ \Pi^{-1}A[z,z^{-1}]$$
$$(\overline{s}\ =\ 1-s\ ,\ s\ =\ (1-z)^{-1}\ ,\ \overline{z}=z^{-1})$$

for any ring with involution A. In particular, the cobordism groups of Seifert complexes over A are identified with the cobordism groups of Blanchfield complexes over $A[z,z^{-1}]$.

An n-dimensional ϵ-symmetric Seifert complex over A is an n-dimensional asymmetric complex (C,λ) such that the A-module chain map

$$(1+T_\epsilon)\lambda\ :\ C^{n-*}\longrightarrow C$$

is a chain equivalence. An ϵ-symmetric Seifert complex $(C,\widehat{\psi})$ is essentially the same as an ϵ-symmetric Poincaré complex with an isometric structure. The ϵ-ultraquadratic L-groups $\widehat{L}_*$ of [237] are the cobordism groups of ϵ-symmetric Seifert complexes over A are just the isometric L-groups of Chap. 31

$$\widehat{L}_*(A,\epsilon)\ =\ LIso^*(A,\epsilon)\ .$$

See Chap. 33 for the identification (already obtained in [237]) of the high-dimensional knot cobordism groups C_* and the isometric L-groups of $\mathbb{Z}$: a choice of Seifert surface $F^n\subset S^{n+1}$ for an $(n-1)$-knot $k:S^{n-1}\subset S^{n+1}$ determines an n-dimensional Seifert complex over $\mathbb{Z}$

$$\sigma^*(k,F)\ =\ (\dot{C}(F),\lambda)$$

and the function

$$C_{n-1}\longrightarrow LIso^n(\mathbb{Z})\ ;\ k\longrightarrow\sigma^*(k,F)$$

is an isomorphism for $n\geq 4$.

An n-dimensional ϵ-symmetric Seifert complex $(C,\widehat{\psi})$ over A is also essentially the same as an $\Omega_+^{-1}A[s]$-contractible $(n+1)$-dimensional ϵ-symmetric Poincaré complex over $A[s]$, with

$$s = f = \widehat{\psi}((1+T_\epsilon)\widehat{\psi})^{-1} : C \longrightarrow C ,$$

so that there is also an identification

$$L\mathrm{Iso}^n(A,\epsilon) = L^n(A[s],\Omega_+,-\epsilon) .$$

There are also identifications with the asymmetric L-groups $L\mathrm{Asy}^*(A,\Pi,\epsilon z)$ of Chap. 26

$$L\mathrm{Iso}^n(A,\epsilon) = L\mathrm{Asy}^n(A,\Pi,\epsilon z) ,$$

with Π the set of A-invertible matrices in $A[z,z^{-1}]$ ($\overline{z} = z^{-1}$). By definition, an n-dimensional ϵ-symmetric Seifert complex (C,λ) is fibred if $\lambda : C^{n-*} \longrightarrow C$ is a chain equivalence; fibred Seifert complexes correspond to fibred knots. The cobordism groups $\widehat{L}_*^{fib}(A,\epsilon)$ of fibred Seifert complexes are identified with the fibred automorphism L-groups $L\mathrm{Aut}^*_{fib}(A,\epsilon)$ of Chap. 28

$$\widehat{L}_*^{fib}(A,\epsilon) = L\mathrm{Iso}^*_{fib}(A,\epsilon) = L\mathrm{Aut}^*_{fib}(A,\epsilon) .$$

An ϵ-symmetric Blanchfield complex over $A[z,z^{-1}]$ is an ϵ-symmetric Poincaré complex which is A-contractible, or equivalently $\Pi^{-1}A[z,z^{-1}]$-contractible. A Blanchfield complex can also be regarded as an ϵ-symmetric Poincaré complex over $A[s,s^{-1},(1-s)^{-1}]$ which is $\Omega_+^{-1}A[s,s^{-1},(1-s)^{-1}]$-contractible. The cobordism groups of Blanchfield complexes are the torsion L-groups

$$L^*(A[z,z^{-1}],\Pi,\epsilon) = L^*(A[s,s^{-1},(1-s)^{-1}],\Omega_+,\epsilon) .$$

Following [237], it will be shown in Chap. 33 that the canonical infinite cyclic cover $\overline{X}$ of the exterior X of an $(n-1)$-knot $k : S^{n-1} \subset S^{n+1}$ determines an $(n+1)$-dimensional Blanchfield complex over $\mathbb{Z}[z,z^{-1}]$

$$\sigma^*(k) = (\dot{C}(\overline{X}),\psi)$$

and that for $n \geq 4$ the function

$$C_{n-1} \longrightarrow L^{n+2}(\mathbb{Z}[z,z^{-1}],P) \; ; \; k \longrightarrow \sigma^*(k)$$

is an isomorphism, with $P = \{p(z) \in \mathbb{Z}[z,z^{-1}] \,|\, p(1) = 1 \in \mathbb{Z}\}$. The inclusion $A[s] \longrightarrow A[s,s^{-1},(1-s)^{-1}]$ induces an isomorphism of L-groups sending Seifert complexes to Blanchfield complexes in Ω_+-torsion L-theory, so that

$$\begin{aligned} L^n(A[s],\Omega_+,\epsilon) &= L\mathrm{Iso}^n(A,-\epsilon) \\ &= L\mathrm{Asy}^n(A,\Pi,-\epsilon z) \\ &= L^n(A[z,z^{-1}],\Pi,\epsilon) \\ &= L^n(A[z,z^{-1},(1-z)^{-1}],\Pi,\epsilon) \\ &= L^n(A[s,s^{-1},(1-s)^{-1}],\Omega_+,\epsilon) \end{aligned}$$

(ignoring the decorations here). See Chap. 33 for a more detailed account of the one-one correspondence between the cobordism groups of Seifert and Blanchfield complexes over $A = \mathbb{Z}$, in the case when these are the high-dimensional knot cobordism groups C_*.

32A. Seifert complexes

Definition 32.1 (Ranicki [237, Chap. 7.8])
(i) An n-dimensional *ϵ-ultraquadratic complex* $(C, \widehat{\psi})$ is a finitely dominated n-dimensional A-module chain complex C together with a chain map over A

$$\widehat{\psi} \,:\, C^{n-*} \longrightarrow C \,.$$

Such a complex is *Poincaré* if the A-module chain map

$$(1 + T_\epsilon)\widehat{\psi} \,:\, C^{n-*} \longrightarrow C$$

is a chain equivalence. An *ϵ-symmetric Seifert complex* is an ϵ-ultraquadratic Poincaré complex.
(ii) An *ϵ-symmetric Seifert form* (L, λ) over A is a f.g. projective A-module L together with an A-module morphism $\lambda : L \longrightarrow L^*$ such that $\lambda + \epsilon\lambda^* : L \longrightarrow L^*$ is an isomorphism.
(iii) The *projective (resp. free) ϵ-ultraquadratic L-group* $\widehat{L}^p_n(A, \epsilon)$ (resp. $\widehat{L}^h_n(A, \epsilon)$) is the cobordism group of finitely dominated (resp. homotopy finite) n-dimensional ϵ-ultraquadratic Poincaré complexes over A. □

Given an n-dimensional ϵ-ultraquadratic (Poincaré) complex $(C, \widehat{\psi})$ define an n-dimensional ϵ-quadratic (Poincaré) complex (C, ψ) by

$$\psi_s \;=\; \begin{cases} \widehat{\psi}_0 \\ 0 \end{cases} :\; C^{n-r-s} \longrightarrow C_r \;\text{ if } \begin{cases} s = 0 \\ s \geq 1 \end{cases},$$

with

$$(1 + T_\epsilon)\psi_0 \;=\; (1 + T_\epsilon)\widehat{\psi} \,:\, C^{n-*} \longrightarrow C \,.$$

Proposition 32.2 (Ranicki [237, Chap. 7.8])
(i) *The homotopy equivalence classes of $2i$-dimensional ϵ-symmetric Seifert complexes $(C, \widehat{\psi})$ over A with $H_r(C) = 0$ for $r \neq i$ are in one-one correspondence with the isomorphism classes of $(-)^i\epsilon$-symmetric Seifert forms (L, λ) over A, with*

$$\lambda : L = H^i(C) \xrightarrow{\widehat{\psi}} H_i(C) = \mathrm{Hom}_A(H^i(C), A) = L^* .$$

(ii) *The ϵ-ultraquadratic L-groups $\widehat{L}^q_*(A,\epsilon)$ for $q = p$ (resp. h) are 4-periodic*

$$\widehat{L}^q_*(A,\epsilon) = \widehat{L}^q_{*+4}(A,\epsilon) ,$$

with $\widehat{L}^q_{2i}(A,\epsilon)$ the Witt group of f.g. projective (resp. f.g. free) $(-)^i\epsilon$-symmetric Seifert forms over A.
(iii) *The forgetful map*

$$\widehat{L}^q_n(A,\epsilon) \longrightarrow L^q_n(A,\epsilon) \; ; \; (C,\widehat{\psi}) \longrightarrow (C,\psi)$$

is onto for n even, and one-one for n odd.
(iv) *The free and projective $\widehat{L}$-groups are related by a Rothenberg-type exact sequence*

$$\dots \longrightarrow \widehat{L}^h_n(A,\epsilon) \longrightarrow \widehat{L}^p_n(A,\epsilon) \longrightarrow \widehat{H}^n(\mathbb{Z}_2;\widetilde{K}_0(A)) \longrightarrow \widehat{L}^h_{n-1}(A) \longrightarrow \dots .$$

□

For $\epsilon = 1$ write

$$\text{1-ultraquadratic} = \text{ultraquadratic} \;,\; \widehat{L}^q_*(A,1) = \widehat{L}^q_*(A) .$$

Proposition 32.3 (i) *The refinements of an n-dimensional ϵ-symmetric Poincaré complex (C,ϕ) over A to an n-dimensional ϵ-ultraquadratic Poincaré (= Seifert) complex $(C,\widehat{\psi})$ are in one-one correspondence with the isometric structures $(f,\delta\phi)$ for (C,ϕ).*
(ii) *The ϵ-ultraquadratic L-groups are isomorphic to the isometric L-groups of Chap. 31, with isomorphisms*

$$\widehat{L}^q_n(A,\epsilon) \xrightarrow{\simeq} LIso^n_q(A,\epsilon) \; ; \; (C,f,\delta\phi,\phi) \longrightarrow (C,f\phi_0) \;\; (n \geq 0, q = h,p) .$$

Proof (i) Given an n-dimensional ϵ-ultraquadratic Poincaré complex $(C,\widehat{\psi})$ define an isometric structure $(f,\delta\phi)$ for the associated n-dimensional ϵ-symmetric Poincaré complex $(C,(1+T_\epsilon)\widehat{\psi})$ by

$$f = \widehat{\psi}((1+T_\epsilon)\widehat{\psi})^{-1} : C \longrightarrow C \;,\; \delta\phi = 0 ,$$

with

$$f(1+T_\epsilon)\widehat{\psi} \simeq (1+T_\epsilon)\widehat{\psi}(1-f^*) \simeq \widehat{\psi} : C^{n-*} \longrightarrow C .$$

Conversely, given an an n-dimensional ϵ-symmetric Poincaré complex (C,ϕ) and an isometric structure $(f,\delta\phi)$ define an ϵ-ultraquadratic Poincaré complex $(C,\widehat{\psi})$ such that $(1+T_\epsilon)\widehat{\psi}=\phi\in Q^n(C,\epsilon)$ by

$$\widehat{\psi} \;=\; f\phi_0 \;:\; C^{n-*} \longrightarrow C \;.$$

(ii) Immediate from (i). □

Remark 32.4 The Witt group $C^\epsilon(A)$ of nonsingular ϵ-symmetric forms over A with an isometric structure considered by Stoltzfus [277] is given by 32.3 (ii) to be

$$C^\epsilon(A) \;=\; \widehat{L}_0(A,\epsilon) \;=\; LIso^0_h(A,\epsilon)\;. \qquad \square$$

32B. Blanchfield complexes

By 10.18 the localization $\Pi^{-1}A[z,z^{-1}]$ is such that the following conditions on a finite f.g. free $A[z,z^{-1}]$-module chain complex C are equivalent:

(i) C is A-contractible via $A[z,z^{-1}]\longrightarrow A;\, z\longrightarrow 1$,
(ii) $\zeta-1: C\longrightarrow C$ is a homology equivalence,
(iii) the induced $\Pi^{-1}A[z,z^{-1}]$-module chain complex $\Pi^{-1}C$ is contractible.

Given an involution on A extend the involution to $A[z,z^{-1}]$, $\Pi^{-1}A[z,z^{-1}]$ by $\overline{z}=z^{-1}$, as in Chap. 24.

Definition 32.5 (i) An *ϵ-symmetric Blanchfield complex* (C,ϕ) is an A-contractible f.g. free ϵ-symmetric Poincaré complex over $A[z,z^{-1}]$.
(ii) An *ϵ-symmetric Blanchfield form* (M,μ) over $A[z,z^{-1}]$ is a nonsingular ϵ-symmetric linking form over $(A[z,z^{-1}],\Pi)$, that is an $(A[z,z^{-1}],\Pi)$-module M with a nonsingular ϵ-symmetric pairing

$$\mu \;:\; M\times M \longrightarrow \Pi^{-1}A[z,z^{-1}]/A[z,z^{-1}]\;. \qquad \square$$

For $\epsilon=1$ write

$$L_*(A[z,z^{-1}],\Pi,1) \;=\; L_*(A[z,z^{-1}],\Pi)\;.$$

Proposition 32.6 (i) *The homotopy equivalence classes of* $(2i+1)$*-dimensional ϵ-symmetric Blanchfield complexes* (C,ϕ) *over* A *with* $H_r(C)=0$ *for* $r\neq i$ *are in one-one correspondence with the isomorphism classes of the* $(-)^{i+1}\epsilon$*-symmetric Blanchfield forms* (M,μ) *over* $A[z,z^{-1}]$, *with*

$$\begin{aligned}\mu \;:\; M \;&=\; H^{i+1}(C) \xrightarrow{\phi_0}\\ H_i(C) \;&=\; \mathrm{Hom}_{A[z,z^{-1}]}(H^{i+1}(C),\Pi^{-1}A[z,z^{-1}]/A[z,z^{-1}]) \;=\; M\widehat{\ }\,.\end{aligned}$$

(ii) *A f.g. free ϵ-symmetric Poincaré complex (C,ϕ) over $A[z,z^{-1}]$ is a Blanchfield complex if and only if (C,ϕ) is $\Pi^{-1}A[z,z^{-1}]$-contractible.*
(iii) *The cobordism group of n-dimensional ϵ-symmetric Blanchfield complexes over $A[z,z^{-1}]$ is the $(n+1)$-dimensional $(A[z,z^{-1}],\Pi)$-torsion ϵ-symmetric L-group $L_h^{n+1}(A[z,z^{-1}],\Pi,\epsilon)$ in the L-theory localization exact sequence*

$$\dots \longrightarrow L_h^{n+1}(A[z,z^{-1}],\epsilon) \longrightarrow L_h^{n+1}(\Pi^{-1}A[z,z^{-1}],\epsilon) \longrightarrow L_h^{n+1}(A[z,z^{-1}],\Pi,\epsilon) \longrightarrow L_h^{n}(A[z,z^{-1}],\epsilon) \longrightarrow \dots .$$

(iv) *The groups $L_h^*(A[z,z^{-1}],\Pi,\epsilon)$ are 4-periodic*

$$L_h^n(A[z,z^{-1}],\Pi,\epsilon) = L_h^{n+2}(A[z,z^{-1}],\Pi,-\epsilon) = L_h^{n+4}(A[z,z^{-1}],\Pi,\epsilon) ,$$

with $L_h^{2i}(A[z,z^{-1}],\Pi,\epsilon)$ the Witt group of $(-)^i\epsilon$-symmetric Blanchfield forms over $A[z,z^{-1}]$.
(v) *Up to natural isomorphism*

$$\mathrm{LIso}_p^n(A,\epsilon) = L_h^n(A[z,z^{-1}],\Pi,\epsilon z) = L_h^n(A[z,z^{-1}],\Pi,-\epsilon) .$$

(vi) *The ϵ-symmetrization maps*

$$1+T_\epsilon : L_n(A[z,z^{-1}],\Pi,\epsilon) \longrightarrow L^n(A[z,z^{-1}],\Pi,\epsilon)$$

are isomorphisms.
Proof (i)+(ii)+(iii)+(iv) This is a special case of 25.4 (ii).
(v) See the proof of 32.24 below.
(vi) Given an A-contractible f.g. free $A[z,z^{-1}]$-module chain complex E let

$$H_n = H_n(\mathrm{Hom}_{A[z,z^{-1}]}(E^*,E)) .$$

By 25.11 there are natural identifications

$$\begin{aligned} Q^n(E,\epsilon) &= \ker(1-T_\epsilon : H_n \longrightarrow H_n) \\ &= \mathrm{coker}(1-T_\epsilon : H_n \longrightarrow H_n) = Q_n(E,\epsilon) , \end{aligned}$$

noting that $t = 1-z \in A[z,z^{-1}]$ is coprime to Π with

$$t\bar{t} = (1-z)(1-z^{-1}) = t+\bar{t} \in A[z,z^{-1}] .$$

In particular, the ϵ-symmetrization maps of Q-groups are isomorphisms

$$1+T_\epsilon : Q_n(E,\epsilon) \longrightarrow Q^n(E,\epsilon) ,$$

and hence so are the ϵ-symmetrization maps of L-groups. □

The following construction is an algebraic analogue of the procedure for obtaining the infinite cyclic cover of an n-knot exterior from a Seifert surface:

Definition 32.7 (Ranicki [237, p. 820])
(i) The *covering* of a finitely dominated n-dimensional ultraquadratic complex $(C,\widehat{\psi})$ over A is the finitely dominated $(n+1)$-dimensional ϵ-symmetric complex over $A[z,z^{-1}]$

$$\beta(C,\widehat{\psi}) \;=\; (E,\theta)$$

defined by

$$\begin{aligned}
E &= \mathcal{C}(T_\epsilon\widehat{\psi}+z\widehat{\psi}: C^{n-*}[z,z^{-1}]\longrightarrow C[z,z^{-1}])\ ,\\
\theta_0 &= \begin{pmatrix} 0 & (-)^{r(n-r+1)}\epsilon(1-z^{-1}) \\ 1-z & 0 \end{pmatrix} :\\
&\qquad E^{n-r+1} \;=\; C^{n-r+1}[z,z^{-1}]\oplus C_r[z,z^{-1}]\\
&\qquad\qquad \longrightarrow E_r = C_r[z,z^{-1}]\oplus C^{n-r+1}[z,z^{-1}]\ ,\\
\theta_s &= 0 \ \ (s\geq 1)
\end{aligned}$$

with projective class

$$[E] \;=\; [C]+(-)^{n+1}[C]^* \in \mathrm{im}(K_0(A)\longrightarrow K_0(A[z,z^{-1}]))\ .$$

(ii) The *covering* of an ϵ-symmetric Seifert form (L,λ) over A is the $(-\epsilon)$-symmetric Blanchfield form over $A[z,z^{-1}]$

$$\beta(L,\lambda) \;=\; (M,\mu)$$

with

$$\begin{aligned}
M &= \mathrm{coker}(z\lambda+\epsilon\lambda^* : L[z,z^{-1}]\longrightarrow L^*[z,z^{-1}])\ ,\\
\mu &: M\times M \longrightarrow \Pi^{-1}A[z,z^{-1}]/A[z,z^{-1}]\ ;\\
&\quad (x,y)\longrightarrow (1-z)(\lambda+\epsilon\lambda^*)(x,(z\lambda+\epsilon\lambda^*)^{-1}(y))\ \ (x,y\in L^*[z,z^{-1}])\ .
\end{aligned}$$

□

In fact, 32.7 (ii) is just the highly-connected special case of 32.7 (i).

Proposition 32.8 *Let $(C,\widehat{\psi})$ be a finitely dominated n-dimensional ϵ-ultraquadratic complex over A, with covering $\beta(C,\widehat{\psi})=(E,\theta)$.*
(i) *$(C,\widehat{\psi})$ is a Seifert complex if and only if (E,θ) is a Blanchfield complex, i.e. $(C,\widehat{\psi})$ is Poincaré if and only if E is A-contractible.*
(ii) *If (E,θ) is A-contractible then it is Poincaré, and is homotopy equivalent to the union $U(\Gamma)$ of the $(n+1)$-dimensional ϵ-isometric Poincaré cobordism*

$$\Gamma \;=\; ((g\ h): C\oplus C\longrightarrow C,(\delta\phi,\phi\oplus-\phi))$$

determined by the isometric structure $(f,\delta\phi)$ on $(C,\phi=(1+T_\epsilon)\widehat{\psi})$ with

$$\begin{aligned} g &= 1 - f = T_\epsilon\widehat{\psi}((1+T_\epsilon)\widehat{\psi})^{-1} : C \longrightarrow C , \\ h &= -f = -\widehat{\psi}((1+T_\epsilon)\widehat{\psi})^{-1} : C \longrightarrow C . \end{aligned}$$

The chain complex E has a canonical (round) finite structure, with respect to which

$$\begin{aligned} \tau(E,\theta) &= \tau(-z : C[z,z^{-1}]\longrightarrow C[z,z^{-1}]) = (0,[C],0,0) \\ &\in Wh_1(A[z,z^{-1}]) = Wh_1(A)\oplus \widetilde{K}_0(A)\oplus \widetilde{\mathrm{Nil}}_0(A)\oplus \widetilde{\mathrm{Nil}}_0(A) . \end{aligned}$$

Proof (i) Immediate from the identity

$$H_*(A\otimes_{A[z,z^{-1}]} E) = H_{*+1}((1+T_\epsilon)\widehat{\psi} : C^{n-*}\longrightarrow C) .$$

(ii) By (i) there is defined a chain equivalence $(1+T_\epsilon)\widehat{\psi} : C^{n-*} \xrightarrow{\simeq} C$. The canonical round finite structure on E is defined using the canonical round finite structure on $\mathcal{C}(1-f+zf : C[z,z^{-1}]\longrightarrow C[z,z^{-1}])$ and the $A[z,z^{-1}]$-module chain equivalence

$$\begin{aligned} 1\oplus(1+T_\epsilon)\widehat{\psi} : E &= \mathcal{C}(T_\epsilon\widehat{\psi}+z\widehat{\psi} : C^{n-*}[z,z^{-1}]\longrightarrow C[z,z^{-1}]) \\ &\xrightarrow{\simeq} \mathcal{C}(g-zh : C[z,z^{-1}]\longrightarrow C[z,z^{-1}]) \\ &= \mathcal{C}(1-f+zf : C[z,z^{-1}]\longrightarrow C[z,z^{-1}]) . \end{aligned}$$ □

Definition 32.9 (i) The covering Blanchfield complex of a projective Seifert complex $(C,\widehat{\psi})$ will be written $\beta^h(C,\widehat{\psi})$, and called the *free covering* of $(C,\widehat{\psi})$.
(ii) A *projective Seifert complex* for an $(n+1)$-dimensional ϵ-symmetric Blanchfield complex (E,θ) over $A[z,z^{-1}]$ is a finitely dominated n-dimensional ϵ-symmetric Seifert complex $(C,\widehat{\psi})$ such that the free covering $\beta^h(C,\widehat{\psi})$ is homotopy equivalent to (E,θ). □

The following result is an algebraic analogue of the transversality construction of a Seifert surface for an $(n+1)$-knot from a fundamental domain of the infinite cyclic cover of the $(n+1)$-knot exterior:

Proposition 32.10 *Blanchfield complexes have projective Seifert complexes. More precisely, every A-contractible based f.g. free $(n+1)$-dimensional ϵ-symmetric Poincaré complex (E,θ) over $A[z,z^{-1}]$ is homotopy equivalent to the free covering $\beta^h(C,\widehat{\psi})$ of a finitely dominated n-dimensional ϵ-ultraquadratic Poincaré complex $(C,\widehat{\psi})$ over A with reduced projective class*

$$[C] = B\tau(E,\theta)\in \widetilde{K}_0(A)$$

the image of $\tau(E,\theta)\in Wh_1(A[z,z^{-1}])$ under the Bass–Heller–Swan projection $B : Wh_1(A[z,z^{-1}])\longrightarrow \widetilde{K}_0(A)$ (= reduced K_0-component of ∂_+ in 5.14 (i)). In particular, the case $n=0$ is that free Blanchfield forms admit projective Seifert forms.

Proof Let

$$\Gamma = ((g\ h) : C \oplus C \longrightarrow D, (\delta\phi, \phi \oplus -\phi))$$

be a finitely balanced $(n+1)$-dimensional symmetric Poincaré cobordism over A such that the union $U(\Gamma)$ is homotopy equivalent to (E, θ), as given by 24.2. Now

$$E \simeq \mathcal{C}(g - zh : C[z, z^{-1}] \longrightarrow D[z, z^{-1}]) ,$$

and

$$H_*(A \otimes_{A[z,z^{-1}]} E) = H_*(g - h : C \longrightarrow D) = 0 .$$

Thus $g - h : C \longrightarrow D$ is an A-module chain equivalence and Γ is isometric. The corresponding isometric structure $(f, \delta\phi)$ for (C, ϕ) with

$$f = -(g-h)^{-1}h : C \longrightarrow C$$

determines by 32.5 (i) a finitely dominated n-dimensional ϵ-symmetric Seifert complex $(C, \widehat{\psi})$ with

$$\widehat{\psi} = f(1+T_\epsilon)\psi_0 = -(g-h)^{-1}g(1+T_\epsilon)\psi_0 : C^{n-*} \longrightarrow C ,$$

and such that there are defined homotopy equivalences

$$\beta^h(C, \widehat{\psi}) \xrightarrow{\simeq} U(\Gamma) \xrightarrow{\simeq} (E, \theta) . \qquad \square$$

Proposition 32.11 (i) *The covering construction of 32.7 defines natural isomorphisms*

$$\beta^h : L\mathrm{Iso}^n_p(A, -\epsilon) \xrightarrow{\simeq} L^n_h(A[z, z^{-1}], \Pi, \epsilon) ; (C, \widehat{\psi}) \longrightarrow \beta^h(C, \widehat{\psi}) .$$

(ii) *The morphism of rings with involution*

$$A[s] \longrightarrow A[z, z^{-1}, (1-z)^{-1}] ; s \longrightarrow (1-z)^{-1}$$

induces isomorphisms in Tate $\mathbb{Z}_2$-cohomology

$$\widehat{H}^n(\mathbb{Z}_2; K_1(A[s], \Omega_+)) = \widehat{H}^n(\mathbb{Z}_2; \mathrm{Iso}_0(A))$$
$$\xrightarrow{\simeq} \widehat{H}^n(\mathbb{Z}_2; K_1(A[z, z^{-1}, (1-z)^{-1}], \Pi)) = \widehat{H}^n(\mathbb{Z}_2; K_1(A[z, z^{-1}], \Pi))$$

and in L-theory

$$L\mathrm{Iso}^n_p(A, -\epsilon) = L^n_h(A[s], \Omega_+, \epsilon) \xrightarrow{\simeq} L^n_h(A[z, z^{-1}, (1-z)^{-1}], \Pi, \epsilon) .$$

The L-theory isomorphism is the composite of the isomorphism β^h *of* (i) *and the isomorphism*

$$L^n_h(A[z, z^{-1}], \Pi, \epsilon) \xrightarrow{\simeq} L^n_h(A[z, z^{-1}, (1-z)^{-1}], \Pi, \epsilon)$$

induced by the inclusion $A[z, z^{-1}] \longrightarrow A[z, z^{-1}, (1-z)^{-1}]$.

Proof (i) The algebraic Poincaré transversality construction of 32.10 defines inverse maps $(\beta^h)^{-1} : (E,\theta) \longrightarrow (C,\widehat{\psi})$.
(ii) Since $s, 1-s \in \Omega_+$ it is possible to identify

$$\Omega_+^{-1}A[s] \;=\; \Omega_+^{-1}A[s,s^{-1},(1-s)^{-1}]\ ,$$

and hence

$$L_h^n(\Omega_+^{-1}A[s],\epsilon) \;=\; L_h^n(\Omega_+^{-1}A[s,s^{-1},(1-s)^{-1}],\epsilon)\ .$$

The inclusion $A[s] \longrightarrow A[s,s^{-1},(1-s)^{-1}]$ induces isomorphisms in Tate $\mathbb{Z}_2$-cohomology and L-theory

$$\widehat{H}^n(\mathbb{Z}_2;K_1(A[s],\Omega_+)) \xrightarrow{\simeq} \widehat{H}^n(\mathbb{Z}_2;K_1(A[s,s^{-1},(1-s)^{-1}],\Omega_+))\ ,$$

$$L_h^n(A[s],\epsilon) \xrightarrow{\simeq} L_h^n(A[s,s^{-1},(1-s)^{-1}],\epsilon)$$

by 25.14 (i). By the 5-lemma there are also induced isomorphisms

$$L_h^n(A[s],\Omega_+,\epsilon) \xrightarrow{\simeq} L_h^n(A[s,s^{-1},(1-s)^{-1}],\Omega_+,\epsilon)\ .$$

The isomorphism of rings with involution

$$A[s,s^{-1},(1-s)^{-1}] \xrightarrow{\simeq} A[z,z^{-1},(1-z)^{-1}]\ ;\ s \longrightarrow (1-z)^{-1}$$

sends Ω_+ to Π, so that there is induced an isomorphism of L-groups

$$L_h^n(A[s,s^{-1},(1-s)^{-1}],\Omega_+,\epsilon) \xrightarrow{\simeq} L_h^n(A[z,z^{-1},(1-z)^{-1}],\Pi,\epsilon)\ . \qquad \square$$

A Blanchfield complex (E,θ) admits many Seifert complexes $(C,\widehat{\psi})$.

Definition 32.12 Finitely dominated n-dimensional ϵ-symmetric Seifert complexes $(C,\widehat{\psi})$, $(C',\widehat{\psi}')$ are *S-equivalent* if the $(n+1)$-dimensional ϵ-symmetric Blanchfield complexes $\beta^h(C,\widehat{\psi})$, $\beta^h(C',\widehat{\psi}')$ over $A[z,z^{-1}]$ are homotopy equivalent. $\square$

The special case $n=0$ of 32.12 defines the S-equivalence relation on Seifert forms (= 0-dimensional Seifert complexes). See 33.11 below for the isotopy classification of simple odd-dimensional knots. by the S-equivalence classes of Seifert forms over $\mathbb{Z}$.

Example 32.13 (i) Homotopy equivalent Seifert complexes are S-equivalent.
(ii) An n-dimensional Seifert complex $(C,\widehat{\psi})$ is S-equivalent to the Seifert complex

$$(C',\widehat{\psi}') \;=\; (C\oplus D\oplus D^{n-*}, \begin{pmatrix} \widehat{\psi} & \beta & 0 \\ \alpha & \gamma & 1 \\ 0 & 0 & 0 \end{pmatrix})\ ,$$

for any finite f.g. free A-module chain complex D and any A-module chain maps

$$\alpha : C^{n-*} \longrightarrow D\ ,\ \beta : D^{n-*} \longrightarrow C\ ,\ \gamma : D^{n-*} \longrightarrow D\ . \qquad \square$$

Proposition 32.14 *The covering construction β^h defines a bijection between the S-equivalence classes of finitely dominated n-dimensional ϵ-symmetric Seifert complexes over A and the homotopy equivalence classes of free $(n+1)$-dimensional ϵ-symmetric Blanchfield complexes over $A[z,z^{-1}]$.* □

32C. Fibred Seifert and Blanchfield complexes

As their name indicates, fibred Seifert and Blanchfield complexes are the Seifert and Blanchfield complex analogues of fibred knots.

Definition 32.15 (i) A finitely dominated n-dimensional ϵ-symmetric Seifert complex $(C,\widehat{\psi})$ over A is *fibred* if $\widehat{\psi}: C^{n-*}\longrightarrow C$ is a chain equivalence, or equivalently if

$$f = \widehat{\psi}((1+T_\epsilon)\widehat{\psi})^{-1} : C \longrightarrow C$$

is a chain equivalence, in which case the *monodromy*

$$h = 1-f^{-1} = -(T_\epsilon\widehat{\psi})\widehat{\psi}^{-1} : C \longrightarrow C .$$

is a fibred chain equivalence, i.e. h and $h-1: C\longrightarrow C$ are both chain equivalences.

(ii) The *fibred ϵ-ultraquadratic L-group* $\widehat{L}^{q,fib}_n(A,\epsilon)$ $(n\geq 0)$ for $q=p$ (resp. h) is the cobordism group of finitely dominated (resp. f.g. free) n-dimensional ϵ-symmetric fibred Seifert complexes over A. □

Proposition 32.16 (i) *The homotopy equivalence classes of the following objects are in one-one correspondence:*

(a) *n-dimensional ϵ-symmetric fibred Seifert complexes over A,*
(b) *fibred self homotopy equivalences of finitely dominated n-dimensional ϵ-symmetric Poincaré complexes over A (28.35),*
(c) *n-dimensional ϵ-symmetric Poincaré complexes over A with an isometric structure (31.4).*

(ii) *The fibred ϵ-ultraquadratic L-groups are isomorphic to the fibred automorphism L-groups (28.35) and also to the fibred isometric L-groups (31.9), with isomorphisms*

$$\widehat{L}^{q,fib}_n(A,\epsilon) \xrightarrow{\simeq} LAut^{q,fib}_n(A,\epsilon) ;$$
$$(C,\widehat{\psi}) \longrightarrow (C,(1+T_\epsilon)\widehat{\psi},-\widehat{\psi}(T_\epsilon\widehat{\psi})^{-1}) ,$$
$$LAut^n_{q,fib}(A,\epsilon) \xrightarrow{\simeq} LIso^n_{q,fib}(A,\epsilon) ;$$
$$(C,\phi,h,\chi) \longrightarrow (C,(h-1)^{-1}\phi_0) \quad (q=s,h,p) .$$

Proof (i) (a) $\Longleftrightarrow$ (b) Given a finitely dominated n-dimensional ϵ-symmetric fibred Seifert complex $(C,\widehat{\psi})$ over A define a fibred self homotopy equivalence

$$h = -\widehat{\psi}(T_\epsilon\widehat{\psi})^{-1} : (C,(1+T_\epsilon)\widehat{\psi}) \longrightarrow (C,(1+T_\epsilon)\widehat{\psi})$$

of the finitely dominated n-dimensional ϵ-symmetric Poincaré complex $(C,(1+T_\epsilon)\widehat{\psi})$ over A.
Conversely, given a fibred self homotopy equivalence

$$(h,\chi) : (C,\phi) \longrightarrow (C,\phi)$$

of a finitely dominated n-dimensional ϵ-symmetric Poincaré complex (C,ϕ) over A the A-module chain equivalence

$$\widehat{\psi} = (h-1)^{-1}\phi_0 : C^{n-*} \longrightarrow C$$

defines a finitely dominated n-dimensional ϵ-symmetric Seifert complex $(C,\widehat{\psi})$ over A.
(a) $\Longleftrightarrow$ (c) This is just the fibred version of 32.3. An ϵ-symmetric Seifert complex $(C,\widehat{\psi})$ is fibred if and only if the corresponding isometric structure $(f,\delta\phi)$ for $(C,(1+T_\epsilon)\widehat{\psi})$ is fibred.
(ii) Immediate from (i). □

Definition 32.17 A nonsingular ϵ-symmetric Seifert form $(M,\widehat{\psi})$ over A is *fibred* if $\widehat{\psi} : M \longrightarrow M^*$ is an isomorphism, in which case the isometric structure is a fibred automorphism

$$f = (\widehat{\psi}+\epsilon\widehat{\psi}^*)^{-1}\widehat{\psi} : M \xrightarrow{\simeq} M . \qquad \square$$

Proposition 32.18 (i) *The $2i$-dimensional fibred Seifert complexes $(C,\widehat{\psi})$ over A with $C_r = 0$ for $r \neq i$ are in one-one correspondence with nonsingular $(-)^i$-symmetric fibred Seifert forms $(M,\widehat{\psi})$ over A, with $M = C^i$.*
(ii) *The nonsingular ϵ-symmetric fibred Seifert forms over A are in one-one correspondence with fibred automorphisms $h : (M,\phi) \longrightarrow (M,\phi)$ of nonsingular ϵ-symmetric forms (M,ϕ) over A.*
(iii) *The fibred automorphism L-group $LAut^{2i}_{fib}(A,\epsilon)$ is the Witt group of f.g. projective nonsingular fibred $(-)^i\epsilon$-symmetric Seifert forms over A.*
Proof (i) Immediate from the definitions.
(ii) The monodromy of a nonsingular ϵ-symmetric fibred Seifert form $(M,\widehat{\psi})$ defines a fibred automorphism of a nonsingular ϵ-symmetric form

$$h = \epsilon(\widehat{\psi}^*)^{-1}\widehat{\psi} : (M,\widehat{\psi}+\epsilon\widehat{\psi}^*) \xrightarrow{\simeq} (M,\widehat{\psi}+\epsilon\widehat{\psi}^*) .$$

Conversely, given a fibred automorphism of an ϵ-symmetric form

$$h : (M,\phi) \longrightarrow (M,\phi)$$

define a fibred ϵ-symmetric Seifert form $(M,\widehat{\psi})$ over A by

$$\widehat{\psi} = \phi(h-1)^{-1} : M \xrightarrow{\simeq} M^* ,$$

with monodromy h and isometric structure

$$f = \phi^{-1}\widehat{\psi} : M \xrightarrow{\simeq} M .$$

(iii) Immediate from (ii). □

And now for fibred Blanchfield complexes.

Definition 32.19 (i) A f.g. free $(n+1)$-dimensional ϵ-symmetric Blanchfield complex (E,θ) over $A[z,z^{-1}]$ is *fibred* if it is homotopy equivalent to the algebraic mapping torus $T(h,\chi)$ of a self homotopy equivalence $(h,\chi) : (C,\phi)\longrightarrow(C,\phi)$ of a f.g. projective n-dimensional ϵ-symmetric Poincaré complex (C,ϕ), which is necessarily fibred (= such that $h-1 : C\longrightarrow C$ is also a chain equivalence).
(ii) The *fibre* of a fibred $(n+1)$-dimensional ϵ-symmetric Blanchfield complex (E,θ) over $A[z,z^{-1}]$ is the finitely dominated n-dimensional ϵ-symmetric fibred Seifert complex $(E^!,\widehat{\psi})$ over A with $E^!$ the A-module chain complex obtained from E by restricting the $A[z,z^{-1}]$-action to A and

$$\widehat{\psi} = (1-\zeta)^{-1}\theta^!_0 : (E^!)^{n-*} \simeq (E^{n+1-*})^! \longrightarrow E^! . \qquad □$$

Let Ω_{fib} be the set defined in Chap. 13 of fibred Fredholm matrices in $A[z,z^{-1}]$. The following conditions on a f.g. free symmetric Poincaré complex (C,ϕ) over $A[z,z^{-1}]$ will be shown to be equivalent:

(i) (C,ϕ) is a fibred Blanchfield complex,
(ii) the $A[z,z^{-1}]$-module chain complex C is a band (= A-finitely dominated) and the A-module chain complex $A\otimes_{A[z,z^{-1}]}C$ is contractible,
(iii) the induced $\Omega_{fib}^{-1}A[z,z^{-1}]$-module chain complex $\Omega_{fib}^{-1}C$ is contractible.

Proposition 32.20 (i) *A f.g. free ϵ-symmetric Poincaré complex (C,ϕ) over $A[z,z^{-1}]$ is a fibred Blanchfield complex if and only if (C,ϕ) is $\Omega_{fib}^{-1}A[z,z^{-1}]$-contractible.*
(ii) *The homotopy equivalence classes of fibred n-dimensional ϵ-symmetric Blanchfield complexes over $A[z,z^{-1}]$ are in one-one correspondence with the homotopy equivalence classes of fibred self homotopy equivalences of finitely dominated $(n-1)$-dimensional ϵ-symmetric Poincaré complexes over A.*
(iii) *The cobordism group of fibred n-dimensional ϵ-symmetric Blanchfield complexes over $A[z,z^{-1}]$ is the $(n-1)$-dimensional projective fibred automorphism L-group*

$$L^{n+1}_h(A[z,z^{-1}],\Omega_{fib},\epsilon) = LAut^{n-1}_{fib,p}(A,\epsilon) ,$$

which is the torsion L-group in the localization exact sequence

$$\dots \longrightarrow L^{n+1}_h(A[z,z^{-1}],\epsilon) \longrightarrow L^{n+1}_h(\Omega_{fib}^{-1}A[z,z^{-1}],\epsilon)$$
$$\longrightarrow LAut^{n-1}_{fib,p}(A,\epsilon) \longrightarrow L^n_h(A[z,z^{-1}],\epsilon) \longrightarrow \dots .$$

The ϵ-symmetric L-groups of $\Omega_{fib}^{-1}A[z,z^{-1}]$ naturally split as

$$L_h^n(\Omega_{fib}^{-1}A[z,z^{-1}],\epsilon) = L_h^n(A,\epsilon)\oplus \widetilde{LAut}_{fib}^{n-2}(A,\epsilon)\ .$$

Proof (i) This is a special case of 25.5.
(ii) This is the fibred version of 32.11 (i).
(iii) The identification of the fibred automorphism L-groups with the torsion L-groups is the fibred analogue of 32.11 (ii). The inclusion $A\longrightarrow\Omega_{fib}^{-1}A[z,z^{-1}]$ is split by the projection $\Omega_{fib}^{-1}A[z,z^{-1}]\longrightarrow A$ given by 13.10, so that $L_h^n(\Omega_{fib}^{-1}A[z,z^{-1}],\epsilon)$ contains $L_h^n(A,\epsilon)$ as a direct summand. The identification of the exterior with $\widetilde{LAut}_{fib}^{n-2}(A,\epsilon)$ follows from (i) and

$$L_h^n(A[z,z^{-1}],\epsilon) = L_h^n(A,\epsilon)\oplus L_p^{n-1}(A,\epsilon)\ . \qquad \square$$

Example 32.21 If A is commutative then the multiplicative subsets

$$P = \{\sum_{j=0}^{d} a_j z^j \mid \sum_{j=0}^{d} a_j \in A^\bullet\}\ ,$$

$$P_{fib} = \{\sum_{j=0}^{d} a_j z^j \mid a_0, a_d, \sum_{j=0}^{d} a_j \in A^\bullet\} \subset A[z,z^{-1}]$$

are such that

$$\begin{aligned}\Pi^{-1}A[z,z^{-1}] &= P^{-1}A[z,z^{-1}]\ ,\\ \Omega_{fib}^{-1}A[z,z^{-1}] &= P_{fib}^{-1}A[z,z^{-1}]\ .\end{aligned} \qquad \square$$

Proposition 32.22 *For any field with involution F every Blanchfield complex over $F[z,z^{-1}]$ fibres, and*

$$L^{n+2}(F[z,z^{-1}],P,\epsilon) = L^{n+2}(F[z,z^{-1}],P_{fib},\epsilon) = LAut_{fib}^n(F,\epsilon)\ .$$

Proof Every Blanchfield complex (E,θ) over $F[z,z^{-1}]$ fibres by 32.20 (i), since

$$P^{-1}F[z,z^{-1}] = \widetilde{P}^{-1}F[z,z^{-1}]\subset F(z)\ .$$

By 32.20 (ii) the homotopy equivalence classes of $(n+1)$-dimensional ϵ-symmetric Blanchfield complexes over $F[z,z^{-1}]$ are in one-one correspondence with the homotopy equivalence classes of fibred self homotopy equivalences of finitely dominated n-dimensional ϵ-symmetric Poincaré complexes over F, and likewise for cobordism classes. $\square$

32D. Based Seifert and Blanchfield complexes

There is also a based version of Blanchfield complex theory.

Definition 32.23 A *simple Blanchfield complex* is an A-contractible based f.g. free ϵ-symmetric Poincaré complex over $A[z,z^{-1}]$ such that

$$\tau(C,\phi) \;=\; 0 \in Wh_1(A[z,z^{-1}]) \;\;,\;\; \tau(A\otimes_{A[z,z^{-1}]} C) \;=\; 0 \in Wh_1(A)\;. \;\square$$

Proposition 32.24 *The cobordism groups $L_s^n(A[z,z^{-1}],\Pi,\epsilon)$ of simple $(n-1)$-dimensional ϵ-symmetric Blanchfield complexes over $A[z,z^{-1}]$ are such that*

$$L\mathrm{Iso}_h^n(A,\epsilon) \;=\; L_h^n(\Pi^{-1}A[z,z^{-1}],\epsilon z) \;=\; L_s^n(A[z,z^{-1}],\Pi,-\epsilon)\;,$$

and fit into the simple ϵ-symmetric L-theory localization exact sequence

$$\begin{aligned}\dots \longrightarrow L_s^n(A[z,z^{-1}],\epsilon) \longrightarrow L_{s'}^n(\Pi^{-1}A[z,z^{-1}],\epsilon) \longrightarrow L_s^n(A[z,z^{-1}],\Pi,\epsilon)\\ \longrightarrow L_s^{n-1}(A[z,z^{-1}],\epsilon) \longrightarrow \dots\;,\end{aligned}$$

with

$$\begin{aligned}s' \;&=\; \ker(Wh_1(\Pi^{-1}A[z,z^{-1}])\longrightarrow Wh_1(A))\\ &=\; \widetilde{\mathrm{End}}_0(A) \subseteq Wh_1(\Pi^{-1}A[z,z^{-1}])\;.\end{aligned}$$

Proof By 26.11 (i) there are defined isomorphisms

$$\begin{aligned}L\mathrm{Asy}_h^n(A,\Pi,\epsilon z) \;&=\; L\mathrm{Iso}_h^n(A,\epsilon) \xrightarrow{\;\simeq\;} L_h^n(\Pi^{-1}A[z,z^{-1}],\epsilon z)\;;\\ &\qquad (C,\lambda) \longrightarrow (\Pi^{-1}C[z,z^{-1}],(1+T_{\epsilon z})\lambda)\;,\\ L\mathrm{Asy}_p^n(A,\Pi,\epsilon z) \;&=\; L\mathrm{Iso}_p^n(A,\epsilon) \xrightarrow{\;\simeq\;} L_h^n(A[z,z^{-1}],\Pi,\epsilon z)\;;\\ &\qquad (C,\lambda) \longrightarrow \partial(C[z,z^{-1}],(1+T_{\epsilon z})\lambda)\;.\end{aligned}$$

Now Π is coprime to $(1-z)$, and the covering isomorphism of 32.12 can be identified with one of the isomorphisms given by 26.11 (v)

$$\begin{aligned}\beta^s \;:\; L\mathrm{Iso}_h^n(A,\epsilon) &\xrightarrow{\;\simeq\;} L_s^n(A[z,z^{-1}],\Pi,-\epsilon)\;;\\ (C,\lambda) &\longrightarrow \partial(C[z,z^{-1}],(1-z^{-1})(1+T_{\epsilon z})\lambda)\;,\\ \beta^h \;:\; L\mathrm{Iso}_p^n(A,\epsilon) &\xrightarrow{\;\simeq\;} L_h^n(A[z,z^{-1}],\Pi,-\epsilon)\;;\\ (C,\lambda) &\longrightarrow \partial(C[z,z^{-1}],(1-z^{-1})(1+T_{\epsilon z})\lambda)\;. \qquad\square\end{aligned}$$

Definition 32.25 A simple ϵ-symmetric Blanchfield complex (E,θ) over $A[z,z^{-1}]$ *fibres* if it is homotopy equivalent to the algebraic mapping torus $T(h,\chi)$ of a simple self homotopy equivalence $(h,\chi):(C,\phi)\longrightarrow(C,\phi)$ of a based f.g. free n-dimensional ϵ-symmetric Poincaré complex (C,ϕ) such that $h-1:C\longrightarrow C$ is also a chain equivalence. $\square$

Remark 32.26 The ϵ-symmetric Poincaré complex (C,ϕ) in 32.25 need not be simple: the torsion is given by

$$\tau(C,\phi) \;=\; B_2\tau_2(E,\theta) \in Wh_1(A)$$

with $\tau_2(E,\theta) \in Wh_2(A[z,z^{-1}])$ the Wh_2-invariant determined by the failure of symmetry in a choice of trivialization of $\tau(E,\theta) \in Wh_1(A[z,z^{-1}])$ and

$$B_2 \;:\; Wh_2(A[z,z^{-1}]) \longrightarrow Wh_1(A)$$

the Wh_2-analogue of $B : Wh_1(A[z,z^{-1}]) \longrightarrow \widetilde{K}_0(A)$. □

The fibering obstructions $\Phi^{\pm}(E)$ of an A-finitely dominated based f.g. free $(n+1)$-dimensional ϵ-symmetric Blanchfield complex (E,θ) over $A[z,z^{-1}]$ are such that

$$\tau(E,\theta) \;=\; \Phi^+(E) + (-)^{n+1}\Phi^-(E)^* \in Wh_1(A[z,z^{-1}]) \;.$$

Thus for simple (E,θ) the two fibering obstructions determine each other by the duality relation

$$\Phi^+(E) \;=\; (-)^n\Phi^-(E)^* \in Wh_1(A[z,z^{-1}]) \;,$$

and E fibres (= simple chain equivalent to the algebraic mapping torus $T^+(h) = \mathcal{C}(1 - zh : C[z,z^{-1}] \longrightarrow C[z,z^{-1}])$ of a simple chain equivalence $h : C \longrightarrow C$ of a based f.g. free A-module chain complex C) if and only if $\Phi^+(E) = 0$.

The based version of 32.20 is given by:

Proposition 32.27 *A simple Blanchfield complex (E,θ) over $A[z,z^{-1}]$ fibres if and only if the based finite f.g. free $A[z,z^{-1}]$-module E is a fibred band, i.e. A-finitely dominated with $\Phi^+(E) = 0 \in Wh_1(A[z,z^{-1}])$.* □

Proposition 32.28 *For any field with involution F*

$$L\mathrm{Aut}^*_{fib}(F,\epsilon) \;=\; L\mathrm{Iso}^*(F,\epsilon) \;=\; L\mathrm{Iso}^*_{fib}(F,\epsilon) \;.$$

Proof Use 32.16 to identify

$$L\mathrm{Aut}^*_{fib}(F,\epsilon) \;=\; L\mathrm{Iso}^*_{fib}(F,\epsilon) \;.$$

The forgetful maps

$$L\mathrm{Aut}^n_{fib}(F,\epsilon) \longrightarrow L\mathrm{Iso}^n(F,\epsilon) \;;\; (C,\phi,h,\chi) \longrightarrow (C,(h-1)^{-1}\phi_0)$$

will now be shown to be isomorphisms, by an explicit construction of the inverse isomorphisms. The covering of an n-dimensional ϵ-symmetric Seifert complex $(D,\widehat{\psi})$ is an $(n+1)$-dimensional ϵ-symmetric Blanchfield complex $\beta(D,\widehat{\psi})$ over $F[z,z^{-1}]$. Now $\beta(D,\widehat{\psi})$ is fibred, so that it is homotopy equivalent to the algebraic mapping torus $T(h,\chi)$ of a fibred self homotopy equivalence $(h,\chi) : (C,\phi) \longrightarrow (C,\phi)$ of an n-dimensional symmetric ϵ-symmetric

Poincaré complex (C, ϕ) over F. The covering construction thus defines the inverse isomorphisms

$$LIso^n(F, \epsilon) \xrightarrow{\simeq} LAut^n_{fib}(F, \epsilon) \; ; \; (D, \widehat{\psi}) \longrightarrow \beta(D, \widehat{\psi}) \; = \; (C, \phi, h, \chi) \; ,$$

with

$$\begin{aligned} H_*(C) &= H_{*+1}(\widehat{\psi} + zT_\epsilon\widehat{\psi} : D^{n-*}[z, z^{-1}] \longrightarrow D[z, z^{-1}]) \; , \\ h &= \zeta \; : \; H_*(C) \longrightarrow H_*(C) \; . \end{aligned}$$

□

Proposition 32.29 *An ϵ-symmetric Seifert complex $(C, \widehat{\psi})$ is S-equivalent to a fibred Seifert complex if and only if $\beta^h(C, \widehat{\psi})$ is a fibred Blanchfield complex.*

Proof If $(C, \widehat{\psi})$ is a fibred Seifert complex then $\beta^h(C, \widehat{\psi})$ is a fibred Blanchfield complex.

Conversely, if $(C, \widehat{\psi})$ is a Seifert complex such that $\beta^h(C, \widehat{\psi})$ is a fibred Blanchfield complex then the fibre of $\beta^h(C, \widehat{\psi})$ is a fibred Seifert complex S-equivalent to $(C, \widehat{\psi})$. □

Proposition 32.30 *An $(n+1)$-dimensional ϵ-symmetric Blanchfield complex (E, θ) over $A[z, z^{-1}]$ fibres if and only it has a fibred projective Seifert complex $(C, \widehat{\psi})$.*

Proof If (E, θ) fibres let $(i^!E, i^!\theta)$ be the finitely dominated n-dimensional symmetric Poincaré complex over A defined in 25.5 (the 'infinite cyclic cover' of (E, θ)). The fibred isometric structure $((1-\zeta)^{-1}, 0)$ for $(i^!E, i^!\theta)$ determines a fibred finitely dominated Seifert complex $(i^!E, \widehat{\psi})$ for (E, θ), with

$$\widehat{\psi} \; = \; (1-\zeta)^{-1}\theta_0 \; : \; (i^!E)^{n-*} \simeq E^{n+1-*} \longrightarrow i^!E \; .$$

Conversely, the free covering $\beta^h(C, \widehat{\psi})$ of a fibred finitely dominated Seifert complex $(C, \widehat{\psi})$ is homotopy equivalent to the algebraic mapping cone $T(h, \chi)$ of the monodromy homotopy equivalence

$$(h, \chi) \; = \; ((T_\epsilon\widehat{\psi})\widehat{\psi}^{-1}, 0) \; : \; (C, (1+T_\epsilon)\widehat{\psi}) \xrightarrow{\simeq} (C, (1+T_\epsilon)\widehat{\psi}) \; ,$$

so that the Blanchfield complex $\beta^h(C, \widehat{\psi})$ is fibred. □

There is also a based version of the covering construction :

Definition 32.31 The *simple covering* of a based f.g. free n-dimensional ϵ-symmetric Seifert complex $(C, \widehat{\psi})$ over A is the simple $(n+1)$-dimensional ϵ-symmetric Blanchfield complex over $A[z, z^{-1}]$

$$\beta^s(C, \widehat{\psi}) \; = \; U(\Gamma) \; = \; (E', \theta')$$

defined by the union of the fundamental symmetric Poincaré cobordism over A

$$\Gamma \; = \; ((1-f \;\; -f) : C \oplus C \longrightarrow C, (\delta\phi, \phi \oplus -\phi)) \; ,$$

with $(f,0)$ the isometric structure on $(C,(1+T_\epsilon)\widehat{\psi})$ given by

$$\begin{aligned} f &= \widehat{\psi}((1+T_\epsilon)\widehat{\psi})^{-1} \;:\; C \longrightarrow C \;, \\ E' &= \mathcal{C}(1-f+zf : C[z,z^{-1}]\longrightarrow C[z,z^{-1}]) \;. \end{aligned}$$ □

The free and simple coverings are related by a homotopy equivalence

$$(h,\chi) \;:\; \beta^h(C,\widehat{\psi}) \xrightarrow{\simeq} \beta^s(C,\widehat{\psi})$$

with

$$\begin{aligned} h = (1\;\;(1+T_\epsilon)\widehat{\psi}) \;:\; & \mathcal{C}(T_\epsilon\widehat{\psi}+z\widehat{\psi} : C^{n-*}[z,z^{-1}]\longrightarrow C[z,z^{-1}]) \\ & \longrightarrow \mathcal{C}(1-f+zf : C[z,z^{-1}]\longrightarrow D[z,z^{-1}]) \;, \end{aligned}$$

such that

$$\tau(h) \;=\; \tau(C,\widehat{\psi}) \in \operatorname{im}(Wh_1(A)\longrightarrow Wh_1(A[z,z^{-1}])) \;.$$

Definition 32.32 A *free Seifert complex* for a simple $(n+1)$-dimensional ϵ-symmetric Blanchfield complex (E,θ) over $A[z,z^{-1}]$ is a based f.g. free n-dimensional ϵ-symmetric Seifert complex $(C,\widehat{\psi})$ such that the simple covering $\beta^s(C,\widehat{\psi})$ is simple homotopy equivalent to (E,θ). □

Remark 32.33 The Seifert complex $(C,\widehat{\psi})$ in 32.32 need not be simple: the torsion is given by

$$\tau(C,\widehat{\psi}) \;=\; B_2\tau_2(E,\theta) \in Wh_1(A)$$

with $\tau_2(E,\theta) \in Wh_2(A[z,z^{-1}])$ the Wh_2-invariant determined by a choice of trivialization of $\tau(E,\theta) \in Wh_1(A[z,z^{-1}])$ and

$$B_2 \;:\; Wh_2(A[z,z^{-1}]) \longrightarrow Wh_1(A)$$

the Wh_2-analogue of the Bass–Heller–Swan projection

$$B \;:\; Wh_1(A[z,z^{-1}]) \longrightarrow \widetilde{K}_0(A) \;.$$ □

Proposition 32.34 *Simple Blanchfield complexes have free Seifert complexes. More precisely, every simple A-contractible based f.g. free $(n+1)$-dimensional symmetric Poincaré complex (E,θ) over $A[z,z^{-1}]$ with*

$$\tau(A\otimes_{A[z,z^{-1}]} E) \;=\; 0 \in Wh_1(A)$$

is simple homotopy equivalent to the simple covering $\beta^s(D,\widehat{\psi})$ of a based f.g. free n-dimensional ultraquadratic Poincaré complex $(D,\widehat{\psi})$.

Proof As for 32.23, but taking account of torsions. □

The based version of 32.24 is given by:

Proposition 32.35 *The simple covering construction of 32.31 defines natural isomorphisms*

$$\beta^s \ : \ L\mathrm{Iso}^n_h(A,\epsilon) \xrightarrow{\simeq} L^s_{n+2}(A[z,z^{-1}],\Pi,\epsilon) \ ; \ (C,\widehat{\psi}) \longrightarrow \beta^s(C,\widehat{\psi}) \ .$$

Proof The algebraic Poincaré transversality construction of 32.23 defines inverse maps $(\beta^s)^{-1} : (E,\theta)\longrightarrow(C,\widehat{\psi})$. □

Definition 32.36 A based f.g. free n-dimensional ϵ-symmetric Seifert complex $(C,\widehat{\psi})$ over A is *simple fibred* if the A-module chain maps

$$\begin{aligned} f &= \widehat{\psi}((1+T_\epsilon)\widehat{\psi})^{-1} \ : C \longrightarrow C \ , \\ 1-f &= T_\epsilon\widehat{\psi}((1+T_\epsilon)\widehat{\psi})^{-1} \ : C \longrightarrow C \end{aligned}$$

are simple chain equivalences. □

The monodromy of a fibred based Seifert complex $(C,\widehat{\psi})$ is a simple chain equivalence

$$h \ = \ (T_\epsilon\widehat{\psi})\widehat{\psi}^{-1} \ = \ 1-f^{-1} \ : \ C \xrightarrow{\simeq} C \ .$$

The based version of 32.30 is given by:

Proposition 32.37 *A simple $(n+1)$-dimensional Blanchfield complex (E,θ) over $A[z,z^{-1}]$ fibres if and only it has a simple fibred free Seifert complex $(C,\widehat{\psi})$.* □

Proposition 32.38 *Let A be an integral domain, with*

$$P \ = \ \{p(z)\,|\,p(1)=1\} \subset A[z,z^{-1}] \ ,$$

and let E be an n-dimensional A-contractible finite f.g. free $A[z,z^{-1}]$-module chain complex with Alexander polynomials

$$\Delta_r(z) \ = \ \sum_{j=0}^{m_r} a_{j,r}z^j \in A[z] \quad (0 \leq r \leq n-1) \ ,$$

and let $E' = E^{n-*}$ be the n-dual n-dimensional A-contractible finite f.g. free $A[z,z^{-1}]$-module chain complex.

(i) *The Alexander polynomials $\Delta'_*(z)$ of E' are given by*

$$\Delta'_{n-r-1}(z) \ = \ z^{m_r}\overline{\Delta_r(z)} \ = \ \sum_{j=0}^{m_r} \bar{a}_{m_r-j,r}z^j \in A[z] \quad (0 \leq r \leq n-1) \ .$$

(ii) *If E is chain equivalent to E' then*

$$\Delta_r(z) \ = \ \Delta'_r(z) \in A[z] \quad (0 \leq r \leq n-1) \ ,$$

so that

$$m_r = m_{n-1-r} \ , \ a_{j,r} = \bar{a}_{m_r-j,n-1-r} \ (0 \leq r \leq n-1) \ .$$

The Reidemeister torsion (16.5) is given by

$$\Delta(E) = \Delta(E') = \prod_{r=0}^{n-1} \Delta_r(1-z)^{(-)^r} \in \widetilde{W}(A) \ ,$$

so that

$$\Delta(E) = \begin{cases} \tau(\tau^*)^{-1} & \text{if } n = 2i \\ \Delta_i(1-z)^{(-)^i}\tau\tau^* & \text{if } n = 2i+1 \end{cases}$$

with

$$\tau = \prod_{2r<n} \Delta_r(1-z)^{(-)^r} \in \widetilde{W}(A) \ .$$

(iii) *The P-primary torsion class morphism of 31.19*

$$\Delta : L^{n+1}(A[z,z^{-1}],P,\epsilon) = L\mathrm{Iso}^{n-1}(A,\epsilon)$$
$$\longrightarrow \widehat{H}^{n+1}(\mathbb{Z}_2;K_1(A[z,z^{-1}],P)) = \widehat{H}^{n-1}(\mathbb{Z}_2;\mathrm{Iso}_0(A)) \ ; \ (E,\theta) \longrightarrow \Delta(E)$$

sends an n-dimensional ϵ-symmetric Blanchfield complex (E,θ) over $A[z,z^{-1}]$ to the Tate $\mathbb{Z}_2$-cohomology class of $\Delta(E)$. By (ii) the map Δ is 0 for even n, and for odd $n = 2i+1$ it is given by the middle-dimensional Alexander polynomial

$$\Delta(E) = \Delta_i(1-z)^{(-)^i} \in \widehat{H}^{n-1}(\mathbb{Z}_2;\mathrm{Iso}_0(A)) = \widehat{H}^{n-1}(\mathbb{Z}_2;\widetilde{W}(A)) \ .$$

Proof (i) It suffices to consider the 1-dimensional case

$$E = \mathcal{C}(1-f+zf : D[z,z^{-1}]\longrightarrow D[z,z^{-1}])$$

with D a f.g. projective A-module and $f \in \mathrm{Hom}_A(D,D)$, and with 1-dual

$$E' = E^{1-*} = \mathcal{C}(1-f^*+z^{-1}f^* : D^*[z,z^{-1}]\longrightarrow D^*[z,z^{-1}]) \ .$$

Let

$$\det(1-f+zf : D[z]\longrightarrow D[z]) = z^k\sum_{j=0}^{m} a_jz^j \in A[z]$$
$$(a_0,a_m \neq 0 \in A, k \geq 0) \ ,$$

so that

$$\det(1-f^*+z^{-1}f^* : D^*[z^{-1}]\longrightarrow D^*[z^{-1}]) = z^{-m-k}\sum_{j=0}^{m} \bar{a}_{m-j}z^j \in A[z^{-1}] \ .$$

The Alexander polynomials of E, E' are given by

$$\Delta_0(z) = \sum_{j=0}^{m} a_j z^j \ , \ \Delta'_0(z) = \sum_{j=0}^{m} \bar{a}_{m-j} z^j \in A[z] \ .$$

(ii) Immediate from (i) and 17.7. □

Example 32.39 Let (E, θ) be a f.g. free $(n+2)$-dimensional Blanchfield complex over $A[z, z^{-1}]$, which by 32.30 is homotopy equivalent to the covering $\beta(C, \widehat{\psi})$ of a finitely dominated $(n+1)$-dimensional Seifert complex $(C, \widehat{\psi})$ over A. Assume that A is an integral domain, so that the Alexander polynomials of (E, θ) are defined (32.38). The Alexander polynomials are determined by the isometric structure

$$f = \widehat{\psi}((1+T)\widehat{\psi})^{-1} : C \longrightarrow C \ ,$$

with

$$\Delta_r(z) = z^{-n_r}\det(1 - f + zf : H_r(C)[z] \longrightarrow H_r(C)[z]) \ \ (r \geq 0) \ ,$$

where

$$n_r = \dim_F H_r(E^+; F) - \dim_F H_r(E; F) \geq 0 \ .$$ □

32E. Minimal Seifert complexes

In 32E. the ground ring A is assumed to be an integral domain.

Definition 32.40 (i) A polynomial

$$q(s) = \sum_{j=0}^{d} a_j s^j \in A[s]$$

is *minimal* if

$$a_0 \neq 0 \in A \ , \ \sum_{j=0}^{d} a_j \neq 0 \in A \ , \ a_d \in A^\bullet \ .$$

(ii) Let $Q_A \subset A[s]$ be the involution-invariant multiplicative subset of all polynomials with leading coefficient a unit in A, and let $Q_{A,min} \subset Q_A$ be the subset of the minimal polynomials, so that $Q_{A,min} \subset A[s]$ is also an involution-invariant multiplicative subset.

(iii) An ϵ-symmetric Seifert complex (C, λ) over A is *minimal* if the chain map

$$s = f = \lambda((1+T_\epsilon)\lambda)^{-1} : C \longrightarrow C$$

(= the isometric structure on $(C, (1+T_\epsilon)\lambda)$) is such that

$$q(f) \simeq 0 : C \longrightarrow C$$

for some $q(s) \in Q_{A,min}$. Similarly for a Seifert form. □

Proposition 32.41 *Let (C,λ) be a Seifert complex over A with covering Blanchfield complex $\beta(C,\lambda)=(E,\theta)$ over $A[z,z^{-1}]$. The natural A-module chain map $g:C\longrightarrow E$ induces injections $g_*:H_*(C)\longrightarrow H_*(E)$ if and only if $f_*,1-f_*:H_*(C)\longrightarrow H_*(C)$ are injections. In particular, this is the case if (C,λ) is minimal.*

Proof The kernel of

$$g_*\ :\ H_r(C)\longrightarrow H_r(E)\ =\ A[s,s^{-1},(1-s)^{-1}]\otimes_{A[s]}H_r(C)$$

consists of the elements $x\in H_r(C)$ such that $(f_*(1-f_*))^N(x)=0\in H_r(C)$ for some $N\geq 0$. It follows from

$$\ker(f_*)+\ker(1-f_*)\subseteq\ker(g_*)$$

that if g_* is injective then so are $f_*,1-f_*$. Conversely, if $f_*,1-f_*$ are injective then so is each $(f_*(1-f_*))^N$, and hence so is g_*.
If (C,λ) is minimal and $q(s)\in Q_{A,min}$ is such that $q(f)\simeq 0$ define

$$p_0(s)\ =\ \frac{q(s)-q(0)}{s}\ ,\ p_1(s)\ =\ \frac{q(1)-q(s)}{1-s}\in A[s]$$

such that

$$sp_0(s)+q(0)\ =\ q(1)-(1-s)p_1(s)\ =\ q(s)\in A[s]$$

with $q(0),q(1)\neq 0\in A$. Now

$$fp_0(f)+q(0)\ =\ q(1)-(1-f)p_1(f)\ =\ q(f)\ \simeq\ 0\ :\ C\longrightarrow C$$

so that $f_*,1-f_*:H_*(C)\longrightarrow H_*(C)$ are injective. □

Remark 32.42 A Seifert surface F for an n-knot $k:S^n\subset S^{n+2}$ is *minimal* if the inclusion $F\subset\overline{X}$ in the canonical infinite cyclic cover of the exterior X induces injections $H_*(F)\longrightarrow H_*(\overline{X})$. By 32.41 this is the case if the Seifert complex $(\dot{C}(F),\lambda)$ over $\mathbb{Z}$ is minimal. Farber [76, 2.4] proved that for $n\geq 4$ every n-knot k has minimal Seifert surfaces $F^{n+1}\subset S^{n+2}$, with $H_*(F)\subseteq H_*(\overline{X})$ prescribed f.g. $\mathbb{Z}[s]$-modules ($*\neq 0$, up to multiplication by some t^i, with self duality if $n+1=2*$) such that

$$\mathbb{Z}[s,s^{-1},(1-s)^{-1}]\otimes_{\mathbb{Z}[s]}H_*(F)\ =\ H_*(\overline{X})\ ,$$

with

$$s\ =\ \lambda((1+T)\lambda)^{-1}\ :\ H_*(F)\longrightarrow H_*(F)\ ,$$

$$s\ =\ (1-\zeta)^{-1}\ :\ H_*(\overline{X})\longrightarrow H_*(\overline{X})\ .$$ □

In 35.12 below it will be proved that every Seifert complex over an integral domain A is cobordant to a minimal Seifert complex. The 0-dimensional case is that every Seifert form is Witt-equivalent to a minimal Seifert form. For the remainder of Chap. 32 only forms will be considered. In 32.45 below it will be proved that every Seifert form over a Euclidean domain is S-equivalent to a minimal Seifert form.

Proposition 32.43 (i) *The characteristic polynomial of an object (P, f) in* $\mathrm{End}(A)$

$$q(s) \;=\; \mathrm{ch}_s(P, f) \;=\; \det(s - f : P[s] \longrightarrow P[s]) \in A[s]$$

is minimal if and only if the endomorphisms $f, 1 - f : P \longrightarrow P$ are injective.
(ii) *The localization of $A[s]$ inverting Q_A is just the Fredholm localization*

$$Q_A^{-1} A[s] \;=\; \Omega_+^{-1} A[s] \;,$$

which fits into a commutative square of rings with involution

$$\begin{array}{ccc} A[s] & \longrightarrow & Q_{A,min}^{-1} A[s] \\ \big\downarrow & & \big\downarrow \\ A[s, s^{-1}, (1-s)^{-1}] & \longrightarrow & Q_A^{-1} A[s] \;. \end{array}$$

The induced morphism of abelian groups

$$Q_{A,min}^{-1} A[s]/A[s] \longrightarrow Q_A^{-1} A[s]/A[s, s^{-1}, (1-s)^{-1}]$$

is injective. This morphism is surjective if and only if $A = F$ is a field, in which case the square is cartesian with

$$Q_{A,min}^{-1} A[s] \;=\; F[s]_{(s,1-s)} \;\;,\;\; Q_A^{-1} A[s] \;=\; F(s) \;.$$

(iii) *Let*

$$P_A \;=\; \{ \sum_{j=-\infty}^{\infty} a_j z^j \mid \sum_{j=-\infty}^{\infty} a_j = 1 \in A \} \subset A[z, z^{-1}] \;.$$

The morphisms

$$Q_A^{-1} A[s]/A[s, s^{-1}, (1-s)^{-1}] \longrightarrow P_A^{-1} A[z, z^{-1}]/A[z, z^{-1}] \;;\; s \longrightarrow (1-z)^{-1} \;,$$
$$P_A^{-1} A[z, z^{-1}]/A[z, z^{-1}] \longrightarrow P_A^{-1} A[z, z^{-1}, (1-z)^{-1}]/A[z, z^{-1}, (1-z)^{-1}]$$

are isomorphisms.

Proof (i) The characteristic polynomial has values

$$q(0) \;=\; \det(-f : P \longrightarrow P) \;\;,\;\; q(1) \;=\; \det(1 - f : P \longrightarrow P) \in A \;.$$

These values are non-zero precisely when f and $1 - f$ are injective.
(ii) The commutative square

$$\begin{array}{ccc} F[s] & \longrightarrow & Q_{F,min}^{-1} F[s] \\ \big\downarrow & & \big\downarrow \\ F[s, s^{-1}, (1-s)^{-1}] & \longrightarrow & Q_F^{-1} F[s] \end{array}$$

is cartesian for any field F, since any polynomial $p(s) \neq 0 \in F[s]$ can be factorized as

$$p(s) = s^j(1-s)^k q(s) \in F[s]$$

with $j,k \geq 0$ and $q(s) \in F[s]$ coprime to $s^j(1-s)^k$. Thus $q(0), q(1) \in F^\bullet$, and there exist $\beta(s), \gamma(s) \in F[s]$ with

$$\beta(s)q(s) + \gamma(s)s^j(1-s)^k = 1 \in F[s] ,$$

giving a partial fraction decomposition

$$\begin{aligned}\frac{1}{p(s)} &= \frac{\beta(s)}{s^j(1-s)^k} + \frac{\gamma(s)}{q(s)} \\ &\in \operatorname{im}(F[s,s^{-1},(1-s)^{-1}] \oplus Q^{-1}_{F,min}F[s] \longrightarrow Q^{-1}_F F[s]) .\end{aligned}$$

If A is not a field and $a \in A \backslash (A^\bullet \cup \{0\})$, then

$$\frac{1}{s(s+a)} \notin \operatorname{im}(A[s,s^{-1},(1-s)^{-1}] \oplus Q^{-1}_{A,min}A[s] \longrightarrow Q^{-1}_A A[s])$$

since otherwise

$$\frac{1}{s(s+a)} = \frac{\beta(s)}{s} + \frac{\gamma(s)}{s+a}$$

for some $\beta(s), \gamma(s) \in A[s]$ and $\beta(0) = a^{-1} \in A$, a contradiction.
(iii) This follows from the isomorphism of rings with involution

$$A[s,s^{-1},(1-s)^{-1}] \xrightarrow{\simeq} A[z,z^{-1},(1-z)^{-1}] \; ; \; s \longrightarrow (1-z)^{-1} .$$

Definition 32.44 The *minimal universal trace map*

$$\chi_{s,min} : Q^{-1}_{A,min}A[s]/A[s] \longrightarrow A$$

is the composite

$$\chi_{s,min} : Q^{-1}_{A,min}A[s]/A[s] \subseteq Q^{-1}_A A[s]/A[s] = \Omega^{-1}_+ A[s]/A[s] \xrightarrow{\chi_s} A$$

with χ_s the universal trace map of 31.20. □

More directly, for any $x \in Q^{-1}_{A,min}A[s]$

$$\chi_{s,min}(x) = a_{-1} \in A$$

is the coefficient of s^{-1} in the power series expansion

$$x = \sum_{j=-\infty}^{\infty} a_j s^j \in Q^{-1}_{A,min}A[s] \subseteq A((s^{-1})) .$$

Proposition 32.45 *Let (L,λ) be an ϵ-symmetric Seifert form over A, corresponding to the isometric structure*

$$f = (\lambda+\epsilon\lambda^*)^{-1}\lambda : L \longrightarrow L$$

on the nonsingular ϵ-symmetric form $(L,\lambda+\epsilon\lambda^*)$ over A, with

$$1-f = (\lambda+\epsilon\lambda^*)^{-1}(\epsilon\lambda^*) : L \longrightarrow L .$$

Also, let $(M,\mu)=\beta(L,\lambda)$ be the covering $(-\epsilon)$-symmetric Blanchfield form over $A[z,z^{-1}]$, with

$$\begin{aligned} M &= \operatorname{coker}(1-f+zf : L[z,z^{-1}]\longrightarrow L[z,z^{-1}]) , \\ \mu &: M\times M \longrightarrow P_A^{-1}A[z,z^{-1}]/A[z,z^{-1}] ; \\ &\qquad (x,y) \longrightarrow (1-z)(\lambda+\epsilon\lambda^*)(x,(1-f+zf)^{-1}(y)) . \end{aligned}$$

(i) *The Blanchfield form is such that for any $x,y\in M$*

$$\begin{aligned} \mu(x,y) &\in Q_{A,min}^{-1}A[s]/A[s] \\ &\subseteq Q_A^{-1}A[s]/A[s,s^{-1},(1-s)^{-1}] = P_A^{-1}A[z,z^{-1}]/A[z,z^{-1}] \end{aligned}$$

with image

$$\mu(x,y) = \sum_{j=-\infty}^{-1}(\lambda+\epsilon\lambda^*)(x,f^{-j-1}(y))s^j \in A[[s^{-1}]] .$$

In particular, μ determines λ by

$$\lambda : L\times L \xrightarrow{i\times si} M\times M \xrightarrow{\mu} Q_{A,min}^{-1}A[s]/A[s] \xrightarrow{\chi_{s,min}} A$$

with $s=(1-z)^{-1}: M\longrightarrow M$.
(ii) *The following conditions are equivalent:*

(a) *$\lambda: L\longrightarrow L^*$ is injective,*
(b) *the natural $A[s]$-module morphism $i: L\longrightarrow M$ is injective,*
(c) *the Seifert form (L,λ) is minimal.*

(iii) *If A is a Euclidean domain then (L,λ) is S-equivalent to a minimal Seifert form.*
(iv) *If A is a field then (L,λ) is S-equivalent to the minimal Seifert form (L',λ') defined by*

$$\begin{aligned} L' &= \operatorname{coker}(z\lambda+\epsilon\lambda^* : L[z,z^{-1}]\longrightarrow L^*[z,z^{-1}]) , \\ \lambda'(x,y) &= \chi_z(\epsilon zx((z\lambda+\epsilon\lambda^*)^{-1}(y))) \in A \ \ (x,y\in L^*[z,z^{-1}]) . \end{aligned}$$

Proof (i) Let

$$\det(1-f+zf : L[z,z^{-1}]\longrightarrow L[z,z^{-1}]) = z^n(\sum_{j=0}^{d} a_j z^j)$$

with

$$a_0, a_d \neq 0 \in A \ , \ \sum_{j=0}^{d} a_j = 1 \in A \ ,$$

so that the Alexander polynomial of (M,μ) is given by 17.2 (iii) to be

$$\begin{aligned}\Delta(z) &= z^{-n}\det(1-f+zf : L[z,z^{-1}]\longrightarrow L[z,z^{-1}]) \\ &= \sum_{j=0}^{d} a_j z^j \in P_A \subset A[z,z^{-1}] \ .\end{aligned}$$

The evaluation of μ on $(x,y) \in M \times M$ is of the type

$$\mu(x,y) = \frac{p(z)}{\Delta(z)} \in P_A^{-1}A[z,z^{-1}]/A[z,z^{-1}]$$

for some $p(z) \in A[z,z^{-1}]$. In terms of $s = 1-z^{-1}$ this is

$$\mu(x,y) = \frac{q(s)}{\mathrm{ch}_s(L,f)} \in Q_A^{-1}A[s]/A[s]$$

with

$$q(s) = (1-s)^{-d}p((1-s)^{-1}) \in Q_A^{-1}A[s] \ ,$$

and denominator

$$(1-s)^m \mathrm{ch}_s(L,f) \in Q_{A,min} \subset Q$$

for some $m \geq 0$, so that

$$\mu(x,y) \in Q_{A,min}^{-1}A[s]/A[s] \subseteq P_A^{-1}A[z,z^{-1}]/A[z,z^{-1}] \ .$$

Work in the completion $A[[s^{-1}]]$ to obtain

$$\begin{aligned}\mu(x,y) &= (1-z)(\lambda+\epsilon\lambda^*)(x,(1-f+zf)^{-1}(y)) \\ &= s^{-1}(\lambda+\epsilon\lambda^*)(x,(1-s^{-1}f)^{-1}y) \\ &= \sum_{j=-\infty}^{-1} (\lambda+\epsilon\lambda^*)(x,f^{-j-1}(y))s^j \in A[[s^{-1}]]\end{aligned}$$

and

$$\chi'_s(\mu(x,y)) = \text{coefficient of } s^{-1} \text{ in } \mu(x,y) = (\lambda+\epsilon\lambda^*)(x,y) \in A \ .$$

(ii) Since $s, 1-s \in A[s]$ are coprime the kernel of

$$i \,:\, L \longrightarrow M \;=\; A[s,s^{-1},(1-s)^{-1}] \otimes_{A[s]} L$$

(with s acting on L by f) is of the form

$$\ker(i) \;=\; K_+ \oplus K_- \;,$$

with

$$\begin{aligned} K_+ &= \{x \in L \,|\, f^k(x) = 0 \text{ for some } k \geq 0\} \;, \\ K_- &= \{x \in L \,|\, (1-f)^k(x) = 0 \text{ for some } k \geq 0\} \;. \end{aligned}$$

(a) $\Longrightarrow$ (b) If λ is injective then so are $\lambda^* : L \longrightarrow L^*, f, 1-f : L \longrightarrow L$, whence $K_+ = K_- = 0$.
(b) $\Longrightarrow$ (a) If $K_+ = 0$ then $f^k : L \longrightarrow L$ is injective for every $k \geq 1$, including $k = 1$.
(c) $\Longrightarrow$ (a) Let

$$q(s) \;=\; \sum_{j=0}^{\infty} a_j s^j \in A[s]$$

be a monic polynomial with $q(0) \neq 0 \in A$ such that $q(f) = 0 : L \longrightarrow L$. If $x \in \ker(\lambda : L \longrightarrow L^*)$ then $f(x) = 0 \in L$, and

$$q(f)(x) \;=\; q(0)x \;=\; 0 \in L \;.$$

Since L is a f.g. projective A-module, it must be the case that $x = 0 \in L$, and hence that f and λ are injective.
(a) $\Longrightarrow$ (c) The characteristic polynomial of f

$$q(s) \;=\; \mathrm{ch}_s(L,f) \in A[s]$$

is a monic polynomial such that

$$q(f) \;=\; 0 \,:\, L \longrightarrow L \;,\;\; q(0) \;=\; \det(-f) \;,\;\; q(1) \;=\; \det(1-f) \neq 0 \in A \;.$$

(iii) If the structure f on (L,λ) is not minimal there exists an element $x \neq 0 \in L$ such that

$$\lambda(x) \;=\; 0 \in L^* \;.$$

Since A is a principal ideal domain $\widetilde{K}_0(A) = 0$, so that L can be taken to be f.g. free, say $L = A^n$, and

$$x \;=\; (a_1, a_2, \ldots, a_n) \in L \;=\; A^n \;.$$

Dividing by the greatest common divisor of $a_1, a_2, \ldots, a_n \in A$ if necessary, it may be assumed that $a_1, a_2, \ldots, a_n$ are coprime, so that $x \in L$ generates a direct summand

$$K \;=\; A\langle x\rangle \subset L \;=\; A^n$$

which is a sublagrangian of $(L, \lambda + \epsilon\lambda^*)$. Since $(L, \lambda + \epsilon\lambda^*)$ is nonsingular there exists an element $x^* \in L$ such that

$$(\lambda + \epsilon\lambda^*)(x, x^*) \;=\; 1 \in A ,$$

generating a direct summand

$$K^* \;=\; A\langle x^* \rangle \subset L \;=\; A^n$$

such that

$$(L, \lambda) \;=\; (K \oplus K^* \oplus L', \begin{pmatrix} 0 & 1 & 0 \\ 0 & \alpha & \beta \\ 0 & -\epsilon\beta^* & \lambda' \end{pmatrix})$$

for some α, β with $(L', \lambda') = (K^{\perp}/K, [\lambda])$. Thus (L, λ) is S-equivalent to (L', λ') with

$$\dim_A(L') \;=\; \dim_A(L) - 2 .$$

If (L', λ') is not minimal repeat the procedure, and so on.
(iv) Immediate from the isomorphism of exact sequences

$$\begin{array}{ccccccccc}
0 & \longrightarrow & L[z, z^{-1}] & \xrightarrow{z\lambda + \epsilon\lambda^*} & L^*[z, z^{-1}] & \longrightarrow & L' & \longrightarrow & 0 \\
 & & {\scriptstyle z\epsilon}\big\downarrow{\scriptstyle \simeq} & & \big\| & & {\scriptstyle \lambda'}\big\downarrow{\scriptstyle \cong} & & \\
0 & \longrightarrow & L[z, z^{-1}] & \xrightarrow{z^{-1}\lambda^* + \epsilon^{-1}\lambda} & L^*[z, z^{-1}] & \longrightarrow & L'^* & \longrightarrow & 0
\end{array}$$

and 28.27. □

Corollary 32.46 *If A is a Euclidean domain every Blanchfield form (M, μ) over $A[z, z^{-1}]$ admits a minimal Seifert form (L, λ) over A.*
Proof The existence of Seifert forms is the 0-dimensional case of 32.10 (for any ring with involution A). Minimal Seifert forms for Euclidean A are given by 32.45 (iii). □

Remark 32.47 The result that every Seifert form over $\mathbb{Z}$ is S-equivalent to a minimal one was first obtained by Trotter [290]. The precise nature of the relationship between Seifert and Blanchfield forms for the case of greatest knot-theoretic significance $A = \mathbb{Z}$ was worked out by Kearton [125] (geometrically) and Trotter [291], [292], Farber [73], [77] (algebraically), including the result that every Blanchfield form over $\mathbb{Z}[z, z^{-1}]$ admits a minimal Seifert form over $\mathbb{Z}$. □

33. Knot theory

This chapter deals with the classification theory of high-dimensional n-knots $k : S^n \subset S^{n+2}$ from the algebraic surgery point of view, extending the treatment in Ranicki [237, Chaps. 7.8, 7.9]. The following Chaps. 34–41 are devoted to various algebraic L-theoretic techniques, which are used in Chap. 42 to describe the computation of the high-dimensional knot cobordism groups C_*.

Definition 33.1 (Kervaire [132])
(i) An *n-knot* $k : S^n \subset S^{n+2}$ is an embedding of S^n in S^{n+2}.
(ii) Two n-knots $k_0, k_1 : S^n \subset S^{n+2}$ are *cobordant* if there exists an embedding $\ell : S^n \times I \subset S^{n+2} \times I$ such that

$$\begin{aligned} \ell(S^n \times I) \cap (S^{n+2} \times \{i\}) &= k_i(S^n) \times \{i\} , \\ \ell(x, i) &= (k_i(x), i) \ \ (i = 0, 1) . \end{aligned}$$

(iii) The *connected sum* of n-knots $k, k' : S^n \subset S^{n+2}$ is the n-knot

$$k\#k' \ : \ S^n\#S^n \ = \ S^n \ \subset \ S^{n+2}\#S^{n+2} \ = \ S^{n+2} .$$

The *n-knot cobordism group* C_n is the abelian group of cobordism classes of n-knots $k : S^n \subset S^{n+2}$, with addition by the connected sum
(iv) An n-knot $k : S^n \subset S^{n+2}$ is *slice* if it is null-cobordant, or equivalently if the $\mathbb{Z}$-homology boundary spine $k(S^n) \subset S^{n+2}$ for (D^{n+3}, S^{n+2}) extends to a $\mathbb{Z}$-homology spine $(D^{n+1}, k(S^n)) \subset (D^{n+3}, S^{n+2})$. □

33A. The Seifert complex of an n-knot

Definition 33.2 (i) A *Seifert surface* for an n-knot $k : S^n \subset S^{n+2}$ is a codimension 1 framed submanifold $F^{n+1} \subset S^{n+2}$ with boundary

$$\partial F \ = \ k(S^n) \subset S^{n+2} ,$$

i.e. a codimension 1 Seifert surface in the sense of 27.3.
(ii) The *Seifert complex* of an n-knot k with respect to a Seifert surface F is the $(n+1)$-dimensional ultraquadratic Poincaré complex over $\mathbb{Z}$

$$\sigma^*(k, F) \ = \ (C, \widehat{\psi})$$

defined in [237, p. 828] using a degree 1 normal map

$$(f,b) : (F^{n+1},\partial F) \longrightarrow (D^{n+3},k(S^n))$$

which is a homeomorphism on ∂F, with

$$H_*(C) = H_{*+1}(D^{n+3},F) = \dot{H}_*(F) . \qquad \square$$

Proposition 33.3 (Kervaire [132], Levine [157], Ranicki [237, 7.9.4]) *The Seifert complex morphisms*

$$\sigma^* : C_n \longrightarrow \widehat{L}_{n+1}(\mathbb{Z}) = LIso^{n+1}(\mathbb{Z}) ; k \longrightarrow \sigma^*(k,F)$$

are isomorphisms for $n \geq 3$.

Proof For even n note that $C_{2*} = 0$ by the simply-connected surgery of [132] and $LIso^{2*+1}(\mathbb{Z}) = 0$ by algebraic surgery as in [237]. For odd $n = 2i-1$ the identification of C_{2i-1} $(i \geq 2)$ with the Witt group of $(-)^i$-symmetric Seifert forms over $\mathbb{Z}$ is due to [157], and $LIso^{2i}(\mathbb{Z})$ was identified with this Witt group in [237]. (The groups C_{2*-1} are infinitely generated – see Chap. 42 for more details.) $\square$

From the point of view of the framed codimension 2 surgery theory of Chap. 27 the high-dimensional n-knot cobordism groups are the $\mathbb{Z}$-homology spine cobordism groups

$$\begin{aligned} C_n &= AB_{n+1}(\text{pt.};\mathcal{F}) = BB_{n+1}(\text{pt.};\mathcal{F}) \\ &= \Gamma_{n+3}(\mathcal{F},-z) = LAsy^{n+3}(\mathbb{Z},P,-z) \ \ (n \geq 3) \end{aligned}$$

with $\mathcal{F} : \mathbb{Z}[z,z^{-1}] \longrightarrow \mathbb{Z}$ the augmentation map $z \longrightarrow 1$, and

$$P = \{ \sum_{j=-\infty}^{\infty} a_j z^j \mid \sum_{j=-\infty}^{\infty} a_j = \pm 1 \} \subset \mathbb{Z}[z,z^{-1}] .$$

Remark 33.4 (i) The morphisms $\widehat{\psi} : H^{n+1-*}(C) \longrightarrow H_*(C)$ are given by

$$\begin{aligned} \widehat{\psi} : H^{n+1-*}(C) &= \dot{H}^{n-*}(F) = \dot{H}_*(F) \\ &\longrightarrow \dot{H}_*(S^{n+2}\backslash F) = \dot{H}^{n+1-*}(F) = \dot{H}_*(F) . \end{aligned}$$

The suspension $S\sigma^*(k,F)$ is the $(n+3)$-dimensional $(P,-z)$-Poincaré asymmetric complex obtained by applying the construction of 27.6 to the $(n+3)$-dimensional manifold with boundary (D^{n+3},S^{n+2}) and the homology framed codimension 2 submanifold $k(S^n) \subset S^{n+2}$ with respect to the codimension 1 Seifert surface $F^{n+1} \subset S^{n+2}$.

(ii) The Seifert form of a $(2i-1)$-knot $k : S^{2i-1} \subset S^{2i+1}$ $(i \geq 1)$ with Seifert surface $F^{2i} \subset S^{2i+1}$ is the $(-)^i$-symmetric Seifert form $(H^i(F)/\text{torsion},\widehat{\psi})$ over $\mathbb{Z}$ determined by the Seifert complex $\sigma^*(k,F)$. See 27.9 (ii) for the surgery construction of a $(2i-1)$-knot $k : S^{2i-1} \subset S^{2i+1}$ with prescribed

$(-)^i$-symmetric Seifert form over $\mathbb{Z}$.
(iii) See Bredon [25] and Kauffman and Neumann [123, Chap. 8] for the geometric construction of maps $C_n \longrightarrow C_{n+4}$ ($n \geq 1$) using the 2-fold cyclic suspension of n-knots (cf. 29.22 (iii)), which are isomorphisms for $n \geq 3$ by Cappell and Shaneson [39].
(iv) The classical knot cobordism group C_1 was defined by Fox and Milnor [82], and shown to be infinitely generated using the Alexander polynomial (cf. 42.1 below). The map $C_1 \longrightarrow C_{4*+1}$ to the high-dimensional $(4*+1)$-knot cobordism group is a surjection, with the invariants of Casson and Gordon [48], [49] detecting non-trivial elements in the kernel – see Litherland [168] for the relationship of these invariants with $L^{2*}(\mathbb{Q}(z)) = LAsy^0(\mathbb{Q})$. □

33B. Fibred n-knots

Definition 33.5 (i) A homology framed knot $K^n \subset S^{n+2}$ is *fibred* if the exterior $X = \mathrm{cl.}(S^{n+2} \backslash (K \times D^2)) \subset S^{n+2}$ fibres over S^1, i.e. if $X = T(h)$ is the mapping torus of the *monodromy* homeomorphism $h : F \longrightarrow F$ of a Seifert surface $F^{n+1} \subset S^{n+2}$ with $h| = \mathrm{id.} : \partial F = K \longrightarrow K$, in which case $(\overline{X}, \zeta) = (F \times \mathbb{R}, h \times T) \simeq (F, h)$ with $T : \mathbb{R} \longrightarrow \mathbb{R}; x \longrightarrow x+1$, and S^{n+2} is an open book with page F and binding K.
(ii) The cobordism group of fibred n-knots $k : S^n \subset S^{n+2}$ is denoted by C_n^{fib}.
(iii) A fibred n-knot $k : S^n \subset S^{n+2}$ with fibre F has a fibred Seifert complex $(\dot{C}(F), \widehat{\psi})$. The *fibred automorphism signature* of k is the cobordism class

$$\sigma^*_{fib}(k, F) = (\dot{C}(F), \widehat{\psi}) \in LIso^{n+1}_{fib}(\mathbb{Z}) = LAut^{n+1}_{fib}(\mathbb{Z}).$$

(iv) A fibred knot $K^{2i-1} \subset S^{2i+1}$ with fibre F^{2i} is *simple* if K and F are $(i-1)$-connected. □

Durfee [64] and Kato [118] identified the isotopy classes of simple fibred knots $K^{2i-1} \subset S^{2i+1}$ for $i \geq 3$ with the isomorphism classes of nonsingular asymmetric forms (L, λ) over $\mathbb{Z}$, with $\lambda + (-)^i\lambda^* : L = H_i(F) \longrightarrow L^* = H_i(F, K)$ the intersection form of the fibre F and $(\lambda^*)^{-1}\lambda = h : L \longrightarrow L$. The isotopy classes of the simple fibred $(2i-1)$-knots $k : S^{2i-1} \subset S^{2i+1}$ for $i \geq 3$ are thus in one-one correspondence with the isomorphism classes of the nonsingular $(-)^i$-symmetric Seifert forms over $\mathbb{Z}$, i.e. the asymmetric forms (L, λ) over $\mathbb{Z}$ with both $\lambda : L \longrightarrow L^*$ and $\lambda + (-)^i\lambda^* : L \longrightarrow L^*$ an isomorphism.

Proposition 33.6 *The fibred automorphism signature map*

$$\sigma^*_{fib} : C_n^{fib} \longrightarrow LIso^{n+1}_{fib}(\mathbb{Z}) \; ; \; k \longrightarrow \sigma^*_{fib}(k, F)$$

is an isomorphism for $n \geq 4$.
Proof For $n = 2i$ work as in Kervaire [132], with $C_{2*}^{fib} = 0$. For $n = 2i-1$ note that every fibred $(2i-1)$-knot is cobordant to a simple one, and use the isotopy classification of simple fibred $(2i-1)$-knots. □

33C. The Blanchfield complex of an n-knot

Let $k : S^n \subset S^{n+2}$ be an n-knot, for any $n \geq 1$. The n-knot exterior

$$(X, \partial X) = (\mathrm{cl.}(S^{n+2} \backslash (k(S^n) \times D^2)), k(S^n) \times S^1)$$

is a compact $(n+2)$-dimensional manifold with a degree 1 normal map to an $(n+2)$-dimensional geometric Poincaré pair

$$(f, b) : (X, \partial X) \longrightarrow (D^{n+3}, k(S^n)) \times S^1$$

which is a $\mathbb{Z}$-homology equivalence on X and the identity on ∂X. The induced group morphism $f_* : \pi_1(X) \longrightarrow \pi_1(D^{n+3} \times S^1) = \mathbb{Z}$ is onto, representing a generator $1 \in H^1(X) = \mathbb{Z}$, and

$$\overline{X} = f^*(D^{n+3} \times \mathbb{R})$$

is the canonical infinite cyclic cover of X.

Definition 33.7 The *Blanchfield complex* of an n-knot $k : S^n \subset S^{n+2}$ is the kernel $\mathbb{Z}$-contractible $(n+2)$-dimensional symmetric Poincaré complex over $\mathbb{Z}[z, z^{-1}]$

$$\sigma^*(k) = (E, \theta)$$

defined in [237, p. 822] with homology the knot $\mathbb{Z}[z, z^{-1}]$-modules

$$H_*(E) = H_{*+1}(\overline{f} : \overline{X} \longrightarrow D^{n+3} \times \mathbb{R}) = \dot{H}_*(\overline{X})$$

and $\mathbb{Z}$-coefficient homology

$$H_*(\mathbb{Z} \otimes_{\mathbb{Z}[z,z^{-1}]} E) = H_{*+1}(f : X \longrightarrow D^{n+3} \times S^1) = 0 . \qquad \square$$

Write the reduced homology and cohomology groups as

$$\dot{H}_*(X) = H_*(X, \mathrm{pt.}) , \quad \dot{H}^*(X) = H^*(X, \mathrm{pt.}) .$$

The $\mathbb{Z}[z, z^{-1}]$-coefficient Poincaré duality isomorphisms

$$H^{n+2-*}(E) \cong H_*(E)$$

are the original Blanchfield duality isomorphisms

$$\begin{aligned} \theta_0 \ : \ & \dot{H}^{n+2-r}(\overline{X}) \xrightarrow{\simeq} \\ & \dot{H}_r(\overline{X}) = \mathrm{Hom}_{\mathbb{Z}[z,z^{-1}]}(\dot{H}^{r+1}(\overline{X}), P^{-1}\mathbb{Z}[z, z^{-1}]/\mathbb{Z}[z, z^{-1}]) , \end{aligned}$$

with

$$\dot{H}^*(\overline{X}) = H_{-*}(\mathrm{Hom}_{\mathbb{Z}[z,z^{-1}]}(\dot{C}(\overline{X}), \mathbb{Z}[z, z^{-1}])) .$$

The adjoints are the nonsingular Blanchfield pairings

$$\dot{H}^{n+2-r}(\overline{X}) \times \dot{H}^{r+1}(\overline{X}) \longrightarrow P^{-1}\mathbb{Z}[z, z^{-1}]/\mathbb{Z}[z, z^{-1}] .$$

Sending an n-knot $k : S^n \subset S^{n+2}$ to its Blanchfield complex $\sigma^*(k)$ defines an isomorphism

$$C_n \xrightarrow{\simeq} L_{n+3}(\mathbb{Z}[z,z^{-1}],P) \; ; \; k \longrightarrow \sigma^*(k)$$

for $n \geq 4$, with $C_{2*} = 0$. The Blanchfield complex of a fibred n-knot k is fibred.

Proposition 33.8 *Let $k : S^n \subset S^{n+2}$ be an n-knot, with knot exterior*

$$X = \text{cl.}(S^{n+2}\backslash(k(S^n) \times D^2)) ,$$

and let $F^{n+1} \subset S^{n+2}$ be a Seifert surface for k.
(i) *The Blanchfield complex of k is homotopy equivalent to the covering of the Seifert complex of k with respect to F*

$$\sigma^*(k) \simeq \beta\sigma^*(k,F) .$$

(ii) *The covering isomorphism (32.9)*

$$\beta : L\mathrm{Iso}^{n+1}(\mathbb{Z}) \xrightarrow{\simeq} L^{n+3}(\mathbb{Z}[z,z^{-1}],P) = C_n$$

sends the cobordism class of the Seifert complex $\sigma^*(k,F)$ to the cobordism class of the Blanchfield complex $\sigma^*(k)$.
(iii) *The homotopy equivalence class of the Blanchfield complex $\sigma^*(k)$ of an n-knot $k : S^n \subset S^{n+2}$ is an isotopy invariant of k.*
(iv) *The S-equivalence class of the Seifert complex $\sigma^*(k)$ of an n-knot k : $S^n \subset S^{n+2}$ is an isotopy invariant of k.*
Similarly for fibred n-knots.
Proof (i) Cutting X along F gives a fundamental domain $(G; F, zF)$ for the canonical infinite cyclic cover of X

$$\overline{X} = \bigcup_{j=-\infty}^{\infty} z^j G .$$

The fundamental $(n+2)$-dimensional symmetric Poincaré cobordism over $\mathbb{Z}[z,z^{-1}]$

$$\Gamma = ((i_+ \; i_-) : \dot{C}(F) \oplus \dot{C}(zF) \longrightarrow \dot{C}(G) \, , (\delta\phi, \phi \oplus -\phi))$$

is such that $i_+ - i_- : \dot{C}(F) \longrightarrow \dot{C}(G)$ is a chain equivalence. The union of Γ is the Blanchfield complex of k

$$U(\Gamma) = \sigma^*(k) ,$$

corresponding to an isometric structure $(f, \delta\phi)$ for the n-dimensional symmetric complex $\sigma^*(F) = (\dot{C}(F), \phi)$ over $\mathbb{Z}$, with

$$f = -(i_+ - i_-)^{-1} i_- : \dot{C}(F) \longrightarrow \dot{C}(F) .$$

The covering of the Seifert complex of an n-knot $k : S^n \subset S^{n+2}$ with respect to a Seifert surface $F^{n+1} \subset S^{n+2}$ is the Blanchfield complex of k

$$\beta\sigma^*(k,F) = \sigma^*(k) .$$

(ii) Immediate from (i).
(iii) The homotopy type of the pair $(X, S^n \times S^1)$ is an isotopy invariant of k.
(iv) Immediate from (iii). □

Similarly for fibred n-knots:

Proposition 33.9 *Let $k : S^n \subset S^{n+2}$ be a fibred n-knot, with knot exterior*

$$X = \text{cl.}(S^{n+2} \backslash (k(S^n) \times D^2))$$

the mapping torus $X = T(h)$ of a monodromy homeomorphism $h : F \longrightarrow F$ on a Seifert surface $F^{n+1} \subset S^{n+2}$ with

$$h| = \text{id.} : \partial F = k(S^n) \longrightarrow \partial F .$$

(i) *The Blanchfield complex of k is homotopy equivalent to the covering of the Seifert complex of k with respect to F*

$$\sigma^*_{fib}(k) \simeq \beta\sigma^*_{fib}(k,F) .$$

(ii) *The covering isomorphism*

$$\beta^{fib} : LIso^{n+1}_{fib}(\mathbb{Z}) \xrightarrow{\simeq} L^{n+3}(\mathbb{Z}[z,z^{-1}], P_{fib}) = C^{fib}_n$$

sends the cobordism class of the fibred Seifert complex $\sigma^*_{fib}(k,F)$ to the cobordism class of the fibred Blanchfield complex $\sigma^*_{fib}(k)$, with $P_{fib} \subset \mathbb{Z}[z,z^{-1}]$ as in 32.21.
Proof As for 33.8. □

33D. Simple n-knots

Definition 33.10 Let $k : S^n \subset S^{n+2}$ be an n-knot, with exterior X.
(i) The n-knot k is *r-simple* if the map $X \longrightarrow S^1$ is r-connected, that is if

$$\pi_i(X) = \pi_i(S^1) \text{ for } i \leq r ,$$

or equivalently if k admits an r-connected Seifert surface $F^{n+1} \subset S^{n+2}$, with

$$\pi_i(F) = 0 \text{ for } i \leq r .$$

(ii) The n-knot k is *simple* if it is $[(n-1)/2]$-simple.
(iii) The n-knot k is *stable* if it is $([n/3]+1)$-simple. □

An n-knot $k : S^n \subset S^{n+2}$ is simple if and only if the Blanchfield complex $\sigma^*(k)$ is $[(n-1)/2]$-connected, corresponding to a nonsingular $(-)^{i+1}$-symmetric linking form over $(\mathbb{Z}[z, z^{-1}], P)$ (= Blanchfield form) for $n = 2i-1$ and a nonsingular $(-)^{i+1}$-symmetric linking formation over $(\mathbb{Z}[z, z^{-1}], P)$ for $n = 2i$. See Ranicki [237, Chap. 3.5] for linking formations, and [237, Chap. 7.9] for more on Blanchfield complexes.

Proposition 33.11 (Levine [159], Kearton [125], Trotter [291], [292])
For $i \geq 2$ the following sets are in one-one correspondence:

(i) *the isotopy classes of simple $(2i-1)$-knots $k : S^{2i-1} \subset S^{2i+1}$,*
(ii) *the S-equivalence classes of $(-)^i$-symmetric Seifert forms over $\mathbb{Z}$,*
(iii) *the isomorphism classes of $(-)^{i+1}$-symmetric Blanchfield forms over $\mathbb{Z}[z, z^{-1}]$.*

Proof See the original references for detailed proofs. The S-equivalence equivalence relation on $(-)^i$-symmetric Seifert forms over $\mathbb{Z}$ (originally defined by Murasugi [206]) is generated by isomorphism and relations of the type

$$(L, \lambda) \sim (L \oplus K \oplus K^*, \begin{pmatrix} \lambda & \beta & 0 \\ \alpha & \gamma & 1 \\ 0 & 0 & 0 \end{pmatrix})$$

which correspond to modifying a Seifert surface by ambient codimension 1 surgery. (See 32.16 for the S-equivalence relation on Seifert complexes.) □

Remark 33.12 (i) For $i \geq 2$ the isotopy classes of simple $2i$-knots $k : S^{2i} \subset S^{2i+2}$ are in one-one correspondence with the isomorphism classes of nonsingular $(-)^{i+1}$-symmetric linking formations over $(\mathbb{Z}[z, z^{-1}], P)$ together with an extra homotopy pairing, by the work of Kearton [126] and Farber [73], [74].
(ii) Schubert [258] proved that every 1-knot has a unique factorization as the connected sum of finitely many irreducible 1-knots. For $n \geq 2$ every n-knot factorizes into finitely many irreducible knots, by Dunwoody and Fenn [63]. In Bayer [16] and Bayer-Fluckiger [17] the classification of simple $(2i-1)$-knots and algebraic results on quadratic forms were used to prove that factorization is not unique for high-dimensional knots, and that cancellation fails.
(iii) Farber [72] obtained an isotopy classification of stable n-knots $k : S^n \subset S^{n+2}$ for $n \geq 5$ in terms of homotopy Seifert pairings.
(iv) For $n \geq 5$ the isotopy class of an n-knot $k : S^n \subset S^{n+2}$ is determined by the complement up to two possibilities, by Browder [28]: if k has exterior X and $\tau : S^n \times S^1 \longrightarrow S^n \times S^1$ is a diffeomorphism representing the generator $\tau \in \pi_1(SO(n+1)) = \mathbb{Z}_2$ the n-knot

$$k_\tau \ : \ S^n \subset S^n \times D^2 \cup_\tau X \ = \ S^{n+2}$$

with the same exterior $X_\tau = X$ may not be equivalent to k. This generalized the result of Gluck [87] for $n = 2$, and was subsequently extended to

$n = 3, 4$ by Lashof and Shaneson [147]. It followed from the classifications that high-dimensional simple and stable n-knots are determined by their complements, since the isotopy invariants could be derived from the n-knot exterior. Richter [251] used Poincaré embedding theory to prove that $[n/3]$-simple n-knots $k : S^n \subset S^{n+2}$ $(n \geq 5)$ are determined by their complements. Cappell and Shaneson [43] and Gordon [93] constructed examples of n-knots $S^n \subset S^{n+2}$ for $n = 3, 4$ which are not determined by their complements. Suciu [280] constructed pairs of inequivalent n-knots $S^n \subset S^{n+2}$ with the same complement, for every $n \geq 3$ with $n \equiv 3$ or $4 (\mathrm{mod}\, 8)$. Gordon and Luecke [95] showed that a classical knot $S^1 \subset S^3$ is determined by its complement.
(v) Let $k : S^n \subset S^{n+2}$ be a 1-simple n-knot, so that the exterior X is such that $\pi_1(X) = \mathbb{Z}$, and there exists a simply-connected Seifert surface $F^{n+1} \subset S^{n+2}$. The n-knot k fibres if and only if it admits a simply-connected Seifert surface $F^{n+1} \subset S^{n+2}$ such that $\sigma^*(k, F)$ is a fibred Seifert complex over $\mathbb{Z}$. For $n \geq 3$ this is equivalent to the Blanchfield complex $\sigma^*(k)$ over $\mathbb{Z}[z, z^{-1}]$ being fibred. □

33E. The Alexander polynomials of an n-knot

The algebraic theory of Alexander polynomials developed in Chap. 17 will now be applied to n-knots.

Definition 33.13 The *Alexander polynomials* of an n-knot $k : S^n \subset S^{n+2}$ are the Alexander polynomials (17.1)

$$\Delta_r(z) \in \mathbb{Z}[z] \ \ (1 \leq r \leq n)$$

of the Blanchfield complex $\sigma^*(k) = (E, \theta)$ over $\mathbb{Z}[z, z^{-1}]$, such that

$$\Delta_{n+1-r}(z) \ = \ z^{m_r}\Delta_r(z^{-1}) \in \mathbb{Z}[z] \ \ (1 \leq r \leq n)$$

by 32.38 (ii), with $m_r = \mathrm{degree}(\Delta_r(z))$. □

By convention

$$\Delta_0(z) \ = \ 1 - z \ \ , \ \ \Delta_r(z) \ = \ 1 \ \ (r \geq n+1) \ .$$

The Blanchfield complex of k is the covering of the Seifert complex over $\mathbb{Z}$ (32.5) with respect to a Seifert surface $F^{n+1} \subset S^{n+2}$

$$\sigma^*(k) \ = \ \beta\sigma^*(k, F)$$

with isometric structure $(C(F)^{n+1-*}, ((1+T)\widehat{\psi})^{-1}\widehat{\psi})$. The Alexander polynomials of k are given by

$$\begin{aligned} \Delta_r(z) \ &= \ \pm z^{-n_r}\det((z\widehat{\psi} + T\widehat{\psi})((1+T)\widehat{\psi})^{-1} : H_r(F)[z] \longrightarrow H_r(F)[z]) \\ &\in \mathbb{Z}[z] \ \ (1 \leq r \leq n+1) \ , \end{aligned}$$

with
$$n_r = \dim_{\mathbb{Q}} H_r(\overline{X}^+;\mathbb{Q}) - \dim_{\mathbb{Q}} H_r(\overline{X};\mathbb{Q}) \geq 0 .$$

Example 33.14 (i) If (L,λ) is a nonsingular ϵ-symmetric Seifert form over an integral domain with involution A then by 17.5 the Alexander polynomial of the 2-dimensional A-contractible finite f.g. free $A[z,z^{-1}]$-module chain complex
$$E \ : \ \ldots \longrightarrow 0 \longrightarrow L[z,z^{-1}] \xrightarrow{z\lambda+\epsilon\lambda^*} L^*[z,z^{-1}] \longrightarrow 0$$
is
$$\Delta_1(z) \ = \ z^{-n}\det((\lambda+\epsilon\lambda^*)^{-1}(z\lambda+\epsilon\lambda^*) : L[z] \longrightarrow L[z]) \in A[z] ,$$
with $n \geq 0$ determined by the condition $\Delta_1(0) \neq 0 \in \mathbb{Z}$. Note that $n = 0$ if and only if $\lambda : L \longrightarrow L^*$ is injective.
(ii) Let $k : S^{2i-1} \subset S^{2i+1}$ be a simple $(2i-1)$-knot. A Seifert surface $M^{2i} \subset S^{2i+1}$ determines a nonsingular $(-)^i$-symmetric Seifert form (L,λ) over $\mathbb{Z}$ with $L = H^i(M)/\text{torsion}$, via the Seifert complex (32.5). By (i) the ith Alexander polynomial of k is given by
$$\Delta_i(z) \ = \ \det((\lambda+(-)^i\lambda^*)^{-1}(z\lambda+(-)^i\lambda^*) : L[z] \longrightarrow L[z]) \in \mathbb{Z}[z] . \quad \square$$

Proposition 33.15 *The Alexander polynomials of a fibred n-knot $k : S^n \subset S^{n+2}$ with exterior $X = T(h : F \longrightarrow F)$ are related to the characteristic polynomials of the monodromy automorphisms*
$$q_r(s) \ = \ \mathrm{ch}_s(H_r(F),h_*) \in \mathbb{Z}[s] \ \ (1 \leq r \leq n)$$
by
$$q_r(s) \ = \ s^{n_r}\Delta_r(1-s^{-1}) \in \mathbb{Z}[s] ,$$
$$\Delta_r(z) \ = \ z^{n_r}q_r(1-z^{-1}) \in \mathbb{Z}[z]$$
with $n_r = \text{degree}\,(q_r(s)) = \text{degree}\,(\Delta_r(z)) = \dim_{\mathbb{Q}} H_r(F;\mathbb{Q})$.
Proof The Seifert complex $(C,\widehat{\psi})$ is fibred, and the monodromy is the isometric structure
$$h \ = \ T\widehat{\psi}((1+T)\widehat{\psi})^{-1} \ : \ C \longrightarrow C .$$
Setting $s = (1-z)^{-1}$ gives
$$s - h \ = \ (1-z)^{-1}(zT\widehat{\psi}+\widehat{\psi})((1+T)\widehat{\psi})^{-1} \ :$$
$$C[z,z^{-1},(1-z)^{-1}] \longrightarrow C[z,z^{-1},(1-z)^{-1}] . \quad \square$$

Proposition 33.16 *Let $k : S^n \subset S^{n+2}$ be an n-knot with exterior X, such that $\pi_1(X) = \mathbb{Z}$. For $n \geq 4$ the following conditions are equivalent:*

(i) *X fibres over S^1,*
(ii) *the infinite cyclic cover $\overline{X}$ is finitely dominated,*
(iii) *the $\mathbb{Z}[z,z^{-1}]$-module chain complex $C(\overline{X})$ is $\mathbb{Z}$-finitely dominated,*

(iv) *the homology* $\mathbb{Z}$*-modules* $H_r(\overline{X})$ $(2 \leq r \leq (n+1)/2)$ *are finitely generated,*

(v) *the extreme coefficients of the Alexander polynomials*

$$\Delta_r(z) \;=\; \sum_{j=0}^{m_r} a_{j,r} z^j \in \mathbb{Z}[z] \quad (2 \leq r \leq (n+1)/2)$$

are $a_{0,r}, a_{m_r,r} = \pm 1 \in \mathbb{Z}$.

Proof (i) $\iff$ (ii) By Browder and Levine [31]. Alternatively, apply the Farrell–Siebenmann fibering obstruction theory, noting that the obstruction takes value in $Wh_1(\mathbb{Z}) = 0$.
(ii) $\iff$ (iii) A special case of 3.5.
(iii) $\iff$ (iv) The finite domination of $\overline{X}$ is equivalent to the finite generation of the homology $\mathbb{Z}$-modules $H_*(\overline{X})$, which by Poincaré duality is equivalent to (iv).
(iv) $\iff$ (v) Immediate from 32.6. □

Remark 33.17 The equivalence 33.16 (i) $\iff$ (v) for $n \geq 4$ was first obtained by Sumners [281, 3.4]. □

34. Endomorphism L-theory

Open books and automorphisms of manifolds were shown in Chaps. 28–30 to be closely related to the L-theory of the Laurent polynomial extensions $A[z,z^{-1}]$ of rings with involution A, with the involution extended by $\overline{z}=z^{-1}$. High-dimensional knots have been shown in Chaps. 31–33 to be closely related to the L-theory of the polynomial extensions $A[s]$ of rings with involution A, with the involution extended by $\overline{s}=1-s$. This chapter deals with the L-theory of $A[x]$ with $\overline{x}=x$, which is somewhat easier to deal with, yet shares many essential features with the L-theories of $A[z,z^{-1}]$ and $A[s]$.

An endometric structure on a symmetric form (M,ϕ) is an endomorphism $f:M\longrightarrow M$ such that

$$\phi f \;=\; f^*\phi \;:\; M \longrightarrow M^* \;.$$

Endomorphism L-theory is the study of symmetric forms and algebraic Poincaré complexes with an endometric structure. The localization exact sequence of Chap. 25 and endomorphism L-theory give L-theory analogues of the K-theory splitting theorems of Chaps. 5, 10

$$\begin{aligned}
K_1(A[z]) &= K_1(A)\oplus \widetilde{\mathrm{Nil}}_0(A)\;,\\
K_1(A[z,z^{-1}]) &= K_1(A[z])\oplus \mathrm{Nil}_0(A)\\
&= K_1(A)\oplus K_0(A)\oplus \widetilde{\mathrm{Nil}}_0(A)\oplus \widetilde{\mathrm{Nil}}_0(A)\;,\\
K_1(\Omega_+^{-1}A[z]) &= K_1(A[z])\oplus \mathrm{End}_0(A)\\
&= K_1(A)\oplus K_0(A)\oplus \widetilde{\mathrm{Nil}}_0(A)\oplus \widetilde{\mathrm{End}}_0(A)\;,\\
K_1(\widetilde{\Omega}_+^{-1}A[z]) &= K_1(A)\oplus \widetilde{\mathrm{End}}_0(A)\;,
\end{aligned}$$

with $\Omega_+,\widetilde{\Omega}_+$ the sets of matrices in $A[z]$ defined in Chap. 10.

34A. Endometric structures

Terminology 34.1 Given a ring with involution A let $A[x]$ denote the polynomial extension of A with the involution extended by

$$\overline{x} = x . \qquad \square$$

Recall that Ω_+ is the set of Fredholm matrices in $A[x]$ (10.4). The localization exact sequence

$$\dots \longrightarrow L_h^n(A[x],\epsilon) \longrightarrow L_h^n(\Omega_+^{-1}A[x],\epsilon) \longrightarrow L_h^n(A[x],\Omega_+,\epsilon) \longrightarrow L_h^{n-1}(A[x],\epsilon) \longrightarrow \dots$$

will now be shown to break up into split exact sequences

$$0 \longrightarrow L_h^n(A[x],\epsilon) \longrightarrow L_h^n(\Omega_+^{-1}A[x],\epsilon) \longrightarrow L_h^n(A[x],\Omega_+,\epsilon) \longrightarrow 0$$

with

$$L_h^n(A[x],\Omega_+,\epsilon) = L\mathrm{End}_p^n(A,\epsilon)$$

the L-group of f.g. projective n-dimensional ϵ-symmetric Poincaré complexes over A with an endometric structure, and similarly in the ϵ-quadratic case. The reduced endomorphism L-groups $\widetilde{L\mathrm{End}}_p^n(A,\epsilon)$ are defined such that

$$L\mathrm{End}_p^n(A,\epsilon) = L_p^n(A,\epsilon) \oplus \widetilde{L\mathrm{End}}_p^n(A,\epsilon) .$$

The splitting theorems

$$L_h^n(\Omega_+^{-1}A[x],\epsilon) = L_h^n(A[x],\epsilon) \oplus L\mathrm{End}_p^n(A,\epsilon) ,$$

$$L_h^n(\widetilde{\Omega}_+^{-1}A[x],\epsilon) = L_h^n(A,\epsilon) \oplus \widetilde{L\mathrm{End}}_p^n(A,\epsilon)$$

are generalizations of the splitting theorems of Ranicki [232], [237] for the L-theory of $A[x]$ and $A[x,x^{-1}]$: the nilpotent and reduced nilpotent L-groups $L\mathrm{Nil}_p^*(A,\epsilon)$, $\widetilde{L\mathrm{Nil}}_p^*(A,\epsilon)$ are the cobordism groups of f.g. projective ϵ-symmetric Poincaré complexes over A with a nilpotent structure, such that

$$L\mathrm{Nil}_p^n(A,\epsilon) = L_p^n(A,\epsilon) \oplus \widetilde{L\mathrm{Nil}}_p^n(A,\epsilon) ,$$

$$L_h^n(A[x,x^{-1}],\epsilon) = L_h^n(A[x],\epsilon) \oplus L\mathrm{Nil}_p^n(A,\epsilon) ,$$

$$L_h^n(A[x],\epsilon) = L_h^n(A,\epsilon) \oplus \widetilde{L\mathrm{Nil}}_p^n(A,\epsilon) .$$

By 10.14 there is defined an isomorphism of categories

$$\mathrm{End}(A) \xrightarrow{\simeq} \mathbb{H}(A[x],\Omega_+) \ ; \ (P,f) \longrightarrow \mathrm{coker}(x-f : P[x] \longrightarrow P[x])$$

for any ring A, and the algebraic K-theory localization exact sequence

$$\ldots \longrightarrow K_1(A[x]) \xrightarrow{i_+} K_1(\Omega_+^{-1}A[x]) \xrightarrow{\partial_+} K_1(A[x],\Omega_+)$$
$$\xrightarrow{j_+} K_0(A[x]) \longrightarrow \ldots$$

splits, with $j_+ = 0$ and ∂_+ split by

$$\begin{aligned}\Delta_+ \ : \ K_1(A[x],\Omega_+) \ &= \ \mathrm{End}_0(A) \longrightarrow K_1(\Omega_+^{-1}A[x]) \ ; \\ [P,f] &\longrightarrow \tau(x - f : \Omega_+^{-1}P[x] \longrightarrow \Omega_+^{-1}P[x]) \ .\end{aligned}$$

A based f.g. free $A[x]$-module chain complex D is $\Omega_+^{-1}A[x]$-contractible if and only if D is A-finitely dominated, in which case

$$\begin{aligned}\tau(\Omega_+^{-1}D) \ &= \ (\Phi(D),[D^!,\xi]) \\ &\in K_1(\Omega_+^{-1}A[x]) \ = \ K_1(A[x]) \oplus \mathrm{End}_0(A)\end{aligned}$$

with $\Phi(D) \in K_1(A[x])$ the fibering obstruction of 5.11, $D^!$ the finitely dominated A-module chain complex defined by D and

$$\xi \ : \ D^! \longrightarrow D^! \ ; \ w \longrightarrow xw \ .$$

Proposition 34.2 (i) *The set Ω_+ of Fredholm matrices in $A[x]$ is involution-invariant, such that the Ω_+-duality involution on $K_1(A[x],\Omega_+)$ corresponds to the involution*

$$* \ : \ \mathrm{End}_0(A) \longrightarrow \mathrm{End}_0(A) \ ; \ (P,f) \longrightarrow (P^*,f^*)$$

on the endomorphism class group $\mathrm{End}_0(A)$.
(ii) *The direct sum system of 10.14*

$$K_1(A[x]) \underset{j_+}{\overset{i_+}{\rightleftarrows}} K_1(\Omega_+^{-1}A[x]) \underset{\Delta_+}{\overset{\partial_+}{\rightleftarrows}} \mathrm{End}_0(A)$$

is a direct sum system of $\mathbb{Z}[\mathbb{Z}_2]$-modules, so that

$$\widehat{H}^*(\mathbb{Z}_2;K_1(\Omega_+^{-1}A[x])) \ = \ \widehat{H}^*(\mathbb{Z}_2;K_1(A[x])) \oplus \widehat{H}^*(\mathbb{Z}_2;\mathrm{End}_0(A)) \ .$$

(iii) *The direct sum system of 10.16*

$$K_1(A) \underset{\widetilde{j}_+}{\overset{\widetilde{i}_+}{\rightleftarrows}} K_1(\widetilde{\Omega}_+^{-1}A[x]) \underset{\widetilde{\Delta}_+}{\overset{\widetilde{\partial}_+}{\rightleftarrows}} \widetilde{\mathrm{End}}_0(A)$$

is a direct sum system of $\mathbb{Z}[\mathbb{Z}_2]$-modules, so that

$$\widehat{H}^*(\mathbb{Z}_2;K_1(\widetilde{\Omega}_+^{-1}A[x])) \ = \ \widehat{H}^*(\mathbb{Z}_2;K_1(A)) \oplus \widehat{H}^*(\mathbb{Z}_2;\widetilde{\mathrm{End}}_0(A)) \ .$$

Proof (i) Given a $k \times k$ Fredholm matrix $\omega = (\omega_{ij})$ in $A[x]$ there is defined a f.g. projective A-module

$$P = \mathrm{coker}(x - \omega : A[x]^k \longrightarrow A[x]^k)$$

with an endomorphism

$$f : P \longrightarrow P \; ; \; y \longrightarrow xy$$

and an $A[x]$-module exact sequence

$$0 \longrightarrow A[x]^k \xrightarrow{x-\omega} A[x]^k \longrightarrow P \longrightarrow 0 \, .$$

The dual $k \times k$ matrix $\omega^* = (\overline{\omega}_{ji})$ is Fredholm, with an $A[x]$-module exact sequence

$$0 \longrightarrow A[x]^k \xrightarrow{x-\omega^*} A[x]^k \longrightarrow P^* \longrightarrow 0$$

where

$$x = f^* : P^* \longrightarrow P^* \, .$$

(ii)+(iii) Immediate from (i). □

Definition 34.3 (i) Given an A-module chain complex C and a chain map $f : C \longrightarrow C$ define the $\mathbb{Z}[\mathbb{Z}_2]$-module chain map

$$\Gamma_f^{end} = f \otimes 1 - 1 \otimes f \; : \; (C \otimes_A C, T_\epsilon) \longrightarrow (C \otimes_A C, T_{-\epsilon}) \, .$$

The $\begin{cases} \epsilon\text{-}symmetric \\ \epsilon\text{-}quadratic \end{cases}$ *endometric Q-groups* of (C, f) are the relative groups

$$\begin{cases} Q^{n+1}_{end}(C,f,\epsilon) = H_{n+1}(\Gamma_f^{end} : \mathrm{Hom}_{\mathbb{Z}[\mathbb{Z}_2]}(W,(C \otimes_A C, T_\epsilon)) \\ \qquad\qquad\qquad \longrightarrow \mathrm{Hom}_{\mathbb{Z}[\mathbb{Z}_2]}(W,(C \otimes_A C, T_{-\epsilon}))) \\ Q_{n+1}^{end}(C,f,\epsilon) = H_{n+1}(\Gamma_f^{end} : W \otimes_{\mathbb{Z}[\mathbb{Z}_2]} (C \otimes_A C, T_\epsilon) \\ \qquad\qquad\qquad \longrightarrow W \otimes_{\mathbb{Z}[\mathbb{Z}_2]} (C \otimes_A C, T_{-\epsilon}))) \end{cases}$$

in the exact sequence

$$\begin{cases} \dots \longrightarrow Q^{n+1}(C,-\epsilon) \longrightarrow Q^{n+1}_{end}(C,f,\epsilon) \\ \qquad\qquad \longrightarrow Q^n(C,\epsilon) \xrightarrow{\Gamma_f^{end}} Q^n(C,-\epsilon) \longrightarrow \dots \\ \dots \longrightarrow Q_{n+1}(C,-\epsilon) \longrightarrow Q_{n+1}^{end}(C,f,\epsilon) \\ \qquad\qquad \longrightarrow Q_n(C,\epsilon) \xrightarrow{\Gamma_f^{end}} Q_n(C,-\epsilon) \longrightarrow \dots \, . \end{cases}$$

(ii) An *endometric structure* $(f, \delta\phi)$ for an n-dimensional ϵ-symmetric complex (C, ϕ) over A is a chain map $f : C \longrightarrow C$ together with a chain

$$\delta\phi \in (W \otimes_{\mathbb{Z}[\mathbb{Z}_2]} (C \otimes_A C))_{n+1}$$

such that

$$f\phi_s - \phi_s f^* = d\delta\phi_s + (-)^r\delta\phi_s d^* + (-)^{n+s-1}(\delta\phi_{s-1} + (-)^s T_{-\epsilon}\delta\phi_{s-1})$$
$$: C^{n-r+s} \longrightarrow C_r \quad (r, s \geq 0, \delta\phi_{-1} = 0) ,$$

representing an element $(\delta\phi, \phi) \in Q^{n+1}_{end}(C, f, \epsilon)$.
Similarly in the ϵ-quadratic case. □

Example 34.4 An *endometric structure* for an ϵ-symmetric form (M, ϕ) over A is an endomorphism $f : M \longrightarrow M$ such that

$$f^*\phi = \phi f : M \longrightarrow M^* .$$

This is just an endometric structure $(f^*, 0)$ on the 0-dimensional ϵ-symmetric complex (C, ϕ) with $C_0 = M^*$, $C_r = 0$ for $r \neq 0$. □

Proposition 34.5 (i) *The $(-\epsilon)$-symmetric Q-groups of an A-finitely dominated $(= \Omega_+^{-1}A[x]$-contractible) finite f.g. free $A[x]$-module chain complex D are the endometric ϵ-symmetric Q-groups of $(D^!, \xi)$*

$$Q^*(D, -\epsilon) = Q^*_{end}(D^!, \xi) .$$

Similarly in the ϵ-quadratic case.
(ii) *The homotopy equivalence classes of finitely dominated n-dimensional ϵ-symmetric Poincaré complexes (C, ϕ) over A with an endometric structure $(f, \delta\phi)$ are in one-one correspondence with the homotopy equivalence classes of f.g. free $(n+1)$-dimensional $\Omega_+^{-1}A[x]$-contractible $(-\epsilon)$-symmetric Poincaré complexes (D, θ) over $A[x]$.*
(iii) *A finitely dominated n-dimensional ϵ-symmetric Poincaré complex (C, ϕ) over A with an endometric structure $(f, \delta\phi)$ determines a f.g. projective n-dimensional ϵ-symmetric $\Omega_+^{-1}A[x]$-Poincaré complex $(C[x], \Phi)$ over $A[x]$ with*

$$\Phi_s = (x - f)\phi_s + \delta\phi_{s-1} : C[x]^{n-r+s} \longrightarrow C[x]_r \quad (r, s \geq 0) .$$

The skew-suspension $(SC[x], \Phi)$ is a f.g. projective $(n+2)$-dimensional $(-\epsilon)$-symmetric $\Omega_+^{-1}A[x]$-Poincaré complex over $A[x]$. The boundary

$$(D, \theta) = \partial(SC[x], \Phi)$$

is a (homotopy) f.g. free $(n+1)$-dimensional $(-\epsilon)$-symmetric Poincaré complex over $A[x]$ corresponding to (C, ϕ) and $(f, \delta\phi)$ in (i), *such that*

$$\tau_{\Omega_+}(D) = [C, f] \in K_1(A[x], \Omega_+) = \mathrm{End}_0(A) ,$$
$$\tau(D, \theta) = \Phi(D) + (-)^n\Phi(D)^* \in K_1(A[x])$$

with respect to the canonical round finite structure on D.

Proof (i) Let (C, f) be a finite chain complex of $(A[x], \Omega_+)$-modules, i.e. a finite f.g. projective A-module chain complex C together a chain map $f : C \longrightarrow C$. The f.g. projective $A[x]$-module chain complex

$$D = \mathcal{C}(x - f : C[x] \longrightarrow C[x])$$

is homology equivalent to (C, f), and every A-finitely dominated f.g. free $A[x]$-module chain complex D is chain equivalent to one of this type, with $(D^!, \xi) \simeq (C, f)$. The $\mathbb{Z}[\mathbb{Z}_2]$-module chain map

$$\mathcal{C}(\Gamma_f^{end} : C \otimes_A C \longrightarrow C \otimes_A C) \longrightarrow D \otimes_{A[x]} D$$

defined by

$$\begin{aligned}\mathcal{C}(\Gamma_f^{end})_r &= (C \otimes_A C)_r \oplus (C \otimes_A C)_{r-1} \\ &\longrightarrow (D \otimes_{A[x]} D)_r = (C[x] \otimes_{A[x]} C[x])_r \oplus (C[x] \otimes_{A[x]} C[x])_{r-1} \\ &\oplus (C[x] \otimes_{A[x]} C[x])_{r-1} \oplus (C[x] \otimes_{A[x]} C[x])_{r-2} \; ; \; (u, v) \longrightarrow (u, v, v, 0)\end{aligned}$$

is a homology equivalence inducing isomorphisms

$$Q^*_{end}(C, f, \epsilon) \cong Q^*(D, -\epsilon) \; ,$$

so that there is defined an exact sequence

$$\dots \longrightarrow Q^{n+1}(D, -\epsilon) \longrightarrow Q^n(C, \epsilon) \xrightarrow{\Gamma_f^{end}} Q^n(C, -\epsilon) \longrightarrow Q^n(D, -\epsilon) \longrightarrow \dots \; .$$

An $(n+1)$-dimensional $(-\epsilon)$-symmetric structure $\theta \in Q^{n+1}(D, -\epsilon)$ on D is thus the same as an n-dimensional ϵ-symmetric structure $\phi \in Q^n(C, \epsilon)$ with a refinement to an endometric structure $(\delta\phi, \phi) \in Q^{n+1}_{end}(C, f, \epsilon)$. Moreover

$$\theta_0 \; : \; H^{n+1-*}(D) \cong H^{n-*}(C) \xrightarrow{\phi_0} H_*(C) \cong H_*(D) \; ,$$

so that (D, θ) is a Poincaré complex over $A[x]$ if and only if (C, ϕ) is a Poincaré complex over A.
(ii)+(iii) Immediate from (i). □

Definition 34.6 (i) Let $U \subseteq \mathrm{End}_0(A)$ be a $*$-invariant subgroup. The *ϵ-symmetric endomorphism L-group* $L\mathrm{End}^n_U(A, \epsilon)$ $(n \geq 0)$ is the cobordism group of finitely dominated n-dimensional ϵ-symmetric Poincaré complexes (C, ϕ) over A with an endometric structure $(f, \delta\phi)$ such that

$$[C, f] \in U \subseteq \mathrm{End}_0(A) \; .$$

(ii) Let

$$U = U_0 \oplus V \subseteq \mathrm{End}_0(A) = K_0(A) \oplus \widetilde{\mathrm{End}}_0(A)$$

for some $*$-invariant subgroups $U_0 \subseteq K_0(A)$, $V \subseteq \widetilde{\mathrm{End}}_0(A)$. The *$\epsilon$-symmetric reduced endomorphism L-group* $\widetilde{L\mathrm{End}}^n_V(A, \epsilon)$ $(n \geq 0)$ is

$$\widetilde{L\mathrm{End}}^n_V(A, \epsilon) = \ker(L\mathrm{End}^n_U(A, \epsilon) \longrightarrow L^n_{U_0}(A, \epsilon); (C, f, \delta\phi, \phi) \longrightarrow (C, \phi)) \; ,$$

so that

$$L\mathrm{End}^n_U(A, \epsilon) = L^n_{U_0}(A, \epsilon) \oplus \widetilde{L\mathrm{End}}^n_V(A, \epsilon) \; .$$

(iii) Similarly for the *ε-quadratic endomorphism L-group* $L\mathrm{End}_n^U(A,\epsilon)$ and the *ε-quadratic reduced endomorphism L-group* $\widetilde{L\mathrm{End}}_n^V(A,\epsilon)$. □

The L-theoretic analogue of 10.14 is:

Proposition 34.7 (i) *For any ∗-invariant subgroup $U \subseteq \mathrm{End}_0(A)$ there are natural identifications of the $(A[x],\Omega_+)$-torsion L-groups and the endomorphism L-groups of A*

$$L^n_U(A[x],\Omega_+,\epsilon) \;=\; L\mathrm{End}^n_U(A,\epsilon)\ .$$

(ii) *For any ∗-invariant subgroups $T \subseteq K_1(A[x])$, $U \subseteq \mathrm{End}_0(A)$ the localization exact sequence of 25.4*

$$\begin{aligned}\ldots \longrightarrow L^n_T(A[x],\epsilon) &\xrightarrow{i_+} L^n_{i_+(T)\oplus\Delta_+(U)}(\Omega_+^{-1}A[x],\epsilon)\\ &\xrightarrow{\partial} L\mathrm{End}^n_U(A,\epsilon) \longrightarrow L^{n-1}_T(A[x],\epsilon) \longrightarrow \ldots\end{aligned}$$

breaks up into split exact sequences, with direct sum systems

$$L^n_{i_+(T)}(A[x],\epsilon) \underset{j_+}{\overset{i_+}{\rightleftarrows}} L^n_{i_+(T)\oplus\Delta_+(U)}(\Omega_+^{-1}A[x],\epsilon) \underset{\Delta_+}{\overset{\partial_+}{\rightleftarrows}} L\mathrm{End}^n_U(A,\epsilon)$$

where

$$\begin{aligned}\Delta_+ \ &: L\mathrm{End}^n_U(A,\epsilon) \longrightarrow L^n_{i_+(T)\oplus\Delta_+(U)}(\Omega_+^{-1}A[x],\epsilon)\ ;\\ &(C,f,\delta\phi,\phi) \longrightarrow \Omega_+^{-1}(\mathcal{C}(x-f:C[x]\longrightarrow C[x]),\Phi)\end{aligned}$$

(Φ as in 34.5 (iii)*).*
(iii) *The ε-symmetric endomorphism L-groups associated to a pair $(U_2, U_1 \subseteq U_2)$ of ∗-invariant subgroups $U_1 \subseteq U_2 \subseteq \mathrm{End}_0(A)$ are related by a Rothenberg-type exact sequence*

$$\begin{aligned}\ldots \longrightarrow \widehat{H}^{n+1}(\mathbb{Z}_2;U_2/U_1) &\longrightarrow L\mathrm{End}^n_{U_1}(A,\epsilon)\\ &\longrightarrow L\mathrm{End}^n_{U_2}(A,\epsilon) \longrightarrow \widehat{H}^n(\mathbb{Z}_2;U_2/U_1) \longrightarrow \ldots\end{aligned}$$

with

$$L\mathrm{End}^n_{U_2}(A,\epsilon) \longrightarrow \widehat{H}^n(\mathbb{Z}_2;U_2/U_1)\ ;\ (C,f,\delta\phi,\phi) \longrightarrow [C,f]\ .$$

Similarly in the ε-quadratic case.
Proof (i) Immediate from 34.5 (ii).
(ii) By 34.5 (iii) the maps ∂_+, Δ_+ are such that $\partial_+\Delta_+ = 1$.
(iii) As for the ordinary L-groups in Ranicki [235, Chap. 10]. □

Recall that $\widetilde{\Omega}_+$ is the set of square matrices $\omega = \sum\limits_{j=0}^{\infty}\omega_j x^j$ in $A[x]$ with ω_0 invertible (10.8). The L-theoretic analogue of 10.16 is:

Proposition 34.8 *For any $*$-invariant subgroups $U_1 \subseteq K_1(A), V \subseteq \widetilde{\mathrm{End}}_0(A)$ there is defined a direct sum system*

$$L^n_{U_1}(A,\epsilon) \underset{\widetilde{j}_+}{\overset{\widetilde{i}_+}{\rightleftarrows}} L^n_{\widetilde{i}_+(U_1)\oplus\widetilde{\Delta}_+(V)}(\widetilde{\Omega}_+^{-1}A[x],\epsilon) \underset{\widetilde{\Delta}_+}{\overset{\widetilde{\partial}_+}{\rightleftarrows}} \widetilde{LEnd}^n_V(A,\epsilon)$$

with

$$\begin{aligned}
\widetilde{i}_+ \ :\ & L^n_{U_1}(A,\epsilon) \longrightarrow L^n_{\widetilde{i}_+(U_1)\oplus\widetilde{\Delta}_+(V)}(\widetilde{\Omega}_+^{-1}A[x],\epsilon) \ ;\\
& (C,\phi) \longrightarrow \widetilde{\Omega}_+^{-1}A[x]\otimes_A (C,\phi)\ ,\\
\widetilde{\partial}_+ \ :\ & L^n_{\widetilde{i}_+(U_1)\oplus\widetilde{\Delta}_+(V)}(\widetilde{\Omega}_+^{-1}A[x],\epsilon) \longrightarrow L^n_{i_-(U_1)\oplus\Delta_-(V)}(\Omega_-^{-1}A[x^{-1}],\epsilon)\\
& \xrightarrow{\partial_-} \widetilde{LEnd}^n_V(A,\epsilon)\ ,\\
\widetilde{j}_+ \ :\ & L^n_{\widetilde{i}_+(U_1)\oplus\widetilde{\Delta}_+(V)}(\widetilde{\Omega}_+^{-1}A[x],\epsilon) \longrightarrow L^n_{U_1}(A,\epsilon)\ ;\\
& \widetilde{\Omega}_+^{-1}(E^+,\theta^+) \longrightarrow A\otimes_{A[x]}(E^+,\theta^+)\ ,\\
\widetilde{\Delta}_+ \ :\ & \widetilde{LEnd}^n_V(A,\epsilon) \longrightarrow L^n_{U_1\oplus V}(\widetilde{\Omega}_+^{-1}A[x],\epsilon)\ ;\\
& (C,f,\delta\phi,\phi) \longrightarrow \widetilde{\Omega}_+^{-1}(C[x],\widetilde{\Phi})\\
& (\widetilde{\Phi}_s \ =\ (1-xf)\phi_s - x\delta\phi_{s-1}\ ,\ s\geq 0\ ,\ \delta\phi_{-1}\ =\ 0)\ .
\end{aligned}$$

Proof As in the proof of 10.16 consider the localization $\Omega_-^{-1}A[x^{-1}]$ of $A[x^{-1}]$ inverting the set Ω_- of Fredholm matrices in $A[x^{-1}]$. By 34.7

$$\begin{aligned}
L^n_{i_-(U_1)\oplus\Delta_-(V)}(\Omega_-^{-1}A[x^{-1}],\epsilon) \ &=\ L^n_{U_1}(A[x],\epsilon)\oplus LEnd^n_V(A,\epsilon)\\
&=\ L^n_{U_1}(A,\epsilon)\oplus \widetilde{LNil}^n_{\{0\}}(A,\epsilon)\oplus L^n_h(A,\epsilon)\oplus \widetilde{LEnd}^n_V(A,\epsilon)\ .
\end{aligned}$$

The cartesian square of rings with involution

$$\begin{array}{ccc}
A[x] & \longrightarrow & \widetilde{\Omega}_+^{-1}A[x]\\
\big\downarrow & & \big\downarrow\\
A[x,x^{-1}] & \longrightarrow & \Omega_-^{-1}A[x^{-1}]
\end{array}$$

induces excision isomorphisms

$$L^*_{U_1}(A[x],X,\epsilon)\ \cong\ L^*_{U_1\oplus V}(\widetilde{\Omega}_+^{-1}A[x],\Omega_-,\epsilon)\ .$$

Thus

$$\begin{aligned}
L^n_{\widetilde{i}_+(U_1)\oplus\widetilde{\Delta}_+(V)}(\widetilde{\Omega}_+^{-1}A[x],\Omega_-,\epsilon) \ &=\ L^n_{U_1}(A[x],X,\epsilon)\ =\ L^n_h(A[x^{-1}],\epsilon)\\
&=\ L^n_h(A,\epsilon)\oplus \widetilde{LNil}^n_{\{0\}}(A,\epsilon)
\end{aligned}$$

and

$$L^n_{\widetilde{i}_+(U_1)\oplus\widetilde{\Delta}_+(V)}(\widetilde{\Omega}_+^{-1}A[x],\epsilon)\ =\ L^n_{U_1}(A,\epsilon)\oplus \widetilde{LEnd}^n_V(A,\epsilon)\ . \qquad \square$$

The splitting theorems of 34.7 and 34.8 can be written as

$$\begin{aligned} L^n_{T\oplus U}(A[x,x^{-1}],\epsilon) &= L^n_T(A[x],\epsilon)\oplus LEnd^n_U(A,\epsilon) \\ L^n_{U_1\oplus V}(\widetilde{\Omega}_+^{-1}A[x],\epsilon) &= L^n_{U_1}(A,\epsilon)\oplus \widetilde{LEnd}^n_V(A,\epsilon)\ . \end{aligned}$$

Terminology 34.9 Write

$$LEnd^n_p(A,\epsilon) = LEnd^n_{\mathrm{Endo}(A)}(A,\epsilon)$$

and similarly in the reduced and ϵ-quadratic cases. □

Proposition 34.10 (i) *The free ϵ-symmetric L-groups of $\Omega_+^{-1}A[x]$ and $\widetilde{\Omega}_+^{-1}A[x]$ are such that*

$$\begin{aligned} L^n_h(\Omega_+^{-1}A[x],\epsilon) &= L^n_h(A[x],\epsilon)\oplus LEnd^n_p(A,\epsilon)\ , \\ L^n_h(\widetilde{\Omega}_+^{-1}A[x],\epsilon) &= L^n_h(A,\epsilon)\oplus \widetilde{LEnd}^n_p(A,\epsilon)\ . \end{aligned}$$

Similarly for the ϵ-quadratic L-groups.
(ii) *The reduced projective odd-dimensional ϵ-quadratic endomorphism L-groups vanish*

$$\widetilde{LEnd}^p_{2*+1}(A,\epsilon) = 0$$

and

$$\begin{aligned} L^h_{2*+1}(\Omega_+^{-1}A[x],\epsilon) &= L^h_{2*+1}(A[x],\epsilon)\oplus L^h_{2*+1}(A,\epsilon)\ , \\ L^h_{2*+1}(\widetilde{\Omega}_+^{-1}A[x],\epsilon) &= L^h_{2*+1}(A,\epsilon)\ , \\ LEnd^p_{2*+1}(A,\epsilon) &= L^p_{2*+1}(A,\epsilon)\ . \end{aligned}$$

Proof (i) These are special cases of 34.7 and 34.8.
(ii) The forgetful maps $\Gamma^h_{2*+1}(\mathcal{F},\epsilon)\longrightarrow L^h_{2*+1}(S,\epsilon)$ are injective for any locally epic morphism of rings with involution $\mathcal{F}:R\longrightarrow S$ (20.19 (ii)). Applying this to the projection

$$\mathcal{F}\ :\ R\ =\ A[x]\longrightarrow S\ =\ A\ ;\ x\longrightarrow 0$$

gives that the forgetful maps

$$\begin{aligned} \Gamma^h_{2*+1}(\mathcal{F}:A[x]\longrightarrow A,\epsilon) &= L^h_{2*+1}(\widetilde{\Omega}_+^{-1}A[x],\epsilon) \\ &= L^h_{2*+1}(A,\epsilon)\oplus\widetilde{LEnd}^p_{2*+1}(A,\epsilon) \\ &\longrightarrow L^h_{2*+1}(A,\epsilon) \end{aligned}$$

are injections, and hence that $\widetilde{LEnd}^p_{2*+1}(A,\epsilon)=0$. □

Example 34.11 Algebraic surgery below the middle dimension was used in Ranicki [235] to prove that the ϵ-quadratic L-groups of a ring with involution A are 4-periodic

$$L_n^U(A,\epsilon) = L_{n+2}^U(A,-\epsilon) = L_{n+4}^U(A,\epsilon) \quad (U \subseteq Wh_i(A))$$

and that the ϵ-symmetric L-groups of a Dedekind ring with involution A are 4-periodic

$$L_U^n(A,\epsilon) = L_U^{n+2}(A,-\epsilon) = L_U^{n+4}(A,\epsilon) .$$

The method extends to prove that the ϵ-quadratic endomorphism L-groups of any ring with involution A are 4-periodic

$$LEnd_n^U(A,\epsilon) = LEnd_{n+2}^U(A,-\epsilon) = LEnd_{n+4}^U(A,\epsilon) \quad (U \subseteq \mathrm{End}_0(A))$$

and that the ϵ-symmetric endomorphism L-groups of a Dedekind ring with involution A are 4-periodic

$$LEnd_U^n(A,\epsilon) = LEnd_U^{n+2}(A,-\epsilon) = LEnd_U^{n+4}(A,\epsilon) .$$

See Chap. 39 below for more on the endomorphism L-theory of fields. □

Remark 34.12 (i) Given an m-dimensional ϵ-symmetric Poincaré complex (C,ϕ) over A define an involution on the ring $H_0(\mathrm{Hom}_A(C,C))$ of chain homotopy classes of chain maps $f : C \longrightarrow C$ by

$$^- : H_0(\mathrm{Hom}_A(C,C)) \longrightarrow H_0(\mathrm{Hom}_A(C,C)) \; ; \; f \longrightarrow \overline{f} = \phi_0 f^*(\phi_0)^{-1} .$$

(ii) The surgery transfer group $L^m(R,A,\epsilon)$ $(m \geq 0)$ of Lück and Ranicki [177] (cf. 27.15, 27.24) is defined for any rings with involution R, A to be the cobordism group of triples (C,ϕ,ρ) with (C,ϕ) a m-dimensional ϵ-symmetric Poincaré complex over A and $\rho : R \longrightarrow H_0(\mathrm{Hom}_A(C,C))^{op}$ a morphism of rings with involution, where it is assumed that either $m = 0$ or that A is a Dedekind ring. The surgery transfer groups act on the quadratic L-groups by products

$$L^m(R,A,\epsilon) \otimes L_n(R \otimes_{\mathbb{Z}} B,\eta) \longrightarrow L_{m+n}(A \otimes_{\mathbb{Z}} B,\epsilon\eta) .$$

(iii) An endometric structure $(f,\delta\phi)$ on an m-dimensional ϵ-symmetric Poincaré complex (C,ϕ) over A determines an element $f \in H_0(\mathrm{Hom}_A(C,C))^{op}$ such that $\overline{f} = f$, so that there is defined a morphism of rings with involution

$$\rho : \mathbb{Z}[x] \longrightarrow H_0(\mathrm{Hom}_A(C,C))^{op} \; ; \; x \longrightarrow f .$$

The morphism

$$LEnd^m(A,\epsilon) \longrightarrow L^m(\mathbb{Z}[x],A,\epsilon) \; ; \; (C,f,\delta\phi,\phi) \longrightarrow (C,\phi,\rho)$$

is an isomorphism for $m = 0$ (by 34.4). The products of [177] give products

$$LEnd^m(A,\epsilon) \otimes L_n(B[x],\eta) \longrightarrow L_{m+n}(A \otimes_{\mathbb{Z}} B,\epsilon \otimes \eta)$$

for any rings with involution A, B and any $m, n \geq 0$. □

By analogy with the round L-groups (20.15, 20.16):

Example 34.17 Suppose that A is commutative, so that (as in 10.10 (iii))

$$\Omega_+^{-1}A[x] = S^{-1}A[x]$$

with $S \subset A[x]$ the multiplicative subset consisting of all the monic polynomials

$$p(x) = \sum_{j=0}^{d} a_j x^j \in A[x] \ \ (a_d = 1) .$$

Let L be an $(A[x], S)$-module, i.e. a f.g. projective A-module (also denoted by L) together with an endomorphism $f : L \longrightarrow L$. If $p(x) \in S$ is such that $p(f) = 0$ (e.g. $p(x) = \mathrm{ch}_x(L, f)$ if L is a f.g. free A-module) there is defined an $A[x]$-module isomorphism

$$\mathrm{Hom}_{A[x]}(L, A[x]/(p(x))) \xrightarrow{\simeq} \mathrm{Hom}_{A[x]}(L, S^{-1}A[x]/A[x]) \ ;$$
$$g \longrightarrow (x \longrightarrow g(x)/p(x)) .$$

Every $g \in \mathrm{Hom}_{A[x]}(L, A[x]/(p(x)))$ can be expressed as

$$g = \sum_{j=0}^{d-1} z^j g_j \ : \ L \longrightarrow A[x]/(p(x)) = \sum_{j=0}^{d-1} x^j A ,$$

with $g_j \in L^* = \mathrm{Hom}_A(L, A)$ and

$$\chi_x\left(\frac{g(x)}{p(x)}\right) = g_{d-1}(x) \in A ,$$

$$g_j(x) = g_{d-1}\Big(\sum_{i=j+1}^{d} a_i f^{i-j-1}(x)\Big) \in A$$
$$(x \in L, 0 \le j \le d-1) .$$

The isomorphism of 34.16 is given in this case by

$$L\widehat{\ } = \mathrm{Hom}_{A[x]}(L, A[x]/(p(x))) \xrightarrow{\simeq} L^* = \mathrm{Hom}_A(L, A) \ ; \ g \longrightarrow g_{d-1} . \ \square$$

35. Primary L-theory

The various types of L-groups defined in Chaps. 26,27,30,32 (asymmetric, automorphism, endomorphism, isometric etc.) have primary analogues in which the endomorphism f is required to be such that $p(f) = 0$ for some polynomial $p(z)$ in a prescribed class of polynomials. The primary L-groups feature in the L-theory analogues of the splitting theorems of Chap. 12

$$\begin{aligned} K_1(S^{-1}A[x]) &= K_1(A[x]) \oplus \mathrm{End}_0^S(A) , \\ K_1(\widetilde{S}^{-1}A[x]) &= K_1(A) \oplus \widetilde{\mathrm{End}}_0^S(A) , \\ \mathrm{End}_0^S(A) &= K_0(A) \oplus \widetilde{\mathrm{End}}_0^S(A) \end{aligned}$$

which will will now be obtained, with $S \subset A[x]$ a multiplicative subset with leading units and $x \in S$, and $\widetilde{S} \subset A[x]$ the reverse multiplicative subset (12.15). The S-primary endomorphism L-groups $L\mathrm{End}_S^*(A,\epsilon)$, $L\widetilde{\mathrm{End}}_S^*(A,\epsilon)$ are defined for an involution-invariant $S \subset A[x]$, such that

$$\begin{aligned} L_h^n(S^{-1}A[x],\epsilon) &= L_h^n(A[x],\epsilon) \oplus L\mathrm{End}_S^n(A,\epsilon) , \\ L_h^n(\widetilde{S}^{-1}A[x],\epsilon) &= L_h^n(A[x],\epsilon) \oplus L\widetilde{\mathrm{End}}_S^n(A,\epsilon) , \\ L\mathrm{End}_S^n(A,\epsilon) &= L_p^n(A,\epsilon) \oplus L\widetilde{\mathrm{End}}_S^n(A,\epsilon) . \end{aligned}$$

There will also be defined the S-primary automorphism, isometric and asymmetric L-groups $L\mathrm{Aut}_S^*(A,\epsilon)$, $L\mathrm{Iso}_S^*(A,\epsilon)$, $L\mathrm{Asy}_S^*(A)$, for appropriate multiplicative sets S of polynomials with coefficients in A. The various S-primary L-groups will be used in the computations of Chap. 39 of the L-groups of function fields. For the sake of brevity only the S-primary L-groups arising in the localization exact sequence for free L-groups will be considered, but it is possible to develop the intermediate versions.

The primary endomorphism, isometric, automorphism, asymmetric and fibred automorphism L-groups will now be considered, in that order.

As in Chaps. 27,30,32 the three distinct ways of extending an involution on a ring A to an involution on the polynomial extensions $A[x]$, $A[x,x^{-1}]$ are distinguished by different names for the variable x :

(i) $A[x]$, with $\overline{x} = x$,
(ii) $A[s]$, with $\overline{s} = 1 - s$,
(iii) $A[z, z^{-1}]$, with $\overline{z} = z^{-1}$.

35A. Endomorphism L-theory

Consider first the polynomial extension ring $A[x]$ with involution $\overline{x} = x$.

Let $S \subset A[x]$ be an involution-invariant multiplicative subset. Recall from 12.12 that for a finite f.g. projective A-module chain complex P an A-module chain map $f : P \longrightarrow P$ is S-primary if and only if $p(f) \simeq 0 : P \longrightarrow P$ for some $p(x) \in S$.

Definition 35.1 (i) The *ϵ-symmetric S-primary endomorphism L-group* $LEnd^n_S(A, \epsilon)$ $(n \geq 0)$ is the cobordism group of finitely dominated n-dimensional ϵ-symmetric Poincaré complexes (C, ϕ) over A with an endometric structure $(f, \delta\phi)$ such that $f : C \longrightarrow C$ is S-primary.
(ii) The *reduced ϵ-symmetric S-primary endomorphism L-group* $\widetilde{LEnd}^n_S(A, \epsilon)$ $(n \geq 0)$ is the cobordism group of null-cobordant f.g. free n-dimensional ϵ-symmetric Poincaré complexes (C, ϕ) over A with an endometric structure $(f, \delta\phi)$ such that $f : C \longrightarrow C$ is S-primary.
Similarly in the ϵ-quadratic case. □

Proposition 35.2 (i) *The absolute and reduced ϵ-symmetric S-primary endomorphism L-groups are related by natural splittings*

$$LEnd^n_S(A, \epsilon) \;=\; L^n_p(A, \epsilon) \oplus \widetilde{LEnd}^n_S(A, \epsilon) \ .$$

(ii) *If $S, T \subset A[x]$ are coprime involution-invariant multiplicative subsets*

$$LEnd^n_{ST}(A, \epsilon) \;=\; LEnd^n_S(A, \epsilon) \oplus LEnd^n_T(A, \epsilon) \ ,$$
$$\widetilde{LEnd}^n_{ST}(A, \epsilon) \;=\; \widetilde{LEnd}^n_S(A, \epsilon) \oplus \widetilde{LEnd}^n_T(A, \epsilon) \ .$$

Similarly in the ϵ-quadratic case. □

By analogy with 12.13:

Proposition 35.3 *Let $S \subset A[x]$ be an involution-invariant multiplicative subset such that each $A[x]/(p(x))$ $(p(x) \in S)$ is a f.g. projective A-module (i.e. if $S \subseteq A[x] \cap A((x^{-1}))^\bullet$). The ϵ-symmetric $(A[x], S)$-torsion L-groups are such that*

$$L^n_h(A[x], S, \epsilon) \;=\; LEnd^n_S(A, \epsilon) \ ,$$

and the L-groups of $S^{-1}A[x]$ fit into the direct sum systems

$$L^n_h(A[x], \epsilon) \underset{j_+}{\overset{i_+}{\rightleftarrows}} L^n_h(S^{-1}A[x], \epsilon) \underset{\Delta_+}{\overset{\partial_+}{\rightleftarrows}} LEnd^n_S(A, \epsilon) \ .$$

Similarly in the ϵ-quadratic case.
Proof As for 34.7, noting that S is a subset of the set of Fredholm matrices Ω_+ over $A[x]$. □

As in 11.1, given a polynomial

$$p(x) = \sum_{j=0}^{d} a_j x^j \in A[x]$$

with leading unit $a_d \in A^\bullet$ define a polynomial

$$\widetilde{p}(x) = (a_d)^{-1} x^d p(x^{-1}) = (a_d)^{-1}\Big(\sum_{j=0}^{d} a_{d-j} x^j\Big) \in A[x]$$

with constant unit $\widetilde{p}(0) = 1 \in A$.

By analogy with 12.20:

Proposition 35.4 *If $S \subset A[x]$ is an involution-invariant multiplicative subset with leading units and $x \in S$ the ϵ-symmetric L-groups of $\widetilde{S}^{-1}A[x]$ fit into the direct sum systems*

$$L^n_h(A,\epsilon) \underset{\widetilde{j}_+}{\overset{\widetilde{i}_+}{\rightleftarrows}} L^n_h(\widetilde{S}^{-1}A[x],\epsilon) \underset{\widetilde{\Delta}_+}{\overset{\widetilde{\partial}_+}{\rightleftarrows}} \widetilde{LEnd}^n_S(A,\epsilon) .$$

Similarly in the ϵ-quadratic case.
Proof As for 34.8, noting that $\widetilde{S}$ is a subset of the set $\widetilde{\Omega}_+$ of the reverses of Fredholm matrices in $A[x]$. □

By analogy with 12.22:

Definition 35.5 (i) Let

$$X = \{x^k \,|\, k \geq 0\} \subset A[x] ,$$

an involution-invariant multiplicative subset with leading units such that

$$X\text{-primary} = \text{nilpotent} ,$$

$$\widetilde{X} = \{1\} , \quad X^{-1}A[x] = A[x,x^{-1}] , \quad \widetilde{X}^{-1}A[x] = A[x] .$$

(ii) The *ϵ-symmetric nilpotent L-group* $L\mathrm{Nil}^n_q(A,\epsilon)$ $(n \geq 0)$ for $q = h$ (resp. p) is the cobordism group of f.g. free (resp. f.g. projective) n-dimensional ϵ-symmetric Poincaré complexes (C,ϕ) over A with an endometric structure $(\nu,\delta\phi)$ such that $\nu : C \longrightarrow C$ is chain homotopy nilpotent. For $q = p$

$$L\mathrm{End}^*_X(A,\epsilon) = L\mathrm{Nil}^*_p(A) .$$

(iii) The *reduced ϵ-symmetric nilpotent L-group* $\widetilde{LNil}_q^n(A,\epsilon)$ $(n \geq 0)$ for $q=h$ (resp. p) is the cobordism group of null-cobordant f.g. free (resp. f.g. projective) n-dimensional ϵ-symmetric Poincaré complexes (C,ϕ) over A with an endometric structure $(\nu,\delta\phi)$ such that $\nu: C \longrightarrow C$ is chain homotopy nilpotent. For $q=h$

$$\widetilde{LEnd}_X^*(A,\epsilon) = \widetilde{LNil}_h^*(A,\epsilon) .$$

Similarly in the ϵ-quadratic case. □

Proposition 35.6 (i) *The absolute and reduced ϵ-symmetric nilpotent L-groups are related by natural splittings*

$$LNil_q^n(A,\epsilon) = L_q^n(A,\epsilon) \oplus \widetilde{LNil}_q^n(A,\epsilon) \quad (q=h,p) .$$

Similarly in the ϵ-quadratic case.

(ii) *The free and projective nilpotent L-groups are related by a Rothenberg-type exact sequence*

$$\dots \longrightarrow \widehat{H}^{n+1}(\mathbb{Z}_2;Nil_0(A)) \longrightarrow LNil_h^n(A,\epsilon) \longrightarrow LNil_p^n(A,\epsilon) \longrightarrow \widehat{H}^n(\mathbb{Z}_2;Nil_0(A)) \longrightarrow \dots .$$

(iii) *The ϵ-symmetric L-groups of $A[x]$ and $A[x,x^{-1}]$ fit into direct sum systems*

$$L_h^n(A[x],\epsilon) \underset{j_+}{\overset{i_+}{\rightleftarrows}} L_h^n(A[x,x^{-1}],\epsilon) \underset{\Delta_+}{\overset{\partial_+}{\rightleftarrows}} LNil_p^n(A,\epsilon) ,$$

$$L_s^n(A[x],\epsilon) \underset{j_+}{\overset{i_+}{\rightleftarrows}} L_s^n(A[x,x^{-1}],\epsilon) \underset{\Delta_+}{\overset{\partial_+}{\rightleftarrows}} LNil_h^n(A,\epsilon) ,$$

$$L_h^n(A,\epsilon) \underset{\widetilde{j}_+}{\overset{\widetilde{i}_+}{\rightleftarrows}} L_h^n(A[x],\epsilon) \underset{\widetilde{\Delta}_+}{\overset{\widetilde{\partial}_+}{\rightleftarrows}} \widetilde{LNil}_p^n(A,\epsilon) ,$$

$$L_h^n(A,\epsilon) \underset{\widetilde{j}_+}{\overset{\widetilde{i}_+}{\rightleftarrows}} L_{\widetilde{K}_1(A)}^n(A[x],\epsilon) \underset{\widetilde{\Delta}_+}{\overset{\widetilde{\partial}_+}{\rightleftarrows}} \widetilde{LNil}_h^n(A,\epsilon)$$

with identifications

$$\begin{aligned}
L_h^n(A[x,x^{-1}],\epsilon) &= L_h^n(A,\epsilon) \oplus L_p^n(A,\epsilon) \oplus \widetilde{LNil}_p^n(A,\epsilon) \oplus \widetilde{LNil}_p^n(A,\epsilon) , \\
L_s^n(A[x,x^{-1}],\epsilon) &= L_s^n(A,\epsilon) \oplus L_h^n(A,\epsilon) \oplus \widetilde{LNil}_h^n(A,\epsilon) \oplus \widetilde{LNil}_h^n(A,\epsilon) , \\
L_h^n(A[x],\epsilon) &= L_h^n(A,\epsilon) \oplus \widetilde{LNil}_p^n(A,\epsilon) , \\
L_{\widetilde{K}_0(A)}^n(A[x],\epsilon) &= L_p^n(A,\epsilon) \oplus \widetilde{LNil}_p^n(A,\epsilon) = LNil_p^n(A,\epsilon) , \\
L_{\widetilde{K}_1(A)}^n(A[x],\epsilon) &= L_h^n(A,\epsilon) \oplus \widetilde{LNil}_h^n(A,\epsilon) = LNil_h^n(A,\epsilon)
\end{aligned}$$

Similarly in the ϵ-quadratic case.

Proof Set $S = X$ in 35.3 and 35.4. □

Remark 35.7 The direct sum decompositions of 35.6 (iii) were obtained in Ranicki [232], [237, Chap. 5], generalizing the results of Karoubi [116]. In [116] it was proved that if $1/2 \in A$

$$L\widetilde{\mathrm{Nil}}_p^*(A,\epsilon) \;=\; 0 \;\;,\;\; L_h^*(A[x],\epsilon) \;=\; L_h^*(A,\epsilon) \;.$$

This follows from the binomial expansion

$$(1-x)^{1/2} \;=\; \sum_{k=0}^{\infty}(-)^k\binom{1/2}{k}x^k \in \mathbb{Z}[1/2][[x]]$$

as follows. Given a nonsingular ϵ-symmetric form (M,ϕ) over A with a nilpotent endometric structure $f : M \longrightarrow M$ there is defined an isomorphism of nonsingular ϵ-symmetric forms over $A[x]$

$$(1-xf)^{1/2} \;=\; \sum_{k=0}^{\infty}(-)^k\binom{1/2}{k}x^k f^k \;:\; (M[x],\phi(1-xf)) \xrightarrow{\;\simeq\;} (M[x],\phi)\;,$$

so that the isomorphism

$$L\mathrm{Nil}_p^0(A,\epsilon) \xrightarrow{\;\simeq\;} L^0_{\widetilde{K}_0(A)}(A[x],\epsilon)\;;\; (M,\phi,f) \longrightarrow (M[x],\phi(1-xf))$$

is onto $L_p^0(A,\epsilon)$. Similarly for the odd-dimensional case. □

Remark 35.8 For $\epsilon = \pm 1$ the reduced ϵ-quadratic nilpotent quadratic L-groups of a ring with involution A can be identified with the UNil-groups of Cappell [38]

$$L\widetilde{\mathrm{Nil}}_n^h(A,\epsilon) \;=\; \mathrm{UNil}_n(\Phi^\epsilon)$$

associated to the pushout square of rings with involution

$$\begin{array}{ccc} A & \longrightarrow & A[\mathbb{Z}_2^\epsilon] \\ \big\downarrow & \Phi^\epsilon & \big\downarrow \\ A[\mathbb{Z}_2^\epsilon] & \longrightarrow & A[D_\infty^\epsilon] \end{array}$$

which expresses the A-group ring of the infinite dihedral group $D_\infty = \mathbb{Z}_2 * \mathbb{Z}_2$ with the involution $\bar{t}_i = \epsilon t_i$ $(i = 1,2)$ on the two generators t_1, t_2 as an amalgamated free product

$$A[D_\infty^\epsilon] \;=\; A[\mathbb{Z}_2^\epsilon] *_A A[\mathbb{Z}_2^\epsilon]\;.$$

See Ranicki [237, p. 743] for the case of a group ring $A = \mathbb{Z}[\pi]$, and Connolly and Ranicki [56] for arbitrary A. □

35B. Isometric L-theory

Consider now the polynomial extension ring $A[s]$ with involution $\bar{s} = 1 - s$.

Let $S \subset A[s]$ be an involution-invariant multiplicative subset.

Definition 35.9 The *ϵ-symmetric S-primary isometric L-group* $LIso_S^n(A,\epsilon)$ ($n \geq 0$) is the cobordism group of finitely dominated n-dimensional ϵ-symmetric Poincaré complexes (C,ϕ) over A with an isometric structure $(f,\delta\phi)$ such that $f : C \longrightarrow C$ is S-primary. □

By analogy with 35.2 (ii):

Proposition 35.10 *If $S,T \subset A[s]$ are coprime involution-invariant multiplicative subsets*

$$LIso_{ST}^n(A,\epsilon) \;=\; LIso_S^n(A,\epsilon) \oplus LIso_T^n(A,\epsilon) \;. \qquad \square$$

By analogy with 35.3:

Proposition 35.11 *Let $S \subset A[s]$ be an involution-invariant multiplicative subset such that each $A[s]/(p(s))$ ($p(s) \in S$) is a f.g. projective A-module (i.e. if $S \subseteq A[s] \cap A((s^{-1}))^\bullet$). The ϵ-symmetric $(A[s],S)$-torsion L-groups are such that*

$$L_h^n(A[s],S,\epsilon) \;=\; LIso_S^n(A,-\epsilon) \;,$$

and the L-groups of $S^{-1}A[s]$ fit into the direct sum systems

$$L_h^n(A[s],\epsilon) \underset{j_+}{\overset{i_+}{\rightleftarrows}} L_h^n(S^{-1}A[s],\epsilon) \underset{\Delta_+}{\overset{\partial_+}{\rightleftarrows}} LIso_S^n(A,-\epsilon) \;. \qquad \square$$

In Chap. 36 $L_h^n(A[s],\epsilon)$ will be identified with the cobordism group of n-dimensional 'almost symmetric' Poincaré complexes over A.

Recall the definition of a minimal isometric structure (32.40).

Proposition 35.12 *Let A be an integral domain with involution, and let $Q, Q_{min} \subset A[s]$ be the involution-invariant multiplicative subsets*

$$Q \;=\; \{\sum_{j=0}^{d} a_j s^j \in A[s] \,|\, a_d = 1 \in A\} \;,$$

$$Q_{min} \;=\; \{p(s) \in Q \,|\, p(0), p(1) \neq 0 \in A\} \subset Q \;.$$

(i) *The Q_{min}-primary isometric L-group $LIso_{Q_{min}}^n(A,\epsilon)$ is the cobordism group of n-dimensional ϵ-symmetric Poincaré complexes over A together with a minimal isometric structure.*
(ii) *The inclusion*

$$Q_{min}^{-1}A[s] \longrightarrow Q^{-1}A[s] \;=\; \Omega_+^{-1}A[s]$$

induces isomorphisms

$$L^n(Q_{min}^{-1}A[s],\epsilon) \xrightarrow{\simeq} L^n(Q^{-1}A[s],\epsilon)\,,$$

$$LIso^n_{Q_{min}}(A,\epsilon) = L^n(A[s],Q_{min},-\epsilon) \xrightarrow{\simeq} LIso^n_Q(A,\epsilon) = L^n(A[s],Q,-\epsilon)\,.$$

The isometric L-group $LIso^n(A,\epsilon) = LIso^n_Q(A,\epsilon)$ *is thus isomorphic to the minimal isometric L-group* $LIso^n_{Q_{min}}(A,\epsilon)$.

Proof (i) An isometric structure is minimal if and only if it is Q_{min}-primary.
(ii) Immediate from 25.11 (ii) since $Q^{-1}A[s]$ is obtained from $Q_{min}^{-1}A[s]$ by inverting the multiplicative subset

$$\{s^j(1-s)^k \mid j,k \geq 0\} \subset Q_{min}^{-1}A[s]\ . \qquad \square$$

Remark 35.13 The case $n=0$ of 35.12 is that for an integral domain with involution A the natural map from the Witt group of minimal nonsingular ϵ-symmetric Seifert forms to the Witt group of nonsingular ϵ-symmetric Seifert forms over A is an isomorphism. $\square$

Example 35.14 Stoltzfus [277, 5.15] computed the high-dimensional knot cobordism groups

$$C_n = LIso^{n+1}(\mathbb{Z}) = LIso^{n+1}_Q(\mathbb{Z})$$

and the fibred knot cobordism groups

$$C_n^{fib}(\mathbb{Z}) = LIso^{n+1}_{fib}(\mathbb{Z}) = LIso^{n+1}_{Q_{fib}}(\mathbb{Z})$$

and also the Brieskorn fibred knot cobordism groups

$$C_n^{Bfib}(\mathbb{Z}) = LIso^{n+1}_{Q_{Bfib}}(\mathbb{Z})$$

with $Q \subset \mathbb{Z}[s]$ as in 35.12 (with $A=\mathbb{Z}$),

$$Q_{fib} = \{p(s) \in Q \mid p(0),p(1) = \pm 1 \in \mathbb{Z}\} \subset Q$$

and $Q_{Bfib} \subset Q_{fib}$ the involution-invariant multiplicative subset (denoted by K in [277, p. 28]) generated by the cyclotomic polynomials $\Phi_m(s)$ for composite (non-prime power) integers m. The odd-dimensional groups are such that

$$C_{2i-1}^{Bfib} \subset C_{2i-1}^{fib} = LAut^{2i}_{fib}(\mathbb{Z}) \subset LAut^{2i}(\mathbb{Z})$$

and the even-dimensional groups are

$$C_{2i}^{Bfib} = C_{2i}^{fib} = C_{2i} = 0\ .$$

A fibred knot $k : S^{2i-1} \subset S^{2i+1}$ with exterior $T(h : F \longrightarrow F)$ represents an element $[k] \in C_{2i-1}^{fib}$ since

$$\mathrm{ch}_s(H_i(F),h_*) \in Q_{fib}$$

(cf. 33.15). The fibred knots $k : \Sigma^{2i-1} \subset S^{2i+1}$ of Brieskorn [26] are such that

$$\mathrm{ch}_s(H_i(F), h_*) \in Q_{Bfib}$$

(cf. Milnor [196, 9.6]), and so represent elements $[k] \in C^{Bfib}_{2i-1}$. □

35C. Automorphism L-theory

Consider now the Laurent polynomial extension ring $A[z, z^{-1}]$ with involution $\overline{z} = z^{-1}$.

Let $S \subset A[z, z^{-1}]$ be an involution-invariant multiplicative subset.

Proposition 35.15 *The following conditions on a self chain equivalence $f : P \longrightarrow P$ of a finite f.g. projective A-module chain complex P are equivalent:*

(i) *the $S^{-1}A[z, z^{-1}]$-module chain map*

$$z - f \ : \ S^{-1}P[z, z^{-1}] \longrightarrow S^{-1}P[z, z^{-1}]$$

is a chain equivalence,

(ii) *f is S_+-primary, with S_+ the multiplicative subset of $A[z]$ defined by*

$$S_+ \ = \ A[z] \cap S \subset A[z] \ ,$$

(iii) *$p(f) \simeq 0 : P \longrightarrow P$ for some $p(z) \in S$.*

Proof Immediate from 12.13 and the identities

$$S \ = \ \bigcup_{j=-\infty}^{\infty} z^j(S_+) \ , \ S^{-1}A[z, z^{-1}] \ = \ S_+^{-1}A[z] \ .$$ □

Definition 35.16 (i) An A-module chain map $f : P \longrightarrow P$ is *S-primary* if it satisfies the equivalent conditions of 35.15.
(ii) The *S-primary automorphism ϵ-symmetric L-group* $LAut^n_S(A, \epsilon)$ ($n \geq 0$) is the cobordism group of finitely dominated n-dimensional ϵ-symmetric Poincaré complexes (C, ϕ) over A with a self homotopy equivalence $(h, \chi) : (C, \phi) \longrightarrow (C, \phi)$ such that $h : C \longrightarrow C$ is S-primary.
(iii) The *reduced S-primary automorphism ϵ-symmetric L-group* is

$$\widetilde{LAut}^n_S(A, \epsilon) \ = \ \ker(LAut^n_S(A, \epsilon) \longrightarrow L^n_p(A, \epsilon)) \ \ (n \geq 0)$$

with

$$LAut^n_S(A, \epsilon) \longrightarrow L^n_p(A, \epsilon) \ ; \ (C, \phi, h, \chi) \longrightarrow (C, \phi) \ .$$

Similarly in the ϵ-quadratic case. □

By analogy with 35.2:

Proposition 35.17 *If $S,T \subset A[z,z^{-1}]$ are coprime involution-invariant multiplicative subsets*

$$LAut^n_{ST}(A,\epsilon) \;=\; LAut^n_S(A,\epsilon) \oplus LAut^n_T(A,\epsilon)\ . \qquad \square$$

By analogy with 35.3:

Proposition 35.18 *Let $S \subset A[z,z^{-1}]$ be an involution-invariant multiplicative subset such that each $A[z,z^{-1}]/(p(z))$ $(p(z) \in S)$ is a f.g. projective A-module (i.e. if $S \subseteq A[z,z^{-1}] \cap A((z))^\bullet \cap A((z^{-1}))^\bullet$).*
(i) *The ϵ-symmetric $(A[z,z^{-1}],S)$-torsion L-groups are such that*

$$L^n_h(A[z,z^{-1}],S,\epsilon) \;=\; LAut^{n-2}_S(A,\epsilon)\ ,$$

so that there is defined an exact sequence

$$\dots \longrightarrow L^n_h(A[z,z^{-1}],\epsilon) \longrightarrow L^n_h(S^{-1}A[z,z^{-1}],\epsilon) \longrightarrow LAut^{n-2}_S(A,\epsilon) \longrightarrow L^{n-1}_h(A[z,z^{-1}],\epsilon) \longrightarrow \dots\ .$$

(ii) *If $1-z \in S$ then up to natural isomorphism*

$$LAut^n_S(A,\epsilon) \;=\; L^n_p(A,\epsilon) \oplus \widetilde{LAut}^n_S(A,\epsilon)$$

and there is defined an exact sequence

$$\dots \longrightarrow L^n_h(A,\epsilon) \longrightarrow L^n_h(S^{-1}A[z,z^{-1}],\epsilon) \longrightarrow \widetilde{LAut}^{n-2}_S(A,\epsilon) \longrightarrow L^{n-1}_h(A,\epsilon) \longrightarrow \dots\ .$$

(iii) *If $S \subset A[z,z^{-1}]$ is such that*

$$S^{-1}A[z,z^{-1}] \;=\; \Omega^{-1}A[z,z^{-1}]$$

is the Fredholm localization of $A[z,z^{-1}]$ then every automorphism of a f.g. projective A-module is S-primary, and

$$LAut^n_S(A,\epsilon) \;=\; LAut^n_p(A,\epsilon) \;=\; L^n_p(A,\epsilon) \oplus \widetilde{Aut}^n_p(A,\epsilon)\ .$$

The reduced S-primary automorphism L-groups in this case are the projective twisted double L-groups (30.19)

$$\widetilde{LAut}^n_S(A,\epsilon) \;=\; DBL^n_p(A,\epsilon)$$

Proof (i) This is just the localization exact sequence of Chap. 25

$$\dots \longrightarrow L^n_h(A[z,z^{-1}],\epsilon) \longrightarrow L^n_h(S^{-1}A[z,z^{-1}],\epsilon) \longrightarrow L^n(A[z,z^{-1}],S,\epsilon) \longrightarrow L^{n-1}_h(A[z,z^{-1}],\epsilon) \longrightarrow \dots$$

using 35.15 to identify

$$L^n(A[z,z^{-1}],S,\epsilon) \;=\; LAut_S^{n-2}(A,\epsilon) \;.$$

(ii) Since $1-z \in S$ the projection $LAut_S^n(A,\epsilon) \longrightarrow L_p^n(A,\epsilon)$ is split by

$$L_p^n(A,\epsilon) \longrightarrow LAut_S^n(A,\epsilon) \;;\; (C,\phi) \longrightarrow (C,\phi,1,0) \;,$$

so that there is defined a direct sum system

$$L_p^n(A,\epsilon) \rightleftarrows LAut_S^n(A,\epsilon) \rightleftarrows \widetilde{LAut}_S^n(A,\epsilon) \;.$$

The exact sequence of (ii) is a direct summand of the exact sequence of (i).
(iii) Apply (ii), noting that $1-z \in S$ is a Fredholm 1×1 matrix. □

Example 35.19 If A is an integral domain and

$$\begin{aligned} S \;&=\; A[z,z^{-1}] \cap A((z))^\bullet \cap A((z^{-1}))^\bullet \\ &=\; \{\sum_{j=m}^{n} a_j z^j \,|\, a_m, a_n \in A^\bullet\} \end{aligned}$$

then the Fredholm localization of $A[z,z^{-1}]$ is just the localization inverting S

$$\Omega^{-1}A[z,z^{-1}] \;=\; S^{-1}A[z,z^{-1}] \;,$$

and every automorphism $f : P \longrightarrow P$ of a f.g. projective A-module is S-primary. The S-primary automorphism L-groups of 35.16 (ii) are just the projective automorphism L-groups of Chap. 27

$$LAut_S^*(A,\epsilon) \;=\; LAut_p^*(A,\epsilon) \;,$$

and the exact sequence of 35.18 (i) is just the exact sequence of 28.17

$$\begin{aligned} \dots \longrightarrow L_h^n(A[z,z^{-1}],\epsilon) \longrightarrow L_h^n(\Omega^{-1}A[z,z^{-1}],\epsilon) \longrightarrow LAut_p^{n-2}(A,\epsilon) \\ \longrightarrow L_h^{n-1}(A[z,z^{-1}],\epsilon) \longrightarrow \dots \;. \end{aligned}$$

□

Example 35.20 For a field with involution F take

$$A \;=\; F \;\;,\;\; S \;=\; F[z,z^{-1}]\backslash\{0\} \subset F[z,z^{-1}]$$

in 35.19 (i).
(i) By 35.18 (iii)

$$\begin{aligned} LAut_S^n(F,\epsilon) \;&=\; LAut^n(F,\epsilon) \;=\; L^n(F,\epsilon) \oplus \widetilde{Aut}^n(F,\epsilon) \;, \\ \widetilde{LAut}^n(F,\epsilon) \;&=\; DBL^n(F,\epsilon) \end{aligned}$$

with $DBL^n(F,\epsilon)$ the ϵ-symmetric twisted double L-group of F (30.19).
(ii) The multiplicative subset

$$P \;=\; \{p(z) \in F[z,z^{-1}] \,|\, p(1) \neq 0 \in F\} \subset F[z,z^{-1}]$$

has the property that a self chain equivalence $h : C \longrightarrow C$ of a finite f.g. free F-module chain complex is P-primary if and only if $1-h : C \longrightarrow C$ is a chain equivalence, i.e. h is fibred. Thus the P-primary ϵ-symmetric L-groups of F are just the fibred automorphism L-groups (28.35)

$$LAut^n_P(F,\epsilon) \;=\; LAut^n_{fib}(F,\epsilon) \;.$$

Now

$$S \;=\; (1-z)^\infty P$$

so that by 35.17

$$\begin{aligned} LAut^n(F,\epsilon) \;&=\; LAut^n_{(1-z)^\infty}(F,\epsilon) \oplus LAut^n_P(F,\epsilon) \\ &=\; L^n(F,\epsilon) \oplus LAut^n_{fib}(F,\epsilon) \end{aligned}$$

with

$$LAut^n_{fib}(F,\epsilon) \;=\; DBL^n(F,\epsilon) \;.$$

By definition, $DBL^n(F,\epsilon)$ is the cobordism group of self homotopy equivalences of null-cobordant n-dimensional ϵ-symmetric Poincaré complexes over F. The identification with the fibred automorphism L-group is via the isomorphism

$$LAut^n_{fib}(F,\epsilon) \xrightarrow{\simeq} DBL^n(F,\epsilon) \;;\; (C,\phi,h,\chi) \longrightarrow (C,\phi,1,0) \oplus (C,-\phi,h,\chi) \;.$$

Combining this with 31.17 gives identifications

$$LAut^n_{fib}(F,\epsilon) \;=\; LIso^n(F,\epsilon) \;=\; DBL^n(F,\epsilon) \;. \qquad \square$$

35D. Asymmetric L-theory

Definition 35.21 Let $S \subset A[z,z^{-1}]$ ($\overline{z} = z^{-1}$) be an involution-invariant multiplicative subset.

(i) An n-dimensional asymmetric Poincaré complex (C,λ) over A is S-*primary* if it satisfies the following equivalent conditions:

(a) the A-module chain equivalence $(T\lambda)\lambda^{-1} : C \longrightarrow C$ is S-primary,
(b) $p((T\lambda)\lambda^{-1}) \simeq 0 : C \longrightarrow C$ for some $p(z) \in S$,
(c) the $S^{-1}A[z,z^{-1}]$-module chain map

$$z\lambda - T\lambda \;:\; S^{-1}C^{n-*}[z,z^{-1}] \longrightarrow S^{-1}C[z,z^{-1}]$$

is a chain equivalence.

(ii) The free (resp. projective) S-*primary asymmetric L-group* $LAsy^n_S(A)$ ($n \geq 0$) is the cobordism group of f.g. projective n-dimensional S-primary asymmetric Poincaré complexes (C,λ) over A.

(iii) For any central unit $\epsilon \in A$ with $\overline{\epsilon} = \epsilon^{-1}$ let $S_\epsilon \subset A[z,z^{-1}]$ be the image of S under the additive automorphism

$$A[z,z^{-1}] \longrightarrow A[z,z^{-1}] \; ; \; \sum_{j=-\infty}^{\infty} a_j z^j \longrightarrow \sum_{j=-\infty}^{\infty} a_j(\epsilon^{-1}z)^j \; ,$$

so that an n-dimensional asymmetric Poincaré complex (C,λ) over A is S_ϵ-primary if the $S^{-1}A[z,z^{-1}]$-module chain map

$$z\lambda - T_\epsilon\lambda \; : \; S^{-1}C^{n-*}[z,z^{-1}] \longrightarrow S^{-1}C[z,z^{-1}]$$

is a chain equivalence. □

Proposition 35.22 *If $S,T \subset A[z,z^{-1}]$ are coprime involution-invariant multiplicative subsets*

$$LAsy^n_{ST}(A,\epsilon) \; = \; LAsy^n_S(A,\epsilon) \oplus LAsy^n_T(A,\epsilon) \; .$$

Proof This is a special case of 25.11 (iii). Here is a detailed proof in the case $n=0$. Given a nonsingular asymmetric form (L,λ) over A let

$$h \; = \; \lambda^{-1}\lambda^* \; : \; L \longrightarrow L \; .$$

For any central polynomial $f(z) \in A[z,z^{-1}]$

$$f(z)^*\lambda \; = \; \overline{f(z)}\lambda \; = \; \lambda f(z) \; : \; L \longrightarrow L^*$$

with

$$z \; = \; (\lambda^{-1}\lambda^*)^* \; = \; \lambda(\lambda^*)^{-1} \; : \; L^* \longrightarrow L^* \; .$$

Suppose that

$$p(h)q(h) \; = \; 0 \; : \; L \longrightarrow L$$

for some $p(z) \in S$, $q(z) \in T$. Since S,T are coprime there exist $a(z), b(z) \in A[z,z^{-1}]$ such that

$$a(z)p(z) + b(z)q(z) \; = \; 1 \in A[z,z^{-1}] \; .$$

Since S,T are involution invariant

$$u\overline{p(z)} \; = \; p(z) \in S \; , \; v\overline{q(z)} \; = \; q(z) \in T$$

for some central units $u,v \in A[z,z^{-1}]^\bullet$. The endomorphisms

$$f \; = \; a(h)p(h) \; , \; g \; = \; b(h)q(h) \; : \; L \longrightarrow L$$

are such that

$$\begin{aligned} &f^2 \; = \; f \; , \; g^2 \; = \; g \; , \; fg \; = \; gf \; = \; 0 \; , \; f+g \; = \; 1 \; : \; L \longrightarrow L \; , \\ &f^*\lambda g \; = \; ua(h)^*\lambda p(h) g \; = \; 0 \; , \; g^*\lambda f \; = \; vb(h)^*\lambda q(h) f \; = \; 0 \; : \; L \longrightarrow L^* \; . \end{aligned}$$

Thus (L,λ) and h split as

$$\begin{aligned}(L,\lambda) &= (L_S,\lambda_S)\oplus(L_T,\lambda_T)\ ,\\ h &= h_S\oplus h_T\ :\ L = L_S\oplus L_T \longrightarrow L = L_S\oplus L_T\end{aligned}$$

with

$$L_S = S^{-1}L = f(L)\ ,\ L_T = T^{-1}L = g(L)$$

(as in 12.7), and $q(h_S)=0$, $p(h_T)=0$. □

Proposition 35.23 *Let $S\subset A[z,z^{-1}]$ be an involution-invariant multiplicative subset such that each $A[z,z^{-1}]/(p(z))$ ($p(z)\in S$) is a f.g. projective A-module, or equivalently such that $p(z)\in A((z))^\bullet\cap A((z^{-1}))^\bullet$ (i.e. if $S\subset\Omega$). For such S with $1-z\in S$*

$$L^n_h(S^{-1}A[z,z^{-1}],\epsilon) = LAsy^n_{h,S_\epsilon}(A)\ .$$

Proof The morphism

$$\begin{aligned}&LAsy^n_{h,S_\epsilon}(A)\longrightarrow L^n_h(S^{-1}A[z,z^{-1}],\epsilon)\ ;\\ &(C,\lambda)\longrightarrow S^{-1}(C[z,z^{-1}],(1-z^{-1})\lambda+(1-z)T_\epsilon\lambda)\end{aligned}$$

is an isomorphism by 25.11. □

Example 35.24 As in 14.24 let A be an integral domain and

$$\begin{aligned}S &= A[z,z^{-1}]\cap A((z))^\bullet\cap A((z^{-1}))^\bullet\\ &= \{\sum_{j=m}^{n} a_jz^j\mid a_m,a_n\in A^\bullet\}\end{aligned}$$

so that

$$\Omega^{-1}A[z,z^{-1}] = S^{-1}A[z,z^{-1}]\ ,$$

and every automorphism $f:P\longrightarrow P$ of a f.g. projective A-module is S-primary. The S-primary asymmetric L-groups of 28.14 are just the free asymmetric L-groups

$$LAsy^*_S(A) = LAsy^*_h(A)\ ,$$

and the localization exact sequence of 22.4 is just the exact sequence of 28.17

$$\begin{aligned}\dots\longrightarrow L^n_h(A[z,z^{-1}])\longrightarrow LAsy^n_h(A)\longrightarrow LAut^{n-2}_p(A)&\\ \longrightarrow L^{n-1}_h(A[z,z^{-1}])\longrightarrow\dots&\end{aligned}$$

(with $LAsy^n_h(A)=L^n_h(\Omega^{-1}A[z,z^{-1}])$). □

Proposition 35.25 *If $S\subset A[z,z^{-1}]$ is an involution-invariant multiplicative subset which is coprime to $(z-\epsilon)^\infty$ then*

$$LAsy^n_S(A) = LAut^n_S(A,-\epsilon)\ \ (n\geq 0)\ .$$

Proof For every $p(z)\in S$ there exist $a(z),b(z)\in A[z,z^{-1}]$ such that

$$a(z)p(z) + b(z)(z - \epsilon) \ = \ 1 \in A[z, z^{-1}] \ .$$

A self chain equivalence $h : C \longrightarrow C$ of an A-module chain complex is S-primary if and only if $p(h) \simeq 0$ for some $p(z) \in S$, in which case $h - \epsilon : C \longrightarrow C$ is also a self chain equivalence, with chain homotopy inverse $(h - \epsilon)^{-1} = b(h)$. An n-dimensional asymmetric Poincaré complex (C, λ) over A is S-primary if and only if the self chain equivalence $(T\lambda)\lambda^{-1} : C \longrightarrow C$ is S-primary, in which case $(C, (1 + T_{-\epsilon})\lambda)$ is an n-dimensional $(-\epsilon)$-symmetric Poincaré complex over A with an S-primary autometric structure $(T\lambda)\lambda^{-1} : C \longrightarrow C$. The morphisms

$$LAsy^n_S(A) \longrightarrow LAut^n_S(A, -\epsilon) \ ; \ (C, \lambda) \longrightarrow (C, (1 + T_{-\epsilon})\lambda, (T\lambda)\lambda^{-1}) \ ,$$
$$LAut^n_S(A, -\epsilon) \longrightarrow LAsy^n_S(A) \ ; \ (C, \phi, h) \longrightarrow (C, (h - \epsilon)^{-1}\phi_0)$$

are inverse isomorphisms. □

Proposition 35.26 *Suppose that A is a commutative ring with involution, and let*

$$P \ = \ \{p(z) \in A[z, z^{-1}] \,|\, p(1) \in A^\bullet\} \ ,$$

so that for a central unit $\epsilon \in A$ with $\bar{\epsilon} = \epsilon^{-1}$

$$P_\epsilon \ = \ \{p(z) \in A[z, z^{-1}] \,|\, p(\epsilon) \in A^\bullet\}$$

is defined as in 35.21 (ii).
The following conditions on an n-dimensional asymmetric Poincaré complex (C, λ) over A are equivalent:

(a) *(C, λ) is P_ϵ-primary,*
(b) *$z\lambda - T_\epsilon\lambda : C^{n-*}[z, z^{-1}] \longrightarrow C[z, z^{-1}]$ is a $P^{-1}A[z, z^{-1}]$-module chain equivalence,*
(c) *$(1 - T_\epsilon)\lambda : C^{n-*} \longrightarrow C$ is an A-module chain equivalence,*
(d) *(C, λ) is a $(-\epsilon)$-symmetric Seifert complex.*

Thus there are defined forgetful maps

$$LAsy^n_{P_\epsilon}(A) \longrightarrow LIso^n(A, -\epsilon) \ ; \ (C, \lambda) \longrightarrow (C, \lambda) \ .$$

Proof An $A[z, z^{-1}]$-module chain complex D is $P^{-1}A[z, z^{-1}]$-contractible if and only if the induced A-module chain complex $A \otimes_{A[z, z^{-1}]} D$ is chain contractible. □

Example 35.27 For a field with involution F every $(-\epsilon)$-symmetric Seifert form over F is S-equivalent to a nonsingular P_ϵ-primary asymmetric form over F, by 32.45 (iii), so that the forgetful maps $LAsy^n_{P_\epsilon}(F) \longrightarrow LIso^n(F, -\epsilon)$ are isomorphisms, and

$$LAsy^n_{P_\epsilon}(F) \ = \ LIso^n(F, -\epsilon) \ .$$

□

Definition 35.28 Let $S \subset A[z,z^{-1},(1-z)^{-1}]$ be an involution-invariant multiplicative subset.
(i) An A-module chain map $h : P \longrightarrow P$ is *S-primary* if it is a fibred chain equivalence such that $p(h) \simeq 0$ for some $p(z) \in S$, or equivalently if the $S^{-1}A[z,z^{-1},(1-z)^{-1}]$-module chain map

$$z - h \,:\, S^{-1}P[z,z^{-1},(1-z)^{-1}] \longrightarrow S^{-1}P[z,z^{-1},(1-z)^{-1}]$$

is a chain equivalence.
(ii) The *S-primary fibred automorphism ϵ-symmetric L-group* $LAut^n_{fib,S}(A,\epsilon)$ ($n \geq 0$) is the cobordism group of finitely dominated n-dimensional ϵ-symmetric Poincaré complexes (C,ϕ) over A with a fibred self homotopy equivalence $(h,\chi) : (C,\phi) \longrightarrow (C,\phi)$ such that $h : C \longrightarrow C$ is S-primary.
Similarly in the ϵ-quadratic case. □

Recall from Chap. 13 that a chain map $h : C \longrightarrow C$ is chain homotopy unipotent if $h - 1 : C \longrightarrow C$ is chain homotopy nilpotent, in which case $h : C \longrightarrow C$ is a chain equivalence.

Definition 35.29 (i) The *ϵ-symmetric unipotent L-group* $LAut^n_{uni}(A,\epsilon)$ ($n \geq 0$) is the cobordism group of f.g. projective n-dimensional ϵ-symmetric Poincaré complexes (C,ϕ) over A with a self homotopy equivalence $(h,\chi) : (C,\phi) \longrightarrow (C,\phi)$ which is chain homotopy unipotent.
(ii) The *reduced ϵ-symmetric unipotent L-group* $\widetilde{LAut}^n_{uni}(A,\epsilon)$ ($n \geq 0$) is the cobordism group of null-cobordant f.g. projective n-dimensional ϵ-symmetric Poincaré complexes (C,ϕ) over A with an endometric structure $(f,\delta\phi)$ such that $f : C \longrightarrow C$ is chain homotopy unipotent. □

Proposition 35.30 (i) *The unipotent L-groups*

$$\begin{aligned} LAut^*_{uni}(A,\epsilon) &= LAut^*_{(1-z)^\infty}(A,\epsilon) \\ &= L^*_p(A,\epsilon) \oplus \widetilde{LAut}^*_{uni}(A,\epsilon) \end{aligned}$$

fit into the localization exact sequence

$$\begin{aligned} \dots \longrightarrow L^n_h(A[z,z^{-1}],\epsilon) \longrightarrow L^n_h(A[z,z^{-1},(1-z)^{-1}],\epsilon) \longrightarrow LAut^{n-2}_{uni}(A,\epsilon) \\ \longrightarrow L^{n-1}_h(A[z,z^{-1}],\epsilon) \longrightarrow \dots . \end{aligned}$$

The reduced unipotent L-groups $\widetilde{LAut}^*_{uni}(A,\epsilon)$ fit into an exact sequence

$$\begin{aligned} \dots \longrightarrow L^n_h(A,\epsilon) \longrightarrow L^n_h(A[z,z^{-1},(1-z)^{-1}],\epsilon) \longrightarrow \widetilde{LAut}^{n-2}_{uni}(A,\epsilon) \\ \longrightarrow L^{n-1}_h(A,\epsilon) \longrightarrow \dots . \end{aligned}$$

(ii) *If $S \subset A[z,z^{-1}]$ is an involution-invariant multiplicative subset which is coprime to $(1-z)^\infty$ and such that each $A[z,z^{-1}]/(p(z))$ ($p(z) \in S$) is a f.g. projective A-module then there is defined a commutative braid of exact sequences*

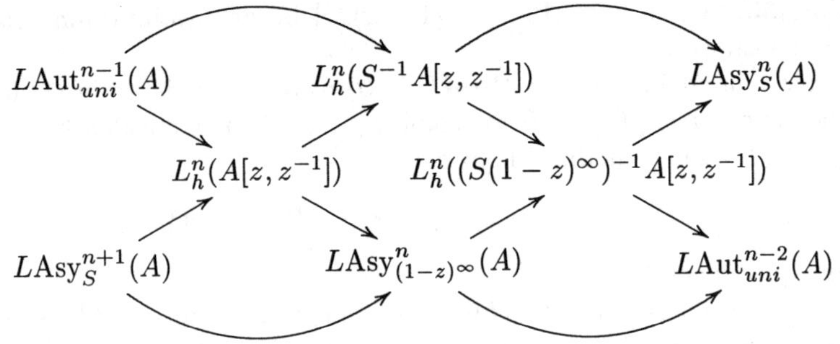

with

$$\begin{aligned}
L^n_h(A[z,z^{-1},(1-z)^{-1}]) &= LAsy^n_{(1-z)^\infty}(A)\ ,\\
LAut^n_{uni}(A) &= LAut^n_{(1-z)^\infty}(A)\ ,\\
L^n((S(1-z)^\infty)^{-1}A[z,z^{-1}]) &= LAsy^n_{S(1-z)^\infty}(A)\\
&= LAsy^n_S(A)\oplus LAsy^n_{(1-z)^\infty}(A)\ .
\end{aligned}$$

Proof (i) The following conditions on a self chain map $f:C\longrightarrow C$ of a finitely dominated A-module chain complex C are equivalent:

(a) $f:C\longrightarrow C$ is $(1-z)$-primary,
(b) $f-1:C\longrightarrow C$ is chain homotopy nilpotent,
(c) $-f:C\longrightarrow C$ is $(1+z)$-primary,
(d) $f:C\longrightarrow C$ is chain homotopy unipotent.

(ii) Combine 25.14, 35.22, 35.23 and 35.25. □

Proposition 35.31 (i) *The assignation $z=(1-s)^{-1}$ and the functions*

$$A[z]\longrightarrow A[s]\ ;$$

$$p(z) = \sum_{j=0}^d a_jz^j \longrightarrow q(s) = s^dp(1-s^{-1}) = \sum_{j=0}^d a_js^{d-j}(s-1)^j\ ,$$

$$A[s]\longrightarrow A[z]\ ;$$

$$q(s) = \sum_{j=0}^d b_js^j \longrightarrow p(z) = (1-z)^dq((1-z)^{-1}) = \sum_{j=0}^d b_j(1-z)^{d-j}$$

determine inverse bijections

$$\{p(z)\in A[z]\,|\,p(1)\neq 0\in A\} \rightleftarrows \{q(s)\in A[s]\,|\,q(0)\neq 0\in A\}$$

such that:

(a) *leading coefficient* $(p(z)) = a_d = (-)^db_0 = (-)^dq(0)$
$= (-)^d$ *constant coefficient* $(q(s))\in A$,

(b) $p(1) \;=\; \sum\limits_{j=0}^{d} a_j \;=\; b_d = \textit{leading coefficient}\,(q(s)) \in A$,

(c) $p(0) \;=\; a_0 \;=\; \sum\limits_{j=0}^{d} b_j \;=\; q(1) \in A$,

(d) *if* $a_0 \neq 0 \in A$ *then* $(-z)^d\overline{p(z)}$ *corresponds to* $\overline{q(s)}$,

(e) *if* $p_1(z), p_2(z) \in A[z]$ *correspond to* $q_1(s), q_2(s) \in A[s]$ *respectively, and*

$$\textit{degree}(p_1(z)p_2(z)) \;=\; \textit{degree}(p_1(z)) + \textit{degree}(p_2(z)) \;,$$

then $p_1(z)p_2(z) \in A[z]$ *corresponds to* $q_1(s)q_2(s) \in A[s]$.

(ii) *Let* $S \subset A[z,z^{-1}]$ *be an involution-invariant multiplicative subset such that* $p(1) \neq 0 \in A$ *for each* $p(z) \in S$, *with* $z \in S$. *Let* $T \subset A[s]$ *be the involution-invariant multiplicative subset of the polynomials* $q(s) \in A[s]$ *corresponding to the polynomials* $p(z) \in S \cap A[z]$, *such that* $q(0) \neq 0 \in A$. *If* $p(1) \in A$ *is a unit for each* $p(z) \in S$ *(or equivalently if the leading coefficient of each* $q(s) \in T$*) is a unit) then*

$$L\mathrm{Iso}^n_T(A,\epsilon) \;=\; L^n_h(A[s],T,-\epsilon) \;=\; L^n_h(A[z,z^{-1}],S,-\epsilon)\;.$$

(iii) *If* S, T *are as in* (ii), *and the constant coefficient of each* $q(s) \in T$ *is a unit (or equivalently, if the leading coefficient of each* $p(z) \in S$ *is a unit) then*

$$\begin{aligned} L\mathrm{Asy}^n_{S_\epsilon}(A) \;&=\; L\mathrm{Aut}^n_S(A,-\epsilon) \;=\; L\mathrm{Aut}^n_{fib,S}(A,-\epsilon)\\ &=\; L^n_h(A[z,z^{-1}],S,\epsilon) \;=\; L^n_h(A[z,z^{-1},(1-z)^{-1}],S,\epsilon)\\ &=\; L\mathrm{Iso}^n_T(A,-\epsilon) \;=\; L^n_h(A[s],T,\epsilon)\;. \end{aligned}$$

Proof (i) Obvious.
(ii) The identification $L\mathrm{Iso}^n_T(A,\epsilon) = L^n_h(A[s],T,-\epsilon)$ is given by 35.11. The multiplicative subsets $S, (1-z)^\infty \subset A[z,z^{-1}]$ are coprime, so that

$$L^n_h(A[z,z^{-1}],S,\epsilon) \;\cong\; L^n_h(A[z,z^{-1},(1-z)^{-1}],S,\epsilon)\;.$$

The isomorphism of rings with involution

$$A[s,s^{-1},(1-s)^{-1}] \longrightarrow A[z,z^{-1},(1-z)^{-1}]\;;\; s \longrightarrow (1-z)^{-1}$$

induces isomorphisms

$$L^n_h(A[s,s^{-1},(1-s)^{-1}],T,\epsilon) \;\cong\; L^n_h(A[z,z^{-1},(1-z)^{-1}],S,\epsilon)\;.$$

The inclusion

$$A[s] \longrightarrow A[s,s^{-1},(1-s)^{-1}]$$

induces isomorphisms

$$L^n_h(A[s],T,\epsilon) \;\cong\; L^n_h(A[s,s^{-1},(1-s)^{-1}],T,\epsilon)$$

by 25.11 (i).
(iii) Since $p(1) \in A^\bullet$ for every $p(z) \in S$, there is defined a morphism of rings with involution

$$S^{-1}A[z,z^{-1}] \longrightarrow A ~;~ z \longrightarrow 1 ~.$$

An A-module chain equivalence $h : C \longrightarrow C$ is S-primary if and only if

$$z - h ~:~ S^{-1}C[z,z^{-1}] \longrightarrow S^{-1}C[z,z^{-1}]$$

is an $S^{-1}A[z,z^{-1}]$-module chain equivalence, in which case $1 - h : C \longrightarrow C$ is an A-module chain equivalence. An n-dimensional asymmetric Poincaré complex (C,λ) is S_ϵ-primary if and only if the $S^{-1}A[z,z^{-1}]$-module chain map

$$z\lambda - T_\epsilon\lambda ~:~ S^{-1}C^{n-*}[z,z^{-1}] \longrightarrow S^{-1}C[z,z^{-1}]$$

is a chain equivalence, in which case $(1 - T_\epsilon)\lambda : C^{n-*} \longrightarrow C$ is an A-module chain equivalence. The functions

$$LAsy^n_{S_\epsilon}(A) \longrightarrow LAut^n_S(A,-\epsilon) ~;~ (C,\lambda) \longrightarrow (C,\lambda - T_\epsilon\lambda,(T_\epsilon\lambda)^{-1}\lambda) ~,$$
$$LAut^n_S(A,-\epsilon) \longrightarrow LAsy^n_{S_\epsilon}(A) ~;~ (C,\phi,h) \longrightarrow (C,(1-h^{-1})^{-1}\phi)$$

define inverse isomorphisms.

Since $q(0) = (-1)^d a_d \in A^\bullet$ and $q(1) = a_0 \in A^\bullet$ for each $q(s) \in T$, there are defined ring morphisms

$$T^{-1}A[s] \longrightarrow A ~;~ s \longrightarrow 0 ~,~ s \longrightarrow 1 ~.$$

(which do not preserve the involutions). An A-module chain map $f : C \longrightarrow C$ is T-primary if and only if

$$s - f ~:~ T^{-1}C[s] \longrightarrow T^{-1}C[s]$$

is a $T^{-1}A[s]$-module chain equivalence, in which case $f, 1 - f : C \longrightarrow C$ are A-module chain equivalences with $q(f) \simeq 0$ for the polynomial $q(s) \in A[s]$ associated to some $p(z) \in S$, and

$$h ~=~ 1 - f^{-1} ~:~ C \longrightarrow C$$

is an S-primary A-module chain equivalence with

$$p(h) ~\simeq~ f^{-d}q(f) ~\simeq~ 0 ~:~ C \longrightarrow C ~.$$

The functions

$$LAut^n_S(A,\epsilon) \longrightarrow LIso^n_T(A,\epsilon) ~;~ (C,\phi,h) \longrightarrow (C,\phi,(1-h)^{-1})$$
$$LIso^n_T(A,\epsilon) \longrightarrow LAut^n_S(A,\epsilon) ~;~ (C,\phi,f) \longrightarrow (C,\phi,1-f^{-1})$$

define inverse isomorphisms. □

36. Almost symmetric L-theory

As already noted in Chap. 31 the high-dimensional knot cobordism groups C_* can be expressed in terms of the L-theory of $\mathbb{Z}[s]$ ($\overline{s} = 1 - s$). The ϵ-symmetric L-theory of $A[s]$ for any ring with involution A is now identified with the L-theory of almost ϵ-symmetric forms over A, generalizing the work of Clauwens [51], and relating it to knot theory and automorphism L-theory.

A nonsingular asymmetric form (L, λ) is 'almost ϵ-symmetric' if

$$1 - (\epsilon\lambda^*)^{-1}\lambda \,:\, L \longrightarrow L$$

is nilpotent, i.e. if (L, λ) is $(z - \epsilon)$-primary in the terminology of Chap. 35. The free L-group $L^0_h(A[s], \epsilon)$ was identified in [51] with the Witt group of nonsingular almost ϵ-symmetric forms over A. The main results of Chap. 36 are the identification of $L^*_h(A[s], \epsilon)$ with the cobordism groups of almost ϵ-symmetric Poincaré complexes over A, and the exact sequence

$$\ldots \longrightarrow L^n_h(A, \epsilon) \longrightarrow L^n_h(A[s], \epsilon) \longrightarrow \widetilde{LAut}^{n-2}_{uni}(A, \epsilon) \longrightarrow L^{n-1}_h(A, \epsilon) \longrightarrow \ldots$$

relating the almost ϵ-symmetric and ϵ-symmetric L-groups of A, with $\widetilde{LAut}^*_{uni}(A, \epsilon)$ the reduced unipotent L-groups. (In 38.22 it will be proved that $\widetilde{LAut}^*_{uni}(A, \epsilon) = 0$ for Dedekind A, so that $L^*_h(A[s], \epsilon) = L^*_h(A, \epsilon)$ for such A).

Let

$$S \;=\; (1 - z)^\infty \;=\; \{\pm z^j(1 - z)^k \,|\, j \in \mathbb{Z}\,,\, k \geq 0\} \subset A[z, z^{-1}]\,,$$

so that

$$S^{-1}A[z, z^{-1}] \;=\; A[z, z^{-1}, (1 - z)^{-1}]\,,$$

and

$$S_\epsilon \;=\; (\epsilon - z)^\infty \;=\; \{\pm \epsilon^j z^j (\epsilon - z)^k \,|\, j \in \mathbb{Z}\,,\, k \geq 0\} \subset A[z, z^{-1}]$$

(in the terminology of 35.21 (iii)). An endomorphism $f : M \longrightarrow M$ of a f.g. projective A-module is S_ϵ-primary if and only if $\epsilon - f : M \longrightarrow M$ is nilpotent. Similarly for chain maps.

Definition 36.1 (i) An asymmetric Poincaré complex (C, λ) is *almost ϵ-symmetric* if the self chain map

$$1 - \lambda(T_\epsilon \lambda)^{-1} \ : \ C \longrightarrow C$$

is chain homotopy nilpotent, or equivalently if (C, λ) is S_ϵ-primary (in the terminology of 35.21 (i)).
(ii) The *free (resp. projective) almost ϵ-symmetric L-groups* $LAsy^*_{q,S_\epsilon}(A)$ ($q = h$ (resp. p)) are the cobordism groups of f.g. free almost ϵ-symmetric Poincaré complexes over A, which are just the free (resp. projective) S_ϵ-primary asymmetric L-groups of 35.21 (ii). □

Example 36.2 A nonsingular asymmetric form (M, ϕ) is *almost ϵ-symmetric* if it is S_ϵ-primary, i.e. if the endomorphism

$$1 - (\epsilon\phi^*)^{-1}\phi \ : \ M \longrightarrow M$$

is nilpotent. The 0-dimensional free (resp. projective) almost ϵ-symmetric L-group $LAsy^0_{h,S_\epsilon}(A)$ (resp. $LAsy^0_{p,S_\epsilon}(A)$) is the Witt group of nonsingular almost ϵ-symmetric forms (M, ϕ) over A, with M f.g. free (resp. projective). □

Proposition 36.3 (i) *The free almost ϵ-symmetric L-groups of A are such that*

$$\begin{aligned} LAsy^n_{h,S_\epsilon}(A) &= L^n_h(A[s], \epsilon) \\ &= L^n_h(A[z, z^{-1}, (1-z)^{-1}], \epsilon) \ . \end{aligned}$$

(ii) *The free almost ϵ-symmetric L-groups $LAsy^*_{h,S_\epsilon}(A)$ fit into a braid of exact sequences*

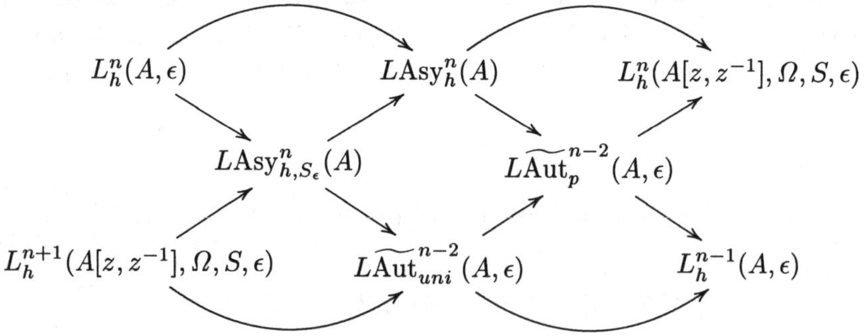

(iii) *If*

either *there exists a central element $t \in A$ such that $t + \bar{t} = 1 \in A$ (e.g. $t = 1/2$)*
or *A is a Dedekind ring*

then

$$L_h^*(A,\epsilon) \;=\; \mathrm{LAsy}^*_{h,S_\epsilon}(A) \;=\; L_h^*(A[s],\epsilon)\;,\quad \widetilde{L\mathrm{Aut}}^*_{uni}(A,\epsilon) \;=\; 0\;.$$

Proof (i) By 25.11 the morphism of rings with involution

$$A[s] \longrightarrow S^{-1}A[z,z^{-1}] \;=\; A[z,z^{-1},(1-z)^{-1}]\;;\; s \longrightarrow (1-z)^{-1}$$

induces isomorphisms of L-groups

$$L_h^*(A[s],\epsilon) \xrightarrow{\;\simeq\;} L_h^*(S^{-1}A[z,z^{-1}],\epsilon) \;=\; \mathrm{LAsy}^*_{h,S_\epsilon}(A)\;,$$

with the inverse isomorphisms explicitly given by

$$\mathrm{LAsy}^n_{h,S_\epsilon}(A) \xrightarrow{\;\simeq\;} L_h^n(A[s],\epsilon)\;;\; (C,\lambda) \longrightarrow (C[s],s\lambda+(1-s)T\lambda)\;.$$

(ii) This is a direct summand of the braid of localization exact sequences

$$\begin{array}{ccccc}
L_h^n(A[z,z^{-1}],\epsilon) & & L_h^n(\Omega^{-1}A[z,z^{-1}],\epsilon) & & L_h^n(A[z,z^{-1}],\Omega,S,\epsilon)\\
& \searrow \nearrow & & \searrow \nearrow & \\
& L_h^n(S^{-1}A[z,z^{-1}],\epsilon) & & L_h^n(A[z,z^{-1}],\Omega,\epsilon) & \\
& \nearrow \searrow & & \nearrow \searrow & \\
L_h^{n+1}(A[z,z^{-1}],\Omega,S,\epsilon) & & L_h^n(A[z,z^{-1}],S,\epsilon) & & L_h^{n-1}(A[z,z^{-1}],\epsilon)
\end{array}$$

(iii) Assume A is such that there exists a central element $t\in A$ with $t+\bar{t}=1$. The inclusion $A\longrightarrow A[s]$ is split by

$$A[s] \longrightarrow A\;;\; s \longrightarrow t\;,$$

so that one of the exact sequences in (ii) breaks up into short exact sequences

$$0 \longrightarrow L_h^n(A,\epsilon) \xrightarrow{\;i\;} L_h^n(A[s],\epsilon) \longrightarrow \widetilde{L\mathrm{Aut}}^{n-2}_{uni}(A,\epsilon) \longrightarrow 0$$

with i split by

$$L_h^n(A[s],\epsilon) \;=\; \mathrm{LAsy}^n_{h,S_\epsilon}(A) \longrightarrow L_h^n(A,\epsilon)\;;\; (C,\lambda) \longrightarrow (C,t\lambda+(1-t)T_\epsilon\lambda)\;,$$

and it suffices to prove that the split injections

$$L_h^n(A,\epsilon) \longrightarrow L_h^n(A[s],\epsilon) \;=\; \mathrm{LAsy}^n_{h,S_\epsilon}(A)$$

are onto. Given an n-dimensional almost ϵ-symmetric Poincaré complex (C,λ) over A there is a chain map

$$\nu \;=\; 1-\lambda(T_\epsilon\lambda)^{-1}\,:\; C \longrightarrow C$$

which is chain homotopy nilpotent. Working as in the proof of Theorem 3 of Clauwens [51] it is possible to find integers $\alpha_1, \alpha_2, \ldots, \alpha_p \in \mathbb{Z}$ and a chain map $\psi : C^{n-*} \longrightarrow C$ over A such that

$$f(s\lambda + (1-s)T_\epsilon\lambda)f^* \simeq t\psi + (1-t)T_\epsilon\psi \ : \ C^{n-*} \longrightarrow C \ ,$$

with

$$f \ = \ 1 + \alpha_1\nu + \alpha_2\nu^2 + \ldots + \alpha_p\nu^p \ : \ C \longrightarrow C$$

an A-module chain equivalence. Then $(C, (1+T_\epsilon)\psi)$ is an n-dimensional ϵ-symmetric Poincaré complex over A with image

$$\begin{aligned}(C[s], (1+T_\epsilon)\psi) \ &= \ (C[s], t\psi + (1-t)T_\epsilon\psi) \\ &= \ (C[s], s\lambda + (1-s)T_\epsilon\lambda) \ = \ (C, \lambda) \\ &\in \operatorname{im}(L^n_h(A,\epsilon) \longrightarrow L^n_h(A[s],\epsilon) \ = \ LAsy^n_{h,S_\epsilon}(A)) \ .\end{aligned}$$

The case of A Dedekind is deferred to Proposition 41.19 below. □

Remark 36.4 (i) The identification of 36.3 (i)

$$L^n_h(A[s]) \ = \ LAsy^n_{h,S}(A)$$

was first obtained by Clauwens [51] for $n = 2i$. An element

$$(M, \lambda) \in LAsy^{2i}_{h,S}(A) \ = \ LAsy^0_{h,S_{(-)^i}}(A)$$

is represented by a f.g. free nonsingular almost $(-)^i$-symmetric form (M, λ) over A. In particular, for a $2i$-dimensional f.g. free ϵ-symmetric Poincaré complex (C, ϕ) over A the image of $(C, \phi) \in L^{2i}_h(A)$ in $LAsy^{2i}_{h,S}(A)$ is the Witt class of the nonsingular almost $(-)^i$-symmetric form

$$\begin{aligned}&(M, \lambda) \\ &= \ (\operatorname{coker}(\begin{pmatrix} d^* & 0 \\ (-)^{i-1}\phi_0 & d \end{pmatrix} : C^{i-1} \oplus C_{i+2} \longrightarrow C^i \oplus C_{i+1}), \begin{bmatrix} \phi_0 & d \\ (-)^i d^* & 0 \end{bmatrix})\end{aligned}$$

obtained in [51], and it is only this Witt class which enters the surgery product formula, as described in (ii) below.

(ii) The L-theory product of Ranicki [235, Chap. 10]

$$L_m(A,\epsilon) \otimes L^n(B,\eta) \longrightarrow L_{m+n}(A \otimes B, \epsilon \otimes \eta) \ ,$$

the L-theory product of Clauwens [50]

$$L_m(A,\epsilon) \otimes L^n(B[s],\eta) \longrightarrow L_{m+n}(A \otimes B, \epsilon \otimes \eta) \ ,$$

and the L-theory product of Lück and Ranicki [177, p. 148]

$$L_m(A[s],\epsilon) \otimes LIso^n(B,\eta) \longrightarrow L_{m+n}(A \otimes_{\mathbb{Z}} B, \epsilon \otimes \eta)$$

are related by a commutative diagram

$$\begin{array}{ccccc}
LIso^m(A,\epsilon)\otimes L^n(B,\eta) & & \longrightarrow & & L_m(A,\epsilon)\otimes L^n(B,\eta) \\
\big\downarrow & \searrow & & \swarrow & \big\downarrow \\
 & & L_m(A,\epsilon)\otimes L^n(B[s],\eta) & & \\
\big\downarrow & \nearrow & & \searrow & \big\downarrow \\
LIso^m(A,\epsilon)\otimes L^n(B[s],\eta) & & \longrightarrow & & L_{m+n}(A\otimes B,\epsilon\otimes\eta)
\end{array}$$

The product

$$LIso^m(A,\epsilon)\otimes L^n(B[s],\eta) \longrightarrow L_{m+n}(A\otimes_{\mathbb{Z}} B,\epsilon\otimes\eta)$$

is given for $m=n=0$ on the level of forms by

$$\begin{array}{l}
LIso^0(A,\epsilon)\otimes LAsy^0(B,\eta) \longrightarrow L_0(A\otimes B,\epsilon\otimes\eta)\ ; \\
\qquad\qquad\qquad\qquad (P,\theta,f)\otimes(Q,\phi) \longrightarrow (P\otimes Q, f\theta\otimes\phi)\ .
\end{array}$$

The product of an m-dimensional normal map $(f,b): M\longrightarrow X$ with $\pi_1(X)=\pi$ and an n-dimensional manifold N with $\pi_1(N)=\rho$ is an $(m+n)$-dimensional normal map

$$(g,c) \;=\; (f,b)\times 1 \;:\; M\times N \longrightarrow X\times N$$

with surgery obstruction the product

$$\sigma_*(g,c) \;=\; \sigma_*(f,b)\otimes\sigma^*(N)\in L_{m+n}(\mathbb{Z}[\pi\times\rho])$$

of the surgery obstruction $\sigma_*(f,b)\in L_m(\mathbb{Z}[\pi])$ and the almost symmetric signature of [50]

$$\sigma^*(N)\in L^n_h(\mathbb{Z}[\rho][s]) \;=\; LAsy^n_{h,S}(\mathbb{Z}[\rho])$$

(= the image of the symmetric signature $\sigma^*(N)\in L^n_h(\mathbb{Z}[\rho])$). The oozing conjecture concerning the image of the assembly map

$$A \;:\; H_*(B\pi;\mathbb{L}_\bullet(\mathbb{Z})) \longrightarrow L_*(\mathbb{Z}[\pi])$$

for a finite group π was solved by Hambleton, Milgram, Taylor and Williams [103] and Milgram [187], [188]. The method of [187], [188] used the almost symmetric signature and the commutative square

$$\begin{array}{ccc}
LIso^m(\mathbb{Z})\otimes L^n(\mathbb{Z}[\pi]) & \longrightarrow & L_m(\mathbb{Z})\otimes L^n(\mathbb{Z}[\pi]) \\
\big\downarrow & & \big\downarrow \\
LIso^m(\mathbb{Z})\otimes L_n(\mathbb{Z}[\pi][s]) & \longrightarrow & L_{m+n}(\mathbb{Z}[\pi])
\end{array}$$

The isometric L-group $LIso^m(\mathbb{Z})=C_{m-1}$ is the cobordism group of knots $S^{m-1}\subset S^{m+1}$ (33.3). □

37. L-theory of fields and rational localization

The computation of the knot cobordism groups reduces to the computation of the L-theory of various fields and integral domains. The main properties of the L-groups of a field with involution F are summarized in 37A. The computations of the L-groups of $\mathbb{Z}$, $\mathbb{Z}_m$, $\widehat{\mathbb{Z}}_m$, $\mathbb{Q}$ and the localization exact sequence are used in 37B to prove a general result: the natural maps $L^*(A) \longrightarrow L^*(\mathbb{Q} \otimes_{\mathbb{Z}} A)$ are isomorphisms modulo 8-torsion for any additively torsion-free ring involution A (e.g. a group ring $A = \mathbb{Z}[\pi]$). In Chap. 42 the version of this result for torsion L-groups will be used to prove that the natural maps

$$C_* = L\mathrm{Iso}^{*+1}(\mathbb{Z}) \longrightarrow L\mathrm{Iso}^{*+1}_{U_{\mathbb{Z}}}(\mathbb{Q})$$

are isomorphisms modulo 8-torsion, for an appropriately defined $U_{\mathbb{Z}}$.

In Chap. 37 the undecorated L-groups L^*, L_* are the projective L-groups L^*_p, L^p_*.

37A. Fields

Given $\epsilon \in F^{\bullet}$ such that $\epsilon\bar{\epsilon} = 1$ let

$$U(F,\epsilon) = \{x \in F^{\bullet} \mid \epsilon\bar{x} = x\}.$$

Proposition 37.1 (Milnor and Husemoller [200, Chap. III], Scharlau [256])
(i) *Every nonsingular ϵ-symmetric form over a field F can be diagonalized.*
(ii) *The abelian group morphism*

$$\mathbb{Z}[U(F,\epsilon)] \longrightarrow L^0(F,\epsilon) \; ; \; x \longrightarrow (F,x)$$

is onto, with kernel generated by elements of the type

$$[x] - [ax\bar{a}] \; , \; [x] + [-x] \; , \; [x] + [y] - [x+y] - [x(x+y)^{-1}y]$$

for any $a \in F^{\bullet}$, $x, y \in U(F,\epsilon)$ with $x + y \neq 0$.
(iii) *The odd-dimensional ϵ-symmetric L-groups of F vanish*

$$L^{2*+1}(F,\epsilon) = 0 \, .$$

(iv) *The ϵ-symmetric L-groups of F are 4-periodic*

$$L^n(F,\epsilon) = L^{n+2}(F,-\epsilon) = L^{n+4}(F,\epsilon) .$$

(v) *If* char$(F) \neq 2$ *and F has the identity involution then*

$$L^{4*+2}(F) = L^{4*}(F,-1) = 0 .$$

(vi) *If* char$(F) = 2$ *then*

$$L^n(F,\epsilon) = L^n(F,-\epsilon) = L^{n+2}(F,\epsilon) .$$

(vii) *If* char$(F) \neq 2$ *and F has a non-identity involution then* $F = F_0(\sqrt{a})$ *is a quadratic Galois extension of the fixed field* $F_0 = \{u \in F \,|\, \overline{u} = u\}$ *for some non-square* $a \in F_0$ *with* $\overline{\sqrt{a}} = -\sqrt{a} \in F$, *the ϵ-symmetric L-groups of F are 2-periodic*

$$L^n(F,\epsilon) = L^n(F,-\epsilon) = L^{n+2}(F,\epsilon)$$

and there are defined isomorphisms

$$L^0(F,\epsilon) \xrightarrow{\simeq} L^0(F,-\epsilon) \; ; \; (M,\phi) \longrightarrow (M,\sqrt{a}\phi) .$$

(viii) *If F is a finite field with the identity involution*

$$L^0(F) = \begin{cases} \mathbb{Z}_2 \oplus \mathbb{Z}_2 & \text{if } \mathrm{char}(F) \equiv 1 (\mathrm{mod}\, 4) \\ \mathbb{Z}_4 & \text{if } \mathrm{char}(F) \equiv 3 (\mathrm{mod}\, 4) \\ \mathbb{Z}_2 & \text{if } \mathrm{char}(F) = 2 . \end{cases}$$

If F is a finite field with a non-identity involution

$$L^0(F) = \mathbb{Z}_2 .$$

(ix) *The symmetric Witt group of* $\mathbb{R}$ *is*

$$L^0(\mathbb{R}) = \mathbb{Z} .$$

The Witt class of a nonsingular symmetric form (M,ϕ) *over* $\mathbb{R}$ *is given by the signature*

$$\sigma(M,\phi) = d_+ - d_- \in \mathbb{Z} ,$$

with d_+ *(resp.* d_-*) the number of positive (resp. negative) entries in a diagonalization of* (M,ϕ).

(x) *Let* $\mathbb{C}^-$ *be* $\mathbb{C}$ *with the complex conjugation. For* $\epsilon \in S^1$ *the ϵ-symmetric Witt group of* $\mathbb{C}^-$ *is*

$$L^0(\mathbb{C}^-,\epsilon) = \mathbb{Z} .$$

The Witt class of a nonsingular ϵ-symmetric form (M,ϕ) *over* $\mathbb{C}^-$ *is given by the signature*

$$\sigma(M,\phi) = d_+ - d_- \in \mathbb{Z} ,$$

with d_+ *(resp.* d_-*) the number of positive (resp. negative) entries in a diagonalization of the symmetric form* $(M,\eta^{-1}\phi)$, *with* $\eta \in S^1$ *such that* $\eta^2 = \epsilon$. *(The choice of* η *determines the sign of* $\sigma(M,\phi)$.*)*

(xi) *Let* $\mathbb{C}^+$ *be* $\mathbb{C}$ *with the identity involution. The symmetric Witt group of* $\mathbb{C}^+$ *is*

$$L^0(\mathbb{C}^+) = \mathbb{Z}_2 .$$

The Witt class of a nonsingular symmetric form (M,ϕ) *over* $\mathbb{C}^+$ *is given by* $\dim_{\mathbb{C}}(M)$ mod 2.
(xii) *The symmetric Witt group of* $\mathbb{Q}$ *is*

$$L^0(\mathbb{Q}) = L^0(\mathbb{Z}) \oplus \bigoplus_{p \text{ prime}} L^0(\mathbb{Z}_p) = \mathbb{Z} \oplus \bigoplus_{\infty} \mathbb{Z}_2 \oplus \bigoplus_{\infty} \mathbb{Z}_4 . \qquad \square$$

The computation of the high-dimensional knot cobordism groups C_* makes use of the computations of the L-groups of algebraic number fields (= finite extensions of $\mathbb{Q}$) and their rings of integers. The L-groups modulo torsion of the fields are given in 37.2 below, and those of the rings of integers in 38.10 below.

Proposition 37.2 (Hasse [108], Landherr [145], Milnor and Husemoller [200, p. 81], Conner [55, II], Scharlau [256, pp. 224, 351])
Let F *be an algebraic number field, with* $F = \mathbb{Q}[x]/(p(x))$ *for an irreducible monic polynomial* $p(x) \in \mathcal{M}_x(\mathbb{Q})$ *of degree* $d = r+2s$, *with* r *real roots and* $2s$ *non-real roots. For each real root* $\alpha \in \mathbb{R}$ *of* $p(x)$ *there is defined an embedding*

$$\rho_\alpha : F \longrightarrow \mathbb{R} ; \ x \longrightarrow \alpha ,$$

and hence an ordering of F. *For each complex root* $\omega \in \mathbb{C}$ *of* $p(x)$ *there is defined an embedding*

$$\sigma_\omega : F \longrightarrow \mathbb{C} ; \ x \longrightarrow \omega .$$

(i) *Nonsingular symmetric forms over* F *with the identity involution are classified by dimension, determinant, the signatures associated to the* r *orderings of* F, *and the Hasse invariants at the local completions of* F. *The Witt group of* F *is of the type*

$$L^0(F) = \mathbb{Z}^r \oplus 8\text{-torsion}$$

with

$$\bigoplus_r \rho_\alpha : L^0(F) \longrightarrow \bigoplus_r L^0(\mathbb{R}) = \mathbb{Z}^r$$

an isomorphism modulo 8-torsion.
(ii) *Suppose that* F *has a non-identity involution with fixed field* F_0, *so that* $F = F_0(\sqrt{a})$ *(as in 37.1 (vii)). Let* s_0 *be the number of the non-real roots* $\omega \in \mathbb{C}$ *of* $p(x)$ *with* $\sigma_\omega(F_0) \subset \mathbb{R}$ *and* $\sigma_\omega(a) < 0$, *so that* $\sigma_\omega : F \longrightarrow \mathbb{C}^-$ *is a morphism of rings with involution. Nonsingular symmetric forms over* F *are classified by dimension, determinant and the signatures associated to the* s_0 *embeddings* $\sigma_\omega : F \longrightarrow \mathbb{C}^-$. *The Witt group of* F *is of the type*

$$L^0(F) = \mathbb{Z}^{s_0} \oplus 8\text{-torsion}$$

with

$$\bigoplus_{s_0} \sigma_\omega : L^0(F) \longrightarrow \bigoplus_{s_0} L^0(\mathbb{C}^-) = \mathbb{Z}^{s_0}$$

an isomorphism modulo 8-torsion. Moreover, $L^2(F) = L^0(F)$. $\square$

Let F be any field with a non-identity involution, and let F_0 be the fixed field of the involution. A symmetric form (M,ϕ) over F determines a quadratic form (M,ψ_0) over F_0, with

$$\psi_0(x)(x) \;=\; \phi(x)(x) \in F_0 \;\; (x \in M)\;.$$

By a theorem of Jacobson nonsingular symmetric forms over F are isomorphic if and only if the corresponding quadratic forms over F_0 are isomorphic; Milnor and Husemoller [200, Appendix 2] used this result to obtain the exact sequence

$$0 \longrightarrow L^0(F) \longrightarrow L_0(F_0) \longrightarrow L_0(F^+)$$

with F^+ denoting F with the identity involution. This sequence will be used in Chap. 40 below, but only in the case when $\mathrm{char}(F) \neq 2$, for which $L_0(F_0) = L^0(F_0)$.

Proposition 37.3 (Lewis [163])
If F is a field with non-identity involution and $\mathrm{char}(F) \neq 2$ *there is defined an exact sequence of L-groups*

$$0 \longrightarrow L^0(F) \longrightarrow L^0(F_0) \longrightarrow L^0(F^+) \longrightarrow L^0(F_0) \longrightarrow L^0(F) \longrightarrow 0\;. \;\square$$

Example 37.4 (i) As in 37.2 let $F = \mathbb{Q}[x]/(p(x))$ be an algebraic number field with a non-identity involution, so that $F = F_0(\sqrt{a})$ is a quadratic extension of the fixed field F_0. Then F_0 is an algebraic number field, say $F_0 = \mathbb{Q}[y]/(p_0(y))$. Let r_0 (resp. r_1) be the number of the real roots $\alpha \in \mathbb{R}$ of $p_0(y)$ for which $\rho_\alpha : F_0 \longrightarrow \mathbb{R}$ has $\rho_\alpha(a) > 0$ (resp. $\rho_\alpha(a) < 0$). Each of the r_0+r_1 real roots of $p_0(y)$ determines a signature map $L^0(F_0) \longrightarrow \mathbb{Z}$. Now $p(x)$ has exactly $r = 2r_0$ real roots, the images of $x \in F$ under the $2r_0$ embeddings

$$F \longrightarrow \mathbb{R}\;;\; u+v\sqrt{a} \longrightarrow \rho_\alpha(u) \pm \rho_\alpha(v)\sqrt{\rho_\alpha(a)} \;\; (u,v \in F_0, \rho_\alpha(a) > 0)\;,$$

which determine $2r_0$ signature maps $L^0(F^+) \longrightarrow \mathbb{Z}$. Also, $p(x)$ has exactly $s_0 = r_1$ conjugate pairs of non-real roots $\omega \in \mathbb{C}$ which determine r_1 signature maps $\sigma_\omega : L^0(F) \longrightarrow \mathbb{Z}$, the images of $x \in F$ under the r_1 conjugate pairs of involution-preserving embeddings

$$F \longrightarrow \mathbb{C}^-\;;\; u+v\sqrt{a} \longrightarrow \rho_\alpha(u) \pm \rho_\alpha(v)\sqrt{\rho_\alpha(a)} \;\; (u,v \in F_0, \rho_\alpha(a) < 0)\;.$$

In this case, 37.4 gives

$$\begin{aligned} L^0(F) &= \mathbb{Z}^{r_1} \oplus 8\text{-torsion}\;, \\ L^0(F_0) &= \mathbb{Z}^{r_0+r_1} \oplus 8\text{-torsion}\;, \\ L^0(F^+) &= \mathbb{Z}^{2r_0} \oplus 8\text{-torsion}\;. \end{aligned}$$

(ii) For $p(x) = x^{q-1} + x^{q-2} + \ldots + x + 1$ ($q \geq 3$ prime) in (i) and the complex conjugation $\overline{x} = x^{-1}$ take $a = -1$, $y = x + x^{-1}$, so that

$$F \;=\; \mathbb{Q}(\zeta)\;,\;\; F_0 \;=\; \mathbb{Q}(\zeta + \zeta^{-1})\;,\;\; r_0 \;=\; 0\;,\;\; r_1 \;=\; (q-1)/2$$

with $\zeta = e^{2\pi i/q}$. $\square$

37B. Rational localization

Following the recollection of the periodicity properties of the algebraic L-groups and the computations of the L-groups of $\mathbb{Z}, \mathbb{Q}$ and related rings, the localization exact sequence of Chap. 25 will be used to investigate the natural maps $L^*(A,\epsilon)\longrightarrow L^*(\mathbb{Q}\otimes_{\mathbb{Z}} A,\epsilon)$, $L^*(A,\Sigma,\epsilon)\longrightarrow L^*(\mathbb{Q}\otimes_{\mathbb{Z}} A,\Sigma,\epsilon)$.

The original algebraic L-groups of $L_*(A)$ of Wall [302] were 4-periodic by construction. The algebraic L-groups $L^*(A,\epsilon)$, $L_*(A,\epsilon)$ defined using algebraic Poincaré complexes have the following 4-periodicity properties.

Proposition 37.5 (Ranicki [235])
(i) *The ϵ-quadratic L-groups of any ring with involution A are 4-periodic*

$$L_n(A,\epsilon) \;=\; L_{n+2}(A,-\epsilon) \;=\; L_{n+4}(A,\epsilon) \;.$$

(ii) *The ϵ-symmetric L-groups of a ring with involution A of global dimension d are such that*

$$L^n(A,\epsilon) \;=\; L^{n+2}(A,-\epsilon) \;=\; L^{n+4}(A,\epsilon)$$

for $n+2\geq 2d$. A Dedekind ring A has global dimension ≤ 1, so that

$$L^n(A,\epsilon) \;=\; L^{n+2}(A,-\epsilon) \;=\; L^{n+4}(A,\epsilon) \;\; (n\geq 0) \;. \qquad \square$$

Proposition 37.6 (Kervaire and Milnor [134], Ranicki [237, 4.4])
(i) *The quadratic L-groups of $\mathbb{Z}$ are given by*

$$L_n(\mathbb{Z}) \;=\; \begin{cases} \mathbb{Z} & \text{if } n\equiv 0 (\bmod 4) \text{ (signature/8)} \\ \mathbb{Z}_2 & \text{if } n\equiv 2 (\bmod 4) \text{ (Arf invariant)} \\ 0 & \text{otherwise .} \end{cases}$$

(ii) *The symmetric L-groups of $\mathbb{Z}$ are given by*

$$L^n(\mathbb{Z}) \;=\; \begin{cases} \mathbb{Z} & \text{if } n\equiv 0 (\bmod 4) \text{ (signature)} \\ \mathbb{Z}_2 & \text{if } n\equiv 1 (\bmod 4) \text{ (deRham invariant)} \\ 0 & \text{otherwise .} \end{cases}$$

(iii) *For any prime p*

$$L^0(\mathbb{Z}_p) \;=\; \begin{cases} \mathbb{Z}_2 & \text{if } p=2 \\ \mathbb{Z}_2\oplus\mathbb{Z}_2 & \text{if } p\equiv 1 (\bmod 4) \\ \mathbb{Z}_4 & \text{if } p\equiv 3 (\bmod 4) \;, \end{cases}$$

and for any integer $k\geq 1$

$$L^0(\mathbb{Z}_{p^k}) \;=\; \begin{cases} L^0(\mathbb{Z}_p) & \text{if } p \text{ is odd, or if } p=2 \text{ and } k=1 \\ L^0(\mathbb{Z}_4) \;=\; \mathbb{Z}_2\oplus\mathbb{Z}_4 & \text{if } p=2 \text{ and } k=2 \\ L^0(\mathbb{Z}_8) \;=\; \mathbb{Z}_2\oplus\mathbb{Z}_8 & \text{if } p=2 \text{ and } k\geq 3 \;. \end{cases}$$

For any integer m with factorization

$$m \;=\; (p_1)^{k_1}(p_2)^{k_2}\dots(p_r)^{k_r}$$

as a product of powers of distinct primes $p_1, p_2, \ldots, p_r$

$$L^0(\mathbb{Z}_m) = \bigoplus_{i=1}^{r} L^0(\mathbb{Z}_{(p_i)^{k_i}}) .$$

(iv) *For any odd prime* p *the projection of the* p*-adic completion of* $\mathbb{Z}$

$$\widehat{\mathbb{Z}}_p = \varprojlim_k \mathbb{Z}_{p^k}$$

onto the residue class field $\widehat{\mathbb{Z}}_p/(p) = \mathbb{Z}_p$ *induces isomorphisms*

$$L^0(\widehat{\mathbb{Z}}_p) \cong L^0(\mathbb{Z}_p)$$

For $p = 2$

$$L^0(\widehat{\mathbb{Z}}_2) \cong L^0(\mathbb{Z}_8) = \mathbb{Z}_2 \oplus \mathbb{Z}_8 .$$

(v) *The symmetric* L*-theory of the profinite completion of* $\mathbb{Z}$

$$\widehat{\mathbb{Z}} = \varprojlim_m \mathbb{Z}_m = \prod_{p \text{ prime}} \widehat{\mathbb{Z}}_p$$

is given by

$$L^n(\widehat{\mathbb{Z}}) = \bigoplus_p L^n(\widehat{\mathbb{Z}}_p) ,$$

with

$$L^0(\widehat{\mathbb{Z}}) = \bigoplus_p L^0(\widehat{\mathbb{Z}}_p) = \bigoplus_\infty \mathbb{Z}_2 \oplus \bigoplus_\infty \mathbb{Z}_4 \oplus \mathbb{Z}_8 .$$ □

Definition 37.7 (i) For any integer m let

$$\begin{aligned} \psi(m) &= \text{exponent}(L^0(\mathbb{Z}_m)) \\ &= \begin{cases} 2 & \text{if } m = d \text{ or } 2d \\ 4 & \text{if } m = 4d, e, 2e \text{ or } 4e \\ 8 & \text{otherwise} \end{cases} \end{aligned}$$

with

$$\begin{aligned} d &= \text{a product of odd primes } p \equiv 1 (\text{mod } 4) \\ e &= \text{a product of odd primes, including at least one } p \equiv 3 (\text{mod } 4) . \end{aligned}$$

(ii) For any integer m let

$$\begin{aligned} \widehat{\psi}(m) &= \text{exponent}(L^0(\widehat{\mathbb{Z}}_m)) \\ &= \begin{cases} 2 & \text{if } m \text{ is a product of odd primes} \equiv 1 (\text{mod } 4) \\ 4 & \text{if } m \text{ is odd and has a prime factor} \equiv 3 (\text{mod } 4) \\ 8 & \text{if } m \text{ is even .} \end{cases} \end{aligned}$$

□

Proposition 37.8 (Ranicki [234, 4.2])
If A is a ring with involution of finite characteristic $m \geq 2$ then the algebraic L-groups $L^*(A,\epsilon)$, $L_*(A,\epsilon)$ are of exponent $\psi(m)$.
Proof The action of $\mathbb{Z}_m$ on A induces an action of the symmetric Witt group $L^0(\mathbb{Z}_m)$ on the algebraic L-groups $L^*(A,\epsilon)$, $L_*(A,\epsilon)$, and $L^0(\mathbb{Z}_m)$ is a ring with 1 of exponent $\psi(m)$. □

By 25.4 there is a quadratic L-theory localization exact sequence

$$\ldots \longrightarrow L^n(A,\epsilon) \longrightarrow L^n_{\widetilde{K}_0(A)}(S^{-1}A,\epsilon) \longrightarrow L^n(A,S,\epsilon) \longrightarrow L^{n-1}(A,\epsilon) \longrightarrow \ldots$$

for any ring with involution A and involution-invariant multiplicative subset $S \subset A$, with $L^*(A,\epsilon)$ the projective L-groups of A, $L^*_{\widetilde{K}_0(A)}(S^{-1}A,\epsilon)$ the $\mathrm{im}(\widetilde{K}_0(A) \longrightarrow \widetilde{K}_0(S^{-1}A))$-intermediate L-groups of $S^{-1}A$, and $L^*(A,S,\epsilon)$ the cobordism groups of $S^{-1}A$-contractible projective Poincaré complexes over A.

Definition 37.9 (i) A ring A is *additively m-torsion-free* for some integer $m \geq 2$ if multiplication by m is an injective map

$$m \,:\, A \longrightarrow A \,;\, a \longrightarrow ma\ ,$$

that is if $\{a \in A \,|\, ma = 0\} = \{0\}$.
(ii) A ring A is *additively torsion-free* if it is additively m-torsion-free for every integer $m \geq 2$. □

Example 37.10 If A is an integral domain of characteristic p and π is a group then $A[\pi]$ is additively m-torsion-free for any integer $m \geq 2$ coprime to p. Furthermore, if $p = 0$ then $A[\pi]$ is additively torsion-free. □

Proposition 37.11 (Ranicki [234, 4.4])
(i) *Let A be a ring with involution which is additively m-torsion-free for some $m \geq 2$, so that*

$$S \;=\; \{m^k \,|\, k \geq 0\} \subset A$$

is a central multiplicative subset of non-zero divisors with localization

$$S^{-1}A \;=\; \mathbb{Z}[1/m] \otimes_{\mathbb{Z}} A\ .$$

The torsion ϵ-symmetric L-groups $L^*(A,S,\epsilon)$ are of exponent $\widehat{\psi}(m)$, and the localization maps $L^n(A,\epsilon) \longrightarrow L^n_{\widetilde{K}_0(A)}(S^{-1}A,\epsilon)$ are isomorphisms modulo $\widehat{\psi}(m)$-torsion.
(ii) *If A is a ring with involution which is additively torsion-free, so that*

$$S \;=\; \mathbb{Z}\backslash\{0\} \subset A$$

is a central multiplicative subset of non-zero divisors with localization

$$S^{-1}A \;=\; \mathbb{Q} \otimes_{\mathbb{Z}} A\ .$$

The torsion ϵ-symmetric L-groups $L^*(A,S,\epsilon)$ are of exponent 8, and the localization maps $L^n(A,\epsilon)\longrightarrow L^n_{\widetilde{K}_0(A)}(S^{-1}A,\epsilon)$ are isomorphisms modulo 8-torsion.
Similarly in the ϵ-quadratic case.
Proof (i) By 25.26 the inclusion of A in the m-adic completion of A

$$\widehat{A} \;=\; \varprojlim_k A/m^kA$$

induces excision isomorphisms in L-theory

$$L^*(A,S,\epsilon) \;\cong\; L^*(\widehat{A},\widehat{S},\epsilon)$$

The symmetric Witt group $L^0(\widehat{\mathbb{Z}}_m)$ is a ring with 1 which acts on $L^*(\widehat{A},\widehat{S},\epsilon)$, and which has additive exponent $\widehat{\psi}(m)$.
(ii) By 25.26 the inclusion of A in the profinite completion of A

$$\widehat{A} \;=\; \varprojlim_m A/mA$$

induces excision isomorphisms in L-theory

$$L^*(A,S,\epsilon) \;\cong\; L^*(\widehat{A},S,\epsilon)$$

The symmetric Witt group $L^0(\widehat{\mathbb{Z}})$ is a ring with 1 which acts on $L^*(\widehat{A},S,\epsilon)$, and which has additive exponent 8. □

Example 37.12 A group ring $A=\mathbb{Z}[\pi]$ is additively torsion-free, so that the inclusion

$$A \;=\; \mathbb{Z}[\pi] \longrightarrow S^{-1}A \;=\; \mathbb{Q}[\pi] \;\; (S=\mathbb{Z}\backslash\{0\})$$

induces isomorphisms modulo 8-torsion in algebraic L-theory, by 37.11 (ii). □

Proposition 37.13 *Let A be a ring with involution which is additively m-torsion-free (resp. torsion-free) and let $S=\mathbb{Z}\backslash\{0\}\subset A$, as in 37.11. For any localization $\Sigma^{-1}A$ of A with $A\longrightarrow\Sigma^{-1}A$ injective the natural morphisms of ϵ-symmetric L-groups*

$$L^*(\Sigma^{-1}A,\epsilon) \longrightarrow L^*_{\widetilde{K}_0(\Sigma^{-1}A)}(S^{-1}(\Sigma^{-1}A),\epsilon)\ ,$$
$$L^*(A,\Sigma,\epsilon) \longrightarrow L^*(S^{-1}A,\Sigma,\epsilon)$$

are isomorphisms modulo $\widehat{\psi}(m)$- (resp. 8-) torsion.
Similarly for the ϵ-quadratic L-groups.
Proof The localization $\Sigma^{-1}A$ is also additively m-torsion-free (resp. torsion-free), and the morphisms induced in algebraic L-theory by

$$\Sigma^{-1}A \longrightarrow S^{-1}(\Sigma^{-1}A) \;=\; \Sigma^{-1}(S^{-1}A)$$

are isomorphisms modulo $\widehat{\psi}(m)$- (resp. 8-) torsion by 37.11 (i) (resp. (ii)). Similarly for the torsion L-groups, noting that the ring $L^0(\widehat{\mathbb{Z}}_m)$ (resp. $L^0(\widehat{\mathbb{Z}})$) also acts on the relative torsion L-groups $L^*(A\longrightarrow S^{-1}A,\Sigma,\epsilon)$. □

38. L-theory of Dedekind rings

The high-dimensional knot cobordism groups C_* are isomorphic to the L-groups of various polynomial extensions over $\mathbb{Z}$. The actual computation of C_* first works out the corresponding L-groups of various polynomial extensions over $\mathbb{Q}$, and then uses L-theoretic localization techniques to pass from $\mathbb{Q}$ to $\mathbb{Z}$. This chapter describes particular features of the general localization exact sequence of Chap. 25 in the case when the ground ring with involution A is Dedekind.

The localization exact sequence relates the algebraic L-groups of a ring with involution A and the localization $S^{-1}A$ of A inverting an involution-invariant multiplicative subset $S \subset A$

$$\ldots \longrightarrow L_p^n(A,\epsilon) \longrightarrow L_S^n(S^{-1}A,\epsilon) \longrightarrow L_p^n(A,S,\epsilon) \longrightarrow L_p^{n-1}(A,\epsilon) \longrightarrow \ldots$$

with $L_p^*(A,S,\epsilon)$ the ϵ-symmetric L-groups of (A,S)-modules, and $L_S^*(S^{-1}A,\epsilon)$ the ϵ-symmetric L-groups of the $S^{-1}A$-modules induced from f.g. projective A-modules.

In Chap. 38 A is a Dedekind ring with involution, with quotient field F. The undecorated L-groups L^*, L_* are the projective L-groups L_p^*, L_*^p.

For a Dedekind ring with involution A and any involution-invariant multiplicative subset $S \subset A$ the splitting of (A,S)-modules into $\mathcal{P}$-primary components and a devissage argument give identifications

$$L^n(A,S,\epsilon) \;=\; \bigoplus_{\mathcal{P}} L^n(A,\mathcal{P}^\infty,\epsilon) \;=\; \bigoplus_{\mathcal{P}} L^n(A/\mathcal{P},u_{\mathcal{P}}\epsilon)$$

with $\mathcal{P} = \overline{\mathcal{P}} \lhd A$ the involution-invariant prime ideals such that $\mathcal{P} \cap S \neq \emptyset$, and $u_{\mathcal{P}} \in A/\mathcal{P}$ appropriate units in the residue class fields $A/\mathcal{P}$. The odd-dimensional torsion L-groups vanish

$$L^{2*+1}(A,S,\epsilon) \;=\; 0\;,$$

so that the localization exact sequence breaks up into exact sequences

$$\begin{aligned}0 \longrightarrow L^{2i}(A,\epsilon) \longrightarrow L_S^{2i}(S^{-1}A,\epsilon) \longrightarrow &\bigoplus_{\mathcal{P}} L^{2i}(A/\mathcal{P},u_{\mathcal{P}}\epsilon)\\ &\longrightarrow L^{2i-1}(A,\epsilon) \longrightarrow L_S^{2i-1}(S^{-1}A,\epsilon) \longrightarrow 0\;.\end{aligned}$$

Chap. 38 is devoted to the localization exact sequence for an arbitrary Dedekind ring with involution A. In Chap. 39 the endomorphism L-groups $L\mathrm{End}^*(F,\epsilon)$ (resp. $L\mathrm{Iso}^*(F,\epsilon)$, $L\mathrm{Aut}^*(F,\epsilon)$) of a field with involution F are expressed in terms of ordinary L-groups, using the localization exact sequence for the Dedekind rings $F[x]$ (resp. $F[s]$, $F[z,z^{-1}]$). These results are generalized to the various endomorphism L-groups of a Dedekind ring with involution A in Chap. 41.

For any ring with involution A and any $*$-invariant subgroup $U \subseteq Wh_i(A)$ $(i=0,1)$ the skew-suspension maps define isomorphisms

$$\overline{S} \;:\; L_n^U(A,\epsilon) \xrightarrow{\simeq} L_{n+2}^U(A,-\epsilon) \;;\; (C,\psi) \longrightarrow (SC,\overline{S}\psi) \quad ((\overline{S}\psi)_j = \psi_j) \;,$$

so that the ϵ-quadratic L-groups are 4-periodic

$$L_*^U(A,\epsilon) \;=\; L_{*+2}^U(A,-\epsilon) \;=\; L_{*+4}^U(A,\epsilon) \;.$$

The ϵ-symmetric L-groups are not 4-periodic in general, but they are 4-periodic for a Dedekind ring A:

Proposition 38.1 (Ranicki [235], [237])
Let A be a Dedekind ring with involution.
(i) *For any $*$-invariant subgroup $U \subseteq Wh_i(A)$ $(i=0,1)$ the skew-suspension maps define isomorphisms*

$$\overline{S} \;:\; L_U^n(A,\epsilon) \xrightarrow{\simeq} L_U^{n+2}(A,-\epsilon) \;;\; (C,\phi) \longrightarrow (SC,\overline{S}\phi) \quad ((\overline{S}\phi)_j = \phi_j) \;,$$

so that the ϵ-symmetric L-groups are 4-periodic

$$L_U^*(A,\epsilon) \;=\; L_U^{*+2}(A,-\epsilon) \;=\; L_U^{*+4}(A,\epsilon) \;.$$

(ii) *Let $S \subset A$ be any involution-invariant multiplicative subset. For any $*$-invariant subgroup $U \subseteq Wh_1(A,S)$ with*

$$I_1(A,S) \;=\; \mathrm{im}(Wh_1(S^{-1}A) \longrightarrow Wh_1(A,S)) \subseteq U$$

and with image $V \subseteq \widetilde{K}_0(A)$ the localization exact sequence is given by

$$\begin{aligned} \dots \longrightarrow L_U^{n+1}(A,S,\epsilon) \longrightarrow L_V^n(A,\epsilon) \longrightarrow L^n(S^{-1}A,\epsilon) \\ \longrightarrow L_U^n(A,S,\epsilon) \longrightarrow L_V^{n-1}(A,\epsilon) \longrightarrow \dots \;. \end{aligned}$$

The ϵ-symmetric L-groups $L_U^*(A,S,\epsilon)$ are 4-periodic for $* \geq 2$. The even-dimensional groups are such that $L_U^{2i}(A,S,\epsilon)$ is the Witt group of nonsingular $(-)^i\epsilon$-symmetric linking forms (M,λ) over (A,S) such that $[M] \in U \subseteq Wh_1(A,S)$. The odd-dimensional groups are such that

$$\begin{aligned} L^{2i-1}(A,S,\epsilon) \;&=\; 0 \;, \\ L_U^{2i-1}(A,S,\epsilon) \;&=\; \mathrm{coker}(L^{2i}(A,S,\epsilon) \longrightarrow \widehat{H}^{2i}(\mathbb{Z}_2;Wh_1(A,S)/U)) \end{aligned}$$

for any $i \geq 1$.

(iii) *Let* $S = A\backslash\{0\} \subset A$, *so that* $S^{-1}A = F$. *Then*

$$L^{2i+1}(F,\epsilon) \;=\; 0 \;\;,\;\; L^{2i+1}(A,S,\epsilon) \;=\; 0$$

and the localization exact sequences for

$$(U,V) \;=\; (Wh_1(A,S), \widetilde{K}_0(A)) \;\;,\;\; (I_1(A,S),\{0\})$$

fit into a commutative braid of exact sequences

$$\begin{array}{ccccccc}
L^{2i}_h(A,\epsilon) & & L^{2i}(F,\epsilon) & & L^{2i}(A,S,\epsilon) & & \widehat{H}^1(\mathbb{Z}_2;\widetilde{K}_0(A)) \\
& \searrow \;\; \nearrow & & \searrow \;\; \nearrow & & \searrow \;\; \nearrow & \\
& L^{2i}(A,\epsilon) & & L^{2i}_h(A,S,\epsilon) & & L^{2i-1}(A,\epsilon) & \\
& \nearrow \;\; \searrow & & \nearrow \;\; \searrow & & \nearrow \;\; \searrow & \\
0 & & \widehat{H}^0(\mathbb{Z}_2;\widetilde{K}_0(A)) & & L^{2i-1}_h(A,\epsilon) & & 0
\end{array}$$

The Witt class $(C,\phi) \in L^{2i-1}(A,\epsilon)$ *of a f.g. projective* $(2i-1)$*-dimensional* ϵ*-symmetric Poincaré complex* (C,ϕ) *over* A *is determined by the Witt class* $(M,\lambda) \in L^{2i}(A,S,\epsilon)$ *of the nonsingular* $(-)^i\epsilon$*-symmetric linking form* (M,λ) *over* (A,S) *with*

$$\begin{aligned}
&M \;=\; \text{torsion}\, H^i(C) \;=\; \ker(H^i(C)\longrightarrow S^{-1}H^i(C)) \;, \\
&\lambda \;:\; M \times M \longrightarrow F/A \;;\; (x,y) \longrightarrow \phi_0(x)(z)/s \\
&(x,y \in \ker(d^* : C^i \longrightarrow C^{i+1}) \;,\; z \in C^{i-1} \;\;,\; s \in S \;,\; d^*z = sy) \;,
\end{aligned}$$

defining an isomorphism

$$L^{2i-1}(A,\epsilon) \xrightarrow{\;\simeq\;} \text{coker}(L^{2i}(F,\epsilon) \longrightarrow L^{2i}(A,S,\epsilon)) \;;\; (C,\phi) \longrightarrow (M,\lambda) \;.$$

Similarly in the free case, with an isomorphism

$$L^{2i-1}_h(A,\epsilon) \xrightarrow{\;\simeq\;} \text{coker}(L^{2i}(F,\epsilon) \longrightarrow L^{2i}_h(A,S,\epsilon)) \;.$$

Proof (i) If (C,ϕ) is an n-dimensional ϵ-symmetric Poincaré complex over A and $n \geq 2$ then a f.g. projective A-module resolution of $H_0(C)$

$$0 \longrightarrow D_1 \longrightarrow D_0 \longrightarrow H_0(C) \longrightarrow 0$$

determines an $(n+1)$-dimensional ϵ-symmetric pair $(C\longrightarrow D,(0,\phi))$ allowing $H_0(C)$ to be killed by surgery on (C,ϕ). See Ranicki [235, Chap. 4] for further details.

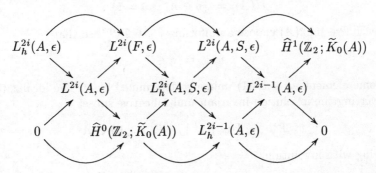

(ii) The proof of 4-periodicity proceeds as for (i), but using F-contractible chain complexes. See [237, 3.6.5, 4.2] for further details. (iii) See Ranicki [233] for the proof of $L^{2*+1}(F,\epsilon) = 0$, and [237, 3.2.5] for the proof that $L^{2*+1}(A,S,\epsilon) = 0$. □

Definition 38.2 Let $\max(A)$ be the set of maximal ideals in A, and let

$$\overline{\max}(A) = \{\mathcal{P} \in \max(A) \mid \mathcal{P} = \overline{\mathcal{P}}\}$$

be the set of involution-invariant maximal ideals. Also, let

$$U(A) = \{u \in A^\bullet \mid u\overline{u} = 1\} .$$

For each $\mathcal{P} \in \overline{\max}(A)$ choose a uniformizer $\pi \in \mathcal{P} \backslash \mathcal{P}^2$, so that

$$u_\pi \overline{\pi} = \pi \in \mathcal{P}$$

for some element $u_\pi \in U(A)$ uniquely determined by π. The localization of A inverting the involution-invariant multiplicative subset

$$\mathcal{P}^\infty = \{(u_\pi)^j \pi^k \mid j \in \mathbb{Z}, k \geq 0\} \subset A$$

is a ring with involution

$$(\mathcal{P}^\infty)^{-1}A = A[1/\pi] .$$

The *$\mathcal{P}$-local ring*

$$A_{\mathcal{P}} = (A\backslash\mathcal{P})^{-1}A$$

is a local ring, with unique maximal ideal $\mathcal{P}A_{\mathcal{P}} \triangleleft A_{\mathcal{P}}$ and residue class field

$$A_{\mathcal{P}}/\mathcal{P}A_{\mathcal{P}} = A/\mathcal{P} .$$ □

Remark 38.3 If $\pi, \pi' \in \mathcal{P}\backslash\mathcal{P}^2$ are two uniformizers for an involution-invariant maximal ideal $\mathcal{P} \in \overline{\max}(A)$ then

$$\pi' = \pi v \in \mathcal{P}$$

for some unit $v \in A^\bullet$. Now

$$u_{\pi'} = v(\overline{v})^{-1}u_\pi \in U(A) ,$$

so that there are defined scaling isomorphisms in $u\epsilon$-symmetric L-theory

$$L^n(A/\mathcal{P}, u_\pi\epsilon) \xrightarrow{\simeq} L^n(A/\mathcal{P}, u_{\pi'}\epsilon) \; ; \; (C,\phi) \longrightarrow (C, v\phi) .$$

Thus the $u_\pi\epsilon$-symmetric L-groups $L^*(A/\mathcal{P}, u_\pi\epsilon)$ only depend on $A, \mathcal{P}, \epsilon$ and are independent of the choice of uniformizer π. Similarly for $u_\pi\epsilon$-quadratic L-groups $L_*(A/\mathcal{P}, u_\pi\epsilon)$. From now on $u_\pi \in U(A)$ will be denoted by $u_{\mathcal{P}}$, even though u_π depends on the choice of uniformizer π. □

Remark 38.4 Every maximal ideal $\mathcal{P} \triangleleft A$ is a f.g. projective A-module, which is isomorphic to the image $\mathrm{im}(p)$ of a projection

$$p = p^2 = \begin{pmatrix} a & b \\ c & 1-a \end{pmatrix} : A \oplus A \longrightarrow A \oplus A$$

for some $a, b, c \in A$ such that $a(1-a) = bc$, as in 17.1 (i). For any involution on A the dual f.g. projective A-module $\mathcal{P}^*$ is isomorphic to $\mathrm{im}(p^*)$, with

$$p^* = \begin{pmatrix} \bar{a} & \bar{c} \\ \bar{b} & 1-\bar{a} \end{pmatrix} : A \oplus A \longrightarrow A \oplus A$$

the dual projection. For the identity involution on A

$$p^* = \begin{pmatrix} a & c \\ b & 1-a \end{pmatrix} = \tau^{-1}(1-p)\tau : A \oplus A \longrightarrow A \oplus A$$

with $\tau = \begin{pmatrix} 0 & -1 \\ 1 & 0 \end{pmatrix}$, so that

$$\mathcal{P} \oplus \mathcal{P}^* \cong A \oplus A$$

and the duality involution on the reduced projective class group is given by

$$* = -1 : \widetilde{K}_0(A) \longrightarrow \widetilde{K}_0(A) ; [\mathcal{P}] \longrightarrow [\mathcal{P}^*] = -[\mathcal{P}] . \qquad \square$$

Proposition 38.5 *Let $\mathcal{P} \in \overline{\max}(A)$, with uniformizer $\pi \in \mathcal{P}$.*
(i) *The cartesian morphism of rings with involution and multiplicative subsets*

$$(A, \mathcal{P}^\infty) \longrightarrow (A_\mathcal{P}, \mathcal{P}^\infty)$$

determines a cartesian square of rings with involution

$$\begin{array}{ccc} A & \longrightarrow & A[1/\pi] \\ \downarrow & & \downarrow \\ A_\mathcal{P} & \longrightarrow & F \end{array}$$

(ii) *The relative algebraic K-groups are such that*

$$\begin{aligned} K_*(A, \mathcal{P}^\infty) &= K_*(A_\mathcal{P}, \mathcal{P}^\infty) = K_{*-1}(A/\mathcal{P}) , \\ Wh_1(A, \mathcal{P}^\infty) &= K_0(A/\mathcal{P}) = \mathbb{Z} , \\ I_1(A, \mathcal{P}^\infty) &= \mathrm{im}(Wh_1(A, \mathcal{P}^\infty) \longrightarrow \widetilde{K}_0(A)) = \{0\} . \end{aligned}$$

(iii) *The relative ϵ-symmetric L-groups are such that*

$$L^*(A, \mathcal{P}^\infty, \epsilon) = L^*(A_\mathcal{P}, \mathcal{P}^\infty, \epsilon) = L^*(A/\mathcal{P}, u_\mathcal{P}\epsilon)$$

with

$$L^{2*+1}(A, \mathcal{P}^\infty, \epsilon) = 0 .$$

Proof (i) Immediate from 17.1.
(ii) See Chap. 18 for devissage in algebraic K-theory.

(iii) See Ranicki [237, 4.2] for devissage in ϵ-symmetric L-theory. Here are the details of the identification

$$L^0(A, \mathcal{P}^\infty, \epsilon) = L^0(A/\mathcal{P}, u_{\mathcal{P}}\epsilon) .$$

For any finite-dimensional $A/\mathcal{P}$-vector space L there is defined an isomorphism

$$\mathrm{Hom}_{A/\mathcal{P}}(L, A/\mathcal{P}) \xrightarrow{\simeq} \mathrm{Hom}_A(L, A[1/\pi]/A) \ ; \ f \longrightarrow (x \longrightarrow f(x)/\pi) ,$$

so that as in 25.20 there is defined a one-one correspondence

$$\{u_{\mathcal{P}}\epsilon\text{-symmetric forms } (L, \phi) \text{ over } A/\mathcal{P}\} \rightleftarrows$$

$$\{\epsilon\text{-symmetric linking forms } (L, \lambda) \text{ over } (A, \mathcal{P}^\infty) \text{ with } \mathcal{P}L = 0\}$$

with $\lambda = \phi/\pi$. The one-one correspondence determines a morphism

$$L^0(A/\mathcal{P}, u_{\mathcal{P}}\epsilon) \longrightarrow L^0(A, \mathcal{P}^\infty, \epsilon) \ ; \ (L, \phi) \longrightarrow (L, \lambda) .$$

In order to construct the inverse start with a nonsingular ϵ-symmetric linking form (M, μ) over $(A, \mathcal{P}^\infty)$. As in 17.2 decompose the $(A, \mathcal{P}^\infty)$-module M as a direct sum

$$M = \sum_{k=1}^{j} M_k$$

with $\mathrm{ann}(M_k) = \mathcal{P}^k$ and M_k a f.g. free $A/\mathcal{P}^k$-module. Write the dual $(A, \mathcal{P}^\infty)$-module as

$$N = M\hat{} = \sum_{k=1}^{j} N_k$$

with $N_k = (M_k)\hat{}$. An A-module morphism $f : M \longrightarrow N$ can be expressed as

$$f = \begin{pmatrix} f' & g \\ h & f_j \end{pmatrix} : M = M' \oplus M_j \longrightarrow N = N' \oplus N_j$$

with

$$f' : M' = \sum_{k=1}^{j-1} M_k \longrightarrow N' = \sum_{k=1}^{j-1} N_k ,$$

$$f_j : M_j \longrightarrow N_j \ , \ g : M_j \longrightarrow N' \ , \ h : M' \longrightarrow N_j .$$

An element $a \in A/\mathcal{P}^j$ is a unit if and only if the projection $[a] \in A/\mathcal{P}$ is non-zero, so that the $A/\mathcal{P}^j$-module morphism $f_j : M_j \longrightarrow N_j$ is an isomorphism if and only if the $A/\mathcal{P}$-module morphism

$$[f] = [f_j] : M/\mathcal{P}^{j-1}M = M_j/\mathcal{P}^{j-1}M_j \longrightarrow N/\mathcal{P}^{j-1}N = N_j/\mathcal{P}^{j-1}N_j$$

is an isomorphism. Thus f is an isomorphism if and only if $f_j : M_j \longrightarrow N_j$ is an isomorphism and

$$f' - g(f_j)^{-1}h \;:\; M' \longrightarrow N'$$

is an isomorphism, in which case

$$f = \begin{pmatrix} f' & g \\ h & f_j \end{pmatrix} = \begin{pmatrix} f' - g(f_j)^{-1}h & g \\ 0 & f_j \end{pmatrix} \begin{pmatrix} 1 & 0 \\ (f_j)^{-1}h & 1 \end{pmatrix}$$
$$: M = M' \oplus M_j \longrightarrow N = N' \oplus N_j \;.$$

In particular, we have that $\mu : M \longrightarrow N = M^\wedge$ is an isomorphism, so that the restriction of (M,μ) to $M_j \subseteq M$ is a nonsingular ϵ-symmetric linking form (M_j,μ_j) over $(A,\mathcal{P}^\infty)$, which can be split off from (M,μ) with an isomorphism

$$(M,\mu) \;\cong\; (M',\mu') \oplus (M_j,\mu_j) \;.$$

Proceeding by induction (as in Milnor [197, p. 95]) there is obtained an isomorphism of the type

$$(M,\mu) \;\cong\; \sum_{k=1}^{j} (M_k,\mu_k) \;.$$

Consider (M_k,μ_k). For any integer $\ell \geq 1$ with $2\ell \geq k \geq \ell$ there is defined a sublagrangian

$$L = \mathcal{P}^\ell M_k \subset M_k$$

with an isomorphism

$$M_k/\mathcal{P}^{2\ell-k}M_k \xrightarrow{\;\cong\;} L^\perp/L \;;\; x \longrightarrow \pi^{k-\ell}x \;.$$

If k is $\begin{cases}\text{even}\\ \text{odd}\end{cases}$ let $\ell_k = \begin{cases} k/2 \\ (k+1)/2 \end{cases}$. The submodule

$$L_k = \mathcal{P}^{\ell_k} M_k \subset M_k$$

is a $\begin{cases}\text{lagrangian}\\ \text{sublagrangian}\end{cases}$ of (M_k,μ_k), such that

$$\begin{cases} L_k = L_k^\perp \\ L_k \subseteq L_k^\perp \end{cases} , \quad (L_k^\perp/L_k, \mu_k^\perp/\mu_k) = \begin{cases} 0 \\ (M_k/\mathcal{P}M_k, [\mu_k]) \;. \end{cases}$$

The terms (M_k,μ_k) with even k are thus hyperbolic, and do not contribute to the Witt class of (M,μ). For odd k the nonsingular ϵ-symmetric linking form $(M_k/\mathcal{P}M_k,[\mu_k])$ over $(A,\mathcal{P}^\infty)$ is defined on a finite dimensional $A/\mathcal{P}$-vector space $M_k/\mathcal{P}M_k$, and so can be regarded as a nonsingular $u_{\mathcal{P}}\epsilon$-symmetric form over $A/\mathcal{P}$. The morphisms

$$L^0(A/\mathcal{P}, u_{\mathcal{P}}\epsilon) \longrightarrow L^0(A,\mathcal{P}^\infty,\epsilon) \;;\; (L,\phi) \longrightarrow (L,\lambda) \;,$$
$$L^0(A,\mathcal{P}^\infty,\epsilon) \longrightarrow L^0(A/\mathcal{P}, u_{\mathcal{P}}\epsilon) \;;\; (M,\mu) \longrightarrow \sum_{k\,\mathrm{odd}} (M_k/\mathcal{P}M_k, [\mu_k])$$

are inverse isomorphisms. □

Remark 38.6 In ϵ-quadratic L-theory

$$L_*(A, \mathcal{P}^\infty, \epsilon) \;=\; L_*(A_{\mathcal{P}}, \mathcal{P}^\infty, \epsilon) \;,$$

but the map

$$L_*(A/\mathcal{P}, u_{\mathcal{P}}\epsilon) \longrightarrow L_*(A, \mathcal{P}^\infty, \epsilon)$$

(determined by a choice of uniformizer $\pi \in \mathcal{P}$) is not an isomorphism in general (Ranicki [237, p. 413]) – for example, the map

$$L_0(\mathbb{Z}_2) \;=\; \mathbb{Z}_2 \longrightarrow L_0(\mathbb{Z}, (2)^\infty) \;=\; \mathbb{Z}_8 \oplus \mathbb{Z}_2$$

is only an injection. □

Given a subset $J \subseteq \overline{\max}(A)$ write $\mathcal{P}_j$ for the ideal indexed by $j \in J$, with uniformizer $\pi_j \in \mathcal{P}_j \backslash \mathcal{P}_j^2$ and $u_{\mathcal{P}_j} = u_j \in U(A)$ such that $u_j \overline{\pi}_j = \pi_j \in \mathcal{P}_j$. Also, define the involution-invariant multiplicative subset

$$S_J \;=\; \{\pi_1^{k_1}\pi_2^{k_2}\dots\pi_m^{k_m} \,|\, k_1, k_2, \dots, k_m \geq 1\} \subset A \;.$$

Proposition 38.7 *For any subset $J \subseteq \overline{\max}(A)$ the ϵ-symmetric L-groups of a Dedekind ring with involution A and the localization $S_J^{-1}A$ are related by an exact sequence*

$$\begin{aligned} 0 \longrightarrow L^{2*}(A, \epsilon) \longrightarrow L^{2*}_{\widetilde{K}_0(A)}(S_J^{-1}A, \epsilon) &\xrightarrow{\;\partial\;} L^{2*}(A, S_J, \epsilon) \\ \longrightarrow L^{2*-1}(A, \epsilon) &\longrightarrow L^{2*-1}_{\widetilde{K}_0(A)}(S_J^{-1}A, \epsilon) \longrightarrow 0 \;, \end{aligned}$$

with

$$L^{2*}(A, S_J, \epsilon) \;=\; \bigoplus_{j \in J} L^{2*}(A/\mathcal{P}_j, u_j\epsilon) \;,$$

$$L^{2*+1}(A, S_J, \epsilon) \;=\; 0$$

by $\mathcal{P}_j$-primary decomposition and devissage.

Proof Every nonsingular $(-)^i\epsilon$-symmetric linking form (M, λ) over (A, S_J) has a direct sum decomposition

$$(M, \lambda) \;=\; \sum_{j \in J}\sum_{k \geq 1}(M_{j,k}, \lambda_{j,k})$$

with $M_{j,k}$ $\mathcal{P}_j$-primary and f.g. free over $A/\mathcal{P}_j^k$. The devissage isomorphisms are given by 38.5 (iii) to be

$$L^{2i}(A, S_J, \epsilon) \xrightarrow{\;\simeq\;} \bigoplus_{j \in J} L^{2i}(A/\mathcal{P}_j, u_j\epsilon) \;;$$

$$(M, \lambda) \longrightarrow \sum_{j \in J}\sum_{k \text{ odd}} (M_{j,k}/\mathcal{P}_j M_{j,k}, [\lambda_{j,k}]) \;. \qquad \square$$

Remark 38.8 The relative ϵ-quadratic L-groups have $\mathcal{P}$-primary decomposition

$$L_*(A, S_J, \epsilon) = \bigoplus_{j\in J} L_*(A, \mathcal{P}_j^\infty, \epsilon)$$

but do not in general have devissage: the maps

$$L_*(A/\mathcal{P}_j, u_j\epsilon) \longrightarrow L_*(A, \mathcal{P}_j^\infty, \epsilon)$$

may not be isomorphisms if $\mathrm{char}(A/\mathcal{P}_j) = 2$ (cf. Remark 38.6). □

Proposition 38.9 *The ϵ-symmetric L-groups of A and the quotient field $F = S^{-1}A$ are related by an exact sequence*

$$0 \longrightarrow L^{2*}(A, \epsilon) \longrightarrow L^{2*}(F, \epsilon) \xrightarrow{\partial} L^{2*}(A, S, \epsilon) \longrightarrow L^{2*-1}(A, \epsilon) \longrightarrow 0$$

with

$$L^{2*}(A, S, \epsilon) = \bigoplus_{\mathcal{P}\in\overline{\max}(A)} L^{2*}(A, \mathcal{P}^\infty, \epsilon) = \bigoplus_{\mathcal{P}\in\overline{\max}(A)} L^{2*}(A/\mathcal{P}, u_{\mathcal{P}}\epsilon)$$

by $\mathcal{P}$-primary decomposition and devissage.

Proof Take

$$J = \overline{\max}(A)\ ,\quad J' = \max(A)\backslash\overline{\max}(A)$$

in 38.8. There is defined a cartesian square of rings with involution

$$\begin{array}{ccc} A & \longrightarrow & S_J^{-1}A \\ \big\downarrow & & \big\downarrow \\ S_{J'}^{-1}A & \longrightarrow & F\ . \end{array}$$

The involution acts freely on J', so that the involution on

$$\mathbb{H}(S_J^{-1}A, S_{J'}) = \mathbb{H}(A, S_{J'})$$

is hyperbolic and by 22.6

$$\begin{aligned} L^*(A, S_{J'}, \epsilon) &= L^*(S_J^{-1}A, S_{J'}, \epsilon) = 0\ , \\ L^*(A, \epsilon) &= L^*(S_{J'}^{-1}A, \epsilon)\ ,\ \ L^*(S_J^{-1}A, \epsilon) = L^*(F, \epsilon)\ . \end{aligned}$$

□

The computation of the Witt group of an algebraic number field F in 37.2 can now be extended to the Witt group of the ring of integers:

Proposition 38.10 (Milnor and Husemoller [200, IV.4])
Let F be an algebraic number field and let $A \subset F$ be the ring of integers, so that A is a Dedekind ring with quotient field F. Let $F = \mathbb{Q}[x]/(p(x))$ for an irreducible monic polynomial $p(x) \in \mathcal{M}_x(\mathbb{Q})$ of degree $d = r + 2s$, with r real roots and $2s$ non-real roots.

(i) *For any involution on F and $\epsilon = \pm 1$ the ϵ-symmetric Witt groups of A and F are related by the localization exact sequence*

$$0 \longrightarrow L^0(A,\epsilon) \longrightarrow L^0(F,\epsilon) \longrightarrow \bigoplus_{\mathcal{P}\in\overline{\max}(A)} L^0(A/\mathcal{P}, u_{\mathcal{P}}\epsilon) \longrightarrow L^{-1}(A,\epsilon) \longrightarrow 0 .$$

The residue class fields $A/\mathcal{P}$ are finite, and $L^0(A/\mathcal{P}, u_{\mathcal{P}}\epsilon)$ is an $L^0(A/\mathcal{P})$-module. Hence each $L^0(A/\mathcal{P}, u_{\mathcal{P}}\epsilon)$ is 4-torsion, and so is $L^{-1}(A)$.
(ii) *For the identity involution on F*

$$\text{dimension}\, L^0(A,\epsilon) \;=\; \text{dimension}\, L^0(F,\epsilon) \;=\; \begin{cases} r & \text{if } \epsilon = 1 \\ 0 & \text{if } \epsilon = -1 \ . \end{cases}$$

Both $L^0(A)$ and $L^0(F)$ are of the form $\mathbb{Z}^r \oplus$ 8-torsion, with one signature in $L^0(\mathbb{R}) = \mathbb{Z}$ for each embedding $F \subset \mathbb{R}$, and

$$L^0(A,-1) \;=\; L^0(F,-1) \;=\; 0 \ .$$

(iii) *For a non-trivial involution on F with fixed field F_0 and $F = F_0(\sqrt{a})$ $(\overline{\sqrt{a}} = -\sqrt{a})$*

$$\text{dimension}\, L^0(A,\epsilon) \;=\; \text{dimension}\, L^0(F,\epsilon) \;=\; r_0 + s \ ,$$

with r_0 the number of real roots of $p(x)$ for which the corresponding embedding $\rho_\xi : F_0 \longrightarrow \mathbb{R}; x \longrightarrow \xi$ has $\rho_\xi(a) < 0$. Both $L^0(A,\epsilon)$ and $L^0(F,\epsilon)$ are of the form $\mathbb{Z}^{r_0+s} \oplus$ 8-torsion. □

Addendum 38.11 The symmetric Witt groups of any Dedekind ring with involution A and quotient field F fit into the commutative braid of exact sequences

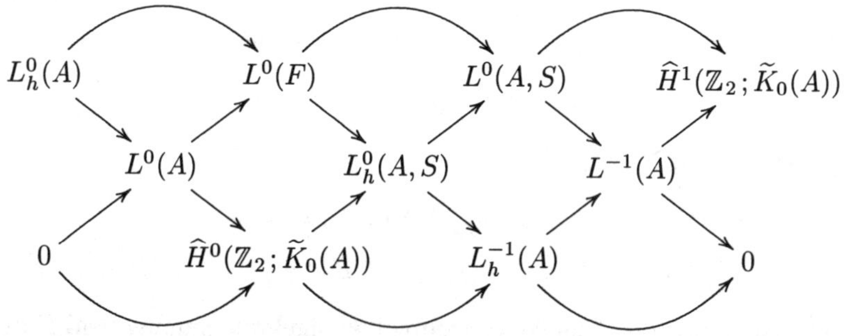

as in 38.1. If A has the identity involution then

$$L^0(A,S) \;=\; \bigoplus_{\mathcal{P}\in\max(A)} L^0(A/\mathcal{P}) \ .$$

Moreover, $* = -1 : \widetilde{K}_0(A) \longrightarrow \widetilde{K}_0(A)$ (38.4), so that

$$\widehat{H}^i(\mathbb{Z}_2;\widetilde{K}_0(A)) = \begin{cases}\{x\in\widetilde{K}_0(A)\,|\,2x=0\} & \text{if } i\equiv 0(\text{mod } 2)\\ \widetilde{K}_0(A)/2\widetilde{K}_0(A) & \text{if } i\equiv 1(\text{mod } 2)\ .\end{cases}$$

If A is the Dedekind ring of algebraic integers in an algebraic number field F (with the identity involution) it now follows from the exact sequence of Milnor and Husemoller [200, IV.3.4]

$$0 \longrightarrow L^0(A) \longrightarrow L^0(F) \longrightarrow \bigoplus_{\mathcal{P}\in\max(A)} L^0(A/\mathcal{P}) \longrightarrow \widetilde{K}_0(A)/2\widetilde{K}_0(A) \longrightarrow 0$$

that

$$L^{-1}(A) = \widehat{H}^1(\mathbb{Z}_2;\widetilde{K}_0(A))\ .$$

In Ranicki [235, p. 417] it is shown that

$$L_h^{-1}(A) = \text{coker}(L^0(A)\longrightarrow\widehat{H}^1(\mathbb{Z}_2;\widetilde{K}_0(A))) = 0\ . \qquad\square$$

Example 38.12 (i) For $F=\mathbb{Q}$, $A=\mathbb{Z}$ with $r=1$, $s=0$ and by 38.10 (ii)

$$\begin{aligned}L^0(\mathbb{Z}) &= \mathbb{Z}\ ,\\ L^0(\mathbb{Q}) &= \mathbb{Z}\oplus\bigoplus_{p \text{ prime}} L^0(\mathbb{F}_p) = \mathbb{Z}\oplus\bigoplus_{\infty}\mathbb{Z}_2\oplus\bigoplus_{\infty}\mathbb{Z}_4\ ,\\ L^0(\mathbb{Z},-1) &= L^0(\mathbb{Q},-1) = 0\ .\end{aligned}$$

(ii) Let

$$F = \mathbb{Q}[z,z^{-1}]/(z^2+2qz+1) = \mathbb{Q}(\sqrt{q^2-1})$$

for some $q\in\mathbb{Q}$ such that $\sqrt{q^2-1}\notin\mathbb{Q}$, with

$$A = \mathbb{Z}[\sqrt{p^2-1}]\ ,\ \overline{\sqrt{q^2-1}} = -\sqrt{q^2-1}\ ,\ r=1\ ,\ s=0\ ,$$

The fixed field of the involution $\overline{z}=z^{-1}$ is $F_0=\mathbb{Q}$, and 38.10 (iii) gives

$$\text{dimension}\, L^0(A,\epsilon) = \text{dimension}\, L^0(F,\epsilon) = r_0+s = \begin{cases}1 & \text{if } q^2<1\\ 0 & \text{if } q^2>1\ .\end{cases}$$

(iii) Let $\zeta_m=e^{2\pi i/m}$, a primitive mth root of 1, with $m\geq 3$. The minimal polynomial of ζ_m is the cyclotomic polynomial

$$\Phi_m(z) = \prod_{(m,k)=1}(z-(\zeta_m)^k)\in\mathbb{Z}[z]$$

of degree

$$\begin{aligned}d &= \phi(m)\ \text{(the Euler function)}\\ &= \text{the number of units } u\in\mathbb{Z}_m^\bullet\ .\end{aligned}$$

(If m is prime then $\phi(m)=m-1$, $\Phi_m(z)=z^{m-1}+z^{m-2}+\ldots+z+1$.) The cyclotomic algebraic number field

$$F = \mathbb{Q}(\zeta_m) = \mathbb{Q}[z,z^{-1}]/(\Phi_m(z))$$

is of degree $\phi(m)$, with

$$r = 0 \ , \ s = \phi(m)/2 \ ,$$

and the ring of integers is

$$A = \mathbb{Z}[z,z^{-1}]/(\Phi_m(z)) = \mathbb{Z}[\zeta_m] \ .$$

For the complex conjugation involution $\overline{z} = z^{-1}$ on F 38.10 (iii) gives

$$\text{dimension}\, L^0(A,\epsilon) = \text{dimension}\, L^0(F,\epsilon) = \phi(m)/2 \ ,$$

with one signature for each conjugate pair of embeddings

$$F = \mathbb{Q}(\zeta_m) \longrightarrow \mathbb{C}^- \ ; \ \zeta_m \longrightarrow (\zeta_m)^u \quad (u \in \mathbb{Z}_m^\bullet) \ . \qquad \square$$

39. L-theory of function fields

The L-theory localization exact sequence will now be used to express the various L-groups $LEnd^*(F,\epsilon)$, $LAut^*(F,\epsilon)$, $LAut^*(F,\epsilon)$, $LAut^*_{fib}(F,\epsilon)$, $LAsy^*(F)$, $LIso^*(F)$ for a field with involution F in terms of the ordinary L-groups of fields with involution associated to F. This expression will allow the computation in the cases $F = \mathbb{Q}, \mathbb{R}, \mathbb{C}$ of particular interest in high-dimensional knot theory.

The basic tools for expressing the L-theory of a function field in terms of the L-theory of the ground field are the localization exact sequence of Chap. 25 and the devissage technique of Chap. 38. For each of the three basic ways of extending the involution on F to an involution on the function field $F(x)$ (= the quotient field of $F[x]$) the ϵ-symmetric L-groups $L^*(F(x),\epsilon)$ are expressed in terms of the L-groups of the residue class fields $F[x]/\mathcal{P}$. As before, the different involutions will be distinguished by the name assigned to the variable x:

(i) $F[x]$, with $\overline{x} = x$,
(ii) $F[s]$, with $\overline{s} = 1 - s$,
(iii) $F[z, z^{-1}]$, with $\overline{z} = z^{-1}$.

In fact, there are defined isomorphisms of fields with involution

$$F(s) \xrightarrow{\simeq} F(x)[s]/(s^2 - s + x) \; ; \; s \longrightarrow s \, ,$$

$$F(s) \xrightarrow{\simeq} F(z) \; ; \; s \longrightarrow (1 - z)^{-1}$$

but it is convenient to deal separately with (i), (ii) and (iii). In all cases only the even-dimensional L-groups need be considered, since the odd-dimensional L-groups of fields are 0, by 37.1 (iii). The L-groups of the function fields for (i), (ii) and (iii) are 2-periodic if $\mathrm{char}(F) = 2$ by 37.1 (vi), and also for (ii) and (iii) if $\mathrm{char}(F) \neq 2$ by 37.1 (vii), since

$$F(x)[\sqrt{a}] \cong F(s) \cong F(z)$$

with $a = 1 - 4x = (2s - 1)^2$, $\overline{\sqrt{a}} = -\sqrt{a}$.

39A. L-theory of $F(x)$ ($\overline{x} = x$)

The polynomial ring $F[x]$ is a Dedekind ring with quotient field $F(x)$. Recall that $\mathcal{M}(F)$ denotes the set of irreducible monic polynomials $p(x) \in F[x]$ (4.9).

Definition 39.1 Let $\overline{\mathcal{M}}_x(F) \subseteq \mathcal{M}(F)$ be the set of irreducible monic polynomials

$$p(x) = a_0 + a_1 x + \ldots + a_{d-1}x^{d-1} + a_d x^d \in F[x] \quad (a_d = 1)$$

which are involution-invariant

$$\overline{p(x)} = p(x) \in F[x] ,$$

that is $\overline{a}_j = a_j \in F$ $(0 \leq j \leq d-1)$. □

The involution-invariant maximal ideals $\mathcal{P} \in \overline{\max}(F[x])$ are the principal ideals $\mathcal{P} = (p(x)) \triangleleft F[x]$ generated by the elements $p(x) \in \overline{\mathcal{M}}_x(F)$, so identify

$$\overline{\max}(F[x]) = \overline{\mathcal{M}}_x(F) .$$

For the identity involution on F

$$\overline{\mathcal{M}}_x(F) = \mathcal{M}(F) .$$

For any involution on F the duality involution

$$\mathrm{End}_0(F) \longrightarrow \mathrm{End}_0(F) \; ; \; (P, f) \longrightarrow (P^*, f^*)$$

corresponds under the identification of 19.3

$$\mathrm{End}_0(F) = \mathbb{Z}[\mathcal{M}_x(F)]$$

to the involution

$$\mathbb{Z}[\mathcal{M}_x(F)] \longrightarrow \mathbb{Z}[\mathcal{M}_x(F)] \; ; \; p(x) \longrightarrow \overline{p(x)} ,$$

so that

$$\widehat{H}^n(\mathbb{Z}_2; \mathrm{End}_0(F)) = \widehat{H}^n(\mathbb{Z}_2; \mathbb{Z})[\overline{\mathcal{M}}_x(F)]$$

with

$$\widehat{H}^n(\mathbb{Z}_2; \mathbb{Z}) = \begin{cases} \mathbb{Z}_2 & \text{if } n \equiv 0 (\mathrm{mod}\ 2) \\ 0 & \text{if } n \equiv 1 (\mathrm{mod}\ 2) . \end{cases}$$

Proposition 39.2 *For any $p(x) \in \overline{\mathcal{M}}_x(F)$, $\epsilon \in U(F)$*

$$L^0(F[x], p(x)^\infty, \epsilon) = L\mathrm{End}^0_{p(x)^\infty}(F, \epsilon) = L^0(E, \epsilon)$$

with $E = F[x]/(p(x))$ the residue class field.

Proof The identification

$$L^0(F[x], p(x)^\infty, \epsilon) = L\mathrm{End}^0_{p(x)^\infty}(F, \epsilon)$$

was already obtained in 35.3, and the identification

$$L^0(F[x], p(x)^\infty, \epsilon) \;=\; L^0(E, \epsilon)$$

is a special case of 38.5 (iii). These identifications will now be made explicit. In view of the devissage result of 38.5 (iii) this need only be done in the $p(x)$-torsion case, so that $L^0(F[x], p(x)^\infty, \epsilon)$ will be regarded as the cobordism group of 1-dimensional $(-\epsilon)$-symmetric Poincaré complexes (C, ϕ) over $F[x]$ such that $p(x)H_*(C) = 0$, and $LEnd^0_{p(x)^\infty}(F, \epsilon)$ will be regarded as the Witt group of nonsingular ϵ-symmetric forms (L, θ) over F with an endometric structure $f : L \longrightarrow L$ such that $p(f) = 0 : L \longrightarrow L$. A finite dimensional F-vector space L with an endomorphism $f : L \longrightarrow L$ such that $p(f) = 0$ is the same as an $(F[x], p(x)^\infty)$-module L with $f = x$, $p(x)L = 0$, and also the same as a finite-dimensional E-vector space. Identify

$$\mathrm{Hom}_E(L, E) \;=\; \mathrm{Hom}_{F[x]}(L, F[x, p(x)^{-1}]/F[x])$$

using the E-vector space isomorphism

$$L^* \;=\; \mathrm{Hom}_E(L, E) \xrightarrow{\simeq} L^\wedge \;=\; \mathrm{Hom}_{F[x]}(L, F[x, p(x)^{-1}]/F[x]) \;;$$
$$g \longrightarrow (y \longrightarrow p(x)^{-1}g(y)) \;.$$

In order to identify L^* with $\mathrm{Hom}_F(L, F)$ proceed as follows. Write

$$p(x) \;=\; a_0 + a_1x + \ldots + a_{d-1}x^{d-1} + a_dx^d \in \overline{\mathcal{M}}_x(F) \;\; (a_d = 1) \;,$$

and let X be an indeterminate over E. The polynomial

$$p(X) \;=\; \sum_{j=0}^{d} a_jX^j \in E[X]$$

has root $x \in E$, with $p(x) = 0 \in E$. The polynomial

$$q(X) \;=\; \frac{p(X) - p(x)}{X - x} \;=\; \sum_{j=0}^{d-1} q_j(x)X^j \in E[X]$$

has coefficients $q_j(x) \in E$ the evaluations at x of the polynomials

$$q_j(X) \;=\; \sum_{i=0}^{d-j-1} a_{i+j+1}X^i \in F[X] \;\; (0 \leq j \leq d-1) \;.$$

Every element $y \in E$ has a unique representation as

$$y \;=\; \sum_{j=0}^{d-1} y_jx^j \in E \;\; (y_j \in F) \;,$$

i.e. $\{1, x, x^2, \ldots, x^{d-1}\}$ is a basis for E as an F-vector space. Now

$$\sum_{j=0}^{d-1} x^j q_j(f)(x-f) = p(x) : L[x] \longrightarrow L[x] ,$$

so that for any E-linear map $g : L \longrightarrow E$ the F-linear maps

$$g_j : L \longrightarrow F \quad (0 \leq j \leq d-1)$$

defined by

$$g(y) = \sum_{j=0}^{d-1} g_j(y) x^j \in E \quad (y \in L)$$

are in fact all determined by g_{d-1}, with

$$g_j = g_{d-1} q_j(f) : L \longrightarrow F .$$

The function

$$\mathrm{Hom}_E(L, E) \longrightarrow \mathrm{Hom}_F(L, F) \; ; \; g \longrightarrow g_{d-1}$$

is thus an E-vector space isomorphism, and there are identifications

$$L\widehat{\ } = \mathrm{Hom}_{F[x]}(L, F[x, p(x)^{-1}]/F[x]) = \mathrm{Hom}_E(L, E) = \mathrm{Hom}_F(L, F) .$$

The one-one correspondence of 25.19

$$\begin{array}{l}\{\text{1-dimensional } (-\epsilon)\text{-symmetric complexes } (C, \phi) \text{ over } F[x] \\ \qquad \text{such that } p(x)H_*(C) = 0\} \rightleftarrows \\ \{\epsilon\text{-symmetric linking forms } (L, \rho) \text{ over } (F[x], p(x)^\infty) \text{ with } p(x)L = 0\}\end{array}$$

is given by

$$\begin{array}{l} f = x : L = H^1(C) \longrightarrow H^1(C) , \\ \rho : H^1(C) \times H^1(C) \longrightarrow F[x, x^{-1}, p(x)^{-1}]/F[x] \; ; \\ \qquad (u, v) \longrightarrow p(x)^{-1} \phi_0(u, w) \\ (u, v \in C^1 , \; w \in C^0 , \; d^*(w) = p(x)(v)) . \end{array}$$

Define also the one-one correspondence

$$\begin{array}{l}\{\text{1-dimensional } (-\epsilon)\text{-symmetric complexes } (C, \phi) \text{ over } F[x] \\ \qquad \text{such that } p(x)H_*(C) = 0\} \rightleftarrows \\ \{\epsilon\text{-symmetric forms } (L, \theta) \text{ over } F \text{ with an} \\ \qquad \text{endometric structure } f : L \longrightarrow L \text{ such that } p(f) = 0\}\end{array}$$

with (L, θ, f) corresponding to (C, ϕ) given by

$$d_C = x - f^* : C_1 = L^*[x] \longrightarrow C_0 = L^*[x] ,$$

$$\phi_0 = \begin{cases} \theta : C^0 = L[x] \longrightarrow C_1 = L^*[x] \\ \theta : C^1 = L[x] \longrightarrow C_0 = L^*[x] , \end{cases}$$

$$\phi_1 = 0 : C^1 = L[x] \longrightarrow C_1 = L^*[x] .$$

The inverse isomorphisms of 38.5 (iii) are given by

$$LEnd^0_{p(x)^\infty}(F, \epsilon) \longrightarrow L^0(F[x], p(x)^\infty, \epsilon) ; (L, \rho) \longrightarrow (C, \phi) ,$$
$$L^0(F[x], p(x)^\infty, \epsilon) \longrightarrow LEnd^0_{p(x)^\infty}(F, \epsilon) ; (C, \phi) \longrightarrow (L, \rho) .$$

The one-one correspondence

$\{\epsilon$-symmetric forms (L, μ) over $E\} \rightleftarrows$

$\{\epsilon$-symmetric linking forms (L, ρ) over $(F[x], p(x)^\infty)$ with $p(x)L = 0\}$

with

$$\rho(u, v) = p(x)^{-1}\lambda(u, v) \in F[x, p(x)^{-1}]/F[x] \ (u, v \in L) .$$

gives the inverse isomorphisms of 35.3

$$L^0(E, \epsilon) \longrightarrow L^0(F[x], p(x)^\infty, \epsilon) ; (L, \lambda) \longrightarrow (L, \rho) ,$$
$$L^0(F[x], p(x)^\infty, \epsilon) \longrightarrow L^0(E, \epsilon) ; (L, \rho) \longrightarrow (L, \lambda) .$$

The composite isomorphism

$$L^0(E, \epsilon) \xrightarrow{\cong} L^0(F[x], p(x)^\infty, \epsilon) \xrightarrow{\cong} LEnd^0_{p(x)^\infty}(F, \epsilon)$$

is given by the one-one correspondence

$\{\epsilon$-symmetric forms (L, λ) over $E\} \rightleftarrows$

$\{\epsilon$-symmetric forms (L, θ) over F with an

endometric structure $f : L \longrightarrow L$ such that $p(f) = 0\}$

with

$$\lambda(u, v) = \theta(u, w) \in E$$
$$(u, v \in L \ , \ w \in L[x] \ , \ (x - f)(w) = p(x)v \in L[x]) .$$

For u, v, w as above

$$w = \sum_{j=0}^{d-1} w_j x^j \in L[x] \ \ (w_j \in L)$$

with $w_{d-1} = v$, so that λ determines θ by

$$\theta(u,v) \;=\; (\text{coefficient of } x^{d-1} \text{ in } \lambda(u,v)) \in F \;. \qquad \square$$

Remark 39.3 As in 39.2 let

$$p(x) \;=\; \sum_{j=0}^{d} a_j x^j \in \overline{\mathcal{M}}_x(F) \;\; (a_d = 1) \;,$$

with $q(X), q_j(X)$ as before.
(i) Given any non-zero involution-preserving F-linear map

$$h \;:\; E \;=\; F[x]/(p(x)) \longrightarrow F$$

there is defined a one-one correspondence

$$\begin{array}{l}\{\epsilon\text{-symmetric forms } (L,\lambda) \text{ over } E\} \;\rightleftarrows \\ \{\epsilon\text{-symmetric forms } (L,\theta) \text{ over } F \text{ with an} \\ \qquad\qquad \text{endometric structure } f : L \longrightarrow L \text{ such that } p(f)=0\}\end{array}$$

with

$$\theta(u,v) \;=\; h(\lambda(u,v)) \in F \;\;,\;\; f(u) \;=\; xu \in L \;,$$

so that there is an isomorphism of Witt groups

$$L^0(E,\epsilon) \xrightarrow{\;\simeq\;} L\mathrm{End}^0_{p(x)^\infty}(F,\epsilon) \;;\; (L,\lambda) \longrightarrow (L,\theta,f) \;.$$

This follows from Milnor [197, Remark 1.4] – the methods developed there for automorphisms of forms (which will be used in 39.20 below) apply equally well to forms with endometric structures.
(ii) It has already been verified that the one-one correspondence used in the proof of 39.2 is determined by the choice

$$h \;:\; E \longrightarrow F \;;\; y \;=\; \sum_{j=0}^{d-1} y_j x^j \longrightarrow y_{d-1} \;\; (y_j \in F)$$

in (i). In (iii) this choice will be related to the trace map used in [197] (in the separable case). The one-one correspondence between ϵ-symmetric forms (L,λ) over E and ϵ-symmetric forms (L,θ) over F with an endometric structure $f : L \longrightarrow L$ such that $p(f) = 0$ in the proof of 39.2 is given by

$$\begin{aligned}\lambda(u,v) \;&=\; \sum_{j=0}^{d-1} \theta(u, q_j(f)(v)) x^j \in E \;\;,\;\; f(u) \;=\; xu \in L \;, \\ \theta(u,v) \;&=\; (\text{coefficient of } x^{d-1} \text{ in } \lambda(u,v)) \in F \;\; (u,v \in L) \;,\end{aligned}$$

noting that $q_{d-1}(x) = 1$ and that $w = \sum_{j=0}^{d-1} q_j(f)(v)x^j \in L[x]$ has image

$$(x - f)(w) \;=\; p(x)(v) \in L[x] \,.$$

(iii) The field E is a degree d extension of F, and the trace map is defined by

$$\mathrm{tr}_{E/F} \;:\; E \longrightarrow F \;;\; y \;=\; \sum_{j=0}^{d-1} y_j x^j \longrightarrow \mathrm{tr}_{E/F}(y) \;=\; \sum_{j=0}^{d-1} y_j (\sum_{k=1}^{d} (x_k)^j)$$

with $x_1, x_2, \ldots, x_d$ the roots of $p(x)$ in an algebraic closure of F. Equivalently, $\mathrm{tr}_{E/F}(y) \in F$ is the trace of the F-linear endomorphism of E

$$E \longrightarrow E \;;\; z \longrightarrow yz$$

regarded as a d-dimensional F-vector space.
Suppose now that $p(x) \in \overline{\mathcal{M}}_x(F)$ is separable, i.e. with $x_1, x_2, \ldots, x_d$ distinct (which is automatically the case if $\mathrm{char}(F) = 0$), so that $\mathrm{tr}_{E/F}$ is onto and it is possible to choose $h = \mathrm{tr}_{E/F}$ in (i). The evaluation of $q(X) \in E[X]$ at x is the derivative of $p(x)$

$$q(x) \;=\; p'(x) \;=\; \sum_{j=0}^{d-1} q_j(x) x^j \;=\; \sum_{j=1}^{d-1} j a_j x^{j-1} \in E \,.$$

with trace

$$\mathrm{tr}_{E/F}(p'(x)) \;=\; \mathrm{discriminant}\, p(x) \;=\; \prod_{i \neq j} (x_i - x_j) \in F^{\bullet} \,,$$

and the elements

$$q_j(x) x^k \;=\; \sum_{i=0}^{d-j-1} a_{i+j+1} x^{i+k} \in E$$

are such that

$$\mathrm{tr}_{E/F}(p'(x)^{-1} q_j(x) x^k) \;=\; \delta_{jk} \in F \;\; (0 \leq j, k \leq d-1)$$

(Lang [146, p. 287]). It follows that the coefficients $y_j \in F$ in the expression of any element $y \in E$ as a polynomial in x of degree $d - 1$

$$y \;=\; \sum_{j=0}^{d-1} y_j x^j \in E$$

are given by

$$y_j \;=\; \mathrm{tr}_{E/F}(p'(x)^{-1} q_j(x) y) \in F \;\; (0 \leq j \leq d-1) \,.$$

In particular, the choice of h in (ii) is given in terms of the trace map by

$$h : E \longrightarrow F ; y = \sum_{j=0}^{d-1} y_j x^j \longrightarrow y_{d-1} = \mathrm{tr}_{E/F}(p'(x)^{-1}y) .$$

(iv) For any choice of h in (i) the composite of the isomorphism

$$L^0(E,\epsilon) \xrightarrow{\simeq} LEnd^0_{p(x)^\infty}(F,\epsilon) ; (L,\lambda) \longrightarrow (L,\theta,f)$$

and the forgetful map

$$LEnd^0_{p(x)^\infty}(F,\epsilon) \longrightarrow L^0(F,\epsilon) ; (L,\theta,f) \longrightarrow (L,\theta)$$

can be expressed as an algebraic surgery transfer map

$$p^! : L^0(E,\epsilon) \longrightarrow L^0(F,\epsilon) ; (L,\lambda) \longrightarrow (L,\theta)$$

in the sense of Lück and Ranicki [175], as follows. The symmetric form (E,θ) over F defined by

$$\theta : E \longrightarrow E^* = \mathrm{Hom}_F(E,F) ; u \longrightarrow (v \longrightarrow h(\overline{u}v))$$

is nonsingular. Write the opposite of the F-endomorphism ring of E as

$$R = \mathrm{Hom}_F(E,E)^{op} ,$$

and let it have the involution

$$^- : R \longrightarrow R ; f \longrightarrow \theta^{-1}f^*\theta ,$$

so that there is defined a morphism of rings with involution

$$p : E \longrightarrow R ; u \longrightarrow (v \longrightarrow \overline{u}v)$$

and $p^!$ is defined as in [175]. □

Proposition 39.4 (i) *The localization exact sequence of 34.7*

$$\dots \longrightarrow L^n(F[x],\epsilon) \xrightarrow{i} L^n(F(x),\epsilon) \xrightarrow{\partial} LEnd^n(F,\epsilon)$$
$$\xrightarrow{j} L^{n-1}(F[x],\epsilon) \longrightarrow \dots$$

breaks up into split exact sequences

$$0 \longrightarrow L^{2k}(F[x],\epsilon) \xrightarrow{i} L^{2k}(F(x),\epsilon) \xrightarrow{\partial} LEnd^{2k}(F,\epsilon) \longrightarrow 0$$

with ∂ split by

$$LEnd^{2k}(F,\epsilon) \longrightarrow L^{2k}(F(x),\epsilon) ; (L,\theta,f) \longrightarrow (L(x),\theta(x-f)) .$$

(ii) *The L-groups of $F[x]$ are such that*

$$L^{2k}(F[x],\epsilon) \;=\; L^{2k}(F,\epsilon) \;=\; L^0(F,(-)^k\epsilon)$$

and the endomorphism L-groups of F are such that

$$\begin{aligned} LEnd^{2k}(F,\epsilon) &= \bigoplus_{p(x)\in\overline{\mathcal{M}}_x(F)} LEnd^{2k}_{p(x)^\infty}(F,\epsilon) \\ &= \bigoplus_{p(x)\in\overline{\mathcal{M}}_x(F)} L^0(F[x]/(p(x)),(-)^k\epsilon)\ . \end{aligned}$$

(iii) *The morphism*

$$\begin{aligned} LEnd^{2k}(F) \;&=\; \bigoplus_{p(x)\in\overline{\mathcal{M}}_x(F)} LEnd^{2k}_{p(x)^\infty}(F) \\ &\longrightarrow \widehat{H}^0(\mathbb{Z}_2;\mathrm{End}_0(F)) \;=\; \mathbb{Z}_2[\overline{\mathcal{M}}_x(F)]\ ; \\ (L,\theta,f) &\longrightarrow [L,f] \;=\; \sum_{p(x)\in\overline{\mathcal{M}}_x(F)} [L_{p(x)^\infty}]p(x) \end{aligned}$$

fits into the exact sequence of 34.14

$$0 \longrightarrow LEnd^{2k}_r(F) \longrightarrow LEnd^{2k}(F) \longrightarrow \mathbb{Z}_2[\overline{\mathcal{M}}_x(F)] \longrightarrow LEnd^{2k-1}_r(F) \longrightarrow 0$$

with $L_{p(x)^\infty}$ *the* $p(x)$*-primary component of* L, $x = f : L \longrightarrow L$ *and*

$$\begin{aligned} [L_{p(x)^\infty}] \;&=\; \sum_{j=0}^{\infty} \dim_E(p(x)^j L/p(x)^{j+1}L) \\ &\in \widehat{H}^0(\mathbb{Z}_2;\mathrm{End}_0^{p(x)^\infty}(F)) \;=\; \widehat{H}^0(\mathbb{Z}_2;K_0(E)) \;=\; \mathbb{Z}_2 \end{aligned}$$

(using 18.2 (ii)) with $E = F[x]/p(x)$ *the quotient field.*

Proof (i) By construction.

(ii) This is the special case $(A,J) = (F[x],\overline{\mathcal{M}}_x(F))$ of 38.7, using 39.2. If (L,θ) is a nonsingular $(-)^k\epsilon$-symmetric form over F with an endometric structure $f : L \longrightarrow L$ then a factorization

$$\text{minimal polynomial}(L,f) \;=\; p_1(x)p_2(x)\dots p_m(x)q(x)\overline{q(x)} \in F[x]$$

with $p_j(x) \in \overline{\mathcal{M}}_x(F)$ coprime determines a decomposition

$$(L,\theta,f) \;=\; \bigoplus_{j=1}^{m}(L_j,\theta_j,f_j) \oplus \text{hyperbolic}$$

with

$$L_j \;=\; L_{p_j(x)}\ ,\ \ p_j(f_j) \;=\; 0 : L_j \longrightarrow L_j \ \ (1 \le j \le m)\ .$$

Thus

$$(L,\theta,f) = \bigoplus_{j=1}^{m}(L_j,\theta_j,f_j) \in LEnd^{2k}(F,\epsilon) = \bigoplus_{p(x)\in\overline{\mathcal{M}}_x(F)} LEnd^{2k}_{p(x)^\infty}(F,\epsilon) .$$

(iii) As in 39.3 (i) choose a non-zero involution-preserving F-linear map $h : E \longrightarrow F$ (which may be taken to be the trace map for separable $p(x)$). The F-linear map $x : E \longrightarrow E$ is a $p(x)$-primary endometric structure on the nonsingular symmetric form (E,θ) over F given by

$$\theta : E \times E \longrightarrow F ; (u,v) \longrightarrow h(v\overline{u}) . \qquad \square$$

Remark 39.5 The split exact sequence

$$0 \longrightarrow L^0(F) \xrightarrow{i} L^0(F(x)) \xrightarrow{\partial} \bigoplus_{p(x)\in\overline{\mathcal{M}}_x(F)} L^0(F[x]/(p(x))) \longrightarrow 0$$

was first obtained by Knebusch [136] and Milnor [198] – see Scharlau [256, 6.3] for a detailed account. See [136] and Knus [138, VIII] for the Witt groups of schemes. $\square$

Example 39.6 If F is a finite field the residue class fields $F[x]/(p(x))$ ($p(x) \in \mathcal{M}(F)$) are also finite, so that

$$\begin{aligned} L^0(F(x),\epsilon) &= L^0(F,\epsilon) \oplus LEnd^0(F,\epsilon) \\ &= L^0(F,\epsilon) \oplus \bigoplus_{p(x)\in\overline{\mathcal{M}}_x(F)} L^0(F[x]/(p(x)),\epsilon) \end{aligned}$$

is an abelian group of exponent 4 (cf. 37.1 (viii)). $\square$

Proposition 39.7 *The involution-invariant irreducible monic polynomials in $F[x]$ of degree 1 are of the type*

$$(x-a) \in \overline{\mathcal{M}}_x(F)$$

with $a \in F$ such that $\overline{a} = a \in F$. Fix one such $a \in F$.
(i) *Up to isomorphism*

$$\begin{aligned} L^0(F[x],(x-a)^\infty,\epsilon) &= LEnd^0_{(x-a)^\infty}(F,\epsilon) \\ &= L^0(F[x]/(x-a),\epsilon) = L^0(F,\epsilon) , \\ L^0(F[x,(x-a)^{-1}],\epsilon) &= L^0(F[x],\epsilon) \oplus L^0(F[x],(x-a)^\infty,\epsilon) \\ &= L^0(F,\epsilon) \oplus L^0(F,\epsilon) . \end{aligned}$$

(ii) *The forgetful map*

$$LEnd^0_{(x-a)^\infty}(F,\epsilon) \longrightarrow L^0(F,\epsilon) ; (L,\theta,f) \longrightarrow (L,\theta)$$

is an isomorphism, with inverse

$$L^0(F,\epsilon) \longrightarrow L\mathrm{End}^0_{(x-a)^\infty}(F,\epsilon)\ ;\ (L,\theta) \longrightarrow (L,\theta,a)\ .$$

Proof (i) Apply 39.4.
(ii) Immediate from (i). □

Example 39.8 Let F be algebraically closed, so that

$$\overline{\mathcal{M}}_x(F) \;=\; \{(x-a)\,|\,a=\overline{a}\in F\}$$

and

$$L^0(F(x),\epsilon) \;=\; L^0(F,\epsilon)\oplus L\mathrm{End}^0(F,\epsilon)$$

with

$$L\mathrm{End}^0(F,\epsilon) \;=\; \bigoplus_{a=\overline{a}\in F} L^0(F[x]/(x-a),\epsilon)$$

$$=\; \bigoplus_{a=\overline{a}\in F} L^0(F,\epsilon)\ .$$

(i) A nonsingular ϵ-symmetric form (L,θ) over F with an endometric structure $f : L\longrightarrow L$ has a direct sum decomposition of the type

$$(L,\theta,f) \;=\; \sum_{a=\overline{a}\in F}(L_a,\theta_a,f_a)\oplus\left(K\oplus K^*, \begin{pmatrix}0 & 1\\ \epsilon & 0\end{pmatrix}, \begin{pmatrix}g & 0\\ 0 & g^*\end{pmatrix}\right)$$

with

$$L_a \;=\; \bigcup_{j=0}^{\infty}\ker((f-a)^j : L\longrightarrow L)\ ,\quad K\oplus K^* \;=\; \sum_{b\neq\overline{b}\in F^\bullet} L_b\ ,$$

$$\mathrm{ch}_x(L_a,f_a) \;=\; (x-a)^{d_a}\ \ (d_a \;=\; \dim_F(L_a))\ .$$

The endomorphism Witt class is given by

$$(L,\theta,f) \;=\; \sum_{a=\overline{a}\in F}(L_a,\theta_a)\in L\mathrm{End}^0(F,\epsilon) \;=\; \bigoplus_{a=\overline{a}\in F} L^0(F,\epsilon)\ .$$

Furthermore, it is possible to decompose each (L_a,θ_a,f_a) as a direct sum

$$(L_a,\theta_a,f_a) \;=\; \sum_{k=1}^{\infty}(L_{a,k},\theta_{a,k},f_{a,k})$$

with

$$\text{minimal polynomial}(L_{a,k},f_{a,k}) \;=\; (x-a)^k\ ,$$

and

$$(L_a,\theta_a) \;=\; \sum_{k\ \mathrm{odd}}(L_{a,k}/(f-a)L_{a,k},[\theta_{a,k}])$$

$$\in L\mathrm{End}^0_{(x-a)^\infty}(F,\epsilon) \;=\; L^0(F,\epsilon)\ .$$

(ii) If $F=\mathbb{C}^+$ then

$$LEnd^0(\mathbb{C}^+) = \sum_{a\in\mathbb{C}} L^0(\mathbb{C}^+) = \sum_{a\in\mathbb{C}} \mathbb{Z}_2 ,$$

with the components generated by $(\mathbb{C}^+, 1, a)$.
(iii) If $F = \mathbb{C}^-$ then

$$LEnd^0(\mathbb{C}^-) = \bigoplus_{a\in\mathbb{R}} L^0(\mathbb{C}^-) = \bigoplus_{a\in\mathbb{R}} \mathbb{Z} .$$

There is one $\mathbb{Z}$-valued signature for each real number $a \in \mathbb{R}$, and the a-component is generated by $(\mathbb{C}^-, 1, a)$. This is the *endomorphism multisignature*, about which more in Chap. 40 below. □

Example 39.9 For $F = \mathbb{R}$

$$\begin{aligned} \mathcal{M}(\mathbb{R}) &= \overline{\mathcal{M}}_x(\mathbb{R}) \\ &= \{(x-a)\,|\,a\in\mathbb{R}\} \sqcup \{(x-b)(x-\bar{b})\,|\,b\in\mathbb{H}^+\} \end{aligned}$$

with $\mathbb{H}^+ = \{u+iv \in \mathbb{C}\,|\,v>0\}$ the upper-half complex plane. The residue class fields are such that there are defined isomorphisms of rings with involution

$$\begin{aligned} &\mathbb{R}[x]/(x-a) \xrightarrow{\simeq} \mathbb{R} \; ; \; x \longrightarrow a , \\ &\mathbb{R}[x]/(x-b)(x-\bar{b}) \xrightarrow{\simeq} \mathbb{C}^+ \; ; \; x \longrightarrow b . \end{aligned}$$

The endomorphism L-groups of $\mathbb{R}$ are given by

$$\begin{aligned} LEnd^0(\mathbb{R},\epsilon) &= \sum_{a\in\mathbb{R}} L^0(\mathbb{R},\epsilon) \oplus \sum_{b\in\mathbb{H}^+} L^0(\mathbb{C}^+,\epsilon) \\ &= \begin{cases} \sum\limits_{a\in\mathbb{R}} \mathbb{Z} \oplus \sum\limits_{b\in\mathbb{H}^+} \mathbb{Z}_2 & \text{if } \epsilon = 1 \\ 0 & \text{if } \epsilon = -1 . \end{cases} \end{aligned}$$

The components of $LEnd^0(\mathbb{R})$ are generated by $(\mathbb{R}, 1, a)$ $(a \in \mathbb{R})$ and $(\mathbb{R}\oplus\mathbb{R}, \begin{pmatrix} 0 & 1 \\ 1 & 0 \end{pmatrix}, \begin{pmatrix} u & -v \\ v & u \end{pmatrix})$ $(b = u+iv \in \mathbb{H}^+)$.

□

Example 39.10 Let $F = \mathbb{Q}$.
(i) If $p(x) \in \mathcal{M}(\mathbb{Q})$ has $r_{p(x)}$ real roots and $2s_{p(x)}$ complex roots then the residue class field $\mathbb{Q}[x]/(p(x))$ is an algebraic number field with $r_{p(x)}$ embeddings in $\mathbb{R}$ and $2s_{p(x)}$ embeddings in $\mathbb{C}$, so that by 38.10 (ii)

$$L^0(\mathbb{Q}[x]/(p(x)),\epsilon) = \begin{cases} \mathbb{Z}^{r_{p(x)}} \oplus \text{8-torsion} & \text{if } \epsilon = 1 \\ 0 & \text{if } \epsilon = -1 . \end{cases}$$

The endomorphism L-groups of $\mathbb{Q}$ are thus given by

$$\begin{aligned} L\mathrm{End}^0(\mathbb{Q},\epsilon) &= \sum_{p(x)\in\mathcal{M}(\mathbb{Q})} L^0(\mathbb{Q}[x]/(p(x)),\epsilon) \\ &= \begin{cases} \sum\limits_{p(x)\in\mathcal{M}(\mathbb{Q})} \mathbb{Z}^{r_{p(x)}} \oplus 8\text{-torsion} & \text{if } \epsilon=1 \\ 0 & \text{if } \epsilon=-1 . \end{cases} \end{aligned}$$

(ii) Let (L,θ,f) be a nonsingular symmetric form (L,θ) over $\mathbb{Q}$ with an endometric structure $f: L\longrightarrow L$. Let $p(x)\in\mathcal{M}(\mathbb{Q})$ be the minimal polynomial of f, with r real roots $\alpha_1,\alpha_2,\dots,\alpha_r\in\mathbb{R}$ and $2s$ non-real roots $\beta_1,\overline{\beta}_1,\beta_2,\dots,\overline{\beta}_s\in\mathbb{C}$, so that

$$p(x) = \prod_{j=1}^{r}(x-\alpha_j)\prod_{k=1}^{s}(x^2-(\beta_k+\overline{\beta}_k)x+\beta_k\overline{\beta}_k)\in\mathbb{R}[x] .$$

The endomorphism Witt class $(L,\theta,f)\in L\mathrm{End}^0(\mathbb{Q})$ is determined up to 8-torsion by the signatures

$$\sigma_j(L,\theta,f) = \sigma(g_j(L,\theta))\in L^0(\mathbb{R}) = \mathbb{Z}\ \ (1\leq j\leq r)$$

with

$$g_j : \mathbb{Q}[x]/(p(x)) \longrightarrow \mathbb{R} ;\ x\longrightarrow \alpha_j .$$

The induced nonsingular symmetric form over $\mathbb{R}$ splits as

$$\mathbb{R}\otimes_{\mathbb{Q}}(L,\theta) = \sum_{j=1}^{r}(L_j,\theta_j)\oplus\text{hyperbolic}$$

with

$$L_j = \{u\in\mathbb{R}\otimes_{\mathbb{Q}}L\,|\,(1\otimes f-\alpha_j)^N(u)=0 \text{ for some } N\geq 0\}$$

the generalized α_j-eigenspace of $1\otimes f:\mathbb{R}\otimes_{\mathbb{Q}}L\longrightarrow\mathbb{R}\otimes_{\mathbb{Q}}L$. Since

$$g_j : \mathbb{Q}[x]/(p(x)) \longrightarrow \mathbb{R}[x]/(p(x)) \cong \prod_r\mathbb{R}\times\prod_t\mathbb{C}^+ \xrightarrow{p_j} \mathbb{R}$$

the jth signature is

$$\sigma_j(L,\theta,f) = (L_j,\theta_j)\in L^0(\mathbb{R}) = \mathbb{Z} .$$

Thus

$$L\mathrm{End}^0(\mathbb{Q}) = \mathbb{Z}^\infty\oplus 8\text{-torsion}$$

with one $\mathbb{Z}$-valued signature for each real algebraic number $\alpha\in\mathbb{R}$ (= real root of a polynomial $p(x)\in\mathbb{Q}[x]$). □

39B. L-theory of $F(s)$ $(\overline{s} = 1 - s)$

Definition 39.11 Let $\overline{\mathcal{M}}_s(F) \subseteq \mathcal{M}(F)$ be the set of irreducible monic polynomials

$$p(s) \;=\; a_0 + a_1 s + \ldots + a_{d-1}s^{d-1} + a_d s^d \in F[s] \;\; (a_d = 1)$$

which are involution-invariant, with

$$u_{p(s)}\overline{p(s)} \;=\; p(s) \in F[s]$$

for some unit $u_{p(s)} \in U(F[s])$, that is

$$\sum_{i=j}^{d} \binom{i}{j} \overline{a}_i \;=\; (-)^{j+d} a_j \in F \;\; (0 \leq j \leq d-1 \;,\; a_d = 1)$$

with $u_{p(s)} = (-)^d \in U(F[s])$. □

The involution-invariant maximal ideals $(p(s)) \in \overline{\max}(F[s])$ are the principal ideals $\mathcal{P} = (p(s)) \lhd F[s]$ generated by the elements $p(s) \in \overline{\mathcal{M}}_s(F)$, so identify

$$\overline{\max}(F[s]) \;=\; \overline{\mathcal{M}}_s(F) \;.$$

As in 31.2 let $\mathrm{Iso}_0(F)$ denote $\mathrm{End}_0(F)$ with the duality involution

$$\mathrm{Iso}_0(F) \longrightarrow \mathrm{Iso}_0(F) \;;\; [P, f] \longrightarrow [P^*, 1 - f^*]$$

Under the identification of 19.3

$$\mathrm{Iso}_0(F) \;=\; \mathbb{Z}[\mathcal{M}_s(F)]$$

this corresponds to the involution

$$\mathbb{Z}[\mathcal{M}_s(F)] \longrightarrow \mathbb{Z}[\mathcal{M}_s(F)] \;;\; p(s) \longrightarrow u_{p(s)}\overline{p(s)} \;,$$

so that

$$\widehat{H}^n(\mathbb{Z}_2; \mathrm{Iso}_0(F)) \;=\; \widehat{H}^n(\mathbb{Z}_2; \mathbb{Z})[\overline{\mathcal{M}}_s(F)] \;.$$

Proposition 39.12 (i) *For any* $p(s) \in \overline{\mathcal{M}}_s(F)$, $\epsilon \in U(F)$

$$\begin{aligned} L\mathrm{Iso}^0_{p(s)^\infty}(F, \epsilon) \;&=\; L^0(F[s], (p(s))^\infty, -\epsilon) \\ &=\; L^0(E, \epsilon) \;=\; L^0(E, -u_{p(s)}\epsilon) \end{aligned}$$

with

$$E = F[s]/(p(s)) \;\;,\;\; u_{p(s)} \;=\; (-)^d \in E^\bullet \;\;,\;\; d \;=\; \mathrm{degree}\,(p(s)) \;.$$

(ii) *The boundary map* ∂ *in the split exact localization exact sequence of 31.10*

$$0 \longrightarrow L^{2k}(F[s],\epsilon) \xrightarrow{i} L^{2k}(F(s),\epsilon) \xrightarrow{\partial} LIso^{2k}(F,-\epsilon) \longrightarrow 0$$

is split by

$$LIso^{2k}(F,-\epsilon) \longrightarrow L^{2k}(F(s),\epsilon) \; ; \; (L,\theta,f) \longrightarrow (L(s),\theta(s-f)) \; .$$

(iii) *The L-groups of $F[s]$ are such that*

$$L^{2k}(F[s],\epsilon) \;=\; L^{2k}(F,\epsilon) \;=\; L^0(F,(-)^k\epsilon)$$

and the isometric L-groups of F are such that

$$\begin{aligned} LIso^{2k}(F,-\epsilon) \;&=\; \bigoplus_{p(s)\in\overline{\mathcal{M}}_s(F)} LIso^{2k}_{p(s)^\infty}(F,-\epsilon) \\ &=\; \bigoplus_{p(s)\in\overline{\mathcal{M}}_s(F)} L^0(F[s]/(p(s)),(-)^k u_{p(s)}\epsilon) \; . \end{aligned}$$

(iv) *The morphism*

$$\begin{aligned} LIso^{2k}(F) \;=\; & \bigoplus_{p(s)\in\overline{\mathcal{M}}_s(F)} LIso^{2k}_{p(s)^\infty}(F) \\ & \longrightarrow \widehat{H}^0(\mathbb{Z}_2;Iso_0(F)) \;=\; \mathbb{Z}_2[\overline{\mathcal{M}}_s(F)] \; ; \\ & (L,\theta,f) \longrightarrow [L,f] \;=\; \sum_{p(s)\in\overline{\mathcal{M}}_s(F)} [L_{p(s)^\infty}]p(s) \end{aligned}$$

fits into the exact sequence of 31.19

$$0 \longrightarrow LIso^{2k}_r(F) \longrightarrow LIso^{2k}(F) \longrightarrow \mathbb{Z}_2[\overline{\mathcal{M}}_s(F)] \longrightarrow LIso^{2k-1}_r(F) \longrightarrow 0$$

with $L_{p(s)^\infty}$ the $p(s)$-primary component of L, $s = f : L \longrightarrow L$ and

$$\begin{aligned} [L_{p(s)^\infty}] \;&=\; \sum_{j=0}^{\infty} \dim_E(p(s)^j L/p(s)^{j+1}L) \\ &\in \widehat{H}^0(\mathbb{Z}_2;Iso_0^{p(s)^\infty}(F)) \;=\; \widehat{H}^0(\mathbb{Z}_2;K_0(E)) \;=\; \mathbb{Z}_2 \end{aligned}$$

(using 18.2 (ii)) with $E = F[s]/p(s)$ the quotient field.

Proof (i) Direct applications of 35.11 and 38.5 (iii) give identifications

$$LIso^0_{p(s)^\infty}(F,\epsilon) \;=\; L^0(E,-u_{p(s)}\epsilon) \;=\; L^0(F[s],(p(s))^\infty,-\epsilon) \; ,$$

using a devissage argument to identify $LIso^0_{p(s)^\infty}(F,\epsilon)$ with the Witt group of nonsingular ϵ-symmetric forms (L,θ) over F with an isometric structure $f : L \longrightarrow L$ such that $p(f) = 0$. Working as in 39.2 there is obtained a one-one correspondence

$$\{(-u_{p(s)}\epsilon)\text{-symmetric forms } (L,\mu) \text{ over } E\} \rightleftarrows$$
$$\{\epsilon\text{-symmetric forms } (L,\theta) \text{ over } F \text{ with an isometric}$$
$$\text{structure } f : L \longrightarrow L \text{ such that } p(f)=0\}$$

with

$$\theta(u,v) \;=\; (\text{coefficient of } s^{d-1} \text{ in } \mu(u,v)) \in F \;\;,\;\; f(u) \;=\; su \in L \;\; (u,v \in L)\;,$$

giving the isomorphism

$$L^0(E,-u_{p(s)}\epsilon) \xrightarrow{\simeq} LIso^0_{p(s)^\infty}(F,\epsilon)\;;\;(L,\mu) \longrightarrow (L,\theta,f)\;.$$

As in Milnor [197] (cf. 39.4 (ii)) use any non-zero involution-preserving F-linear map $h : E \to F$ to define a one-one correspondence

$$\{\epsilon\text{-symmetric forms } (L,\lambda) \text{ over } E\} \rightleftarrows$$
$$\{\epsilon\text{-symmetric forms } (L,\theta) \text{ over } F \text{ with an isometric structure}$$
$$f : L \longrightarrow L \text{ such that } p(f)=0\}$$

with

$$\theta(u,v) \;=\; h(\lambda(u,v)) \in F \;\;,\;\; f(u) \;=\; su \in L \;\; (u,v \in L)\;.$$

This gives an isomorphism of Witt groups

$$L^0(E,\epsilon) \xrightarrow{\simeq} LIso^0_{p(s)^\infty}(F,\epsilon)\;;\;(L,\lambda) \longrightarrow (L,\theta,f)\;.$$

(In the separable case it is possible to choose $h = \mathrm{tr}_{E/F} : E \longrightarrow F$ as in [197], and λ, μ are related by

$$\lambda(u,v) \;=\; p'(s)\mu(u,v) \in E \;\; (u,v \in L)$$

with

$$p'(s) \;=\; \sum_{j=1}^{d} ja_j s^{j-1} \in F[s]$$

such that

$$p'(s) + u_{p(s)}\overline{p'(s)} \;=\; 0 \in F[s]\;,$$
$$\mathrm{tr}_{E/F}(p'(s)) \;=\; \text{discriminant}\,(p(s)) \neq 0 \in F$$

and $p'(s) \neq 0 \in E$.)

(ii) By construction.

(iii) This is the special case $(A,J) = (F[s],\mathcal{M}_s(F))$ of 38.7. If (L,θ) is a nonsingular $(-)^k\epsilon$-symmetric form over F with an isometric structure $f : L \longrightarrow L$ then the factorization of the characteristic polynomial as a product of coprime factors

$$\mathrm{ch}_s(L,f) \;=\; p_1(s)p_2(s)\ldots p_m(s)q(s)\overline{q(s)} \in F[s]$$

with each $p_j(s) \in \overline{\mathcal{M}}_s(F)$ determines a decomposition

$$(L,\theta,f) = \bigoplus_{j=1}^{m}(L_j,\theta_j,f_j) \in LIso^{2k}(F,\epsilon) = \bigoplus_{p(s)\in\overline{\mathcal{M}}_s(F)} LIso^{2k}_{p(s)^\infty}(F,\epsilon)$$

with $\mathrm{ch}_s(L_j,f_j) = p_j(s)$ $(1 \leq j \leq m)$.
(iv) As for 39.4 (iii). □

Example 39.13 For a finite field F the residue class fields $F[s]/(p(s))$ ($p(s) \in \max(F[s])$) are finite, so that

$$\begin{aligned} L^{2k}(F(s),\epsilon) &= L^{2k}(F,\epsilon) \oplus LIso^{2k}(F,-\epsilon) \\ &= L^{2k}(F,\epsilon) \oplus \bigoplus_{p(s)\in\mathcal{M}_s(F)} L^{2k}(F[s]/(p(s)),u_{p(s)}\epsilon) \end{aligned}$$

is an abelian group of exponent 8. □

Proposition 39.14 *The involution-invariant irreducible monic polynomials in $F[s]$ of degree 1 are of the type*

$$(s-a) \in \overline{\mathcal{M}}_s(F)$$

with $a \in F$ such that $a + \overline{a} = 1 \in F$, and $u_{(s-a)} = -1$. For any such $a \in F$

$$\begin{aligned} L^0(F[s],(s-a)^\infty,\epsilon) &= LIso^0_{(s-a)^\infty}(F,\epsilon) \\ &= L^0(F[s]/(s-a),-\epsilon) = L^0(F,-\epsilon)\ , \\ L^0(F[s,(s-a)^{-1}],\epsilon) &= L^0(F[s],\epsilon) \oplus L^0(F[s],(s-a)^\infty,\epsilon) \\ &= L^0(F,\epsilon) \oplus L^0(F,-\epsilon)\ . \end{aligned}$$

Proof Apply 39.12. □

Example 39.15 Let F be algebraically closed, so that

$$\mathcal{M}_s(F) = \{(s-a)\,|\,a+\overline{a} = 1 \in F\}$$

and

$$L^0(F(s),\epsilon) = L^0(F,\epsilon) \oplus LIso^0(F,-\epsilon)$$

with

$$\begin{aligned} LIso^0(F,\epsilon) &= \bigoplus_{a+\overline{a}=1\in F} L^0(F[s]/(s-a),\epsilon) \\ &= \bigoplus_{a+\overline{a}=1\in F} L^0(F,\epsilon)\ . \end{aligned}$$

(i) A nonsingular ϵ-symmetric form (L,θ) over F with an isometric structure $f : L \longrightarrow L$ has a direct sum decomposition of the type

$$(L,\theta,f) \;=\; \sum_{a+\overline{a}=1\in F}(L_a,\theta_a,f_a)\oplus(K\oplus K^*,\begin{pmatrix}0&1\\ \epsilon&0\end{pmatrix},\begin{pmatrix}g&h\\0&1-g^*\end{pmatrix})$$

with

$$L_a \;=\; \bigcup_{j=0}^{\infty}\ker((f-a)^j:L\longrightarrow L)\;,\quad K\oplus K^* \;=\; \sum_{b+\overline{b}\neq 1\in F}L_b\;,$$

$$\mathrm{ch}_s(L_a,f_a) \;=\; (s-a)^{d_a}\quad (d_a \;=\; \dim_F(L_a))\;.$$

The isometric Witt class is given by

$$(L,\theta,f) \;=\; \sum_{a+\overline{a}=1\in F}(L_a,\theta_a)\in LIso^0(F,\epsilon) \;=\; \bigoplus_{a+\overline{a}=1\in F}L^0(F,\epsilon)\;.$$

Furthermore, it is possible to decompose each (L_a,θ_a,f_a) as a direct sum

$$(L_a,\theta_a,f_a) \;=\; \sum_{k=0}^{\infty}(L_{a,k},\theta_{a,k},f_{a,k})$$

with

$$\text{minimal polynomial}(L_{a,k},f_{a,k}) \;=\; (s-a)^k\;,$$

and

$$\begin{aligned}(L_a,\theta_a) \;&=\; \sum_{k\,\mathrm{odd}}(L_{a,k}/(f-a)L_{a,k},[\theta_{a,k}])\\ &\in LIso^0_{(s-a)^\infty}(F,\epsilon) \;=\; L^0(F,\epsilon)\;.\end{aligned}$$

(ii) If $F=\mathbb{C}^+$ then

$$LIso^0(\mathbb{C}^+) \;=\; L^0(\mathbb{C}^+) \;=\; \mathbb{Z}_2\;,$$

generated by $(\mathbb{C}^+,1,1/2)$.
(iii) If $F=\mathbb{C}^-$ then

$$LIso^0(\mathbb{C}^-) \;=\; \bigoplus_{a\in\mathbb{C},\mathrm{Re}(a)=1/2}L^0(\mathbb{C}^-) \;=\; \bigoplus_{a\in\mathbb{C},\mathrm{Re}(a)=1/2}\mathbb{Z}\;.$$

There is one $\mathbb{Z}$-valued signature for each complex number a with $\mathrm{Re}(a)=1/2$, and the a-component is generated by $(\mathbb{C}^-,1,a)$. This is the *isometric multisignature*, about which more in Chap. 40. □

Remark 39.16 Let F have the identity involution, so that

$$\overline{\mathcal{M}}_s(F) \;=\; \{\sum_{j=0}^{d}a_js^j\in\mathcal{M}(F)\,|\,\sum_{i=j}^{d}\binom{i}{j}a_i=(-)^{i+j}a_j\in F\;(0\le j\le d)\}\;.$$

(i) The fixed subfield of the non-trivial involution $\overline{s}=1-s$ on $F(s)$ is

$$F(s)_0 \;=\; F(s(1-s))\ ,$$

and there is defined an isomorphism of fields

$$F(x) \xrightarrow{\simeq} F(s)_0\ ;\ x \longrightarrow s(1-s)\ .$$

The computations of $L^*(F(x))$, $L^*(F(s))$ given by 39.5, 39.12 in the case $\mathrm{char}(F) \neq 2$ are related by the exact sequence of 37.3

$$\begin{aligned} 0 \longrightarrow L^0(F(s)) \longrightarrow L^0(F(s)_0) \longrightarrow L^0(F(s)^+) \\ \longrightarrow L^0(F(s)_0) \longrightarrow L^0(F(s)) \longrightarrow 0 \end{aligned}$$

where $F(s)^+ = F(s)$ with the identity involution. This sequence can be written as

$$\begin{aligned} 0 \longrightarrow L^0(F(s)) \longrightarrow L^0(F(x)) \longrightarrow L^0(F(x)) \\ \longrightarrow L^0(F(x)) \longrightarrow L^0(F(s)) \longrightarrow 0\ . \end{aligned}$$

(ii) If $\mathrm{char}(F) \neq 2$ and $p(s) \in \overline{\mathcal{M}}_s(F)$ is of degree d then

$$\overline{(-)^d p(1/2)} \;=\; (-)^d p(1/2) \;=\; p(1/2) \in F\ ,$$

so that either d is even or $p(s) = s - 1/2$. □

Example 39.17 Let $F = \mathbb{R}$.

(i) The involution-invariant irreducible monic polynomials in $\mathbb{R}[s]$ are given by

$$\overline{\mathcal{M}}_s(\mathbb{R}) \;=\; \{(s-1/2)\} \sqcup \{(s-b)(s-\overline{b}) \mid b \in \mathbb{H}^+\,, \mathrm{Re}(b) = 1/2\}\ .$$

The residue class fields are such that there are defined isomorphisms of rings with involution

$$\begin{aligned} \mathbb{R}[s]/(s-1/2) &\xrightarrow{\simeq} \mathbb{R}\ ;\ s \longrightarrow 1/2\ , \\ \mathbb{R}[s]/(s-b)(s-\overline{b}) &\xrightarrow{\simeq} \mathbb{C}^-\ ;\ s \longrightarrow b\ . \end{aligned}$$

The isometric L-groups of $\mathbb{R}$ are thus given by

$$L\mathrm{Iso}^0(\mathbb{R}, \epsilon) \;=\; \begin{cases} L^0(\mathbb{R}) \oplus \sum\limits_{c>0} L^0(\mathbb{C}^-) = \mathbb{Z} \oplus \sum\limits_{c>0} \mathbb{Z} & \text{if } \epsilon = 1 \\ \sum\limits_{c>0} L^0(\mathbb{C}^-) \;=\; \sum\limits_{c>0} \mathbb{Z} & \text{if } \epsilon = -1 \end{cases}$$

with $b = 1/2 + ic \in \mathbb{H}^+$ $(c > 0)$, and the components are generated by

$$\begin{cases} (\mathbb{R}, 1, 1/2),\ (\mathbb{R} \oplus \mathbb{R}, \begin{pmatrix} 1 & 0 \\ 0 & 1 \end{pmatrix}, \begin{pmatrix} 1/2 & -c \\ c & 1/2 \end{pmatrix}) \\ (\mathbb{R} \oplus \mathbb{R}, \begin{pmatrix} 0 & 1 \\ -1 & 0 \end{pmatrix}, \begin{pmatrix} 1/2 & -c \\ c & 1/2 \end{pmatrix})\ . \end{cases}$$

(ii) For every real number $\alpha > 1/4$ there is a signature isomorphism

$$LIso^0_{(s^2-s+\alpha)}(\mathbb{R},\epsilon) \xrightarrow{\simeq} L^0(\mathbb{C}^-,\epsilon) = \mathbb{Z} ; \ (L,\theta,f) \longrightarrow \sigma(L,\phi)$$

as follows: by devissage every element of $LIso^0_{(s^2-s+\alpha)}(\mathbb{R},\epsilon)$ is represented by a nonsingular ϵ-symmetric form (L,θ) over $\mathbb{R}$ with an isometric structure $f : L \longrightarrow L$ such that $f^2 - f + \alpha = 0 : L \longrightarrow L$, and (L,ϕ) is the nonsingular ϵ-symmetric form over $\mathbb{C}^-$ given by

$$i = \frac{f - 1/2}{\sqrt{\alpha - 1/4}} : L \longrightarrow L ,$$

$$\phi : L \times L \longrightarrow \mathbb{C} ; \ (u,v) \longrightarrow \theta(u,v) - i\theta(u,iv) . \qquad \square$$

Example 39.18 Let $F = \mathbb{Q}$.
(i) If $p(s) \in \overline{\mathcal{M}}_s(\mathbb{Q})$ has degree $d_{p(s)}$ then the residue class field

$$E = \mathbb{Q}[s]/(p(s))$$

is an algebraic number field with involution which is a degree $d_{p(s)}$ extension of $\mathbb{Q}$. The involution is non-trivial if $p(s) \neq s - 1/2$, in which case $d_{p(s)}$ is even, say $d_{p(s)} = 2m$, and $p(s)$ splits over $\mathbb{R}$ as a product

$$p(s) = \prod_{j=1}^{m}(s^2 - s + \alpha_j)\prod_{k=1}^{n}[(s^2 - s + \beta_k)(s^2 - s + \overline{\beta}_k)] \in \mathbb{R}[s]$$

with $\alpha_j \neq 1/4 \in \mathbb{R}$, and $\beta_k \in \mathbb{C}$ such that $\mathrm{Re}(\beta_k) \neq 1/2$, $\mathrm{Im}(\beta_k) \neq 0$. Now $\mathbb{R}[s]/(p(s))$ is a product of rings with involution

$$\mathbb{R}[s]/(p(s)) = \prod_{j=1}^{m}\mathbb{R}[s]/(s^2 - s + \alpha_j) \times \prod_{k=1}^{n}\mathbb{R}[s]/[(s^2 - s + \beta_k)(s^2 - s + \overline{\beta}_k)] .$$

For any $\alpha \neq 1/4 \in \mathbb{R}$

$$\mathbb{R}[s]/(s^2 - s + \alpha) \cong \begin{cases} (\mathbb{R} \times \mathbb{R})^T & \text{if } \alpha < 1/4 \\ \mathbb{C}^- & \text{if } \alpha > 1/4 \end{cases}$$

where $(\mathbb{R} \times \mathbb{R})^T$ denotes $\mathbb{R} \times \mathbb{R}$ with the hyperbolic involution

$$T : \mathbb{R} \times \mathbb{R} \longrightarrow \mathbb{R} \times \mathbb{R} ; \ (u,v) \rightarrow (v,u) ,$$

so that

$$L^0(\mathbb{R}[s]/(s^2 - s + \alpha),\epsilon) = \begin{cases} L^0((\mathbb{R} \times \mathbb{R})^T,\epsilon) = 0 & \text{if } \alpha < 1/4 \\ L^0(\mathbb{C}^-,\epsilon) = \mathbb{Z} & \text{if } \alpha > 1/4 . \end{cases}$$

For any $\beta \in \mathbb{C}$ with $\mathrm{Re}(\beta) \neq 1/2$, $\mathrm{Im}(\beta) \neq 0$.

$$\mathbb{R}[s]/[(s^2 - s + \beta)(s^2 - s + \overline{\beta})] \cong (\mathbb{C} \times \mathbb{C})^T ,$$

so that

$$L^0(\mathbb{R}[s]/[(s^2-s+\beta)(s^2-s+\overline{\beta})],\epsilon) \;=\; 0\ .$$

It now follows from 38.10 (iii) that if $p(s)\neq s-1/2$

$$L^0(E,\epsilon) \;=\; \mathbb{Z}^{t_{p(s)}}\oplus 8\text{-torsion}$$

with $t_{p(s)}$ the number of the factors $(s^2-s+\alpha_j)$ in $p(s)$ with $\alpha_j > 1/4$. Working as in 39.12 the isometric L-groups of $\mathbb{Q}$ are given by

$$\begin{aligned} L\mathrm{Iso}^0(\mathbb{Q},\epsilon) \;&=\; \sum_{p(s)\in\overline{\mathcal{M}}_s(\mathbb{Q})} L^0(\mathbb{Q}[s]/(p(s)),(-)^{d_{p(s)}+1}\epsilon)\\ &=\; \begin{cases} \mathbb{Z}\oplus \displaystyle\sum_{p(s)\neq(s-1/2)} \mathbb{Z}^{t_{p(s)}}\oplus 8\text{-torsion} & \text{if } \epsilon=1\\ \displaystyle\sum_{p(s)\neq(s-1/2)} \mathbb{Z}^{t_{p(s)}}\oplus 8\text{-torsion} & \text{if } \epsilon=-1\ . \end{cases}\end{aligned}$$

(ii) Let (L,θ) be a nonsingular ϵ-symmetric form over $\mathbb{Q}$ with an isometric structure $f: L\longrightarrow L$ such that $1/2$ is not an eigenvalue of f. The minimal polynomial of f can be expressed as a product

$$p(s)q(s)\overline{q(s)}\in\mathbb{Q}[s]$$

with

$$p(s) \;=\; \prod_{j=1}^{m}(s^2-s+\alpha_j)\prod_{k=1}^{n}[(s^2-s+\beta_k)(s^2-s+\overline{\beta}_k)]\in\mathbb{R}[s]$$

as above, numbered in such a way that $\alpha_j > 1/4$ for $j\leq t$ and $\alpha_j < 1/4$ for $j>t$. Thus

$$\mathbb{R}\otimes_{\mathbb{Q}}(L,\theta,f) \;=\; \bigoplus_{j=1}^{m}(L_j,\theta_j,f_j)\oplus\text{hyperbolic}\ ,$$

with (L_j,θ_j,f_j) $(s^2-s+\alpha_j)$-primary and

$$L_j \;=\; \{u\in\mathbb{R}\otimes_{\mathbb{Q}}L\,|\,(1\otimes f^2-1\otimes f+\alpha_j)^N(u)=0 \text{ for some } N\geq 0\}\ .$$

The isometric Witt class $(L,\theta,f)\in L\mathrm{Iso}^0(\mathbb{Q},\epsilon)$ is determined up to 8-torsion by the signatures

$$\begin{aligned}\sigma_j(L,\theta,f) \;&=\; (L_j,\theta_j,f_j) \;=\; \sigma(g_j(L,\theta))\\ &\in L\mathrm{Iso}^0_{(s^2-s+\alpha_j)}(\mathbb{R},\epsilon) \;=\; L^0(\mathbb{C}^-,\epsilon) \;=\; \mathbb{Z}\ \ (1\leq j\leq t)\end{aligned}$$

(using 39.17 (ii)) with

$$g_j\ :\ \mathbb{Q}[s]/(p(s))\longrightarrow\mathbb{C}^-\ ;\ s\longrightarrow\lambda_j \;=\; \frac{1}{2}\pm\sqrt{\frac{1}{4}-\alpha_j}\ .$$

The two possible values of λ_j are the two roots of $s^2 - s + \alpha_j = 0$, which are complex conjugate algebraic numbers with $\mathrm{Re}(\lambda_j) = \frac{1}{2}$. Thus

$$LIso^0_R(\mathbb{Q}, \epsilon) \;=\; LIso^0(\mathbb{Q}, -1) \;=\; \mathbb{Z}^\infty \oplus 8\text{-torsion}$$

with $R = \{p(s) \in \mathbb{Q}[s] \,|\, p(1/2) \neq 0\}$. There is one $\mathbb{Z}$-valued signature for each algebraic number $\lambda \in \mathbb{C}$ with $\mathrm{Re}(\lambda) = 1/2$, $\mathrm{Im}(\lambda) > 0$. □

39C. L-theory of $F(z)$ $(\overline{z} = z^{-1})$

Definition 39.19 Let $\overline{\mathcal{M}}_z(F) \subseteq \mathcal{M}(F)$ be the set of irreducible monic polynomials

$$p(z) \;=\; a_0 + a_1 z + \ldots + a_{d-1} z^{d-1} + a_d z^d \in F[z, z^{-1}] \;\; (a_d = 1)$$

which are involution-invariant, such that

$$u_{p(z)}\overline{p(z)} \;=\; p(z) \in F[z, z^{-1}]$$

for some unit $u_{p(z)} \in U(F[z, z^{-1}])$, that is

$$\overline{a}_j \;=\; \overline{a}_0 a_{d-j} \in F \;\; (0 \leq j \leq d \;,\; a_d = 1)$$

with $a_0 \in U(F)$ and

$$u_{p(z)} \;=\; z^d a_0 \in U(F[z, z^{-1}]) \;.$$

□

The involution-invariant maximal ideals $p(z) \in \overline{\max}(F[z, z^{-1}])$ are the principal ideals $\mathcal{P} = (p(z)) \triangleleft F[z, z^{-1}]$ generated by the elements $p(z) \in \overline{\mathcal{M}}_z(F)$, so identify

$$\overline{\max}(F[z, z^{-1}]) \;=\; \overline{\mathcal{M}}_z(F) \;.$$

The duality involution

$$\mathrm{Aut}_0(F) \longrightarrow \mathrm{Aut}_0(F) \;;\; [P, f] \longrightarrow [P^*, (f^*)^{-1}]$$

corresponds under the identification of 19.3

$$\mathrm{Aut}_0(F) \;=\; \mathbb{Z}[\mathcal{M}_z(F)]$$

to the involution

$$\mathbb{Z}[\mathcal{M}_z(F)] \longrightarrow \mathbb{Z}[\mathcal{M}_z(F)] \;;\; p(z) \longrightarrow u_{p(z)}\overline{p(z)} \;,$$

so that

$$\widehat{H}^n(\mathbb{Z}_2; \mathrm{Aut}_0(F)) \;=\; \widehat{H}^n(\mathbb{Z}_2; \mathbb{Z})[\overline{\mathcal{M}}_z(F)] \;.$$

Proposition 39.20 (i) *For any $p(z) \in \overline{\mathcal{M}}_z(F)$, $\epsilon \in U(F)$*

$$\begin{aligned} LAut^0_{p(z)^\infty}(F,\epsilon) &= L^0(F[z,z^{-1}],(p(z))^\infty,-\epsilon) \\ &= L^0(E,\epsilon) = L^0(E,-u_{p(z)}\epsilon) \end{aligned}$$

with $E = F[z,z^{-1}]/(p(z))$, $u_{p(z)} = p(0)z^d \in E^\bullet$ *(*d *= degree* $(p(z))$ *).*
(ii) *Given*

$$p(z) = \sum_{j=0}^{d} a_j z^j \neq (z-1) \in \overline{\mathcal{M}}_z(F) \ \ (a_d = 1)$$

let

$$q(s) = p(1)^{-1}s^d p(1-s^{-1}) = p(1)^{-1}\sum_{j=0}^{d} a_j s^{d-j}(s-1)^j \in \overline{\mathcal{M}}_s(F)$$

(35.31). For any $\epsilon \in U(F)$

$$\begin{aligned} LAut^0_{p(z)^\infty}(F,\epsilon) &= LIso^0_{q(s)^\infty}(F,\epsilon) \\ &= L^0(E,-u_{p(z)}\epsilon) = L^0(E,\epsilon) \\ &= L^0(D,-u_{q(s)}\epsilon) = L^0(D,\epsilon) \end{aligned}$$

with $D = F[s]/(q(s))$, $u_{q(s)} = (-)^d$.
(iii) *The* ϵ*-symmetric automorphism and fibred automorphism L-groups of F are such that*

$$\begin{aligned} LAut^n(F,\epsilon) &= L^n(F,\epsilon) \oplus LAut^n_{fib}(F,\epsilon) \\ &\qquad (= 0 \text{ for } n \text{ odd}) . \end{aligned}$$

(iv) *The automorphism L-groups of F are such that*

$$\begin{aligned} LAut^{2k}(F,\epsilon) &= \bigoplus_{p(z)\in\overline{\mathcal{M}}_z(F)} LAut^{2k}_{p(z)^\infty}(F,\epsilon) \\ &= \bigoplus_{p(z)\in\overline{\mathcal{M}}_z(F)} L^0(F[z,z^{-1}]/(p(z)),(-)^k\epsilon) . \end{aligned}$$

(v) *The localization exact sequence of 28.17*

$$\begin{aligned} 0 \longrightarrow L^{2k}(F[z,z^{-1}],\epsilon) \xrightarrow{\ i\ } L^{2k}(F(z),\epsilon) &\xrightarrow{\ \partial\ } LAut^{2k}(F,-\epsilon) \\ &\xrightarrow{\ j\ } L^{2k-1}(F[z,z^{-1}],\epsilon) \longrightarrow 0 \end{aligned}$$

is isomorphic to

$$0 \longrightarrow L^0(F,(-)^k\epsilon) \xrightarrow{\begin{pmatrix} 1 \\ 0 \end{pmatrix}} L^0(F,(-)^k\epsilon) \oplus LAut^0_{fib}(F,(-)^{k+1}\epsilon)$$
$$\xrightarrow{\begin{pmatrix} 0 & -j_{fib} \\ 0 & 1 \end{pmatrix}} L^0(F,(-)^{k+1}\epsilon) \oplus LAut^0_{fib}(F,(-)^{k+1}\epsilon)$$
$$\xrightarrow{(1 \;\; j_{fib})} L^0(F,(-)^{k+1}\epsilon) \longrightarrow 0$$

with

$$j_{fib} \;:\; LAut^0_{fib}(F,(-)^{k+1}\epsilon) \longrightarrow L^0(F,(-)^{k+1}\epsilon) \;;\; (L,\theta,f) \longrightarrow (L,\theta)$$

and

$$\begin{aligned} LAut^0_{fib}(F,(-)^{k+1}\epsilon) &= \bigoplus_{p(z)\neq(z-1)\in\overline{\mathcal{M}}_z(F)} LAut^0_{p(z)^\infty}(F,(-)^{k+1}\epsilon) \\ &= \bigoplus_{p(z)\neq(z-1)\in\overline{\mathcal{M}}_z(F)} L^0(F[z,z^{-1}]/(p(z)),(-)^{k+1}\epsilon) \;. \end{aligned}$$

In particular, there is defined an isomorphism

$$(i \;\; \Delta^{fib}) \;:\; L^0(F,(-)^k\epsilon) \oplus LAut^0_{fib}(F,(-)^{k+1}\epsilon) \xrightarrow{\simeq} L^0(F(z),(-)^k\epsilon)$$

with

$$\begin{aligned} &\Delta^{fib} \;:\; LAut^0_{fib}(F,(-)^{k+1}\epsilon) \longrightarrow L^0(F(z),(-)^k\epsilon) \;; \\ &(L,\theta,f) \longrightarrow (L(z),(1-z^{-1})\theta(1-f)^{-1} + (-)^k\epsilon(1-z)(1-f^*)^{-1}\theta^*) \;. \end{aligned}$$

(vi) *If there exists an element $a \neq \overline{a} \in F^\bullet$ then the ϵ-symmetric L-groups of F are 2-periodic*

$$L^*(F,\epsilon) \;=\; L^*(F,-\epsilon) \;.$$

(vii) *The ϵ-symmetric L-groups of $F(z)$ are 2-periodic*

$$L^*(F(z),\epsilon) \;=\; L^*(F(z),-\epsilon) \;.$$

(viii) *If $p(z) \neq (z-1),(z+1) \in \overline{\mathcal{M}}_z(F)$ then the $(p(z))$-primary automorphism L-groups of F are 2-periodic*

$$L^*(F[z,z^{-1}],(p(z))^\infty,\epsilon) \;=\; L^*(F[z,z^{-1}],(p(z))^\infty,-\epsilon) \;.$$

(ix) *The morphism*

$$\begin{aligned} LAut^{2k}(F) \;=\; & \bigoplus_{p(z)\in\overline{\mathcal{M}}_z(F)} LAut^{2k}_{p(z)^\infty}(F) \\ & \longrightarrow \widehat{H}^0(\mathbb{Z}_2; Aut_0(F)) \;=\; \mathbb{Z}_2[\overline{\mathcal{M}}_s(F)]\ ; \\ & (L,\theta,f) \longrightarrow [L,f] \;=\; \sum_{p(z)\in\overline{\mathcal{M}}_s(F)} [L_{p(z)^\infty}]p(z) \end{aligned}$$

fits into the exact sequence of 28.37

$$0 \longrightarrow LAut^{2k}_r(F) \longrightarrow LAut^{2k}(F) \longrightarrow \mathbb{Z}_2[\overline{\mathcal{M}}_s(F)] \longrightarrow LAut^{2k-1}_r(F) \longrightarrow 0$$

with $L_{p(z)^\infty}$ *the* $p(z)$*-primary component of* L, $s = f : L \longrightarrow L$ *and*

$$\begin{aligned} [L_{p(z)^\infty}] \;=\; & \sum_{j=0}^{\infty} \dim_E(p(z)^j L/p(z)^{j+1}L) \\ & \in \widehat{H}^0(\mathbb{Z}_2; Aut_0^{p(z)^\infty}(F)) \;=\; \widehat{H}^0(\mathbb{Z}_2; K_0(E)) \;=\; \mathbb{Z}_2 \end{aligned}$$

(using 18.2 (ii)) with $E = F[z,z^{-1}]/p(z)$ *the quotient field.*

Proof (i) By definition, $LAut^0_{p(z)^\infty}(F,\epsilon)$ is the Witt group of nonsingular ϵ-symmetric forms (L,θ) over F with a $(p(z))$-primary automorphism $f : (L,\theta) \longrightarrow (L,\theta)$. The 1-dimensional $(-\epsilon)$-symmetric Poincaré complex (C,ϕ) defined over $F[z,z^{-1}]$ by

$$\begin{aligned} d_C \;&=\; z - f^* \;:\; C_1 \;=\; L^*[z,z^{-1}] \longrightarrow C_0 \;=\; L^*[z,z^{-1}]\ , \\ \phi_0 \;&=\; \begin{cases} -z\theta \;:\; C^0 \;=\; L[z,z^{-1}] \longrightarrow C_1 \;=\; L^*[z,z^{-1}] \\ \theta \;:\; C^1 \;=\; L[z,z^{-1}] \longrightarrow C_0 \;=\; L^*[z,z^{-1}]\ , \end{cases} \\ \phi_1 \;&=\; 0 \;:\; C^1 \;=\; L[z,z^{-1}] \longrightarrow C_1 \;=\; L^*[z,z^{-1}] \end{aligned}$$

is $p(z)^\infty$-primary. Conversely, if (C,ϕ) is a $p(z)^\infty$-primary 1-dimensional $(-\epsilon)$-symmetric Poincaré complex over $F[z,z^{-1}]$ there is defined a nonsingular ϵ-symmetric form (L,θ) over F with an automorphism

$$f \;=\; z \;:\; (L,\theta) \longrightarrow (L,\theta)$$

such that

$$\begin{aligned} \phi_0 \;:\; L = H^1(C) \xrightarrow{\;\simeq\;} H_0(C) \cong Hom_{F[z,z^{-1}]}(L, F[z,z^{-1},p(z)^{-1}]/F[z,z^{-1}])\ ; \\ u \longrightarrow (v \longrightarrow \frac{\phi(u,w)}{p(z)^k}) \\ (u,v \in L\,,\ w \in L[z,z^{-1}]\,,\ (z-f)(w) = p(z)^k(v))\,. \end{aligned}$$

The morphisms

$$\begin{aligned} & LAut^0_{p(z)^\infty}(F,\epsilon) \longrightarrow L^0(F[z,z^{-1}],p(z)^\infty,-\epsilon)\ ;\ (L,\theta,f) \longrightarrow (C,\phi)\ , \\ & L^0(F[z,z^{-1}],p(z)^\infty,-\epsilon) \longrightarrow LAut^0_{p(z)^\infty}(F,\epsilon)\ ;\ (C,\phi) \longrightarrow (H^1(C),\phi_0,f) \end{aligned}$$

are inverse isomorphisms. The identification

$$L^0(E, -u_{p(z)}\epsilon) = L^0(F[z, z^{-1}], p(z)^\infty, -\epsilon)$$

is a special case of 38.5 (iii), with a one-one correspondence

$$\begin{array}{l}\{(-u_{p(z)}\epsilon)\text{-symmetric forms } (L,\mu) \text{ over } E\} \rightleftarrows \\ \{(-\epsilon)\text{-symmetric linking forms } (L,\rho) \\ \qquad\qquad \text{over } (F[z,z^{-1}], p(z)^\infty) \text{ with } p(z)L = 0\}\end{array}$$

given by

$$\rho(u,v) = \mu(u,v)p(z)^{-1} \in F[z,z^{-1},p(z)^{-1}]/F[z,z^{-1}] \ \ (u,v \in L) .$$

Direct applications of 35.18 and 38.5 (iii) give identifications

$$LAut^0_{p(z)^\infty}(F,\epsilon) = L^0(E,-u_{p(z)}\epsilon) = L^0(F[z,z^{-1}],(p(z))^\infty,-\epsilon) ,$$

using a devissage argument to identify $LAut^0_{p(z)^\infty}(F,\epsilon)$ with the Witt group of nonsingular ϵ-symmetric forms (L,θ) over F with an automorphism $f : (L,\theta) \longrightarrow (L,\theta)$ such that $p(f) = 0$. The isomorphism of Witt groups

$$L^0(E,-u_{p(z)}\epsilon) \xrightarrow{\simeq} LAut^0_{p(z)^\infty}(F,\epsilon) \ ; \ (L,\mu) \longrightarrow (L,\theta,f)$$

is determined by the one-one correspondence

$$\begin{array}{l}\{(-u_{p(z)}\epsilon)\text{-symmetric forms } (L,\mu) \text{ over } E\} \rightleftarrows \\ \{\epsilon\text{-symmetric forms } (L,\theta) \text{ over } F \text{ with an} \\ \qquad\qquad \text{automorphism } f:(L,\theta)\longrightarrow(L,\theta) \text{ such that } p(f)=0\}\end{array}$$

given by

$$\begin{array}{rcl}\theta(u,v) & = & (\text{coefficient of } z^{d-1} \text{ in } \mu(u,v)) \in F , \\ f(u) & = & zu \in L \ \ (u,v \in L) ,\end{array}$$

working as in 39.2. As in Milnor [197] (cf. 39.4 (ii)) use any non-zero involution-preserving F-linear map $h : E \to F$ to define a one-one correspondence

$$\begin{array}{l}\{\epsilon\text{-symmetric forms } (L,\lambda) \text{ over } E\} \rightleftarrows \\ \{\epsilon\text{-symmetric forms } (L,\theta) \text{ over } F \text{ with an} \\ \qquad\qquad \text{automorphism } f:(L,\theta)\longrightarrow(L,\theta) \text{ such that } p(f)=0\}\end{array}$$

with

$$\theta(u,v) = h(\lambda(u,v)) \in F \ , \ f(u) = zu \in L \ \ (u,v \in L) .$$

This gives an isomorphism of Witt groups

$$L^0(E,\epsilon) \xrightarrow{\simeq} LAut^0_{p(z)^\infty}(F,\epsilon) \; ; \; (L,\lambda) \longrightarrow (L,\theta,f) \; .$$

(In the separable case it is possible to choose $h = \mathrm{tr}_{E/F} : E \longrightarrow F$ as in [197], and λ, μ are related by

$$\lambda(u,v) \;=\; q(z)\mu(u,v) \in E \;\; (u,v \in L)$$

with

$$q(z) \;=\; zp'(z) \;=\; \sum_{j=1}^{d} ja_j z^j \in F[z,z^{-1}]$$

such that

$$\begin{aligned} q(z) + u_{p(z)}\overline{q(z)} \;&=\; dp(z) \in F[z,z^{-1}] \; , \\ q(z) \;&=\; -u_{p(z)}\overline{q(z)} \in E \; , \\ \mathrm{tr}_{E/F}(p'(z)) \;&=\; \text{discriminant}\,(p(z)) \neq 0 \in F \end{aligned}$$

and $q(z) \neq 0 \in E$.)
(ii) The identification

$$LAut^0_{p(z)^\infty}(F,-\epsilon) \;=\; LIso^0_{(q(s))^\infty}(F,-\epsilon)$$

is the special case $(A,S) = (F,(p(z))^\infty)$ of 35.31 (iii). Moreover, there is defined an isomorphism of fields with involution

$$F[z,z^{-1}]/(p(z)) \longrightarrow F[s]/(q(s)) \; ; \; z \longrightarrow 1 - s^{-1} \; ,$$

with inverse

$$F[s]/(q(s)) \longrightarrow F[z,z^{-1}]/(p(z)) \; ; \; s \longrightarrow (1-z)^{-1} \; .$$

(iii) The involution-invariant multiplicative subsets

$$P \;=\; \{p(z)\,|\,p(1) \in F^\bullet\} \;\; , \;\; (z-1)^\infty \subset F[z,z^{-1}]$$

are coprime and such that the localizations $P^{-1}F[z,z^{-1}]$, $F[z,z^{-1},(z-1)^{-1}]$ fit into a cartesian square of rings with involution

$$\begin{array}{ccc} F[z,z^{-1}] & \longrightarrow & F[z,z^{-1},(z-1)^{-1}] \\ \downarrow & & \downarrow \\ P^{-1}F[z,z^{-1}] & \longrightarrow & F(z) \end{array}$$

It follows that

$$\begin{aligned} LAut^n(F,\epsilon) \;&=\; L^{n+2}(F[z,z^{-1}],(z-1)^\infty P,\epsilon) \\ &=\; L^{n+2}(F[z,z^{-1}],(z-1)^\infty,\epsilon) \oplus L^{n+2}(F[z,z^{-1}],P,\epsilon) \\ &=\; LAut^n_{uni}(F,\epsilon) \oplus LAut^n_{fib}(F,\epsilon) \end{aligned}$$

and by 38.5 (iii)

$$\mathrm{LAut}^n_{uni}(F,\epsilon) \;=\; L^n(F,\epsilon)\ .$$

(iv) This is the special case $(A,J)=(F[z,z^{-1}],\overline{\mathcal{M}}_z(F))$ of 38.10, using (i).
(v) Apply (iv), or alternatively use the isomorphism of fields with involution

$$F(s) \xrightarrow{\simeq} F(z)\ ;\ s \longrightarrow (1-z)^{-1}$$

to interpret the direct sum decomposition of 39.12

$$L^0(F(s),(-)^k\epsilon) \;=\; L^0(F,(-)^k\epsilon)\oplus L\mathrm{Iso}^0(F,(-)^{k+1}\epsilon)$$

as a direct sum decomposition

$$L^0(F(z),(-)^k\epsilon) \;=\; L^0(F,(-)^k\epsilon)\oplus L\mathrm{Aut}^0_{fib}(F,(-)^{k+1}\epsilon)\ .$$

(vi) The unit $u=a-\overline{a}\in F^\bullet$ is such that $\overline{u}=-u$, so that there are defined isomorphisms

$$L^n(F,\epsilon) \xrightarrow{\simeq} L^n(F,-\epsilon)\ ;\ (C,\phi) \longrightarrow (C,u\phi)\ .$$

(vii) Apply (vi) with $a\in F^\bullet$ replaced by $z\in F(z)^\bullet$, to obtain a unit

$$u \;=\; z-z^{-1}\in F(z)^\bullet$$

such that $\overline{u}=-u$.
(viii) For any $p(z)\in\overline{\mathcal{M}}_z(F)$

$$L^*(F[z,z^{-1}],(p(z))^\infty,\epsilon) \;=\; L^*(F[z,z^{-1}]/(p(z)),u_{p(z)}\epsilon)$$

by the devissage isomorphism of 38.5 (iii), and by (i)

$$L^*(F[z,z^{-1}]/(p(z)),u_{p(z)}\epsilon) \;=\; L^*(F[z,z^{-1}]/(p(z)),-\epsilon)\ .$$

If $p(z)\neq z-1,z+1$ then

$$v \;=\; z-z^{-1} \;=\; z^{-1}(z-1)(z+1)\notin(p(z))\ ,$$

so that $v\in(F[z,z^{-1}]/(p(z)))^\bullet$ is a unit such that $\overline{v}=-v$ and there are defined isomorphisms

$$\begin{aligned}L^n(F[z,z^{-1}]/(p(z)),u_{p(z)}\epsilon) &\xrightarrow{\simeq} L^n(F[z,z^{-1}]/(p(z)),-u_{p(z)}\epsilon)\ ;\\ &(C,\phi)\longrightarrow(C,v\phi)\ .\end{aligned}$$

(ix) As for 39.4 (iii). □

Proposition 39.21 *The involution-invariant irreducible monic polynomials in $F[z,z^{-1}]$ of degree 1 are of the type*

$$(z-a)\in\overline{\mathcal{M}}_z(F)$$

with $a \in F$ such that $\overline{a}=a^{-1} \in F$, and $u_{(z-a)} = -za$. For any such $a \in F$

$$\begin{aligned} L^n(F[z,z^{-1}],(z-a)^{\infty},\epsilon) &= LAut^{n-2}_{(z-a)^{\infty}}(F,\epsilon) \\ &= L^n(F[z,z^{-1}]/(z-a),-z\epsilon a) \\ &= L^n(F,-a^2\epsilon) \ = \ L^n(F,-\epsilon) \ , \end{aligned}$$

$$L^n(F[z,z^{-1},(z-a)^{-1}],\epsilon) \ = \ L^n(F,\epsilon) \ .$$

The inclusion $F \longrightarrow F[z,z^{-1},(z-a)^{-1}]$ thus induces an isomorphism in ϵ-symmetric L-theory. If $U(F)$ has more than 2 elements there exists $b \neq a \in U(F)$, and the inverse is induced by the morphism of rings with involution

$$F[z,z^{-1},(z-a)^{-1}] \longrightarrow F \ ; \ z \longrightarrow b \ .$$

Proof Apply 39.20 (i). □

Example 39.22 Let F be algebraically closed, so that

$$\overline{\mathcal{M}}_z(F) \ = \ \{(z-a)\,|\,a \in U(F)\}$$

and

$$\begin{aligned} LAut^0(F,\epsilon) &= \bigoplus_{a \in U(F)} L^0(F[z,z^{-1}]/(z-a),\epsilon) \\ &= \bigoplus_{a \in U(F)} L^0(F,\epsilon) \ . \end{aligned}$$

(i) A nonsingular ϵ-symmetric form (L,θ) over F with an automorphism $f : (L,\theta) \longrightarrow (L,\theta)$ has a direct sum decomposition of the type

$$(L,\theta,f) \ = \ \sum_{a \in U(F)} (L_a,\theta_a,f_a) \oplus (K \oplus K^*, \begin{pmatrix} 0 & 1 \\ \epsilon & 0 \end{pmatrix}, \begin{pmatrix} g & h \\ 0 & (g^*)^{-1} \end{pmatrix})$$

with

$$L_a \ = \ \bigcup_{j=0}^{\infty} \ker((f-a)^j : L \longrightarrow L) \ , \ K \oplus K^* \ = \ \sum_{b \neq \overline{b}^{-1} \in F^\bullet} L_b \ ,$$

$$\mathrm{ch}_z(L_a,f_a) \ = \ (z-a)^{d_a} \ (d_a \ = \ \dim_F(L_a)) \ .$$

The automorphism Witt class is given by

$$(L,\theta,f) \ = \ \sum_{a \in U(F)} (L_a,\theta_a) \in LAut^0(F,\epsilon) \ = \ \bigoplus_{a \in U(F)} L^0(F,\epsilon) \ .$$

Furthermore, it is possible to decompose each (L_a,f_a,θ_a) as a direct sum

$$(L_a,\theta_a,f_a) \ = \ \sum_{k=0}^{\infty} (L_{a,k},\theta_{a,k},f_{a,k})$$

with

$$\text{minimal polynomial}(L_{a,k}, f_{a,k}) = (z-a)^k ,$$

and

$$(L_a, \theta_a, f_a) = \sum_{k \text{ odd}} (L_{a,k}/(f-a)L_{a,k}, [\theta_{a,k}])$$
$$\in LAut^0_{(z-a)^\infty}(F,\epsilon) = L^0(F,\epsilon) .$$

(ii) If $F = \mathbb{C}^+$ then $U(F) = \{+1, -1\}$ and

$$LAut^0(\mathbb{C}^+) = L^0(\mathbb{C}^+) \oplus L^0(\mathbb{C}^+) = \mathbb{Z}_2 \oplus \mathbb{Z}_2 ,$$

with the components generated by $(\mathbb{C}^+, 1, 1)$ and $(\mathbb{C}^+, 1, -1)$.
(iii) If $F = \mathbb{C}^-$ then $U(F) = S^1$ and

$$LAut^0(\mathbb{C}^-) = \bigoplus_{a \in S^1} L^0(\mathbb{C}^-) = \bigoplus_{a \in S^1} \mathbb{Z} ,$$

with the components generated by $(\mathbb{C}^-, 1, a)$. (This is the *automorphism multisignature*, about which more in Chap. 40). □

Example 39.23 Let $F = \mathbb{R}$. The involution-invariant irreducible monic polynomials in $\mathbb{R}[z, z^{-1}]$ are given by

$$\overline{\mathcal{M}}_z(\mathbb{R}) = \{(z-1)\} \sqcup \{(z+1)\} \sqcup \{p_\theta(z) \,|\, 0 < \theta < \pi\}$$

with

$$p_\theta(z) = z^2 - 2z\cos\theta + 1 = (z - e^{i\theta})(z - e^{-i\theta}) \in \overline{\mathcal{M}}_z(\mathbb{R}) .$$

The residue class fields are such that there are defined isomorphisms of rings with involution

$$\mathbb{R}[z, z^{-1}]/(z \pm 1) \xrightarrow{\simeq} \mathbb{R} ; \ z \longrightarrow \mp 1 ,$$
$$\mathbb{R}[z, z^{-1}]/(p_\theta(z)) \xrightarrow{\simeq} \mathbb{C}^- ; \ z \longrightarrow e^{i\theta} .$$

By 39.21

$$LAut^0(\mathbb{R}, \epsilon) = \bigoplus_{p(z) \in \overline{\mathcal{M}}_z(\mathbb{R})} L^0(\mathbb{R}[z, z^{-1}]/(p(z)), \epsilon)$$
$$= \begin{cases} \mathbb{Z} \oplus \mathbb{Z} \oplus \bigoplus\limits_{0<\theta<\pi} \mathbb{Z} & \text{if } \epsilon = 1 \\ \bigoplus\limits_{0<\theta<\pi} \mathbb{Z} & \text{if } \epsilon = -1 , \end{cases}$$

with components generated by

$$\begin{cases} (\mathbb{R}, 1, \pm 1)\ ,\ (\mathbb{R} \oplus \mathbb{R}, \begin{pmatrix} 1 & 0 \\ 0 & 1 \end{pmatrix}, \begin{pmatrix} \cos\theta & -\sin\theta \\ \sin\theta & \cos\theta \end{pmatrix}) \\ (\mathbb{R} \oplus \mathbb{R}, \begin{pmatrix} 0 & 1 \\ -1 & 0 \end{pmatrix}, \begin{pmatrix} \cos\theta & -\sin\theta \\ \sin\theta & \cos\theta \end{pmatrix})\ . \end{cases}$$

□

Example 39.24 Let $F = \mathbb{Q}$.
(i) If $p(z) \in \overline{\mathcal{M}}_z(\mathbb{Q})$ has degree $d_{p(z)}$ then the residue class field

$$E = \mathbb{Q}[z, z^{-1}]/(p(z))$$

is an algebraic number field with involution which is a degree $d_{p(z)}$ extension of $\mathbb{Q}$. The involution is non-trivial if $p(z) \neq z-1, z+1$, in which case $d_{p(z)}$ is even, say $d_{p(z)} = 2m$, and $p(z)$ splits over $\mathbb{R}$ as a product

$$p(z) = (z^2 - 2\beta_1 z + 1)(z^2 - 2\beta_2 z + 1)\dots(z^2 - 2\beta_m z + 1) \in \mathbb{R}[z, z^{-1}]$$

with

$$\mathbb{R}[z, z^{-1}]/(p(z)) = \prod_{k=1}^{m} \mathbb{R}[z, z^{-1}]/(z^2 - 2\beta_k z + 1) \quad (\beta_k \neq \pm 1 \in \mathbb{R})$$

a product of rings with involution. For $\beta \neq \pm 1 \in \mathbb{R}$

$$\mathbb{R}[z, z^{-1}]/(z^2 - 2\beta z + 1) \cong \begin{cases} (\mathbb{R} \times \mathbb{R})^T & \text{if } |\beta| > 1 \\ \mathbb{C}^- & \text{if } |\beta| < 1 \end{cases}$$

where $(\mathbb{R} \times \mathbb{R})^T$ denotes $\mathbb{R} \times \mathbb{R}$ with the hyperbolic involution $T : (u, v) \to (v, u)$, so that

$$L^0(\mathbb{R}[z, z^{-1}]/(z^2 - 2\beta z + 1), \epsilon) = \begin{cases} L^0((\mathbb{R} \times \mathbb{R})^T, \epsilon) = 0 & \text{if } |\beta| > 1 \\ L^0(\mathbb{C}^-, \epsilon) = \mathbb{Z} & \text{if } |\beta| < 1 \end{cases}.$$

It now follows from 38.10 (iii) that if $p(z) \neq z - 1, z + 1$

$$L^0(E, \epsilon) = \mathbb{Z}^{t_{p(z)}} \oplus 8\text{-torsion}$$

with $t_{p(z)}$ the number of the factors $(z^2 - 2\beta_k z + 1)$ in $p(z)$ with $|\beta_k| < 1$. Working as in 39.12 the automorphism L-groups of $\mathbb{Q}$ are given by

$$\begin{aligned} LAut^0(\mathbb{Q}, \epsilon) &= \sum_{p(z) \in \overline{\mathcal{M}}_z(\mathbb{Q})} L^0(\mathbb{Q}[z, z^{-1}]/(p(z)), (-)^{d_{p(z)}}\epsilon) \\ &= \begin{cases} \mathbb{Z} \oplus \mathbb{Z} \oplus \displaystyle\sum_{p(z) \neq (z-1),(z+1)} \mathbb{Z}^{t_{p(z)}} \oplus 8\text{-torsion} & \text{if } \epsilon = 1 \\ \displaystyle\sum_{p(z) \neq (z-1),(z+1)} \mathbb{Z}^{t_{p(z)}} \oplus 8\text{-torsion} & \text{if } \epsilon = -1\ . \end{cases} \end{aligned}$$

(ii) Let $h : (L, \theta) \longrightarrow (L, \theta)$ be an automorphism of a nonsingular ϵ-symmetric form (L, θ) over $\mathbb{Q}$ such that $h - 1, h + 1 : L \longrightarrow L$ are also automorphisms. The minimal polynomial of f can be expressed as a product

$$p(z)q(z)\overline{q(z)} \in \mathbb{Q}[z,z^{-1}]\ ,$$

with

$$p(z) = (z^2-2\beta_1 z+1)(z^2-2\beta_2 z+1)\dots(z^2-2\beta_m z+1) \in \mathbb{R}[z]$$

as above, numbered in such a way that $|\beta_k|<1$ for $k\le t$ and $|\beta_k|>1$ for $k>t$. Thus

$$\mathbb{R}\otimes_{\mathbb{Q}}(L,\theta,h) = \bigoplus_{k=1}^{m}(L_k,\theta_k,h_k)\oplus \text{hyperbolic}\ ,$$

with (L_k,θ_k,h_k) $(z^2-2\beta_k z+1)$-primary. The automorphism Witt class $(L,\theta,h)\in LAut^0(\mathbb{Q},\epsilon)$ is determined up to 8-torsion by the signatures

$$\sigma_k(L,\theta,h) = \sigma(g_k(L,\theta))\in L^0(\mathbb{C}^-,\epsilon) = \mathbb{Z}\ \ (1\le k\le t)$$

with

$$g_k\ :\ \mathbb{Q}[z,z^{-1}]/(p(z)) \longrightarrow \mathbb{C}^-\ ;\ z\longrightarrow \mu_k = \beta_k\pm\sqrt{(\beta_k)^2-1}\ .$$

The two possible values of μ_k are the two roots of $z^2-2\beta_k z+1=0$, which are complex conjugate algebraic numbers with $|\mu_k|=1$. Since

$$g_k\ :\ \mathbb{Q}[z,z^{-1}]/(p(z)) \longrightarrow \mathbb{R}[z,z^{-1}]/(p(z)) \cong \prod_t \mathbb{C}^-\times\prod_{m-t}(\mathbb{R}\times\mathbb{R})^T \xrightarrow{p_k} \mathbb{C}^-$$

the kth signature is

$$\sigma_k(L,\theta) = (L'_k,\theta'_k)\in L^0(\mathbb{C}^-,\epsilon) = \mathbb{Z}$$

where (L'_k,θ'_k) is the restriction of $\mathbb{C}\otimes_{\mathbb{Q}}(L,\theta)$ to the generalized μ_k-eigenspace of $1\otimes h:\mathbb{C}\otimes_{\mathbb{Q}}L\longrightarrow\mathbb{C}\otimes_{\mathbb{Q}}L$

$$L'_k = \{u\in\mathbb{C}\otimes_{\mathbb{Q}}L\,|\,(1\otimes h-\mu_k)^N(u)=0 \text{ for some } N\ge 0\}\ .$$

Thus

$$LAut^0_{\widehat{P}}(\mathbb{Q},\epsilon) = LAut^0(\mathbb{Q},-1) = \mathbb{Z}^\infty\oplus 8\text{-torsion}$$

with $\widehat{P}=\{p(z)\in\mathbb{Q}[z,z^{-1}]\,|\,p(1),p(-1)\neq 0\}$. There is one $\mathbb{Z}$-valued signature for each algebraic number $\mu\in S^1\subset\mathbb{C}$ such that $\mathrm{Im}(\mu)>0$.
(iii) As in 39.18 let $R=\{p(s)\in\mathbb{Q}[s]\,|\,p(1/2)\neq 0\}$, so that there is defined an isomorphism

$$LAut^0_{\widehat{P}}(\mathbb{Q},\epsilon) \xrightarrow{\simeq} LIso^0_R(\mathbb{Q},\epsilon)\ ;\ (L,\theta,h)\longrightarrow(L,\theta,f)$$

with $f=(1-h)^{-1}:L\longrightarrow L$. If $h:L\longrightarrow L$ has minimal polynomial $p(z)\in\mathbb{Q}[z]$ with factorization

$$p(z) = (z^2-2\beta_1 z+1)(z^2-2\beta_2 z+1)\dots(z^2-2\beta_m z+1) \in \mathbb{R}[z]$$

then $f : L \longrightarrow L$ has minimal polynomial

$$q(s) = s^{2m}p(1-s^{-1}) \in \mathbb{Q}[s]$$

with factorization

$$q(s) = (s^2 - s + \alpha_1)(s^2 - s + \alpha_2)\dots(s^2 - s + \alpha_m) \in \mathbb{R}[s]$$

where

$$\alpha_k = \frac{1}{2(1-\beta_k)} \quad (1 \leq k \leq m) .$$

(See 41.22 below for a generalization of the relationship between $p(z)$ and $q(s)$.) The equations $s^2 - s + \alpha = 0$ with $\alpha < 1/4$ correspond to the equations $z^2 - 2\beta z + 1 = 0$ with $|\beta| < 1$ under the identifications

$$s = (1-z)^{-1} , \quad \alpha = \frac{1}{2(1-\beta)} .$$

The automorphism multisignature components for $(L, \theta, h) \in \mathbb{Z}$ are thus given by (ii) for the coefficients $\beta_1, \beta_2, \dots, \beta_t \in \mathbb{R}$ in the factorization of $p(z)$ such that $|\beta_k| < 1$ coincide with the isometric multisignature components for (L, θ, f) given by 39.18 for the coefficients $\alpha_1, \alpha_2, \dots, \alpha_t \in \mathbb{R}$ in the factorization of $q(s)$ such that $\alpha_k < 1/4$ coincide

$$\sigma_k(L, \theta, h) = \sigma_k(L, \theta, f) \in \mathbb{Z} \quad (1 \leq k \leq t) . \qquad \square$$

39D. The asymmetric *L*-theory of *F*

The asymmetric Witt group $LAsy^0(F)$ was defined in 25.3.

Definition 39.25 Let $p(z) \in \overline{\mathcal{M}}_z(F)$.
(i) A nonsingular asymmetric form (L, λ) over F is *$(p(z))$-primary* if the automorphism $(\lambda^*)^{-1}\lambda : L \longrightarrow L$ is $(p(z))$-primary, i.e. if it has characteristic polynomial a power of $p(z)$

$$\text{ch}_z(L, (\lambda^*)^{-1}\lambda) = p(z)^k .$$

(ii) The *$(p(z))$-primary asymmetric Witt group* $LAsy^0_{p(z)^\infty}(F)$ is the Witt group of nonsingular $(p(z))$-primary asymmetric forms over F. $\square$

The splitting of ϵ-symmetric automorphism L-theory

$$LAut^0(F, \epsilon) = \bigoplus_{p(z) \in \overline{\mathcal{M}}_z(F)} LAut^0_{p(z)^\infty}(F, \epsilon)$$

will now be extended to asymmetric L-theory

$$LAsy^0(F) \;=\; \bigoplus_{p(z)\in\overline{\mathcal{M}}_z(F)} LAsy^0_{p(z)^\infty}(F) \;,$$

and $LAsy^0_{p(z)^\infty}(F)$ will be identified with the appropriate u-symmetric Witt group of the residue class field $F[z,z^{-1}]/(p(z))$.

Proposition 39.26 (Riehm [252, p. 47], Warshauer [304, p. 83])
Every nonsingular asymmetric form (L,λ) over F splits as a sum

$$(L,\lambda) \;=\; \sum_{p(z)\in\overline{\mathcal{M}}_z(F)} (L_{p(z)},\lambda_{p(z)}) \oplus (L',\lambda')$$

according to the $p(z)^\infty$-primary decomposition of the automorphism $\lambda^{-1}\lambda^* : L \longrightarrow L$, with

$$L' \;=\; \sum_{p(z)\neq u_{p(z)}\overline{p(z)}\in\mathcal{M}(F)} L_{p(z)}$$

and (L',λ') metabolic.

Proof Immediate from the special case $A = F$ of 35.22, using the unique factorization in $F[z,z^{-1}]$. □

The asymmetric Witt group $LAsy^0(F)$ was defined in 25.3.

Proposition 39.27 *Let $\epsilon \in U(F)$.*
(i) *The asymmetric Witt group of F is isomorphic to the ϵ-symmetric Witt group of $F(z)$, with an isomorphism*

$$LAsy^0(F) \xrightarrow{\simeq} L^0(F(z),\epsilon)\;;\;(L,\lambda) \longrightarrow (L(z),(1-z^{-1})\lambda+\epsilon(1-z)\lambda^*)$$

such that the composite

$$L^0(F,\epsilon) \longrightarrow LAsy^0(F) \cong L^0(F(z),\epsilon)$$

is the injection induced by the inclusion $F \longrightarrow F(z)$.
(ii) *The asymmetric Witt group of F decomposes as the sum of $(p(z))$-primary asymmetric Witt groups*

$$LAsy^0(F) \;=\; \sum_{p(z)\in\overline{\mathcal{M}}_z(F)} LAsy^0_{p(z)^\infty}(F) \;.$$

(iii) *For $p(z) \in \overline{\mathcal{M}}_z(F)$ of degree d the ideal $(p(\epsilon^{-1}z)) \in \overline{\max}(F[z,z^{-1}])$ is generated by $\epsilon^d p(\epsilon^{-1}z) \in \overline{\mathcal{M}}_z(F)$, and*

$$\begin{aligned} LAsy^0_{p(\epsilon^{-1}z)^\infty}(F) \;&=\; LAsy^0_{(\epsilon^d p(\epsilon^{-1}z))^\infty}(F) \\ &= \begin{cases} L^0(F,\epsilon) & \text{if } p(z)=z-1 \\ LAut^0_{p(z)^\infty}(F,-\epsilon) \;=\; L^0(F[z,z^{-1}]/(p(z)),-\epsilon) & \text{if } p(z)\neq z-1. \end{cases}\end{aligned}$$

In particular, if $p(z) = z-a$ then $\epsilon p(\epsilon^{-1}z) = z-\epsilon a$ and the natural map

$$L^0(F,\epsilon a) \longrightarrow LAsy^0_{(z-\epsilon a)^\infty}(F)\ ;\ (L,\lambda) \longrightarrow (L,\lambda)$$

from the ϵ-symmetric Witt group of F to the $(z-\epsilon a)$-primary asymmetric (= almost ϵa-symmetric) Witt group of F is an isomorphism. If $\mathrm{char}(F) \neq 2$ *the inverse isomorphism is given by*

$$LAsy^0_{(z-\epsilon a)^\infty}(F) \longrightarrow L^0(F,\epsilon a)\ ;\ (L,\lambda) \longrightarrow (L,\frac{1}{2}(\lambda+\epsilon a\lambda^*))\ .$$

If $a \neq 1$ then the composite of the isomorphism

$$L^0(F,\epsilon a) \cong LAsy^0_{(z-\epsilon a)^\infty}(F) \xrightarrow{\simeq} LAut^0_{(z-\epsilon a)^\infty}(F,-\epsilon)\ ;$$
$$(L,\lambda) \longrightarrow (L,\lambda-\epsilon\lambda^*,\epsilon a)$$

and the forgetful isomorphism

$$LAut^0_{(z-\epsilon a)^\infty}(F,-\epsilon) \xrightarrow{\simeq} L^0(F,-\epsilon)\ ;\ (L,\theta,f) \longrightarrow (L,\theta)$$

is the isomorphism

$$L^0(F,\epsilon a) \xrightarrow{\simeq} L^0(F,-\epsilon)\ ;\ (L,\lambda) \longrightarrow (L,(1-\overline{a})\lambda)\ .$$

(iv) *The localization exact sequence*

$$0 \longrightarrow L^{4*}(F[z,z^{-1}],\epsilon) \xrightarrow{i} L^{4*}(F(z),\epsilon) \xrightarrow{\partial} LAut^{4*-2}(F,\epsilon)$$
$$\xrightarrow{j} L^{4*-1}(F[z,z^{-1}],\epsilon) \longrightarrow 0$$

is isomorphic to

$$0 \longrightarrow L^0(F,\epsilon) \xrightarrow{\begin{pmatrix}1\\0\end{pmatrix}} L^0(F,\epsilon) \oplus \bigoplus_{p(z)\neq(z-1)\in\overline{\mathcal{M}}_z(F)} L^0(F[z,z^{-1}]/(p(z)),-\epsilon)$$

$$\xrightarrow{\begin{pmatrix}0 & -\bigoplus\limits_{p(z)} j_{p(z)}\\ 0 & 1\end{pmatrix}} L^0(F,-\epsilon) \oplus \bigoplus_{p(z)\neq(z-1)\in\overline{\mathcal{M}}_z(F)} L^0(F[z,z^{-1}]/(p(z)),-\epsilon)$$

$$\xrightarrow{\begin{pmatrix}1 & \bigoplus\limits_{p(z)} j_{p(z)}\end{pmatrix}} L^0(F,-\epsilon) \longrightarrow 0$$

with

$$j_{p(z)}\ :\ L^0(F[z,z^{-1}]/(p(z)),-\epsilon)\ =\ LAut^0_{p(z)^\infty}(F,-\epsilon) \longrightarrow L^0(F,-\epsilon)\ ;$$
$$(L,\phi,f) \longrightarrow (L,\phi)\ .$$

(v) *Given a nonsingular asymmetric form* (L,λ) *over* F *and an element* $a\in U(F)$ *the following conditions are equivalent:*

(a) *the* $(-\epsilon a)$*-symmetric form* $(L,\lambda-\epsilon a\lambda^*)$ *over* F *is nonsingular,*

(b) a *is not a root of the characteristic polynomial*

$$\begin{aligned}\mathrm{ch}_z(L,(\epsilon\lambda^*)^{-1}\lambda) &= \det(z-(\epsilon\lambda^*)^{-1}\lambda : L[z,z^{-1}]\longrightarrow L[z,z^{-1}])\\ &\in F[z,z^{-1}] .\end{aligned}$$

The conditions are satisfied if (L,λ) *is* $p(\epsilon^{-1}z)$*-primary and* $a\in U(F)$ *is such that the polynomials* $p(z),(z-a)\in F[z,z^{-1}]$ *are coprime (i.e. if* $p(a)\neq 0\in F$*). For* $a\neq 1$ *these conditions are also equivalent to:*

(c) *the* ϵ*-symmetric form over* $F[z,z^{-1}]_{(z-a)}$

$$(L[z,z^{-1}]_{(z-a)},(1-z^{-1})\lambda+\epsilon(1-z)\lambda^*)$$

is nonsingular,

(d) *the* ϵ*-symmetric form over* F

$$(L,(1-\overline{a})\lambda+\epsilon(1-a)\lambda^*)$$

is nonsingular.

(vi) *For* $a\neq 1\in U(F)$ *the* ϵ*-symmetric Witt group of* $F[z,z^{-1}]_{(z-a)}$ *is such that there is defined an isomorphism*

$$\begin{aligned}\sum_{p(z)\neq(z-a)\in\overline{\mathcal{M}}_z(F)} LAsy^0_{p(\epsilon^{-1}z)^\infty}(F) &\xrightarrow{\simeq} L^0(F[z,z^{-1}]_{(z-a)},\epsilon) ;\\ (L,\lambda) &\longrightarrow (L[z,z^{-1}]_{(z-a)},(1-z^{-1})\lambda+\epsilon(1-z)\lambda^*) .\end{aligned}$$

The morphism of rings with involution

$$\tau_a : F[z,z^{-1}]_{(z-a)}\longrightarrow F ; z\longrightarrow a$$

induces a morphism of ϵ*-symmetric Witt groups*

$$\begin{aligned}\tau_a : L^0(F[z,z^{-1}]_{(z-a)},\epsilon) \cong \sum_{p(z)\neq(z-a)} LAsy^0_{p(\epsilon^{-1}z)^\infty}(F) &\longrightarrow L^0(F,\epsilon) ;\\ (L,\lambda) &\longrightarrow (L,(1-\overline{a})\lambda+\epsilon(1-a)\lambda^*)\end{aligned}$$

which is given on the components for $p(z)=z-b$ $(b\neq a\in U(F))$ *by the isomorphism*

$$LAsy^0_{p(\epsilon^{-1}z)^\infty}(F) = L^0(F,\epsilon b) \xrightarrow{\simeq} L^0(F,\epsilon) ; (L,\lambda)\longrightarrow(L,v\lambda)$$

with $v=(1-a)(\overline{b}-\overline{a})\in F^\bullet$ *such that* $\overline{v}=vb$.

Proof (i) This is just the special case $A=F$, $n=0$ of the identification of 27.33

$$LAsy^n(A) = L^n(\Omega^{-1}A[z,z^{-1}],\epsilon) .$$

(ii) The splitting is immediate from 38.25.

(iii) Consider first the case $p(z) = z-1$. The following conditions on a nonsingular asymmetric form (L,λ) over F are equivalent:

(a) (L,λ) is $(z-\epsilon)$-primary,

(b) the endomorphism $1-(\epsilon\lambda^*)^{-1}\lambda : L \longrightarrow L$ is nilpotent,

(c) (L,λ) is almost ϵ-symmetric (35.1).

The morphisms

$$LAsy^0_{(z-\epsilon)^\infty}(F) \longrightarrow L^0(F[z,z^{-1},(z-1)^{-1}],\epsilon) ;$$
$$(L,\lambda) \longrightarrow (L[z,z^{-1},(z-1)^{-1}],(1-z^{-1})\lambda+\epsilon(1-z)\lambda^*) ,$$
$$L^0(F,\epsilon) \longrightarrow L^0(F[s],\epsilon) = L^0(F[z,z^{-1},(z-1)^{-1}],\epsilon)$$

are isomorphisms by 35.3 (i). Thus the forgetful map

$$L^0(F,\epsilon) \longrightarrow LAsy^0_{(z-\epsilon)^\infty}(F) ; (L,\theta) \longrightarrow (L,\theta)$$

is an isomorphism.

Now suppose given $p(z) \in \overline{\mathcal{M}}_z(F)$ with $p(z) \neq (z-1)$. Given a nonsingular asymmetric form (L,λ) over F define a $(-\epsilon)$-symmetric form (L,θ) over F with an automorphism $f : (L,\theta) \longrightarrow (L,\theta)$ by

$$\theta = \lambda - \epsilon\lambda^* , \quad f = (\epsilon\lambda^*)^{-1}\lambda .$$

The asymmetric form (L,λ) is $(p(\epsilon^{-1}z))$-primary if and only if the automorphism $f : L \longrightarrow L$ is $(p(z))$-primary, in which case (L,θ) is nonsingular and $1-f : L \longrightarrow L$ is an automorphism. The nonsingular $(p(\epsilon^{-1}z))$-primary asymmetric forms (L,λ) over F are thus in one-one correspondence with the $(p(z))$-primary automorphisms of nonsingular $(-\epsilon)$-symmetric forms over F, and there are defined inverse isomorphisms

$$LAsy^0_{p(\epsilon^{-1}z)^\infty}(F) \longrightarrow LAut^0_{p(z)^\infty}(F,-\epsilon) ;$$
$$(L,\lambda) \longrightarrow (L,\lambda-\epsilon\lambda^*,(\epsilon\lambda^*)^{-1}\lambda) ,$$
$$LAut^0_{p(z)^\infty}(F,-\epsilon) \longrightarrow LAsy^0_{p(\epsilon^{-1}z)^\infty}(F) ;$$
$$(L,\theta,f) \longrightarrow (L,\theta(1-f^{-1})^{-1}) .$$

(This is the special case $(A,S) = (F,(p(z))^\infty)$ of 35.31 (iii).) Identify

$$LAut^0_{p(z)^\infty}(F,-\epsilon) = L^0(F[z,z^{-1}],(p(z))^\infty,\epsilon) = L^0(F[z,z^{-1}]/(p(z)),-\epsilon)$$

as in 38.20.

(iv) Working as in 28.12 it may be verified that the composite

$$LAsy^0(F) \cong L^0(F(z),\epsilon) \xrightarrow{\partial} LAut^0(F,-\epsilon)$$

sends the Witt class of a nonsingular asymmetric form (L, λ) over F to the Witt class of the automorphism $f : (M, \phi) \longrightarrow (M, \phi)$ of the nonsingular $(-\epsilon)$-symmetric form (M, ϕ) over F defined by

$$\phi = \begin{pmatrix} 0 & 1 \\ -\epsilon & 0 \end{pmatrix} : M = L \oplus L^* \longrightarrow M^* = L^* \oplus L ,$$

$$f = \begin{pmatrix} 1 + (\epsilon\lambda^*)^{-1}\lambda & -(\epsilon\lambda^*)^{-1} \\ \lambda & 0 \end{pmatrix} : M = L \oplus L^* \longrightarrow M = L \oplus L^* .$$

If $\epsilon\lambda^* = \lambda$ then

$$K = \mathrm{im}(\begin{pmatrix} 1 \\ \lambda \end{pmatrix} : L \longrightarrow L \oplus L^*) \subset L$$

is a lagrangian of (M, ϕ) such that $f| = 1 : K \longrightarrow K$, so that

$$(M, \phi, f) = 0 \in LAut^0(F, -\epsilon) .$$

If (L, λ) is $(p(z))_\epsilon$-primary with $p(z) \neq (z-1)$ then $(L, \lambda - \epsilon\lambda^*)$ is a nonsingular $(-\epsilon)$-symmetric form over F and there is defined an isomorphism

$$\begin{pmatrix} 1 & 1 \\ \epsilon\lambda^* & \lambda \end{pmatrix} : (L, -(\lambda - \epsilon\lambda^*), (\epsilon\lambda^*)^{-1}\lambda) \oplus (L, \lambda - \epsilon\lambda^*, 1) \xrightarrow{\simeq} (M, \phi, f) ,$$

so that

$$\begin{aligned}(M, \phi, f) &= (L, -(\lambda - \epsilon\lambda^*), (\epsilon\lambda^*)^{-1}\lambda) \oplus (L, \lambda - \epsilon\lambda^*, 1) \\ &\in LAut^0_{p(z)^\infty}(F, -\epsilon) \oplus LAut^0_{(z-1)^\infty}(F, -\epsilon) \subseteq LAut^0(F, -\epsilon) .\end{aligned}$$

(vi)+(vii) Combine 38.20, 38.25 and work as in (v), but using the localization exact sequence

$$\begin{aligned}0 \longrightarrow L^0(F[z, z^{-1}], \epsilon) &\longrightarrow L^0(F[z, z^{-1}]_{(z-a)}, \epsilon) \\ &\longrightarrow \bigoplus_{p(z) \neq (z-a)} LAut^0_{p(z)^\infty}(F, -\epsilon) \\ &\longrightarrow L^{-1}(F[z, z^{-1}], \epsilon) \longrightarrow 0 .\end{aligned}$$

□

Example 39.28 For a finite field F the residue class fields $F[z, z^{-1}]/(p(z))$ $(p(z) \in \mathcal{M}(F))$ are finite, so that the automorphism L-group of F

$$LAut^0(F, \epsilon) = \bigoplus_{p(z) \in \overline{\mathcal{M}}_z(F)} L^0(F[z, z^{-1}]/(p(z)), -\epsilon)$$

is also an abelian group of exponent 8, and hence so is the asymmetric Witt group

$$LAsy^0(F) = L^0(F) \oplus \bigoplus_{p(z) \neq (z-1) \in \overline{\mathcal{M}}_z(F)} L^0(F[z, z^{-1}]/(p(z)), -1) .$$

□

The multiplicative subset

$$P = \{p(z) \,|\, p(1) \in F^{\bullet}\} \subset F[z,z^{-1}]$$

is involution-invariant, and such that

$$P^{-1}F[z,z^{-1}] = F[z,z^{-1}]_{(z-1)} .$$

A self chain equivalence $f : C \longrightarrow C$ of a finite f.g. free F-module chain complex is fibred (i.e. $1-f : C \longrightarrow C$ is a chain equivalence) if and only if the $P^{-1}F[z,z^{-1}]$-module chain map

$$z - f \, : \, P^{-1}C[z,z^{-1}] \longrightarrow P^{-1}C[z,z^{-1}]$$

is a chain equivalence. As in Chap. 35, define also the multiplicative subset

$$P_{\epsilon} = \{p(z) \,|\, p(\epsilon) \in F^{\bullet}\} \subset F[z,z^{-1}] .$$

An n-dimensional asymmetric Poincaré complex (C,λ) over F is P_{ϵ}-primary if and only if $(1+T_{\epsilon})\lambda : C^{n-*} \longrightarrow C$ is a chain equivalence.

Proposition 39.29 *For any field with involution F there are natural identifications*

$$\begin{aligned} L\mathrm{Aut}^n_{fib}(F,\epsilon) &= L^{n+2}(F[z,z^{-1}],P,\epsilon) \\ &= L\mathrm{Iso}^n(F,\epsilon) = L\mathrm{Iso}^n_{fib}(F,\epsilon) = L\mathrm{Asy}^n_{P_{-\epsilon}}(F) . \end{aligned}$$

Proof Every $P^{-1}F[z,z^{-1}]$-contractible $(n+1)$-dimensional ϵ-symmetric Poincaré complex (E,θ) over $F[z,z^{-1}]$ is homotopy equivalent to the algebraic mapping torus $T(h,\chi)$ of a fibred self homotopy equivalence $(h,\chi) : (C,\phi) \longrightarrow (C,\phi)$ of an n-dimensional ϵ-symmetric Poincaré complex (C,ϕ) over F.

A self homotopy equivalence $(h,\chi) : (C,\phi) \longrightarrow (C,\phi)$ of an n-dimensional ϵ-symmetric Poincaré complex (C,ϕ) over F is fibred if and only if the algebraic mapping torus $T(h,\chi)$ over $F[z,z^{-1}]$ is $P^{-1}F[z,z^{-1}]$-contractible.

The combination of these two statements proves that the algebraic mapping torus defines isomorphisms

$$L\mathrm{Aut}^n_{fib}(F,\epsilon) \xrightarrow{\simeq} L^{n+2}(F[z,z^{-1}],P,\epsilon) \; ; \; (C,\phi,f,\chi) \longrightarrow T(f,\chi) .$$

See 31.13 for the identifications

$$L^{n+2}(F[z,z^{-1}],P,\epsilon) = L\mathrm{Iso}^n(F,\epsilon) .$$

Use the covering construction of 32.7 to define an isomorphism

$$\beta \, : \, L\mathrm{Iso}^n(F,\epsilon) \longrightarrow L^{n+2}(F[z,z^{-1}],P,\epsilon) \; ; \; (C,\widehat{\psi}) \longrightarrow \beta(C,\widehat{\psi}) .$$

(See 39.33 below for an explicit inverse β^{-1}). There are also defined inverse isomorphisms

$$\begin{aligned}
&LAut^n_{fib}(F,\epsilon) \longrightarrow LIso^n(F,\epsilon)\ ;\ (C,\phi,f) \longrightarrow (C,(1-f)^{-1}\phi_0)\ ,\\
&LIso^n(F,\epsilon) \longrightarrow LAut^n_{fib}(F,\epsilon)\ ;\ (C,\widehat{\psi}) \longrightarrow (C,(1+T_\epsilon)\widehat{\psi},-(T_\epsilon\widehat{\psi})\widehat{\psi}^{-1})\ .
\end{aligned}$$

(This is the special case $(A,S)=(F,F[z,z^{-1}]\backslash\{0\})$ of 35.31 (iii)). See 35.27 for the identification

$$LAsy^n_{P_{-\epsilon}}(F) \ =\ LIso^n(F,\epsilon)\ . \qquad \square$$

39E. The automorphism L-theory of F

The L-theory localization exact sequence of Chap. 25 combined with the identifications

$$\begin{aligned}
&\widetilde{K}_0(F[z,z^{-1}]) \ =\ 0\ ,\\
&I_1(F[z,z^{-1}],P) \ =\ I_1(F[z,z^{-1}],S) \ =\ 0\ ,\\
&J_1(F[z,z^{-1}],P) \ =\ K_1(F[z,z^{-1}],P) \ =\ \mathbb{Z}[\mathcal{M}_z(F)\backslash\{(z-1)\}]\ ,\\
&J_1(F[z,z^{-1}],S) \ =\ K_1(F[z,z^{-1}],S) \ =\ \mathbb{Z}[\mathcal{M}_z(F)]\ ,\\
&L^*(F[z,z^{-1}],P,\epsilon) \ =\ L^*(F[z,z^{-1}],P,\epsilon) \ =\ LAut^*_{fib}(F,-\epsilon)\ ,\\
&L^{2*+1}(P^{-1}F[z,z^{-1}],\epsilon) \ =\ L^{2*+1}(F[z,z^{-1}],P,\epsilon) \ =\ 0\ ,\\
&L^*(F[z,z^{-1}],S,\epsilon) \ =\ L^*(F[z,z^{-1}],(z-1)^\infty,\epsilon)\oplus L^*(F[z,z^{-1}],P,\epsilon)\\
&\qquad =\ LAut^*(F,-\epsilon)\\
&\qquad =\ L^*(F,-\epsilon)\oplus LAut^*_{fib}(F,-\epsilon)\ ,\\
&L^{2*+1}(F,\epsilon) \ =\ L^{2*+1}(F(z),\epsilon)\\
&\qquad =\ L^{2*+1}(F[z,z^{-1}],P,\epsilon) \ =\ L^{2*+1}(F[z,z^{-1}],S,\epsilon)\\
&\qquad =\ LAut^{2*+1}(F,-\epsilon) \ =\ LAut^{2*+1}_{fib}(F,-\epsilon) \ =\ 0
\end{aligned}$$

will now be used to express the L-groups $L^*(F(z),\epsilon)$, $LAut^*(F,\epsilon)$ and $LAut^*_{fib}(F,\epsilon)$ in terms of the L-groups of F and the residue class fields $F[z,z^{-1}]/(p(z))$, extending the results already obtained in 39C above.

Proposition 39.30 (i) *For any field with involution F*

$$L^{2k}(F(z),\epsilon) \;=\; L^{2k}(F,\epsilon)\oplus \sum_{p(z)\neq(z-1)\in\overline{\mathcal{M}}_z(F)} L^{2k}(F[z,z^{-1}]/(p(z)),-\epsilon)\ ,$$

$$\begin{aligned} LAut^{2k}_{fib}(F,\epsilon) \;&=\; L^{2k}(F[z,z^{-1}],P,-\epsilon)\\ &=\; \sum_{p(z)\neq(z-1)\in\overline{\mathcal{M}}_z(F)} L^{2k}(F[z,z^{-1}]/(p(z)),\epsilon)\ ,\end{aligned}$$

$$\begin{aligned} &LAut^{2k}(F,\epsilon)\\ &\quad=\; L^{2k}(F,\epsilon)\oplus \sum_{p(z)\neq(z-1)\in\overline{\mathcal{M}}_z(F)} L^{2k}(F[z,z^{-1}]/(p(z))\ ,\epsilon)\ ,\end{aligned}$$

$$\begin{aligned} L^{2k+1}(F(z),\epsilon) \;&=\; L^{2k+1}(F[z,z^{-1}],P,\epsilon)\\ &=\; LAut^{2k-1}(F,\epsilon) \;=\; LAut^{2k-1}_{fib}(F,\epsilon) \;=\; 0\ .\end{aligned}$$

(ii) *The ϵ-symmetric L-groups of $P^{-1}F[z,z^{-1}]$ are given by*

$$\begin{aligned} &L^{2k}(P^{-1}F[z,z^{-1}],\epsilon)\\ &=\; L^0(F,(-)^k\epsilon)\oplus \ker\Big(j_{fib}: LAut^0_{fib}(F,(-)^{k+1}\epsilon)\longrightarrow L^0(F,(-)^{k+1}\epsilon)\Big)\\ &=\; L^0(F,(-)^k\epsilon)\\ &\qquad \oplus\ker\Big(\bigoplus_{p(z)} j_{p(z)}: \sum_{p(z)\neq(z-1)\in\overline{\mathcal{M}}_z(F)} L^0(F[z,z^{-1}]/(p(z)),(-)^{k+1}\epsilon)\\ &\qquad\qquad\qquad\qquad\longrightarrow L^0(F,(-)^{k+1}\epsilon)\Big)\ ,\end{aligned}$$

$$\begin{aligned} &L^{2k+1}(P^{-1}F[z,z^{-1}],\epsilon)\\ &=\; \mathrm{coker}\Big(j_{fib}: LAut^0_{fib}(F,(-)^{k+1}\epsilon)\longrightarrow L^0(F,(-)^{k+1}\epsilon)\Big)\\ &=\; \mathrm{coker}\Big(\bigoplus_{p(z)} j_{p(z)}: \sum_{p(z)\neq(z-1)\in\overline{\mathcal{M}}_z(F)} L^0(F[z,z^{-1}]/(p(z)),(-)^{k+1}\epsilon)\\ &\qquad\qquad\qquad\qquad\longrightarrow L^0(F,(-)^{k+1}\epsilon)\Big)\ ,\end{aligned}$$

with $j_{fib}, j_{p(z)}$ as in 39.20, 39.27.
(iii) *If there exists an element $a\neq 1\in U(F)$ (e.g. if* $\mathrm{char}(F)\neq 2$ *with $a=-1$) then*

$$\begin{aligned}
&L^{2k}(P^{-1}F[z,z^{-1}],\epsilon)\\
&= L^0(F,(-)^k\epsilon)\oplus \sum_{p(z)\neq(z-1),(z-a)\in\overline{\mathcal{M}}_z(F)} L^0(F[z,z^{-1}]/(p(z)),(-)^{k+1}\epsilon)\ ,\\
&L^{2k+1}(P^{-1}F[z,z^{-1}],\epsilon) = 0\ .
\end{aligned}$$

(iv) *If* $\mathrm{char}(F)\neq 2$ *then the involution-invariant multiplicative subsets*

$$\begin{aligned}
(z+1)^\infty &= \{z^j(z+1)^k\,|\,j\in\mathbb{Z},k\geq 0\}\ ,\\
\widehat{P} &= \{p(z)\,|\,p(1),p(-1)\in F^\bullet\}\subset F[z,z^{-1}]
\end{aligned}$$

are coprime, with

$$(z+1)^\infty\widehat{P} = P\ ,$$

so that

$$\begin{aligned}
L^n(F[z,z^{-1}],P,\epsilon) &= L^n(F[z,z^{-1}],(z+1)^\infty,\epsilon)\oplus L^n(F[z,z^{-1}],\widehat{P},\epsilon)\\
&= L^n(F,-\epsilon)\oplus L^n(F[z,z^{-1}],\widehat{P},\epsilon)\ .
\end{aligned}$$

The automorphism L-group $L^n(F[z,z^{-1}],\widehat{P},\epsilon)=\mathrm{LAut}^n_{\widehat{P}}(F,-\epsilon)$ *is the cobordism group of self homotopy equivalences* $(h,\chi):(C,\phi)\longrightarrow(C,\phi)$ *of n-dimensional* $(-\epsilon)$*-symmetric Poincaré complexes* (C,ϕ) *over F such that* $h-h^{-1}:C\longrightarrow C$ *is also a chain equivalence. The unit* $z-z^{-1}\in(\widehat{P}^{-1}F[z,z^{-1}])^\bullet$ *is such that there is defined an isomorphism of exact sequences*

$$\begin{array}{ccccccc}
\cdots\to & L^n(\widehat{P}^{-1}F[z,z^{-1}],\epsilon) & \to & L^n(F(z),\epsilon) & \to & L^n(F[z,z^{-1}],\widehat{P},\epsilon) & \to\cdots\\
& \downarrow{\scriptstyle z-z^{-1}}\ \cong & & \downarrow{\scriptstyle z-z^{-1}}\ \cong & & \downarrow{\scriptstyle z-z^{-1}}\ \cong & \\
\cdots\to & L^n(\widehat{P}^{-1}F[z,z^{-1}],-\epsilon) & \to & L^n(F(z),-\epsilon) & \to & L^n(F[z,z^{-1}],\widehat{P},-\epsilon) & \to\cdots
\end{array}$$

and

$$\begin{aligned}
L^*(F[z,z^{-1}],\widehat{P},\epsilon) &= \sum_{p(z)\neq(z-1),(z+1)\in\overline{\mathcal{M}}_z(F)} L^*(F[z,z^{-1}]/(p(z)),\epsilon)\\
&= \sum_{p(z)\neq(z-1),(z+1)\in\overline{\mathcal{M}}_z(F)} L^*(F[z,z^{-1}]/(p(z)),-\epsilon)\\
&= L^*(F[z,z^{-1}],\widehat{P},-\epsilon)\ ,\\
L^*(F(z),\epsilon) &= L^*(F(z),-\epsilon)\\
&= L^*(F[z,z^{-1}],P,\epsilon)\oplus L^*(F,\epsilon)\\
&= L^*(P^{-1}F[z,z^{-1}],-\epsilon)\oplus L^*(F,\epsilon)\ ,
\end{aligned}$$

$$LAut^*_{fib}(F,\epsilon) = L^*(F[z,z^{-1}],P,-\epsilon) = L^*(P^{-1}F[z,z^{-1}],\epsilon)$$
$$= L^*(F,\epsilon) \oplus \sum_{p(z)\neq(z-1),(z+1)\in\overline{\mathcal{M}}_z(F)} L^*(F[z,z^{-1}]/(p(z)),\epsilon)$$

$$LAut^*(F,\epsilon) = L^*(F,\epsilon) \oplus L^*(F,\epsilon)$$
$$\oplus \sum_{p(z)\neq(z-1),(z+1)\in\overline{\mathcal{M}}_z(F)} L^*(F[z,z^{-1}]/(p(z)),\epsilon)\ .$$

Proof (i) Let $J = \overline{\mathcal{M}}_z(F)\backslash\{(z-1)\}$, and let $S_J \subset F[z,z^{-1}]$ be as in 39.10, so that the inclusion

$$S_J^{-1}F[z,z^{-1}] \longrightarrow P^{-1}F[z,z^{-1}]$$

induces isomorphisms

$$L^*(S_J^{-1}F[z,z^{-1}],\epsilon) \xrightarrow{\simeq} L^*(P^{-1}F[z,z^{-1}],\epsilon)$$

and

$$L^{2*}(F[z,z^{-1}],P,\epsilon) = L^{2*}(F[z,z^{-1}],S_J,\epsilon)$$
$$= \sum_{p(z)\in J} L^{2*}(F[z,z^{-1}]/(p(z)),-\epsilon)\ .$$

Write

$$G = F[z,z^{-1},(z-1)^{-1}]\ ,$$

and substitute

$$L^{2*-1}(F(z),\epsilon) = 0\ ,\ L^{2*+1}(F[z,z^{-1}],P,\epsilon) = 0\ ,$$
$$L^*(F[z,z^{-1}],(z-1)^\infty,\epsilon) = L^*(F,-\epsilon)\ ,$$
$$L^*(G,\epsilon) = L^*(F,\epsilon)$$

in the commutative braid of exact sequences

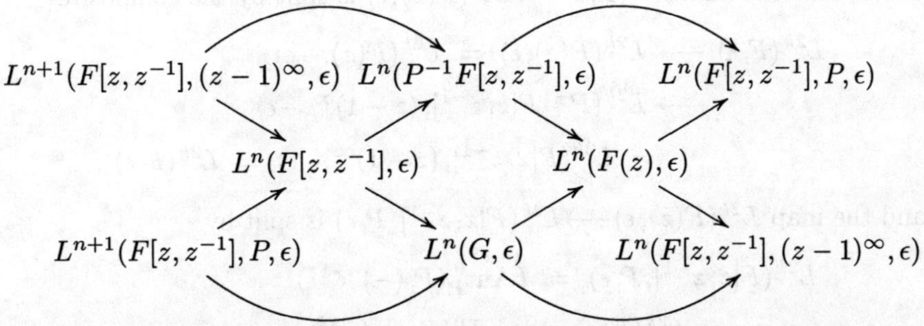

to obtain the expressions for $L^*(P^{-1}F[z,z^{-1}],\epsilon)$, $L^*(F(z),\epsilon)$.
(ii) Immediate from 39.27 and (i).
(iii) Immediate from (ii), on noting that $j_{(z-a)^\infty}$ is an isomorphism.
(iv) As $\mathrm{char}(F) \neq 2$ the connecting map in the exact sequence of (i)

$$\begin{aligned} L^{2k}(F[z,z^{-1}],P,\epsilon) \ &= \ \mathrm{LAut}^{2k}_{fib}(F,-\epsilon) \\ &\longrightarrow L^{2k-1}(F[z,z^{-1}],\epsilon) \ = \ L^{2k}(F,-\epsilon) \ ; \ (L,\theta,f) \longrightarrow (L,\theta) \end{aligned}$$

is a surjection, with a splitting

$$L^{2k}(F,-\epsilon) \longrightarrow \mathrm{LAut}^{2k}_{fib}(F,-\epsilon) \ ; \ (L,\theta) \longrightarrow (L,\theta,-1) \ ,$$

so that

$$L^{2k-1}(P^{-1}F[z,z^{-1}],\epsilon) \ = \ 0 \ .$$

(See 39.31 below for an example with $\mathrm{char}(F)=2$ where this group is non-zero.) The exact sequence

$$0 \longrightarrow L^{2k}(F,\epsilon) \longrightarrow L^{2k}(P^{-1}F[z,z^{-1}],\epsilon) \longrightarrow L^{2k}(F[z,z^{-1}],\widehat{P},\epsilon) \longrightarrow 0$$

splits: the map $L^{2k}(F,\epsilon)\longrightarrow L^{2k}(P^{-1}F[z,z^{-1}],\epsilon)$ is split by the composite

$$\begin{aligned} L^{2k}(P^{-1}F[z,z^{-1}],\epsilon) &\longrightarrow L^{2k}(F(z),\epsilon) \ = \ L^{2k}(F(z),-\epsilon) \\ &\longrightarrow L^{2k}(P^{-1}F[z,z^{-1}],(z-1)^{\infty},-\epsilon) \\ &= \ L^{2k}(F[z,z^{-1}],(z-1)^{\infty},-\epsilon) \ = \ L^{2k}(F,\epsilon) \ , \end{aligned}$$

and the map $L^{2k}(P^{-1}F[z,z^{-1}],\epsilon)\longrightarrow L^{2k}(F[z,z^{-1}],\widehat{P},\epsilon)$ is split by

$$\begin{aligned} L^{2k}(F[z,z^{-1}],\widehat{P},\epsilon) \ &= \ \mathrm{LAut}^{0}_{\widehat{P}}(F,(-)^{n+1}\epsilon) \\ &\longrightarrow L^{2k}(P^{-1}F[z,z^{-1}],\epsilon) \ = \ L^{0}(P^{-1}F[z,z^{-1}],(-)^{k}\epsilon) \ ; \\ (M,\phi,f) &\longrightarrow (M(z),(1-z)\phi(f-1)^{-1}(1-z^{-1}f)) \ . \end{aligned}$$

The exact sequence given by (i)

$$0 \longrightarrow L^{2k}(F,\epsilon) \longrightarrow L^{2k}(F(z),\epsilon) \longrightarrow L^{2k}(F[z,z^{-1}],\widehat{P},\epsilon) \longrightarrow 0$$

also splits: the map $L^{2k}(F,\epsilon)\longrightarrow L^{2k}(F(z),\epsilon)$ is split by the composite

$$\begin{aligned} L^{2k}(F,\epsilon) &\longrightarrow L^{2k}(F(z),\epsilon) \ = \ L^{2k}(F(z),-\epsilon) \\ &\longrightarrow L^{2k}(P^{-1}F[z,z^{-1}],(z-1)^{\infty},-\epsilon) \\ &= \ L^{2k}(F[z,z^{-1}],(z-1)^{\infty},-\epsilon) \ = \ L^{2k}(F,\epsilon) \ , \end{aligned}$$

and the map $L^{2k}(F(z),\epsilon)\longrightarrow L^{2k}(F[z,z^{-1}],P,\epsilon)$ is split by

$$\begin{aligned} L^{2k}(F[z,z^{-1}],P,\epsilon) \ &= \ \mathrm{LAut}^{0}_{P}(F,(-)^{n+1}\epsilon) \\ &\longrightarrow L^{2k}(F(z),\epsilon) \ = \ L^{0}(F(z),(-)^{k}\epsilon) \ ; \\ (M,\phi,f) &\longrightarrow (M(z),(1-z)\phi(1-f^{-1})(1-z^{-1}f)) \ . \end{aligned}$$ □

Example 39.31 If F is an algebraically closed field of characteristic 2 with the identity involution then

$$\begin{aligned}
\overline{\mathcal{M}}_z(F) &= \{(z-1)\}\ ,\\
L^0(F) &= L^0(P^{-1}F[z,z^{-1}]) \;=\; L^0(F(z)) \;=\; L\mathrm{Asy}^0(F)\\
&= L\mathrm{Aut}^0(F) \;=\; L^1(P^{-1}F[z,z^{-1}]) \;=\; \mathbb{Z}_2\ ,\\
L^0(F[z,z^{-1}],P) &= L\mathrm{Aut}^0_{fib}(F) \;=\; 0\ . \qquad \square
\end{aligned}$$

Remark 39.32 Let F have the identity involution. The fixed subfield of the non-trivial involution $\overline{z}=z^{-1}$ on $F(z)$ is

$$F(z)_0 \;=\; F(z+z^{-1})\ ,$$

and there is defined an isomorphism of fields

$$F(x) \xrightarrow{\;\simeq\;} F(z)_0\ ;\ x \longrightarrow z+z^{-1}\ .$$

The computations of $L^*(F(x))$, $L^*(F(z))$ given by 39.5, 39.30 in the case $\mathrm{char}(F)\neq 2$ are related by the exact sequence of 37.3

$$\begin{aligned}
0 \longrightarrow L^0(F(z)) \longrightarrow L^0(F(x)) \longrightarrow L^0(F(x))&\\
\longrightarrow L^0(F(x)) \longrightarrow L^0(F(z)) \longrightarrow 0\ .& \qquad \square
\end{aligned}$$

Proposition 39.33 (i) *Every free Blanchfield complex over $F[z,z^{-1}]$ fibres.*
(ii) *The cobordism group $L^{n+1}(F[z,z^{-1}],P,\epsilon)$ of free n-dimensional ϵ-symmetric Blanchfield complexes is isomorphic to the cobordism group of P-primary homotopy equivalences $(h,\chi):(C,\phi)\longrightarrow(C,\phi)$ of f.g. free $(n-2)$-dimensional ϵ-symmetric Poincaré complexes (C,ϕ) over F.*
(iii) *The L-groups of $(F[z,z^{-1}],P)$ are such that*

$$L^{n+1}(F[z,z^{-1}],P) \;=\; \sum_{p(z)\neq(z-1)\in\overline{\mathcal{M}}_z(F)} L^{n+1}(F[z,z^{-1}]/(p(z)),-1)\ ,$$

$$(=0 \text{ for } n \text{ even})\ .$$

(iv) *The inverse of the covering isomorphism of 32.11*

$$\beta\ :\ L\mathrm{Iso}^n(F) \cong L^{n+2}(F[z,z^{-1}],P)$$

is given by

$$\beta^{-1}\ :\ L^{n+2}(F[z,z^{-1}],P) \xrightarrow{\;\simeq\;} L\mathrm{Iso}^n(F)\ ;\ (E,\theta) \longrightarrow (i^!E,\widehat{\psi})\ .$$

(v) *Every Seifert complex $(C,\widehat{\psi})$ over F is cobordant to a fibred Seifert complex.*
(vi) *The isometric L-groups of F are such that*

$$\begin{aligned}
LIso^{2*}(F) &= L_{2*+2}(F[z,z^{-1}],P) = L^{2*+2}(F[z,z^{-1}],P) \\
&= \sum_{p(z)\neq(z-1)\in\overline{\mathcal{M}}_z(F)} L^{2*}(F[z,z^{-1}]/(p(z)),-1)\ , \\
LIso^{2*+1}(F) &= 0\ .
\end{aligned}$$

The cobordism class of a 2i-dimensional Seifert complex $(C,\widehat{\psi})$ *over* F *is the Witt class of the nonsingular* $(-)^i$*-symmetric linking form* (L,λ) *over* $(F[z,z^{-1}],P)$ *defined by*

$$\begin{aligned}
L &= \mathrm{coker}(\widehat{\psi}+(-)^i z\widehat{\psi}^* : H^i(C)[z,z^{-1}]\longrightarrow H_i(C)[z,z^{-1}])\ , \\
\lambda(x,y) &= (1-z^{-1})(\widehat{\psi}+(-)^i z\widehat{\psi}^*)^{-1}(x)(y) \in P^{-1}F[z,z^{-1}]/F[z,z^{-1}]\ ,
\end{aligned}$$

with one component for each irreducible polynomial $p_j(z)\in F[z,z^{-1}]$ *appearing in a factorization of the Alexander polynomial*

$$\begin{aligned}
\Delta(z) &= \det\big((\widehat{\psi}+(-)^i\widehat{\psi}^*)^{-1}(\widehat{\psi}+(-)^i z\widehat{\psi}^*) : \\
&\qquad\qquad H^i(C)[z,z^{-1}]\longrightarrow H^i(C)[z,z^{-1}]\big) \\
&= \Big(\prod_{j=1}^{m} p_j(z)^{a_j}\Big) q(z)\overline{q(z)} \in F[z,z^{-1}]
\end{aligned}$$

for some coprime polynomials $p_j(z),q(z)\in F[z,z^{-1}]$ *such that*

$$\overline{(p_j(z))} = (p_j(z))\ ,\quad (q(z)) \neq \overline{(q(z))} \lhd F[z,z^{-1}]$$

with $p_j(z)$ *irreducible. If* $(C,\widehat{\psi})$ *is fibred then* $L=H^i(C)$*, and* (L,λ) *corresponds to the monodromy automorphism of the nonsingular* $(-)^i$*-symmetric form* $(L,\widehat{\psi}+(-)^i\widehat{\psi}^*)$ *over* F

$$h = (-)^{i+1}(\widehat{\psi}^*)^{-1}\widehat{\psi} : (L,\widehat{\psi}+(-)^i\widehat{\psi}^*) \xrightarrow{\simeq} (L,\widehat{\psi}+(-)^i\widehat{\psi}^*)\ .$$

Proof (i)+(ii)+(iii) Immediate from 31.24 and 39.30.
(iv) Every free $(n+1)$-dimensional Blanchfield complex (E,θ) over $F[z,z^{-1}]$ fibres, and is homotopy equivalent to the covering $\beta(i^!E,\widehat{\psi})$ of the fibred n-dimensional Seifert complex $(i^!E,\widehat{\psi})$ over F.
(v) Immediate from (ii).
(vi) Identify $LIso^*(F)=L_{*+2}(F[z,z^{-1}],P)$, and apply 39.30. □

Example 39.34 (i) For an algebraically closed field F 39.15 gives

$$LIso^{2*}(F) = \sum_{a\in U(F)\backslash\{1\}} L^{2*}(F)\ ,$$

(ii) For $F=\mathbb{C}^-$

$$\begin{aligned} LIso^{2*}(\mathbb{C}^-) &= \bigoplus_{(z-e^{i\theta})\neq(z-1)\in\overline{\mathcal{M}}_z(\mathbb{C}^-)} L^{2*}(\mathbb{C}^-[z,z^{-1}]/(z-e^{i\theta}),-1) \\ &= \sum_{0<\theta\leq\pi} L^{2*}(\mathbb{C}^-[z,z^{-1}]/(z-e^{i\theta}),-1) = \sum_{0<\theta\leq\pi} \mathbb{Z} . \qquad \square \end{aligned}$$

For any field F the multiplicative subset

$$S = F[z,z^{-1}]\backslash\{0\} \subset F[z,z^{-1}]$$

is such that the category $\mathbb{H}(F[z,z^{-1}],S)$ of $(F[z,z^{-1}],S)$-modules is isomorphic to the category $\mathrm{Aut}(F)$ of finite dimensional F-vector spaces with an automorphism. For a field with involution F the category of nonsingular ϵ-symmetric linking forms over $(F[z,z^{-1}],S)$ is isomorphic to the category of nonsingular ϵ-dimensional symmetric forms over F with an autometric structure (i.e. an automorphism). A $(2i+1)$-dimensional symmetric Poincaré complex (C,ϕ) over the Laurent polynomial extension $F[z,z^{-1}]$ determines by 38.1 (iii) a nonsingular $(-)^{i+1}$-symmetric linking form (T,λ) over $(F[z,z^{-1}],S)$, which corresponds to the automorphism

$$\zeta : (T,\phi_0) \longrightarrow (T,\phi_0)$$

of the nonsingular $(-)^i$-symmetric form (T,ϕ_0) over F with

$$\begin{aligned} T = T_i(C) &= F[z,z^{-1}]\text{-torsion}\, H_i(C) \\ &= \ker(H^{i+1}(C)\longrightarrow F(z)\otimes_{F[z,z^{-1}]} H_i(C)39, \\ \lambda = \phi_0 : T &\xrightarrow{\simeq} T\hat{~} = \mathrm{Hom}_{F[z,z^{-1}]}(T,F(z)/F[z,z^{-1}]) = T^* \\ &= \mathrm{Hom}_F(T_i(C),F) . \end{aligned}$$

Definition 39.35 The *monodromy* of a $(2i+1)$-dimensional symmetric Poincaré complex (C,ϕ) over $F[z,z^{-1}]$ is the nonsingular $(-)^i$-symmetric form over F with autometric structure

$$\mathcal{H}(C,\phi) = (T_i(C),\phi_0,\zeta) . \qquad \square$$

Remark 39.36 (i) If $(C,\phi) = T(f,\chi)$ is the algebraic mapping torus of a homotopy equivalence $(f,\chi) : (D,\theta)\longrightarrow(D,\theta)$ of a $2i$-dimensional symmetric Poincaré complex (D,θ) over F the monodromy is given by

$$\mathcal{H}(C,\phi) = (H^i(D),\theta_0,f) .$$

More generally, if (C,ϕ) is a $(2i+1)$-dimensional symmetric Poincaré band over $F[z,z^{-1}]$ (3.2) then

$$\mathcal{H}(C,\phi) = (H^i(C^!),\phi_0^!,\zeta) ,$$

with $(C^!,\phi_0^!)$ the $2i$-dimensional symmetric Poincaré complex over F defined by the restriction of the $F[z,z^{-1}]$-action to $F\subset F[z,z^{-1}]$.

(ii) Let X be a $(2i+1)$-dimensional geometric Poincaré complex with an infinite cyclic cover $\overline{X}$, so that there is defined a symmetric $(2i+1)$-dimensional symmetric Poincaré complex $\sigma^*(X;F) = (C(\overline{X}),\phi)$ over $F[z,z^{-1}]$. The F-coefficient monodromy of X in the sense of Neumann [214] is just the monodromy of $\sigma^*(X;F)$

$$\mathcal{H}(X;F) \;=\; \mathcal{H}(\sigma^*(X;F)) \;.$$

(iii) Working as in [214, Chaps. 2, 11] it is possible to express the monodromy $\mathcal{H}(C,\phi)$ in terms of the locally finite homology of C, as follows. The Laurent polynomial extension $F[z,z^{-1}]$ fits into an exact sequence

$$0 \longrightarrow F[z,z^{-1}] \longrightarrow F((z)) \oplus F((z^{-1})) \longrightarrow F[[z,z^{-1}]] \longrightarrow 0$$

with $F((z))$, $F((z^{-1}))$ the Novikov completions (6.4) and $F[[z,z^{-1}]]$ the $F[z,z^{-1}]$-module of formal Laurent series $\sum\limits_{j=-\infty}^{\infty} a_j z^j$, without any restrictions on the coefficients $a_j \in F$. Given a finite f.g. free $F[z,z^{-1}]$-module chain complex C define the *locally finite* $F[z,z^{-1}]$-module chain complex

$$C^{lf} \;=\; F[[z,z^{-1}]] \otimes F[z,z^{-1}]C \;,$$

and define also

$$C^{((lf))} \;=\; F((z)) \otimes_{F[z,z^{-1}]} C \;,$$
$$C^{((lf^{-1}))} \;=\; F((z^{-1})) \otimes_{F[z,z^{-1}]} C \;.$$

The short exact sequence of $F[z,z^{-1}]$-module chain complexes

$$0 \longrightarrow C \longrightarrow C^{((lf))} \oplus C^{((lf^{-1}))} \longrightarrow C^{lf} \longrightarrow 0$$

induces an exact sequence in homology

$$\dots \longrightarrow H_{r+1}(C^{lf}) \xrightarrow{\partial} H_r(C)$$
$$\longrightarrow H_r(C^{((lf))}) \oplus H_r(C^{((lf^{-1}))}) \longrightarrow H_r(C^{lf}) \longrightarrow \dots \;.$$

The torsion $F[z,z^{-1}]$-module of $H_r(C)$ is

$$\begin{aligned} T_r(C) &= \{x \in H_r(C) \,|\, p(z)(x) = 0 \text{ for some } p(z) \in S\} \\ &= \mathrm{im}(\partial : H_{r+1}(C^{lf}) \longrightarrow H_r(C)) \\ &= \ker(H_r(C) \longrightarrow H_r(C^{((lf))}) \oplus H_r(C^{((lf^{-1}))})) \;. \end{aligned}$$

It follows from 8.3 that the following conditions on C are equivalent :

(a) $\dim_F(H_*(C)) < \infty$,
(b) C is F-finitely dominated,
(c) $H_*(C^{((lf))}) = H_*(C^{((lf^{-1}))}) = 0$,
(d) $\partial : H_*(C^{lf}) \cong H_{*-1}(C)$,

(e) $T_*(C) = H_*(C)$,
(f) $\Delta(z)H_*(C) = 0$ for some $\Delta(z) \neq 0 \in F[z,z^{-1}]$.

(iv) If $(C,\widehat{\psi})$ is a $2i$-dimensional fibred symmetric Seifert complex over a field F then the covering $\beta(C,\widehat{\psi}) = (E,\theta)$ (31.31) is a $(2i+1)$-dimensional fibred symmetric Blanchfield complex over $F[z,z^{-1}]$ with

$$H_{*+1}(E^{lf}) \;=\; H_*(E) \;=\; H_*(C)\ ,$$

and monodromy

$$\mathcal{H}\beta(C,\psi) \;=\; (H_i(C),(1+T)\widehat{\psi},-(T\widehat{\psi})\widehat{\psi}^{-1})\ .$$

(v) Let W be an open $(2i+1)$-dimensional manifold with a proper map $p: W \longrightarrow \mathbb{R}$ which is transverse regular at $0 \in \mathbb{R}$, and let $j: M \longrightarrow W$ be the inclusion of the closed codimension 1 submanifold $M = p^{-1}(0) \subset W$. For any coefficient field F the subspace

$$\begin{aligned} T \;=\; T_i(W;F) \;&= \mathrm{im}(j_*[M]\cap - : H^{lf}_{i+1}(W;F) = H^i(W;F) \longrightarrow H_i(W;F)) \\ &\subseteq \mathrm{im}(j_* : H_i(M;F) \longrightarrow H_i(W;F)) \end{aligned}$$

is finite dimensional. The $(-)^i$-symmetric pairing

$$\begin{aligned} &H^i(W;F) \times H^i(W;F) \longrightarrow F\ ; \\ &\qquad (x,y) \longrightarrow \langle x \cup y, j_*[M]\rangle \;=\; \langle j^*x \cup j^*y,[M]\rangle \end{aligned}$$

(on the potentially infinite dimensional F-vector space $H^i(W;F)$) determines a nonsingular $(-)^i$-symmetric form (T,ϕ_0) over F such that the Witt class

$$(T,\phi_0) \;=\; \sigma^*(M;F) \in L^{2i}(F) \;=\; L^0(F,(-)^i)$$

is a proper homotopy invariant of W. Novikov [216] used this proper homotopy invariance in the case $i = 2k$, $F = \mathbb{R}$ to prove that the kth component of the $\mathcal{L}$-genus $\mathcal{L}_k(M) \in H^{4k}(M;\mathbb{Q})$ is a homotopy invariant for closed $(4k+1)$-dimensional manifolds M^{4k+1}. The proof of the topological invariance of the rational Pontrjagin classes in Novikov [217] used a different (albeit related) method. Gromov [100] obtained a new proof of topological invariance, using a twisted-coefficient version of the invariant of [216]. See Ranicki [247] for the relationship between the method of [100], the original proof in [217], and the lower L-theory method of Ranicki [244], [245]. If $W = \overline{X}$ is the infinite cyclic cover of a closed $(2i+1)$-dimensional manifold X with $\pi_1(X) = \pi \times \mathbb{Z}$ (as in (ii)) then (T,ϕ_0) is the pairing used to define the monodromy $\mathcal{H}(X;F) = (T,\phi_0,\zeta)$, with $C = C(\overline{X};F)$ such that $H_*(C^{lf}) = H^{lf}_*(\overline{X};F)$. □

40. The multisignature

A multisignature is a collection of integer-valued signature invariants. For example, the torsion-free part of the symmetric Witt group $L^0(F)$ of an algebraic number field F is detected by the multisignature consisting of the signatures associated to involution-preserving embeddings $F \longrightarrow \mathbb{C}$ (37.2). Similarly, the torsion-free parts of the quadratic L-groups $L_{2*}(\mathbb{Z}[\pi])$ for a finite group π are detected by the signatures associated to irreducible real representations of π (cf. 40.26 below). In these examples the multisignatures have a finite number of components.

Multisignatures can also be used to detect the torsion-free parts of the high-dimensional knot cobordism groups $C_{2*-1} = L\mathrm{Iso}^{2*}(\mathbb{Z})$, the asymmetric L-groups $L\mathrm{Asy}^{2*}(\mathbb{Z})$ and the automorphism L-groups $L\mathrm{Aut}^{2*}(\mathbb{Z})$. These multisignatures have an infinite number of components, essentially one for each complex conjugate pair of algebraic integers on the line $\mathrm{Re} = 1/2$.

The multisignature of a form over $\mathbb{Z}$ is the multisignature of the induced form over $\mathbb{C}$. The multisignatures can also be defined for other rings with involution (e.g. $\mathbb{Q}$, $\mathbb{R}$, algebraic number fields), but in each case they are the restrictions of the complex multisignatures. In this chapter only multisignatures over $\mathbb{C}$ will be considered.

Symmetric forms over $\mathbb{C}$ with the complex conjugation involution are traditionally called hermitian forms, although they will be called symmetric forms here. The even-dimensional symmetric L-groups of $\mathbb{C}$ are detected by the signature isomorphisms

$$\sigma \ : \ L^{2j}(\mathbb{C}) \xrightarrow{\simeq} \mathbb{Z} \, .$$

The even-dimensional isometric Witt groups of $\mathbb{C}$ are detected by the *isometric multisignature* isomorphism

$$\sigma_*^{Iso} \ : \ L\mathrm{Iso}^{2j}(\mathbb{C}) \xrightarrow{\simeq} \mathbb{Z}[R_{1/2}]$$

defined in 40A below, where $R_{1/2} \subset \mathbb{C}$ is the line $\mathrm{Re} = 1/2$. More precisely, the Witt class

$$(L, \lambda) \in L\mathrm{Iso}^{2j}(\mathbb{C}) \ = \ L\mathrm{Iso}^0(\mathbb{C}, (-)^j)$$

of a $(-)^j$-symmetric Seifert form (L,λ) over $\mathbb{C}$ is determined by the formal linear combination of signatures

$$\sigma_*^{Iso}(L,\lambda) \;=\; \sum_{a\in R_{1/2}} \sigma(L_a,\lambda_a+(-)^j\lambda_a^*).a \in \mathbb{Z}[R_{1/2}]$$

with (L_a,λ_a) the restriction of (L,λ) to the generalized a-eigenspace of the endomorphism

$$f \;=\; (\lambda+(-)^j\lambda^*)^{-1}\lambda \;:\; L \longrightarrow L\ ,$$

that is

$$L_a \;=\; \{x\in L\,|\,(f-a)^N(x)=0\in L \text{ for some } N\geq 1\}\ .$$

Similarly, the even-dimensional asymmetric Witt groups of $\mathbb{C}$ are detected by the *asymmetric multisignature* isomorphism

$$\sigma_*^{Asy} \;:\; LAsy^{2j}(\mathbb{C}) \xrightarrow{\;\simeq\;} \mathbb{Z}[S^1]$$

defined in 40B below, and the even-dimensional automorphism Witt groups of $\mathbb{C}$ are detected by the *automorphism multisignature* isomorphism

$$\sigma_*^{Aut} \;:\; LAut^{2j}(\mathbb{C}) \xrightarrow{\;\simeq\;} \mathbb{Z}[S^1]$$

defined in 40D below. In fact, the isometric multisignature is the restriction of the automorphism multisignature to the fibred automorphism L-groups

$$\sigma_*^{Iso} \;=\; \sigma_*^{Aut}| \;:\; LIso^{2*}(\mathbb{C}) \;=\; LAut_{fib}^{2*}(\mathbb{C}) \xrightarrow{\;\simeq\;} \mathbb{Z}[R_{1/2}] \;\cong\; \mathbb{Z}[S^1\backslash\{1\}]\ .$$

The asymmetric and automorphism multisignatures fit together with the signature isomorphisms $\sigma : L^{2j}(\mathbb{C}) \cong \mathbb{Z}$ to define an isomorphism of exact sequences

$$\begin{array}{ccccccccc}
0 \longrightarrow & L^{2j+2}(\mathbb{C}) & \longrightarrow & LAsy^{2j+2}(\mathbb{C}) & \longrightarrow & LAut^{2j}(\mathbb{C}) & \longrightarrow & L^{2j}(\mathbb{C}) & \longrightarrow 0 \\
 & \sigma\downarrow\cong & & \sigma_*^{Asy}\downarrow\cong & & \sigma_*^{Aut}\downarrow\cong & & \sigma\downarrow\cong & \\
0 \longrightarrow & \mathbb{Z} & \xrightarrow{\;1\;} & \mathbb{Z}[S^1] & \xrightarrow{\;\partial\;} & \mathbb{Z}[S^1] & \xrightarrow{\;e\;} & \mathbb{Z} & \longrightarrow 0
\end{array}$$

with

$$\partial(a) \;=\; a-1\ ,\quad e(a) \;=\; 1\ \ (a\in S^1)\ .$$

The general results of Chaps. 39–41 will be used in Chap. 42 to show that the morphisms induced by $\mathbb{Z}\subset\mathbb{C}$

$$C_{2j-1} \;=\; LIso^{2j}(\mathbb{Z}) \longrightarrow LIso^{2j}(\mathbb{C}) \;=\; \mathbb{Z}[R_{1/2}]$$

are such that

$$C_{2j-1} \longrightarrow \mathrm{im}(C_{2j-1}) \;=\mathbb{Z}^\infty$$

has 4-torsion kernel.

Definition 40.1 (i) The *signature* of a $2j$-dimensional symmetric complex (C,ϕ) over $\mathbb{C}$ is the signature of the $(-)^j$-symmetric form $(H^j(C),\phi_0)$ over $\mathbb{C}$

$$\sigma(C,\phi) \;=\; \sigma(H^j(C),\phi_0) \in \mathbb{Z}\ .$$

(ii) The *signature* of a $2j$-dimensional symmetric Poincaré pair over $\mathbb{C}$

$$(D,C) \;=\; (f:C\longrightarrow D,(\phi_D,\phi_C))$$

is the signature of the quotient $2j$-dimensional symmetric complex over $\mathbb{C}$ $(D/C,\phi_D/\phi_C)$

$$\sigma(D,C) \;=\; \sigma(H^j(D/C),(\phi_D/\phi_C)_0) \in \mathbb{Z}\ . \qquad \square$$

The $2j$-dimensional symmetric L-groups of $\mathbb{C}$ are given by

$$L^{2j}(\mathbb{C}) \;=\; L^0(\mathbb{C},(-)^j) = \mathbb{Z}\ .$$

By 37.1 (x) the signature defines isomorphisms

$$\sigma\ :\ L^{2j}(\mathbb{C}) \xrightarrow{\;\cong\;} \mathbb{Z}\ ;\ (C,\phi) \longrightarrow \sigma(C,\phi)$$

with generator $(S^j\mathbb{C},1) \in L^{2j}(\mathbb{C})$ for j even, and $(S^j\mathbb{C},-i) \in L^{2j}(\mathbb{C})$ for j odd.

40A. Isometric multisignature

Given an ϵ-symmetric Seifert form (L,λ) over $\mathbb{C}$ let $\mathbb{C}[s]$ act on L by the isometric structure

$$f \;=\; (\lambda+\epsilon\lambda^*)^{-1}\lambda\ :\ L\longrightarrow L$$

for the nonsingular ϵ-symmetric form $(L,\lambda+\epsilon\lambda^*)$ over $\mathbb{C}$. The Seifert form decomposes into the $(s-a)$-primary components given by 39.15

$$(L,\lambda) \;=\; \sum_{a\in R_{1/2}} (L_a,\lambda_a) \oplus \text{hyperbolic}$$

with

$$R_{1/2} \;=\; \{a\in\mathbb{C}\,|\,\mathrm{Re}(a)=1/2\}\ ,$$

$$L_a \;=\; \bigcup_{N=0}^{\infty} \ker((f-a)^N : L\longrightarrow L)\ ,$$

$$\mathrm{ch}_s(L_a,f_a) \;=\; (s-a)^{d_a}\quad (d_a \;=\; \dim_{\mathbb{C}}(L_a))\ .$$

Definition 40.2 (i) The *isometric multisignature* of an ϵ-symmetric Seifert form (L,λ) over $\mathbb{C}$ is the formal linear combination

$$\sigma_*^{Iso}(L,\lambda) = \sum_{a\in R_{1/2}} \sigma_a^{Iso}(L,\lambda).a \in \mathbb{Z}[R_{1/2}]$$

with

$$\sigma_a^{Iso}(L,\lambda) = \sigma(L_a,\lambda_a+\epsilon\lambda_a^*) \in \mathbb{Z}\ .$$

(ii) The *isometric multisignature* of a $2j$-dimensional Seifert complex (C,λ) over $\mathbb{C}$ is the isometric multisignature of the $(-)^j$-symmetric Seifert form $(H^j(C),\lambda)$

$$\sigma_*^{Iso}(C,\lambda) = \sigma_*^{Iso}(H^j(C),\lambda) \in \mathbb{Z}[R_{1/2}]\ .$$

(iii) The *multisignature* of a $(2j-1)$-knot $k: S^{2j-1}\subset S^{2j+1}$ is the isometric multisignature

$$\sigma_*(k) = \sigma_*^{Iso}(S^jH_j(\overline{X};\mathbb{C}),\lambda) \in \mathbb{Z}[R_{1/2}]$$

of the $2j$-dimensional Seifert complex $(S^jH_j(\overline{X};\mathbb{C}),\lambda)$ over $\mathbb{C}$, with $\overline{X}$ the canonical infinite cyclic cover of the knot exterior X and

$$\lambda : H_j(\overline{X};\mathbb{C})^* \cong H^j(\overline{X};\mathbb{C}) \xrightarrow[\simeq]{[\overline{X}]\cap -} H_j(\overline{X};\mathbb{C}) \xrightarrow[\simeq]{(1-\zeta)^{-1}} H_j(\overline{X};\mathbb{C})\ .$$ □

Proposition 40.3 *The isometric multisignature defines isomorphisms*

$$\sigma_*^{Iso} : L\mathrm{Iso}^{2j}(\mathbb{C}) \xrightarrow{\simeq} \mathbb{Z}[R_{1/2}]$$

inverse to

$$\mathbb{Z}[R_{1/2}] \xrightarrow{\simeq} L\mathrm{Iso}^{2j}(\mathbb{C})\ ;\ a \longrightarrow \begin{cases} (S^j\mathbb{C},a) & \text{if } j\equiv 0(\bmod 2) \\ (S^j\mathbb{C},ia) & \text{if } j\equiv 1(\bmod 2)\ . \end{cases}$$

Proof This is a special case of 39.15. □

Remark 40.4 (i) The multisignature of a $(2j-1)$-knot $k: S^{2j-1}\subset S^{2j+1}$ was first defined by Milnor [194]. The finite-dimensionality of $H^j(\overline{X};\mathbb{C})$ (as a complex vector space) and the Poincaré duality

$$[\overline{X}]\cap - : H^j(\overline{X};\mathbb{C}) \cong H_j(\overline{X};\mathbb{C})$$

were shown to follow from $H_*(X;\mathbb{C})\cong H_*(S^1;\mathbb{C})$. See 39.36 (iii) for a general discussion of such finite-dimensionality results.
(ii) The multisignature of knots defines morphisms

$$i_j : C_{2j-1} = L\mathrm{Iso}^{2j}(\mathbb{Z}) \longrightarrow L\mathrm{Iso}^{2j}(\mathbb{C}) = \mathbb{Z}[R_{1/2}]$$

which were shown by Milnor [195] and Levine [158] to map $C_{2j-1}=L\mathrm{Iso}^{2j}(\mathbb{Z})$ onto a countably infinitely generated subgroup $\mathrm{im}(i_j)=\mathbb{Z}^\infty\subset\mathbb{Z}[R_{1/2}]$ with countably infinite 2- and 4-torsion kernel. See Chap. 42 for a more detailed account. □

40B. Asymmetric multisignature

Given a nonsingular asymmetric form (L,λ) over $\mathbb{C}$ let $\mathbb{C}[z,z^{-1}]$ act on L by the automorphism

$$h \;=\; \lambda^{-1}\lambda^* \;:\; L \longrightarrow L \;.$$

The asymmetric form decomposes into the $(z-a)$-primary components given by 39.26

$$(L,\lambda) \;=\; \bigoplus_{a\in S^1}(L_a,\lambda_a)\oplus \text{hyperbolic} \;,$$

with

$$L_a \;=\; \{x\in L\,|\,(h-a)^N(x)=0 \text{ for some } N\geq 1\} \;.$$

Definition 40.5 (i) The *asymmetric multisignature* of a nonsingular asymmetric form (L,λ) over $\mathbb{C}$ is the formal linear combination

$$\sigma^{Asy}_*(L,\lambda) \;=\; \sum_{a\in S^1}\sigma^{Asy}_a(L,\lambda).a\in\mathbb{Z}[S^1]$$

with

$$\sigma^{Asy}_a(L,\lambda) \;=\; \begin{cases}\sigma\Big(L_a,\dfrac{\lambda_a+\lambda_a^*}{2}\Big) & \text{if } a\neq -1\\[2ex] \sigma\Big(L_a,\dfrac{\lambda_a-\lambda_a^*}{2i}\Big) & \text{if } a=-1\;.\end{cases}$$

(ii) The *asymmetric multisignature* of a $2j$-dimensional asymmetric Poincaré complex (C,λ) over $\mathbb{C}$ is

$$\begin{aligned}\sigma^{Asy}_*(C,\lambda) \;&=\; \sum_{a\in S^1}\sigma^{Asy}_a(C,\lambda).a\\ &=\; \sum_{a\in S^1}\sigma^{Asy}_{(-)^ja}(H^j(C),\lambda).a\in\mathbb{Z}[S^1]\end{aligned}$$

with

$$\sigma^{Asy}_a(C,\lambda) \;=\; \sigma^{Asy}_{(-)^ja}(H^j(C),\lambda)\in\mathbb{Z}\;.$$

(Note that (i) is just the case $j=0$). □

Proposition 40.6 *The asymmetric multisignature defines isomorphisms*

$$\sigma^{Asy}_* \;:\; LAsy^{2j}(\mathbb{C}) \xrightarrow{\;\simeq\;} \mathbb{Z}[S^1]\;;\;(C,\lambda)\longrightarrow\sigma^{Asy}_*(C,\lambda)$$

inverse to

$$\mathbb{Z}[S^1] \xrightarrow{\;\simeq\;} LAsy^{2j}(\mathbb{C})\;;$$

$$b\longrightarrow\begin{cases}(S^j\mathbb{C},i(1-b)^{-1}) & \textit{if } b\neq 1 \textit{ and } j\equiv 0(\mathrm{mod}\,2)\\ (S^j\mathbb{C},1) & \textit{if } b=1 \textit{ and } j\equiv 0(\mathrm{mod}\,2)\\ (S^j\mathbb{C},(1-b)^{-1}) & \textit{if } b\neq 1 \textit{ and } j\equiv 1(\mathrm{mod}\,2)\\ (S^j\mathbb{C},i) & \textit{if } b=1 \textit{ and } j\equiv 1(\mathrm{mod}\,2)\;.\end{cases}$$

Proof By 39.27 the asymmetric L-group splits as a direct sum

$$\begin{aligned} LAsy^{2j}(\mathbb{C}) &= \sum_{a\in S^1} LAsy^{2j}_{(z-a)^\infty}(\mathbb{C}) \\ &= \sum_{a\in S^1} LAsy^{0}_{(z-(-)^j a)^\infty}(\mathbb{C}) \end{aligned}$$

according to the decomposition of nonsingular asymmetric Poincaré complexes over $\mathbb{C}$ into $(z-a)$-primary components, corresponding to the decomposition given by 39.26 of nonsingular asymmetric forms over $\mathbb{C}$ into $(z-(-)^j a)$-primary components. The $(z-a)$-primary component of the multisignature

$$\sigma^{Asy}_a \; : \; LAsy^{2j}_{(z-a)^\infty}(\mathbb{C}) \;=\; LAsy^{0}_{(z-(-)^j a)^\infty}(\mathbb{C}) \longrightarrow \mathbb{Z}$$

is an isomorphism, since for $a \neq (-)^{j+1}$ σ^{Asy}_a is the composite of the isomorphisms

$$LAsy^{0}_{(z-(-)^j a)^\infty}(\mathbb{C}) \xrightarrow{\simeq} L^0(\mathbb{C}) \; ; \; (L,\lambda) \longrightarrow (L, \frac{1}{2}(\lambda+\lambda^*)) \; ,$$

$$L^0(\mathbb{C}) \xrightarrow{\simeq} \mathbb{Z} \; ; \; (M,\mu) \longrightarrow \sigma(M,\mu) \; ,$$

and for $a = (-)^{j+1}$ the composite of σ^{Asy}_a and the isomorphism

$$L^0(\mathbb{C},-1) \xrightarrow{\simeq} LAsy^{0}_{(z+1)^\infty}(\mathbb{C}) \; ; \; (M,\mu) \longrightarrow (M,\mu)$$

is the isomorphism

$$L^0(\mathbb{C},-1) \xrightarrow{\simeq} \mathbb{Z} \; ; \; (L,\lambda) \longrightarrow \sigma(L,\lambda) \; . \qquad \square$$

Proposition 40.7 *The isometric and asymmetric multisignatures are related by a commutative square of isomorphisms*

$$\begin{array}{ccc} LIso^{2j}(\mathbb{C}) & \xrightarrow{\cong} & LAsy^{2j+2}_P(\mathbb{C}) \\ \cong \big\downarrow \sigma^{Iso}_* & & \cong \big\downarrow \sigma^{Asy}_* \\ \mathbb{Z}[R_{1/2}] & \xrightarrow{\cong} & \mathbb{Z}[S^1\backslash\{1\}] \end{array}$$

with $P = \{p(z) \,|\, p(1) = 1\} \subset \mathbb{C}[z,z^{-1}]$ *and*

$$\mathbb{Z}[R_{1/2}] \xrightarrow{\simeq} \mathbb{Z}[S^1\backslash\{1\}] \; ; \; a \longrightarrow 1 - a^{-1} \; .$$

Proof A $2j$-dimensional Seifert complex (C,λ) over $\mathbb{C}$ is homotopy equivalent to a fibred complex (C',λ'), with $\lambda' : (C')^{2j-*} \longrightarrow C'$ a chain equivalence. Indeed, by 32.45 (iv) it is possible to take

$$C' = \mathcal{C}(z\lambda - T\lambda : C^{2j-*}[z,z^{-1}] \longrightarrow C[z,z^{-1}])$$

with

$$\lambda' : H^j(C') \times H^j(C') \longrightarrow \mathbb{C} \; ; \; (x,y) \longrightarrow \chi_z(zx((z\lambda - T\lambda)^{-1}(y))) \; .$$

Now (SC',λ') is a $(2j+2)$-dimensional P-primary asymmetric Poincaré complex over $\mathbb{C}$, and the function

$$L\mathrm{Iso}^{2j}(\mathbb{C}) \longrightarrow L\mathrm{Asy}_P^{2j+2}(\mathbb{C}) \; ; \; (C,\lambda) \longrightarrow (SC',\lambda')$$

is an isomorphism by 35.27. For any $a \in R_{1/2}$ the $2j$-dimensional Seifert complex $(S^j\mathbb{C},a)$ with isometric multisignature

$$\sigma_*^{Iso}(S^j\mathbb{C},a) = 1.a \in \mathbb{Z}[R_{1/2}]$$

is sent by this isomorphism to the $(2j+2)$-dimensional P-primary asymmetric Poincaré complex $(S^{j+1}\mathbb{C},a)$ with asymmetric multisignature

$$\sigma_*^{Asy}(S^{j+1}\mathbb{C},a) = 1.(1-a^{-1}) \in \mathbb{Z}[S^1\backslash\{1\}] \; . \qquad \square$$

40C. The ω-signatures

The isometric and asymmetric multisignatures are cobordism invariants which are closely related to the 'ω-signatures'. The ω-signatures are not cobordism invariants in general, but they do have a nice geometric interpretation as signatures of branched covers (cf. 40.15).

Definition 40.8 Let $\omega \in S^1$.
(i) The *ω-signature* of an asymmetric form (L,λ) over $\mathbb{C}$ is

$$\tau_\omega(L,\lambda) = \sigma(L,(1-\overline{\omega})\lambda + (1-\omega)\lambda^*) \in \mathbb{Z} \; .$$

(ii) The *ω-signature* of a $2j$-dimensional asymmetric complex (C,λ) over $\mathbb{C}$ is

$$\tau_\omega(C,\lambda) = \tau_\omega(H^j(C),\lambda) \in \mathbb{Z} \; . \qquad \square$$

Note that $\tau_1 = 0$.

Example 40.9 The (-1)-signature of an asymmetric form (L,λ) over $\mathbb{C}$ is the signature of the symmetric form $(L,\lambda+\lambda^*)$ over $\mathbb{C}$

$$\tau_{-1}(L,\lambda) = \sigma(L,\lambda+\lambda^*) \in \mathbb{Z} \; . \qquad \square$$

In general, the ω-signature $\tau_\omega(L,\lambda) \in \mathbb{Z}$ of a nonsingular asymmetric form (L,λ) over $\mathbb{C}$ is an invariant of the Witt class $(L,\lambda) \in L\mathrm{Asy}^0(\mathbb{C})$ only if ω is not an eigenvalue of $(\lambda^*)^{-1}\lambda : L \longrightarrow L$ (or equivalently, if $\lambda + \omega\lambda^* : L \longrightarrow L^*$ is an isomorphism), in which case the ω-signature can be expressed as a linear combination of multisignature components:

Proposition 40.10 *Let $a = e^{i\theta}, \omega = e^{i\phi} \in S^1$, with $\omega \neq 1$.*
(i) *If $a \neq \omega$ the ω-symmetric signature is an invariant of the $(z-a)$-primary asymmetric Witt group of $\mathbb{C}$, defining an isomorphism*

$$\tau_\omega \ : \ \mathrm{LAsy}^0_{(z-a)^\infty}(\mathbb{C}) \xrightarrow{\simeq} \mathbb{Z} \ ; \ (L,\lambda) \longrightarrow \tau_\omega(L,\lambda)$$

which sends the generator $(\mathbb{C}, e^{i\theta/2}) \in \mathrm{LAsy}^0_{(z-a)^\infty}(\mathbb{C})$ to

$$\begin{aligned}\tau_\omega(\mathbb{C}, e^{i\theta/2}) &= \sigma(\mathbb{C}, \cos(\theta/2) - \cos(\theta/2 - \phi)) \\ &= \mathrm{sign}(\sin(\phi/2)\sin((\phi-\theta)/2)) \ .\end{aligned}$$

The generator has asymmetric multisignature components

$$\sigma^{Asy}_b(\mathbb{C}, e^{i\theta/2}) = \begin{cases} \mathrm{sign}(\cos(\theta/2)) & \text{if } b = a \neq -1, \\ \mathrm{sign}(\sin(\theta/2)) & \text{if } b = a = -1, \\ 0 & \text{if } b \neq a. \end{cases}$$

(ii) *Let (L,λ) be a nonsingular asymmetric form over $\mathbb{C}$, and let $\omega_1, \omega_2, \ldots, \omega_n \in S^1$ be the eigenvalues of $(\lambda^*)^{-1}\lambda : L \longrightarrow L$. For any $a \in S^1$ let*

$$\tau_{a^+}(L,\lambda) = \lim_{h\to 0+} \tau_{ae^{ih}}(L,\lambda) \ , \ \tau_{a^-}(L,\lambda) = \lim_{h\to 0^-} \tau_{ae^{ih}}(L,\lambda) \in \mathbb{Z} \ .$$

The function

$$S^1 \longrightarrow \mathbb{Z} \ ; \ \omega \longrightarrow \tau_\omega(L,\lambda)$$

is constant on each component of $S^1 \backslash \{1, \omega_1, \omega_2, \ldots, \omega_n\}$, with

$$\tau_{a^+}(L,\lambda) - \tau_{a^-}(L,\lambda) = \begin{cases} 2\sigma^{Asy}_{\omega_k}(L,\lambda) & \text{if } a = \omega_k \neq 1 \\ 0 & \text{if } a \in S^1\backslash\{1, \omega_1, \omega_2, \ldots, \omega_n\} \ , \end{cases}$$

and

$$\tau_\omega(L,\lambda) = \mathrm{sign}(\mathrm{im}(\omega))(\sum_{a\neq -1 \in S^1} \mathrm{sign}(\phi - \theta)\sigma^{Asy}_a(L,\lambda) - \sigma^{Asy}_{-1}(L,\lambda)) \in \mathbb{Z}$$
$$(-\pi < \theta, \phi < \pi)$$

for any $\omega \in S^1\backslash\{-1, 1, \omega_1, \omega_2, \ldots, \omega_n\}$. For $\omega = -1 \notin \{\omega_1, \omega_2, \ldots, \omega_n\}$

$$\tau_{-1}(L,\lambda) = \sum_{a\in S^1} \sigma^{Asy}_a(L,\lambda) \in \mathbb{Z}$$

with $\sigma^{Asy}_{-1}(L,\lambda) = 0$.
In particular, $\tau_\omega(L,\lambda) \in \mathbb{Z}$ is an invariant of the Witt class $(L,\lambda) \in \mathrm{LAsy}^0(\mathbb{C})$ for $\omega \in S^1\backslash\{1, \omega_1, \omega_2, \ldots, \omega_n\}$.
Proof (i) The ω-signature map is the special case $F = \mathbb{C}$ of the isomorphism obtained in 39.27 (vi)

$$\tau_\omega \ : \ \mathrm{LAsy}^0_{(z-a)^\infty}(\mathbb{C}) \xrightarrow{\simeq} L^0(\mathbb{C}) = \mathbb{Z} \ ; \ (L,\lambda) \longrightarrow \sigma(L, (1-\overline{\omega})\lambda + (1-\omega)\lambda^*)$$

as the composite of the split injection

$$\begin{aligned} LAsy^0_{(z-a)^\infty}(\mathbb{C}) &\longrightarrow L^0(\mathbb{C}[z,z^{-1}]_{(z-\omega)}) ; \\ (L,\lambda) &\longrightarrow (L[z,z^{-1}]_{(z-\omega)}, (1-z^{-1})\lambda + (1-z)\lambda^*) \end{aligned}$$

and the morphism of symmetric Witt groups induced by the morphism of rings with involution

$$\tau_\omega : \mathbb{C}[z,z^{-1}]_{(z-\omega)} \longrightarrow \mathbb{C} ; z \longrightarrow \omega .$$

(ii) Decompose (L,λ) into primary components

$$(L,\lambda) = \bigoplus_{a\in S^1} (L_a,\lambda_a) \oplus \bigoplus_{b\in \mathbb{C}^\bullet, |b|<1} (L_b \oplus L_{\bar{b}^{-1}}, \begin{pmatrix} 0 & \lambda_{\bar{b}^{-1}} \\ \lambda_b & 0 \end{pmatrix}) ,$$

as above. For $a \neq -1 \in S^1$

$$(L_a,\lambda_a) = \sigma(L_a,\lambda_a+\lambda_a^*)(\mathbb{C},e^{i\theta/2}) \in LAsy^0_{(z-a)^\infty}(\mathbb{C}) = \mathbb{Z}$$

with

$$\mathrm{ch}_z(L_a,\lambda_a) = (z-a)^{\dim_{\mathbb{C}}(L_a)}$$

and for $\omega \neq a$ the contribution of (L_a,λ_a) to the ω-signature is given by (i) to be

$$\begin{aligned} \tau_\omega(L_a,\lambda_a) &= \sigma(L_a,\lambda_a+\lambda_a^*)\tau_\omega(\mathbb{C},e^{i\theta/2}) \\ &= \sigma(L_a,\lambda_a+\lambda_a^*)\mathrm{sign}(\sin(\phi/2)\sin((\phi-\theta)/2)) \\ &= \mathrm{sign}(\mathrm{im}(\omega))\mathrm{sign}(\phi-\theta)\sigma_a^{Asy}(L_a,\lambda_a) \in \mathbb{Z} . \end{aligned}$$

For $\omega \neq a = -1 \in S^1$ it follows from

$$\begin{aligned} (L_{-1},\lambda_{-1}) &= \sigma(L_{-1},\lambda_{-1})(\mathbb{C},i) \in LAsy^0_{(z+1)^\infty}(\mathbb{C}) = \mathbb{Z} , \\ \tau_\omega(\mathbb{C},i) &= -\mathrm{sign}(\mathrm{im}(\omega)) \in \mathbb{Z} \end{aligned}$$

that the contribution of (L_{-1},λ_{-1}) is $-\mathrm{sign}(\mathrm{im}(\omega))\sigma(L_{-1},\lambda_{-1})$. The hyperbolic summands $L_b \oplus L_{\bar{b}^{-1}} \subseteq L$ ($b \in \mathbb{C}^\bullet\backslash S^1$) do not contribute to the ω-signatures. □

Definition 40.11 (i) The *ω-signatures* of a $(2j-1)$-knot $k : S^{2j-1} \subset S^{2j+1}$ are the ω-signatures

$$\tau_\omega(k) = \tau_\omega(H^j(\overline{X};\mathbb{C}),\lambda) \in \mathbb{Z} \quad (\omega \neq 1 \in S^1)$$

of the nonsingular asymmetric form $(H^j(\overline{X};\mathbb{C}),\lambda)$ over $\mathbb{C}$ given by the $(-)^j$-symmetric Seifert form

$$\lambda : H^j(\overline{X};\mathbb{C}) \xrightarrow[\simeq]{[\overline{X}]\cap -} H_j(\overline{X};\mathbb{C}) \xrightarrow[\simeq]{(1-\zeta)^{-1}} H_j(\overline{X};\mathbb{C}) \cong H^j(\overline{X};\mathbb{C})^*$$

with $\overline{X}$ the canonical infinite cyclic cover of the knot exterior X (as in 40.2).
(ii) The *signature* of a $(2j-1)$-knot $k : S^{2j-1} \subset S^{2j+1}$ is the (-1)-signature

$$\sigma(k) = \tau_{-1}(k) \in \mathbb{Z} . \qquad \square$$

Proposition 40.12 (i) *Let* $k : S^{2j-1} \subset S^{2j+1}$ *be a* $(2j-1)$*-knot, and let* $\omega_1, \omega_2, \ldots, \omega_n \in S^1$ *be the roots of the Alexander polynomial*

$$\begin{aligned} \Delta_j(z) = \det((\lambda + (-)^j\lambda^*)^{-1}(z\lambda + (-)^j\lambda^*) : \\ H^j(\overline{X};\mathbb{C})[z] \longrightarrow H^j(\overline{X};\mathbb{C})[z]) \in \mathbb{C}[z] \end{aligned}$$

(which are the eigenvalues of $\zeta : H^j(\overline{X};\mathbb{C}) \longrightarrow H^j(\overline{X};\mathbb{C})$*).*
The ω*-signatures* $\tau_\omega(k) \in \mathbb{Z}$ *are invariants of the knot cobordism class* $k \in C_{2j-1}$ *for any* $\omega \in S^1 \backslash \{1, \omega_1, \ldots, \omega_n\}$.
(ii) *The signature is a cobordism invariant*

$$\sigma = \tau_{-1} : C_{2j-1} = L\mathrm{Iso}^{2j}(\mathbb{Z}) \longrightarrow \mathbb{Z} .$$

(iii) *If* $\omega \in S^1$ *is not an algebraic number then the* ω*-signature is a cobordism invariant*

$$\tau_\omega : C_{2j-1} = L\mathrm{Iso}^{2j}(\mathbb{Z}) \longrightarrow \mathbb{Z} .$$

Proof (i) Immediate from 40.10.
(ii) It follows from $\Delta_j(1) = 1 \in \mathbb{Z}$ that $\Delta_j(-1) = 1 \in \mathbb{Z}_2$, and hence that $\Delta_j(-1) \neq 0 \in \mathbb{Z}$, so that (i) applies.
(iii) The roots of Alexander polynomials are algebraic numbers, so that (i) applies. $\square$

Remark 40.13 (i) The signature $\sigma(k) \in \mathbb{Z}$ of a $(2j-1)$-knot $k : S^{2j-1} \subset S^{2j+1}$ is just the signature of k in the sense of Trotter [290], Murasugi [206] and Milnor [195], such that

$$\sigma(k) = \tau_{-1}(k) = \tau_{-1}(L,\lambda) = \sigma(L, \lambda + \lambda^*) \in \mathbb{Z}$$

with (L,λ) any $(-)^j$-symmetric Seifert form over $\mathbb{Z}$ for k. The signature was first proved to be a knot cobordism invariant in [195]. Kauffman and Taylor [124] proved the cobordism invariance of the signature of a 1-knot $k : S^1 \subset S^3$ by identifying it with the signature of a double cover of D^4 branched over a push-in of a Seifert surface $F^2 \subset D^4$ for k (cf. 40.15 for $\omega = -1$).
(ii) Suppose that (L,λ) is a nonsingular asymmetric form over $\mathbb{Z}$ which is also a (-1)-symmetric Seifert form (i.e. such that $\lambda - \lambda^* : L \longrightarrow L^*$ is an isomorphism). The eigenvalues of $(\lambda^*)^{-1}\lambda : L \longrightarrow L$ are the inverses of the roots of the Alexander polynomial

$$\Delta(z) = \det((\lambda - \lambda^*)^{-1}(z\lambda - \lambda^*) : L[z] \longrightarrow L[z]) \in \mathbb{Z}[z]$$

with $\Delta(1) = 1$. If $\omega \in S^1$ is a pth root of unity (p prime) then $\Delta(\omega^{-1}) \neq 0$, since otherwise $(1 + z + \ldots + z^{p-1}) \,|\, \Delta(z)$ and $p \,|\, \Delta(1)$. The ω-signature

$\tau_\omega(L,\lambda) \in \mathbb{Z}$ is thus a Witt invariant of (L,λ). Tristram [289] used the ω-signature map

$$\tau_\omega \;:\; LAsy^0_P(\mathbb{Z}) \;=\; LIso^0(\mathbb{Z},-1) \longrightarrow \mathbb{Z}\;;\;(L,\lambda) \longrightarrow \tau_\omega(L,\lambda)$$

to obtain cobordism invariants of links, and to map the classical knot cobordism group C_1 onto $\mathbb{Z}^\infty$.
(iii) Levine [157], [158] identified $C_{2j-1} = LIso^{2j}(\mathbb{Z})$ and used the ω-signatures of Seifert forms over $\mathbb{Z}$ to map C_{2j-1} $(i \geq 2)$ onto $\mathbb{Z}^\infty$ with 2- and 4-torsion kernel.
(iv) The expression in 40.10 (ii) above of the ω-signature in terms of the multisignature components is due to Matumoto [185]. See also Kearton [128] and Litherland [168].
(v) Casson and Gordon [48], [49] obtained their invariants of the classical knot group C_1 using the ω-signatures and the specific Rochlin-type 4-dimensional properties of the cyclic covers of D^4 branched over knots $S^1 \subset S^3$.
(vi) Viro [295], Durfee and Kauffman [65], Cappell and Shaneson [41] and Kauffman [129, 5.6] used the ω-signatures to express the signatures of the branched cyclic covers of a high-dimensional knots $S^{4i+1} \subset S^{4i+3}$ $(i \geq 0)$ in terms of Seifert forms, as outlined in 40.15 below. □

Remark 40.14 Here are some specific examples of ω-signatures which are not invariants of the asymmetric Witt class.
(i) Given any $\omega \in \mathbb{C}^\bullet$, $a \in \mathbb{C}$ define a nonsingular asymmetric form over $\mathbb{C}$

$$(L,\lambda) \;=\; (\mathbb{C}\oplus\mathbb{C}, \begin{pmatrix} 0 & \omega \\ 1 & a \end{pmatrix})$$

with

$$(L,\lambda) \;=\; 0 \in LAsy^0(\mathbb{C}) \;\;,\;\; (\lambda^*)^{-1}\lambda \;=\; \begin{pmatrix} \omega & 0 \\ -\overline{a} & \overline{\omega}^{-1} \end{pmatrix}.$$

If $\omega \in S^1$ and $(1-\overline{\omega})a + (1-\omega)\overline{a} \neq 0 \in \mathbb{R}$ then

$$\begin{aligned}\tau_\omega(L,\lambda) \;&=\; \sigma(\mathbb{C}\oplus\mathbb{C}, \begin{pmatrix} 0 & 0 \\ 0 & (1-\overline{\omega})a + (1-\omega)\overline{a} \end{pmatrix})\\ &=\; \sigma(\mathbb{C}, (1-\overline{\omega})a + (1-\omega)\overline{a})\\ &=\; \mathrm{sign}((1-\overline{\omega})a + (1-\omega)\overline{a}) \neq 0 \in \mathbb{Z}\,.\end{aligned}$$

(ii) Setting $\omega = -1$ in (i), it is clear that the signature $\sigma(L,\lambda+\lambda^*)$ (40.9) is not an invariant of the asymmetric Witt class $(L,\lambda) \in LAsy^0(\mathbb{C})$. In particular, the nonsingular asymmetric form over $\mathbb{C}$

$$(L,\lambda) = (\mathbb{C}\oplus\mathbb{C}, \begin{pmatrix} 0 & -1 \\ 1 & 1 \end{pmatrix})$$

has

$$(L,\lambda) = 0 \in LAsy^0(\mathbb{C}) ,$$

$$\tau_{-1}(L,\lambda) = \sigma(L,\lambda+\lambda^*) = \sigma(\mathbb{C}\oplus\mathbb{C},\begin{pmatrix} 0 & 0 \\ 0 & 2 \end{pmatrix}) = 1 \neq 0 \in \mathbb{Z} .$$

(iii) If (L,λ) is a nonsingular asymmetric form over $\mathbb{C}$ such that $\lambda+\lambda^* : L \longrightarrow L^*$ is an isomorphism then

$$L_{-1} = 0 \ , \ \sigma^{Asy}_{-1}(L,\lambda) = 0 .$$

The (-1)-signature of such (L,λ) is the sum of all the multisignature components

$$\tau_{-1}(L,\lambda) = \sigma(L,\lambda+\lambda^*) = \sum_{a\in S^1} \sigma^{Asy}_a(L,\lambda) \in \mathbb{Z} ,$$

and is thus a Witt invariant. □

Remark 40.15 (i) Given a nonsingular asymmetric form (L,λ) over $\mathbb{C}$ and an integer $a \geq 2$ define the symmetric form over $\mathbb{C}$

$$\begin{aligned}\psi^a(L,\lambda) &= (L[z,z^{-1}],(1-z^{-1})\lambda+(1-z)\lambda^*)/(\sum_{j=0}^{a-1} z^j) \\ &= \Big(\bigoplus_{a-1} L, \begin{pmatrix} \lambda+\lambda^* & -\lambda & 0 & \dots & 0 & 0 \\ -\lambda^* & \lambda+\lambda^* & -\lambda & \dots & 0 & 0 \\ 0 & -\lambda^* & \lambda+\lambda^* & \dots & 0 & 0 \\ \vdots & \vdots & \vdots & \ddots & \vdots & \vdots \\ 0 & 0 & 0 & \dots & \lambda+\lambda^* & -\lambda \\ 0 & 0 & 0 & \dots & -\lambda^* & \lambda+\lambda^* \end{pmatrix}\Big)\end{aligned}$$

with the automorphism

$$\zeta = \begin{pmatrix} 0 & 0 & 0 & \dots & 0 & -1 \\ 1 & 0 & 0 & \dots & 0 & -1 \\ 0 & 1 & 0 & \dots & 0 & -1 \\ \vdots & \vdots & \vdots & \ddots & \vdots & \vdots \\ 0 & 0 & 0 & \dots & 0 & -1 \\ 0 & 0 & 0 & \dots & 1 & -1 \end{pmatrix} : \psi^a(L,\lambda) \longrightarrow \psi^a(L,\lambda)$$

such that

$$\mathrm{ch}_z(\bigoplus_{a-1} L,\zeta) = (1+z+z^2+\dots+z^{a-1})^d$$

where $d = \dim_{\mathbb{C}}(L)$. The eigenvalues of $\zeta : L \longrightarrow L$ are the ath roots of 1 other than 1 itself, that is the powers

$$\omega^j = e^{2\pi ij/a} \quad (1 \leq j \leq a-1)$$

of $\omega = e^{2\pi i/a}$, and the eigenspaces are the images of the inclusions

$$f_j = \begin{pmatrix} 1+\omega^j+\ldots+(\omega^j)^{a-3}+(\omega^j)^{a-2} \\ 1+\omega^j+\ldots+(\omega^j)^{a-3} \\ \vdots \\ 1+\omega^j \\ 1 \end{pmatrix} : L \longrightarrow \bigoplus_{a-1} L .$$

Thus there is defined an isomorphism of symmetric forms over $\mathbb{C}$

$$(f_1\ f_2\ \ldots\ f_{a-1}) : \bigoplus_{j=1}^{a-1}(L,(1-\overline{\omega}^j)\lambda+(1-\omega^j)\lambda^*) \xrightarrow{\simeq} \psi^a(L,\lambda)$$

and the signature of $\psi^a(L,\lambda)$ is the sum of the ω^j-signatures of (L,λ)

$$\sigma(\psi^a(L,\lambda)) = \sum_{j=1}^{a-1}\tau_{\omega^j}(L,\lambda)\in\mathbb{Z} .$$

(ii) If (L,λ) is the Seifert form (over $\mathbb{C}$) of a $(4i+1)$-knot $S^{4i+1}\subset S^{4i+3}$ with $L=H_{2i+1}(F;\mathbb{C})$ for a Seifert surface $F^{4i+2}\subset S^{4i+3}$, then $\psi^a(L,\lambda)$ is the intersection form of the a-fold branched cover of D^{4i+4}, branched over a copy of F pushed into D^{4i+4} as in 26.10. Let ω be a primitive ath root of 1. The ω^j-signatures $\tau_{\omega^j}(L,\lambda)\in\mathbb{Z}$ determine the signature $\sigma(\psi^a(L,\lambda))\in\mathbb{Z}$ as in (i). The Alexander polynomial

$$\Delta(z) = \det((\lambda-\lambda^*)^{-1}(z\lambda-\lambda^*) : L[z,z^{-1}]\longrightarrow L[z,z^{-1}])\in\mathbb{C}[z,z^{-1}]$$

is such that

$$\Delta(1) = 1 ,\quad \Delta(z)\in\mathbb{Z}[z,z^{-1}] ,$$

so that

$$\Delta(\omega^j)\neq 0\in\mathbb{C}\ \ (1\leq j\leq a-1) .$$

The ω^j-signatures $\tau_{\omega^j}(L,\lambda)$ are thus Witt class invariants (cf. 40.13 (ii)), and hence so is $\sigma(\psi^a(L,\lambda))$. In particular, for $a=2$ this is the signature $\sigma(L,\lambda+\lambda^*)\in\mathbb{Z}$. □

40D. Automorphism multisignature

Next, the multisignature of an automorphism of an ϵ-symmetric form over $\mathbb{C}$.

Definition 40.16 (i) Let $f:(M,\phi)\longrightarrow(M,\phi)$ be an automorphism of a nonsingular ϵ-symmetric form (M,ϕ) over $\mathbb{C}$, with $\epsilon=\pm1$. Use the decomposition

$$(M,\phi,f) \; = \; \sum_{a\in S^1}(M_a,\phi_a,f_a) \oplus \text{hyperbolic}$$

(as in 39.22) with

$$M_a \; = \; \{x\in M \,|\, (f-a)^N(x)=0 \text{ for some } N\geq 1\} \;\;,\;\; \phi_a \; = \; \phi|_{M_a}$$

to define the *automorphism multisignature* of (M,ϕ,h) to be the formal linear combination

$$\sigma_*^{Aut}(M,\phi,f) \; = \; \sum_{a\in S^1}\sigma_a^{Aut}(M,\phi,f).a \in \mathbb{Z}[S^1]$$

of the signatures

$$\sigma_a^{Aut}(M,\phi,f) \; = \; \sigma(M_a,\phi_a)\in\mathbb{Z} \;\; (a\in S^1)$$

of the ϵ-symmetric forms (M_a,ϕ_a) over $\mathbb{C}$. The signature of (M,ϕ) is the sum of the multisignature components

$$\sigma(M,\phi) \; = \; \sum_{a\in S^1}\sigma_a^{Aut}(M,\phi,f)\in\mathbb{Z}\,.$$

(ii) The *automorphism multisignature* of a self homotopy equivalence $h: E\longrightarrow E$ of a $2j$-dimensional symmetric Poincaré complex E over $\mathbb{C}$ is the multisignature of the induced automorphism

$$h^* \; : \; (H^j(E),\phi_E) \longrightarrow (H^j(E),\phi_E)$$

of the $(-)^j$-symmetric form on $H^j(E)$

$$\sigma_*^{Aut}(E,h) \; = \; \sigma_*^{Aut}(H^j(E),\phi_E,h^*)\in\mathbb{Z}[S^1]\,.$$

Again, the signature is the sum of the multisignature components

$$\sigma(E) \; = \; \sum_{a\in S^1}\sigma_a^{Aut}(E,h)\in\mathbb{Z}\,.$$

□

Proposition 40.17 (Neumann [212])
The automorphism multisignature defines an isomorphism

$$\sigma_*^{Aut} \; : \; \mathrm{LAut}^{2j}(\mathbb{C}) \xrightarrow{\simeq} \mathbb{Z}[S^1]\;;\; (E,h)\longrightarrow \sigma_*^{Aut}(E,h)\;,$$

with $(S^j\mathbb{C},1,a)$ *(resp.* $(S^j\mathbb{C},-i,a)$*) the generator of the* $(z-a)$*-primary component for* k *even (resp. odd).*

Proof This is the special case $F=\mathbb{C}$ of 39.30, noting that

$$\max(\mathbb{C}[z,z^{-1}]) \; = \; \{(z-a)\,|\,a\in\mathbb{C}^\bullet\}\,,$$
$$\overline{\max}(\mathbb{C}[z,z^{-1}]) \; = \; \{(z-a)\,|\,a\in S^1\}$$

(cf. 39.22, 39.34).

□

By analogy with 40.7:

Proposition 40.18 *The isometric and automorphism multisignatures are related by a commutative square of isomorphisms*

$$\begin{array}{ccc} L\mathrm{Iso}^{2j}(\mathbb{C}) & \xrightarrow{\cong} & L\mathrm{Aut}^{2j+2}_{fib}(\mathbb{C}) \\ \cong \Big\downarrow \sigma^{Iso}_* & & \cong \Big\downarrow \sigma^{Aut}_* \\ \mathbb{Z}[R_{1/2}] & \xrightarrow{\cong} & \mathbb{Z}[S^1\backslash\{1\}] \end{array}$$

with

$$L\mathrm{Iso}^{2j}(\mathbb{C}) \xrightarrow{\cong} L\mathrm{Aut}^{2j}(\mathbb{C}) \ ; \ (C,\lambda) \longrightarrow (C,(1+T)\lambda,\lambda(T\lambda)^{-1}) \ ,$$

$$\mathbb{Z}[R_{1/2}] \xrightarrow{\cong} \mathbb{Z}[S^1\backslash\{1\}] \ ; \ a \longrightarrow 1-a^{-1} \ . \qquad \square$$

The cobordism classes of the following types of algebraic Poincaré structures are in one-one correspondence, for any ring with involution A:

(i) $(2j+2)$-dimensional asymmetric Poincaré complexes over A,
(ii) $(2j+2)$-dimensional symmetric $\Omega^{-1}A[z,z^{-1}]$-Poincaré complexes over $A[z,z^{-1}]$,
(iii) $(2j+2)$-dimensional symmetric Poincaré pairs over $A[z,z^{-1}]$ with band boundary.

The corresponding identifications

$$L\mathrm{Asy}^{2j+2}_h(A) \ = \ L^{2j+2}_h(\Omega^{-1}A[z,z^{-1}])$$

and exact sequence

$$0 \longrightarrow L^{2j+2}_h(A) \longrightarrow L\mathrm{Asy}^{2j+2}_h(A) \longrightarrow L\mathrm{Aut}^{2j}_p(A) \longrightarrow L^{2j}_h(A) \longrightarrow 0$$

were obtained in Chap. 28. These identifications will now be used for $A=\mathbb{C}$ to define the asymmetric multisignature for (ii) and (iii), and to examine the relationship with the automorphism multisignature.

Definition 40.19 Let $(D,C,\phi)=(f:C\longrightarrow D,(\phi_D,\phi_C))$ be a $(2j+2)$-dimensional symmetric Poincaré pair over $\mathbb{C}[z,z^{-1}]$ with band boundary, i.e. such that $(C,\phi_C)=T(h)$ is the algebraic mapping torus of a self homotopy equivalence $h:(E,\phi_E)\longrightarrow(E,\phi_E)$ of a $2j$-dimensional symmetric Poincaré complex (E,ϕ_E) over $\mathbb{C}$.
The *asymmetric multisignature* of (D,E,h) is the formal sum

$$\sigma^{Asy}_*(D,E,h) \ = \ \sum_{a\in S^1} \sigma^{Asy}_a(D,E,h)a \in \mathbb{Z}[S^1]$$

with components

$$\sigma_a^{Asy}(D,E,h) \;=\; \sigma_a^{Aut}(H^j(E),\phi_E,h^*) \in \mathbb{Z} \;\; (a \neq 1)$$

and 1-component

$$\begin{aligned}\sigma_1^{Asy}(D,E,h) &= \sigma(\mathbb{C}\otimes_{\mathbb{C}[z,z^{-1}]}(D,C)) + \sigma_1^{Asy}(E,h)\\ &= \sigma(H^{j+1}(\mathbb{C}\otimes_{\mathbb{C}[z,z^{-1}]}(D/C)), 1\otimes(\phi_D/\phi_C))\\ &\qquad + \sigma_1^{Aut}(H^j(E),\phi_E(h^*-(h^*)^{-1}),h^*) \in \mathbb{Z}\end{aligned}$$

with $(H^{j+1}(\mathbb{C}\otimes_{\mathbb{C}[z,z^{-1}]}(D/C)), 1\otimes(\phi_D/\phi_C))$ and $(H^j(E),\phi_E(h^*-(h^*)^{-1}))$ $(-)^{j+1}$-symmetric forms over $\mathbb{C}$ (which are singular in general). □

Every $(2j+2)$-dimensional symmetric Poincaré complex B over $\mathbb{C}(z)$ is homotopy equivalent to the localization $\mathbb{C}(z)\otimes_{\mathbb{C}[z,z^{-1}]}C$ of a $(2j+2)$-dimensional symmetric $\mathbb{C}(z)$-Poincaré complex C over $\mathbb{C}[z,z^{-1}]$, corresponding to the $(2j+2)$-dimensional symmetric Poincaré pair over $\mathbb{C}[z,z^{-1}]$ $(C^{2j+3-*},\partial C)$ with $\partial C = T(\zeta)$ the algebraic mapping torus of the self homotopy equivalence $\zeta : \partial C^! \longrightarrow \partial C^!$ of a $2j$-dimensional symmetric Poincaré complex $\partial C^!$ over $\mathbb{C}$.

Definition 40.20 The *asymmetric multisignature* of a $(2j+2)$-dimensional symmetric Poincaré complex B over $\mathbb{C}(z)$ is defined by

$$\sigma_*^{Asy}(B) \;=\; \sigma_*^{Asy}(C^{2j+3-*},\partial C^!,\zeta) \in \mathbb{Z}[S^1]$$

using any $(2j+2)$-dimensional symmetric $\mathbb{C}(z)$-Poincaré complex C over $\mathbb{C}[z,z^{-1}]$ such that $B \simeq \mathbb{C}(z)\otimes_{\mathbb{C}[z,z^{-1}]} C$. This is just the asymmetric multisignature of 40.19 for the $(2j+2)$-dimensional symmetric Poincaré pair $(C^{2j+3-*},T(\zeta))$ over $\mathbb{C}[z,z^{-1}]$. □

Example 40.21 If D is a $(2j+2)$-dimensional symmetric Poincaré complex over $\mathbb{C}[z,z^{-1}]$ the asymmetric multisignature of the induced $(2j+2)$-dimensional symmetric Poincaré complex over $\mathbb{C}(z)$

$$B \;=\; \mathbb{C}(z)\otimes_{\mathbb{C}[z,z^{-1}]} D$$

is just the signature of the induced $(2j+2)$-dimensional symmetric Poincaré complex over $\mathbb{C}$

$$\sigma_*^{Asy}(B) \;=\; \sigma(\mathbb{C}\otimes_{\mathbb{C}[z,z^{-1}]} D).1 \in \mathbb{Z} \subset \mathbb{Z}[S^1]\,.$$ □

Proposition 40.22 (i) *The asymmetric multisignature of 40.20 is a cobordism invariant, defining an isomorphism*

$$\sigma_*^{Asy} \;:\; L^{2j+2}(\mathbb{C}(z)) \xrightarrow{\;\simeq\;} \mathbb{Z}[S^1]\;;\; B \longrightarrow \sigma_*^{Asy}(B)\,.$$

The 1-component defines a morphism

$$\sigma_1^{Asy} \;:\; L^{2j+2}(\mathbb{C}(z)) \longrightarrow \mathbb{Z}\;;\; B \longrightarrow \sigma_1^{Asy}(B)$$

which splits the injection

$$L^{2j+2}(\mathbb{C}[z,z^{-1}]) \;=\; \mathbb{Z} \longrightarrow L^{2j+2}(\mathbb{C}(z))$$

induced by the inclusion $\mathbb{C}[z,z^{-1}]\longrightarrow\mathbb{C}(z)$.

(ii) *The composite of the isomorphism of 39.27*

$$LAsy^{2j+2}(\mathbb{C}) \xrightarrow{\simeq} L^{2j+2}(\mathbb{C}(z)) \; ; \; (C,\lambda) \longrightarrow (C(z),(1-z^{-1})\lambda+(1-z)T\lambda)$$

and the asymmetric multisignature isomorphism

$$\sigma_*^{Asy} \; : \; L^{2j+2}(\mathbb{C}(z)) \xrightarrow{\simeq} \mathbb{Z}[S^1]$$

of (i) *agrees up to sign with the asymmetric multisignature isomorphism of 40.6*

$$\sigma_*^{Asy} \; : \; LAsy^{2j+2}(\mathbb{C}) \xrightarrow{\simeq} \mathbb{Z}[S^1] \; .$$

(iii) *The asymmetric and automorphism multisignature maps define isomorphisms of exact sequences*

$$\begin{array}{ccccccccc}
0 \longrightarrow & L^{2j+2}(\mathbb{C}) & \longrightarrow & L^{2j+2}(\mathbb{C}(z)) & \longrightarrow & LAut^{2j}(\mathbb{C}) & \longrightarrow & L^{2j}(\mathbb{C}) & \longrightarrow 0 \\
 & \sigma \big\downarrow \cong & & \sigma_*^{Asy} \big\downarrow \cong & & \sigma_*^{Aut} \big\downarrow \cong & & \sigma \big\downarrow \cong & \\
0 \longrightarrow & \mathbb{Z} & \xrightarrow{1} & \mathbb{Z}[S^1] & \xrightarrow{\partial} & \mathbb{Z}[S^1] & \xrightarrow{e} & \mathbb{Z} & \longrightarrow 0
\end{array}$$

and

$$\begin{array}{ccccccccc}
0 \longrightarrow & L^{2j+2}(\mathbb{C}) & \longrightarrow & LAsy^{2j+2}(\mathbb{C}) & \longrightarrow & LAut^{2j}(\mathbb{C}) & \longrightarrow & L^{2j}(\mathbb{C}) & \longrightarrow 0 \\
 & \sigma \big\downarrow \cong & & \sigma_*^{Asy} \big\downarrow \cong & & \sigma_*^{Aut} \big\downarrow \cong & & \sigma \big\downarrow \cong & \\
0 \longrightarrow & \mathbb{Z} & \xrightarrow{1} & \mathbb{Z}[S^1] & \xrightarrow{\partial} & \mathbb{Z}[S^1] & \xrightarrow{e} & \mathbb{Z} & \longrightarrow 0
\end{array}$$

with

$$\begin{aligned}
\partial \; &: \; \mathbb{Z}[S^1] \longrightarrow \mathbb{Z}[S^1] \; ; \; a \longrightarrow a-1 \; , \\
e \; &: \; \mathbb{Z}[S^1] \longrightarrow \mathbb{Z} \; ; \; a \longrightarrow 1 \; .
\end{aligned}$$

Proof (i) This is immediate from the special case $F = \mathbb{C}$ of 39.30, once it has been verified that the 1-component of the multisignature is in fact a cobordism invariant. Suppose then that (D,C) is a $(2j+2)$-dimensional symmetric Poincaré pair over $\mathbb{C}[z,z^{-1}]$ with band boundary $C = T(h)$, such that there exists a $(2j+3)$-dimensional symmetric Poincaré pair $(R,\partial R)$ with boundary

$$\partial R = D \cup_{T(h)} -T(g)$$

for some extension of h to a self homotopy equivalence $(g,h) : (Q,P) \to (Q,P)$ of a $(2j+1)$-dimensional symmetric Poincaré pair (Q,P) over $\mathbb{C}$. By the Novikov additivity of signature

$$\sigma(\partial R) \; = \; \sigma(D,C) - \sigma(T(g)) \; = \; 0 \in \mathbb{Z} \; .$$

Write

$$(M,\phi,h) \;=\; (H^j(P),\phi_P,h^*)$$

and let $L\subset M$ be the h-invariant lagrangian of (M,ϕ) defined by

$$L \;=\; \mathrm{im}(H^j(Q)\longrightarrow H^j(P))\ .$$

Apply the formula of Wall [301] for the nonadditivity of the signature exactly as in Neumann [214, p. 164], to obtain

$$\sigma(T(g)) \;=\; -\sigma(N,\psi)\in\mathbb{Z}$$

with (N,ψ) the (potentially singular) $(-)^{j+1}$-symmetric form over $\mathbb{C}$ defined by

$$(N,\psi) \;=\; (\{x\in M\,|\,(1-h)(x)\in L\},\phi(h-h^{-1})|)\ ,$$

such that h restricts to an automorphism $h|:(N,\psi)\longrightarrow(N,\psi)$. The $(z-a)$-primary decomposition of $h:M\longrightarrow M$ determines $(z-a)$-primary decompositions

$$(M,\phi,h) \;=\; \sum_{a\in S^1}(M_a,\phi_a,h_a)\oplus(M',\phi',h')\ ,$$

$$(N,\psi) \;=\; \sum_{a\in S^1}(N_a,\psi_a)\oplus(N',\psi')$$

with

$$M' \;=\; \sum_{a\neq\overline{a}}M_a\ ,\quad N' \;=\; \sum_{a\neq\overline{a}}N_a$$

and (N',ψ') hyperbolic. As in [214], the $(z-a)$-primary components of (N,ψ) for $a\neq 1\in S^1$ have $\psi_a=0$, and (N_1,ψ_1) is a subform of $(M_1,\phi_1(h_1-h_1^{-1}))$ with $N_1^\perp\subseteq N_1$, so that

$$\begin{aligned}\sigma(N,\psi) \;&=\; \sigma(N_1,\psi_1)\\ &=\; \sigma(M_1,\phi_1(h_1-h_1^{-1})) \;=\; \sigma_1^{Aut}(M,\phi(h-h^{-1}),h)\in\mathbb{Z}\end{aligned}$$

and hence

$$\begin{aligned}\sigma_1^{Asy}(D,C) \;&=\; \sigma(D,C)+\sigma_1^{Aut}(M,\phi(h-h^{-1}),h)\\ &=\; \sigma(T(g))+\sigma(N,\psi) \;=\; 0\in\mathbb{Z}\ .\end{aligned}$$

(ii) By 40.6 the asymmetric L-group $LAsy^{2j+2}(\mathbb{C})$ is generated by the cobordism classes of the $(2j+2)$-dimensional asymmetric Poincaré complexes over $\mathbb{C}$ of the type $(S^{j+1}\mathbb{C},\lambda)$ ($\lambda\in S^1$), so it suffices to prove that the multisignature of $(S^{j+1}\mathbb{C},\lambda)$ agrees up to sign with the multisignature of the $(2j+2)$-dimensional symmetric Poincaré complex over $\mathbb{C}(z)$

$$B_{\lambda,(-)^{j+1}} \;=\; (S^{j+1}\mathbb{C}(z),\phi_{\lambda,(-)^{j+1}})$$

defined by

$$\phi_{\lambda,(-)^{j+1}} \;=\; (1-z^{-1})\lambda+(-)^{j+1}(1-z)\overline{\lambda}\;:\;\mathbb{C}(z)\longrightarrow\mathbb{C}(z)^*\ .$$

Let

$$a = (-)^{j+1}\overline{\lambda}^{-1}\lambda = (-)^{j+1}\lambda^2 \in S^1 .$$

By definition (40.5 (ii)), the asymmetric multisignature of $(S^{j+1}\mathbb{C},\lambda)$ is

$$\begin{aligned}\sigma^{Asy}_*(S^{j+1}\mathbb{C},\lambda) &= (-)^{j+1}\sigma^{Asy}_*(\mathbb{C},\lambda)\\ &= \begin{cases}\sigma(\mathbb{C},\frac{1}{2}(\lambda+\overline{\lambda})).a & \text{if } a\neq(-)^j\\ \sigma(\mathbb{C},\frac{1}{2i}(\lambda-\overline{\lambda})).a & \text{if } a=(-)^j\end{cases}\\ &= \pm a\in\mathbb{Z}[S^1] .\end{aligned}$$

By construction

$$B_{\lambda,(-)^{j+1}} = \mathbb{C}(z)\otimes_{\mathbb{C}[z,z^{-1}]}(C,\phi)$$

is induced from $(C,\phi)=(S^{j+1}\mathbb{C}[z,z^{-1}],\phi_{\lambda,(-)^{j+1}})$, with

$$\sigma(\mathbb{C}\otimes_{\mathbb{C}[z,z^{-1}]}(C,\phi)) = \sigma(\mathbb{C},0) = 0\in\mathbb{Z} ,$$

so that

$$\sigma^{Asy}_*(B_{\lambda,(-)^{j+1}}) = \sigma^{Aut}_1(\partial C^!,\partial\phi^!_0(\zeta-\zeta^{-1}),\zeta).1+\sum_{b\neq 1}\sigma^{Aut}_b(\partial C^!,\partial\phi^!_0,\zeta).b$$

$$\in\mathbb{Z}[S^1] .$$

As in the proof of 39.27 (iv)

$$(\partial C^!,\partial\phi^!_0,\zeta) = \left(\mathbb{C}\oplus\mathbb{C},\begin{pmatrix}0 & 1\\ (-)^j & 0\end{pmatrix},\begin{pmatrix}1+(-)^{j+1}\overline{\lambda}^{-1}\lambda & (-)^j\overline{\lambda}^{-1}\\ \lambda & 0\end{pmatrix}\right)$$

$$\begin{cases}\text{has lagrangian } \{(x,\lambda(x))\,|\,x\in\mathbb{C}\}\subset\mathbb{C}\oplus\mathbb{C} & \text{if } a=1\\ \cong(\mathbb{C},\lambda+(-)^j\overline{\lambda},1)\oplus(\mathbb{C},-(\lambda+(-)^j\overline{\lambda}),a) & \text{if } a\neq 1 .\end{cases}$$

The $(1-z)$-primary component of the $(-)^{j+1}$-symmetric form over $\mathbb{C}$

$$(\partial C^!,\partial\phi^!_0(\zeta-\zeta^{-1})) = \left(\mathbb{C}\oplus\mathbb{C},\begin{pmatrix}\lambda+(-)^{j+1}\overline{\lambda} & -1+(-)^j\overline{\lambda}\lambda^{-1}\\ (-)^j-\overline{\lambda}^{-1}\lambda & \overline{\lambda}^{-1}+(-)^{j+1}\lambda^{-1}\end{pmatrix}\right)$$

with respect to the automorphism ζ is given up to isomorphism by

$$\sigma^{Aut}_1(\partial C^!,\partial\phi^!_0(\zeta-\zeta^{-1}),\zeta) = \begin{cases}(\mathbb{C},\lambda)\oplus(\mathbb{C},0) & \text{if } a=1\\ (\mathbb{C},0) & \text{if } a\neq 1.\end{cases}$$

The asymmetric multisignature of $B_{\lambda,(-)^{j+1}}$ is thus

$$\begin{aligned}\sigma^{Asy}_*(B_{\lambda,(-)^{j+1}}) &= \sigma^{Aut}_1(\partial C^!,\partial\phi^!_0(\zeta-\zeta^{-1}),\zeta).1+\sum_{b\neq 1}\sigma^{Aut}_b(\partial C^!,\partial\phi^!_0,\zeta).b\\ &= \begin{cases}\sigma(\mathbb{C},\lambda).1 & \text{if } a=1\\ \sigma(\mathbb{C},-(\lambda+(-)^j\overline{\lambda})).a & \text{if } a\neq 1\end{cases}\\ &= \pm a\in\mathbb{Z}[S^1] .\end{aligned}$$

(iii) Combine 40.6 and (i). □

Definition 40.23 (i) The *signature* of a $(2j+2)$-dimensional geometric Poincaré pair $(N, \partial N)$ is the signature of the $(-)^{j+1}$-symmetric intersection form $(H_{j+1}(N;\mathbb{C}), \phi_N)$ over $\mathbb{C}$

$$\sigma(N) = \sigma(H_{j+1}(N;\mathbb{C}), \phi_N) \in \mathbb{Z} .$$

(The form is singular in general. The invariant is 0 for even k).
(ii) The *automorphism multisignature* of a self homotopy equivalence $h : P \longrightarrow P$ of a $2j$-dimensional geometric Poincaré complex P is the automorphism multisignature of the induced automorphism

$$h^* : (H^j(P;\mathbb{C}), \phi_P) \longrightarrow (H^j(P;\mathbb{C}), \phi_P)$$

of the $(-)^j$-symmetric intersection form over $\mathbb{C}$

$$\sigma_*^{Aut}(P, h) = \sigma_*^{Aut}(H^j(P;\mathbb{C}), \phi_P, h^*) \in \mathbb{Z}[S^1] .$$

The signature of P is the sum of the multisignature components

$$\sigma(P) = \sum_{a \in S^1} \sigma_a^{Aut}(P, h) \in \mathbb{Z} .$$

(iii) Let $(N, \partial N)$ be a $(2j+2)$-dimensional geometric Poincaré pair which is equipped with a map

$$(p, \partial p) : (N, \partial N) \longrightarrow S^1$$

such that $\partial p : \partial N = T(h) \longrightarrow S^1$ is the projection of the mapping torus of a self homotopy equivalence $h : P \longrightarrow P$ of a $2j$-dimensional geometric Poincaré complex P. The *asymmetric multisignature* of (N, P, h) is the asymmetric multisignature of the $(2j+2)$-dimensional symmetric Poincaré pair $(C(\overline{N};\mathbb{C}), T(h : C(P;\mathbb{C}) \longrightarrow C(P;\mathbb{C})))$ over $\mathbb{C}[z, z^{-1}]$

$$\begin{aligned} \sigma_*^{Asy}(N, P, h) &= \sigma_*^{Asy}(C(\overline{N};\mathbb{C}), C(P;\mathbb{C}), h) \\ &= \sum_{a \in S^1} \sigma_a^{Asy}(N, P, h).a \in \mathbb{Z}[S^1] \end{aligned}$$

with components

$$\begin{aligned} &\sigma_a^{Asy}(N, P, h) \\ &= \begin{cases} \sigma(N) + \sigma_1^{Aut}(H^j(P;\mathbb{C}), \phi_P(h^* - (h^*)^{-1}), h^*) & \text{if } a = 1 \\ \sigma_a^{Aut}(P, h) & \text{if } a \neq 1 . \end{cases} \end{aligned}$$

□

Proposition 40.24 (i) *Let $(N, \partial N)$ be a $(2j+2)$-dimensional geometric Poincaré pair with $\partial N = T(h : P \longrightarrow P)$. The asymmetric multisignature of (N, P, h) is the asymmetric multisignature of the $(2j+2)$-dimensional asymmetric Poincaré complex $(C(N, P;\mathbb{C}), \lambda)$ obtained as in 27.6*

$$\sigma^{Asy}_*(N,P,h) \;=\; \sigma^{Asy}_*(C(N,P;\mathbb{C}),\lambda) \in \mathbb{Z}[S^1] .$$

The automorphism multisignature of (P,h) *is given by*

$$\sigma^{Aut}_*(P,h) \;=\; \partial\sigma^{Asy}_*(N,P,h) \in \mathbb{Z}[S^1]$$

with $\partial : \mathbb{Z}[S^1] \longrightarrow \mathbb{Z}[S^1]; a \longrightarrow a-1$.

(ii) *The asymmetric and automorphism multisignature maps define a natural transformation of exact sequences*

$$\begin{array}{ccccccccc}
\Delta_{2j+1} & \longrightarrow & \Omega_{2j+2}(S^1) & \longrightarrow & AB_{2j} & \longrightarrow & \Delta_{2j} & \longrightarrow & \Omega_{2j+1}(S^1)\\
\downarrow{\scriptstyle \sigma^{Aut}_*} & & \downarrow{\scriptstyle \sigma} & & \downarrow{\scriptstyle \sigma^{Asy}_*} & & \downarrow{\scriptstyle \sigma^{Aut}_*} & & \downarrow{\scriptstyle \sigma}\\
L\mathrm{Aut}^{2j+1}(\mathbb{C}) & \longrightarrow & L^{2j+2}(\Lambda) & \longrightarrow & L^{2j+2}(\mathbb{C}(z)) & \longrightarrow & L\mathrm{Aut}^{2j}(\mathbb{C}) & \longrightarrow & L^{2j+1}(\Lambda)\\
\downarrow{\scriptstyle \cong} & & \downarrow{\scriptstyle \cong} & & \downarrow{\scriptstyle \cong} & & \downarrow{\scriptstyle \cong} & & \downarrow{\scriptstyle \cong}\\
0 & \longrightarrow & \mathbb{Z} & \xrightarrow{\;1\;} & \mathbb{Z}[S^1] & \xrightarrow{\;\partial\;} & \mathbb{Z}[S^1] & \xrightarrow{\;e\;} & \mathbb{Z}
\end{array}$$

with $\Lambda = \mathbb{C}[z,z^{-1}]$ *and* $e : \mathbb{Z}[S^1] \longrightarrow \mathbb{Z}; a \longrightarrow 1$. □

Corollary 40.25 (Neumann [214])
Let $(N,\partial N)$ *be a* $(2j+2)$*-dimensional manifold with boundary which is a fibre bundle over* S^1, *with*

$$(N,\partial N) \;=\; T((g,\partial g) : (Q,\partial Q) \longrightarrow (Q,\partial Q)) .$$

The asymmetric multisignature of $(N,\partial Q,\partial g)$ *is*

$$\sigma^{Asy}_*(N,\partial Q,\partial g) \;=\; \sum_{a\in S^1} \sigma^{Asy}_a(N,\partial Q,\partial g).a \;=\; 0 \in \mathbb{Z}[S^1] .$$

In particular

$$\begin{aligned}\sigma^{Asy}_1(N,\partial Q,\partial g) \;&=\; \sigma(N) + \sigma^{Aut}_1(H^j(P;\mathbb{C}),\phi_P(h^*-(h^*)^{-1}),h^*)\\ &=\; 0 \in \mathbb{Z} .\end{aligned}$$

Proof The double of N defines a null-cobordism of $(N,\partial Q,\partial g)$, so that

$$(N,\partial Q,\partial g) \;=\; 0 \in AB_{2j}$$

and the asymmetric multisignature vanishes by 40.22. □

Remark 40.26 (i) The various multisignatures considered in this chapter are infinite versions of the multisignature used to compute $L^0_p(\mathbb{C}[\pi])$ for finite groups π, which only has a finite number of components corresponding to the self dual irreducible complex representations of π. See Wall [302, Chap. 13A] (and also Ranicki [237, Chap. 22]) for more on the multisignature for finite π, including the result that $L_{2*}(\mathbb{Z}[\pi])$ is detected modulo 2-primary torsion

by the real multisignature.
(ii) If $\pi = \mathbb{Z}_m$ is a cyclic group of order m there is defined an isomorphism of rings with involution

$$\mathbb{C}[\mathbb{Z}_m] \;=\; \mathbb{C}[z,z^{-1}]/(z^m-1) \xrightarrow{\simeq} \prod_m \mathbb{C} \;;\; z \longrightarrow (1,\omega,\omega^2,\dots,\omega^{m-1})$$

with $\omega = e^{2\pi i/m} \in S^1$ a primitive mth root of 1. A projective nonsingular symmetric form (L,λ) over $\mathbb{C}[\mathbb{Z}_m]$ can be regarded as a nonsingular symmetric form $(L,\lambda^!)$ over $\mathbb{C}$ together with an automorphism $\zeta : (L,\lambda^!)\longrightarrow(L,\lambda^!)$ such that $\zeta^m = 1 : L\longrightarrow L$, with

$$\lambda^!(u,v) \;=\; \Big(\text{coefficient of 1 in } \lambda(u,v) \in \mathbb{C}[\pi]\Big) \in \mathbb{C} \;\; (u,v\in L)\;.$$

Now

$$(L,\lambda^!) \;=\; \bigoplus_{j=0}^{m-1}(L_j,\lambda^!_j)$$

with

$$L_j \;=\; \{u \in L \,|\, (\zeta-\omega^j)^N(u) = 0 \text{ for some } N \geq 0\}\;,$$

the generalized ω^j-eigenspace of ζ. The jth projection

$$f_j \;:\; \mathbb{C}[\mathbb{Z}_m] \longrightarrow \mathbb{C} \;;\; z \longrightarrow \omega^j$$

is a morphism of rings with involution such that

$$f_j(L_k,\lambda^!_k) \;=\; \begin{cases} (L_k,\lambda^!_k) & \text{if } j = k \\ 0 & \text{if } j \neq k\;. \end{cases}$$

The Witt class is determined by the signatures

$$\sigma_j \;=\; (L_j,\lambda^!_j) \in L^0(\mathbb{C}) \;=\; \mathbb{Z}\;,$$

with

$$(L,\lambda) \;=\; (\sigma_0,\sigma_1,\dots,\sigma_{m-1}) \in L^0_p(\mathbb{C}[\mathbb{Z}_m]) \;=\; LAut^0_{(z^m-1)}(\mathbb{C}) \;=\; \mathbb{Z}[\mathbb{Z}_m]\;.$$

The symmetric Witt group $L^0_p(\mathbb{C}[\mathbb{Z}_m]) = \mathbb{Z}[\mathbb{Z}_m]$ can thus be regarded as a subgroup of the automorphism Witt group $LAut^0(\mathbb{C}) = \mathbb{Z}[S^1]$. □

41. Coupling invariants

In Chap. 41 A is a Dedekind ring with involution, and $F = S^{-1}A$ is the quotient field, with $S = A\backslash\{0\} \subset A$. The undecorated L-groups L^*, L_* are the projective L-groups L^*_p, L^p_*.

The results of Chap. 39 for the endomorphism L-groups of a field F are now extended to the endomorphism L-groups of a Dedekind ring A, using the ordinary L-theory of the appropriate localizations of the polynomial extensions:

(i) $A[x]$, with $\overline{x} = x$,
(ii) $A[s]$, with $\overline{s} = 1 - s$,
(iii) $A[z, z^{-1}]$, with $\overline{z} = z^{-1}$.

Note that if A is not a field these polynomial extensions have global dimension 2, so that they are not themselves Dedekind rings (although some of the localizations may be so, e.g. if $S = A\backslash\{0\} \subset A$ is inverted). The case most directly relevant to the computation of the high-dimensional knot cobordism groups C_{2*+1} is (ii) with $A = \mathbb{Z}$, as will be discussed in Chap. 42 below. However, the other cases are also of interest, for example in open books and the bordism of automorphisms of manifolds.

The endomorphism class group of A splits as

$$\mathrm{End}_0(A) = \bigoplus_{p(x)} \mathrm{End}_0^{p(x)^\infty}(A)$$

with the sum running over all the monic irreducible polynomials $p(x) \in A[x]$ (14.16), and

$$\mathrm{End}_0^{p(x)^\infty}(A) = \begin{cases} \mathrm{Nil}_0(A) = K_0(A) & \text{if } p(x) = x \\ \mathrm{End}_0^{p(x)^\infty}(F) = \mathbb{Z} & \text{if } p(x) \neq x. \end{cases}$$

As already indicated in Chap. 14 the situation is considerably more complicated for L-theory, involving the coupling invariants used by Stoltzfus [277] to unravel the knot cobordism groups. The *coupling endomorphism L-groups* $CL\mathrm{End}^*(A, \epsilon)$ will now be defined, to fit into exact sequences

$$\ldots \longrightarrow \bigoplus_{p(x)} LEnd^n_{p(x)^\infty}(A,\epsilon) \longrightarrow LEnd^n(A,\epsilon) \longrightarrow CLEnd^n(A,\epsilon)$$
$$\longrightarrow \bigoplus_{p(x)} LEnd^{n-1}_{p(x)^\infty}(A,\epsilon) \longrightarrow \ldots$$

for any central unit $\epsilon \in A$ such that $\overline{\epsilon} = \epsilon^{-1}$, with the sums running over all the involution-invariant monic irreducible polynomials $p(x) \in A[x]$. Similarly for the isometric, automorphism and asymmetric L-groups. The isometric L-theory of $A = \mathbb{Z}$ is particularly significant, since the high-dimensional knot cobordism groups are given by $C_n = LIso^{n+1}(\mathbb{Z})$.

41A. Endomorphism L-theory

Definition 41.1 (i) Let

$$T_A = \{\sum_{j=0}^{d} a_j x^j \,|\, \overline{a}_j = a_j \in A\,,\, a_d = 1\} \subset A[x]$$

be the multiplicative subset of the involution-invariant monic polynomials, with

$$\sum_{j=0}^{d} \overline{a}_j x^j = \sum_{j=0}^{d} a_j x^j \in A[x]\ ,$$

and let $T_A^{-1}A[x]$ be the localization of $A[x]$ inverting T_A.
(ii) Let $\mathcal{M}_x(A) \subset T_A$ be the set of the involution-invariant monic polynomials $p(x) \in A[x]$ which are irreducible. □

Proposition 41.2 (i) *The endomorphism L-groups of A are related to the T_A-primary endomorphism L-groups of F by a localization exact sequence*

$$\ldots \longrightarrow LEnd^n(A,\epsilon) \longrightarrow LEnd^n_{T_A}(F,\epsilon)$$
$$\longrightarrow LEnd^n(A,S,\epsilon) \longrightarrow LEnd^{n-1}(A,\epsilon) \longrightarrow \ldots$$

with $LEnd^n(A,S,\epsilon)$ the cobordism groups of $S^{-1}A$-contractible $(n-1)$-dimensional f.g. projective ϵ-symmetric Poincaré complexes over A with an endometric structure.
(ii) *The endomorphism L-groups of A are 4-periodic*

$$LEnd^n(A,\epsilon) = LEnd^{n+2}(A,-\epsilon) = LEnd^{n+4}(A,\epsilon)\ .$$

The T_A-primary endomorphism L-groups of F are 4-periodic and such that:

(a) *$LEnd^{2i}_{T_A}(F,\epsilon)$ is the Witt group of F-nonsingular $(-)^i\epsilon$-symmetric forms (L,λ) over A with an endometric structure $f : L \longrightarrow L$,*

(b) $LEnd^{2i+1}_{T_A}(F,\epsilon) = 0$.

(iii) *The T_A-primary endomorphism ϵ-symmetric L-groups of F split as*

$$LEnd^n_{T_A}(F,\epsilon) \;=\; \bigoplus_{p(x)\in\overline{\mathcal{M}}_x(A)} LEnd^n_{p(x)^\infty}(F,\epsilon)\ .$$

If $p(x)\in\overline{\mathcal{M}}_x(A)$ factorizes in $F[x]$ as a product of coprime monic polynomials

$$p(x) \;=\; p_1(x)p_2(x)\dots p_k(x)q(x)\overline{q(x)} \in F[x]$$

with each $p_j(x)\in\overline{\mathcal{M}}_x(F)$ involution-invariant and irreducible the

$$LEnd^n_{p(x)^\infty}(F,\epsilon) \;=\; \bigoplus_{j=1}^{k} L^n(F[x]/(p_j(x)),\epsilon)\ .$$

(iv) *The endomorphism torsion L-groups split as*

$$LEnd^n(A,S,\epsilon) \;=\; \bigoplus_{\mathcal{P}\in\overline{\max}(A)} LEnd^n(A,\mathcal{P}^\infty,\epsilon) = \bigoplus_{\mathcal{P}\in\overline{\max}(A)} LEnd^n(A/\mathcal{P},u_{\mathcal{P}}\epsilon)$$

and

$$LEnd^{2*+1}(A,S,\epsilon) \;=\; 0\ .$$

Proof (i) Each monic polynomial $p(x)\in T_A\subset A[x]$ is a Fredholm 1×1 matrix in $A[x]$, with $A[x]/(p(x))$ a f.g. free A-module of rank degree$(p(x))$. The localization of $A[x]$ inverting T_A is thus a subring

$$T_A^{-1}A[x]\subseteq \Omega_+^{-1}A[x]$$

of the Fredholm localization of $A[x]$ inverting the set Ω_+ of the Fredholm matrices in $A[x]$ (10.4). (For the identity involution on A $T_A^{-1}A[x]=\Omega_+^{-1}A[x]$ by 8.7 and 10.10 (iv).) If (C,ϕ) is an n-dimensional ϵ-symmetric Poincaré complex over A with an endometric structure $f: C\longrightarrow C$ then

$$p(f)\;\simeq\;0\;:\;C\longrightarrow C$$

for some $p(x)\in T_A$, so that by 34.8

$$L^n_{\widetilde{K}_0(A)}(\widetilde{T}_A^{-1}A[x],\epsilon) \;=\; L^n_{\widetilde{K}_0(A)}(\widetilde{\Omega}_+^{-1}A[x],\epsilon) \;=\; LEnd^n(A,\epsilon)$$

with

$$\widetilde{T}_A \;=\; \{\sum_{j=0}^{d} b_jx^j \mid \overline{b}_j = b_j\in A, b_0=1\}\subset A[x]$$

the multiplicative subset consisting of the involution-invariant polynomials with constant coefficient 1. The stated localization exact sequence is just the localization exact sequence of 25.4

$$\ldots \longrightarrow L^n_{\widetilde{K}_0(A)}(\widetilde{T}_A^{-1}A[x],\epsilon) \longrightarrow L^n(\widetilde{T}_A^{-1}F[x],\epsilon)$$
$$\longrightarrow L^n(\widetilde{T}_A^{-1}A[x],S,\epsilon) \longrightarrow L^{n-1}_{\widetilde{K}_0(A)}(\widetilde{T}_A^{-1}A[x],\epsilon) \longrightarrow \ldots .$$

(ii) The 4-periodicity of the endomorphism L-groups is obtained by surgery below the middle dimension, as for the ordinary ϵ-symmetric L-groups $L^n(A,\epsilon)$ in 38.1 (i). Here are the details.

(a) Every element of $LEnd^{2i}_{T_A}(F,\epsilon)$ is represented by an endometric structure $g : S^{-1}M \longrightarrow S^{-1}M$ for the form $S^{-1}(M,\phi)$ over $S^{-1}A = F$ induced from an F-nonsingular $(-)^i\epsilon$-symmetric form (M,ϕ) over A with

$$p(g) = 0 : S^{-1}M \longrightarrow S^{-1}M$$

for some

$$p(x) = \sum_{j=0}^{d} a_j x^j \in T_A \subset A[x] \quad (\overline{a}_j = a_j, a_d = 1) .$$

Working as in Stoltzfus [277, 2.4] (cf. 17.5) define a g-invariant submodule

$$L = M + g(M) + \ldots + g^{d-1}(M) \subset S^{-1}M ,$$

and let

$$f = g| : L \longrightarrow L \ , \ \lambda = \overline{s}^d\phi s^d| : L \longrightarrow L^*$$

with $s \in S$ such that $sg(M) \subseteq M$. Now L is a torsion-free f.g. module over a Dedekind ring A, so that it is f.g. projective, and (L,λ) is an F-nonsingular $(-)^i\epsilon$-symmetric form over A with an endometric structure $f : L \longrightarrow L$ such that

$$(S^{-1}(M,\phi),g) = S^{-1}(L,\lambda,f) \in LEnd^{2i}_{T_A}(F,\epsilon) .$$

(b) $LEnd^{2i+1}_{T_A}(F,\epsilon)$ is just the subgroup of

$$LEnd^{2i+1}(F,\epsilon) = \bigoplus_{p(x)\in\overline{\mathcal{M}}_x(F)} LEnd^{2i+1}_{p(x)}(F,\epsilon) = L^{2i+1}(F(x),\epsilon) = 0 ,$$

indexed by $\overline{\mathcal{M}}_x(A) \subseteq \overline{\mathcal{M}}_x(F)$.

(iii) Immediate from 35.2 (ii) and the unique factorization of polynomials in $F[x]$.

(iv) As for the corresponding results for the ordinary L-groups

$$L^n(A,S,\epsilon) = \bigoplus_{\mathcal{P}\in\overline{\max}(A)} L^n(A,\mathcal{P}^\infty,\epsilon) = \bigoplus_{\mathcal{P}\in\overline{\max}(A)} L^n(A/\mathcal{P},u_{\mathcal{P}}\epsilon) ,$$

$$L^{2*+1}(A,S,\epsilon) = 0$$

of 38.5 and 38.9. □

Proposition 41.3 *Let $p(x) \in \overline{\mathcal{M}}_x(A)$.*
(i) *The localization exact sequence*

$$\dots \longrightarrow L^n(A[x],\epsilon) \longrightarrow L^n_{\widetilde{K}_0(A[x])}(A[x,p(x)^{-1}],\epsilon) \longrightarrow L^n(A[x],p(x)^\infty,\epsilon) \longrightarrow L^{n-1}(A[x],\epsilon) \longrightarrow \dots$$

splits, with

$$L^n_{\widetilde{K}_0(A[x])}(A[x,p(x)^{-1}],\epsilon) = L^n(A[x],\epsilon) \oplus L^n(A[x],p(x)^\infty,\epsilon).$$

(ii) *The $(p(x))$-primary torsion ϵ-symmetric L-groups of $A[x]$ are such that*

$$L^n(A[x],p(x)^\infty,\epsilon) = \mathrm{LEnd}^n_{p(x)^\infty}(A,\epsilon) = \mathrm{LEnd}^n_{p(x)}(A,\epsilon)$$

with $\mathrm{LEnd}^n_{p(x)}(A,\epsilon)$ the cobordism group of f.g. projective n-dimensional ϵ-symmetric Poincaré complexes (C,ϕ) over A with an endometric structure $(f,\delta\phi)$ such that $p(f) \simeq 0 : C \longrightarrow C$.
(iii) *The $(p(x))$-primary ϵ-symmetric endomorphism L-groups of A and F are related by a localization exact sequence*

$$\dots \longrightarrow \mathrm{LEnd}^n_{p(x)}(A,\epsilon) \longrightarrow \mathrm{LEnd}^n_{p(x)}(F,\epsilon) \longrightarrow \mathrm{LEnd}^n_{p(x)}(A,S,\epsilon) \longrightarrow \mathrm{LEnd}^{n-1}_{p(x)}(A,\epsilon) \longrightarrow \dots$$

with $\mathrm{LEnd}^n_{p(x)}(A,S,\epsilon)$ the cobordism group of F-contractible f.g. projective $(n-1)$-dimensional ϵ-symmetric Poincaré complexes (C,ϕ) over A with an endometric structure $(f,\delta\phi)$ such that $p(f) \simeq 0 : C \longrightarrow C$, and:

(a) *$\mathrm{LEnd}^{2i}_{p(x)}(A,S,\epsilon)$ is the Witt group of F-nonsingular $(-)^i\epsilon$-symmetric linking forms (L,λ) over (A,S) with an endometric structure $f : L \longrightarrow L$ such that $p(f) = 0$, and*

$$\begin{aligned}\mathrm{LEnd}^{2i}_{p(x)}(A,S,\epsilon) &= \bigoplus_{\mathcal{P}\in\overline{\max}(A)} \ \bigoplus_{q(x)\in\overline{\mathcal{M}}_x(A/\mathcal{P}),q(x)|p(x)} \mathrm{LEnd}^0_{q(x)}(A/\mathcal{P},(-)^i\epsilon) \\ &= \bigoplus_{\mathcal{P}\in\overline{\max}(A)} \ \bigoplus_{q(x)\in\overline{\mathcal{M}}_x(A/\mathcal{P}),q(x)|p(x)} L^0((A/\mathcal{P})[x]/(q(x)),(-)^i\epsilon)\ ,\end{aligned}$$

(b) *$\mathrm{LEnd}^{2i+1}_{p(x)}(A,S,\epsilon) = 0$.*

(iv) *If $A[x]/(p(x))$ is a Dedekind ring (e.g. if A is a field, or if $p(x)$ has degree 1) then*

$$\mathrm{LEnd}^n_{p(x)}(A,\epsilon) = L^n(A[x]/(p(x)),\epsilon)\ .$$

(v) *The ϵ-symmetric L-groups of $A[x]$ are such that*

$$L^n_h(A[x],\epsilon) = L^n_h(A,\epsilon)\ ,\ L^n_{\widetilde{K}_0(A)}(A[x],\epsilon) = L^n(A,\epsilon)\ .$$

Proof (i) This is a special case of 35.4.
(ii) The skew-suspension maps define isomorphisms

$$\overline{S} : L\mathrm{End}^n_{p(x)}(A,\epsilon) \xrightarrow{\simeq} L\mathrm{End}^{n+2}_{p(x)}(A,-\epsilon) ,$$
$$\overline{S} : L\mathrm{End}^n_{p(x)\infty}(A,\epsilon) \xrightarrow{\simeq} L\mathrm{End}^{n+2}_{p(x)\infty}(A,-\epsilon)$$

by algebraic surgery above the middle dimension (as in the ordinary L-theory case of Ranicki [235, Prop. 4.6]), so it suffices to prove that the natural maps

$$L\mathrm{End}^n_{p(x)}(A,\epsilon) \longrightarrow L\mathrm{End}^n_{p(x)\infty}(A,\epsilon) ; \ (C,f,\delta\phi,\phi) \longrightarrow (C,f,\delta\phi,\phi)$$

are isomorphisms for $n = 0, 1$. This will be done using the appropriate devissage argument.
Only the case $n = 0$ will be worked out in detail; the case $n = 1$ is similar, with formations instead of forms. An element

$$(M,\phi,f) \in L\mathrm{End}^0_{p(x)\infty}(A,\epsilon)$$

is represented by a f.g. projective nonsingular ϵ-symmetric form (M,ϕ) over A together with an endomorphism $f : M \longrightarrow M$ such that

$$f^*\phi = \phi f : M \longrightarrow M^* , \quad p(f)^N = 0 : M \longrightarrow M$$

for some $N \geq 1$. If $N \geq 2$ reduce to the case $N = 1$, as follows. Consider the submodule

$$K = p(f)^{N-1}(M) \subset M .$$

The A-submodule

$$L = \{x \in M \,|\, ax \in K \text{ for some } a \in A\backslash\{0\}\} \subseteq M$$

contains K, with L/K the torsion submodule of M/K. The quotient A-module M/L is projective, so that L is a direct summand of M. For any $x_1, x_2 \in L$ there exist $a_1, a_2 \in A\backslash\{0\}$, $y_1, y_2 \in M$ such that

$$\begin{aligned} a_1x_1 &= p(f)^{N-1}(y_1) , \quad a_2x_2 = p(f)^{N-1}(y_2) \in K , \\ a_2\phi(x_1,x_2)\overline{a}_1 &= \phi(p(f)^{N-1}(y_1), p(f)^{N-1}(y_2)) \\ &= \phi(y_1, p(f)^{2N-2}(y_2)) \\ &= 0 \in A \quad (\text{since } 2N-2 \geq N) , \\ \phi(x_1,x_2) &= 0 \in A . \end{aligned}$$

Thus L is a sublagrangian of (M,ϕ) and

$$(M,\phi,f) = (L^\perp/L, [\phi], f) \in \mathrm{im}(L\mathrm{End}^0_{p(x)}(A,\epsilon) \longrightarrow L\mathrm{End}^0_{p(x)\infty}(A,\epsilon)) .$$

(iii) Let

$$p(x) \;=\; \sum_{j=0}^{d} a_j x^j \in \overline{\mathcal{M}}_x(A) \quad (a_d = 1)$$

with reduction

$$\widetilde{p}(x) \;=\; x^d p(x^{-1}) \;=\; \sum_{j=0}^{d} a_{d-j} x^j \in A[x] \;.$$

The ϵ-symmetric L-groups of $A[x, \widetilde{p}(x)^{-1}]$ are such that

$$\begin{aligned} L^n_{\widetilde{K}_0(A)}(A[x,\widetilde{p}(x)^{-1}],\epsilon) \;&=\; L^n(A,\epsilon) \oplus \widetilde{L\mathrm{End}}^n_{p(x)}(A,\epsilon) \\ &=\; L\mathrm{End}^n_{p(x)}(A,\epsilon) \end{aligned}$$

by the special case $S = \{x^j p(x)^k \,|\, j,k \geq 0\} \subset A[x]$ of 35.4. In particular, there is defined an isomorphism

$$\begin{aligned} L\mathrm{End}^0_{p(x)}(A,\epsilon) \xrightarrow{\;\simeq\;} L^0_{\widetilde{K}_0(A)}(A[x,\widetilde{p}(x)^{-1}],\epsilon) \;; \\ (M,\phi,f) \longrightarrow (M[x,\widetilde{p}(x)^{-1}],\phi(1-xf)) \;. \end{aligned}$$

The stated localization exact sequence is just the localization exact sequence of 25.4

$$\begin{aligned} \dots \longrightarrow L^n_{\widetilde{K}_0(A)}(A[x,\widetilde{p}(x)^{-1}],\epsilon) \longrightarrow L^n(F[x,\widetilde{p}(x)^{-1}],\epsilon) \\ \longrightarrow L^n(A[x,\widetilde{p}(x)^{-1}],S,\epsilon) \longrightarrow L^{n-1}_{\widetilde{K}_0(A)}(A[x,\widetilde{p}(x)^{-1}],\epsilon) \longrightarrow \dots \;. \end{aligned}$$

The ring

$$R \;=\; A[x]/(p(x))$$

is an integral domain which is an order with quotient field

$$E \;=\; F[x]/(p(x)) \;.$$

As in 41.3 choose a non-zero involution-preserving F-linear map $h : E \longrightarrow F$, e.g. the trace map $\mathrm{tr}_{E/F}$ if $p(x)$ is separable. The inverse different (alias codifferent) of R

$$\Delta \;=\; \{v \in E \,|\, h(Rv) \subseteq A\}$$

is equipped with a pairing

$$\Delta \times R \longrightarrow A \;;\; (u,v) \longrightarrow h(\overline{u}v)$$

such that the adjoint A-module morphisms

$$\begin{aligned} \Delta \longrightarrow \mathrm{Hom}_A(R,A) \;;\; u \longrightarrow (v \longrightarrow h(\overline{u}v)) \;, \\ R \longrightarrow \mathrm{Hom}_A(\Delta,A) \;;\; v \longrightarrow (u \longrightarrow h(\overline{u}v)) \end{aligned}$$

are isomorphisms. Use h to identify $LEnd^{2i}_{p(x)}(A,\epsilon)$ with the Witt group of nonsingular $(-)^i\epsilon$-symmetric Δ-valued forms on f.g. torsion-free R-modules (cf. Milnor [197], Knebusch and Scharlau [137] and Stoltzfus [277, 4.6], [278, 2.13]). The other even-dimensional endomorphism L-groups in the localization exact sequence

$$\begin{aligned} 0 \longrightarrow LEnd^{2i}_{p(x)}(A,\epsilon) \longrightarrow LEnd^{2i}_{p(x)}(F,\epsilon) \longrightarrow LEnd^{2i}_{p(x)}(A,S,\epsilon) \\ \longrightarrow LEnd^{2i-1}_{p(x)}(A,\epsilon) \longrightarrow 0 \end{aligned}$$

are given by

$$\begin{aligned} LEnd^{2i}_{p(x)}(F,\epsilon) &= L^0(E,(-)^i\epsilon)\ , \\ LEnd^{2i}_{p(x)}(A,S,\epsilon) &= \bigoplus_{\mathcal{Q}\in\overline{\max}(\Delta)} L^0(\Delta/\mathcal{Q},(-)^i\epsilon)\ . \end{aligned}$$

The involution-invariant maximal ideals of Δ are of the form

$$\mathcal{Q} = (\mathcal{P}, q(x)) \lhd \Delta$$

with $\mathcal{P} \in \overline{\max}(A)$ and $q(x) \in A[x]$ a monic polynomial such that the reduction in $(A/\mathcal{P})[x]$ is involution-invariant, irreducible and divides the reduction of $p(x)$, with

$$\Delta/\mathcal{Q} = (A/\mathcal{P})[x]/(q(x))\ .$$

(iv) If $R = A[x]/(p(x))$ is a Dedekind ring it is possible to identify

$$\begin{aligned} &\{\text{f.g. projective } R\text{-modules } M\} \\ &\quad = \{\text{f.g. projective } A\text{-modules } M \text{ with an endomorphism} \\ &\qquad\qquad f : M \longrightarrow M \text{ such that } p(f) = 0 : M \longrightarrow M\}\ , \end{aligned}$$

and also

$$\begin{aligned} &\{\epsilon\text{-symmetric forms } (M,\phi) \text{ over } R\} \\ &\quad = \{\epsilon\text{-symmetric forms } (M,\phi) \text{ over } A \text{ with an endomorphic} \\ &\qquad\qquad \text{structure } f : M \longrightarrow M \text{ such that } p(f) = 0 : M \longrightarrow M\} \end{aligned}$$

with λ, ϕ, f related by

$$\begin{aligned} \lambda(y,z) &= \phi(y,w) \in R \\ (y,z \in M, w \in M[x]&, (x-f)(w) = p(x)z \in M[x])\ . \end{aligned}$$

The function

$$L^n(R,\epsilon) \longrightarrow LEnd^n_{p(x)}(A,\epsilon)\ ;\ (C,\lambda) \longrightarrow (C,x,0,\phi)$$

is an isomorphism already on the level of algebraic Poincaré complexes.

(v) By 35.6 (iii) and the special case $p(x) = x$ of (iii)

$$\begin{aligned} L^n_{\widetilde{K}_0(A)}(A[x],\epsilon) &= LNil^n(A,\epsilon) \\ &= LEnd^n_{(x)^\infty}(A,\epsilon) = LEnd^n_x(A,\epsilon) = L^n(A,\epsilon)\ . \end{aligned}$$

For the free case apply the 5-lemma to the morphism of exact sequences

$$\begin{array}{ccccccccc} \longrightarrow & L^n_h(A,\epsilon) & \longrightarrow & L^n(A,\epsilon) & \longrightarrow & \widehat{H}^n(\mathbb{Z}_2;\widetilde{K}_0(A)) & \longrightarrow & L^{n-1}_h(A,\epsilon) & \longrightarrow \\ & \downarrow & & \downarrow\cong & & \| & & \downarrow & \\ \longrightarrow & L^n_h(A[x],\epsilon) & \longrightarrow & L^n_{\widetilde{K}_0(A)}(A[x],\epsilon) & \longrightarrow & \widehat{H}^n(\mathbb{Z}_2;\widetilde{K}_0(A)) & \longrightarrow & L^{n-1}_h(A[x],\epsilon) & \longrightarrow \end{array}$$

□

Remark 41.4 (i) The endomorphism L-groups which occur in 41.3 (iv) are given in the terminology of 28.15 by

$$LEnd^n_{p(x)}(A,\epsilon) = L^n(A[x]/(p(x)),A,\epsilon)$$

(cf. 34.12). For $n=0$ this is the Witt group of nonsingular ϵ-symmetric forms (M,ϕ) over A with a morphism of rings with involution

$$A[x]/(p(x)) \longrightarrow \mathrm{Hom}_A(M,M)^{op}\ .$$

(ii) For separable $p(x)\in\overline{\mathcal{M}}_x(A)$ the trace map

$$\mathrm{tr}_{E/F}\ :\ E = F[x]/(p(x)) \longrightarrow F$$

is such that

$$\mathrm{tr}_{E/F}\left(\frac{x^k}{p'(x)}\right) = \begin{cases} 0 & \text{if } k=0,1,2,\dots,d-2 \\ 1 & \text{if } k=d-1 \end{cases}$$

(cf. 39.4), so that the inverse different of $R=A[x]/(p(x))$ is

$$\Delta = Rp'(x)^{-1}\subset E\ .$$

(iii) See Levine [160, 28.2] for the necessary and sufficient condition on an integral irreducible monic polynomial $p(x)\in\mathcal{M}_x(\mathbb{Z})$ for the quotient ring $\mathbb{Z}[x]/(p(x))$ to be Dedekind. For example, suppose that $p(x)=x^2+ax+b\in\mathbb{Z}[x]$ is a quadratic polynomial with discriminant $D=a^2-4b\neq 0$. The quotient ring $\mathbb{Z}[x]/(x^2+ax+b)$ is Dedekind unless $p^2|D$ for some prime p, and p divides at most one of a,b, and $D\equiv 0$ or $4 (\mathrm{mod}\ 16)$ if $p=2$. □

Remark 41.5 The analogue of 41.3 is false for the ϵ-quadratic endomorphism L-groups, since the corresponding devissage fails in ϵ-quadratic L-theory (cf. 38.6). The natural map

$$LEnd^{p(x)}_n(A,\epsilon) \longrightarrow LEnd^{p(x)^\infty}_n(A,\epsilon)$$

is injective, but need not be surjective – for example, in the case $A=\mathbb{Z}$, $p(x)=x$, $n=0$, $\epsilon=-1$ the map

$$\begin{aligned} L_0(\mathbb{Z},-1) &= L\mathrm{End}_0^x(\mathbb{Z},-1) \\ &\longrightarrow L\mathrm{End}_0^{x^\infty}(\mathbb{Z},-1) = L\mathrm{Nil}_0(\mathbb{Z},-1) = L_0(\mathbb{Z}[x],-1) \end{aligned}$$

has infinitely generated cokernel (cf. Remark 35.8 and Cappell [37]). □

Definition 41.6 The *coupling endomorphism L-groups* $CL\mathrm{End}^n(A,\epsilon)$ are the relative L-groups in the exact sequences

$$\dots \longrightarrow \bigoplus_{p(x)} L\mathrm{End}^n_{p(x)^\infty}(A,\epsilon) \longrightarrow L\mathrm{End}^n(A,\epsilon) \longrightarrow CL\mathrm{End}^n(A,\epsilon)$$
$$\longrightarrow \bigoplus_{p(x)\in\overline{\mathcal{M}}_x(A)} L\mathrm{End}^{n-1}_{p(x)^\infty}(A,\epsilon) \longrightarrow \dots .$$

□

Thus $CL\mathrm{End}^n(A,\epsilon)$ is the cobordism group of n-dimensional ϵ-symmetric Poincaré pairs $(C\longrightarrow D,(\delta\phi,\phi))$ over A with an endometric structure (f_D,f_C) which is decoupled on the boundary, i.e. such that

$$(C,\phi,f_C) = \bigoplus_{p(x)} (C,\phi,f_C)_{p(x)}$$

with $(C,\phi,f_C)_{p(x)}$ $(p(x))$-primary.

Proposition 41.7 (i) *The coupling endomorphism L-groups fit into a commutative braid of exact sequences*

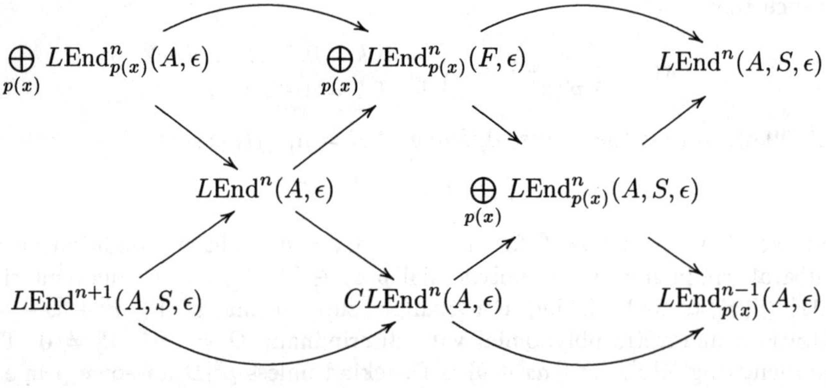

with

$$L\mathrm{End}^{2*+1}(A,S,\epsilon) = CL\mathrm{End}^{2*+1}(A,\epsilon) = 0 .$$

(ii) *The endomorphism L-groups of A fit into an exact sequence*

$$0 \longrightarrow \bigoplus_{p(x)} L\mathrm{End}^{2i}_{p(x)}(A,\epsilon) \longrightarrow L\mathrm{End}^{2i}(A,\epsilon) \longrightarrow \bigoplus_{p(x)} L\mathrm{End}^{2i}_{p(x)}(A,S,\epsilon)$$
$$\longrightarrow L\mathrm{End}^{2i}(A,S,\epsilon) \oplus \bigoplus_{p(x)} L\mathrm{End}^{2i-1}_{p(x)}(A,\epsilon) \longrightarrow 0 .$$

(iii) *If A is additively torsion-free the morphisms*

$$\bigoplus_{p(x)} LEnd^{2i}_{p(x)}(A,\epsilon) \longrightarrow LEnd^{2i}(A,\epsilon)\ ,$$

$$LEnd^{2i}(A,\epsilon) \longrightarrow \bigoplus_{p(x)} LEnd^{2i}_{p(x)}(F,\epsilon)$$

are isomorphisms modulo 8-torsion.

Proof (i) The braid is derived (by working on the chain level) from the morphism of exact sequences

$$\begin{array}{ccccccccc}
0 \longrightarrow & CLEnd^{2i}(A,\epsilon) & \longrightarrow & \bigoplus\limits_{p(x)} LEnd^{2i}_{p(x)}(A,S,\epsilon) & \longrightarrow & LEnd^{2i}(A,S,\epsilon) & \longrightarrow 0 \\
& \Big\| & & \Big\downarrow \cong & & \Big\downarrow \cong & \\
0 \longrightarrow & CLEnd^{2i}(A,\epsilon) & \longrightarrow & \bigoplus\limits_{p(x)} LEnd^{2i}_{p(x)}(\widehat{A},\widehat{S},\epsilon) & \longrightarrow & LEnd^{2i}(\widehat{A},\widehat{S},\epsilon) & \longrightarrow 0
\end{array}$$

and $LEnd^{2*+1}(A,S,\epsilon) = 0$ by 41.2 (iv). Furthermore, there is defined an exact sequence

$$\bigoplus_{p(x)} LEnd^{2i+2}_{p(x)^\infty}(A,S,\epsilon) \xrightarrow{\ \partial\ } LEnd^{2i+2}(A,S,\epsilon) \longrightarrow CLEnd^{2i+1}(A,\epsilon)$$

$$\longrightarrow \bigoplus_{p(x)} LEnd^{2i+1}_{p(x)^\infty}(A,S,\epsilon)$$

with

$$LEnd^{2i+2}(A,S,\epsilon) = \bigoplus_{\mathcal{P}\in\overline{\max}(A)} LEnd^{2i+2}(A/\mathcal{P},u_{\mathcal{P}}\epsilon)$$

by 41.2 (iv). The boundary map ∂ is onto, since if (M,ϕ) is a nonsingular $u_{\mathcal{P}}\epsilon$-symmetric form over $A/\mathcal{P}$ with an endometric structure $f : M \longrightarrow M$ then

$$(M,\phi,f) = \bigoplus_{q(x)\in\overline{\mathcal{M}}_x(A/\mathcal{P})} (M,\phi,f)_{q(x)}$$

with $(M,\phi,f)_{q(x)}$ $(q(x))$-primary, and each $q(x)$ can be lifted to an element $\widetilde{q}(x) \in \overline{\mathcal{M}}_x(A)$. Since $LEnd^{2i+1}_{p(x)^\infty}(A,S,\epsilon) = 0$ (41.3 (iv)), it follows that $CLEnd^{2i+1}(A,\epsilon) = 0$.

(ii) This is the Mayer–Vietoris exact sequence obtained from (i), using $LEnd^{2i+1}(A,S,\epsilon) = 0$ and $CLEnd^{2i+1}(A,\epsilon) = 0$.

(iii) Exactly as for ordinary L-theory in 37.8, with the natural morphism of rings with involution $A \longrightarrow \widehat{A}$ to the S-adic completion inducing an excision isomorphism of exact sequences

$$\begin{array}{ccccccc}
\to \bigoplus\limits_{p(x)} LEnd^n_{p(x)}(A,\epsilon) & \to & \bigoplus\limits_{p(x)} LEnd^n_{p(x)}(F,\epsilon) & \to & \bigoplus\limits_{p(x)} LEnd^n_{p(x)}(A,S,\epsilon) & \to \\
\downarrow & & \downarrow \cong & & \downarrow & \\
\longrightarrow LEnd^n(A,\epsilon) & \longrightarrow & LEnd^n_{T_A}(F,\epsilon) & \longrightarrow & LEnd^n(A,S,\epsilon) & \longrightarrow
\end{array}$$

and the symmetric Witt ring $L^0(\widehat{\mathbb{Z}}) = \mathbb{Z}_8 \oplus \mathbb{Z}_2$ of the profinite completion $\widehat{\mathbb{Z}}$ of $\mathbb{Z}$ acting on the endomorphism L-groups of $(\widehat{A}, \widehat{S})$ (cf. 37.13). □

41B. Isometric L-theory

Definition 41.8 (i) Let

$$U_A = \{\sum_{j=0}^{d} a_j s^j \,|\, \overline{a}_j = (-)^{d-j} \sum_{k=j}^{d} \binom{k}{j} a_k \,,\, a_d = 1\} \subset A[s]$$

be the multiplicative subset of the involution-invariant monic polynomials, with

$$\sum_{j=0}^{d} \overline{a}_j (1-s)^j = (-)^d \sum_{j=0}^{d} a_j s^j \in A[s] \,,$$

and let $U_A^{-1}A[s]$ be the localization of $A[s]$ inverting U_A.
(ii) Let $\overline{\mathcal{M}}_s(A) \subset U_A$ be the set of the involution-invariant monic polynomials $p(s) \in A[s]$ which are irreducible. □

Proposition 41.9 (i) *The isometric L-groups of A are related to the U_A-primary isometric L-groups of F by a localization exact sequence*

$$\begin{aligned} \dots \longrightarrow LIso^n(A,\epsilon) \longrightarrow LIso^n_{U_A}(F,\epsilon) & \\ \longrightarrow LIso^n(A,S,\epsilon) \longrightarrow LIso^{n-1}(A,\epsilon) \longrightarrow \dots \end{aligned}$$

with $LIso^n(A,S,\epsilon)$ the cobordism groups of $(n-1)$-dimensional F-contractible f.g. projective ϵ-symmetric Poincaré complexes over A with an isometric structure.
(ii) *The isometric L-groups of A are 4-periodic*

$$LIso^n(A,\epsilon) = LIso^{n+2}(A,-\epsilon) = LIso^{n+4}(A,\epsilon) \,.$$

The U_A-primary isometric L-groups of F are 4-periodic and such that:

(a) *$LIso^{2i}_{U_A}(F,\epsilon)$ is the Witt group of F-nonsingular $(-)^i\epsilon$-symmetric forms (L,λ) over A with an isometric structure $f : L \longrightarrow L$,*
(b) $LIso^{2i+1}_{U_A}(F,\epsilon) = 0$.

(iii) *The U_A-primary isometric ϵ-symmetric L-groups of F split as*

$$LIso^n_{U_A}(F,\epsilon) \;=\; \bigoplus_{p(s)\in\overline{\mathcal{M}}_s(A)} LIso^n_{p(s)^\infty}(F,\epsilon)\ .$$

If $p(s)\in\overline{\mathcal{M}}_s(A)$ factorizes in $F[s]$ as a product of coprime monic polynomials

$$p(s) \;=\; p_1(s)p_2(s)\dots p_k(s)q(s)\overline{q(s)} \in F[s]$$

with each $p_j(s)\in\overline{\mathcal{M}}_s(F)$ involution-invariant and irreducible then

$$LEnd^n_{p(s)^\infty}(F,\epsilon) \;=\; \bigoplus_{j=1}^{k} L^n(F[s]/(p_j(s)),(-)^{d_j}\epsilon)\ .$$

where $d_j = \text{degree}(p_j(s))$.
(iv) *The isometric torsion L-groups split as*

$$LIso^n(A,S,\epsilon) = \bigoplus_{\mathcal{P}\in\overline{\max}(A)} LIso^n(A,\mathcal{P}^\infty,\epsilon) = \bigoplus_{\mathcal{P}\in\overline{\max}(A)} LIso^n(A/\mathcal{P},u_{\mathcal{P}}\epsilon)$$

and

$$LIso^{2*+1}(A,S,\epsilon) \;=\; 0\ .$$

Proof As for 41.2. □

Proposition 41.10 *Let $p(s)\in\overline{\mathcal{M}}_s(A)$.*
(i) *The localization exact sequence*

$$\dots \longrightarrow L^n(A[s],\epsilon) \longrightarrow L^n_{\widetilde{K}_0(A[s])}(A[s,p(s)^{-1}],\epsilon) \longrightarrow L^n(A[s],p(s)^\infty,\epsilon)$$
$$\longrightarrow L^{n-1}(A[s],\epsilon) \longrightarrow \dots$$

splits, with

$$L^n_{\widetilde{K}_0(A[s])}(A[s,p(s)^{-1}],\epsilon) \;=\; L^n(A[s],\epsilon)\oplus L^n(A[s],p(s)^\infty,\epsilon)\ .$$

(ii) *The $(p(s))$-primary torsion ϵ-symmetric L-groups of $A[s]$ are such that*

$$L^n(A[s],p(s)^\infty,\epsilon) \;=\; LIso^n_{p(s)^\infty}(A,-\epsilon) \;=\; LIso^n_{p(s)}(A,-\epsilon)$$

with $LIso^n_{p(s)}(A,-\epsilon)$ the cobordism group of f.g. projective n-dimensional $(-\epsilon)$-symmetric Poincaré complexes (C,ϕ) over A with an isometric structure $(f,\delta\phi)$ such that $p(f)\simeq 0 : C\longrightarrow C$.
(iii) *The $(p(s))$-primary ϵ-symmetric isometric L-groups of A and F are related by a localization exact sequence*

$$\dots \longrightarrow LIso^n_{p(s)}(A,\epsilon) \longrightarrow LIso^n_{p(s)}(F,\epsilon)$$
$$\longrightarrow LIso^n_{p(s)}(A,S,\epsilon) \longrightarrow LIso^{n-1}_{p(s)}(A,\epsilon) \longrightarrow \dots$$

with $LIso^n_{p(s)}(A,S,\epsilon)$ *the cobordism group of* F*-contractible f.g. projective* $(n-1)$*-dimensional* ϵ*-symmetric Poincaré complexes* (C,ϕ) *over* A *with an isometric structure* $(f,\delta\phi)$ *such that* $p(f)\simeq 0 : C\longrightarrow C$*, and:*

(a) $LIso^{2i}_{p(s)}(A,S,\epsilon)$ *is the Witt group of* F*-nonsingular* $(-)^i\epsilon$*-symmetric linking forms* (L,λ) *over* (A,S) *with an isometric structure* $f : L\longrightarrow L$ *such that* $p(f)=0$*, and*

$$\begin{aligned} & LIso^{2i}_{p(s)}(A,S,\epsilon) \\ &= \bigoplus_{\mathcal{P}\in\overline{\max}(A)} \ \bigoplus_{q(s)\in\overline{\mathcal{M}}_s(A/\mathcal{P}),q(s)|p(s)} LIso^0_{q(s)}(A/\mathcal{P},(-)^i\epsilon) \\ &= \bigoplus_{\mathcal{P}\in\overline{\max}(A)} \ \bigoplus_{q(s)\in\overline{\mathcal{M}}_s(A/\mathcal{P}),q(s)|p(s)} L^0((A/\mathcal{P})[s]/(q(s)),(-)^i\epsilon) , \end{aligned}$$

(b) $LIso^{2i+1}_{p(s)}(A,S,\epsilon)=0$.

(iv) *If* $A[s]/(p(s))$ *is a Dedekind ring (e.g. if* A *is a field, or if* $p(s)$ *has degree 1) then*

$$LIso^n_{p(s)}(A,\epsilon) \ = \ L^n(A[s]/(p(s)),\epsilon) \ .$$

(v) *The* ϵ*-symmetric* L*-groups of* $A[s]$ *are such that*

$$L^n_h(A[s],\epsilon) \ = \ L^n_h(A,\epsilon) \ , \ L^n_{\widetilde{K}_0(A)}(A[s],\epsilon) \ = \ L^n(A,\epsilon) \ .$$

Proof (i) This is a special case of 35.11.
(ii)+(iii)+(iv) As in the proof of 41.3 (ii) it suffices to prove that the natural map

$$LIso^n_{p(s)}(A,\epsilon) \longrightarrow LIso^n_{p(s)^\infty}(A,\epsilon) \ ; \ (C,f,\delta\phi,\phi) \longrightarrow (C,f,\delta\phi,\phi)$$

is an isomorphism for $n=0$. An element

$$(M,\phi,f)\in LIso^0_{p(s)^\infty}(A,\epsilon)$$

is represented by a f.g. projective nonsingular ϵ-symmetric form (M,ϕ) over A together with an endomorphism $f : M\longrightarrow M$ such that

$$(1-f^*)\phi \ = \ \phi f \ : \ M \longrightarrow M^* \ , \ p(f)^N \ = \ 0 \ : \ M \longrightarrow M$$

for some $N\geq 1$. If $N\geq 2$ reduce to the case $N=1$ as in the proof of 41.3, noting that $p(f) : M\longrightarrow M$ is such that

$$p(f)^*\phi \ = \ \overline{p}(f^*)\phi \ = \ \phi\overline{p}(1-f) \ = \ u^{-1}\phi p(f) \ : \ M \longrightarrow M^* \ .$$

The rest is as for 41.3 (ii)+(iii)+(iv)+(v). □

Remark 41.11 (i) The isometric L-groups which occur in 41.10 (iv) are given in the terminology of 28.15 by

$$LIso^n_{p(s)}(A,\epsilon) \ = \ L^n(A[s]/(p(s)),A,\epsilon) \ .$$

For $n = 0$ this is the Witt group of nonsingular ϵ-symmetric forms (M,ϕ) over A with a morphism of rings with involution

$$A[s]/(p(s)) \longrightarrow \text{Hom}_A(M,M)^{op} \ .$$

(ii) In general, $R = A[s]/(p(s))$ is an order in its quotient field $E = F[s]/(p(s))$. As in 41.4 (ii) choose a non-zero involution-preserving F-linear map $h : E \longrightarrow F$ (e.g. the trace map $\text{tr}_{E/F}$ if $p(s)$ is separable) and identify $LIso^{2i}_{p(s)}(A,\epsilon)$ with the Witt group of nonsingular $(-)^i\epsilon$-symmetric Δ-valued forms on f.g. torsion-free $A[s]/(p(s))$-modules, with

$$\Delta \ = \ \{v \in E \,|\, h(vR) \subseteq A\}$$

the inverse different. The other terms in the localization exact sequence

$$\begin{aligned}0 \longrightarrow LIso^{2i}_{p(s)}(A,\epsilon) \longrightarrow LIso^{2i}_{p(s)}(F,\epsilon) \longrightarrow LIso^{2i}_{p(s)}(A,S,\epsilon) \\ \longrightarrow LIso^{2i-1}_{p(s)}(A,\epsilon) \longrightarrow 0\end{aligned}$$

are given by

$$\begin{aligned} LIso^{2i}_{p(s)}(F,\epsilon) \ &= \ L^0(E,(-)^i\epsilon) \ , \\ LIso^{2i}_{p(s)}(A,S,\epsilon) \ &= \ \bigoplus_{\mathcal{P}\in\overline{\max}(\Delta)} L^0(\Delta/\mathcal{P},(-)^i\epsilon) \ . \end{aligned}$$

□

Definition 41.12 The *coupling isometric L-groups* $CLIso^n(A,\epsilon)$ are the relative L-groups in the exact sequences

$$\begin{aligned}\dots \longrightarrow \bigoplus_{p(s)} LIso^n_{p(s)}(A,\epsilon) \longrightarrow LIso^n(A,\epsilon) \longrightarrow CLIso^n(A,\epsilon) \\ \longrightarrow \bigoplus_{p(s)} LIso^{n-1}_{p(s)}(A,\epsilon) \longrightarrow \dots \ .\end{aligned}$$

with $p(s) \in \overline{\mathcal{M}}_s(A)$. □

Thus $CLIso^n(A,\epsilon)$ is the cobordism group of n-dimensional ϵ-symmetric Poincaré pairs $(C \longrightarrow D,(\delta\phi,\phi))$ over A with an isometric structure (f_D,f_C) which is decoupled on the boundary, i.e. such that

$$(C,\phi,f_C) \ = \ \bigoplus_{p(s)} (C,\phi,f_C)_{p(s)}$$

with $(C,\phi,f_C)_{p(s)}$ $(p(s))$-primary.

Proposition 41.13 (i) *The coupling isometric L-groups fit into a commutative braid of exact sequences*

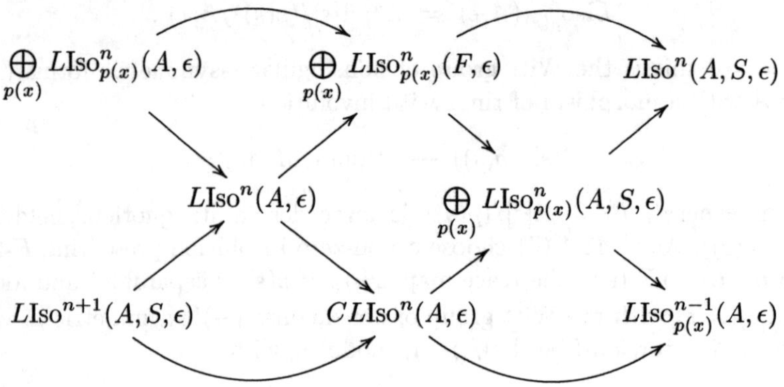

with

$$LIso^{2*+1}(A,S,\epsilon) \;=\; CLIso^{2*+1}(A,\epsilon) \;=\; 0\ .$$

(ii) *The isometric L-groups of A fit into an exact sequence*

$$0 \longrightarrow \bigoplus_{p(s)} LIso^{2i}_{p(s)}(A,\epsilon) \longrightarrow LIso^{2i}(A,\epsilon) \longrightarrow \bigoplus_{p(s)} LIso^{2i}_{p(s)}(A,S,\epsilon)$$

$$\longrightarrow LIso^{2i}(A,S,\epsilon) \oplus \bigoplus_{p(s)} LIso^{2i-1}_{p(s)}(A,\epsilon) \longrightarrow 0\ .$$

(iii) *If A is additively torsion-free the morphisms*

$$\bigoplus_{p(s)} LIso^{2i}_{p(s)}(A,\epsilon) \longrightarrow LIso^{2i}(A,\epsilon)\ ,$$

$$LIso^{2i}(A,\epsilon) \longrightarrow \bigoplus_{p(s)} LIso^{2i}_{p(s)}(F,\epsilon)$$

are isomorphisms modulo 8-torsion. □

Remark 41.14 The odd-dimensional knot cobordism groups

$$C_{2i-1} \;=\; LIso^{2i}(\mathbb{Z})$$

fit into a commutative braid of exact sequences

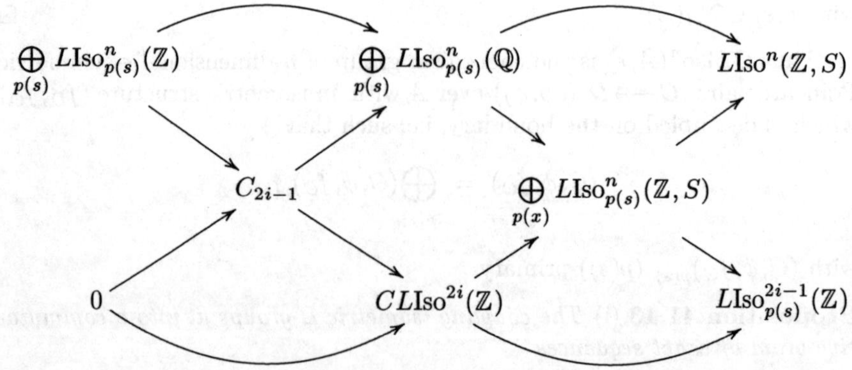

with exact sequences

$$0 \longrightarrow C_{2i-1} \longrightarrow \bigoplus_{p(s)} LIso^{2i}_{p(s)}(\mathbb{Q}) \longrightarrow LIso^{2i}(\mathbb{Z},S) \longrightarrow 0 ,$$

$$\begin{aligned}0 \longrightarrow \bigoplus_{p(s)} LIso^{2i}_{p(s)}(\mathbb{Z}) \longrightarrow C_{2i-1} \longrightarrow \bigoplus_{p(s)} LIso^{2i}_{p(s)}(\mathbb{Z},S) \\ \longrightarrow LIso^{2i}(\mathbb{Z},S) \oplus \bigoplus_{p(s)} LIso^{2i-1}_{p(s)}(\mathbb{Z}) \longrightarrow 0\end{aligned}$$

as in Stoltzfus [277, p. 25], with $p(s) \in \overline{\mathcal{M}}_s(\mathbb{Z})$. Levine [158] showed that the odd-dimensional knot cobordism groups C_{2i-1} have an infinite number of elements of order 4 (cf. 42.1 below). By contrast, it is proved in [277, p. 58] that the groups $LIso^{2i}_{p(s)}(\mathbb{Z})$ do not have elements of order 4 for i even, with the difference picked up by the coupling invariants in $CLIso^{2i}(\mathbb{Z})$. □

41C. Automorphism L-theory

Definition 41.15 (i) Let

$$V_A \;=\; \{\sum_{j=0}^{d} a_j z^j \,|\, a_0 \in U(A), a_d = 1, \overline{a}_j = \overline{a}_0 a_{d-j}\} \subset A[z,z^{-1}]$$

be the multiplicative subset of the involution-invariant monic polynomials, with

$$\sum_{j=0}^{d} \overline{a}_j z^{-j} \;=\; \overline{a}_0 z^{-d}(\sum_{j=0}^{d} a_j z^j) \in A[z,z^{-1}] ,$$

and let $V_A^{-1}A[z,z^{-1}]$ be the localization of $A[z,z^{-1}]$ inverting V_A.
(ii) Let $\overline{\mathcal{M}}_z(A) \subset V_A$ be the set of the involution-invariant monic polynomials $p(z) \in A[z,z^{-1}]$ which are irreducible. □

Proposition 41.16 (i) *The automorphism L-groups of A are related to the V_A-primary automorphism L-groups of F by a localization exact sequence*

$$\begin{aligned}\dots \longrightarrow LAut^n(A,\epsilon) \longrightarrow LAut^n_{V_A}(F,\epsilon) \\ \longrightarrow LAut^n(A,S,\epsilon) \longrightarrow LAut^{n-1}(A,\epsilon) \longrightarrow \dots\end{aligned}$$

with $LAut^n(A,S,\epsilon)$ the cobordism groups of $(n-1)$-dimensional F-contractible f.g. projective ϵ-symmetric Poincaré complexes over A with an autometric structure.
(ii) *The automorphism L-groups of A are 4-periodic*

$$LAut^n(A,\epsilon) \;=\; LAut^{n+2}(A,-\epsilon) \;=\; LAut^{n+4}(A,\epsilon) .$$

The V_A-primary automorphism L-groups of F are 4-periodic and such that:

(a) $LAut^{2i}_{T_A}(F,\epsilon)$ *is the Witt group of F-nonsingular $(-)^i\epsilon$-symmetric forms (L,λ) over A with an F-automorphism $h:(L,\lambda)\longrightarrow(L,\lambda)$, and*

$$LAut^{2i}_{p(z)}(A,S,\epsilon) \;=\; \bigoplus_{\mathcal{P}\in\overline{\max}(A)} LAut^{2i}_{p(z)}(A/\mathcal{P},u_{\mathcal{P}}\epsilon)\ ,$$

(b) $LAut^{2i+1}_{T_A}(F,\epsilon)=0$.

(iii) *The V_A-primary automorphism ϵ-symmetric L-groups of F split as*

$$LAut^n_{V_A}(F,\epsilon) \;=\; \bigoplus_{p(z)\in\overline{\mathcal{M}}_z(A)} LAut^n_{p(z)^\infty}(F,\epsilon)\ .$$

If $p(z)\in\overline{\mathcal{M}}_z(A)$ factorizes in $F[z,z^{-1}]$ as a product of coprime monic polynomials

$$p(z) \;=\; p_1(z)p_2(z)\dots p_k(z)q(z)\overline{q(z)} \in F[z,z^{-1}]$$

with each $p_j(z)\in\overline{\mathcal{M}}_z(F)$ involution-invariant and irreducible the

$$LAut^n_{p(z)^\infty}(F,\epsilon) \;=\; \bigoplus_{j=1}^{k} L^n(F[z,z^{-1}]/(p_j(z)),\epsilon)\ .$$

(iv) *The automorphism torsion L-groups split as*

$$LAut^n(A,S,\epsilon) \;=\; \bigoplus_{\mathcal{P}\in\overline{\max}(A)} LAut^n(A,\mathcal{P}^\infty,\epsilon) \;=\; \bigoplus_{\mathcal{P}\in\overline{\max}(A)} LAut^n(A/\mathcal{P},u_{\mathcal{P}}\epsilon)$$

and

$$LAut^{2*+1}(A,S,\epsilon) \;=\; 0\ .$$

Similarly for the asymmetric L-groups.

Proof As for 41.2. The sequence in (i) may be identified with the localization exact sequence of 25.4

$$\dots\longrightarrow L^n_{\widetilde{K}_0(A)}(A[z,z^{-1}],V_A,-\epsilon)\longrightarrow L^n(F[z,z^{-1}],V_A,-\epsilon)$$
$$\longrightarrow L^n(A[z,z^{-1}],SV_A,-\epsilon)\longrightarrow L^{n-1}_{\widetilde{K}_0(A)}(A[z,z^{-1}],V_A,-\epsilon)\longrightarrow\dots\ ,$$

noting that $V_A\subseteq\Omega$ is such that

$$L^n_{\widetilde{K}_0(A)}(A[z,z^{-1}],V_A,-\epsilon) \;=\; L^n_{\widetilde{K}_0(A)}(A[z,z^{-1}],\Omega,-\epsilon) \;=\; LAut^n(A,\epsilon)\ .\ \square$$

Proposition 41.17 *Let $p(z)\in\overline{\mathcal{M}}_z(A)$.*

(i) *The $(p(z))$-primary ϵ-symmetric L-groups fit into a localization exact sequence*

$$\ldots \longrightarrow L^n_{\widetilde{K}_0(A)}(A[z,z^{-1}],\epsilon) \longrightarrow L^n_{\widetilde{K}_0(A)}(A[z,z^{-1},p(z)^{-1}],\epsilon) \longrightarrow L^n(A[z,z^{-1}],p(z)^\infty,\epsilon) \longrightarrow L^{n-1}_{\widetilde{K}_0(A)}(A[z,z^{-1}],\epsilon) \longrightarrow \ldots$$

(ii) *The $(p(z))$-primary torsion ϵ-symmetric L-groups of $A[z,z^{-1}]$ are such that*

$$L^n(A[z,z^{-1}],p(z)^\infty,\epsilon) = LAut^n_{p(z)^\infty}(A,-\epsilon) = LAut^n_{p(z)}(A,-\epsilon)$$

with $LAut^n_{p(z)}(A,-\epsilon)$ *the cobordism group of f.g. projective n-dimensional $(-\epsilon)$-symmetric Poincaré complexes (C,ϕ) over A with an automorphism $(f,\delta\phi)$ such that $p(f) \simeq 0 : C \longrightarrow C$.*
(iii) *The $(p(z))$-primary ϵ-symmetric automorphism L-groups of A and F are related by a localization exact sequence*

$$\ldots \longrightarrow LAut^n_{p(z)}(A,\epsilon) \longrightarrow LAut^n_{p(z)}(F,\epsilon) \longrightarrow LAut^n_{p(z)}(A,S,\epsilon) \longrightarrow LAut^{n-1}_{p(z)}(A,\epsilon) \longrightarrow \ldots$$

with $LAut^n_{p(z)}(A,S,\epsilon)$ *the cobordism group of F-contractible f.g. projective $(n-1)$-dimensional ϵ-symmetric Poincaré complexes (C,ϕ) over A with an automorphism $(f,\delta\phi)$ such that $p(f) \simeq 0 : C \longrightarrow C$, and:*

(a) $LAut^{2i}_{p(z)}(A,S,\epsilon)$ *is the Witt group of F-nonsingular $(-)^i\epsilon$-symmetric linking forms (L,λ) over (A,S) with an automorphism $f : (L,\lambda) \longrightarrow (L,\lambda)$ such that $p(f) = 0$, and*

$$\begin{aligned} LAut^{2i}_{p(z)}(A,S,\epsilon) &= \bigoplus_{\mathcal{P}\in\overline{\max}(A)} \bigoplus_{q(z)\in\overline{\mathcal{M}}_z(A/\mathcal{P}),q(z)|p(z)} LAut^0_{q(z)}(A/\mathcal{P},(-)^i\epsilon) \\ &= \bigoplus_{\mathcal{P}\in\overline{\max}(A)} \bigoplus_{q(z)\in\overline{\mathcal{M}}_z(A/\mathcal{P}),q(z)|p(z)} L^0((A/\mathcal{P})[z,z^{-1}]/(q(z)),(-)^i\epsilon)\ , \end{aligned}$$

(b) $LAut^{2i+1}_{p(z)}(A,S,\epsilon) = 0$.

(iv) *If $A[z,z^{-1}]/(p(z))$ is a Dedekind ring (e.g. if A is a field, or if $p(z)$ has degree 1) then*

$$LAut^n_{p(z)}(A,\epsilon) = L^n(A[z,z^{-1}]/(p(z)),\epsilon)\ .$$

(v) *The ϵ-symmetric L-groups of $A[z,z^{-1}]$ are such that*

$$L^n_h(A[z,z^{-1}],\epsilon) = L^n_h(A,\epsilon)\ ,\ L^n_{\widetilde{K}_0(A)}(A[z,z^{-1}],\epsilon) = L^n(A,\epsilon)\ .$$

Proof (i) As in the proof of 41.3 it suffices to prove that the natural map

$$LAut^n_{p(z)}(A,\epsilon) \longrightarrow LAut^n_{p(z)^\infty}(A,\epsilon)\ ;\ (C,h,\delta\phi,\phi) \longrightarrow (C,h,\delta\phi,\phi)$$

is an isomorphism for $n = 0$. An element

$$(M,\phi,h) \in LAut^0_{p(z)^\infty}(A,\epsilon)$$

is represented by a f.g. projective nonsingular ϵ-symmetric form (M,ϕ) over A together with an automorphism $h : (M,\phi) \longrightarrow (M,\phi)$ such that

$$p(h)^N = 0 : M \longrightarrow M$$

for some $N \geq 1$. If $N \geq 2$ reduce to the case $N = 1$ as in the proof of 41.3, noting that $p(h) : M \longrightarrow M$ is such that

$$p(h)^*\phi = \overline{p}(h^*)\phi = \phi\overline{p}(h^{-1}) = u^{-1}\phi p(h) : M \longrightarrow M^* .$$

(ii)+(iii) As for 41.3 (ii)+(iii). □

Remark 41.18 (i) The automorphism L-groups which occur in 41.17 (iii) are given in the terminology of 28.15 by

$$LAut^n_{p(z)}(A,\epsilon) = L^n(A[z,z^{-1}]/(p(z)),A,\epsilon) .$$

For $n = 0$ this is the Witt group of nonsingular ϵ-symmetric forms (M,ϕ) over A with a morphism of rings with involution

$$A[z,z^{-1}]/(p(z)) \longrightarrow \mathrm{Hom}_A(M,M)^{op} .$$

(ii) In general, $A[z,z^{-1}]/(p(z))$ is an order in its quotient field

$$E = F[z,z^{-1}]/(p(z)) ,$$

and for separable $p(z)$ it is possible to use the trace map $\mathrm{tr}_{E/F} : E \longrightarrow F$ as in 41.4 (ii) and 41.11 (ii) to identify $LAut^{2i}_{p(z)}(A,\epsilon)$ with the Witt group of nonsingular $(-)^i\epsilon$-symmetric Δ-valued forms on f.g. torsion-free $A[z,z^{-1}]/(p(z))$-modules, with Δ the inverse different. The other terms in the localization exact sequence

$$\begin{aligned}0 \longrightarrow LAut^{2i}_{p(z)}(A,\epsilon) \longrightarrow LAut^{2i}_{p(z)}(F,\epsilon) \longrightarrow LAut^{2i}_{p(z)}(A,S,\epsilon)\\ \longrightarrow LAut^{2i-1}_{p(z)}(A,\epsilon) \longrightarrow 0\end{aligned}$$

are given by

$$\begin{aligned}LAut^{2i}_{p(z)}(F,\epsilon) &= L^0(E,(-)^i\epsilon) ,\\ LAut^{2i}_{p(z)}(A,S,\epsilon) &= \bigoplus_{\mathcal{P}\in\overline{\max}(\Delta)} L^0(\Delta/\mathcal{P},(-)^i\epsilon) .\end{aligned}$$

□

Proposition 41.19 *Let*

$$p(z) = z - a \in \overline{\mathcal{M}}_z(A)$$

for some $a \in U(A)$ (i.e. a unit $a \in A^\bullet$ with $\overline{a} = a^{-1}$). In this case the localization exact sequence of 41.17 (i) splits, with

$$\begin{aligned} L^n_{\widetilde{K}_0(A)}(A[z,z^{-1}],\epsilon) &= L^n(A,\epsilon)\oplus L^{n-1}(A,\epsilon)\ , \\ L^n(A[z,z^{-1}],(z-a)^\infty,\epsilon) &= LAut^{n-2}_{(z-a)^\infty}(A,\epsilon) = L^{n-2}(A,\epsilon)\ , \\ L^n_{\widetilde{K}_0(A)}(A[z,z^{-1},(z-a)^{-1}],\epsilon) &= L^n(A,\epsilon)\ . \end{aligned}$$

In particular, the special case $a = 1$ shows that the unipotent ϵ-symmetric groups of A (35.29) are given by

$$\begin{aligned} LAut^n_{uni}(A,\epsilon) &= L^n_{(z-1)^\infty}(A,\epsilon) = L^n(A,\epsilon)\ , \\ \widetilde{LAut}^n_{uni}(A,\epsilon) &= 0\ . \end{aligned}$$

Proof The case $p(z) = z - a$ of 41.17 (ii) gives

$$L^n(A[z,z^{-1}],(z-a)^\infty,\epsilon) = LAut^{n-2}_{(z-a)}(A,\epsilon) = L^{n-2}(A,\epsilon)\ .$$

Write $\Lambda = A[z,z^{-1},(z-a)^{-1}]$. The application of the 5-lemma to the morphism of exact sequences

$$\begin{array}{ccccccc} \dots\longrightarrow L^n(A,\epsilon)\oplus L^{n-1}(A,\epsilon) & \xrightarrow{(1\ 0)} & L^n(A,\epsilon) & \xrightarrow{0} & L^{n-2}(A,\epsilon) & \longrightarrow\dots \\ \downarrow\cong & & \downarrow & & \downarrow\cong & \\ \dots\longrightarrow L^n_{\widetilde{K}_0(A)}(A[z,z^{-1}],\epsilon) & \longrightarrow & L^n_{\widetilde{K}_0(A)}(\Lambda,\epsilon) & \longrightarrow & LAut^{n-2}_{(z-a)^\infty}(A,\epsilon) & \longrightarrow\dots \end{array}$$

gives

$$L^n_{\widetilde{K}_0(A)}(\Lambda,\epsilon) = L^n(A,\epsilon)\ . \qquad \square$$

Definition 41.20 The *coupling automorphism L-groups* $CLAut^n(A,\epsilon)$ are the relative L-groups in the exact sequences

$$\begin{aligned} \dots\longrightarrow \bigoplus_{p(z)} LAut^n_{p(z)}(A,\epsilon) &\longrightarrow LAut^n(A,\epsilon)\longrightarrow CLAut^n(A,\epsilon) \\ &\longrightarrow \bigoplus_{p(z)\in\overline{\mathcal{M}}_z(A)} LAut^{n-1}_{p(z)}(A,\epsilon)\longrightarrow\dots\ . \qquad \square \end{aligned}$$

Thus $CLAut^n(A,\epsilon)$ is the cobordism group of n-dimensional ϵ-symmetric Poincaré pairs $(C\longrightarrow D,(\delta\phi,\phi))$ over A with an autometric structure (f_D,f_C) which is decoupled on the boundary, i.e. such that

$$(C,\phi,f_C) = \bigoplus_{p(z)}(C,\phi,f_C)_{p(z)}$$

with $(C,\phi,f_C)_{p(z)}$ $(p(z))$-primary.

Proposition 41.21 (i) *The coupling automorphism L-groups fit into a commutative braid of exact sequences*

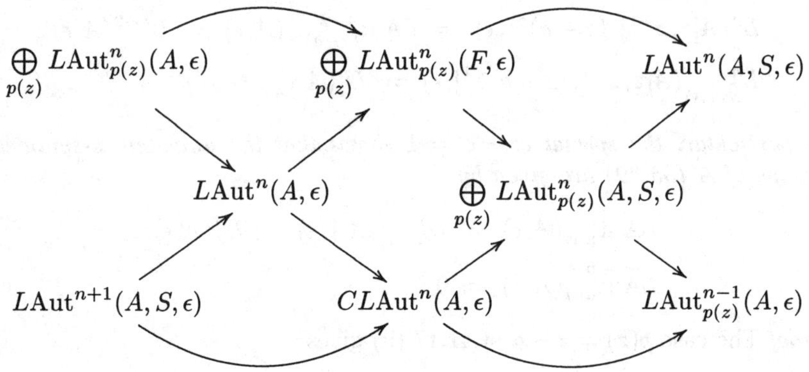

with

$$LAut^{2*+1}(A,S,\epsilon) \;=\; CLAut^{2*+1}(A,\epsilon) \;=\; 0 \;.$$

(ii) *The automorphism L-groups of A fit into an exact sequence*

$$\begin{aligned} 0 \longrightarrow \bigoplus_{p(z)} LAut^{2i}_{p(z)}(A,\epsilon) \longrightarrow LAut^{2i}(A,\epsilon) \longrightarrow \bigoplus_{p(z)} LAut^{2i}_{p(z)}(A,S,\epsilon) \\ \longrightarrow LAut^{2i}(A,S,\epsilon) \oplus \bigoplus_{p(z)} LAut^{2i-1}_{p(z)}(A,\epsilon) \longrightarrow 0 \;. \end{aligned}$$

(iii) *If A is additively torsion-free the morphisms*

$$\bigoplus_{p(z)} LAut^{2i}_{p(z)}(A,\epsilon) \longrightarrow LAut^{2i}(A,\epsilon) \;,$$

$$LAut^{2i}(A,\epsilon) \longrightarrow \bigoplus_{p(z)} LAut^{2i}_{p(z)}(F,\epsilon)$$

are isomorphisms modulo 8-torsion.
Similarly for the asymmetric L-groups. □

Proposition 41.22 (i) *Let*

$$p(z) \;=\; \sum_{j=0}^{d} a_j z^j \in A[z,z^{-1}] \;\;,\;\; q(s) \;=\; \sum_{j=0}^{d} b_j s^j \in A[s] \;,$$

be polynomials which are related by the one-one correspondence of 35.31 (i), with $p(0),p(1) \neq 0 \in A$ *and*

$$p(z) = (1-z)^d q((1-z)^{-1}) = \sum_{j=0}^{d} b_j (1-z)^{d-j} ,$$

$$q(s) = s^d p(1-s^{-1}) = \sum_{j=0}^{d} a_j s^{d-j}(s-1)^j .$$

The polynomial $p(z)$ is irreducible (resp. involution-invariant) if and only if the polynomial $q(s)$ is irreducible (resp. involution-invariant).
(ii) *If $p(z) \in A[z,z^{-1}]$, $q(s) \in A[s]$ are related as in* (i), *involution-invariant, irreducible and $p(1) = 1 \in A$ (or equivalently $q(s)$ is monic) then*

$$LIso^0_{q(s)}(A,\epsilon) = L^0_h(A[z,z^{-1}], p(z)^\infty, -\epsilon) .$$

Furthermore, if $p(z)$ is monic (or equivalently $q(0) = (-)^d \in A$) then

$$\begin{aligned} LIso^0_{q(s)}(A,\epsilon) &= L^0_h(A[z,z^{-1}], p(z)^\infty, -\epsilon) \\ &= LAut^0_{p(z)}(A,\epsilon) \\ &= LAut^0_{fib,p(z)}(A,\epsilon) \\ &= LAsy^0_{p(-\epsilon^{-1}z)}(A) . \end{aligned}$$

Proof Take $S = (p(z))^\infty \subset A[z,z^{-1}]$, $T = (q(s))^\infty \subset A[s]$ in 35.31 (ii). □

42. The knot cobordism groups

The cobordism groups C_n of n-knots $k : S^n \subset S^{n+2}$ were first defined in the 1960's. The computation in the high dimensions $n \geq 3$ was completed by the 1970's. The object of this final chapter is to give a brief description of the algebraic structure of the odd-dimensional groups

$$\begin{aligned} C_{2j-1} &= \widehat{L}_0(\mathbb{Z},(-)^j) \;=\; L\mathrm{Iso}^0(\mathbb{Z},(-)^j) \\ &= L\mathrm{Asy}^0(\mathbb{Z},P,(-)^j) \;=\; L_0(\mathbb{Z}[z,z^{-1}],P,(-)^{j+1}) \quad (j \geq 2) \end{aligned}$$

with $P = \{p(z) \,|\, p(1) = \pm 1\} \subset \mathbb{Z}[z,z^{-1}]$.

Proposition 42.1 (Kervaire [132], Milnor [195], Levine [157], [158], Stoltzfus [277])
(i) *The high-dimensional knot cobordism groups C_n $(n \geq 3)$ are 4-periodic*

$$C_n \;=\; C_{n+4}\;.$$

(ii) *The even-dimensional groups vanish*

$$C_{2*} \;=\; 0\;.$$

(iii) *The odd-dimensional groups C_{2*+1} are countably infinitely generated, with both C_{4*} and C_{4*+2} of the type*

$$\bigoplus_{\infty}\mathbb{Z} \oplus \bigoplus_{\infty}\mathbb{Z}_2 \oplus \bigoplus_{\infty}\mathbb{Z}_4\;.$$

Proof As in [132] apply the high-dimensional surgery method of Kervaire and Milnor [134] to prove that C_n is the cobordism group of simple n-knots, and to extend the computation $L_{2*+1}(\mathbb{Z}) = 0$ to $C_{2*} = 0$.

The first odd-dimensional knot cobordism invariant was obtained for classical knots $k : S^1 \subset S^3$ by Fox and Milnor [82] from Reidemeister torsion modulo squares, and used to show that C_1 is infinitely generated. This invariant can be used as in [132, III.12] to show that all the odd-dimensional knot cobordism groups C_{2j-1} are infinitely generated. Here are the details.

Given a $(2j-1)$-knot $k : S^{2j-1} \subset S^{2j+1}$ with knot exterior X write the jth Alexander polynomial as

$$\begin{aligned}\Delta_j(z) &= \det((z-\zeta)/(1-\zeta) : H_j(\overline{X};\mathbb{Q})[z] \longrightarrow H_j(\overline{X};\mathbb{Q})[z]) \\ &= \sum_{i=0}^{d} a_i z^i \in P \subset \mathbb{Z}[z] \subset \mathbb{Q}[z] \ \ (\sum_{i=0}^{d} a_i = 1\ ,\ a_d = 1)\ ,\end{aligned}$$

with $d = \dim_{\mathbb{Q}} H_j(\overline{X};\mathbb{Q})$. In terms of a Seifert form (L,λ) for k

$$\Delta_j(z) = \det(1 - f + zf : L[z] \longrightarrow L[z]) \in \mathbb{Z}[z]$$

with

$$\begin{aligned} f &= (\lambda + (-)^j\lambda^*)^{-1}\lambda : L \longrightarrow L\ ,\\ H_j(\overline{X};\mathbb{Q}) &= \operatorname{coker}(1 - f + zf : L[z,z^{-1}] \longrightarrow L[z,z^{-1}])\ .\end{aligned}$$

Define the involution-invariant monic polynomial

$$\begin{aligned}\widetilde{\Delta}_j(s) &= s^d \Delta_j(1 - s^{-1}) = \sum_{i=0}^{d} a_i (s-1)^i s^{d-i} \\ &= s^d \det(1 - s^{-1} f : L[s] \longrightarrow L[s]) \in \mathbb{Z}[s]\ ,\end{aligned}$$

and factorize $\widetilde{\Delta}_j(s)$ as a product

$$\widetilde{\Delta}_j(s) = \left(\prod_{r=1}^{t} p_r(s)^{n_r}\right) q(s)\overline{q(s)} \in \mathbb{Z}[s]$$

with each $p_r(s) \in \overline{\mathcal{M}}_s(\mathbb{Z})$ an involution-invariant irreducible monic polynomial. The Reidemeister torsion map in knot cobordism is given by

$$\begin{aligned}\Delta : C_{2j-1} &= L^{2j+2}(\mathbb{Z}[z,z^{-1}],P) = L\mathrm{Iso}^{2j}(\mathbb{Z}) \\ &\longrightarrow \widehat{H}^0(\mathbb{Z}_2; K_1(\mathbb{Z}[z,z^{-1}],P)) = \widehat{H}^0(\mathbb{Z}_2;\mathrm{Iso}_0(\mathbb{Z})) = \mathbb{Z}_2[\overline{\mathcal{M}}_s(\mathbb{Z})]\ ;\end{aligned}$$

$$k \longrightarrow \widetilde{\Delta}_j(s) = [L,f] = \sum_{r=1}^{t} n_r [p_r(s)]\ .$$

The image of the Reidemeister torsion map

$$\Delta : C_{2j-1} \longrightarrow \mathbb{Z}_2[\overline{\mathcal{M}}_s(\mathbb{Z})]$$

was characterized in [157]: Δ is onto for odd j, and for even j the image of Δ is generated by the classes of the involution-invariant monic polynomials $\widetilde{\Delta}(s) \in \mathbb{Z}[s]$ such that $2^d \widetilde{\Delta}(1/2)$ is a square, where $d = \operatorname{degree}(\widetilde{\Delta}(s))$. In both cases the image of Δ is countably infinitely generated, namely $\bigoplus_{\infty} \mathbb{Z}_2$.

The isometric multisignature (Chap. 40A) provides integer-valued knot cobordism invariants, as in [195] and [157],[158]. It follows from 41.13 (iii) that the natural maps

$$C_{2j-1} = L\mathrm{Iso}^{2j}(\mathbb{Z}) \longrightarrow L\mathrm{Iso}^{2j}_{U_{\mathbb{Z}}}(\mathbb{Q})$$

are injections with 8-torsion cokernel, where $U_{\mathbb{Z}} \subset \mathbb{Q}[s]$ is the multiplicative subset of involution-invariant monic polynomials with integer coefficients. If (M,ϕ) is a nonsingular $(-)^j$-symmetric form over $\mathbb{Q}$ with an isometric structure $f : M \longrightarrow M$ such that $\mathrm{ch}_s(M,f) \in U_{\mathbb{Z}}$, say

$$\mathrm{ch}_s(M,f) = \det(s - f : M[s] \longrightarrow M[s]) = \sum_{j=0}^{m} b_j s^j \in \mathbb{Z}[s]$$

$$(b_m = 1 \ , \ m = \dim_{\mathbb{Q}}(M)) \ ,$$

then

$$\det(1 - 2f : M \longrightarrow M) = \sum_{j=0}^{m} b_j 2^{m-j} = 2(\sum_{j=0}^{m-1} b_j 2^{m-j-1}) + 1 \in \mathbb{Z}$$

is an odd integer, so that $1 - 2f : M \longrightarrow M$ is an automorphism. The groups $LIso^{2j}_{U_{\mathbb{Z}}}(\mathbb{Q})$ are 2-periodic (i.e. independent of j), with an isomorphism

$$LIso^{2j}_{U_{\mathbb{Z}}}(\mathbb{Q}) \longrightarrow LIso^{2j+2}_{U_{\mathbb{Z}}}(\mathbb{Q}) \ ; \ (M,\phi,f) \longrightarrow (M,\phi(1-2f),f) \ .$$

Define the sets

$$R_{1/2} = \{a \in \mathbb{C} \,|\, \mathrm{Re}(a) = 1/2\} \ ,$$

$$Q_{1/2} = \{a \in R_{1/2} \,|\, a \text{ is an algebraic integer, } \mathrm{Im}(a) > 0\}$$

and the subgroup

$$I = \mathbb{Z}[\{1.a + 1.\overline{a} \,|\, a \in Q_{1/2}\}] \subset \mathbb{Z}[R_{1/2}] \ .$$

By 40.3 the isometric signature defines an isomorphism

$$\sigma^{Iso}_* \ : \ LIso^{2j}(\mathbb{C}) \xrightarrow{\ \simeq\ } \mathbb{Z}[R_{1/2}] \ .$$

By the primary decomposition of isometric L-theory and devissage (39.12)

$$\begin{aligned} LIso^{2j}_{U_{\mathbb{Z}}}(\mathbb{Q}) &= \bigoplus_{p(s) \in \overline{\mathcal{M}}_s(\mathbb{Z})} LIso^{2j}_{p(s)}(\mathbb{Q}) \\ &= \bigoplus_{p(s) \in \overline{\mathcal{M}}_s(\mathbb{Z})} L^0(\mathbb{Q}[s]/(p(s)),(-)^j) \ . \end{aligned}$$

The image of the natural map

$$i \ : \ LIso^{2j}_{U_{\mathbb{Z}}}(\mathbb{Q}) \longrightarrow LIso^{2j}(\mathbb{C}) = \mathbb{Z}[R_{1/2}]$$

is such that $\mathrm{im}(i) \subseteq I$, with $\ker(i)$ and $I/\mathrm{im}(i)$ 8-torsion. The isometric multisignature maps

$$\sigma^{Iso}_* \ : \ C_{2j-1} = LIso^{2j}(\mathbb{Z}) \longrightarrow I \cong \mathbb{Z}[Q_{1/2}] \cong \bigoplus_{\infty} \mathbb{Z}$$

are thus isomorphisms modulo 2-primary torsion, and

$$C_{2j-1}/\text{torsion} = \bigoplus_{\infty} \mathbb{Z} \ .$$

See [158] for the original proof (based on the methods of Milnor [197]) that there is only countable 2- and 4-torsion in C_{2j-1}, so that

$$\text{torsion} = \bigoplus_{\infty} \mathbb{Z}_2 \oplus \bigoplus_{\infty} \mathbb{Z}_4 .$$

The precise nature of the torsion in C_{2*-1} was computed in [277] by means of the coupling invariants. □

Example 42.2 Specific infinite linearly independent sets in C_{2j-1} were constructed in Kervaire [132, III.12] and Milnor [195], as follows.
(i) For odd j define for each integer $m \geq 1$ the skew-symmetric Seifert form over $\mathbb{Z}$

$$(L_m, \lambda_m) = (\mathbb{Z} \oplus \mathbb{Z}, \begin{pmatrix} m & 0 \\ -1 & 1 \end{pmatrix}) ,$$

with Alexander polynomial

$$\begin{aligned} \Delta_j(z)_m &= \det((\lambda_m - \lambda_m^*)^{-1}(z\lambda_m - \lambda_m^*) : L_m[z, z^{-1}] \longrightarrow L_m[z, z^{-1}]) \\ &= \begin{vmatrix} z & 1-z \\ m(z-1) & 1 \end{vmatrix} \\ &= m - (2m-1)z + mz^2 \in P \subset \mathbb{Z}[z, z^{-1}] . \end{aligned}$$

The roots of $\Delta_j(z)_m$ are a complex conjugate pair of algebraic numbers on the unit circle

$$z_1, z_2 = \cos\theta_m + i\sin\theta_m \in S^1$$

with

$$\cos\theta_m = \frac{2m-1}{2m} , \quad \sin\theta_m = \pm\frac{\sqrt{4m-1}}{2m} .$$

The Reidemeister torsion is determined by the irreducible involution-invariant monic polynomial

$$\begin{aligned} \Delta(s)_m &= s^2 \Delta_j(1 - s^{-1})_m \\ &= m - s + s^2 \in U_{\mathbb{Z}} \subset \overline{\mathcal{M}}_s(\mathbb{Z}) \subset \mathbb{Z}[s] \end{aligned}$$

with roots

$$s_k = (1 - z_k)^{-1} \quad (k = 1, 2)$$

a complex conjugate pair of algebraic integers

$$\begin{aligned} s_1, s_2 &= \frac{1 + i\cot(\theta_m/2)}{2} \\ &= \frac{1 \pm i\sqrt{4m-1}}{2} \in Q_{1/2} \subset R_{1/2} \subset \mathbb{C} . \end{aligned}$$

The subset $\{\Delta(s)_m \,|\, m \geq 1\} \subset \mathbb{Z}_2[\overline{\mathcal{M}}_s(\mathbb{Z})]$ is $\mathbb{Z}_2$-linearly independent, so that $\{(L_m, \lambda_m) \,|\, m \geq 1\} \subset C_{2j-1}$ is $\mathbb{Z}$-linearly independent. Now

$$\mathbb{C}\otimes_{\mathbb{Z}}(L_m,\lambda_m+\lambda_m^*)\ \cong\ (\mathbb{C},1)\oplus(\mathbb{C},1)\ ,$$

so that the isometric multisignature of $\mathbb{C}\otimes_{\mathbb{Z}}(L_m,\lambda_m)$ is

$$\sigma_*^{Iso}(\mathbb{C}\otimes_{\mathbb{Z}}(L_m,\lambda_m))\ =\ 1.s_1+1.s_2\in I\subset\mathbb{Z}[R_{1/2}]$$

with I as in 42.1.

(ii) For even j define for each integer $m\geq 1$ the symmetric Seifert form over $\mathbb{Z}$

$$(L'_m,\lambda'_m)\ =\ (\mathbb{Z}\oplus\mathbb{Z}\oplus\mathbb{Z}\oplus\mathbb{Z},\begin{pmatrix}0&0&m&m\\1&0&1&1\\-m&-1&0&0\\-m&-1&1&0\end{pmatrix})$$

with Alexander polynomial

$$\begin{aligned}\Delta'_j(z)_m\ &=\ \det((\lambda'_m+\lambda'^*_m)^{-1}(z\lambda'_m+\lambda'^*_m):L'_m[z,z^{-1}]\longrightarrow L'_m[z,z^{-1}])\\ &=\ m-(2m-1)z^2+mz^4\in P\subset\mathbb{Z}[z,z^{-1}]\ .\end{aligned}$$

The Reidemeister torsion is determined by the irreducible involution-invariant monic polynomial

$$\begin{aligned}\Delta'(s)_m\ &=\ s^4\Delta'_j(1-s^{-1})_m\\ &=\ m-4ms+(4m+1)s^2-2s^3+s^4\in\overline{\mathbb{M}}_s(\mathbb{Z})\subset\mathbb{Z}[s]\ .\end{aligned}$$

The subset $\{\Delta'(s)_m\,|\,m\geq 1\}\subset\mathbb{Z}_2[\overline{\mathbb{M}}_s(\mathbb{Z})]$ is $\mathbb{Z}_2$-linearly independent, so that $\{(L'_m,\lambda'_m)\,|\,m\geq 1\}\subset C_{2j-1}$ is $\mathbb{Z}$-linearly independent. The roots of $\Delta'_j(z)_m$ are two complex conjugate pairs of algebraic numbers on the unit circle

$$z_1,z_2,z_3,z_4\ =\ \cos\theta_m/2+i\sin\theta_m/2\in S^1\backslash\{1\}$$

(with θ_m as in (i)) and

$$\mathbb{C}\otimes_{\mathbb{Z}}(L'_m,\lambda'_m+\lambda'^*_m)\ \cong\ (\mathbb{C},1)\oplus(\mathbb{C},1)\oplus(\mathbb{C},-1)\oplus(\mathbb{C},-1)\ .$$

The roots of $\Delta'(s)_m$ are algebraic integers

$$s_k\ =\ (1-z_k)^{-1}\ =\ \frac{1+i\cot(\theta_m/4)}{2}\in Q_{1/2}\subset R_{1/2}\ \ (k=1,2,3,4)$$

and the isometric multisignature of $\mathbb{C}\otimes_{\mathbb{Z}}(L'_m,\lambda'_m)$ is

$$\sigma_*^{Iso}(\mathbb{C}\otimes_{\mathbb{Z}}(L'_m,\lambda'_m))\ =\ 1.s_1+1.s_2-1.s_3-1.s_4\in I\subset\mathbb{Z}[R_{1/2}]$$

with I as in 42.1. □

Exercise 42.3 Verify that the even-dimensional automorphism L-groups $LAut^{2*}(\mathbb{Z})$, the asymmetric L-groups $LAsy^{2*}(\mathbb{Z})$ and the twisted double symmetric L-groups

$$DBL^{2*}(\mathbb{Z})\ =\ \ker(LAut^{2*}(\mathbb{Z})\longrightarrow L^{2*}(\mathbb{Z}))$$

are all countably infinitely generated, of the type

$$\bigoplus_\infty\mathbb{Z}\oplus\bigoplus_\infty\mathbb{Z}_2\oplus\bigoplus_\infty\mathbb{Z}_4\ .$$

As stated, this can be worked out by the method of Levine [157], [158] for $C_{2*-1} = L\mathrm{Iso}^{2*}(\mathbb{Z})$, using the computations in Chap. 39 of $L\mathrm{Aut}^{2*}(\mathbb{Q})$, $L\mathrm{Asy}^{2*}(\mathbb{Q})$ and

$$DBL^{2*}(\mathbb{Q}) = L\mathrm{Aut}^{2*}_{fib}(\mathbb{Q}) = L\mathrm{Iso}^{2*}(\mathbb{Q}) .$$

The torsion can then be unraveled by the method of Stoltzfus [277], using the automorphism coupling invariants of Chap. 41C. In fact, there is defined a commutative braid of exact sequences

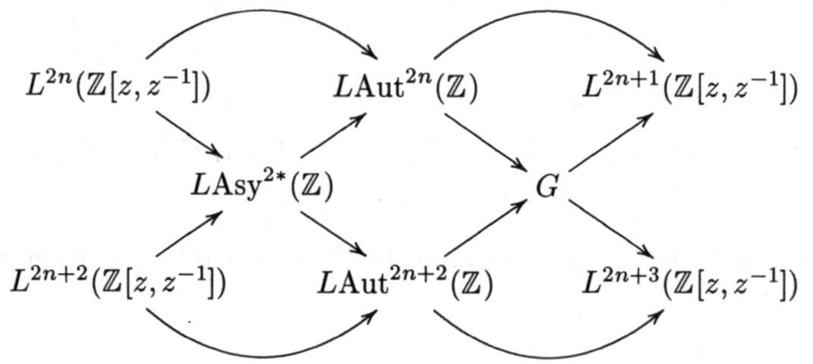

with

$$L^m(\mathbb{Z}[z,z^{-1}]) = L^m(\mathbb{Z}) \oplus L^{m-1}(\mathbb{Z}) ,$$

$$L^m(\mathbb{Z}) = \begin{cases} \mathbb{Z} \text{ (signature)} \\ \mathbb{Z}_2 \text{ (deRham invariant)} \\ 0 \end{cases} \text{if } m \equiv \begin{cases} 0 \\ 1 \\ 2 \text{ or } 3 \end{cases} \pmod 4 ,$$

$$L\mathrm{Aut}^{2n}(\mathbb{Z}) = L^{2n}(\mathbb{Z}) \oplus DBL^{2n}(\mathbb{Z}) ,$$

$$\begin{aligned} G &= DBL^{2n}(\mathbb{Z}) \oplus L^{2n+3}(\mathbb{Z}[z,z^{-1}]) = DBL^{2n+2}(\mathbb{Z}) \oplus L^{2n+1}(\mathbb{Z}[z,z^{-1}]) \\ &= DBL^0(\mathbb{Z}) = DBL^2(\mathbb{Z}) \oplus L^0(\mathbb{Z}) \oplus L^1(\mathbb{Z}) , \end{aligned}$$

$$\begin{aligned} L\mathrm{Asy}^{2*}(\mathbb{Z}) &= L^{2n}(\mathbb{Z}) \oplus \ker(DBL^{2n+2}(\mathbb{Z}) \longrightarrow L^{2n+3}(\mathbb{Z})) \\ &= \ker(DBL^{2n}(\mathbb{Z}) \longrightarrow L^{2n+1}(\mathbb{Z})) \oplus L^{2n+2}(\mathbb{Z}) \\ &= \ker(DBL^0(\mathbb{Z}) \longrightarrow L^1(\mathbb{Z})) = DBL^2(\mathbb{Z}) \oplus L^0(\mathbb{Z}) . \end{aligned}$$

The groups $L\mathrm{Aut}^{2*}(\mathbb{Z})$, $DBL^{2*}(\mathbb{Z})$ and $L\mathrm{Asy}^{2*}(\mathbb{Z})$ only differ in the summands $L^0(\mathbb{Z}) = \mathbb{Z}$ and $L^1(\mathbb{Z}) = \mathbb{Z}_2$, all being of the type

$$\bigoplus_{\infty} \mathbb{Z} \oplus \bigoplus_{\infty} \mathbb{Z}_2 \oplus \bigoplus_{\infty} \mathbb{Z}_4 . \qquad \square$$

Appendix

The history and applications of open books by H. E. Winkelnkemper

A1. Open book theorems, analogues, motivations and historical remarks

Let V be a compact differentiable $(n-1)$-manifold with $\partial V \neq \emptyset$ and $h : V \to V$ a diffeomorphism, which restricts to the identity on ∂V; by forming the mapping torus V_h , which has $\partial V \times S^1$ as boundary, and identifying $(x,t) \sim (x,t')$ on ∂V_h for each $x \in \partial V, t, t' \in S^1$, we obtain a closed differentiable n-manifold M. If we look at a piece of the image N of $\partial V \times S^1$ under the identification map, then M looks like an open book, as in the figure.

Definition A closed manifold is an *open book* if it is diffeomorphic to M. The fibers $V \times \{t\}$ $(t \in S^1)$ of V_h are codimension 1 submanifolds of M whose images are the *pages* of the open book, and the image of the closed codimension 2 submanifold N of M is called the *binding* of the open book.

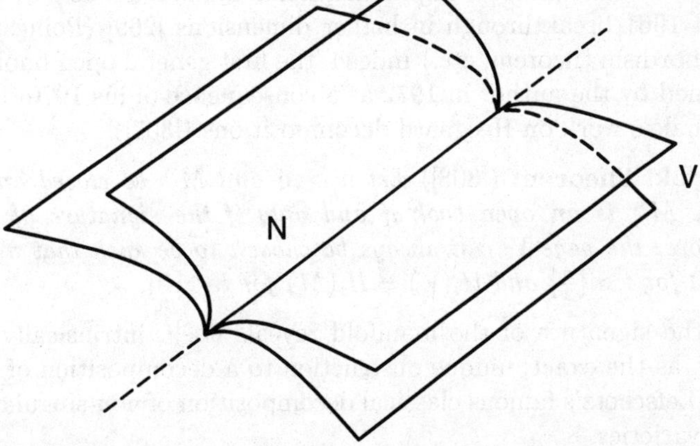

For an open book M every point $x \in M \backslash N$ lies on one and only one page and N is the boundary of each page.

An open book is represented by a page V and a monodromy self diffeomorphism $h : V \to V$ which restricts to the identity on ∂V. If $h_* : H_*(V) \to H_*(V)$ is the identity, we say that the open book decomposition has no monodromy.

Open book decompositions describe an arbitrary closed manifold in terms of lower dimensional ones. Mapping tori are open books with $\partial V = \emptyset$.

Example Let V be any compact manifold with $\partial V \neq \emptyset$; in $V \times I$ identify each interval (x, t), $x \in \partial V, t \in I$, to a point $(x, \frac{1}{2})$ obtaining a manifold with boundary, W. Let $N \subset \partial W$ be the image of $\partial V \times I$ under the identification, which divides ∂W into $(\partial W)_+$ and $(\partial W)_-$; if

$$h \ : \ (\partial W, (\partial W)_+, (\partial W)_-) \to (\partial W, (\partial W)_+, (\partial W)_-)$$

is a diffeomorphism of triples, then $W \cup_h W$ has an open book decomposition with binding N.

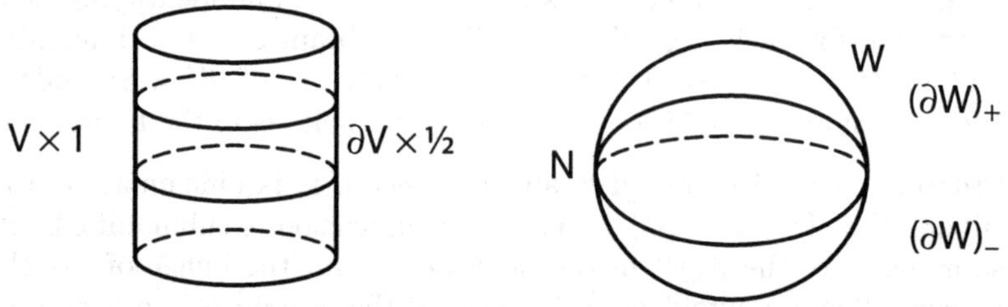

At first glance, open books are just refined Heegaard decompositions ('twisted doubles') whose existence for high-dimensional manifolds was part of Smale's celebrated 1961 breakthrough in higher dimensions [269] (Poincaré conjecture, h-cobordism theorem, etc.) Indeed, the first general open book theorem was obtained by the author in 1972 as a consequence of his 1970 thesis completing Smale's work on Heegaard decompositions ([307]).

Open Book Theorem ([308]) *Let $n > 6$ and M^n be closed and simply-connected. M^n is an open book if and only if the signature of M^n is 0. Furthermore, the page V can always be chosen to be such that $\pi_1(V) = 0$, $H_i(V) = 0$ for $i > [\frac{n}{2}]$ and $H_i(V) \cong H_i(M)$ for $i < [\frac{n}{2}]$.*

Thus the signature of the manifold reveals itself, intrinsically and geometrically, as the exact, unique obstruction to a decomposition of M^n, very similar to Lefschetz's famous classical decomposition of non-singular complex algebraic varieties.

For other, but equivalent, definitions of open books (which were briefly known also as "fibered knots", "spinnable structures", "Alexander decompositions") see Tamura [283], H. B. Lawson [151], Hector and Hirsch [110], Kauffman and Neumann [123], Quinn [227]; for their classification in many cases, using Seifert forms, see Kato [118].

As usual when a general definition is made it turned out that in particular cases these decompositions had appeared in the literature already; the most notable being Milnor's fibration theorem: it gives open book decompositions of odd-dimensional spheres S^{2n+1}, whose monodromy is very interesting ([196]). Also in 1972 I. Tamura [283] used these methods instead of Smale's decompositions to obtain book decompositions of a very large class of manifolds.

Soon afterward, the author discovered that a theorem of Alexander, [3], usually quoted and used in the non-intrinsic form[12] as "every link in S^3 can be written as a braid" leads in fact to the most general open book theorem for any closed, orientable 3-manifold; furthermore, the page can be chosen to be a 2-disk with holes or have a connected binding.

However, this intrinsic interpretation of Alexander's theorem was certainly not well-known among the leading researchers in low-dimensional topology.

As evidence of this see the list in Section A2 of the simple proofs of important theorems as well as new ones in 3-manifold theory obtained after 1972 using Alexander's theorem as an open book theorem.

For modern proofs of Alexander's theorem (as an open book theorem) and refinements see González-Acuña [91], Rolfsen [253], Myers [208], Lyon [180], Harer [106], [107], Montesinos and Morton [204] and in the non-orientable case, Berstein and Edmonds [18].

A2. Applications of Alexander's theorem in dimension 3 as an open book theorem, and Artin presentations

We list some applications to 3-manifold theory of Alexander's theorem:

(i) In 1973 González-Acuña [91] (at the author's request) showed[13] that one can augment Alexander's theorem by requiring that the binding be connected: "every 3-manifold M^3 contains a fibered knot" and J. Simon realized that Bing's characterization of S^3 ("a 3-manifold M^3 is homeomorphic to S^3 if and only if every simple closed curve lies in a topological disk D^3 in M^3" [20]) is an immediate corollary, when combined with Schoenflies' theorem. Indeed, the boundary of Bing's 3-disk D^3 contains the connected binding and also bounds a 3-disk in the complement of D^3. This complement is fibered, and is thus covered by $\mathbb{R}^3$, etc. (See also Rolfsen [253, p. 341], Myers [208].)

(ii) Alexander's theorem gives an immediate proof of the existence of a codimension 1 foliation on any closed 3-manifold due to Lickorish [165]: indeed

[12] Most notably as a starting point of the celebrated work of V. Jones [115].
[13] Using basic work of Lickorish [164].

a simple procedure (H. B. Lawson [150], [151]) allows one to alter the obvious fibering foliation of $M^3\backslash$(binding) so that its boundary becomes a union of leaves; pasting on solid tori with Reeb foliations then gives the required foliation.

(iii) Similarly, Alexander's theorem was used by Thurston and Winkelnkemper [288] to give a short proof of a conjecture of Chern "every M^3 has a contact form", (i.e. a 1-form ω such that $\omega \wedge d\omega \neq 0$ everywhere). This had been shown first by Lutz and Martinet [182].

(iv) In a short paper J. Simon [268] used Alexander's theorem to show that, among all homotopy 3-spheres, S^3 is characterized by its fibered knot theory. This is a converse of a theorem of Neuwirth as well as a sharpening of a theorem of A. Connor.

(v) More recently, Bavard [15] used Alexander's theorem to extend examples of Gromov's from S^3 to any M^3 : given any $\epsilon > 0$, there exists a Riemannian metric on M^3 with sectional curvature $K \leq 1$, diameter $\leq \epsilon$ and volume $\leq \epsilon$. Thus (unlike dimension 2) the existence of a Riemannian metric of almost negative curvature on a 3-dimensional manifold M^3 has no bearing on the topology of M^3, even if one also requires the volume to be small.

For the following applications we use an open book decomposition of a closed, orientable 3-manifold M^3, where the page is the compact 2-disk with n holes, denoted by Ω_n. In 1975 González-Acuña [92] defined the notion of an Artin presentation of $\pi_1(M^3)$, using the HNN construction (Lyndon and Schupp [179]). By a theorem of Nielsen (Birman [21, Chap. 2], Zieschang [313, p. 3]) the monodromy homeomorphism $h : \Omega_n \to \Omega_n$ in the open book decomposition can be substituted in a natural way by an Artin presentation.

In particular, we can say: although the fundamental group $\pi_1(M^3)$ alone does not in general determine the closed, orientable 3-manifold M^3 up to homeomorphisms, a presentation of a certain type of $\pi_1(M^3)$, indeed does so.

Let F_n denote the free group generated by $x_1, x_2, \ldots, x_n$.

Definition (i) A presentation of a group

$$r \;=\; \langle x_1, x_2, \ldots, x_n \,|\, r_1, r_2, \ldots, r_n \rangle$$

is called an *Artin presentation* if

$$x_1 x_2 \ldots x_n \;=\; (r_1^{-1} x_1 r_1)(r_2^{-1} x_2 r_2) \ldots (r_n^{-1} x_n r_n) \in F_n \;.$$

(ii) The set of Artin presentations with n generators is denoted by $\mathcal{R}_n$. For $r \in \mathcal{R}_n$ let $A(r)$ be the exponent sum $n \times n$-matrix of r, let $\pi(r)$ be the group presented by r, and let $M^3(r)$ be the closed, orientable 3-manifold determined by r. It is not difficult to see that an arbitrary integer $n \times n$-matrix is an $A(r)$ for some $r \in \mathcal{R}_n$ if and only if it is symmetric ([311]).

From the work of Milnor [191] we have the important result:

If det $A(r) = \pm 1$ *and* $\pi(r)$ *is a finite group then* $\pi(r)$ *is either trivial or isomorphic to the binary icosahedral group of order 120*

$$I(120) = \langle a, b \,|\, a^3 = b^5 = (ab)^2 \rangle \ .$$

Examples Besides the obvious Artin presentations $\langle x_1, \ldots, x_n \,|\, 1, \ldots, 1 \rangle$ (of F_n) and $\epsilon_n = \langle x_1, \ldots, x_n \,|\, x_1, \ldots, x_n \rangle$ (of the trivial group), we have:

(1) $n = 2$; the only Artin presentations are given by

$$r_1 = x_1^{a-b}(x_1x_2)^b \ , \ r_2 = x_2^{c-b}(x_1x_2)^b \ .$$

Here

$$A(r) = \begin{pmatrix} a & b \\ b & c \end{pmatrix} \ .$$

If say $A = \begin{pmatrix} 1 & -2 \\ -2 & 3 \end{pmatrix}$, then $\pi(r) = I(120)$, and $M^3(r)$ is the Poincaré homology 3-sphere.

(2) $n = 8$; set

$$\begin{aligned}
X &= x_7(x_6x_7x_8)^{-1}x_5x_6x_7x_8x_7^{-1} \ , \\
r_1 &= x_1^2x_2 \ , \\
r_2 &= x_2x_1x_2x_4^{-1}x_3x_4 \ , \\
r_3 &= x_3x_4(x_1x_2)^{-1}r_2 \ , \\
r_4 &= x_4x_3x_4X \ , \\
r_5 &= x_5x_6x_7x_8(x_3x_4x_7)^{-1}r_4 \ , \\
r_6 &= x_6^2Xx_7 \ , \\
r_7 &= x_7(x_6X)^{-1}r_6 \ , \\
r_8 &= x_8^2x_7^{-1}Xx_7 \ .
\end{aligned}$$

Here $A(r) = E_8$, the well-known unimodular, even, positive definite 8×8 matrix used by Milnor to construct his exotic 7-sphere[14], and $\pi(r) = I(120)$.

For more examples, see [311].

(vi) In [92] González-Acuña obtained the fundamental:

Theorem *An arbitrary abstract group is isomorphic to the fundamental group of a closed, orientable 3-manifold, if and only if it has an Artin presentation.*

A much more elaborate *algorithmic* characterization had been given by Neuwirth [215].

---

[14]See, e.g. Lectures on Modern Mathematics, II, T. L. Saaty, ed., Wiley (1964), p. 174.

(vii) In [92] González-Acuña used a theorem of Waldhausen [298] (Burde and Zieschang [34, p. 41]) to give a purely algebraic equivalent of the Poincaré conjecture, which is simpler than those of Birman, Traub, et al (see Birman [21]).

(viii) With the fundamental theorem of Gordon and Luecke [95], one can use Artin presentations to test (on the computer) whether knots in homotopy 3-spheres have exotic peripheral structures or whether they satisfy Property P : in [311] the author obtains the following explicit criterion for testing the Poincaré conjecture. First observe:

Given an Artin presentation $r = \langle x_1, \ldots, x_n \,|\, r_1, \ldots, r_n \rangle$ and an integer j $(1 \leq j \leq n)$ we obtain another Artin presentation, j-red r, called the *j-reduction of* r, by simply removing r_j and setting $x_j = 1$ everywhere else (and renumbering, of course).

Criterion *The Poincaré conjecture implies: if $\pi(r) = 1$ and $\pi(j\text{-red}\, r) = 1$, then the group G_j presented by*

$$\langle x_1, \ldots, x_n \,|\, r_1, \ldots, r_{j-1}, x_j r_j = r_j x_j, r_{j+1}, \ldots, r_n \rangle$$

is isomorphic to $\mathbb{Z}$.

Remark This is not true for just any presentation of deficiency 0; consider

$$H \;=\; \langle x, y \,|\, w_1, w_2 \rangle \;,$$

where

$$w_1 \;=\; x^3 y x^{-2} y^{-1} \;\;,\;\; w_2 \;=\; y^3 x y^{-2} x^{-1} \;.$$

Then $H = 1$ and the 1-reduction also presents the trivial group, but the group

$$H_1 \;=\; \langle x, y \,|\, x w_1 = w_1 x, w_2 \rangle$$

is non-abelian, since, after adding the relation $x^2 = 1$, the commutator $[x, y^2]$ has order 5.

For more criteria, some of which are related to gauge theory and 3+1 $TQFT$'s (= topological quantum field theories), see [311].

We stress that no knowledge of topology or geometry is needed for testing these criteria on a computer.

Since 3-dimensional open books are not discussed in this book we end this section A2 by making some remarks justifying that at least Artin presentation theory (i.e. the case where the pages are planar) is as purely algebraic and arithmetic as possible, in the spirit of the development of the high-dimensional theory in the main body of the book.

A main advantage of Artin presentation theory, which detaches it and pushes it away from all other 3-manifold theories, comes from the discovery in [310] of the following fundamental fact: an Artin presentation r does not just determine the closed, orientable 3-manifold $M^3(r)$, but also, *in a canonical way*, a well-defined, smooth, simply-connected 4-manifold $W^4(r)$, whose boundary $\partial W^4(r) = M^3(r)$ and *whose intersection form is given by the matrix* $A(r)$: namely the cobordism constructed in section A3 (ii) below.[15] Thus each Artin presentation r representing a given M^3 literally materializes as a cobordism[16] of M^3.

The purely discrete coupling $r \to A(r)$ is thus strengthened by the smooth 4-dimensional gauge theory (and 3+1 $TQFT$'s). Conversely, Artin presentation theory becomes a first approximation to 4-dimensional gauge theory in the same philosophical sense as, according to Artin himself [9, p. 491], braid theory is a first approximation to knot theory.

For example, let $\mathcal{D}$ denote the set of unimodular, symmetric, integer matrices prevented by Donaldson's theorem from representing the quadratic form of a a closed, smooth, simply-connected 4-manifold. From a theorem of Taubes [286, p. 366] showing among other things that Donaldson's theorem holds even if one allows homotopy 3-spheres as boundaries we obtain:

Theorem *If $A(r) \in \mathcal{D}$, then $\pi(r)$ cannot be the trivial group; in fact it has a non-trivial representation into the Lie group $SU(2)$.*

Thus for the groups $\pi_1(M^3)$, $\det A(r) \neq \pm 1$ is not the only abelian condition preventing $\pi_1(M^3)$ from being trivial.

Notice the philosophical similarity with Bohm-Aharonov phenomena: we obtain non-trivial homotopy where, a priori, intuitively, none should exist (see [12], [114], [181]).

Does gauge theory, via this theorem, disturb the conjectured equivalence of the two basic decision problems: the simply-connectedness for $M^3(r)$ and the homeomorphism problem for S^3?

If so, this could already disprove the Poincaré conjecture without having to give an explicit counterexample – see Haken [102, p. 147].

Similar arguments can be made with topological quantum field theories in the sense of [11]: given r, any 3+1 $TQFT$ will associate to r not only a vector space $Z(r)$ but also, via $W^4(r)$, a well-defined (vacuum) vector $z(r) \in Z(r)$; furthermore this is done in a completely discrete fashion. Which is the most natural $TQFT$ here?

This again leads to more computer criteria for testing several important conjectures in 3,4-manifold theory (see [311]).

[15] These facts are easy consequences of the van Kampen theorem and the Mayer–Vietoris sequence.

[16] In particular, we obtain a canonical cobordism (i.e. one obtained without using Thom transversality) for any M^3. This was first done by Lickorish [164] in a different way.

A3. Applications to cobordism theory, SK-theory, etc.

(i) Already in [307] it was shown that Heegaard decompositions (i.e. twisted doubles) have applications to the problem of fibering over S^1 in a cobordism class:

Proposition *If $M^n = W \cup_h W$ where $h : \partial W \to \partial W$ is a homeomorphism, then M^n is canonically cobordant to the mapping torus of h, $(\partial W)_h$.*

Proof Let C be a collar of ∂W in W and extend h to a homeomorphism $H : C \to C$; consider $W \times I$ and identify $C \times 0$ with $C \times 1$ via H.

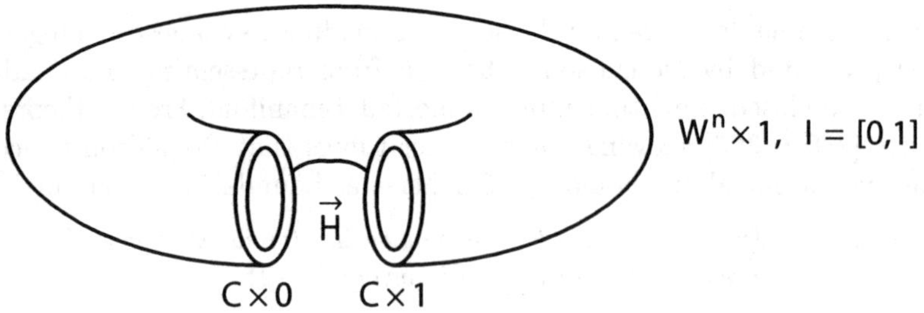

We have obtained a cobordism between $(\partial W)_h$ (on the "outside") and $W^n \cup_h W^n = M^n$ (on the "inside") – see [307]. Hence it follows easily from the open book theorem that a smooth or oriented cobordism class contains a manifold fibered over S^1 if and only if its signature $= 0$; furthermore, the fiber will be cobordant to 0, and will have similar properties to that of the cobordism class; for example, all this is true in the spin category, as well as the stably parallelizable category, etc. (See also Neumann [209]).

Question *Do this canonical cobordism and its cyclic covers induce any universal, categorical relations on certain axiomatic TQFT's ([11], [149], [229])?*

(ii) Another canonical cobordism (which however does need the full open book structure) is the following:

Proposition (see [310]) *Let M^n have an open book decomposition whose page V^{n-1} embeds in S^{n-1}; then there exists a canonical cobordism of M^n, denoted by $W(h)$, which is k-connected if $H_i(V) = 0$ for $i > k$.*
Proof Let cV denote the closure of the complement of V in S^{n-1}; extend h to S^{n-1} and then to all of D^n obtaining a homeomorphism $H : D^n \to D^n$; the mapping torus of H, call it D^n_H, has $cV \times S^1$ canonically embedded in its boundary ∂D^n_H and $W(h)$ is obtained by pasting $cV \times D^2$ onto this part of the boundary of D^n_H by the identity; the k-connectedness follows by Alexander duality.

Remark In the smooth case, in order to obtain H, we have to add the corresponding exotic sphere to M^n.

A theorem of Hirsch [112] can be used to show that a large class of stably parallelizable n-manifolds (i.e. π-manifolds) have open book decompositions whose page embeds in S^{n-1}.

(iii) J. C. Hausmann [109] extended the work of Hatcher and T. Lawson on inertial h-cobordisms (i.e. h-cobordism of a manifold with itself) by inductively relating such a cobordism of an open book to one of its binding.

(iv) The cutting and pasting of Heegaard decompositions (twisted doubles) can be generalized and given a systematic treatment via a Grothendieck construction (Jänich, Karras et. al. [117]) to obtain the so called *SK-invariants* (SK from "Schneiden und Kleben") and SK-groups, $SK_n(X)$, of singular manifolds in a topological space X. For the relative version see Koshikawa [139]. For example, the index of a smooth elliptic operator on an orientable manifold is an SK-invariant. In Chap. 8 of [117] it is shown how open books facilitate their study considerably.

A4. Applications to bordism of automorphisms

In the fall of 1965, in Princeton, William Browder gave the author the problem of cobordisms of diffeomorphisms, as a thesis problem: if $h_1 : M_1^n \to M_1^n$ and $h_2 : M_2^n \to M_2^n$ are self diffeomorphisms, when does there exist a cobordism W^{n+1} between M_1^n and M_2^n and a self diffeomorphism $H : W^{n+1} \to W^{n+1}$ such that H restricts to h_1 and h_2 on M_1, respectively M_2. In the usual way one obtains abelian groups in each dimension which in the orientable, orientation preserving case are denoted by Δ_n.

The following questions are already nontrivial:

(a) Is any diffeomorphism of S^n cobordant to 0 in Δ_n?

(b) If so, is it still true if one requires the cobording diffeomorphism $H : M_0^n \to M_0^n$ to be, say, homotopic to the identity of M_0?

Using the well-known 1 to 1 correspondence between diffeomorphism classes of $h : S^n \to S^n$ and homotopy $(n+1)$-spheres $\sum^{n+1}$, it is easy to see: if $H : M_0^{n+1} \to M_0^{n+1}$ cobords $h : S^n \to S^n$ and $\sum^{n+1}$ is determined by h, then the connected sum $M\#\sum$ is diffeomorphic to M, where M is the closed smooth manifold $M_0^{n+1} \cup_{id} D^{n+1}$. One then says $\sum$ lies in the inertia group, $I(M)$, of M. (If, in addition, the diffeomorphism is homotopic to the 'identity': $M\#\sum \to M$, one says $\sum$ lies in the special inertia group, $I_0(M)$, of M).

The above questions can then be reformulated as follows:

(a) For each n, does there exist a smooth, closed, orientable M^n whose inertia group is the group Θ^n of [309], i.e. as large as possible?
(b) Same question with the special inertia group.

The author succeeded in answering (a) affirmatively (the answer to (b), somewhat surprisingly, is negative [309]) by using Smale's Heegaard decompositions as follows: given an arbitrary homotopy $2m$-sphere, $\sum$, let W_0 be a smooth, simply-connected cobordism of $\sum$, then $W = W_0 \cup (\text{cone on}) \sum$ is a closed $(2m+1)$-manifold, smooth except at the cone; since in Smale's proof the 2 hemispheres of the Heegaard decomposition are obtained as regular neighborhoods, N, N^*, of the m-skeletons of a triangulation of W and its dual, the non-smooth cone can be avoided and one obtains an "h-cobordism" of the form:

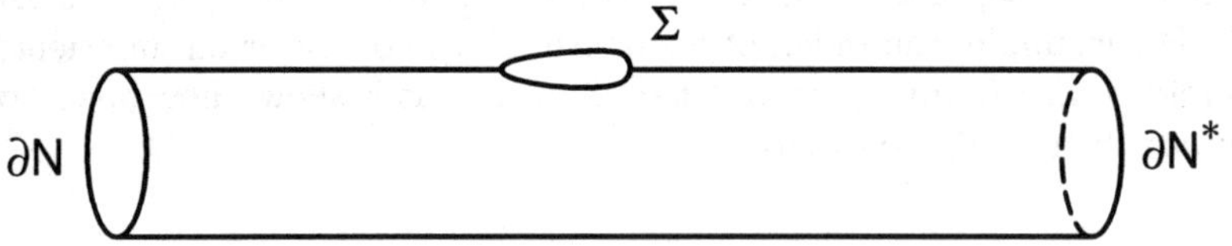

Since ∂N and ∂N^* can be shown to be diffeomorphic, (see [307]) $\sum$ lies in the inertia group $I(\partial N)$.

Of course, since Smale's proof only, a priori, applied to $(2m+1)$-manifolds M where $H_m(M)$ was torsion-free, to solve problem (a) for all dimensions, one had to settle the existence of twisted doubles in all dimensions, which, aided by the engulfing techniques of Levitt [162], the author proceeded to do in the thesis [307]. In a very elementary manner it was shown there that, unlike in many cobordism problems, the groups Δ_{4n} are not finitely generated. (The author gave up trying to compute them).

In 1971, López de Medrano [172] noticed the similarity with knot cobordism computations of Fox, Milnor and Levine and, in the case of $(n-1)$-connected $2n$-manifolds, introduced, as a new invariant in Δ_{2n}, the isometric structure of a diffeomorphism, taking values in an infinitely generated Witt group; in particular he showed that the groups Δ_{2n} were also infinitely generated.

It thus came as quite a surprise when in 1976 M. Kreck [141] succeeded in giving a complete description of all Δ_n, $n \geq 4$, by just using López de Medrano's invariant and the obvious ones determined by the ordinary cobordism classes of M and the mapping torus of h. (The groups Δ_2, Δ_3 were computed by Bonahon [24], Melvin [186] respectively).

In 1978, Quinn [227] identified Δ_n with the bordism groups of open books $BB_n(\text{pt.})$ appearing in his surgery theory of open book decompositions, and used an exact sequence relating $BB_n(\text{pt.})$ to the oriented bordism groups Ω_n to again compute Δ_n for $n \geq 5$. In the same paper Quinn found the exact obstruction for any manifold M, (not necessarily a simply-connected one) to be an open book, (see [227]). T. Lawson had settled the odd-dimensional case in 1976, [152]. Stoltzfus [279] found the explicit relation between the invariants used by Kreck and Quinn. Recently Ranicki [248] (and in the main body of the book) gave his more algebraic L-theory version of Quinn's $BB_n(X)$ groups as surgery groups of a localized ring.

A5. Applications to foliations.

Open books just in the form of Milnor's fibration theorem were first used by Blaine Lawson [150] in his famous 1971 breakthrough (the first since Reeb foliated S^3 in the 1940's) in *explicitly* constructing codimension 1 foliations of the spheres S^{2^k+3} $(k = 1, 2, \ldots)$.

The relevance of open books to this type of construction is given by Lawson's Lemma 1 of [150, p. 495] : in the complement of an open neighborhood of the binding one can always easily change the obvious foliation given by the pages to one where the boundary of the complement becomes a compact leaf; then one is done if one finds a similar foliation for (binding)$\times D^2$. For example, this is why Lickorish's theorem [165] in dimension 3 followed easily from Alexander's theorem.

Lawson's method was extended by others and culminated with the work of Freedman and Tamura : if $n > 2$, then on any closed, $(n-1)$-connected $(2n+1)$-manifold one can construct an explicit codimension 1 foliation in the above manner, [284].

The case $n = 2$ had been settled by A'Campo [1]. Mizutani and Tamura [203] settled the case of $(n-1)$-connected $2n$-manifolds with vanishing signature and Euler characteristic. Tamura and Verjovsky also used these methods to find, among other things explicit foliations on S^7, S^{15} of codimension 3, 7, respectively, (see [285], [294]).

Since Lawson's Steele Prize memoir [151] with its exhaustive bibliography is hard to improve upon we refer the reader to it, as well as to the more recent books of Hector and Hirsch [110] and Godbillon [89].

A6. Applications in differential geometry.

In differential geometry, besides Bavard's theorem in section A2, we have:

(i) H. Gluck [88] used open books to construct a non-singular geodesible vector field on any odd-dimensional closed manifold, i.e. a non-singular vector field such that there exists a Riemannian metric on M so that all orbits are geodesics.

(ii) F. Morgan [205, Thm. 4.15] proved among other things: if a codimension 2 submanifold, B, in certain closed manifolds, bounds an infinite number of minimal submanifolds, then M is an open book with binding B and the bounding minimal submanifolds are among the pages; thus, for example, if the signature of M is non-zero, such a B in M can only bound a finite number of minimal submanifolds.

Open books were also used to study the uniqueness of minimal submanifolds by Hardt and Rosenberg [105].

A7. Miscellaneous applications.

Among the various other applications we single out the following:

(i) L. Kauffman [121] characterizes homotopy n-spheres by their monodromy and uses it to prove a knot cobordism periodicity theorem for fibered knots.

(ii) If $h : V \to V$ defines an open book, then so, of course do the iterates $h^k : V \to V$. J. Stevens [275] generalized the work of Durfee and Kauffman [65], showing that the corresponding open books M_k and M_{k+d} have the same rational homology if the eigenvalues of the monodromy are all dth roots of unity.

(iii) L. Rudolph, [255], attaches a non-zero vector field to an open books decomposition of a 3-manifold and then, in the case of homology 3-spheres, expresses them with linking number formulae and shows they do not depend on the Seifert form and hence are independent of the algebraic monodromy.

(iv) In [260], J. Seade associates open book decompositions to holomorphic vector fields, and uses them to study new types of singularities by analogy with Milnor's use of functions.

(v) Also in the complex case, new studies of complex open books are developed in papers [35], [36], [259].

(vi) In [178], R. Lutz generalized the definition of open books to 'fibrations noueés' and used these to study contact structures invariant under locally free Lie group actions.

In the bibliography the reader will find other applications not discussed here.

References

[1] N. A'Campo, **Feuilletages de codimension 1 sur les variétés de dimension 5**, Comm. Math. Helv. 47, 514–525 (1972)
[2] C. Adams, **The knot book**, W.H.Freeman (1994)
[3] J.W. Alexander, **A lemma on systems of knotted curves**, Proc. Nat. Acad. Sci. U.S.A. 9, 93–95 (1923)
[4] ––, **Topological invariants of knots and links**, Trans. A.M.S. 30, 275–306 (1928)
[5] G. Almkvist, **Endomorphisms of finitely generated projective modules over a commutative ring**, Arkiv för Mat. 11, 263–301 (1973)
[6] ––, **The Grothendieck ring of the category of endomorphisms**, J. of Algebra 28, 375–388 (1974)
[7] ––, ***K*-theory of endomorphisms**, J. of Algebra 55, 308–340 (1978)
[8] E. Artin, **Zur Isotopie zweidimensionaler Flächen in $\mathbb{R}^4$**, Abh. Math. Seminar Univ. Hamburg 4, 174–177 (1925)
[9] ––, **The collected papers of E. Artin**, S. Lang, J. Tate, eds., Addison-Wesley (1965)
[10] M.F. Atiyah, **On the work of Serge Novikov**, Proc. 1970 Nice I.C.M., Gauthier–Villars, Vol. 1, 11–13 (1971)
[11] ––, **Topological quantum field theories**, Publ. Math. I.H.E.S. 68, 175–186 (1989)
[12] ––, **The geometry and physics of knots**, Cambridge (1990) (review by R. Bott in Bull. A.M.S. 26 (1992) p. 182)
[13] H. Bass, **Algebraic *K*-theory**, Benjamin (1968)
[14] ––, A. Heller and R. Swan, **The Whitehead group of a polynomial extension**, Publ. Math. I.H.E.S. 22, 61–80 (1964)
[15] C. Bavard, **Courbure presque negative en dimension 3**, Comp. Math. 63, 223–236 (1987)
[16] E. Bayer, **Factorisation is not unique for higher-dimensional knots**, Comm. Math. Helv. 55, 583–592 (1980)
[17] E. Bayer-Fluckiger, **Definite Hermitian forms and the cancellation of simple knots**, Arch. Math. 40, 182–185 (1983)
[18] I. Berstein and A.L. Edmonds, **On the construction of branched coverings of low–dimensional manifolds**, Trans. A.M.S. 247, 87–124 (1979)
[19] R. Bieri and B. Eckmann, **Finiteness properties of duality groups**, Comm. Math. Helv. 49, 74–83 (1974)
[20] R.H. Bing, **Necessary and sufficient conditions that a 3-manifold be S^3**, Ann. of Maths. 68, 17–37 (1958)
[21] J. Birman, **Braids, links and mapping class groups**, Ann. of Maths. Studies 82, Princeton (1974)
[22] ––, **Poincaré's conjecture and the homeotopy group of a closed, orientable 2-manifold**, J. Austr. Math. Soc. 17, 214–221 (1974)

[23] R. C. Blanchfield, **Intersection theory of manifolds with operators with applications to knot theory**, Ann. of Maths. 65, 340–356 (1957)
[24] F. Bonahon, **Cobordism of automorphisms of surfaces**, Ann. Sci. École Norm. Sup. (4) 16, 237–270 (1983)
[25] G. Bredon, **Regular $O(n)$-manifolds, suspension of knots, and knot periodicity**, Bull. A. M. S. 79, 87–91 (1973)
[26] E. Brieskorn, **Beispiele zur Differentialtopologie von Singularitäten**, Invent. Math. 2, 1–14 (1966)
[27] W. Browder, **Structures on $M \times \mathbb{R}$**, Proc. Camb. Phil. Soc. 61, 337–345 (1965)
[28] ——, **Diffeomorphisms of 1-connected manifolds**, Trans. A. M. S. 128, 155–163 (1967)
[29] ——, **Surgery and the theory of differentiable transformation groups**, Proc. Conf. on Transformation Groups (New Orleans, 1967), Springer (1969)
[30] ——, **Surgery on simply-connected manifolds**, Ergebnisse der Mathematik und ihrer Grenzgebiete 65, Springer (1972)
[31] —— and J. Levine, **Fibering manifolds over the circle**, Comm. Math. Helv. 40, 153–160 (1966)
[32] —— and G. R. Livesay, **Fixed point free involutions on homotopy spheres**, Tohôku J. Math. 25, 69–88 (1973)
[33] K. Brown, **Homological criteria for finiteness**, Comm. Math. Helv. 50, 129–135 (1975)
[34] G. Burde and H. Zieschang, **Knots**, de Gruyter (1985)
[35] O. Calvo–Andrade, **Foliations with Kupka component on algebraic manifolds**, Bol. Soc. Bras. de Mat. (to appear)
[36] —— and M. Soares, **On Chern numbers of Kupka components**, Ann. Inst. Fourier 44, 1219–1236 (1994)
[37] S. Cappell, **Splitting obstructions for hermitian forms and manifolds with $\mathbb{Z}_2 \subset \pi_1$**, Bull. A. M. S. 79, 909–913 (1973)
[38] ——, **Unitary nilpotent groups and hermitian K-theory**, Bull. A. M. S. 80, 1117–1122 (1974)
[39] —— and J. Shaneson, **Topological knots and knot cobordism**, Topology 12, 33–40 (1973)
[40] —— and ——, **The codimension two placement problem, and homology equivalent manifolds**, Ann. of Maths. 99, 277–348 (1974)
[41] —— and ——, **Branched cyclic covers**, in Knots, Groups and 3-Manifolds (Papers Dedicated to the Memory of R. H. Fox), Ann. of Maths. Study 84, 165–173, Princeton (1975)
[42] —— and ——, **Piecewise linear embeddings and their singularities**, Ann. of Maths. 103, 163–228 (1976)
[43] —— and ——, **There exist inequivalent knots with the same complement**, Ann. of Maths. 103, 349–353 (1976)
[44] —— and ——, **Totally spineless manifolds**, Ill. J. Math. 21, 231–239 (1977)
[45] —— and ——, **Singularities and immersions**, Ann. of Maths. 105, 539–552 (1977)
[46] —— and ——, **An introduction to embeddings, immersions and singularities in codimension two**, Proc. Symp. in Pure Math. XXXII (Stanford 1976), vol. 2, 129–149, A. M. S. (1978)
[47] —— and ——, **Singular spaces, characteristic classes, and intersection homology**, Ann. of Maths. 134, 325–374 (1991)
[48] A. J. Casson and C. Mc. Gordon, **On slice knots in dimension three**, Pure Maths. Proc. Symp. in Pure Math. XXXII (Stanford 1976), vol. 2, 39–53, A. M. S. (1978)

[49] –– and ––, **Cobordism of classical knots**, in A la recherche de la topologie perdue, Progress in Maths. 62, 181–197, Birkhäuser (1986)
[50] F. Clauwens, **Surgery on products**, Indagat. Math. 41, I.,II. 121–144 (1979)
[51] ––, **The K-theory of almost symmetric forms**, Mathematical Centre Tracts 115, Amsterdam, 41–49 (1979)
[52] M. Cohen, **A course in simple homotopy theory**, Graduate Texts in Mathematics 10, Springer (1973)
[53] P. M. Cohn, **Free rings and their relations**, Academic Press (1971)
[54] ––, **Rings of fractions**, Amer. Math. Monthly 78, 596–615 (1971)
[55] P. E. Conner, **Notes on the Witt classification of hermitian innerproduct spaces over a ring of algebraic integers**, U. Texas Press (1979)
[56] F. X. Connolly and A. A. Ranicki, **On the computation of Unil_*** (to appear)
[57] P. R. Cromwell and I. J. Nutt, **Embedding knots and links in an open book, II. Bounds on arc index**, Math. Proc. Cambr. Phil. Soc. 11, 309–319 (1996)
[58] R. H. Crowell, **The group G'/G'' of a knot group G**, Duke Math. J. 30, 349–354 (1963)
[59] –– and R. H. Fox, **Introduction to knot theory**, Graduate Texts in Mathematics 57, Springer (1977)
[60] A. W. M. Dress, **Induction and structure theorems for orthogonal representations of finite groups**, Ann. of Maths. 102, 291–325 (1975)
[61] –– and C. Siebeneicher, **Symmetric powers of cyclic sets and the definition of A. Weil's zeta-functions**, Proc. Symp. Pure Math. 47, 407–411 (1987)
[62] –– and ––, **On the number of solutions of certain linear diophantine equations**, Hokkaido Math. J. 19, 385–401 (1990)
[63] M. Dunwoody and R. Fenn, **On the finiteness of higher knot sums**, Topology 26, 337–343 (1987)
[64] A. Durfee, **Fibered knots and algebraic singularities**, Topology 13, 47–59 (1974)
[65] –– and L. Kauffman, **Periodicity of branched cyclic covers**, Math. Ann. 218, 157–174 (1975)
[66] A. Edmonds and J. Ewing, **Remarks on the cobordism group of surface diffeomorphisms**, Math. Ann. 259, 497–504 (1982)
[67] D. Eisenbud, **Commutative algebra with a view toward algebraic geometry**, Graduate Texts in Mathematics 150, Springer (1994)
[68] M. Epple, **Branch points of algebraic functions and the beginnings of modern knot theory**, Historia Mathematica 22, 371–401 (1995)
[69] ––, **Geometric aspects in the development of knot theory**, History of Topology, Chapter 11, Elsevier (1998)
[70] ––, **Geschichte der Knotentheorie**, Vieweg (1998)
[71] D. Erle, **Die quadratische Form eines Knotens und ein Satz über Knotenmannigfaltigkeiten**, J. reine u. angew. Math. 236, 174–218 (1969)
[72] M. Sh. Farber, **Isotopy types of knots of codimension two**, Trans. A. M. S. 261, 185–209 (1980)
[73] ––, **The classification of simple knots**, Uspekhi Mat. Nauk 38, 59–106 (1983) English translation : Russian Math. Surv. 38, 63–117 (1982)
[74] ––, **An algebraic classification of some even-dimensional spherical knots**, Trans. A. M. S. 281, I. 507–527, II. 529–570 (1984)
[75] ––, **Exactness of the Novikov inequalities**, Functional analysis and its applications 19, 49–59 (1985)

[76] ——, **Minimal Seifert manifolds and the knot finiteness theorem**, Israel J. Math. 66, 179–215 (1989)
[77] ——, **Hermitian forms on link modules**, Comm. Math. Helv. 66, 189–236 (1991)
[78] F. T. Farrell, **The obstruction to fibering a manifold over a circle**, Indiana Univ. J. 21, 315–346 (1971)
[79] ——, **The obstruction to fibering a manifold over a circle**, Proc. 1970 I. C. M. Nice, Vol. 2, 69–72 (1971)
[80] —— and W. C. Hsiang, **Manifolds with $\pi_1 = G \times_\alpha T$**, Amer. J. Math. 95, 813–845 (1973)
[81] S. Ferry, A. A. Ranicki and J. Rosenberg, **A history and survey of the Novikov conjecture**, Proc. 1993 Oberwolfach Conf. on the Novikov Conjectures, Index Theorems and Rigidity, Vol. 1, L.M.S. Lecture Notes 226, 7–66, Cambridge (1995)
[82] R. H. Fox and J. Milnor, **Singularities of 2-spheres in 4-space and cobordism of knots**, Osaka J. Math. 3, 257–267 (1966)
[83] F. Frankl and L. Pontrjagin, **Ein Knotensatz mit Anwendung auf die Dimensionstheorie**, Math. Ann. 102, 785–789 (1930)
[84] M. Freedman, **Surgery on codimension two submanifolds**, Memoirs A. M. S. 191 (1977)
[85] ——, **The disk theorem for four-dimensional manifolds**, Proc. 1983 I.C.M., Warsaw, 647–663 (1983)
[86] —— and F. Quinn, **The topology of 4-manifolds**, Princeton (1990)
[87] H. Gluck, **The embedding of two-spheres in the four-sphere**, Bull. A. M. S. 67, 586–589 (1961)
[88] ——, **Geodesible vector fields**, (preprint) (1979)
[89] C. Godbillon, **Feuilletages. Etudes geometriques**, Birkhäuser (1991)
[90] O. Goldman, **Determinants in projective modules**, Nagoya Math. J. 18, 27–36 (1961)
[91] F. González–Acuña, **3-dimensional open books**, Lecture Notes, University of Iowa (1974)
[92] ——, Lecture Notes, UNAM, Mexico (1975)
[93] C. Mc. Gordon, **Knots in the 4-sphere**, Comm. Math. Helv. 39, 585–596 (1977)
[94] ——, **Some aspects of classical knot theory**, Proc. 1977 Plans-sur-Bex Conf. Knot Theory, Lecture Notes in Mathematics 685, 1–65, Springer (1978)
[95] —— and J. Luecke, **Knots are determined by their complements**, Jour. A. M. S. 2, 371–415 (1989)
[96] D. Grayson, **Higher algebraic K-theory: II (after D. Quillen)**, Proc. 1976 Algebraic K-theory Conf., Evanston, Lecture Notes in Mathematics 551, 217–240, Springer (1976)
[97] ——, **The K-theory of endomorphisms**, J. of Algebra 48, 439–446 (1977)
[98] ——, **K-theory and localization of noncommutative rings**, J. Pure Appl. Algebra 18, 125–127 (1980)
[99] ——, **Exact sequences in algebraic K-theory**, Ill. J. Math. 31, 598–617 (1987)
[100] M. Gromov, **Positive curvature, macroscopic dimension, spectral gaps and higher signatures**, Functional Analysis on the Eve of the 21st Century, In Honor of the Eightieth Birthday of I.M.Gelfand, Vol. II, Progress in Mathematics 132, 1–213, Birkhäuser (1996)
[101] A. Haefliger, **Differentiable embeddings of S^n in S^{n+q} for $q > 2$**, Ann. of Maths. 83, 402–436 (1966)

[102] W. Haken, **Various aspects of the 3-dimensional Poincaré problem**, Topology of Manifolds (Proc. Inst., Univ. of Georgia, Athens, Ga., 1969), 140–152, Markham (1970)
[103] I. Hambleton, R. J. Milgram, L. Taylor and B. Williams, **Surgery with finite fundamental group**, Proc. L.M.S. (3) 56, 349–379 (1988)
[104] ––, A. A. Ranicki and L. Taylor, **Round L-theory**, J. Pure and App. Alg. 47, 131–154 (1987)
[105] R. Hardt and H. Rosenberg, **Open book structures and uniqueness of minimal submanifolds**, Ann. Inst. Fourier 40, 701–708 (1990)
[106] J. Harer, **Representing elements of $\pi_1 M^3$ by fibered knots**, Math. Proc. Cambridge Phil. Soc. 92, 133–138 (1982)
[107] ––, **How to construct all fibered knots and links**, Topology 21, 263–280 (1982)
[108] H. Hasse, **Äquivalenz quadratischer Formen in einem beliebigen algebraischen Zahlkörper**, J. reine u. angew. Math. 153, 158–162 (1924)
[109] J. C. Hausmann, **Open books and h-cobordisms**, Comm. Math. Helv. 55, 263–346 (1980)
[110] G. Hector and U. Hirsch, **Introduction to the geometry of foliations**, Aspects of Math., Vieweg, Part A (1986), Part B (1987)
[111] G. Higman, **The units of group rings**, Proc. L.M.S. (2) 46, 231–248 (1940)
[112] M. Hirsch, **Embeddings and compressions of polyhedra and smooth manifolds**, Topology 4, 361–369 (1966)
[113] B. Hughes and A. A. Ranicki, **Ends of complexes**, Cambridge Tracts in Mathematics 123, Cambridge (1996)
[114] Y. Imry and R. A. Webb, **Quantum interference and the Aharonov–Bohm effect**, Scientific American, (April 1989)
[115] V. F. R. Jones, **Hecke algebra representations of braid groups and link polynomials**, Ann. of Maths. 126, 335–388 (1987)
[116] M. Karoubi, **Localisation de formes quadratiques**, Ann. Éc. Norm. Sup (4) I. 7, 359–404 (1974), II. 8, 99–155 (1975)
[117] U. Karras, M. Kreck, W. D. Neumann and E. Ossa, **Cutting and pasting of manifolds; SK groups**, Publish or Perish (1973)
[118] M. Kato, **A classification of simple spinnable structures on a 1-connected Alexander manifold**, J. Math. Soc. Japan 26, 454–463 (1974)
[119] –– and Y. Matsumoto, **Simply-connected surgery of submanifolds in codimension two, I.**, J. Math. Soc. Japan 24, 586–608 (1972)
[120] L. H. Kauffman, **Products of knots**, Bull. A. M. S. 80, 1104–1107 (1974)
[121] ––, **Branched coverings, open books and knot periodicity**, Topology 13, 143–160 (1974)
[122] ––, **On knots**, Ann. of Maths. Study 115, Princeton (1987)
[123] –– and W. D. Neumann, **Products of knots, branched fibrations and sums of singularities**, Topology 16, 369–393 (1977)
[124] –– and L. Taylor, **Signature of links**, Trans. A. M. S. 216, 351–365 (1976)
[125] C. Kearton, **Blanchfield duality and simple knots**, Trans. A. M. S. 202, 141–160 (1975)
[126] ––, **Cobordism of knots and Blanchfield duality**, J. L. M. S. 10, 406–408 (1975)
[127] ––, **An algebraic classification of some even-dimensional knots**, Topology 15, 363–373 (1976)
[128] ––, **Signatures of knots and the free differential calculus**, Quart. J. Math. (2) 30, 157–182 (1979)
[129] ––, **Some non-fibred 3-knots**, Bull. L. M. S. 15, 365–367 (1983)

[130] J. L. Kelley and E. H. Spanier, **Euler characteristics**, Pacific J. Math. 26, 317–339 (1968)
[131] M. Kervaire, **On higher dimensional knots**, Differential and combinatorial topology, A symposium in honor of Marston Morse, Princeton, 105–119 (1965)
[132] ——, **Les noeuds de dimensions supérieures**, Bull. Soc. Math. France 93, 225–271 (1965)
[133] ——, **Knot cobordism in codimension two**, Proc. 1970 Amsterdam Manifolds Conf., Lecture Notes in Mathematics 197, 83–105, Springer (1971)
[134] —— and J. Milnor, **Groups of homotopy spheres I.**, Ann. Math. 77, 504–537 (1963)
[135] —— and C. Weber, **A survey of multidimensional knots**, Proc. 1977 Plans-sur-Bex Conf. Knot Theory, Lecture Notes in Mathematics 685, 61–134, Springer (1978)
[136] M. Knebusch, **Grothendieck und Wittringe von nichtausgearteten symmetrischen Bilinearformen**, Sitzb. Heidelberger Akad. Wiss. Math. Naturwiss. Kl., 89–157 (1969)
[137] —— and W. Scharlau, **Quadratische Formen und quadratische Reziprozitätsgesetze über algebraischen Zahlkörpern**, Math. Z. 121, 346–368 (1971)
[138] M. Knus, **Quadratic and hermitian forms over rings**, Grundlehren der mathematischen Wissenschaften 294, Springer (1991)
[139] H. Koshikawa, ***SK* groups of manifolds with boundary**, Kyushu J. Math. 49, 47–57 (1995)
[140] D. Kotschick, **Gauge theory is dead! Long live gauge theory!**, Notices of the A. M. S. 42, 335–338 (1995)
[141] M. Kreck, **Bordism of diffeomorphisms and related topics**, Lecture Notes in Mathematics 1069, Springer (1984)
[142] K. Y. Lam, A. A. Ranicki and L. Smith, **Jordan normal form projections**, Arch. Math. 50, 113–117 (1988)
[143] T. Y. Lam, **Serre's conjecture**, Lecture Notes in Mathematics 635, Springer (1978)
[144] K. Lamotke, **Die Homologie isolierter Singularitäten**, Math. Z. 143, 27–44 (1975)
[145] W. Landherr, **Äquivalenz Hermitescher Formen über einen beliebigen algebraischen Zahlkörper**, Abh. Math. Sem. Univ. Hamburg 11, 245–248 (1936)
[146] S. Lang, **Algebra**, Third Edition, Addison Wesley (1993)
[147] R. Lashof and J. Shaneson, **Classification of knots in codimension two**, Bull. A. M. S. 75, 171–175 (1969)
[148] F. Latour, **Existence de 1-formes fermées non singuliéres dans une classe de cohomologie de de Rham**, Publ. Math. I. H. E. S. 80, 135–194 (1995)
[149] R. Lawrence, **An introduction to topological field theory**, The interface of knots and physics (San Francisco, CA, 1995), Proc. Sympos. Appl. Math. 51, A. M. S., 89–128 (1996)
[150] H. B. Lawson, **Codimension-one foliations of spheres**, Ann. of Maths. 94, 494–503 (1971)
[151] ——, **Foliations**, Bull. A. M. S. 80, 369–418 (1974)
[152] T. Lawson, **Open book decompositions for odd dimensional manifolds**, Topology 17, 189–192 (1977)
[153] A. Y. Lazarev, **Novikov homologies in knot theory**, Mat. Zam. 51, 53–57 (1992), English translation : Math. Notes 51, 259–262 (1992)
[154] J. Levine, **Unknotting spheres in codimension 2**, Topology 4, 9–16 (1965)

[155] ––, **A classification of differentiable knots**, Ann. of Maths. 82, 15–50 (1965)
[156] ––, **Polynomial invariants of knots of codimension two**, Ann. of Maths. 84, 537–554 (1966)
[157] ––, **Knot cobordism in codimension two**, Comm. Math. Helv. 44, 229–244 (1969)
[158] ––, **Invariants of knot cobordism**, Invent. Math. 8, 98–110 (1969)
[159] ––, **An algebraic classification of some knots of codimension two**, Comm. Math. Helv. 45, 185–198 (1970)
[160] ––, **Algebraic structure of knot modules**, Lecture Notes in Mathematics 772, Springer (1980)
[161] –– and K. Orr, **A survey of surgery and knot theory** (preprint) (1997)
[162] N. Levitt, **Applications of engulfing**, Ph.D. Thesis, Princeton Univ., 1967
[163] D. W. Lewis, **New improved exact sequences of Witt groups**, J. of Algebra 74, 206–210 (1982)
[164] W. B. R. Lickorish, **A representation of orientable combinatorial 3-manifolds**, Ann. of Maths. 76, 531–540 (1962)
[165] ––, **A foliation for 3-manifolds**, Ann. of Maths. 82, 414–420 (1965)
[166] ––, **A finite set of generators for the homeotopy group of a 2-manifold** (corrigendum), Proc. Cambridge Phil. Soc. 62, 674–681 (1966)
[167] ––, **An introduction to knot theory**, Graduate Texts in Mathematics 175, Springer (1997)
[168] R. A. Litherland, **Cobordism of satellite knots**, Four-Manifold Theory, Contemp. Math. 35, 327–362 (1984)
[169] C. Livingston, **Knot theory**, Carus Mathematical Monographs 24, M.A.A. (1993)
[170] S. López de Medrano, **Invariant knots and surgery in codimension 2**, Proc. 1970 Nice I. C. M., Gauthier–Villars, Vol. 2, 99–112 (1971)
[171] ––, **Involutions on manifolds**, Ergebnisse der Mathematik und ihrer Grenzgebiete 59, Springer (1971)
[172] ––, **Cobordism of diffeomorphisms of $(k-1)$-connected $2k$-manifolds**, Proc. Second Conf. on Compact Transformation Groups, Lecture Notes in Mathematics 298, 217–227, Springer (1972)
[173] W. Lück, **The transfer maps induced in the algebraic K_0- and K_1-groups by a fibration I.**, Math. Scand. 59, 93–121 (1986)
[174] ––, **The universal functorial Lefschetz invariant**, (preprint) (1998)
[175] –– and A. A. Ranicki, **Surgery transfer**, Proc. 1987 Göttingen Conf. on Alg. Topology, Lecture Notes in Mathematics 1361, 167–246, Springer (1988)
[176] –– and ––, **Chain homotopy projections**, J. of Algebra 120, 361–391 (1989)
[177] –– and ––, **Surgery obstructions of fibre bundles**, J. of Pure and App. Alg. 81, 139–189 (1992)
[178] R. Lutz, **Sur la geometrie des structures de contact invariants**, Ann. Inst. Fourier (Grenoble) 29, 283–306 (1979)
[179] R. Lyndon and P. Schupp, **Combinatorial group theory**, Ergebnisse der Mathematik und ihrer Grenzgebiete 89, Springer (1977)
[180] H. C. Lyon, **Torus knots in the complements of links and surfaces**, Michigan Math. J. 27, 39–46 (1980)
[181] Y. I. Manin, **Mathematics and physics**, Birkhäuser (1981)
[182] J. Martinet, **Formes de contact sur les variétés de dimension 3**, Lecture Notes in Mathematics 209, 142–163, Springer (1971)
[183] M. Mather, **Counting homotopy types of manifolds**, Topology 4, 93–94 (1965)

[184] Y. Matsumoto, **Knot cobordism groups and surgery in codimension two**, J. Fac. Sci. Tokyo (IA) 20, 253–317 (1973)

[185] T. Matumoto, **On the signature invariants of a non-singular complex sesquilinear form**, J. Math. Soc. Japan 29, 67–71 (1977)

[186] P. Melvin, **Bordism of diffeomorphisms**, Topology 18, 173–175 (1980)

[187] R. J. Milgram, **Orientations for Poincaré duality spaces and applications**, Algebraic Topology: Proc., Arcata, 1986, Lecture Notes in Mathematics 1370, 293–324, Springer (1989)

[188] ——, **Surgery with finite fundamental group**, Pacific J. Math. 151, I. 65–115, II. 117–150 (1991)

[189] —— and A. A. Ranicki, **The L-theory of Laurent polynomial extensions and genus 0 function fields**, J. reine u. angew. Math. 406, 121–166 (1990)

[190] J. Milnor, **On manifolds homeomorphic to the 7-sphere**, Ann. of Math. 64, 399–405 (1956)

[191] ——, **Groups which act on S^n without fixed points**, Amer. J. of Math. 79, 623–630 (1957)

[192] ——, **A procedure for killing the homotopy groups of differentiable manifolds**, Proc. Symp. on Pure Math. 3 (Differential Geometry), 39–55, A. M. S. (1961)

[193] ——, **A duality theorem for Reidemeister torsion**, Ann. of Maths. 76, 137–147 (1962)

[194] ——, **Whitehead torsion**, Bull. A. M. S. 72, 358–426 (1966)

[195] ——, **Infinite cyclic covers**, Proc. 1967 Conf. on the topology of manifolds, Prindle, Weber and Schmidt, 115–133 (1968)

[196] ——, **Singular points of complex hypersurfaces**, Ann. of Maths. Studies 61, Princeton (1968)

[197] ——, **On isometries of inner product spaces**, Invent. Math. 8, 83–97 (1969)

[198] ——, **Algebraic K-theory and quadratic forms**, Invent. Math. 9, 318–344 (1970)

[199] ——, **Introduction to algebraic K-theory**, Ann. of Maths. Studies 72, Princeton (1971)

[200] —— and D. Husemoller, **Symmetric bilinear forms**, Ergebnisse der Mathematik und ihrer Grenzgebiete 73, Springer (1973)

[201] A. S. Mishchenko, **Homotopy invariants of non–simply connected manifolds. III. Higher signatures**, Izv. Akad. Nauk SSSR, ser. mat. 35, 1316–1355 (1971)

[202] G. Mislin, **Wall's obstruction for nilpotent spaces**, Ann. of Math. 103, 547–556 (1976)

[203] T. Mizutani and I. Tamura, **Foliations of even-dimensional manifolds**, Manifolds – Tokyo 1973 (Proc. Internat. Conf., Tokyo, 1973), Univ. Tokyo Press, 189–194 (1975)

[204] J. M. Montesinos–Amilibia and H. R. Morton, **Fibered links from closed braids**, Proc. L.M.S. 62, 167–201 (1991)

[205] F. Morgan, **On the finiteness of the number of stable minimal hypersurfaces with fixed boundary**, Indiana J. of Math. 37. 779–833 (1986)

[206] K. Murasugi, **On a certain numerical invariant of link types**, Trans. A. M. S. 117, 387–422 (1965)

[207] ——, **Knot theory and its applications**, Birkhäuser (1996)

[208] R. Myers, **Open book decompositions of 3-manifolds**, Proc. A. M. S. 72, 397–402 (1978)

[209] W. D. Neumann, **Fibering over the circle within a cobordism class**, Math. Ann. 192, 191–192 (1971)

[210] ––, **Cyclic suspension of knots and periodicity of signature for singularities**, Bull. A. M. S. 80, 977–981 (1974)
[211] ––, **Manifold cutting and pasting groups**, Topology 14, 237–244 (1974)
[212] ––, **Equivariant Witt rings**, Bonner Math. Schriften 100 (1977)
[213] ––, **Multiplicativity of the signature**, J. Pure and App. Alg. 13, 19–31 (1978)
[214] ––, **Signature related invariants of manifolds – I. Monodromy and γ-invariants**, Topology 18, 147–172 (1979)
[215] L. Neuwirth, **Some algebra for 3-manifolds**, Topology of Manifolds (Proc. Inst., Univ. of Georgia, Athens, Ga., 1969), 179–184, Markham (1970)
[216] S. P. Novikov, **Rational Pontrjagin classes. Homeomorphism and homotopy type of closed manifolds I.**, Izv. Akad. Nauk SSSR, ser. mat. 29, 1373–1388 (1965), English translation : A. M. S. Transl. (2) 66, 214–230 (1968)
[217] ––, **Manifolds with free abelian fundamental group and applications (Pontrjagin classes, smoothings, high–dimensional knots)**, Izv. Akad. Nauk SSSR, ser. mat. 30, 208–246 (1966), English translation : A. M. S. Transl. (2) 67, 1–42 (1969)
[218] ––, **The algebraic construction and properties of hermitian analogues of K-theory for rings with involution, from the point of view of the hamiltonian formalism. Some applications to differential topology and the theory of characteristic classes**, Izv. Akad. Nauk SSSR, ser. mat. 34, 253–288, 478–500 (1970)
[219] ––, **Analogues hermitiens de la K-théorie**, Proc. 1970 Nice I. C. M., Gauthier–Villars, Vol. 2, 39–45 (1971)
[220] ––, **The hamiltonian formalism and a multivalued analogue of Morse theory**, Uspekhi Mat. Nauk 37, 3–49 (1982) English translation : Russian Math. Surv. 37, 1–56 (1982)
[221] W. Pardon, **Local surgery and the theory of quadratic forms**, Bull. A. M. S. 82, 131–133 (1976)
[222] A. V. Pazhitnov, **On the Novikov complex for rational Morse forms**, Ann. Fac. Sci. Touluse Math. (6) 4, 297–338 (1995)
[223] ––, **Surgery on the Novikov complex**, K-theory 4, 323–412 (1996)
[224] E. K. Pedersen and A. A. Ranicki, **Projective surgery theory**, Topology 19, 239–254 (1980)
[225] D. Quillen, **Higher algebraic K-theory**, Proc. 1972 Battelle Seattle Algebraic K-theory Conf., Vol. I, Lecture Notes in Mathematics 341, 85–147, Springer (1973)
[226] F. Quinn, **Surgery on Poincaré and normal spaces**, Bull. A. M. S. 78, 262–267 (1972)
[227] ––, **Open book decompositions, and the bordism of automorphisms**, Topology 18, 55–73 (1979)
[228] ––, **Homotopically stratified sets**, Journal of A. M. S. 1, 441–499 (1988)
[229] ––, **Lectures on axiomatic topological quantum field theory**, Geometry and quantum field theory (Park City, UT, 1991), IAS/Park City Math. Ser. 1, A. M. S., 323–453 (1995)
[230] A. A. Ranicki, **Algebraic L–theory I. Foundations**, Proc. L.M.S. (3) 27, 101–125 (1973)
[231] ––, **Algebraic L–theory II. Laurent extensions**, Proc. L.M.S. (3) 27, 126–158 (1973)
[232] ––, **Algebraic L–theory IV. Polynomial extensions**, Comm. Math. Helv. 49, 137–167 (1974)

[233] ——, **On the algebraic L-theory of semisimple rings**, J. Algebra 50, 242–243 (1978)
[234] ——, **Localization in quadratic L-theory**, Proc. 1978 Waterloo Conf. Algebraic Topology, Lecture Notes in Mathematics 741, 102–157, Springer (1979)
[235] ——, **The algebraic theory of surgery I. Foundations**, Proc. L.M.S. (3) 40, 87–192 (1980)
[236] ——, **The algebraic theory of surgery II. Applications to topology**, Proc. L.M.S. (3) 40, 193–287 (1980)
[237] ——, **Exact sequences in the algebraic theory of surgery**, Mathematical Notes 26, Princeton (1981)
[238] ——, **The algebraic theory of finiteness obstruction**, Math. Scand. 57, 105–126 (1985)
[239] ——, **The algebraic theory of torsion I. Foundations**, Proc. 1983 Rutgers Topology Conf., Lecture Notes in Mathematics 1126, 199–237, Springer (1985)
[240] ——, **Algebraic and geometric splittings of the K- and L-groups of polynomial extensions**, Proc. Symp. on Transformation Groups, Poznań 1985, Lecture Notes in Mathematics 1217, 321–364, Springer (1986)
[241] ——, **The algebraic theory of torsion II. Products**, K-theory 1, 115–170 (1987)
[242] ——, **The L-theory of twisted quadratic extensions**, Can. J. Math. 39, 345–364 (1987)
[243] ——, **Additive L-theory**, K-theory 3, 163–194 (1989)
[244] ——, **Lower K- and L-theory**, L. M. S. Lecture Notes 178, Cambridge (1992)
[245] ——, **Algebraic L-theory and topological manifolds**, Cambridge Tracts in Mathematics 102, Cambridge (1992)
[246] ——, **Finite domination and Novikov rings**, Topology 34, 619–632 (1995)
[247] ——, **On the Novikov conjecture**, Proc. 1993 Oberwolfach Conf. on the Novikov Conjectures, Rigidity and Index Theorems, Vol. 1, L. M. S. Lecture Notes 226, 272–337, Cambridge (1995)
[248] ——, **The bordism of automorphisms of manifolds from the algebraic L-theory point of view**, Prospects in Topology; proceedings of a conference in honor of William Browder, Annals of Mathematics Studies 138, 314–327, Princeton (1996)
[249] —— (ed.), **The Hauptvermutung Book**, K-Monographs in Mathematics 1, Kluwer (1996)
[250] K. Reidemeister, **Knotentheorie**, Ergebnisse der Mathematik und ihrer Grenzgebiete 1, Springer (1932) (reprinted 1974)
[251] W. Richter, **High-dimensional knots with $\pi_1 \cong \mathbb{Z}$ are determined by their complements in one more dimension than Farber's range**, Proc. A. M. S. 120, 285–294 (1994)
[252] C. Riehm, **The equivalence of bilinear forms**, J. Algebra 31, 45–66 (1974)
[253] D. Rolfsen, **Knots and links**, Publish or Perish (1976)
[254] J. Rosenberg, **Algebraic K-theory and its Applications**, Graduate Texts in Mathematics 147, Springer (1994)
[255] L. Rudolph, **Mutually braided open books and new invariants of fibered links**, in Braids, Santa Cruz, 1996. Contemp. Math. 78, A. M. S. (1988)
[256] W. Scharlau, **Quadratic and hermitian forms**, Grundlehren der mathematischen Wissenschaften 270, Springer (1985)
[257] A. H. Schofield, **Representation of rings over skew fields**, L. M. S. Lecture Notes 92, Cambridge (1985)
[258] H. Schubert, **Die eindeutige Zerlegbarkeit eines Knotens in Primknoten**, Sitzb. Heidelberger Akad. Wiss. Math. Naturwiss. Kl., 57–104 (1949)

[259] J. Seade, **Fibred links and a construction of real singularities via complex geometry**, Bol. Soc. Bras. Mat. 27, 199–215 (1996)
[260] ––, **Open book decompositions associated to holomorphic vector fields**, Bol. Soc. Mat. Mex. 3, 323–335 (1997)
[261] H. Seifert, **Verschlingungsinvarianten**, Sitzb. Preuß. Akad. Wiss. 26, 811–823 (1933)
[262] ––, **Über das Geschlecht von Knoten**, Math. Ann. 110, 571–592 (1934)
[263] J. Shaneson, **Wall's surgery obstruction groups for $G \times \mathbb{Z}$**, Ann. of Maths. 90, 296–334 (1969)
[264] L. Siebenmann, **The structure of tame ends**, Notices AMS 66T–G7, 861 (1966)
[265] ––, **A torsion invariant for bands**, Notices AMS 68T–G7, 811 (1968)
[266] ––, **A total Whitehead torsion obstruction to fibering over the circle**, Comm. Math. Helv. 45, 1–48 (1970)
[267] J. R. Silvester, **Introduction to algebraic K-theory**, Chapman and Hall (1981)
[268] J. Simon, **Fibered knots in homotopy 3–spheres**, Proc. A. M. S. 58, 325–328 (1976)
[269] S. Smale, **On the structure of manifolds**, Amer. J. of Math. 84, 387–399 (1962)
[270] J. Smith, **Complements of codimension two submanifolds III. Cobordism theory**, Pacific J. Math. 94, 423–484 (1981)
[271] ––, **Non-simply-connected twisted tensor products**, (preprint) (1997)
[272] J. Stallings, **On fibering certain 3-manifolds**, Proc. 1961 Georgia Conf. on the Topology of 3-manifolds, Prentice-Hall, 95–100 (1962)
[273] ––, **On topologically unknotted spheres**, Ann. of Maths. 77, 490–503 (1963)
[274] B. Stenström, **Rings of quotients**, Springer (1975)
[275] J. Stevens, **Periodicity of branched cyclic covers of manifolds with open book decomposition**, Math. Ann. 273, 227–239 (1986)
[276] J. Stienstra, **Operations in the higher K-theory of endomorphisms**, Proc. 1981 Waterloo Topology Conf., Can. Math. Soc. Conf. Proc. 2 (Vol. 1), 59–115 (1982)
[277] N. Stoltzfus, **Unraveling the integral knot concordance group**, Memoirs A. M. S. 192 (1977)
[278] ––, **Algebraic computations of the integral concordance and double null concordance group of knots**, Proc. 1977 Plans-sur-Bex Conf. Knot Theory, Lecture Notes in Mathematics 685, 274–290, Springer (1978)
[279] ––, **The algebraic relationship of Quinn's invariant for open book decomposition bordism and the isometric structure**, Appendix to [141], Lecture Notes in Mathematics 1069, 115–141, Springer (1984)
[280] A. Suciu, **Inequivalent frame-spun knots with the same complement**, Comm. Math. Helv. 67, 47–63 (1992)
[281] D. W. Sumners, **Polynomial invariants and the integral homology of coverings of knots and links**, Invent. Math. 15, 78–90 (1972)
[282] P. G. Tait, **On knots**, Trans. R. S. E. 28, 145–190 (1877)
[283] I. Tamura, **Spinnable structures on differentiable manifolds**, Proc. Japan Acad. 48, 293–296 (1972)
[284] ––, **Foliations and spinnable structures on manifolds**, Ann. Inst. Fourier 23, 197–214 (1973)
[285] ––, **Foliations of total spaces of sphere bundles over spheres**, J. Math. Soc. Japan 24, 698–700 (1972)

[286] C. H. Taubes, **Gauge theory on asymptotically periodic 4-manifolds**, J. Diff. Geo. 25, 363–430 (1987)
[287] E. Thomas and J. Wood, **On manifolds representing homology classes in codimension 2**, Invent. Math. 25, 63–89 (1974)
[288] W. P. Thurston and H. E. Winkelnkemper, **On the existence of contact forms**, Proc. A. M. S. 52, 345–347 (1975)
[289] A. G. Tristram, **Some cobordism invariants for links**, Proc. Camb. Phil. Soc. 66, 251–264 (1969)
[290] H. F. Trotter, **Homology of group systems with applications to knot theory**, Ann. of Maths. 76, 464–498 (1962)
[291] ––, **On S-equivalence of Seifert matrices**, Invent. Math. 20, 173–207 (1973)
[292] ––, **Knot modules and Seifert matrices**, Proc. 1977 Plans-sur-Bex Conf. Knot Theory, Lecture Notes in Mathematics 685, 291–299, Springer (1978)
[293] V. G. Turaev, **Reidemeister torsion in knot theory**, Uspekhi Mat. Nauk 41, 97–147 (1986). English translation: Russian Math. Surv. 41, 119–182 (1986)
[294] A. Verjovsky, **A note on foliations of spheres**, Bol. Soc. Mat. Mexicana 18, 36–37 (1973)
[295] O. J. Viro, **Branched coverings of manifolds with boundary and link invariants**, Izv. Akad. Nauk SSSR, ser. mat. 37, 1241–1258 (1973)
[296] P. Vogel, **Localization in algebraic L-theory**, Proc. 1979 Siegen Topology Conf., Lecture Notes in Mathematics 788, 482–495, Springer (1980)
[297] ––, **On the obstruction group in homology surgery**, Publ. Math. I. H. E. S. 55, 165–206 (1982)
[298] F. Waldhausen, **Heegaard-Zerlegungen der 3-Sphäre**, Topology 7, 195–203 (1968)
[299] ––, **Algebraic K-theory of generalized free products**, Ann. of Maths. 108, 135–256 (1978)
[300] C. T. C. Wall, **Finiteness conditions for CW complexes**, Ann. of Maths. 81, 56–69 (1965)
[301] ––, **Non-additivity of the signature**, Invent. Math. 7, 269–274 (1969)
[302] ––, **Surgery on compact manifolds**, Academic Press (1970)
[303] ––, **On the axiomatic foundations of the theory of Hermitian forms**, Proc. Camb. Phil. Soc. 67, 243–250 (1970)
[304] M. Warshauer, **The Witt group of degree k maps and asymmetric inner product spaces**, Lecture Notes in Mathematics 914, Springer (1982)
[305] C. Weibel, **K-theory and analytic isomorphisms**, Invent. Math. 61, 177–197 (1980)
[306] S. Weinberger, **On fibering four- and five-manifolds**, Israel J. Math. 59, 1–7 (1987)
[307] H. E. Winkelnkemper, **Equators of manifolds and the action of Θ^n**, Ph.D. Thesis, Princeton University, 1970
[308] ––, **Manifolds as open books**, Bull. A. M. S. 79, 45–51 (1973)
[309] ––, **On the action of Θ^n I.**, Trans. A. M. S. 206, 339–346 (1975)
[310] ––, **An algebraic condition implied by the weak Poincaré conjecture**, Abstracts of the A. M. S. (93T–57–19) Jan. 1993
[311] ––, **Artin presentations, I.** (to appear)
[312] E. C. Zeeman, **Unknotting combinatorial balls**, Ann. of Maths. 78, 501–526 (1963)
[313] H. Zieschang, **Finite groups of mapping classes of surfaces**, Lecture Notes in Mathematics 875, Springer (1981)

Index

– class group, $\mathrm{Iso}_0(A)$ 398
– fibred L-groups, $L\mathrm{Iso}^*_{fib,U}(A,\epsilon)$ 402
– reduced L-groups, $\widetilde{L\mathrm{Iso}}^*_U(A,\epsilon)$ 403
– structure on ϵ-symmetric complex 399
– structure on ϵ-symmetric form 401
isotopic xvii

knot
– n-dimensional, (M^{n+2},N^n,k) xv
– n-knot, $k:S^n\subset S^{n+2}$ xv, 443
– classical, $k:S^1\subset S^3$ xv
– cobordism xvii, 443
– cobordism groups, C_* 443
– concordance xvii
– connected sum 443
– fibred xxiii, 445
– homology framed xix, 303
– simple xxvi, 448
– slice 443
– stable 448

L-groups
– (Λ,ϵ)-asymmetric, $L\mathrm{Asy}^*_q(A,\Lambda,\epsilon)$ 290
– (Σ_0,ϵ)-asymmetric, $L\mathrm{Asy}^*_q(A,\Sigma_0,\epsilon)$ 293
– S-primary asymmetric, $L\mathrm{Asy}^*_S(A)$ 477
– S-primary automorphism, $L\mathrm{Aut}^*_S(A,\epsilon)$, $L\mathrm{Aut}^S_*(A,\epsilon)$ 474
– S-primary endomorphism, $L\mathrm{End}^*_S(A,\epsilon)$, $L\mathrm{End}^S_*(A,\epsilon)$ 468
– S-primary fibred automorphism, $L\mathrm{Aut}^*_{fib,S}(A,\epsilon)$, $L\mathrm{Aut}^{fib,S}_*(A,\epsilon)$ 481
– S-primary isometric, $L\mathrm{Iso}^*_S(A,\epsilon)$ 472
– U-intermediate twisted double, $DBL^*_U(A,\epsilon)$ 390
– U-intermediate, $L^*_U(A,\epsilon)$, $L^U_*(A,\epsilon)$ 211
– ϵ-quadratic, $L_*(A,\epsilon)$ 211
– ϵ-symmetric, $L^*(A,\epsilon)$ 211
– ϵ-ultraquadratic, $\widehat{L}^p_*(A,\epsilon)$, $\widehat{L}^h_*(A,\epsilon)$ 417
– almost ϵ-symmetric, $L\mathrm{Asy}^*_{q,S_\epsilon}(A)$ 486
– asymmetric, $L\mathrm{Asy}^*_q(A)$ 344
– automorphism, $L\mathrm{Aut}^*_q(A,\epsilon)$ 334
– automorphism, $L\mathrm{Aut}^*_U(A,\epsilon)$, $L\mathrm{Aut}^U_*(A,\epsilon)$ 333
– endomorphism, $L\mathrm{End}^*_U(A,\epsilon)$, $L\mathrm{End}^U_*(A,\epsilon)$ 458
– fibred ϵ-ultraquadratic, $\widehat{L}^{q,fib}_*(A,\epsilon)$ 425
– fibred automorphism, $L\mathrm{Aut}^*_{q,fib}(A,\epsilon)$ 350
– fibred isometric, $L\mathrm{Iso}^n_{fib,U}(A,\epsilon)$ 402
– isometric, $L\mathrm{Iso}^*_U(A,\epsilon)$ 402
– nilpotent endomorphism, $L\mathrm{Nil}^*_q(A,\epsilon)$, $L\mathrm{Nil}^q_*(A,\epsilon)$ 469
– projective twisted double, $DBL^*_p(A,\epsilon)$ 388
– reduced S-primary automorphism, $\widetilde{L\mathrm{Aut}}^*_S(A,\epsilon)$, $\widetilde{L\mathrm{Aut}}^S_*(A,\epsilon)$ 474
– reduced S-primary endomorphism, $\widetilde{L\mathrm{End}}^*_S(A,\epsilon)$, $\widetilde{L\mathrm{End}}^S_*(A,\epsilon)$ 468
– reduced asymmetric, $\widetilde{L\mathrm{Asy}}^*_q(A,\Lambda,\epsilon)$ 293
– reduced isometric, $\widetilde{L\mathrm{Iso}}^n_U(A,\epsilon)$ 403
– reduced nilpotent, $\widetilde{L\mathrm{Nil}}^*_q(A,\epsilon)$, $L\mathrm{Nil}^q_*(A,\epsilon)$ 469
– reduced unipotent, $\widetilde{L\mathrm{Aut}}^*_{uni}(A,\epsilon)$ 481
– round automorphism, $L\mathrm{Aut}^*_r(A,\epsilon)$, $L\mathrm{Aut}^r_*(A,\epsilon)$ 350
– round endomorphism, $L\mathrm{End}^*_r(A,\epsilon)$, $L\mathrm{End}^r_*(A,\epsilon)$ 463
– round isometric, $L\mathrm{Iso}^*_r(A,\epsilon)$, $L\mathrm{Iso}^r_*(A,\epsilon)$ 410
– round, $L^*_r(A,\epsilon)$, $L^r_*(A,\epsilon)$ 214
– Schneiden und Kleben, $SKL^*(A,\epsilon)$ 395
– torsion, $L^U_*(A,\Sigma,\epsilon)$, $L^*_U(A,\Sigma,\epsilon)$ 269
lagrangian 206
– in asymmetric form 290, 344
Laurent polynomial extension, $A[z,z^{-1}]$ 15
leading
– coefficients 28
– units 109
linking form 279
– metabolic 279
– nonsingular 279
localization
– $P^{-1}A[z,z^{-1}]$ 428
– $P^{-1}_{fib}A[z,z^{-1}]$ 428
– $S^{-1}A$ 20
– $T^{-1}_A A[x]$ 584
– $U^{-1}_A A[s]$ 594
– $V^{-1}_A A[z,z^{-1}]$ 599

projective
- class 4
- class group, $K_0(A)$ 3
- module 3
- reduced class group, $\widetilde{K}_0(A)$ 4
- Seifert complex 422
push-in Seifert surface 311

Q-groups
- absolute, $Q^*(C,\epsilon)$, $Q_*(C,\epsilon)$ 207
- autometric, $Q^*_{aut}(C,h,\epsilon)$, $Q^{aut}_*(C,h,\epsilon)$ 330
- double relative, $Q^*(f,g,\epsilon)$, $Q_*(f,g,\epsilon)$ 261
- endometric, $Q^*_{end}(C,f,\epsilon)$, $Q^{end}_*(C,f,\epsilon)$ 456
- isometric, $Q^*_{iso}(C,f,\epsilon)$ 399
- relative, $Q^*(f,\epsilon)$, $Q_*(f,\epsilon)$ 210
quadratic
- LN-groups, LN_* 230
- LS-groups, $LS_*(\Phi)$ xxxiv, 228
- construction 210
- kernel, $(\mathcal{C}(f^!),\psi)$ xxviii, 210
- signature, Λ-coefficient, $\sigma^\Lambda_*(f,b)$ 216
- signature, $\sigma_*(f,b)$ 214

r-simple n-knot 448
rational Witt vector
- group, $W(A)$ 138
- group, reduced, $\widetilde{W}(A)$ 139
- ring, $W(A)$ 102
reduced
- asymmetric L-groups, $\widetilde{LAsy}^*_q(A,\Lambda,\epsilon)$ 293
- endomorphism K-groups, $\widetilde{End}_*(A)$ 80
- endomorphism L-groups, $\widetilde{LEnd}^*_V(A,\epsilon)$, $\widetilde{LEnd}^V_*(A,\epsilon)$ 458
- endomorphism class group, $\widetilde{End}_0(A)$ 80
- nilpotent K-groups, $\widetilde{Nil}_*(A)$ 30
- nilpotent L-groups, $\widetilde{LNil}^*_q(A,\epsilon)$, $\widetilde{LNil}^q_*(A,\epsilon)$ 469
- nilpotent class group, $\widetilde{Nil}_0(A)$ 31
- projective class 4
- S-primary automorphism L-groups, $\widetilde{LAut}^*_S(A,\epsilon)$, $\widetilde{LAut}^S_*(A,\epsilon)$ 474
- S-primary endomorphism K-groups, $\widetilde{End}^S_*(A)$ 105
- S-primary endomorphism L-groups, $\widetilde{LEnd}^*_S(A,\epsilon)$, $\widetilde{LEnd}^S_*(A,\epsilon)$ 468
Reidemeister torsion 159, 191
- absolute, Δ 159
- relative, Δ 160
rel ∂ self homotopy equivalence 385
relative
- algebraic mapping torus, $t(h,\chi)$ 386
- geometric mapping torus, $t(h)$ 352
reverse
- characteristic polynomial
-- noncommutative, $\widetilde{ch}_z(P,f)$ 140
- characteristic polynomial, $\widetilde{ch}_z(P,f)$ 100
- multiplicative subset, $\widetilde{S}$ 111
- polynomial, $\widetilde{p}(z)$ 97
ribbon handle 361
right denominator set 74
round
- A-finite structure 7
- L-groups, $L^*_r(A,\epsilon)$, $L^r_*(A,\epsilon)$ 214
- automorphism L-groups, $LAut^*_r(A,\epsilon)$, $LAut^r_*(A,\epsilon)$ 350
- chain complex 7
- endomorphism L-groups, $LEnd^*_r(A,\epsilon)$, $LEnd^r_*(A,\epsilon)$ 463
- finite CW complex 7
- finite structure 7
- isometric L-groups, $LIso^*_r(A,\epsilon)$, $LIso^r_*(A,\epsilon)$ 410
- simple L-groups, $L^*_{rs}(A,\epsilon)$, $L^{rs}_*(A,\epsilon)$ 214

S-adic completion, $\widehat{A}_S$ 26
S-equivalence
- Seifert complexes 424
- Seifert forms 449
S-primary
- asymmetric L-groups, $LAsy^*_S(A)$ 477
- asymmetric complex 477
- automorphism 128
- automorphism K-groups, $Aut^S_*(A)$ 128
- automorphism L-groups, $LAut^*_S(A,\epsilon)$, $LAut^S_*(A,\epsilon)$ 474
- automorphism L-groups, $LAut^*_{fib,T}(A,\epsilon)$, $LAut^{fib,T}_*(A,\epsilon)$ 481
- chain complex 109
- chain map 109
- endomorphism 105

Printing: Mercedesdruck, Berlin
Binding: Buchbinderei Lüderitz & Bauer, Berlin